Harnessing Autodesk® Civil 3D

Harnessing Autodesk® Civil 3D

PHILLIP J. ZIMMERMAN

autodesk Press

THOMSON
DELMAR LEARNING™

Australia · Canada · Mexico · Singapore · Spain · United Kingdom · United States

Harnessing Autodesk® Civil 3D
by Phillip Zimmerman

Autodesk Press Staff

Vice President, Technology and Trades ABU:
David Garza

Director of Learning Solutions:
Sandy Clark

Senior Acquisitions Editor:
James Gish

Senior Development Editor:
John Fisher

Marketing Director:
Deborah S. Yarnell

Channel Manager:
Kevin Rivenburg

Marketing Coordinator:
Mark Pierro

Production Director:
Patty Stephan

Production Manager:
Stacy Masucci

Technology Project Manager:
Kevin Smith

Technology Project Specialist:
Chris Catalina

Editorial Assistant:
Niamh Matthews

Printed in Canada
1 2 3 4 5 TC 08 07 06

For more information contact
Delmar Learning
Executive Woods
5 Maxwell Drive, PO Box 8007
Clifton Park, NY 12065-8007
Or find us on the World Wide Web at
www.delmarlearning.com

Library of Congress Cataloging-in-Publication Data:

Zimmerman, Phillip J.
Harnessing Autodesk Civil 3D / Phillip Zimmerman.
p. cm.
Includes index.
ISBN 1-4180-1488-5
1. Civil engineering--Computer programs. 2. Surveying--Computer programs. 3. Autodesk Civil 3D. I. Title.
TA345.Z553 2006
620'.00420285--dc22

2006034881

ISBN-13: 978-1-4180-1488-9
ISBN-10: 1-4180-1488-5

CONTENTS

INTRODUCTION

Civil 3D is the successor to the Autodesk Land Desktop product line (Land Desktop, Civil Design, and Survey). Like Autodesk Land Desktop (LDT), the program is a powerful drafting tool and unlike Autodesk (LDT), Civil 3D is a dynamic engineering environment. Civil 3D does not create data based on drawing entity snapshots. The dynamic engineering environment is the single fundamental change that radically alters the cadd design environment. Changes in data produce changes in documentation.

Autodesk (LDT) with its hidden data and definitions disappear in Civil 3D. Prospector's Toolspace provides open access to object data and its properties. Prospector associates an object with its data and provides an interface to the object's data. In Civil 3D, data is only a few clicks away.

The Settings Toolspace presents a unique way of developing a company "look" that is consistent irrespective of which user develops a design. This is where the journey starts, with the styles that Civil 3D depends on to accomplish its drafting and documentation of your design. Where Land Desktop was a take-it or draft it environment, Civil 3D is a-la-carte. It is like returning to the time when Land Desktop started shipping; you either live with the Civil 3D's shipping content or you spend some time developing new content. It is a new beginning for the Civil Industry using Autodesk tools.

The tool sets of Civil 3D include points, surfaces, parcels, alignments, profiles, sections, pipes, and survey. To use Civil 3D's tool sets requires a fundamental knowledge of the Civil Design and Survey processes. This book addresses the use of and provides a basic understanding of the tool sets. This book, however, is NOT an engineering or surveying textbook nor is it a tips and tricks book. You can peruse the news groups for those needs. This book provides explanations of and exercises helping in the development of a basic understanding of the tools within Civil 3D. Many examples are from people using the software while other examples contain difficulties to demonstrate capabilities.

Autodesk (LDT)'s approach is search a menu for a command to do a task. Civil 3D's is choosing the correct heading, pressing the right mouse button, and selecting a command from a shortcut menu.

When working in Civil 3D you have four potential tasks: creating objects, analyzing their values, editing them if wrong, and labeling their values. Tools from some of these tasks are called from the menus, but most are called from Prospector.

It will become an ingrained habit; left mouse button: right mouse button, and select from a shortcut menu.

Working through the exercises, you will encounter issues with the software. You must remember, Civil 3D is a work in progress. Service packs have solved several bugs or issues, but each service pack has broken features that were functioning before the patch. I have not had a single version of Civil 3D complete all of the tasks in this book. That is why each unit has a drawing for the exercise; it recovers you from a corrupt file or starts you at that point in the unit without having to do the previous exercise(s). Even with these issues, I can not imagine going back to LDT. Civil 3D has too much potential.

Drawing files are included on the book's CD for use in exercises. PDF files of the last two chapters of text, chapters 13 and 14, are also on the CD.

ACKNOWLEDGEMENTS

The examples for the exercises in Chapter 2 are from surveys done by Jerry Bartels. The manual survey of Chapter 13 is from Gail and Gil Evans at Balsamo/Olson Group Inc. The network loop traverse of Chapter 14 is from Bill Laster. The lot exercise for Chapter 3 is from a plat developed by Intech Consultants, Inc. (Tom Fahrenbok, Scott and Joe).

Thanks to the staff at Thompson Publishing for patiently waiting for the manuscript's completion. I want to thank my dear mom (Lorraine), sisters (Sharon and Dawn), and patient friends for their support and help. Also, I want to thank SageCAD and Imaginit, Carl, Len, Angela, Andrea, V, Todd, Brent, Jeremy, Mike Choquette, and many other friends for giving me advice and opportunities. I want to thank Mike Martinez for joyfully and enthusiastically rereading the manuscript for technical errors. Sadly, no matter how many times we reread the manuscript, there will be errors. Let me know where they are so I can fix them.

New Beginnings

INTRODUCTION

This chapter introduces Autodesk Civil 3D 2007, its interface, and its organization. Civil 3D presents the user with a new environment that addresses traditional civil design problems. Civil 3D's focus is creating and documenting a design, the same as any other civil engineering application. The tools in Civil 3D are similar to those found in other civil programs, particularly Autodesk Land Desktop. This also holds true for the process of moving through the design in Civil 3D. However, the difference is Civil 3D's interactive environment and assumed dependencies among civil design elements (points, surfaces, alignments, profiles, etc.). These related design objects make Civil 3D unique among civil design applications.

OBJECTIVES

This chapter focuses on the following topics:

- Civil 3D Objects
- Prospector Overview
- Prospector and Its Preview Area
- Settings
- Data Compatibility and Transfer with Land Desktop 2007

OVERVIEW

Autodesk Civil 3D addresses the same civil engineering design issues as Autodesk Land Desktop (LDT), Inroads, or any other civil engineering application. However, Civil 3D contains new tools that create, evaluate, edit, and annotate a design solution. While Civil 3D's design environment is new, the application still uses the familiar design elements found in any civil design application (points, surfaces, alignments, profiles, etc.). The most radical difference between Civil 3D and other applications is the dynamic environment. Civil 3D understands and implements relationships and dependencies among design elements. If a change to an element that has dependencies is requested, Civil 3D updates all dependent elements to reflect the change. For example, changing a surface causes contours, profiles, corridors, and sections to update, reflecting their adjustment to the change made to the surface.

The impetus for developing Civil 3D was the perceived and real shortcomings of the Autodesk Land Desktop/Civil Design family of products. Many users working with the products had issues with inflexible labeling options, limited handling of special design situations, no major road design and rehabilitation tools, limited design documentation tools, and a passive design environment.

To address these issues, the civil applications group at Autodesk developed Civil 3D. Civil 3D's design/drafting environment is completely new and presents the user with an interactive drafting space, many new road design tools, flexible labeling methods, and several new options for implementing standards. Something so new or so different produces a "fear of the unknown" and raises many questions. Besides learning and becoming comfortable with this new design environment, other major hurdles to using Civil 3D are understanding how to document a design, creating content for a Civil 3D implementation, and learning application behaviors and terminology.

Most company standards focus only on linetypes, layers, pen weights, etc. A Civil 3D implementation focuses on these same standards, but its styles extend an implementation to crafting a unified company look to design documents. Styles are flexible enough to create a unique look to profiles, sections, parcels, and other design solution elements. A Civil 3D implementation should define a "look" for each design element and document prepared for review or submittal. When implementing Civil 3D, some of the questions to ask are: What does a profile or section look like? What types of annotation do they have? In a set sheet, where is the annotation located? This microscopic look at each page of a submission set is necessary to understand how to implement Civil 3D in any office. The power of styles is in creating a correct document as the designer creates or changes the design solution.

UNIT 1

The first unit reviews the anatomy of the Prospector and demonstrates its dynamic qualities, the actions that occur in the object and preview areas, and its extensive use of icons to signal dependencies, out-of-date status, etc.

UNIT 2

The second unit covers the Settings panel, the types of styles available for an object, and how to navigate the Settings tree and branch structure.

UNIT 3

Civil 3D implements design relationships and dependencies with custom objects. These objects act as fundamental building blocks for design solutions. Linked to these objects are styles that produce a look, implement standards, and define formats for annotation for each object's information. This dependent and relational object environment implemented with styles allows Civil 3D to dynamically manage civil design data and annotation in a project.

These objects and their behaviors are the topic of the third unit of this chapter.

UNIT 4

The last unit looks into transferring data between Autodesk Land Desktop and Civil 3D. Civil 3D reads data from the LDT project data structure or imports data from a LandXML file (from LDT or other applications). Autodesk Land Desktop 2007 has routines that read and extract data from a Civil 3D drawing and that store the extracted data in an LDT project structure.

UNIT 1: PROSPECTOR

The heart of Civil 3D is a Toolspace with two tabs, the Prospector, and Settings (see Figure 1.1). The two tabs of this single Toolspace manage all of the objects, styles, references, and data values for a drawing and/or project. The General menu contains the command to display the Toolspace, Show Toolspace, and other options. A user can float the Toolspace and let it Auto-hide when not needed, or dock it as a part of the overall interface.

For the remainder of this book, the term "Prospector" identifies the Prospector panel and "Settings" identifies the Settings panel of this Toolspace.

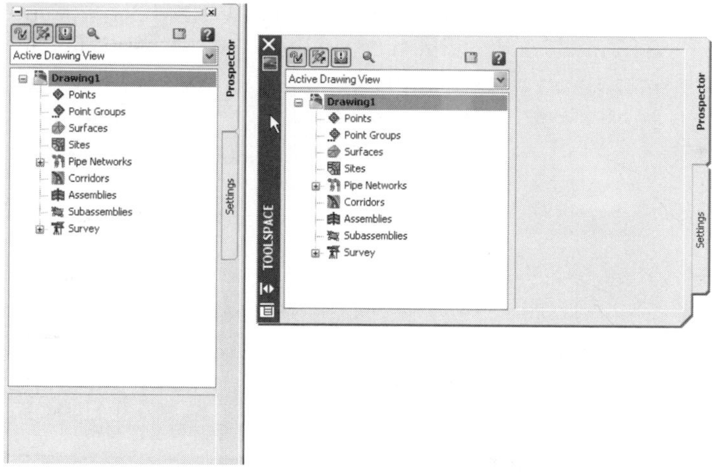

Figure 1.1

Prospector uses a simple hierarchical structure to organize objects, display their essential data, and show relationships among objects. Prospector identifies each object type as a heading with a branch specific to its information and data. Prospector dynamically manages the objects in the drawing, their listings, and makes their data available to other objects. When wanting to view an object's information, the user interacts with an object's listing in Prospector.

A second Prospector area is preview. This area displays a list or an image of a selected Prospector entry (see Figure 1.2). When selecting a Prospector or branch heading, a list appears in the preview area (for example, selecting the Prospector's Surfaces heading displays a list of surfaces). When selecting from the Prospector's Surfaces list, the preview area displays a surface image. When the Toolspace is docked, the preview area is at the Prospector's bottom, and when floating the preview is to one side of Prospector.

Prospector's data management and preview are dynamic. Prospector updates information and responds to changes and additions to objects and their data. The preview area then updates with a new list or image.

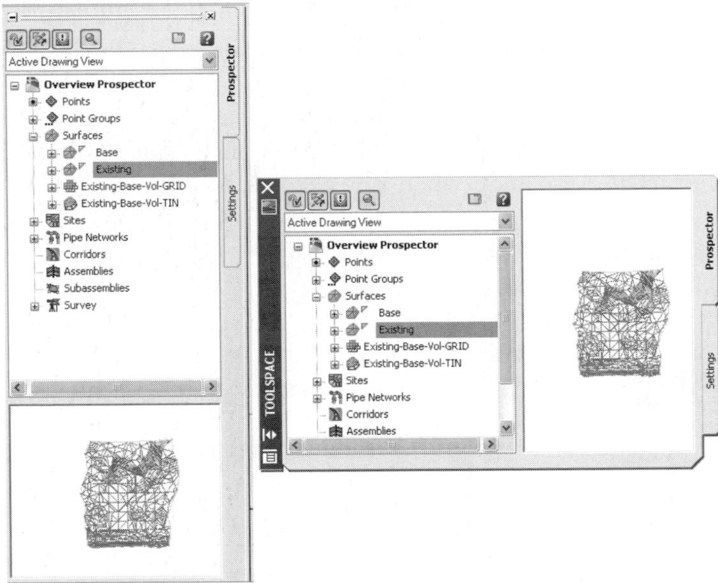

Figure 1.2

OBJECT TYPES

In Prospector each object type has a heading and an icon (see Figure 1.2). The object types are Points, Point Groups, Surfaces, Sites, Pipe Networks, Corridors, Assemblies, Subassemblies, and Survey. Alignments, Profiles, Sample Lines, and Cross Sections are members of the Sites branch. Civil 3D defines a site as the outermost boundary containing alignments and parcels. A site can range from a subdivision boundary or a parcel, to a single alignment (i.e., a site is a container for these objects).

When expanding an object type's branch, the first entry is a list of named instances (occurrences) for that object type. In Figure 1.2, for example, Base and Existing are two instances of the surface object type. Prospector automatically updates this list as objects are added or deleted from a drawing. Depending on the object type, each instance has its own branch containing its data or other critical values. For example, the Sites object type contains a list of sites (Site 1, Site 2, etc.). Each instance of a site has it own list of alignments, profiles, sections, and parcels.

Prospector displays the data of an object at some point down the instance's branch. However, there are times when the data appears in the preview area instead of Prospector. For example, the Definition branch of a surface contains a list of data types assigned to a surface, but shows the data for a selected type in the preview area, not the branch. Some objects such as alignments, profiles, assemblies, and subassembies use a Properties dialog box instead of a branch in Prospector to display their data.

Prospector uses right mouse button (RMB) shortcut menus to call commands, view instances, or generate external files. Each heading (object type, object instance, and so on) have a shortcut menu unique to its location in the branch (see Figure 1.3). The choices on the shortcut menu vary with the type of object and where the user is in the branch. For example, the Surfaces heading shortcut menu has commands to create a surface, import a surface, export a LandXML file, or refresh the surface's instance list (the left side of Figure 1.3). The shortcut menu for a named surface in the instance list includes build options, snapshots, zooming, etc. (the center of Figure 1.3). The shortcut menu for a specific surface component or data element has commands that create, edit, and delete entries (the right side of Figure 1.3).

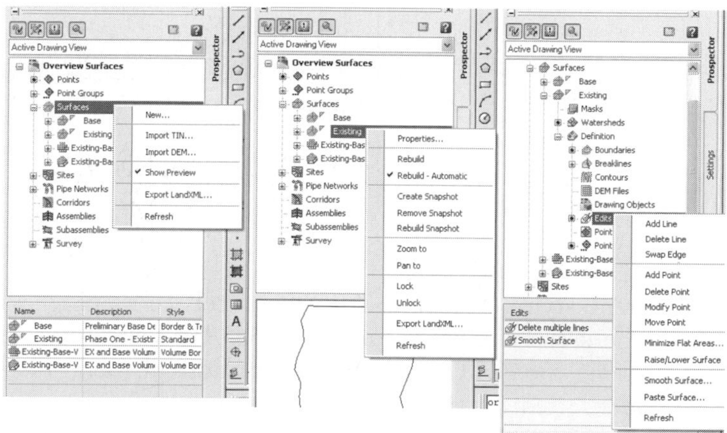

Figure 1.3

PROSPECTOR PREVIEW

When selecting an object type heading, Prospector responds with a list of object instances. When selecting an object instance, Prospector previews a representation of the selected instance (see Figure 1.4). The selected heading determines what displays in the preview area (for example, selecting Sites shows a list of sites, selecting Site 1 shows an image of the site, selecting Parcels shows a list of parcels, and selecting a parcel previews the shape of the parcel).

When selecting a text item from a preview list, the user can edit the entry or press the right mouse button to view a shortcut menu specific to the selected item. After completing changes in the preview area, these changes show in Prospector and the drawing. For example, if the user selects the Surfaces heading, the preview area lists the surface names,

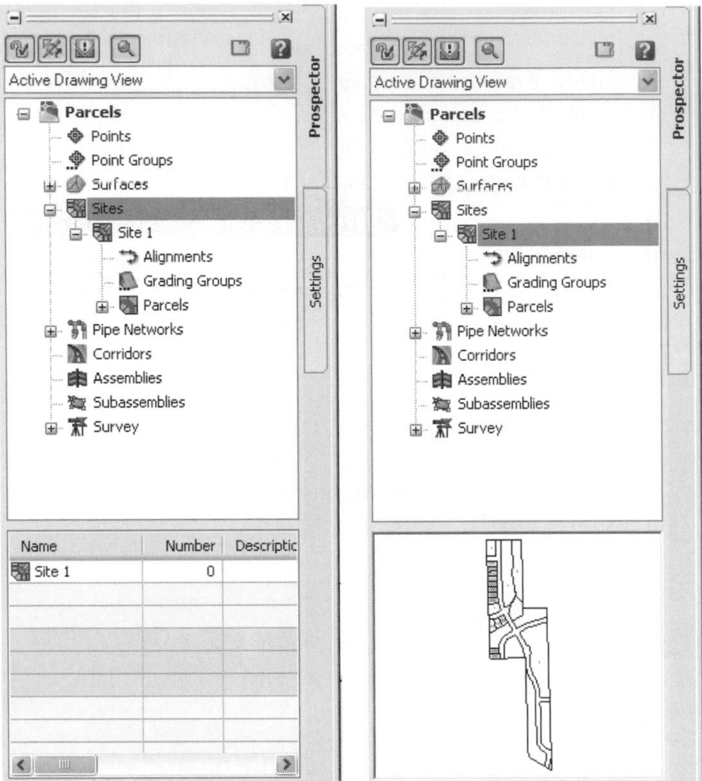

Figure 1.4

descriptions, and current styles. When clicking in a cell containing a surface's name, description, or style, the user can edit its current value. After clicking a cell containing a surface name and pressing the right mouse button, a shortcut menu displays that is the same as if the user had selected a surface name from Prospector's surface instance list (the center shortcut menu in Figure 1.3). When selecting a style name in the preview area, a Select Style dialog box appears for that object type. The user can change the currently assigned surface style by selecting another style from the styles list (see Figure 1.5). Again, changes made to a preview entry update Prospector and the drawing.

CIVIL 3D OBJECTS

In the Autodesk Land Desktop (LDT) series of products, few, if any, objects interact with each other (the exception being a grading object and a surface). None of the roadway design line work could react to a change made to any one of its dependent objects (a profile reacting to changes in a surface's elevations). It was left to the user to verify what changes needed to be made and as to how to synchronize the external data with the entities. In LDT, routines create snapshots of the on-screen line work and store them as data in external project folders. The drawing's representation of the data is valid only if nothing

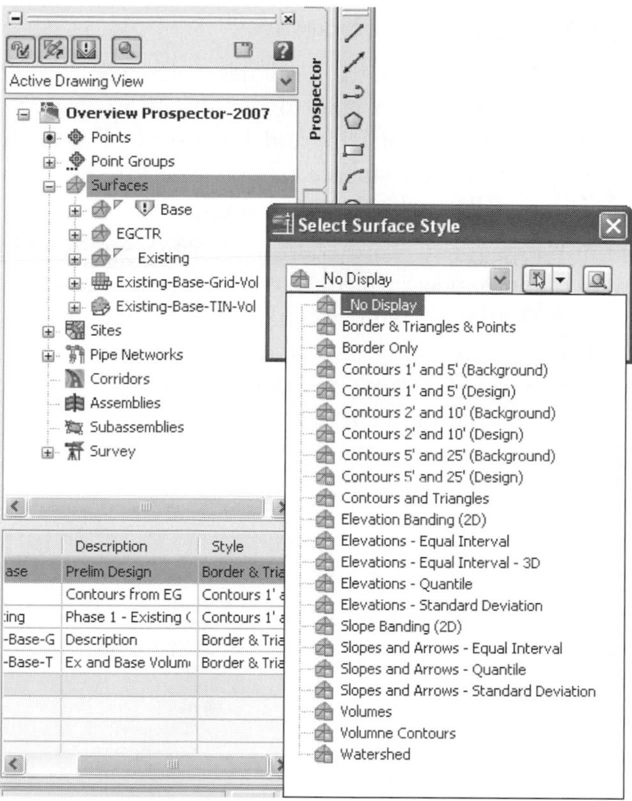

Figure 1.5

changes. If the line work is edited or redrawn, all of the related project data files need updating. In many cases, the final design is the starting point of an LDT project, for fear of data becoming out of sync after numerous changes and handoffs between drafters and engineers.

Civil 3D uses a data schema that gives objects knowledge about their relationships and dependencies with other objects. This knowledge allows them to respond to changes by any object in their group of relationships. So, if a surface's elevations change and an alignment with a profile view crosses the area of change, the profile updates to show the new surface elevations along the alignment's path. If changing the location of an alignment, the profile changes its elevations and either lengthens or shortens its path to show the effects of the alignment change. These types of relationships and dependencies are programmed into the objects of Civil 3D.

OBJECT DEPENDENCIES AND ICONS

Civil 3D identifies the object type, status, and dependencies with icons in Prospector (see Figure 1.6). For surfaces, the green icon to the left of a surface's name indicates the type of surface. The icons to the left of Base and Existing in Figure 1.6 identify them as Triangular Irregular Network or TIN surfaces. The icons for the remaining two surfaces identify them as a Grid and TIN volume surfaces.

The triangle icon pointing diagonally to the left of a surface's name indicates that another object in the drawing references that surface's data. When an object is referenced, it cannot be deleted until the reference is removed. An out-of-date icon (a shield with an exclamation mark), as shown in Figure 1.6, indicates something to the Existing-Base-Grid-Vol surface has changed. To remove the out-of-date markers, the user must rebuild the two volume surfaces. Rebuilding a surface accommodates the change(s) and removes the out-of-date status icon from the surface.

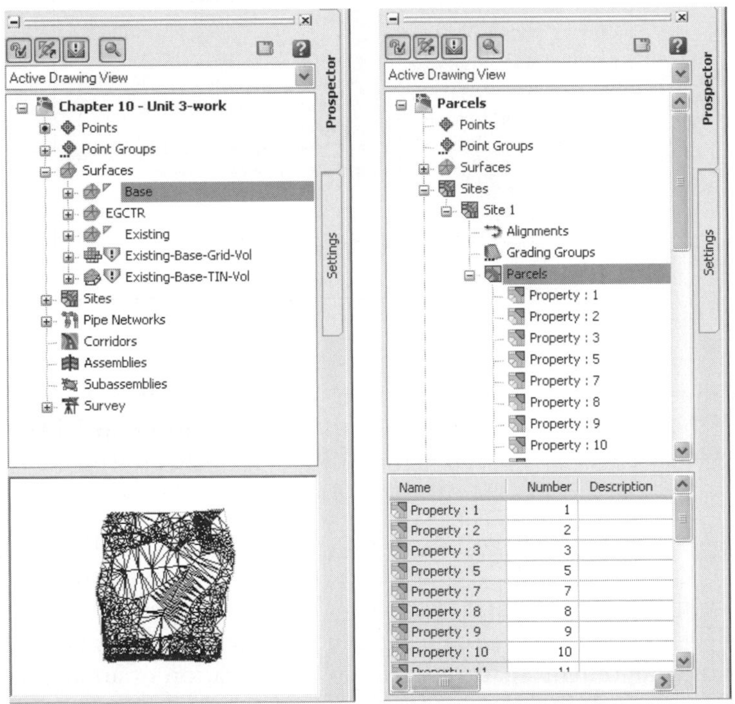

Figure 1.6

PROSPECTOR: PROJECT MANAGEMENT

Civil 3D implements project- and data-sharing capabilities through Prospector, and Autodesk Vault. Project Management is visible only in Prospector's Master view. The user sets the Master view by selecting it from the view list above the name of the drawing (see Figure 1.7).

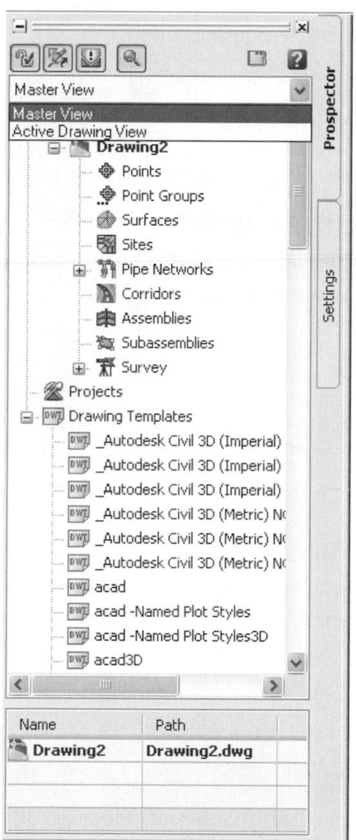

Figure 1.7

The Projects section within Prospector assists in object-level management of a project's data. The main method of control is requiring users to check in or check out data from a project. This allows a single user to have editing control over the data. Other users can reference (consume only) the original data and, if they edit the data, Prospector notifies them that the data is now out-of-sync relative to the project's entries. If the data the user is using changes in Vault, the user is notified that the data has changed and can synchronize his or her data to the new data. Again, Prospector uses icons to indicate the changing status of data in the current drawing and in the project.

Civil 3D Project management is discussed in detail in Chapter 12.

CIVIL 3D TEMPLATES

Civil 3D ships with several content templates. When using Prospector's Master view, it contains a list of template files the user can use to start a new drawing. Rather than selecting New from the File menu, in Prospector's Master View the user can select a template from a list of templates, press the right mouse button, and select Create New Drawing (see Figure 1.8).

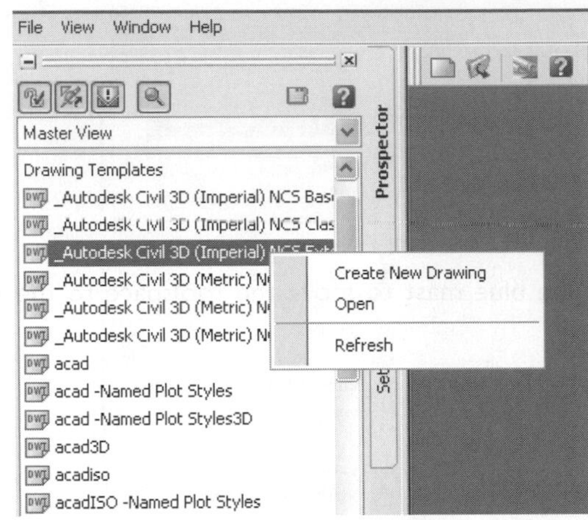

Figure 1.8

This textbook works with the Autodesk Civil 3D (Imperial) NCS Extended template file. This content template assigns needed layer names and styles for most objects. Several of the exercises will modify, create new styles, and define new layers in addition to those in the original template file.

EXERCISE 1-1

After completing this exercise, you will:

- Become familiar with the Toolspace and Prospector.
- Observe how Prospector dynamically manages objects.
- Change an alignment's path and view the changes to the road model.
- Dissolve and add parcels to a site design and view the changes to the Parcels data tree.

This exercise works with drawings from the CD that accompanies this textbook. The drawings are in the Chapter 1 folder of the CD.

TOOLSPACE BASICS

1. If not in Civil 3D, start the application by double-clicking the desktop icon.
2. If the Prospector Toolspace is not showing, display it by selecting **Show Toolspace...** from the General menu.
3. Close any open drawings and do not save them.
4. Select the drop-list arrow under the Toolspace icons and change the view to **Master View**.
5. Expand the Drawing Templates branch, select the template _**Autodesk Civil 3D (Imperial) NCS Extended**, press the right mouse button, and select **Create New Drawing** from the shortcut menu.

6. If the Toolspace is not floating, click and hold the left mouse button down on the double gray lines at the top of the Toolspace, and move it to the center of the screen.
7. Right mouse click the blue mast of the Toolspace, and in the shortcut menu toggle **OFF** Allow Docking.
8. Again, right mouse click the blue mast of the Toolspace and in the shortcut menu, if toggled on, toggle **OFF** Auto-hide.
9. Click and hold the blue mast to move the Toolspace to the right and left sides of the screen.

   The blue mast of the Toolspace switches from side to side. The preview area of Prospector is to one side or the other of the Toolspace.
10. Click the blue mast, press the right mouse button, and in the shortcut menu toggle **ON** Auto-hide.

    The Toolspace hides by rolling up under the mast.
11. Click the blue mast, press the right mouse button, and in the shortcut menu toggle **ON** Allow Docking, and dock the Toolspace on the left side of the screen.
12. If necessary, click the **Prospector** tab to view its panel.
13. If necessary, expand the drawing's object type hierarchy by clicking on the expand tree icon (plus sign) to the left of the drawing's name.
14. Adjust the size of the preview area by placing the cursor over the boundary between Prospector and the preview area, pressing and holding down the left mouse button, and sliding it up or down to size it.
15. Select the drop-list arrow under the Toolspace icons and change the view to **Active Drawing View**.
16. From the File menu, select **Close** and do not save the drawing.

EXERCISES

PROSPECTOR OBJECT MANAGEMENT

Prospector is the control center for Civil 3D. Prospector dynamically updates its object list as the user manipulate's the objects in the drawing. Prospector also manages the object's data and makes the data available either in an object type branch or the preview area.

1. Open the drawing *Overview Prospector*. It is on the CD in the *Chapter 1* folder that accompanies this textbook.
2. Make sure Prospector is showing by selecting the Toolspace's **Prospector** tab.

Points

1. In Prospector select the **Points** heading.

The square icon with a black dot indicates there are points in the drawing. When selecting an object heading, Prospector displays a list of object instances in the preview area (see Figure 1.9). In this case, Prospector lists all of the points in the drawing in the preview area. You can edit any value in the preview area.

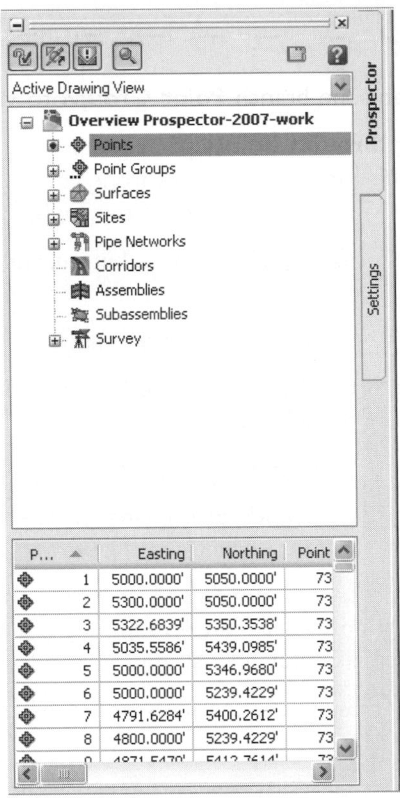

Figure 1.9

Each heading in the data tree has a specific shortcut menu (see Figure 1.3). The number of commands on the shortcut menu depends on the object's complexity or the user's location in the Prospector data tree.

2. With **Points** still highlighted, press the right mouse button and select **Create...** from the shortcut menu.

   This displays the Create Points toolbar.

3. Close the Create Points toolbar by clicking on the red "**X**" in the upper-right corner of the toolbar.

4. In the Prospector's preview area, double-click a point number.

   By clicking any entry in the preview area, you can edit or change the style applied to the object or change an object's value.

Point Groups

1. In Prospector, click the heading **Point Groups** and notice the list of point groups in the preview area (in the lower portion of Prospector).

   Each point group can have a point and/or a label style attached to it. Consequently, to the members of the group, this assignment overrides any point and label styles originally assigned.

2. In Prospector, expand the branch **Point Groups** to view the point group list.
3. Click the **Existing Ground Points** point group.

   This displays a list of the group's points in the preview area.

4. With **Existing Ground Points** still highlighted, press the right mouse button and select **Edit Points...** from the shortcut menu.

   This displays a panorama containing a Point Editor vista for all points in the point group. You can edit and change their values in the Point Editor.

5. Close the panorama by clicking on the "**X**" or green check mark in the upper-right of the panorama.

   The property of the Point Groups heading is the display order of all point groups in the drawing (see Figure 1.10). The display of point groups in the drawing begins with the list's lowest point group and continues to the top.

   The Display Order property allows you to "hide" or change point markers and labels so you can view or display different combinations of point data. In the current drawing, the No Show point group turns off label visibility if it is the last drawn point group.

EXERCISES

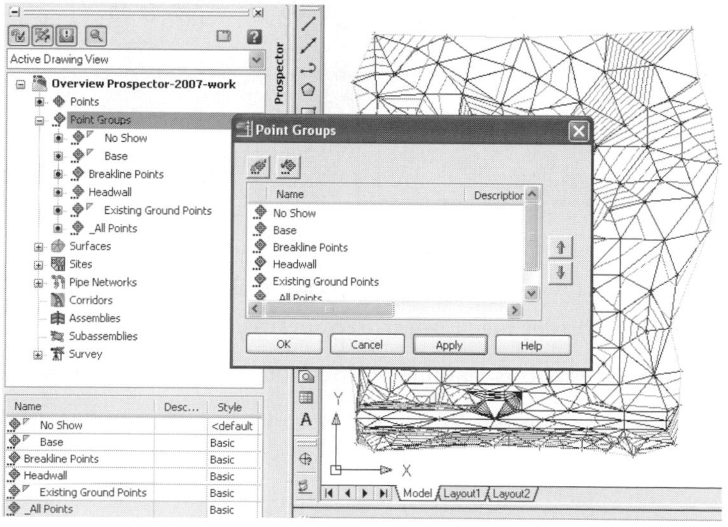

Figure 1.10

6. In Prospector, click the **Point Groups** heading, press the right mouse button, and from the shortcut menu select **Properties....**

   No Show is the top group and suppresses all point labels. To see different point groups, change their position relative to the No Show group using the up and down arrow buttons at the right side of the dialog box. The point groups you want to see should be above No Show.

   If the Event Viewer displays, close it by clicking the green check mark in the upper-right of the panorama.

7. Click **Existing Ground Points** and move it to the top position.
8. If necessary, click **No Show** and move it to the second position.
9. Click the **OK** button to view the Existing Ground Points point group on the screen.

   The drawing displays all of the points in the Existing Ground Points point group and their assigned point and label styles (see Figure 1.11).

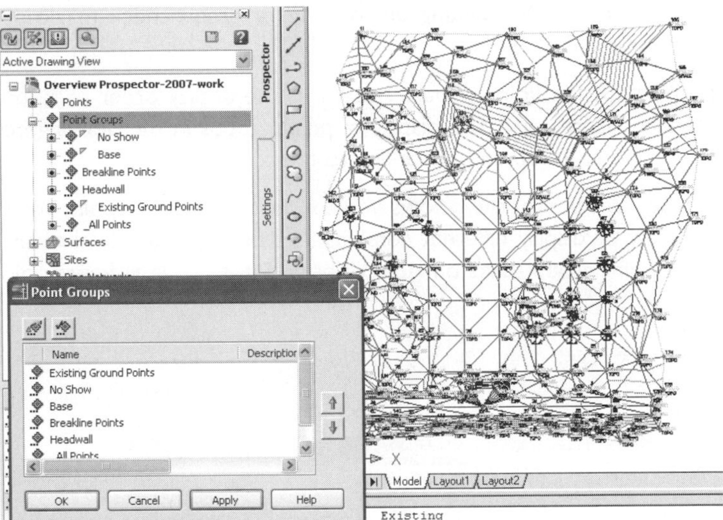

Figure 1.11

10. Again, in Prospector, click the heading **Point Groups**, press the right mouse button, and from the shortcut menu select **Properties....**
11. Select the point group **Breakline Points** and move it to the top position.
12. Select the point group **No Show** and move it to the second position.

13. Click the **OK** button to view the Breakline Points point group on the screen.

    This combination displays Breakline Points point group (see Figure 1.12). By isolating these points and the triangulation of the surface you can review the "successful" triangulation of linear objects on the surface represented by the displayed points.

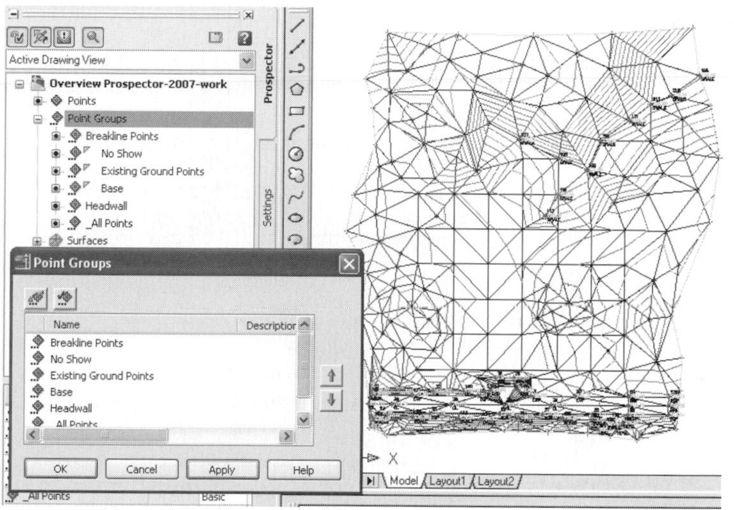

Figure 1.12

14. Close the current drawing and do not save the changes.

CIVIL 3D OBJECT REACTIONS AND DEPENDENCIES

All Civil 3D objects react to changes affecting their display or the information they show. When changing data that an object depends on, or changing the object itself, all objects with dependencies react and accommodate the change.

Alignments and Profiles

1. From the File menu, select **Open...** and select the *Overview Road* drawing from the *Chapter 1* folder of the CD that accompanies this textbook.

   When changing the end of an alignment, the profile accommodates the edit by displaying new elevations and alignment length.

2. Use the AutoCAD Zoom and Pan commands to view the right side of the profile view to see its elevation annotation.
3. If necessary, click the **Prospector** tab to make it current.
4. In Prospector, expand the branch **Surfaces** until viewing the surface **EG**.
5. Expand the **EG** branch until viewing the data entries under the **Definition** heading.
6. From the list, select the heading **Edits**, press the right mouse button, and select **Raise/Lower Surface** (see Figure 1.13).

EXERCISES

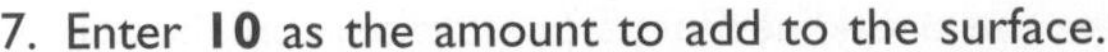

7. Enter **10** as the amount to add to the surface.

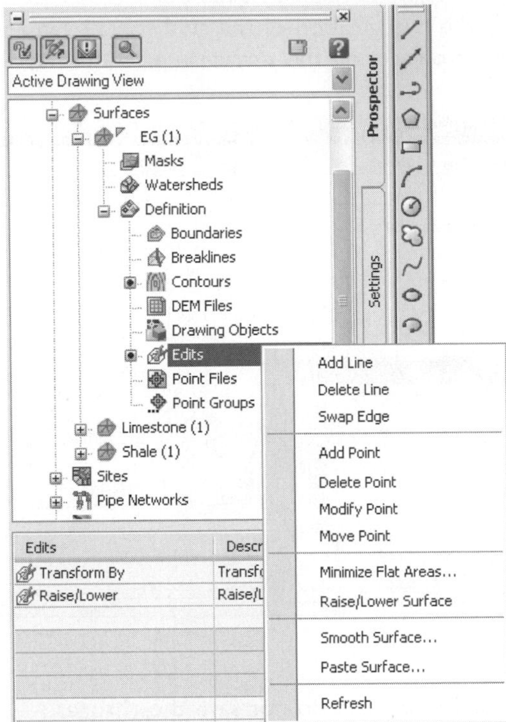

Figure 1.13

The edit raises the surface by 10 feet. The profile view reacts to the change by updating its elevation annotation to reflect the new surface elevations. A surface and a profile view are linked by the alignment and profiles. If something should happen to either the surface or the alignment, the profile and profile view react and change to correctly show the new situation.

8. Select **Close** from the File menu and do not save any changes to the drawing.

Surface and Contours

When a user changes a surface's data, the surface reacts to the change by updating its definition. If the current style displays contours, the contours change to show the new surface definition.

If the Event Viewer displays, close it by clicking the green check mark in the upper-right of the panorama.

1. From the File menu select **Open...** and select the *Overview Surfaces* drawing from the *Chapter 1* folder of the CD that accompanies this textbook.
2. In the drawing, click any existing surface contour, press the right mouse button, and from the shortcut menu select **Surface Properties...**.

EXERCISES

3. In the dialog box, click on the **Information** tab. To the right of Surface Style click on the drop-list arrow to view a style list, and from the list select **Contours and Triangles** to make it the current surface style (see Figure 1.14).
4. Click the **OK** button to exit the dialog box.

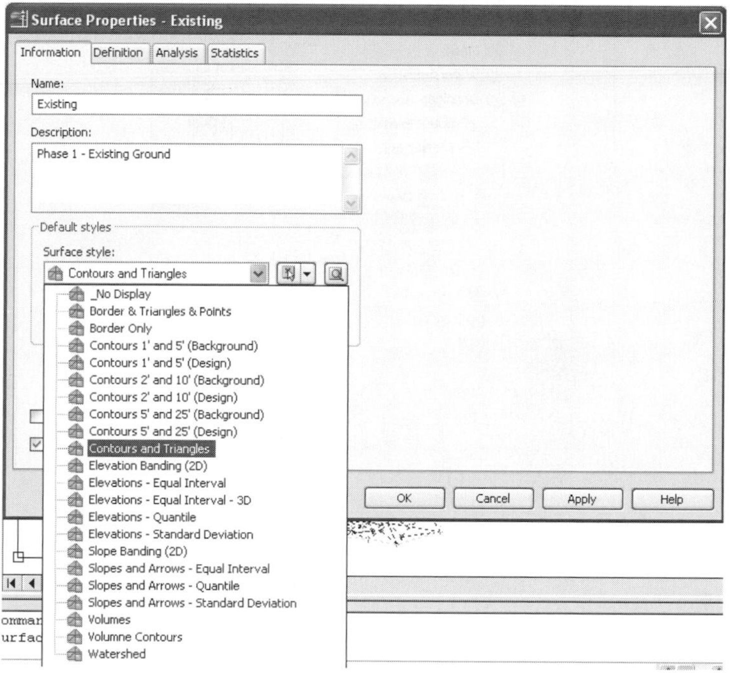

Figure 1.14

The surface should now show the contour and triangle surface components. The next step zooms to the point you are going to modify. After you modify the point, the contours will change to reflect the new surface elevation.

5. At the top of Prospector, click the heading **Points** to list all of the points in the preview area.
6. In the preview area, scroll through the point list until point number **71** is in view.
7. In the preview area, click the point icon to the left of point **71**, press the right mouse button, and select **Zoom to** from the shortcut menu.

   The point is now centered on the screen.
8. In Prospector, click **Surfaces** and expand the **Existing** branch until the Definition heading and its list of entries is in view.

   Your Prospector should look like Figure 1.15.

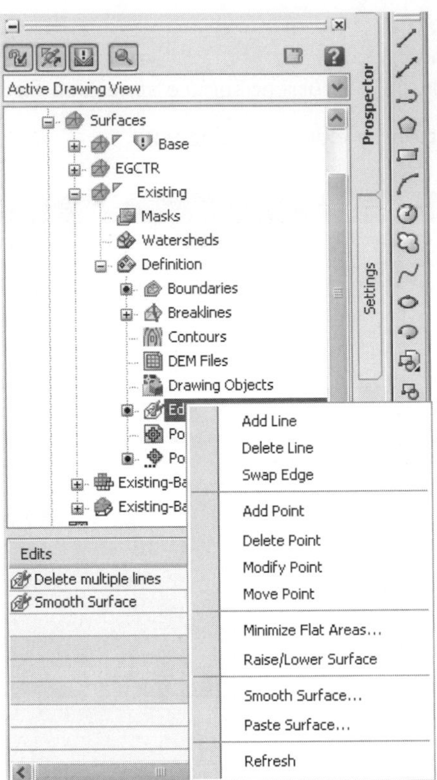

Figure 1.15

9. In the Definition tree, select **Edits**, press the right mouse button, and from the shortcut menu select **Modify Point**.

10. In the drawing, click the intersection of the triangles at the center of the screen, press the **Enter** key, assign a new elevation of **728** to the point, and press the **Enter** key twice to exit the command.

    This edit changes the surface data. The contours change to show the new surface elevation and the Prospector's preview panel shows the edit added to the surface edits list (see Figure 1.16). The surfaces dependent on Existing become out of date because of the change made to Existing.

    If the previous surface edit is speculative, you can temporarily remove the edit from a surface. You do this in the Definition panel of the Surface Properties dialog box. If you want to permanently remove the edit, you can delete it either in the Surface Properties dialog box (Remove) or delete it from the surface edits list of the Prospector's preview panel.

11. In the drawing, click on a contour or triangle leg, press the right mouse button, and from the shortcut menu select **Surface Properties...**.

EXERCISES

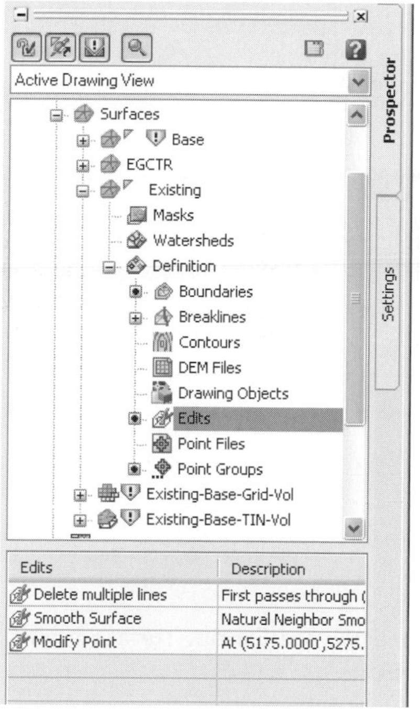

Figure 1.16

12. In the Surface Properties dialog box, click on the **Definition** tab; then at the bottom of the panel in Operation Type select an edit from the list, and press the down arrow key to scroll until you locate the Modify Point operation.

    The operation has a check mark indicating it is an active edit for the surface.

13. Toggle **OFF** the Modify Point edit and click the **OK** button to exit the dialog box.

    A warning dialog box displays.

14. In the Warning dialog box, click the **Yes** button to rebuild the surface, and exit the Surface Properties dialog box (see Figure 1.17).

    The surface shows the modified point removed from the surface data by changing the location of the surface contours.

15. Return to the Surface Properties dialog box (see Step 11). In the Operation Type area of the Definition panel, toggle **ON** the Modify Point edit, and click the **OK** button to exit the dialog box.

16. In the Warning dialog box, click the **Yes** button to rebuild the surface and exit the Surface Properties dialog box.

    The edit reappears in the surface.

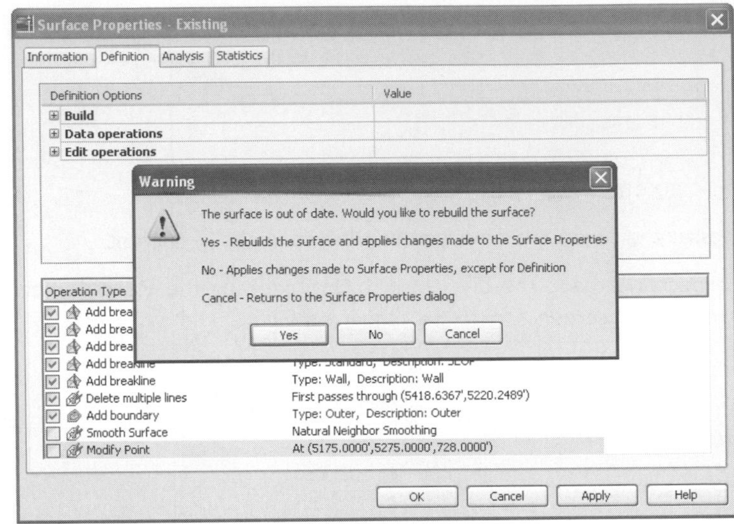

Figure 1.17

17. If necessary, in Prospector, the Existing surface branch, from the list below the Definition heading select **Edits**.

    This displays the existing surface edits in Prospector's preview area.

18. In the preview area, click the **Modify Point** entry, press the right mouse button, and select **Delete...** from the shortcut menu to permanently remove the edit from the surface.

    A Remove From Definition dialog box displays (see Figure 1.18).

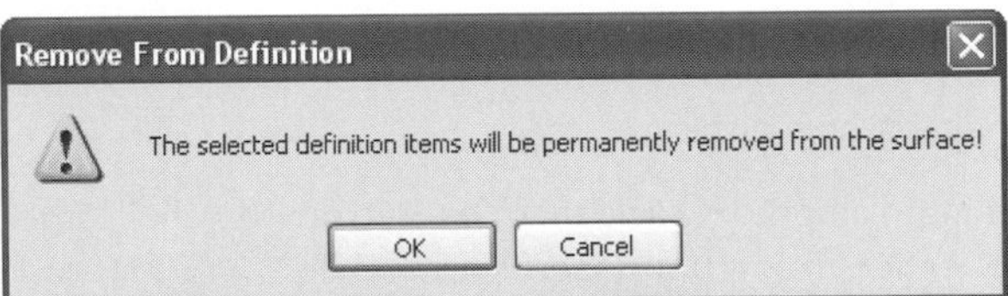

Figure 1.18

19. In the Remove From Definition dialog box click the **OK** button to accept the deletion.

    The edit is permanently removed from the surface.

20. Select **Close** from the File menu to exit the drawing. Do not save your work.

DYNAMIC DATA MANAGEMENT

Prospector dynamically manages an object's list in a drawing. Each list entry is an instance (occurrence) of that object type in the drawing. When adding or removing drawing objects, Prospector automatically updates the object's list.

Prospector Preview

The Parcels list for a site manages all site parcels, identifies their type, and displays their identifier (usually a parcel number). When defining a parcel, Prospector adds the new parcel to the list and labels it in the drawing.

1. In Civil 3D, open the *Overview Parcels* drawing located in the *Chapter 1* folder of the CD that accompanies this textbook.
2. If necessary, click on the **Prospector** tab to make it current.
3. In Prospector, expand the Sites branch until viewing the **Parcels** heading and its list of parcels. Your screen should be similar to Figure 1.19.

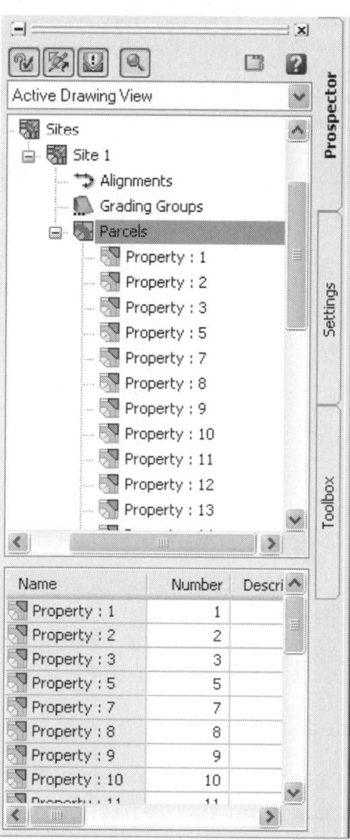

Figure 1.19

The property of each parcel includes an analysis about its boundary (see Figure 1.20).

4. In Parcels, select a parcel from the list, press the right mouse button, and from the shortcut menu select **Properties....**
5. Click on the Analysis tab and view both the **Inverse** and **Mapcheck** reports for the selected parcel.

EXERCISES

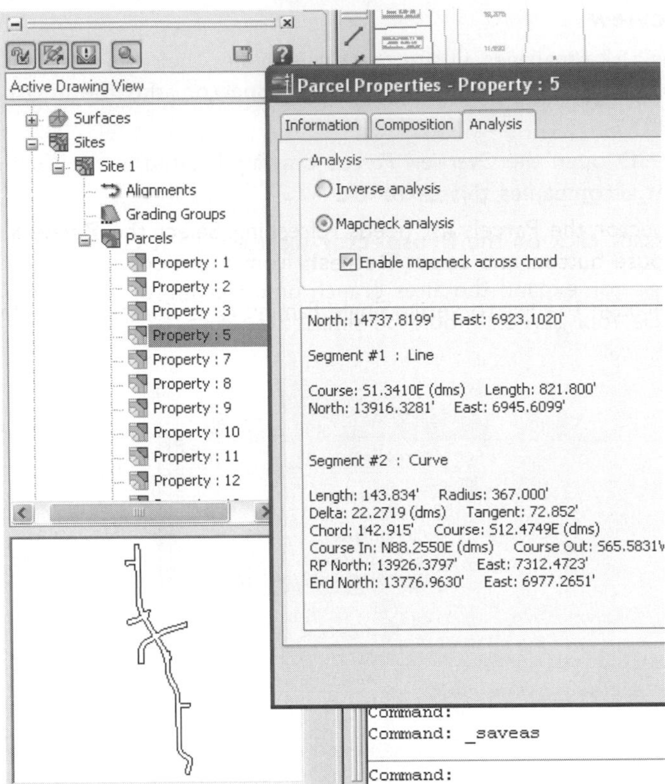

Figure 1.20

6. Click the **OK** button to exit the Parcel Properties dialog box.

   Reports Manager has several parcel reports.

7. From the General menu, select **Reports Manager...**; in the Reports Manager Toolspace, expand the Reports Manager and the Parcel branches.

   A list of parcel reports displays.

8. From the list of parcel reports, select **Inverse_Report**, press the right mouse button, and select **Execute...** from the shortcut menu.
9. In the Export to LandXML dialog box, unselect everything except for two parcels, and click the **OK** button.
10. Internet Explorer displays with a Parcel Inverse Report.
11. Close Internet Explorer.

Adding and Deleting Parcels

Whenever adding or deleting parcels, Prospector modifies its Parcels list.

1. If necessary, click on the **Prospector** tab to make it current.

2. From the Parcels list, select **Single Family: 17**, press the right mouse button, and select **Zoom to** from the shortcut menu.
3. Start the **AutoCAD Erase** command, and in the drawing select the north/south side yard line dividing parcels 4 and 17 and press the **Enter** key.

   The two parcels merge into one.
4. In Prospector, the Parcels list needs refreshing. Select the **Parcels** heading, press the right mouse button, and select **Refresh** from the menu.
5. Scroll through the list to verify Single Family: 17 is no longer on the list.
6. From the File menu select **Close** and do not save the changes to this drawing.

EXERCISES

SUMMARY

- Prospector is the command center and data manager of Civil 3D.
- Civil 3D objects are dynamic and update their display when edited.
- Prospector is Civil 3D's object manager.
- Prospector adds and removes objects from its lists as the user creates and deletes objects in the drawing.
- Prospector displays and manages each object's status (out of date, reference, locked, etc.).
- Civil 3D displays dependencies and references in the Prospector data panel.
- Civil 3D does not allow the user to delete objects that have dependencies on their data.
- Selecting a heading displays a list or image in the preview area.

UNIT 2: SETTINGS

The main function of the Settings hierarchy is managing all of the settings and styles in a drawing. At this highest level the settings affect all of the styles and values that are lower in the hierarchy. For example, the values in the Edit Settings dialog box (located at the very top of Settings) affect drawing scale, coordinate systems, label abbreviations, and layers in a drawing.

In Figure 1.21 the Object Layers panel sets the base layer name for each object type, if using a modifier, and the modifier's value. If there is more than one instance of an object type, the user should define a modifier for the base layer. A modifier can be a prefix or suffix to the base layer name with the value of the modifier being the instance's name (for example, Existing and Base are two TIN surface names that can append to the base layer name). An asterisk (*) uses the name of the object as the modifier value. There also needs to be a spacing character between the base layer name and the modifier (such as a dash (-) or an underscore (_)). If a surface's name is Existing, and the user sets the modifier to Suffix and sets its value to -* (a dash and an asterisk), the resulting name of the surface layer is C-TOPO-Existing.

The Ambient Settings panel of the Edit Drawing Settings dialog box sets basic rules for units, rounding, precision, format, etc. For example, to use cubic feet as the unit for volumes, in the Volume section of the Ambient Settings panel change the Unit value, and all listings and calculations will use cubic feet as the base unit (see Figure 1.22). The settings for reporting and displaying a grade or slope are set by the Grade/Slope Section settings. Any slope or grade will use 2 units of precision, rounded normally; will have a format showing a percent sign; and will use a negative sign if negative.

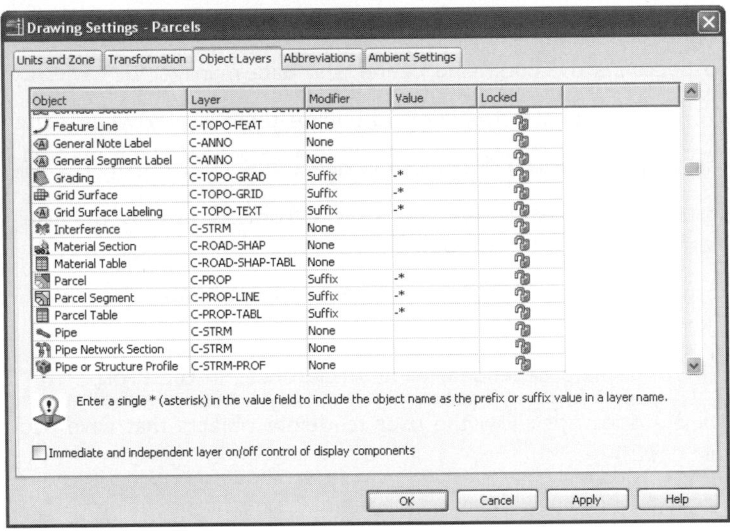

Figure 1.21

Any change made to settings in an object type's branch affects only those settings and styles below to which the change is made. For example, when creating a surface slope label style "Run over Rise," this particular style uses a ratio slope format and is different from the format set in the Ambient Settings panel. The change affects only the style and nothing else in the Surface branch. The reason is that this style is at the end of the Label Styles branch of the Surface branch (see Figure 1.23).

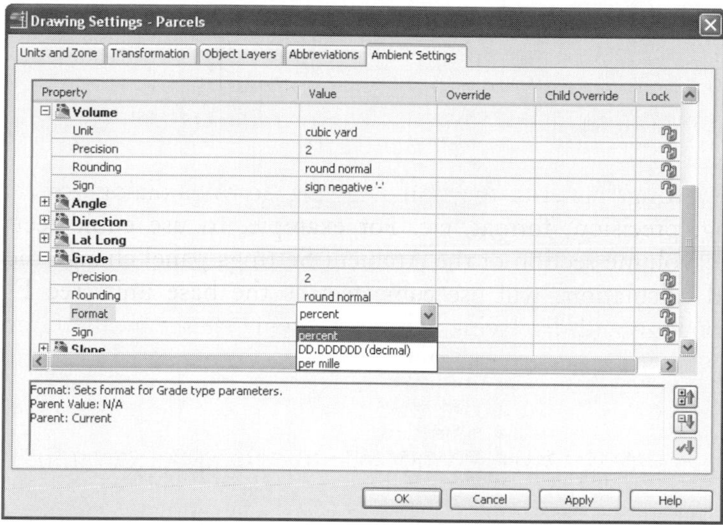

Figure 1.22

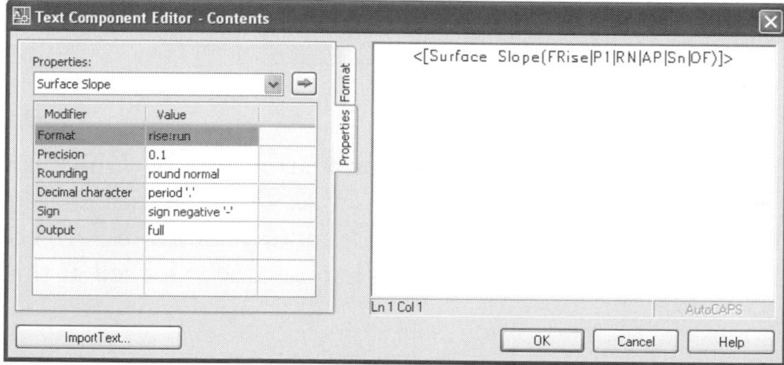

Figure 1.23

The Settings panel displays all settings and styles for each object type. Each object type is a branch heading and the branch contains the settings and styles for that object type. When selecting an object type while pressing the right mouse button, Civil 3D presents a shortcut menu allowing the user to change certain feature (object) values, label style defaults, and so on (see Figure 1.24).

EDIT LABEL STYLE DEFAULTS

The values in this dialog box affect the basic behavior of all labels in a drawing (see Figure 1.25). In the figure, some of the settings are overridden by "lower" styles (down arrows in the Child Override column). The reason for style overrides is that the lower styles may behave, label, or report values differently and may require changes to the values set at this level.

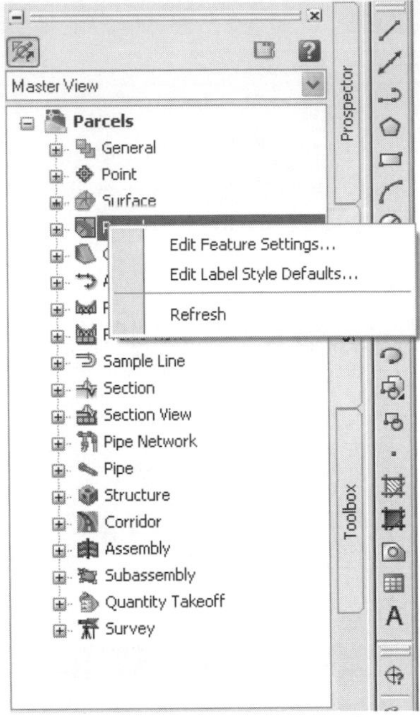

Figure 1.24

Users can negate overrides by clicking the down arrow in the Child Override column, which changes the icon to a down arrow with an X. The X indicates that the override is reset to the values in the Edit Label Style Defaults dialog box. A second method of controlling overrides is by locking the value. When a value is locked at this level, no lower style can change the value.

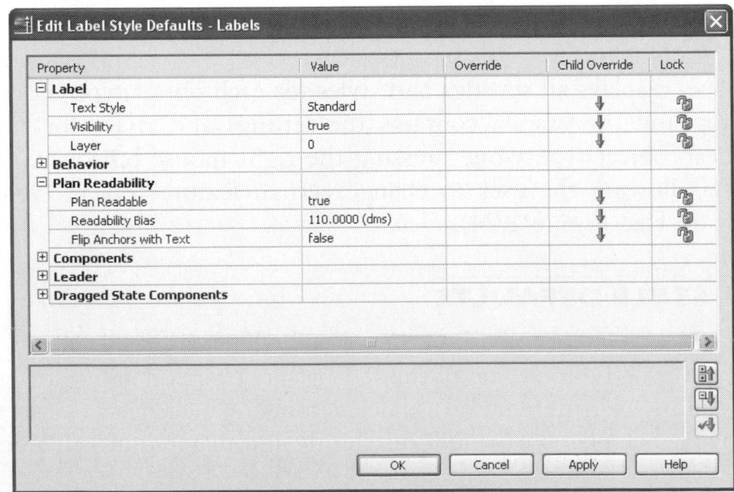

Figure 1.25

EDIT LANDXML SETTINGS

The LandXML Settings dialog box affects data importing and exporting to and from civil applications using LandXML files (see Figure 1.26). LandXML files transfer design elements (surfaces, alignments, points, etc.) between applications without loss of fidelity. The LandXML civil data schema allows Autodesk Land Desktop, Civil 3D, and other applications to transfer data without losing information quality. Civil 3D also uses LandXML files as report data.

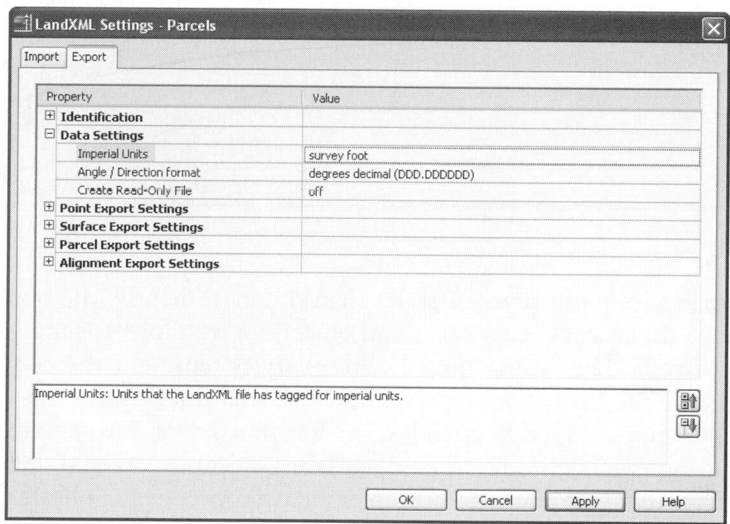

Figure 1.26

EDIT FEATURE SETTINGS

In Settings, when selecting an object type heading while pressing the right mouse button, Civil 3D displays a shortcut menu containing a call to the Edit Feature Settings dialog box. The Edit Feature Settings dialog box combines the settings of the Ambient Settings panel of Edit Drawing Settings with specific settings for the selected object type (see Figure 1.27). In the figure, there are four sections affecting points: Default Styles, Default Name Format, Update Points, and Point Identity. The number of sections appearing in an Edit Feature Settings dialog box depends on the complexity of the object.

In Settings, when expanding an object type's branch, Civil 3D lists the major style types as headings. The major style types include objects, labels, tables, and command settings. The number of types depends on the complexity of the object. Each heading has a shortcut menu containing commands appropriate to the selected heading (see Figure 1.28). For example, selecting the Description Key Sets heading displays a shortcut menu displaying commands to create a new Description Key Set or to view a set's properties. Selecting a named Description Key Set while pressing the right mouse button displays a menu containing commands to view the key set's properties, to copy it, to delete it, or to edit its keys.

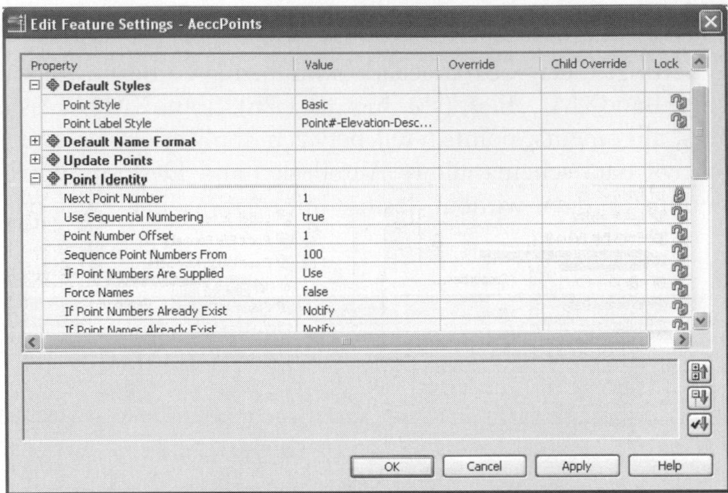

Figure 1.27

Some object types have more types of styles than others to handle the object's complexity or to help manage the object's data. The point object, for example, has marker styles, labels, tables, and commands. The point object (marker) styles represent the coordinate location and/or symbols, and the label styles represent the information a user wants to display about a point. The Description Key Sets styles are for translating raw descriptions into full descriptions, sorting points onto layers, and assigning a point marker style to points. Other styles define external data references or user-defined properties for points. The settings in the Commands branch set values and assign styles used by commands that create object instances (i.e., points, surfaces, etc.).

When selecting a specific style name while pressing the right mouse button, a shortcut menu displays containing commands to Edit..., Delete..., or Copy... the selected style (see Figure 1.29).

To add new styles to an existing template or drawing, drag and drop them between open files. When dragging and dropping a style definition into a file where the style already exists, Civil 3D issues a warning and prompts the user to rename or overwrite the existing definition.

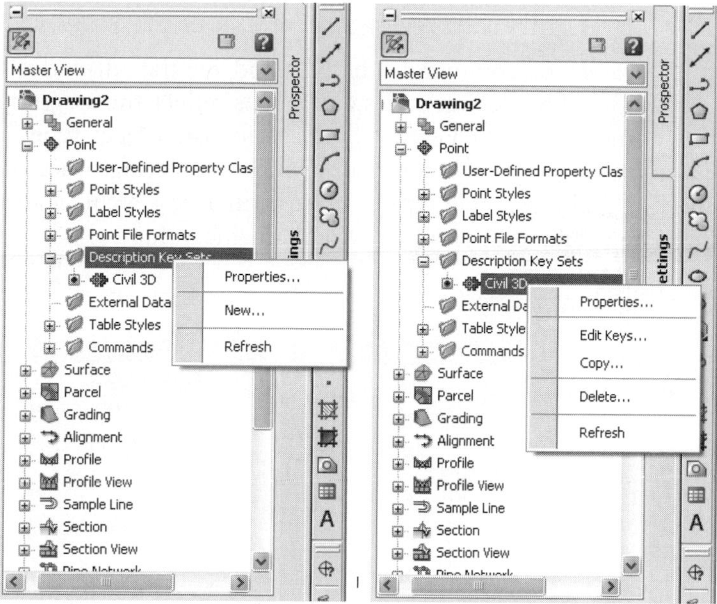

Figure 1.28

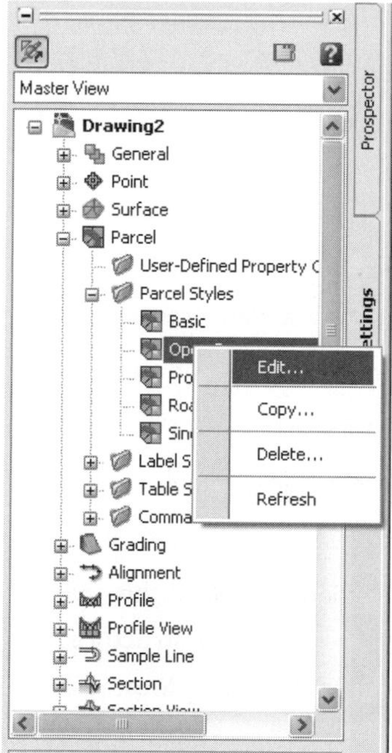

Figure 1.29

EDIT COMMAND SETTINGS

Each object type has a set of commands that depend on default styles and settings to function (see Figure 1.30). The default styles and values reflect the assignments and values of Edit Drawing Settings and Edit Feature Settings for the object type. You can change the default styles and settings for each command without affecting other commands or the feature settings. In this and other chapters, exercises will access a command settings dialog box, evaluate its settings, and if appropriate change some values.

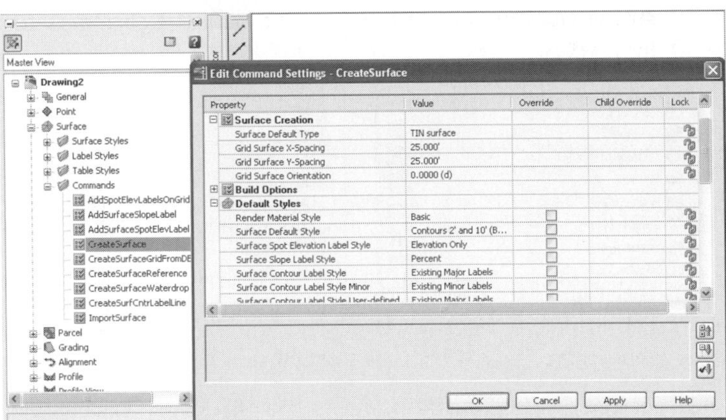

Figure 1.30

EXERCISE 1-2

After completing this exercise, you will:

- Review the Prospector data panel.
- Expand and review the Surface data tree.
- Review Prospector icons for status and dependency.
- Review a panorama and its vistas.
- Become familiar with the Settings panel.

PROSPECTOR SETTINGS

The Settings panel provides management for all of the styles and settings within a drawing. This exercise familiarizes you with the hierarchy structure and types of styles.

1. Open the *Overview Surfaces II* drawing located in the *Chapter 1* folder on the CD that accompanies this textbook.
2. Click on the Setting tab to view its contents.

   The Edit Drawing Settings..., Edit Label Style Defaults..., and Edit LandXML Settings... dialog boxes occupy the top of Settings.

3. At the top of Settings, click the name of the drawing, press the right mouse button, and from the shortcut menu select **Edit Drawing Settings....**
4. In the Edit Drawing Settings dialog box, click the **Units and Zone** tab to view its contents.
5. Click the **Transformation** tab to view its values.
6. Click the **Object Layers** tab and scroll through the layer list to review the layer names settings.
7. Click the **Abbreviations** tab and expand the different segments to view the base abbreviations Civil 3D uses.

   You can change any abbreviation listed in the panel.
8. Click the **Ambient Settings** tab to view its settings.

   All of these values set the general tone for the drawing. Clicking a padlock icon locks the value so that no lower style can override or change its locked value.
9. Click the **OK** button to exit the Edit Drawing Settings dialog box.
10. In Settings, click the **Surface** heading, press the right mouse button, and from the shortcut menu select **Edit Feature Settings....**

    The Edit Feature Settings dialog box sets the default styles and naming values for Surfaces.
11. Click the **OK** button to exit the dialog box.
12. In Settings, click the **Surface** heading, press the right mouse button, and from the shortcut menu select **Edit Label Style Defaults....**

    The Edit Label Style Defaults dialog box sets several default values for all surface label styles.
13. Click the **OK** button to exit the dialog box.
14. In Settings, expand the **Surface** branch by clicking the plus sign to the left of Surface.

    The Surface heading has four branches; Surface Styles, Label Styles, Table Styles, and Commands.
15. Click the expand tree icon to the left of Surface Styles to view the list of styles.
16. Click the style **Border & Triangles & Points**, press the right mouse button, and from the shortcut menu select **Edit....**

    This displays the Surface Style dialog box. This is where you manipulate the values to create the style.
17. Click on the tabs and expand some of the sections in the different panels to view the types of values you can set for a style.

18. Click the **OK** button to exit the dialog box.
19. Explore some of the remaining style trees and view some of their styles.
20. Close the drawing and do not save the changes.

SUMMARY

- The Settings panel manages all styles and commands.
- Using the values in the Settings panel promotes standards and eases implementation by using a hierarchical structure.
- In the Settings hierarchy, the higher a setting is, the more styles and values it affects in the drawing.
- Settings manages, creates, and modifies object and label styles.

EXERCISES

UNIT 3: THE CIVIL 3D ENVIRONMENT

Autodesk Civil 3D uses new terms to describe the drafting environment, its objects, and their behaviors (e.g., Prospector, vistas, baselines, assemblies, styles, etc.). Civil 3D sites, baselines, feature lines, and assembly objects are familiar civil concepts and design elements, but these elements are known by more traditional names in other civil design packages (parcels, alignments, grading objects, templates, etc.). The new terms indicate that these familiar design elements have new or expanded capabilities.

Civil 3D uses styles and settings to graphically display the design, set design limits (criteria), produce reports, and create design documentation. Civil 3D ships with basic content and style definitions in its template files. However, these styles may not be right for a user's specific tasks, or they may not meet a user's CAD drafting standards. The biggest initial cost to implementing Civil 3D will be the time spent in setting up and modifying styles. As with any Autodesk product, there are several methods for creating this content, and each strategy has consequences. Harnessing the interplay of the dynamic design environment and the role styles play in creating and finishing a design are the greatest challenges facing implementers and users of Civil 3D.

Civil 3D styles and settings are contents of a template file. If satisfied with the template content as it was shipped, the user can immediately begin to produce a design document. If the content does not reflect specific standards, the user must create new styles, modify their settings, and modify existing styles to create content for the template file. When using this new template file, all of the modified content is present in the drawing.

The interface that controls, displays, or edits information is consistent for all object types. The object and label styles use similar dialog boxes that have a set structure. By knowing the anatomy and behavior of one object or label style dialog box, a user will know the basics for most of the remaining object type and label styles. What changes is the information available about the object or for the label.

IMPLEMENTATION

A single AutoCAD template file is the ideal implementation of Civil 3D. This file defines all object, label, and miscellaneous styles producing design documentation. This single file contains the office standards that in LDT took several folders and files. When starting a new Civil 3D drawing with this template, all the objects, styles, and settings are there, ready for use. The only remaining settings are the model space plotting scale and a coordinate zone.

A Civil 3D implementation uses a combination of layers and styles. Layers can be used with traditional AutoCAD methods for displaying and hiding drawing elements. Layers are also necessary when implementing AutoCAD Xrefs or sharing data with those who do not have Civil 3D. Styles also allow control of data visibility. Users implement Civil 3D as a combination of styles with layers.

Adding style definitions to a template file refers to opening the template file and defining the new styles. To add styles from an existing drawing to a template or another drawing, open both files and drag and drop the new styles to the template or other drawing.

The biggest issue with a layer implementation is the layers displaying the component or characteristics of an object. If a drawing defines a modifier for an object type, for example the suffix -* for surfaces, Civil 3D creates layers for each surface from the base layer name (C-TOPO-EXISTING and C-TOPO-DESIGN). If a user assigns both surfaces the same object style, the surfaces use the same layer list. For example, assigning the Border & Contour style to two surfaces causes them to use the same layers. The user cannot turn off the major contours layer for one surface, because it is the same layer for both surfaces.

In Settings, the three top settings dialog boxes influence all of the settings and styles below them. These dialog boxes are Edit Drawing Settings, Edit Label Style Defaults, and Edit LandXML Settings. The values in these dialog boxes set the tone for the entire drawing. Any lower style or setting can override these values, unless they are locked at the top level. With a setting locked in any of these dialog boxes, a lower style cannot change the setting's value.

EDIT DRAWING SETTINGS

The Object Layers panel defines the base layer name for each object type with the option of adding a modifier (prefix or suffix) to the layer name when having more than one instance of an object type in the drawing (see Figure 1.21).

One modifier type uses the object instance name and appends it to the beginning or end of the base layer name. Civil 3D uses an asterisk (*) to indicate using the object's name. Using the object's name is the preferred implementation method. One additional consideration is the spacing of the object's name and the base layer. A dash (-) or underscore (_) is the usual spacing character. For example, setting a suffix modifier to a surface's name and spacing the modifier with a dash creates the layer C-TOPO-EG; the base layer is followed by the surface name separated by a dash. A second method is entering an explicit value as the modifier (e.g., C-TOPO-SURFACE). In this case, -SURFACE is the modifier value.

When a drawing has two or more object instances, using an object layer modifier is a necessity. Using surfaces as an example, if a drawing has two surfaces, EG and DESIGN, and is not defining a modifier, both surfaces use the same layer. When objects occupy the same layer, the user cannot independently control their visibility. If the user adds a modifier to the object layer name, Civil 3D creates a new layer for each object from the values of the modifier column. A drawing having two surfaces, EG and DESIGN, would have the layers C-TOPO-EG and C-TOPO-DESIGN (see Figure 1.31).

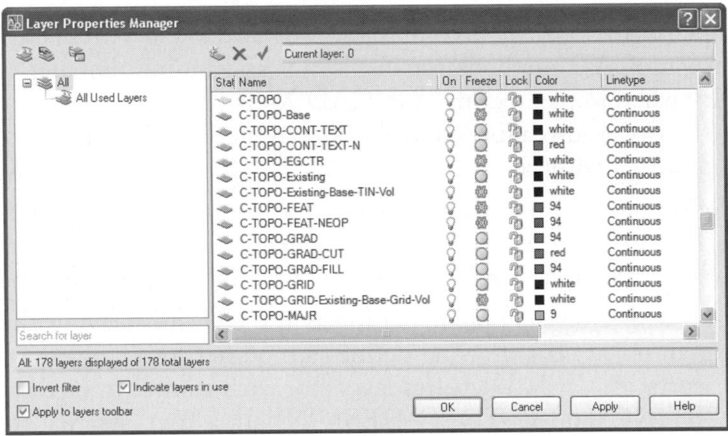

Figure 1.31

SETTING THE DRAWING ENVIRONMENT

When selecting the name of the drawing at the top of the Settings panel while pressing the right mouse button, Civil 3D displays a shortcut menu listing the three settings dialog boxes: Edit Drawing Settings, Edit Label Style Defaults, and Edit LandXML Settings. Because of the hierarchical structure of Settings, the values and settings of these three editors affect all facets of a drawing. When starting a new drawing, the first step should be checking the values of these three dialog boxes.

EDIT DRAWING SETTINGS

When selecting Edit Drawing Settings, a dialog box with five tabs appears (see Figure 1.32). These tabs affect several areas of the drafting environment.

Units and Zone Panel

The top left of the Units and Zone panel sets the base units for linear and angular measurements. There are three angular measurement values for angles; degrees, grads, or radians. Also, there are two important toggles in this area. The first is the Scale objects inserted from other drawings. All newer AutoCAD drawings (2007) contain base unit information. When a user sets the base units of the current drawing to imperial and inserts

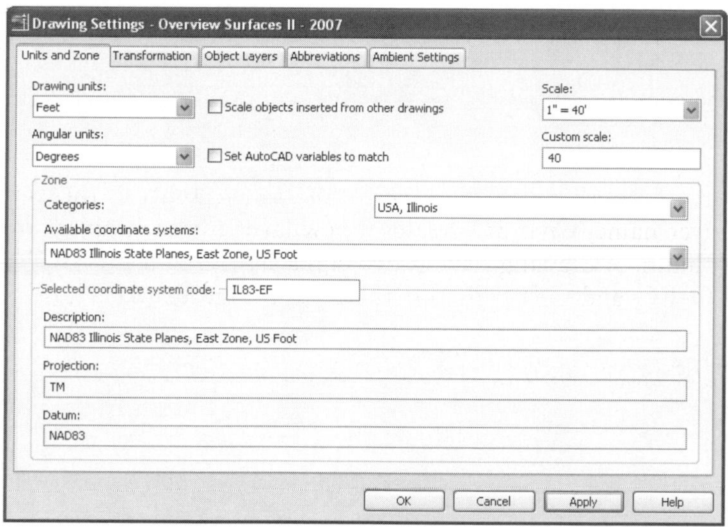

Figure 1.32

a metric drawing, Civil 3D will convert the incoming drawing objects to the current drawing's base units. The X, Y, and Z values of the metric drawing are converted into feet. So, a contour with the elevation of 780 meters becomes 2559.0551 feet (780 / 0.3048). The original coordinate X and Y values also change by the same factor.

The top right of the dialog box sets the model space plotting scale. A "standard" scale can be selected here, or a user can enter in a custom scale factor in the space below the scale drop-list entry.

The remainder of the panel sets and reports coordinate zone values, if set for the drawing. When selecting the drop-list arrow to the right of Categories, Civil 3D displays the currently supported coordinate systems. This section is important when working with state plane or latitude and longitude values.

Transformation Panel

When a user wants to set a transformation to the coordinates in the drawing, this panel is used. (see Figure 1.33). The panel is active only if there is a zone set in the Units and Zone panel. Users are allowed to define corrections for distances at sea level or a transformation to a grid.

Object Layers Panel

The Object Layers panel sets the base layer name for each object type (see Figure 1.34). The left side of the panel lists the object type. The three columns to the right of the object type set the layer name, state if the layer has a modifier, state if the modifier is a prefix or suffix, and give the value of the modifier. Any value can be added to a layer as a prefix or suffix. When using an * (asterisk) as the modifier value, Civil 3D uses the name of an object as the layer modifier. For example, a surface named EG modifies the C-TOPO base layer to C-TOPO-EG when the modifier is set to suffix and the value of the modifier is

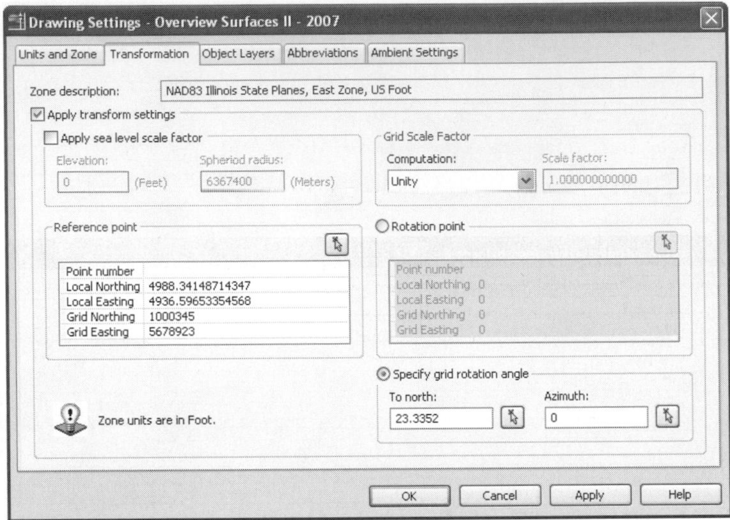

Figure 1.33

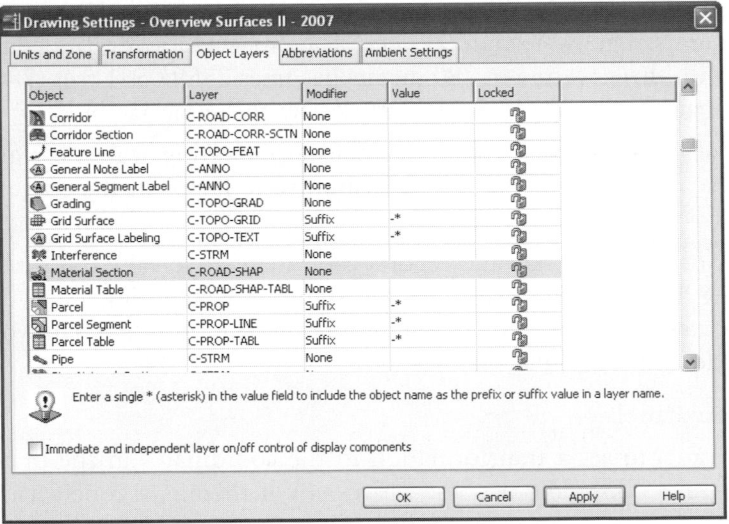

Figure 1.34

a dash asterisk (-*). The last column locks the layer name so it cannot be changed by any lower style of that object type.

Abbreviations Panel

The Abbreviations panel affects alignment, superelevation, and profile labeling (see Figure 1.35). Many of the labels for these object types have regionally accepted values and may need to change from their initial values.

The values in the Alignment Geometry Point Entity Data are a set of abbreviations with a complex format string. The format string defines how Civil 3D displays the geometry point's values. When clicking in a cell containing a format string, an ellipsis (three dots) appears at the right of the entry. Clicking the ellipsis calls the Text Component Editor, and it is within this editor that users edit the abbreviation and its format string. All label styles use this format string unless a lower style changes the format string.

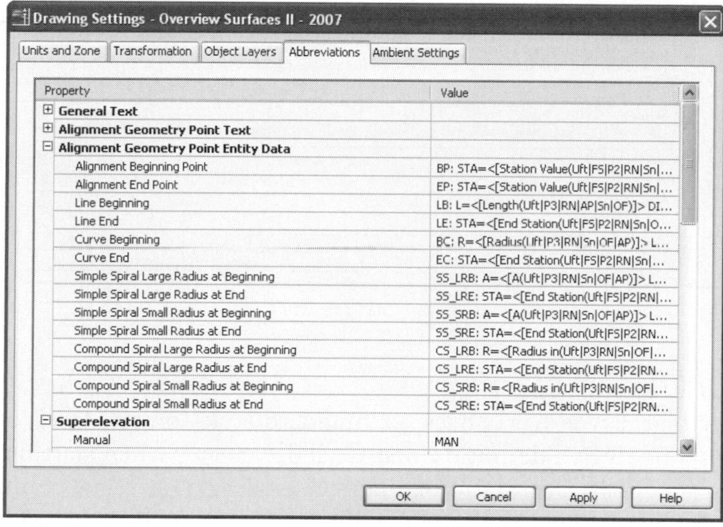

Figure 1.35

Ambient Settings Panel

The Ambient Settings panel affects a multitude of values in a drawing. Values in each section of this panel affect the units, precision, and rounding; set formats for entering and reporting values; and show how do denote a negative value, coordinates and distances, and so on (see Figure 1.36).

The Value column contains the actual setting. A value can be an entered value, a toggle, or a selection from a list of choices.

The Child Override column indicates if any lower styles change the current value of an entry. A down arrow indicates that a value is changed at a lower style in the Settings tree. The user can click on the arrow and reset the value back to the higher-level value.

The Lock column indicates if a lower style can change the current value. If locked, a lower style cannot change the value in the dialog box.

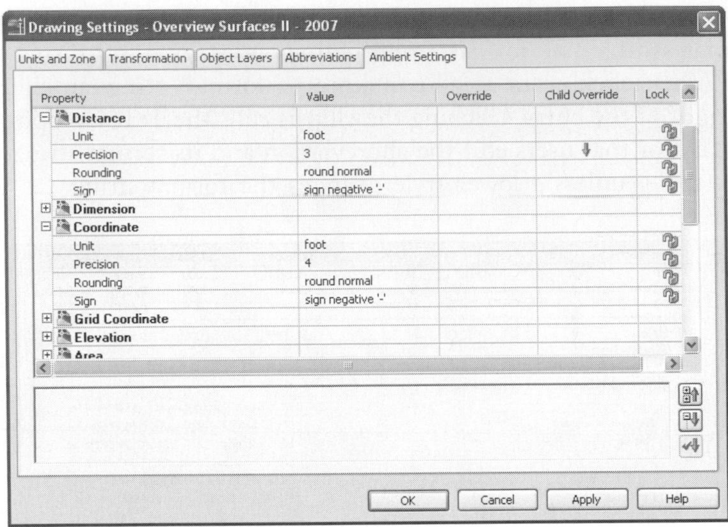

Figure 1.36

STYLES

By default every object type has a Basic or Standard style (object, label, etc.). A Basic or Standard style is a hardwired definition that Civil 3D creates when starting a new drawing without a content template. A Basic or Standard style is a starting point, and Civil 3D expects users to copy and modify it to implement new styles. If using a Civil 3D content template in which there are situations unresolved by the current styles list, the user can define new styles at any time to address these situations.

Each Civil 3D object can display its information on the screen. The information is either a component or characteristic of the object. Civil 3D uses a style to display each component or characteristic, or it can display groups of components or characteristics. A surface object, for example, has triangles, points, and a border. These components are essential to correctly view and edit a surface under development. Surface slopes and elevations are characteristics of a surface and they are essential to developing a site's design solution. A style can change the focus of the information on the screen by how it groups the components and characteristics of the object.

STYLE LAYERS

The intent of a layer is to control the visibility of the object's components and characteristics from within AutoCAD's Layer Properties Manager. The Layer Properties Manager provides a list of layers for the styles present in the drawing, and turning on and off the layers determines what is visible on the screen. This is a typical AutoCAD strategy for controlling the display of information.

The Display panel of an object style assigns a layer name and properties for each object's component or characteristic. The left side of any object style's Display panel lists the object's components and characteristics. The number of entries varies by object type: a more complex object has more entries than a simple object. Compare the layer lists in Figure 1.37 and Figure 1.38.

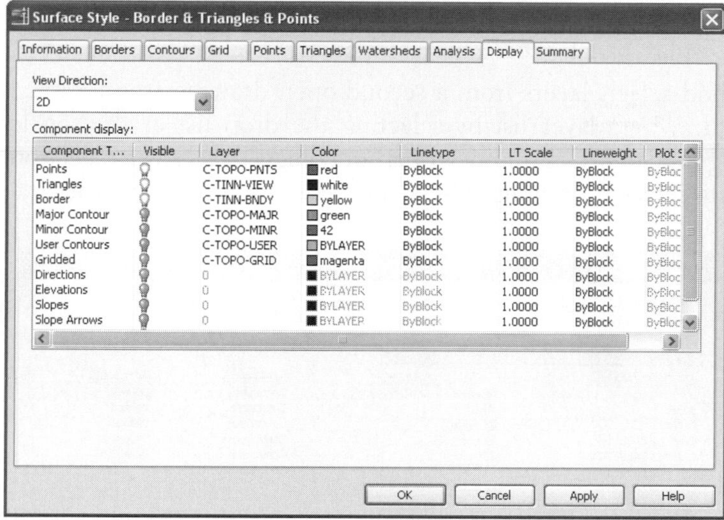

Figure 1.37

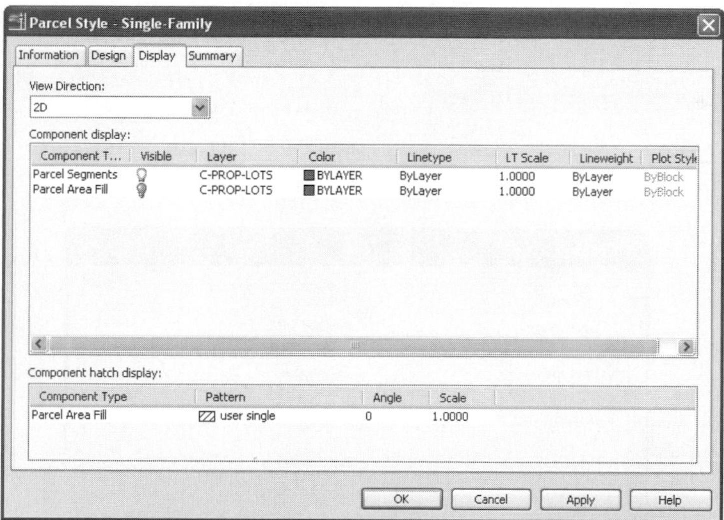

Figure 1.38

There are three methods for assigning a layer to a component or characteristic. The first method uses layers in the current drawing. In the object style's Display panel, clicking on a layer name causes the Layer Selection dialog box to appear, and the user can then select a layer from the list for the current drawing (see Figure 1.39). When returning to the style's Display panel, the component lists the selected layer.

The second method creates a new layer. When selecting a layer name in the Display panel, the Layer Selection dialog box appears. At the top right of the dialog box is a New... button that calls the Create Layer dialog box (see Figure 1.40). In this dialog box, users define the new layer. After returning to the Layer Selection box, users can select the new layer and return to the Display panel, where the component or characteristic lists the new layer.

The third method selects layers from a second open drawing (see Figure 1.39). Users add to the current drawing's layer list by selecting the drop list at the top left of the Layer Selection dialog box to select another open drawing. The Layer Selection list changes to reflect the newly selected drawing. Users then select a layer from the drawing's layer list.

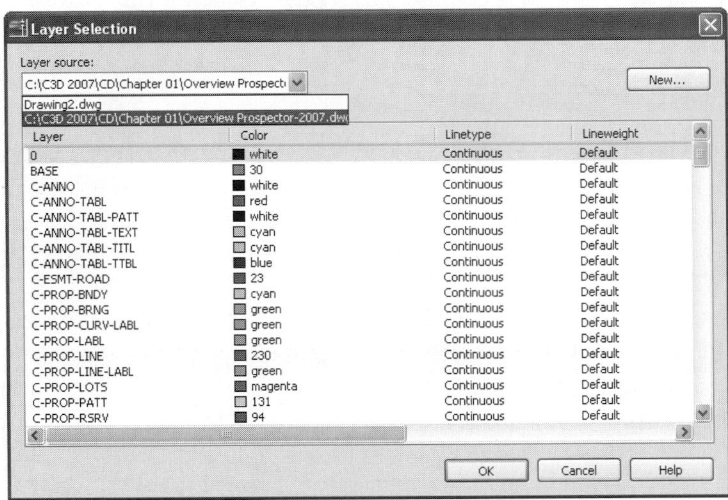

Figure 1.39

Create Layer

Properties	Values
Layer name	V-NODE-PINE
Color	110
Linetype	Continuous
Lineweight	Default
Locked	No
On	Yes
Freeze	No
Plot Style	Color_110
Plot	Yes

OK Cancel Help

Figure 1.40

Civil 3D views all data from two directions: 2D and 3D. A 2D view is from directly overhead, and a 3D view is from any other angle. The list of components and characteristics is the same for 2D and 3D viewing.

Though the Display panel lists all of the available components and characteristics for an object type, a style may display only one or a few components and/or characteristics. A style's purpose determines the components' and/or characteristics' visibility. For example, a surface analysis style focuses on the elevations of a surface. A surface editing/review style shows a surface's border, triangles, and points (see Figure 1.37).

To the right of each component and characteristic is the name and properties of the layer assigned to it. The layer properties of a style must contain the AutoCAD keyword of Bylayer to allow Layer Properties Manager to control the layer properties.

STYLE TYPES

Object styles have a special purpose or function (such as analysis, grouping, or submission documentation). Continuing with the surface objects example, a certain designer has an interest in surface slopes and elevations before starting the design process. To better understand surface slopes, some analysis and ranging of the slope values is necessary. By creating a style that groups slope values into ranges and shows down-slope arrows, the style displays in a more meaningful way the slopes and their spatial distribution. When using an elevation analysis style, the result is a better understanding of their distribution over a site. Surface styles have color schemes that colorize analytical results. Figure 1.41 shows examples of output from different types of analysis styles.

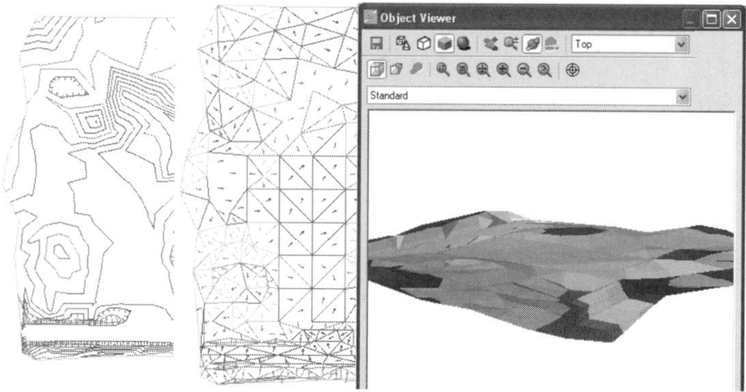

Figure 1.41

OBJECT STYLES

An object style directly controls the display of an object in the drawing. Every drawing has a Basic object style. This style has minimal settings and displays a minimal set of object components. Civil 3D ships with several template files containing object styles. These Autodesk object styles serve as a demonstration of the types of styles a user can define for a Civil 3D implementation. Each chapter in this textbook will expand, modify, and create new styles as they are needed to document a design solution. The number of styles and their complexity will vary depending on the object type. For example, the surface object is complex and has numerous potential styles, whereas parcels have less complexity and fewer styles.

Autodesk Civil 3D uses the same dialog box structure for all object types. Fundamental settings, location of values, and lists of components and characteristics are in the same panels from object to object. Being familiar with one object's dialog box means knowing how to navigate the next object's dialog box.

ASSIGNING OBJECT STYLES

When creating a new object, Civil 3D assigns a default object style set in Edit Feature Settings or the Command Settings dialog boxes for that object type. If changing the style of an existing object is desired, change it in the Information panel of the object's Properties dialog box. Assign a new style by selecting a style from a list of available object styles. When exiting the dialog box, the new style changes how the object appears in the drawing (see Figure 1.42).

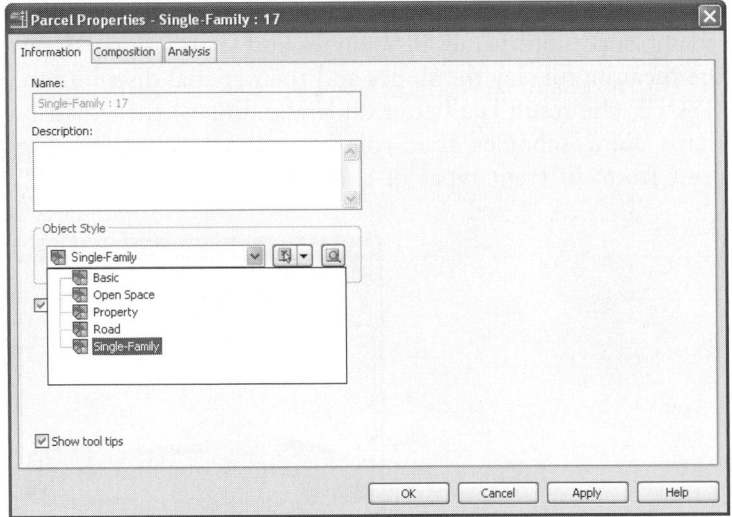

Figure 1.42

LABEL STYLES

A label style directly controls object annotation and its display of data on the screen. Every drawing has a Basic label style. Basic styles have minimal settings and annotate a minimal set of object components. Civil 3D ships with several template files containing label styles. These label styles serve as a demonstration of the types of annotation a user can define for a Civil 3D implementation. Each chapter in this textbook will expand, modify, and create new styles as they are needed to document a design solution. The number of styles and their complexity will vary depending on the object.

When starting a new drawing and using a content template, all of the styles in the template become part of the new drawing. When a style name displays a triangle icon to its upper-left corner, it means that an object in the drawing uses (references) that style.

Label styles are generally single-purpose labels, meaning they annotate a specific facet of the object or its data (see Figure 1.43). The types of label styles for a parcel are Area, Line, and Curve (see the left side of Figure 1.43). The types of labels for an alignment include Station, Station Offset, Line, Curve, Spiral, and Tangent Intersection (see the right side of Figure 1.43).

Each object has its own set of labeling styles. Each object has specific properties that can be part of a style. All of the labels have the same general behavior and use the Text Component Editor interface to create new label styles and to modify existing label styles. When a user becomes familiar with the label editor and its behavior, he or she knows how to edit and create labels for all object types.

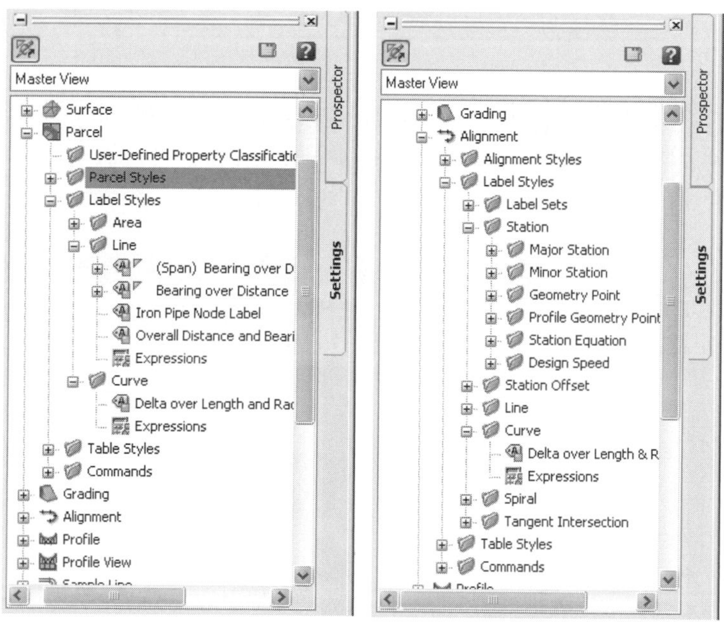

Figure 1.43

Autodesk Civil 3D uses the same dialog boxes for all label styles. The fundamental settings, location of values, and lists of components and characteristics are in the same panels for each label type.

Every label has an anchoring point. In Figure 1.44, the anchor point is the parcel's centroid. The location of the label is "justified" to the anchoring point (located in the middle center of the label is the centroid of the parcel).

To view the label's content definition, click in the value cell of Contents, and click on the ellipsis appearing in the cell. Doing this presents the Text Component Editor and the content

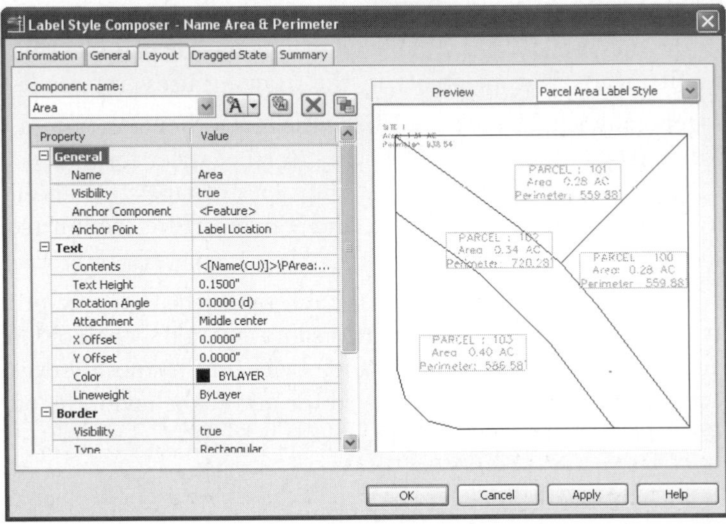

Figure 1.44

of the label (see Figure 1.45). All styles use both Label Style Composure and Text Component Editor. The Layout tab of the Label Style Composure defines a label's text component(s) and the Text Component Editor defines and formats each label component's text.

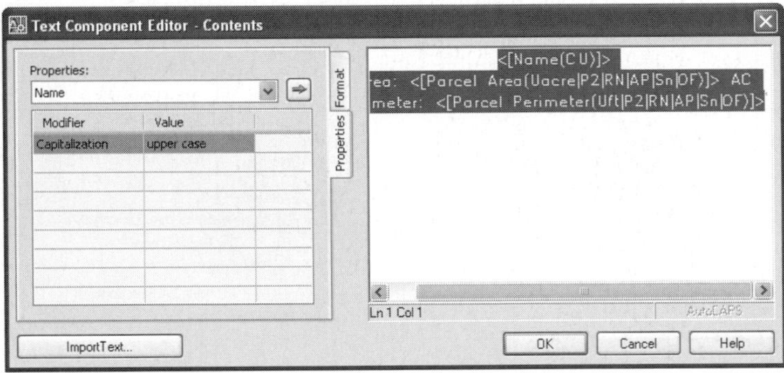

Figure 1.45

TABLE STYLES

Some Civil 3D objects use a table as a method of displaying or documenting their information. The object types using tables are points, surfaces, parcels, and alignments. Each object type has a Basic table style. In this and other chapters, exercises will access these dialog boxes, evaluate their settings, and, if appropriate, change some of their values.

THE PANORAMA WINDOW

An important window in Civil 3D is the panorama. The structure of this window is similar to a spreadsheet (in that it is cell based) and contains a vista (editor, Event Viewer, etc.). In Figure 1.46, the panorama displays a Point Editor vista. Each object type can use a panorama to display its data, and many of the chapter exercises will interact with, edit, and review data in various vistas.

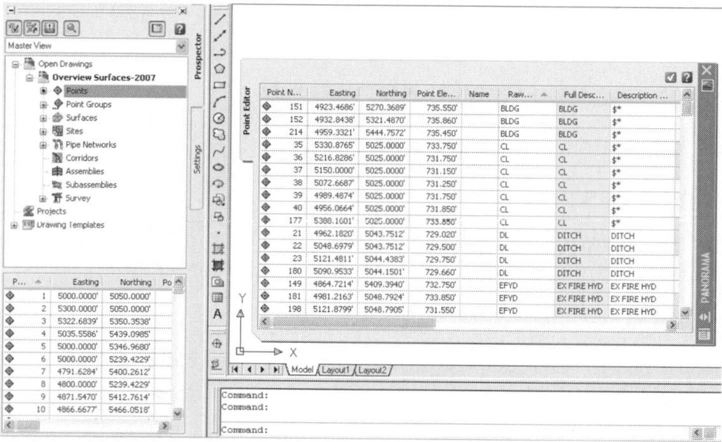

Figure 1.46

EVENT VIEWER

The Event Viewer lists errors, warnings, and information during the execution of a command. Icons at the left of the text message indicate the severity of the message. A red X indicates a failure, a yellow triangle indicates a warning that is not fatal, and an exclamation mark denotes an information message (see Figure 1.47).

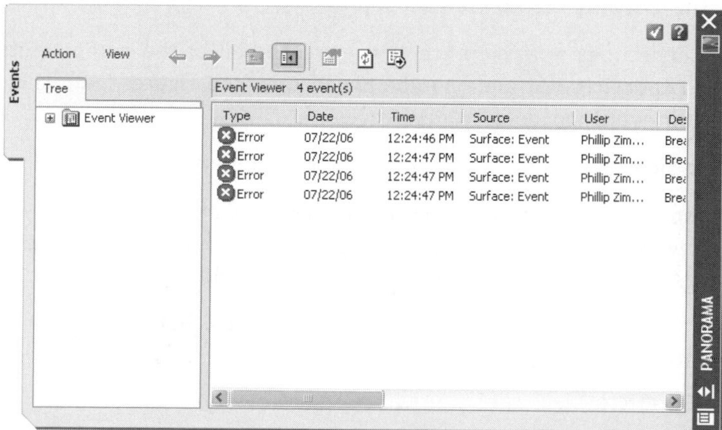

Figure 1.47

EXERCISE 1-3

After completing this exercise, you will:

- View styles from content templates.
- Review the settings and values in Edit Drawing Settings.
- Save a layer-based template file.
- Open an object Properties dialog box.
- Review the anatomy of an object style.
- Review the anatomy of a label style.
- Review the anatomy of a table style.

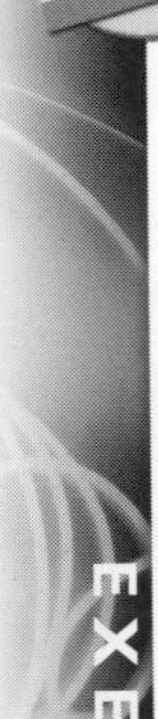

DRAWING TEMPLATES

Civil 3D comes with several template files that you can use and modify as needed. Templates are essential to establishing standard layers, labels, and design documentation.

Template Drawings

1. If you are not in Civil 3D, start the application and close the beginning drawing. If you are in Civil 3D, close the current drawing and do not save it.
2. From the File menu, select **New...**, and from the list of templates select the **_Autodesk Civil 3D (Imperial) NCS Extended** template.
3. Click the **Settings** tab to make it current.
4. In Settings, expand the Surface branch until you view the list of styles under the Surface Styles heading.
5. Click the **Surface Styles** heading, press the right mouse button, and select **New...** from the shortcut menu.
6. Click the **Display** tab to view its settings.

   A new style does not reference any layers, and all of the components have been assigned specific color and linetype properties (see Figure 1.48). This New Style is a seed from which to develop new styles.
7. In the dialog box, to the right of Border, click **layer 0** (zero).

   This displays the Layer Selection dialog box, allowing you to use layers from the current drawing. If more than one drawing is open, you can select layers from other drawings or create new layers (see Figure 1.39 and Figure 1.40).
8. At the top right of the dialog box click the **New...** button to view the Create Layer dialog box.
9. Click the **OK** button until you exit the Surface Style dialog box.

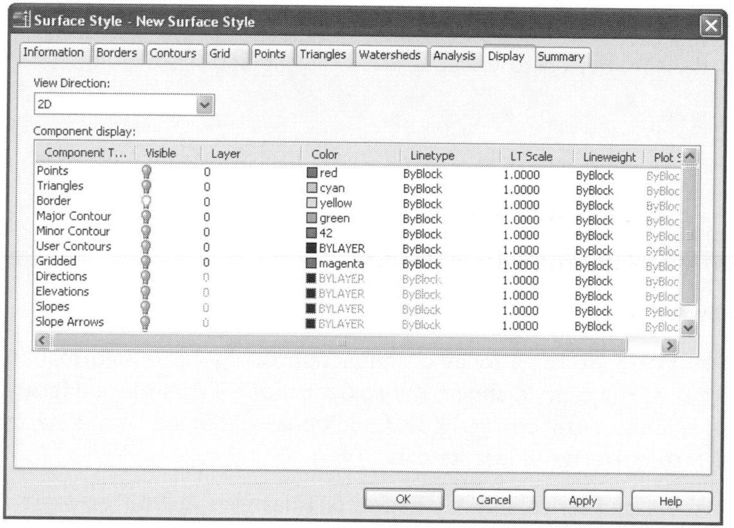

Figure 1.48

10. In Settings, from the list of Surface Styles, select **Contours and Triangles**, press the right mouse button, select **Edit...** from the shortcut menu, and select the **Display** tab to view its contents.
11. Exit the Contours and Triangles style.
12. Close the current drawing and do not save the changes.

EDIT DRAWING SETTINGS

The Edit Drawing Setting, Edit Label Style Defaults, and LandXML Settings dialog boxes occupy the top of the settings tree. Their values influence all the styles below them.

1. Open the *Overview Surfaces II* drawing from the Chapter 1 folder of the CD that accompanies this textbook.
2. If necessary, click the **Settings** tab to make it current.
3. In Settings, click on the name of the drawing located at the top of the panel, press the right mouse button, and from the shortcut menu select **Edit Drawing Settings....**
4. In Edit Drawing Settings, click the **Units and Zone** tab to view the values found in the panel.
5. Click the **Transformation** tab to view its values.
6. Click the **Object Layers** tab to view its settings.
7. Scroll through the layer list and review the settings for layer names.
8. Click the **Abbreviations** tab and expand the different segments to view the base abbreviations Civil 3D uses. You can change any of them in this dialog box.

9. Click the **Ambient Settings** tab and review settings in different sections.
10. Click the **OK** button to exit the Edit Drawing Settings dialog box.

OBJECT STYLE

Each object instance has an object style. A style shows user-specified components or characteristics. Civil 3D has several object styles that affect some aspect of the attached object.

1. Remain in the *Overview Surfaces II* drawing.
2. In Settings, expand the Surface branch until you view the list of styles under the Surface Styles heading.

   Each surface style has a focus on a function or display of information. The Border & Triangles & Points style shows the structure of a TIN surface. Here you can review if linear features show correctly or if additional editing or data is necessary for the surface to correctly reflect its data.
3. From the list of styles, select **Border & Triangles & Points**, press the right mouse button, and from the shortcut menu select **Edit....**
4. Click the **Borders**, **Points**, and **Triangles** tabs to view their contents.
5. Click the **Display** tab to view what components this style displays
6. Exit the dialog box by clicking the **Cancel** button.
7. In Settings, in the Surface Styles list, select from the list **Contours 1' and 5' (Design)**, press the right mouse button, and from the shortcut menu select **Edit....**
8. Click the **Borders** and **Contours** tabs to view their contents.
9. Click the **Display** tab to view what components this style displays.
10. Exit the dialog box by clicking the **Cancel** button.
11. In Settings, expand the Parcel branch until you view a list of styles under the Parcel Styles heading.
12. From the list of styles, select **Single-Family**, press the right mouse button, and from the shortcut menu select **Edit....**
13. Click each tab and review their settings and values.

    The Display tab settings define what components a parcel displays
14. Exit the dialog box by clicking the **Cancel** button.

LABEL STYLE

1. In Settings, expand the Parcel branch until you view the list of styles under the Area heading of Label Styles.
2. From the list select **Name Area & Perimeter**, press the right mouse button, and from the shortcut menu select **Edit....**

3. Click the **Information** tab to view its settings and values.

   The Information tab displays name, description, and authorship credits for the style.

4. Click the **Layout** tab to view its settings and values.
5. In the Text section, in Contents, click in the Value cell to display an ellipsis and click the ellipsis to display the Text Component Editor.
6. Click in the Parcel Area format on the right side of the dialog box.

   This action displays the format string components on the left side of the dialog box. The left side shows in detail what is in shorthand on the right (i.e., Uacre means area is in acre units, P2 means the area is to two decimal places, etc.). You can change a value by clicking in the value cell, dropping a list of options, and selecting a new value from the list.

7. In the dialog box for Precision, click in the value cell to display a drop-list arrow, click the drop-list arrow, and select from the list a new precision value.
8. Exit all of the dialog boxes by clicking on the **Cancel** buttons.
9. Close and exit the drawing without saving it.

EXERCISES

SUMMARY

- Layer-based drawings have layers for each object type as well as layers for object components and characteristics.
- The Edit Drawing Settings, Edit Label Style Defaults, and Edit LandXML Settings dialog boxes influence all of the styles and settings in a drawing.
- You can lock values in the Edit Drawing Settings, Edit Label Style Defaults, and Edit LandXML Settings dialog boxes.
- The Edit Drawing Settings, Edit Label Style Defaults, and Edit LandXML Settings dialog boxes indicate that a lower style has changed the value.
- Object styles emphasize a subset of an object's components or characteristics.
- Label styles label a subset of an object's properties.

UNIT 4: CIVIL 3D AND LAND DESKTOP

Civil 3D works in tandem with LDT and the Civil Design companion. Autodesk gives each application the ability to exchange data, enabling them to complement their strengths and make up for their weaknesses. The LandXML file, direct data reading, and data extraction commands transfer points, point groups, description keys, surfaces, alignments, pipe networks, and profile sampling data between the two programs.

LANDXML SETTINGS

Civil 3D exports and imports LandXML data files from LDT and other civil applications. Setting the proper units for importing and exporting data is critical to successfully using LandXML. The user sets these values in the Edit LandXML Settings dialog box. The first value to set is in the Import panel, Data Conversion section (see Figure 1.49). If toggled on, it makes Civil 3D convert US Foot to International Foot values on the file.

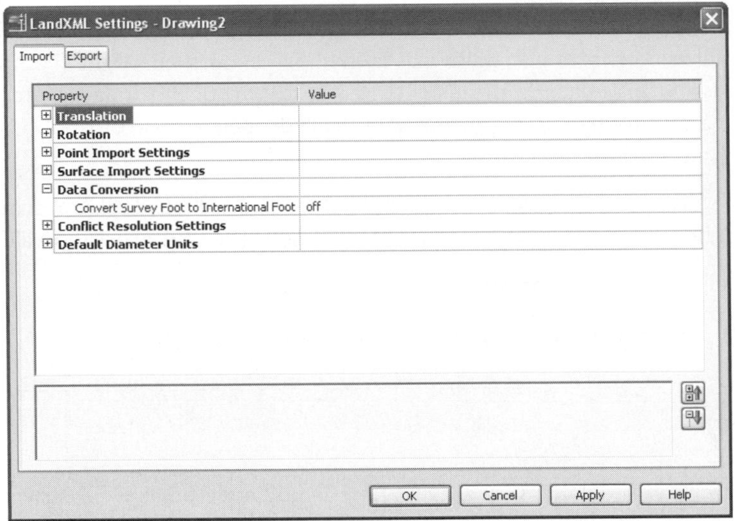

Figure 1.49

The second value to set is the units for exported data. This is set in the Export, Data Settings section. This value sets the unit type for the file: International or survey foot (see Figure 1.50).

LANDXML IMPORT AND EXPORT

Under the File menu, Import flyout has the Import LandXML... command. The command prompts the user to select a file and displays a dialog box listing the types of data in the file (see Figure 1.51). At this time all the data or a subset of data from the file can be selected.

After importing a LandXML file, the information appears in Prospector and on the screen. The exception to this is profile data. When importing profile data, the import creates only entries in the Prospector's Sites branch. Users have to create a profile view using the profile data to view them on the screen, and will have to create or re-create the remainder of the roadway design elements (assembly, corridor, section sample group, and section views) to complete a roadway design.

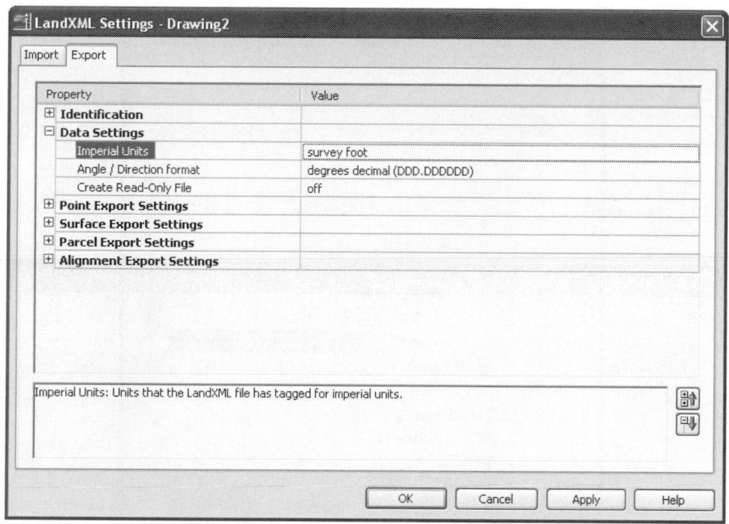

Figure 1.50

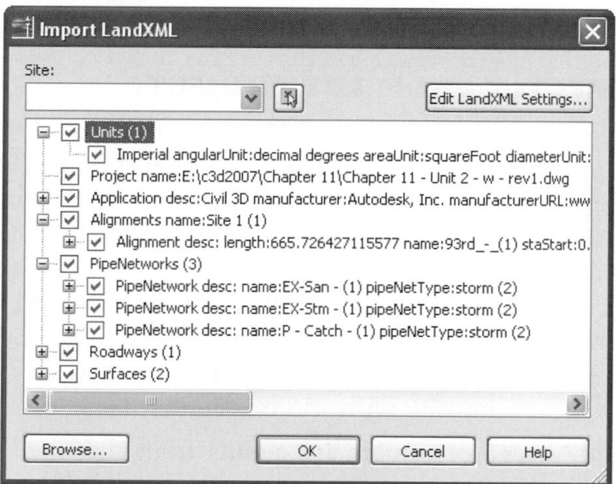

Figure 1.51

When Exporting to LandXML, users select data by toggling on the data in the Export LandXML (see Figure 1.52).

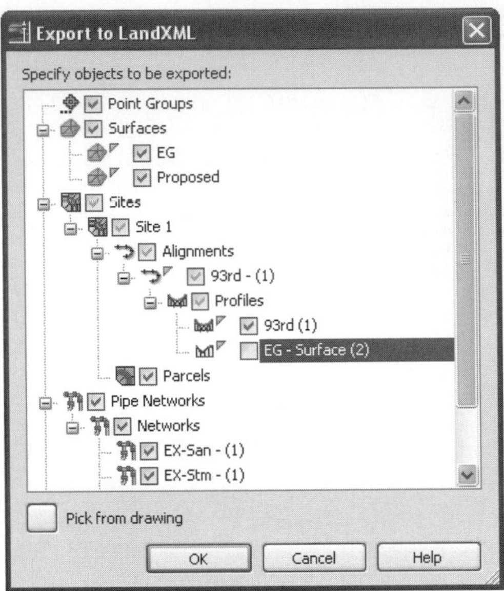

Figure 1.52

IMPORTING DIRECTLY FROM LDT PROJECTS

A second method of transferring data from LDT 2007 to Civil 3D is by directly reading data from an LDT project data structure. The File menu, Import flyout, and the Import Data from Land Desktop... command displays a dialog box that lists the data of a selected project. To use the dialog box, first identify the LDT project folder and then the project name. After selecting the project, the dialog box populates with the project's data (see Figure 1.53). After selecting the data, click the OK button and the data in placed in the drawing. The user can read project data back to LDT Release 2. If attempting to read earlier versions, users may encounter incompatible file formats. This can be overcome by overwriting the data in the drawing with the transferred data.

This method currently does not import any points from the project into the Civil 3D drawing. The best method of importing points into a Civil 3D drawing is by importing an ASCII file.

If the LDT project contains pipe runs, the routine issues a warning about having the proper parts list definition (see Figure 1.54).

After importing from an LDT project, the transferred information appears in Prospector and in the drawing. The exception to this is profile data. When importing profile data, the current implementation creates only entries in Prospector. Users have to create a profile view using the profile data to view them on the screen, and will have to create or re-create the remainder of the roadway design from Civil 3D objects (assembly, corridor, section sample group, and section views) to complete a roadway design.

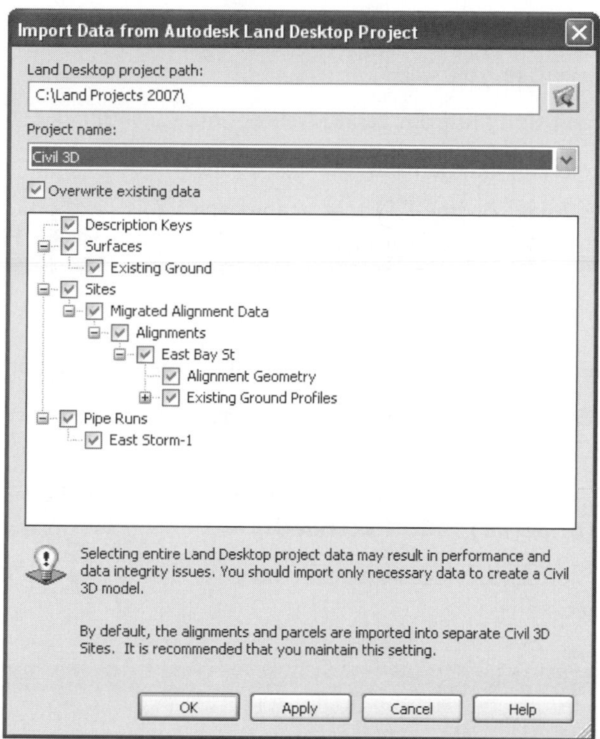

Figure 1.53

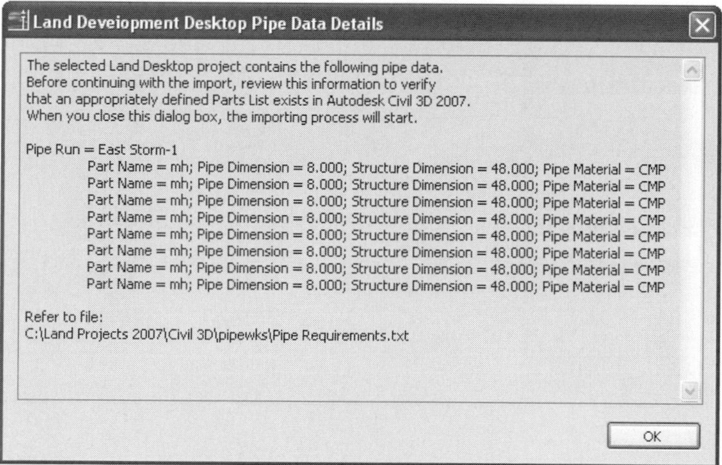

Figure 1.54

EXERCISE 1-4

After completing this exercise, you will:

- Import a LandXML file from Land Desktop into Civil 3D.
- Read data directly from a Land Desktop project folder.

This exercise uses files found on the CD that accompanies this textbook. You need to copy the Autodesk Land Desktop Project, Civil 3D, from the CD to the Civil 3D Projects folder on your computer.

EXERCISE SETUP

1. Using Windows Explorer, locate the Civil 3D Projects folder on your hard drive.
2. Copy the Civil 3D project from the CD that accompanies this textbook, placing it in the Civil 3D Projects folder (see Figure 1.55).
3. In Civil 3D, from the File menu select **New...** and select the template **_Autodesk Civil 3D (Imperial) NCS Extended**.
4. From the File menu, select **Autocad Save As...** and save the drawing as **Import and Export**.

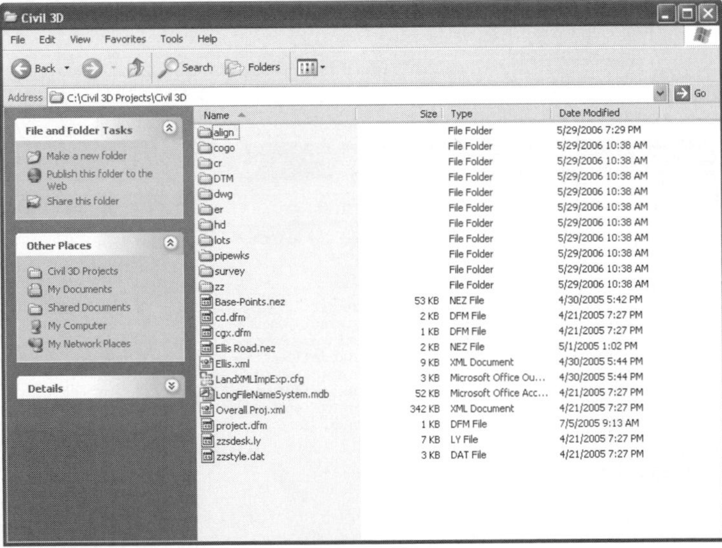

Figure 1.55

IMPORTING A LANDXML FILE

A LandXML file contains several different types of data (surface, alignment, parcel, pipe networks, and profile data). When viewing the file in the Import LandXML dialog box, it lists all of the data within the file. You can select what items you want to import by toggling them on or off.

1. From the File menu, Import flyout, select **Import LandXML...**.

2. In the file open dialog box, browse to the folder *Civil 3D Projects\Civil 3D*, select **Overall Proj.xml**, and click the **Open** button to display the Import LandXML dialog box.

   The Import LandXML dialog box displays, listing the file's data. It is here you select the data to import (see Figure 1.51).

3. Leave all of the toggles on and click the **OK** button to import the data.
4. Click the **Prospector** tab to make it current.
5. Expand the Sites data tree until viewing the alignments from the LandXML file.

   After reading in the points, surface, and alignment data, Civil 3D creates the objects in the current drawing and lists them under the appropriate object type (see Figure 1.56).

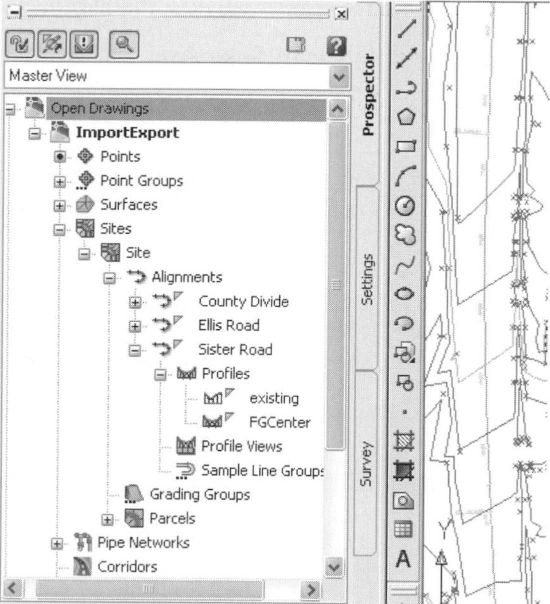

Figure 1.56

6. In Prospector, click the **Points** heading and view imported points in Prospector's preview area.
7. Expand the Surface branch until viewing the surface list.
8. Close the Import and Export drawing, do not save the changes, and reopen it.

READING FROM A LAND DESKTOP PROJECT

1. From the File menu, select **Import Data from Land Desktop....**

   This displays the Import Data from Autodesk Land Desktop Project dialog box, which sets the projects and named project folder (see Figure 1.53). After setting the project folder and identifying the project, the dialog box displays all of data you can transfer from the project.

2. In the Import Data from Autodesk Land Desktop Project dialog box, set the project folder to the **Civil 3D Projects** and the project to **Civil 3D**.
3. In the dialog box, click the **OK** button to import the project data.
4. If necessary, click the **Prospector** tab to make it current.
5. In Prospector, expand the Surfaces branch to view the surface.
6. In Prospector, expand the Sites branch to view the alignments and profiles entries.
7. From the File menu, select **Autocad Save As...** and save the drawing as Overview Import.
8. Exit Civil 3D.

Each of the following chapters explores in greater detail an object type and its styles.

SUMMARY

- LandXML files provide an effective method of transferring data between Civil 3D, Autodesk Land Desktop, and other civil engineering applications.
- Civil 3D reads Land Desktop project data and imports the data directly into a drawing.

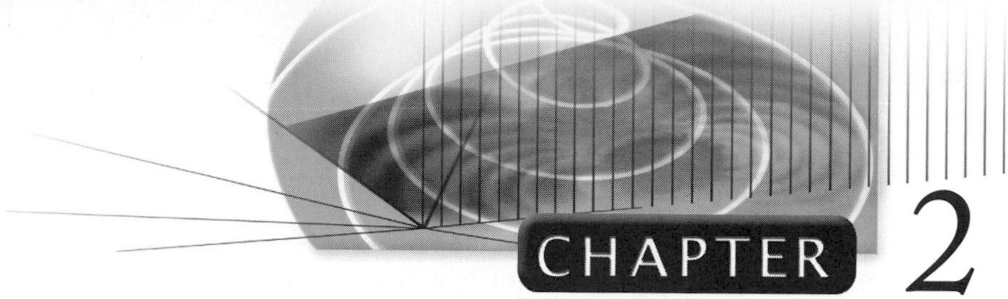

Points

INTRODUCTION

Many civil design projects start with coordinate data. Coordinate data can come from surveyors, public records, or a Web site, among others. It is necessary to import and modify these coordinates into usable point data. Methods for importing and modifying coordinate data include defining standard symbols with descriptions, transforming coordinates to other grid coordinate systems, organizing points into groups based on function, description, and common names, and labeling their coordinates and information as a part of a submitted document.

OBJECTIVES

This chapter focuses on the following topics:

- Importing Points from External Files
- Defining Point Styles
- Defining Point Label Styles
- Civil 3D Transparent Commands
- Description Key Set
- Point Groups

OVERVIEW

Coordinate data serves a dual purpose in civil designs. First, it communicates the current state of a site to the office. Second, it communicates the design from the office to the field. Many of the coordinates representing the current site's state become points for a plat of survey or a surface. A surface influences the strategies for access roads, parcel layout, water control, grading, and so on. Civil 3D introduces new capabilities, and changes some of the methods for working with points. This chapter reviews the implementation of points in Civil 3D.

Points represent several types of data. How these points appear in the drawing and following submittal documents determines what marker and label styles the user assigns them. There are three basic point categories that determine how points should appear in a drawing. The first type is a point that does not require any action other than placing it into the drawing with a simple marker and label style. An example of this type of point is a ground shot from a topographic survey. A second type of point requires a symbol as the marker style. The symbol will be part of the final product, but the point label may not be visible. This means the marker and label styles should use different layers in the drawing. Examples of this type of point are points representing manholes, signs, power poles, and so on. The third type of point is one that is a part of a lineation (a boundary, roadway centerline, edge-of-pavement, etc.) and it is found on a line or arc in the drawing. Civil 3D will use these types of points as surface breaklines, grading feature lines, or line work for a submittal document.

Civil 3D manipulates, represents, and organizes points from the Prospector's Points and Point Groups areas or from the Points menu. The point tools are flexible, comprehensive, and provide necessary management tools. The Points menu contains routines affecting points from creation to annotation (see Figure 2.1). The command for creating points is also in Prospector's Points shortcut menu.

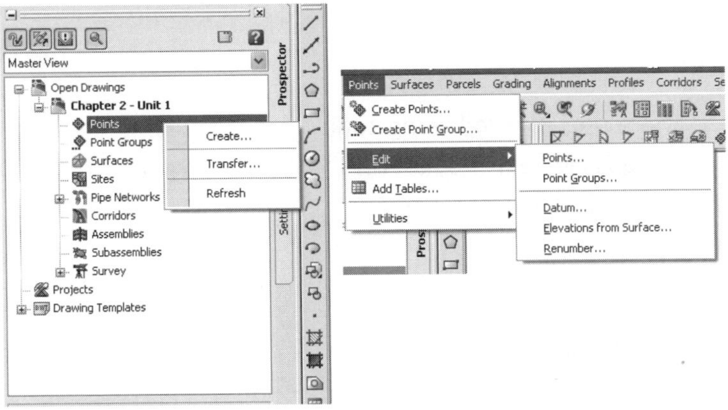

Figure 2.1

The number of routines creating point data reflects the importance of points to existing conditions and design solutions. The point commands create points from lists, external files, AutoCAD entities, surface elevations, roadway corridors, by interpolating between existing points, from slopes and intersections (distance and direction), and more. The Create Points toolbar organizes its icons by what data source creates the resulting points (see Figure 2.2).

A point represents coordinate data in a drawing. Civil 3D uses a point object consisting of a marker (the coordinate location) and label. Traditionally, the label portion displays a point's number, elevation, and description. Civil 3D styles (marker and label) define the actual look of a point in two and three dimensions. Users need to develop a basic understanding of style definitions and behaviors when defining and assigning styles to points in a drawing. The assignment of marker and label styles are from default point creation settings (Edit Feature Settings for points). In addition to default label settings, a

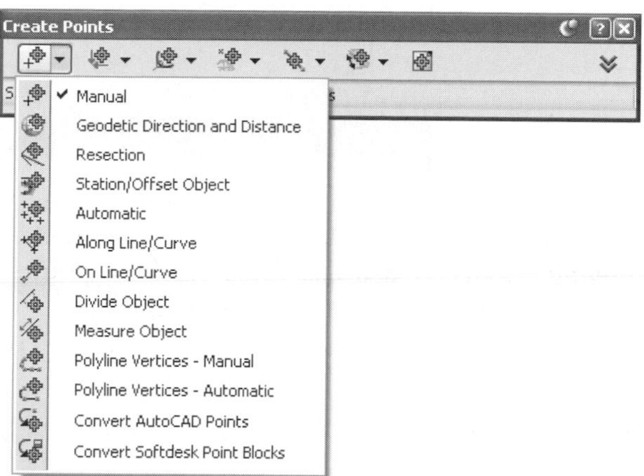

Figure 2.2

Description Key Set, if used, affects the assigned point styles, or a point group can override the originally assigned styles.

If organizing points into point groups, point group properties can change how groups display on the screen. When using point groups, there are two choices. First, the group retains the original marker and label style. In this situation, display order makes no difference. In the second choice, the group overrides one or both styles. In this situation, display order makes a difference in how a point displays on the screen.

If a point belongs to several point groups with each group having overridden point and/or label styles, it is the point group that was drawn last that determines how the point displays on the screen. The Point Groups Properties dialog box displays a drawing's draw order list (lowest being the first drawn). In Figure 2.3, all points belong to the No Show group, and the override hides the markers and labels. However, since the Breakline Points group is the last drawn (located at the top of the list) only those points in the group show in the drawing.

When using the Standard or Basic point style, the marker and label are on the same layer. A point (marker) style can define a symbol for the point's coordinate, a behavior (two- and three-dimensional views, at elevation, etc.), and can assign layers to the marker and label.

If using a Description Key Set, the set assigns the point and label styles for each matching point. If desired, a set can assign a marker layer, and it is up to the marker style to define a (separate) layer for the point label.

Prospector allows users to lock individual or groups of points. Locking points or groups prevents the AutoCAD Move command from changing the coordinates of selected points. Locking is easily undone, and the current implementation of a project does improve on safeguarding point coordinates.

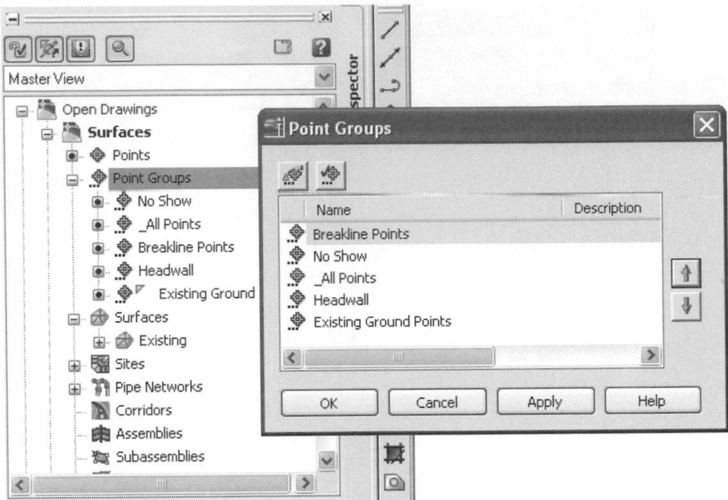

Figure 2.3

A point in the drawing represents an instance of a point object. If using the AutoCAD Erase command to erase a point from the drawing, Civil 3D removes the point from the overall point list and any point groups it belongs to. Erase causes a permanent deletion.

Points as a part of a project are discussed in Chapter 12 of this textbook.

Here are some rules about the relationships among points and Civil 3D:

- Points must be present in a Civil 3D drawing.
- Generic AutoCAD routines edit only the AutoCAD properties of a point object.
- If allowed, the AutoCAD Move command will move a point to a new location and Civil 3D will update its coordinates.

UNIT 1

The point object's default point and label styles, import options, point group naming, and point numbering settings are in the point's Edit Feature Settings dialog box of the Settings panel. These settings assign default values found in the settings area of commands that create points. When reviewing the settings for each point command, the dialog box shows a combined set of values from the Edit Feature Settings dialog box (the diamond icon entries) and command parameter settings (the square icon entries). These combined settings affect several aspects in creating new points (next point number, styles to use, data entry prompts for elevation and description, and other settings). These settings and style definitions are the focus of the first unit of this chapter.

UNIT 2

The second unit of this chapter reviews routines that create points from printed lists, external ASCII files, or calculated points (such as the Intersection icons of the Create

Points toolbar). Other routines in the Create Points toolbar are discussed in their appropriate context in later chapters of this textbook.

UNIT 3

The subject of the third unit of this chapter is the point analysis and listing tools. Points are viewed in Prospector's preview area or in a Point Editor vista within a panorama. The Point Utilities flyout menu of the Points menu contains routines to view, export, and preview project point information. The report capabilities of Prospector allow users to review critical point values and to create user-formatted point reports.

UNIT 4

The editing of points, including changing their number or location, is the subject of the fourth unit of this section. The point editing routines are in a shortcut menu in the Points area of Prospector, and the user can directly edit point values in the preview area, or edit a selection set of points from a right mouse button shortcut menu.

UNIT 5

As previously mentioned, some points will have a symbol as a marker. A Description Key Set is the easiest method of assigning a marker style for new points and includes a symbol. The symbol usually is part of the final document while the point label may not be. This means the point and point label should be on separate layers in the drawing. The Description Key Set, point groups, and Table styles are the subjects of the last unit of this section.

UNIT 1: POINT SETTINGS AND STYLES

A point object's creation and display on the screen is the result of user-defined settings and styles. The routines that create points look at these settings and their values (including default styles) in the Point branch of Settings. This branch's settings determine what the next point number is, what symbols to use, what colors to display, and other behaviors. Values within a point marker and label style also affect many aspects of a point's behavior (2D and 3D).

EDIT FEATURE SETTINGS

When selecting the Point heading under Settings, press the right mouse button and select Edit Feature Settings, and Civil 3D presents a dialog box that is a combination of Ambient Settings from Edit Drawing Settings and new entries specific to point objects. The dialog box values set initial values for all of the styles and commands below this point.

DEFAULT STYLES

The first section for the point feature is the name of the default point marker and label styles. Civil 3D uses these entries when there is no active Description Key Set or point groups with marker and label styles overrides (see Figure 2.4).

Edit Feature Settings - AeccPoints
Property
Value
Override
Child Override
Lock
General
Labeling
Default Styles
Point Style
Basic
Point Label Style
Point#-Elevation-Desc...
Default Name Format
Update Points
Point Identity
Unitless
Distance
Dimension
Coordinate
Grid Coordinate
OK
Cancel
Apply
Help

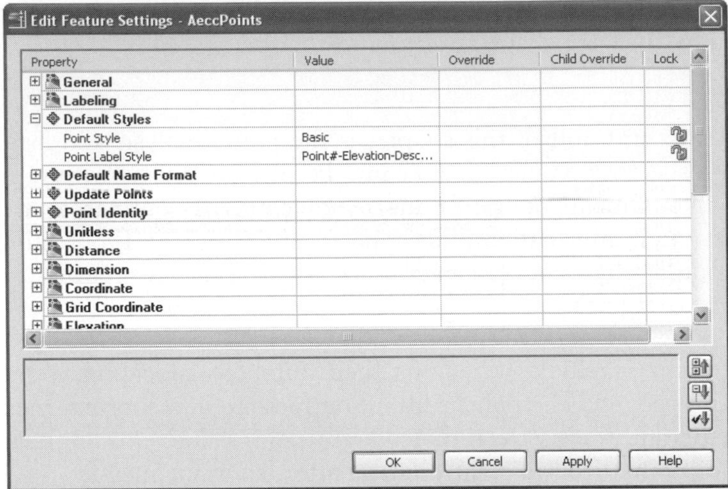

Figure 2.4

DEFAULT NAME FORMAT

This section sets the naming convention for point groups (see Figure 2.5). By default, a point group name is simply the words "Point Group" with a sequential number counter that follows. When defining a point group, a user can assign a more descriptive name. When clicking in the point name cell, an ellipsis (...) appears. When clicking on the ellipsis, the Name Template dialog box displays. In this dialog box, users can change the name prefix (Point Group -) as well as the number style, beginning counter number, and increment values (see Figure 2.6).

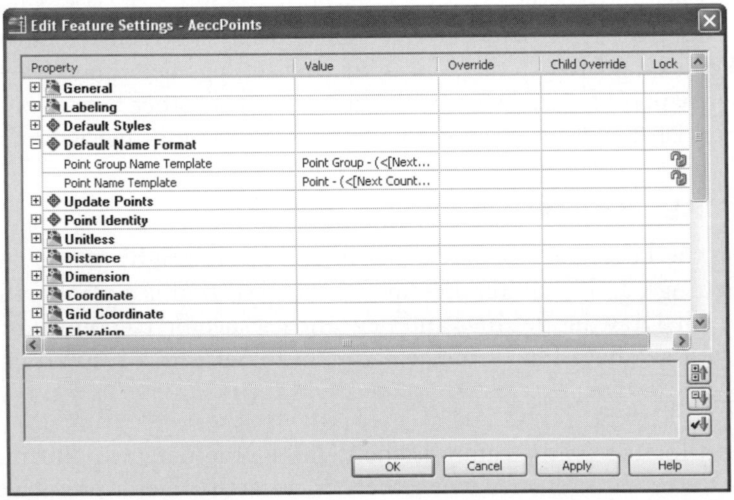

Edit Feature Settings - AeccPoints
Property
Value
Override
Child Override
Lock
General
Labeling
Default Styles
Default Name Format
Point Group Name Template
Point Group - (<[Next...
Point Name Template
Point - (<[Next Count...
Update Points
Point Identity
Unitless
Distance
Dimension
Coordinate
Grid Coordinate
OK
Cancel
Apply
Help

Figure 2.5

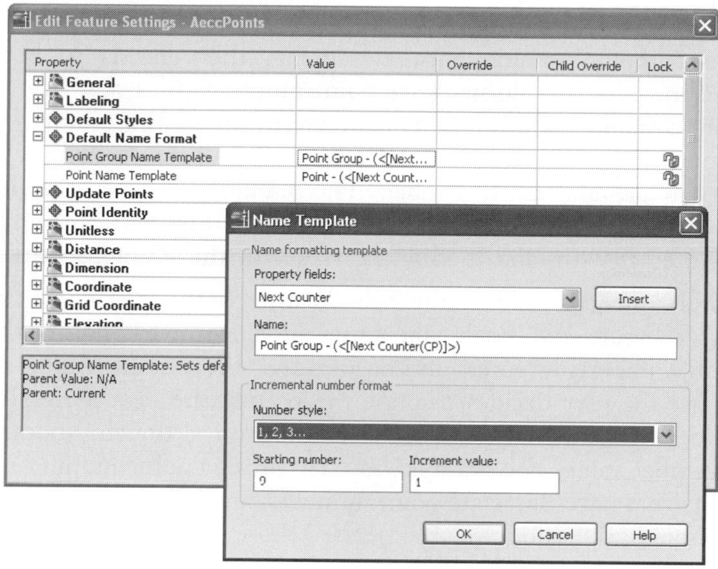

Figure 2.6

UPDATE POINTS

The Allow Checked-In Points to be Modified toggle allows a user to prevent points checked in from a project to be modified from the original project values (see Figure 2.7). However, by simply changing this value to true, a user can modify the point values and then check them back in by updating the project.

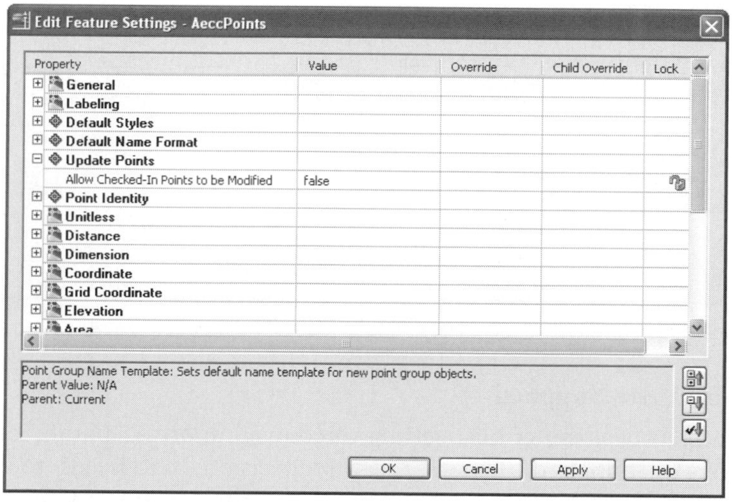

Figure 2.7

IMPORT OPTIONS

Whenever a user imports points into a drawing, there needs to be set conditions concerning handling files that do not have point numbers, files that have point numbers, or files that may contain duplicate point numbers of preexisting points (see Figure 2.8). The Point Number Offset and Sequence Point Numbers From values are used if referred to by other settings below these two values. The remaining entries have multiple options that display when the user clicks the drop-list arrow (see Figure 2.9). All routines creating points look to, interact with, and modify the values in the Point Identity section.

The Next Point Number is held in the Edit Feature Settings dialog box or the Create Points toolbar (see Figure 2.8). Each time a user adds or imports points, this number may change. If a user is importing points and decides that their numbering is incorrect, or for some other reason the user decides to erase the points, the next available point number does not reset to reflect the erasure. Erasing points from a drawing does not change the current point number value. For example, say the current point number is 500. The user imports 20 points that have duplicate point numbers, and those 20 points are assigned the numbers 500 through 519. The user then erases those 20 points from the drawing. The current point number remains 520. If the user wants to reuse numbers 500 through 519, he or she must edit the current point number value in Edit Feature Settings.

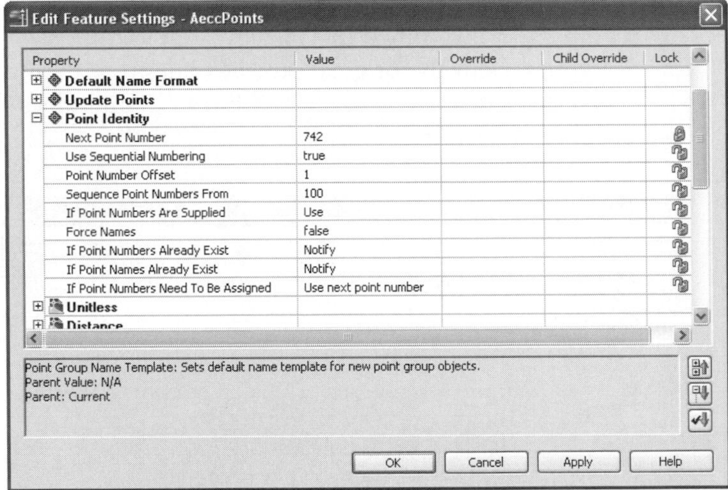

Figure 2.8

If Point Numbers Are Supplied

Of all the point settings, the critical setting affecting importing points is the value of If Point Numbers Are Supplied. This is the first decision that needs to be made when importing points. If Point Numbers Are Supplied has three options: Use, Ignore, and Add an offset. Each option has its own set of choices. Some of the choices are complex because they depend on other settings in the Edit Feature Settings for points.

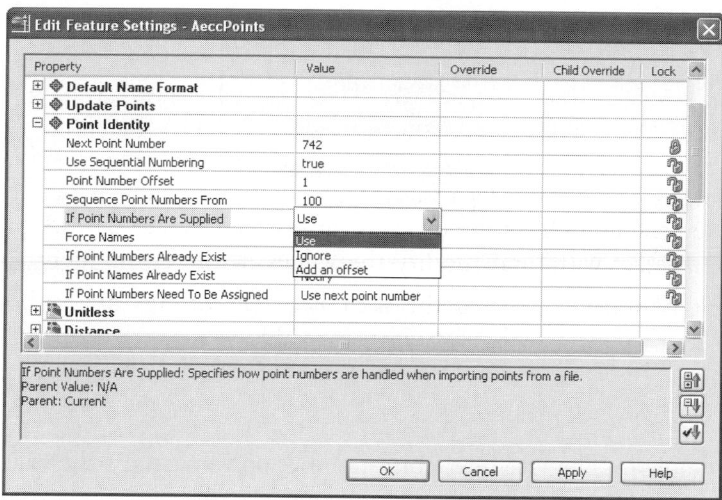

Figure 2.9

Use When If Point Numbers Are Supplied is set to Use, the next decision is what to do with file point numbers, some of which may be duplicate numbers. When importing and encountering a duplicate point number, the routine checks the value of If Point Numbers Already Exist to determine its next step.

Ignore When If Point Numbers Are Supplied is set to Ignore, the next decision to make is how to assign new point numbers to the points in the file. The import routine looks at the value of If Point Numbers Need To Be Assigned.

Add an Offset When If Point Numbers Are Supplied is set to Add an Offset, the next decision to make is what offset to add (see Figure 2.9). The import routine looks at the value of Point Number Offset. The potential problem with the Add an Offset option is that it may produce a duplicate point number. When this happens, the import routine displays the Duplicate Point Number dialog box and asks for a solution to this situation.

If Point Numbers Already Exist

If Point Numbers Already Exist has four options to choose from: Notify, Renumber, Merge, and Overwrite. Each option handles the situation differently, and some look to other settings in the Edit Features Settings dialog box to determine how to resolve a duplicate point number.

Notify This option displays the Duplicate Point Number dialog box and asks the user to resolve the situation (see Figure 2.10). The resolution can be applied to all duplicate points by toggling on the Apply to all duplicate point numbers option. If this toggle is off, each time a duplicate point number is encountered, this dialog box displays and the user must resolve the current situation. The Duplicate Point Number dialog box has five possible solutions: Use next point number, Add offset, Sequence from, Merge, and Overwrite.

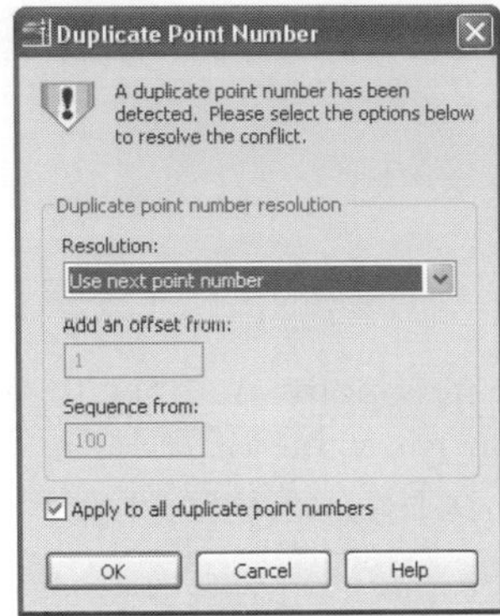

Figure 2.10

Use next point number The Use next point number option assigns the value of Next Point Number to the duplicate point.

Add offset and Sequence from The Add offset and Sequence from options can produce a new point number that is also a duplicate of another existing point in the drawing. For example, if the user chooses the offset option and if the offset is 1 and the file point number is 100, the Add offset option creates a new point number of 101. If that point number is also a duplicate of a point in the drawing, the point coming in is assigned the next number. The same can happen using the sequencing option. If the new point number is a still a duplicate number, the point is assigned the next available point number.

Merge The merge option blends a drawing point's data with the values from a source file. If the point entry of the file does not contain a value, the current drawing point value remains after the import. For example, say the source file has point number, northing, and easting data. This source file data will replace the northing and easting values of the same point number in the drawing, but will not change any other values for the point.

Source file values for point 101:

```
101, 5000.0000, 6000.0000
```

Values for point 101 in the drawing:

```
101, 100.0000, 100.0000, 723.84, IPF
```

With Merge on and after importing the file, the new values for point 101 are:

```
101, 5000.0000, 6000.0000, 723.84, IPF
```

Overwrite The last option is to Overwrite. This option replaces a duplicate point number's values in the drawing with the values found in the file. When using the Overwrite option, there may be a loss of data. An example of this option is:

Source file values for point 401:

```
401, 5010.0000, 6050.0000
```

Values for 401 in the drawing:

```
401, 5100.0000, 6500.0000, 723.84, IPF
```

With Overwrite on and after importing the file, the new values for point 401 are:

```
401, 5100.0000, 65000.0000, NULL, NULL
```

Renumber This option assigns a new point number to the incoming point. The value of the new point number depends on the value of If Point Numbers Need To Be Assigned. When If Point Numbers Need To Be Assigned is set to Use Next Point Number, the import routine assigns the value on Next Point Number as the new point number. If the value of If Point Numbers Need To Be Assigned is set to Sequence from, the import routine looks to the current value of Sequence Point Numbers From. The potential problem with this option is it may produce a duplicate point number. When this happens, the import routine displays the Duplicate Point Number dialog box and asks for a solution to this situation.

If Point Numbers Need To Be Assigned

When If Point Numbers Need To Be Assigned is set to Use Next Point Number, the import routine uses the current value of Next Point Number. If the value of If Point Numbers Need To Be Assigned is set to Sequence from, the import routine uses the current value of Sequence Point Numbers From. A potential problem with this option is it may produce a duplicate point number. When this happens, the import routine displays the Duplicate Point Number dialog box and asks for a solution to this situation.

EDIT LABEL STYLE DEFAULTS

This dialog box sets the general behavior of all point labels. See the discussion of this dialog box in Chapter 1, Unit 3 of this textbook.

POINT STYLES

The Standard and Basic point styles are simple marker definitions. A marker is a Civil 3D object that locates a point's coordinates in a drawing and has two possible views: two and three dimensions.

The Point Styles list in the Point branch of the Settings panel lists all of the currently defined styles for a drawing (see Figure 2.11).

The Point Style dialog box contains five panels, three of which affect some aspect of the marker.

INFORMATION

The Information panel displays the name, description, and who created or modified the style definition (see Figure 2.12).

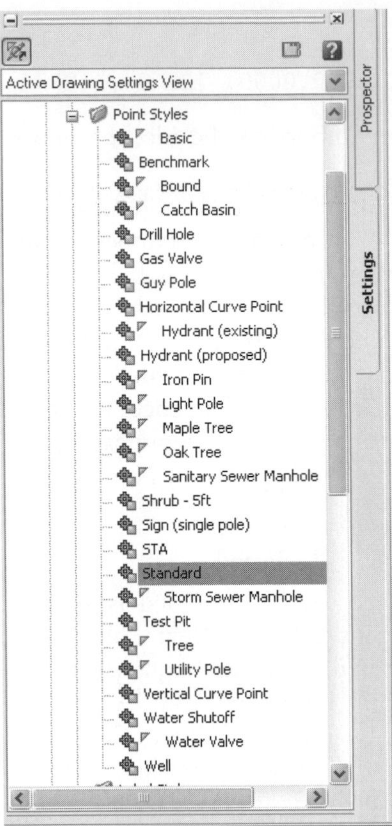

Figure 2.11

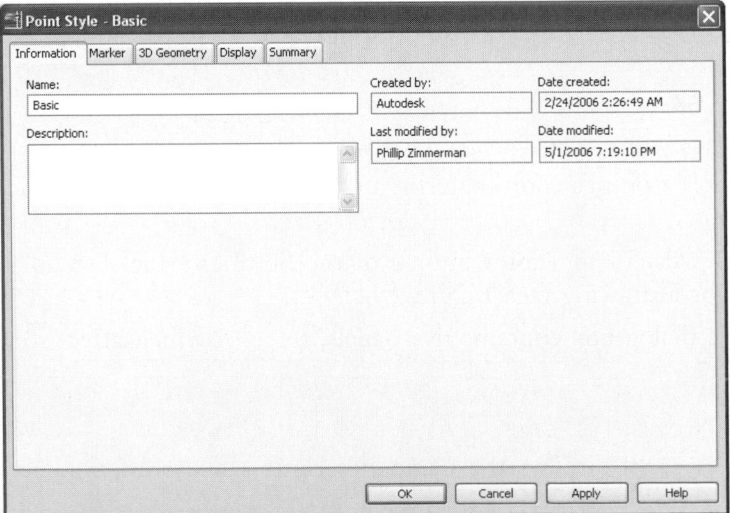

Figure 2.12

MARKER

The Marker panel defines the appearance of the coordinate marker in the drawing (see Figure 2.13). The default the marker is an X from the Custom marker style list. However, markers can be a combination of values in this section of the dialog box.

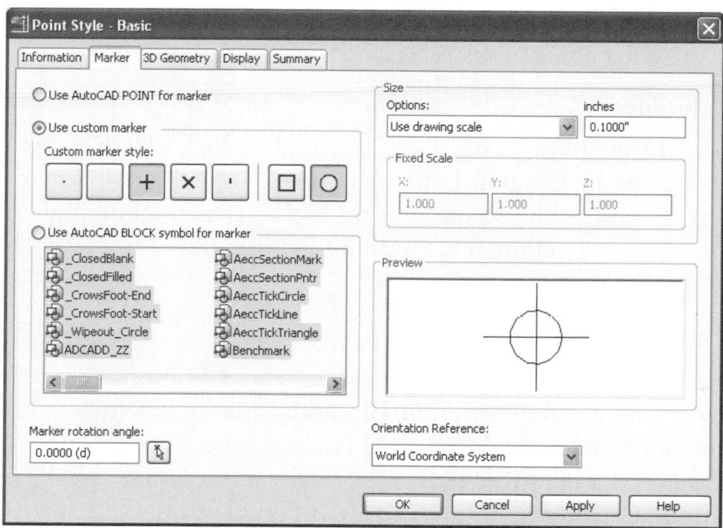

Figure 2.13

The left side of the Marker panel shows three options for defining a marker. The first option is to Use AutoCAD POINT (node) for marker. This uses the system variable of pdmode and sizes the node with pdsize. See the Civil 3D help entry for these two AutoCAD system variables.

The second option creates a marker from the choices of the Use custom marker area. A custom marker looks similar to an AutoCAD node, but it is an object rather than an AutoCAD node. A custom marker has a base shape (the five leftmost icons) to which users can add two more shapes (a square and/or circle). The base shapes include a node, nothing, a plus symbol (+), an X, or a vertical line.

The third option is selecting an AutoCAD block from the Use AutoCAD BLOCK symbol for marker section. This option assigns a symbol (AutoCAD block) as the coordinate marker. One of the markers, _Wipeout_Circle, contains a wipeout to hide any line work connecting to the point. Other markers are AutoCAD multi-view blocks (a block containing a two- and three-dimensional representation) that show a two-dimensional symbol for plan viewing and a three-dimensional symbol for any other view of the point. All block definitions must be present in the drawing to be used as a symbol.

Marker Rotation Angle

This is a rotation angle for the marker in a drawing. This is the only way to rotate a marker.

Size

The upper right of the Marker panel controls the marker size (see Figure 2.13). By default, the marker size is the result of multiplying the drawing scale by the symbol size. For example, the Use drawing scale method of sizing a marker reads the drawing scale (1" = 40") and the marker size (0.6). By multiplying 40 by 0.6, the resulting marker size is 24 units. Other methods for setting thc marker are: Use fixed scale, Use size in absolute units, and Use size relative to screen. The Fixed Scale option specifies a scaling factor for X, Y, or Z.

Orientation Reference

By default, Civil 3D orients a marker relative to the World UCS. This setting has two additional options. The first option is that the rotation angle of the marker is relative to the object it is attached to. The second option is that the rotation angle is relative to the current AutoCAD view direction. This option allows markers to always be plan readable in any viewport.

3D GEOMETRY

When placing points in a drawing, there are three choices for vertical placement (see Figure 2.14). The default, Use Point Elevation, places a point in the drawing at the elevation of the point. A second choice, Flatten Points to Elevation, assigns a user-specified elevation to the point. For example, setting the Point Display Mode to Flatten Points to Elevation and entering 0 (zero) as the point elevation will create objects with an elevation of zero when selecting the point with an AutoCAD object snap. The point will always list the correct elevation, but selecting it with object snaps will always produce objects with a zero elevation. If a user uses a Civil 3D Transparent command and selects the point, the resulting entity will have the elevation of the point. This method is how users define a 3D polyline using Civil 3D point elevations for vertex elevations.

The third option is Exaggerate Points by Scale Factor. This option scales the elevation of a point and places it at a new elevation in the drawing. For example, setting the option to Exaggerate Points by Scale Factor and setting the scale factor to 2.0 places a point whose elevation is 325.25 at the AutoCAD elevation of 650.50 in the drawing.

DISPLAY

The Basic style defines a layer for a point's marker (default layer 0) and associated label. The Display panel can also set the layer state (ON/OFF, and for two- and three-dimensional views), and layer properties (see Figure 2.15).

To change the point marker and/or point label layers, select the layer name value in the layer column. A Layer Selection dialog box displays, listing layers from which the user can select a layer for the component, create a ncw layer in the drawing and assign it to the point style, or switch to another open drawing and use its layer definitions.

If using a Description Key Set, an entry in the key set can assign the layer for the point style (marker). When implementing Civil 3D, a user will have to decide how to assign layers to the point and label styles.

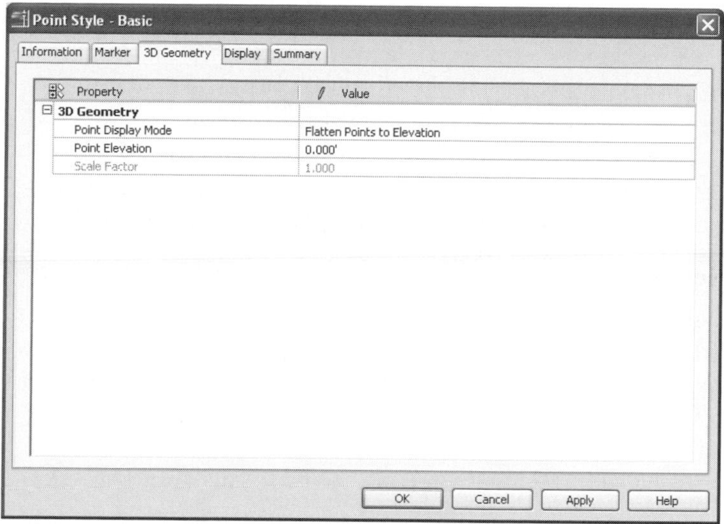

Figure 2.14

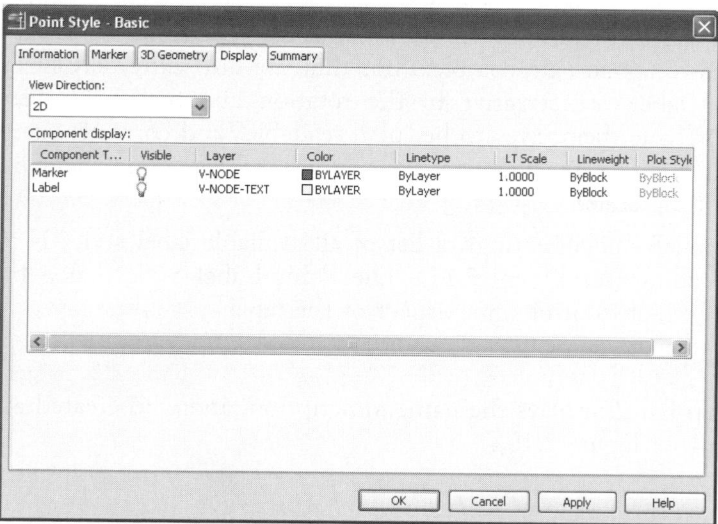

Figure 2.15

SUMMARY

The Summary panel reviews all of the settings in the style (see Figure 2.16).

LABEL STYLE

The Point#-Elevation-Description label style is a basic label definition for a point. A label style displays a point's information adjacent to a marker and has two possible views: two and three dimensions. A label style can contain up to 14 point properties.

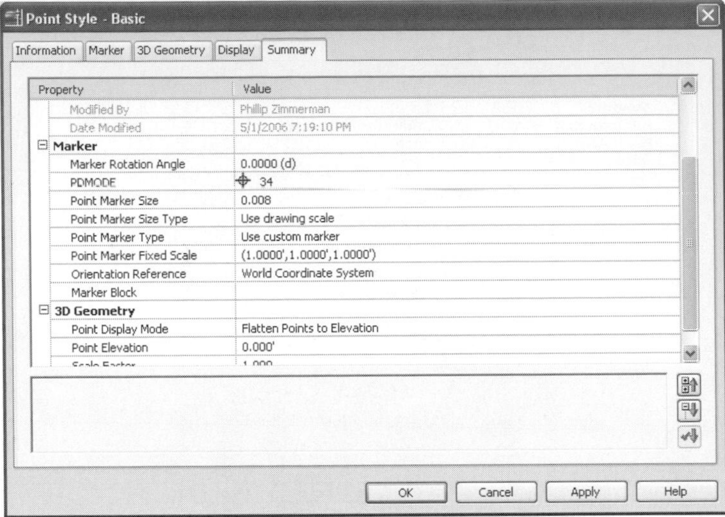

Figure 2.16

A label style places text and point information at user-specified locations around a point. The label styles for points are complex, but they are also extremely flexible and easy to customize. The labels are sensitive to the rotation and scale of a viewport and will re-orientate or rescale themselves to be "plan readable" and correctly sized (i.e., they will read from left to right across the view and the printed page, and they will be the correct size for the plotting scale).

In the Point branch under Settings, a list of all available label styles is found under the Label Styles heading (see Figure 2.17). The Point Label Style dialog box contains five panels, three of which control some aspect of the label.

INFORMATION

The Information panel displays the name, description, and who created and modified the style definition (see Figure 2.18).

GENERAL

The General panel reflects many of the values from the Edit Label Style defaults at the top of Settings (see Figure 2.19). The Label section identifies the text style for the point label, its overall visibility, and if necessary a specific layer when using this style. Often, it is an associated point style that assigns a layer for the label. If Visibility is set to false, all labels using this style do not show. The Behavior section of the panel sets the initial orientation of the label. The last group of settings sets the orientation of the label and affects its plan readability. The Readability Bias setting triggers the change in orientation of a label.

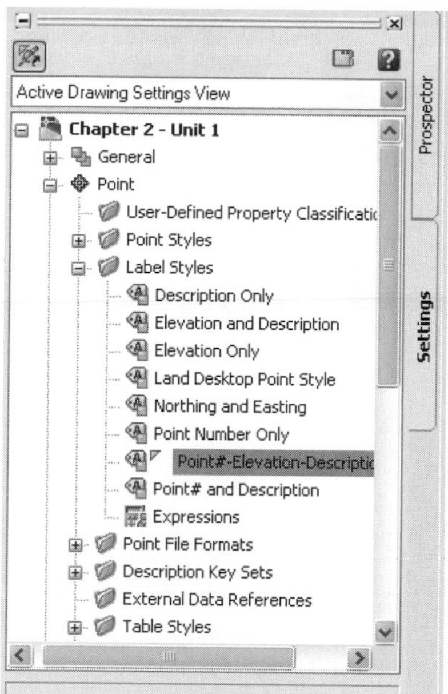

Figure 2.17

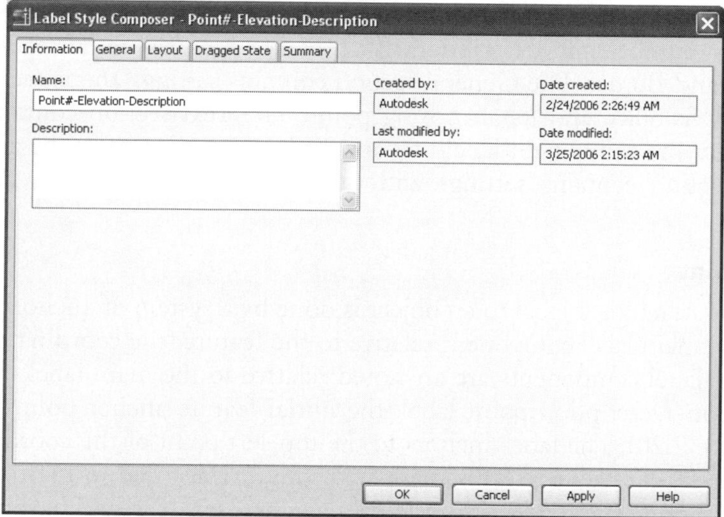

Figure 2.18

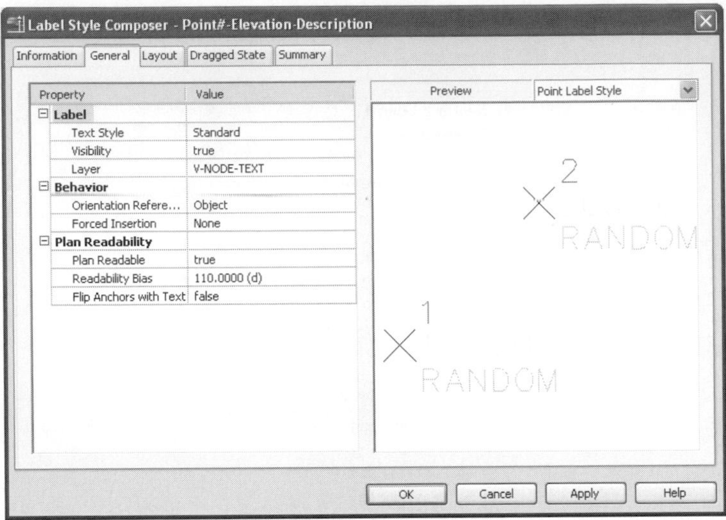

Figure 2.19

LAYOUT

The Layout panel displays the settings and values for each component of a label (see Figure 2.20). The Point#-Elevation-Description point label has three components: elevation, point number, and description. Each component has three sections of settings: General, Text, and Border. The General section contains settings and values for the named component, its visibility, and its anchoring point. The Text section contains settings and values that affect the component label text and its location relative to the anchoring point. The Border section contains settings and values that create a border around the text component.

Anchoring Point

The method of attaching a label to an object is done by a system of anchoring points. Civil 3D anchors the main label component relative to the feature (the coordinates of the point). The remaining label components are anchored relative to the main label element. For the Point#-Elevation-Description point label, the initial feature anchor point is the elevation label (see Figure 2.20). The label anchors to the top-left point of the coordinates (feature). The text of the label attaches to this point with the text justified to the top left. To move the label off the point's left side, the style uses an X offset. It would have been simpler to use the feature's middle-right anchor point to be the component's left-middle attachment point. In this case, no offset would have to be used.

The main component's attachment to the feature sets the location for the remaining components (i.e., point number and description). When viewing the anchoring and attachment information, these refer to the elevation label for the Anchor Component, not the feature (see Figure 2.21). The Point Number label's bottom left attaches to the top left

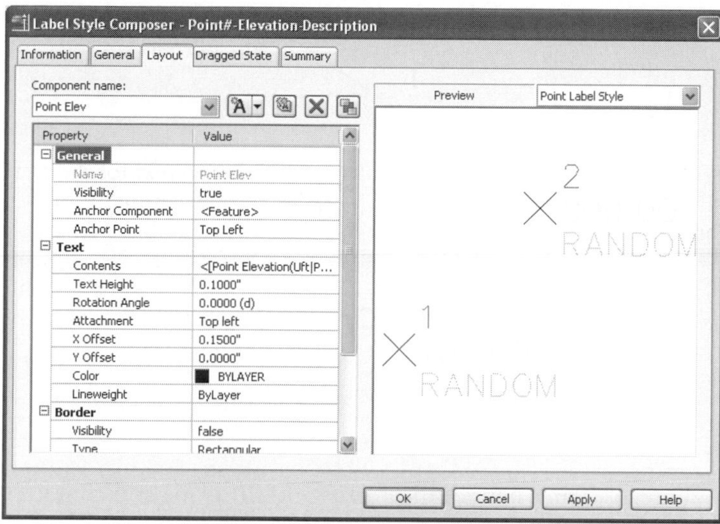

Figure 2.20

of the elevation label. The point description attaches to the bottom left of the elevation label and is top-left-justified from that point (see Figure 2.21).

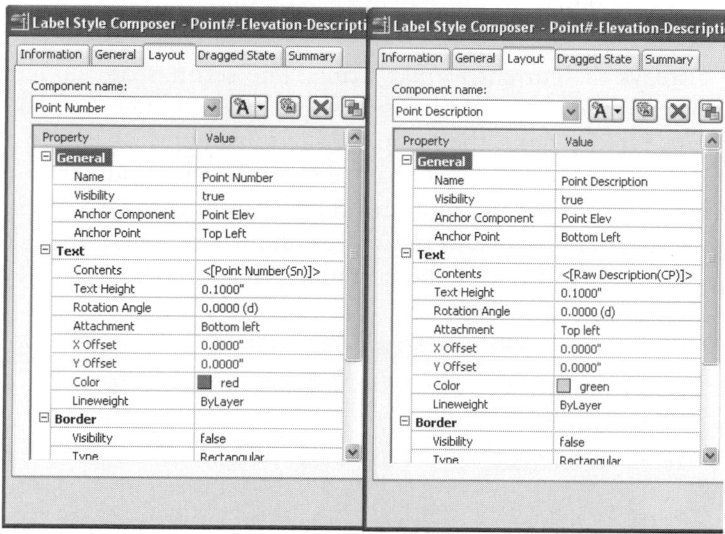

Figure 2.21

Ideally, the anchoring of a point and its labels should be the following: The right-middle of the feature is the middle-left attachment point for the elevation. The elevation's top-left anchor point is the bottom-left attachment point for the point number. The elevation's bottom-left anchor point is the top-left attachment point for the point description (see Figure 2.22).

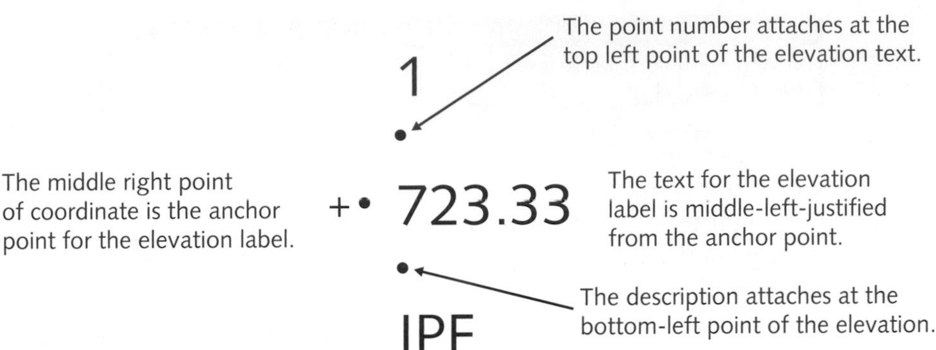

Figure 2.22

Contents

The Contents value is a string containing formatting information for the text component. Each component name in the label has its own text format string. This allows each entry to have its own color, precision, base units, and so on (see Figure 2.23). Users create and edit the format string in the Text Component Editor dialog box. Access this dialog box by clicking in the Value cell for Contents in the Text section of the Layout panel. When the user clicks in the cell, an ellipsis displays at the right side of the cell. When clicking on the ellipsis, the Text Component Editor – Label Text dialog box appears, containing the current format string for the selected component.

The right side displays the current format string. The settings on the left side create the format string shown on the right. The left side shows the modifiers and their values for the current component. On the right, the format string must appear between the < (less than) and > (greater than) signs. A vertical bar separates each modifier. For example, Uft indicates the elevation is in feet, P2 indicates the elevation is to two places, and so on. To change the value of a modifier, click in the Value column of the modifier, change the value, and select the blue arrow to transfer the current settings on the left to the right side.

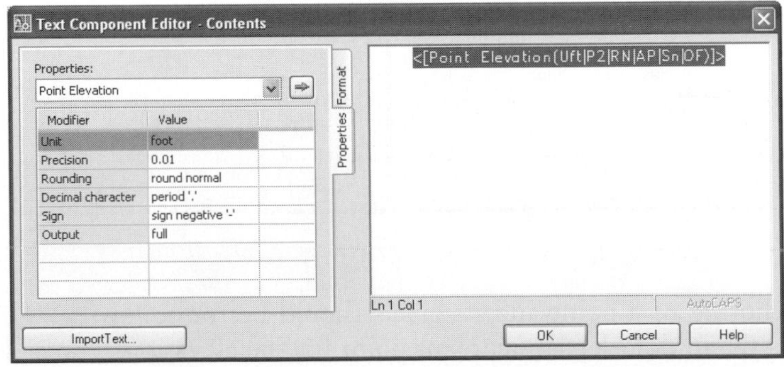

Figure 2.23

Format string values vary depending on the selected point property and its modifier types. When selecting a property from the list, the Text Component Editor provides all the necessary modifiers and settings for a valid format string. An Elevation format includes units (foot or metric), precision, rounding, the decimal character, sign, and output. Precision can vary from an integer (no decimal places) to seven decimal places. A decimal point can either be a period or a comma. To indicate the sign value, Civil 3D allows the user to use a minus sign (-) or brackets to denote negativity, a plus sign, a drop sign (no sign), a left or right parenthesis, or the option to display a value to the right of, left of, all of, or only the decimal character.

To modify an existing format string, highlight the format string on the right side. This action displays on the modifiers on the left side and values of the format string on the right side. Next, make changes to the values on the left side. Finally, click the blue right arrow at the top center of the dialog box to transfer the current settings from the left side to the highlighted format string on the right side. The format string on the right now represents the updated values from the left side.

To add a new property to the label, select its position in the format string area on the right side, then select an item from the Property drop-down list at the upper left of the dialog box. Default values for the property appear on the left. Modify their values and click the blue right arrow to place the values on the left in the format area on the right. When exiting the dialog box the format string becomes the value for the Text Contents of the Current Component name.

EXERCISES

POINT DISPLAY MANAGEMENT

The AutoCAD Erase command permanently deletes a point from the screen and the drawing. This creates an issue of how to prevent points from displaying on the screen while still having them active in a drawing. One method is to turn off the layers used by the point marker and label styles. A second method of hiding markers and labels is by defining a point group that hides the display of a marker and/or its label.

EXERCISE 2-1

When you complete this exercise, you will:

- Become familiar with the Edit Feature Settings for points.
- Become familiar with the label style defaults.
- Be able to create new point styles.
- Become familiar with the point label style.

EXERCISE SETUP

The first task is to set up the drawing environment for the point exercises. The new drawing will use a template file that contains several point and label styles. After reviewing the settings, you will create additional point and label styles.

1. From the File menu, select **New...** and browse to and select the *Chapter 2 – Unit 1.dwt* file from the *Chapter 2* folder of the CD that accompanies this textbook.
2. From the File menu select the **Save As...** command and save the drawing as **Points-1**.

   Review the values in the Edit Drawing Settings dialog box.
3. Select the **Settings** tab to make it current.
4. In Settings, select the drawing name at the top, press the right mouse button, and from the shortcut menu select **Edit Drawing Settings....**
5. Select the **Units and Zone** tab; if necessary, change the drawing scale to **1"=40'** and the zone to **No Datum, No Projection**.
6. Select the **Object Layers** tab. Locate the Point Group and Point Table entries, change the modifier to **Suffix**, and set its value to **-*** (a dash followed by an asterisk).
7. Select the **Ambient Settings** tab. Expand the Direction section, click in the value box for format, and if necessary change the format to **DD.MMSSSS (decimal dms)**.
8. Click the **OK** button to exit the dialog box.

EDIT FEATURE SETTINGS

The values found in the Edit Feature Settings dialog box affect many aspects of points in a drawing. The Default Styles section controls the assignment of point and label styles and the Name Format section controls the naming of point groups and point names.

1. In Settings, select the **Point** heading, press the right mouse button, and from the shortcut menu select **Edit Feature Settings....**
2. Expand the Default Styles section; if necessary, change the point style to **Basic** and the label style to **Point#-Elevation-Description**.
3. Expand the Default Name Format section containing the name and sequential counters for both Point Groups and Point Name.
4. Expand the Update Points section and make sure it is set to false.

   The Update Points section controls the modification of checked-in points. If this is set to true, you would be able to edit points that should not be changed.
5. Click the **OK** button to exit the dialog box.

The Point Identity section controls and lists important point values. The Next Point Number lists the next available point number. The remainder of the section affects Civil 3D point numbers as they are imported into the drawing (see Figure 2.8).

EDIT LABEL STYLE DEFAULTS

The values in this dialog box set the general behavior for all point label styles. The point's Edit Label Style Defaults dialog box allows you to override the current drawing settings. The changes made in the dialog box affect only the labels in the Point branch of Settings.

1. In Settings, select the Point heading, press the right mouse button, and from the shortcut menu select **Edit Label Style Defaults....**
2. Expand each section that applies to points and review their values.
3. Click the **OK** button to exit the dialog box.

POINT (MARKER) STYLES

Civil 3D identifies a point's location in the coordinate system with a marker. This marker can be a simple AutoCAD node, a custom Civil 3D object, or a drawing block (symbol) (see Figure 2.24).

1. In Settings, expand the Point branch until you view the list of point styles.
2. From the Point Styles list, select the **Basic** style, press the right mouse button, and from the shortcut menu select **Edit....**

   The Information tab contains the name and other data about the style.

3. Select the **Marker** tab to view its values.

   Currently, the marker is a custom object, an X, whose size is a product of the drawing scale and the value 0.1, and its orientation is relative to the world coordinate system (see Figure 2.24).

4. Change the marker to a plus sign (+) and toggle on the circle.

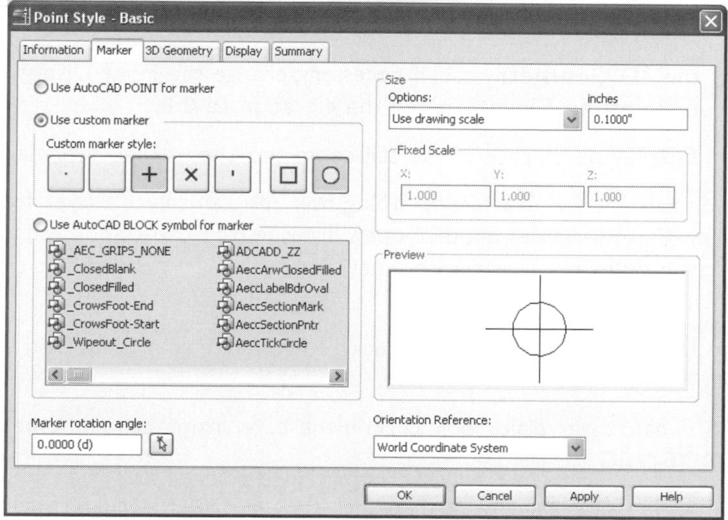

Figure 2.24

5. Select the **3D Geometry** tab; if necessary, change the Point Display Mode to **Flatten Points to Elevation** and the Point Elevation value to **0** (zero).

   The effect of these settings produces a zero elevation object when selecting a point with an AutoCAD object snap. When selecting this point with a Civil 3D Transparent command, the resulting object will have the elevation of the point.

6. Select the **Display** tab to view its contents.

   The Display tab indicates that the marker and label are on the default point layer, V-NODE and V-NODE-TEXT.

7. Click the **Summary** tab to view its contents.

   This tab displays all of the settings from the previous panels and graphically shows what the node will look like in the drawing.

8. Click the **OK** button to exit the dialog box.

CREATE POINT STYLES

The first new point style is for a maple tree. This style uses a maple tree symbol as the marker. Use Figure 2.25 and the information in Table 2.1 for the definition of the point style.

1. In Settings, from the list of point styles, select **Basic**, press the right mouse button, and from the shortcut menu select **Copy...**.

2. Select the **Information** tab; enter **Maple** as the name, give it a short description, and then select the **Marker** tab.

3. In the Marker panel, set the marker to **Use AutoCAD BLOCK symbol for Marker**, and select the **Maple** marker from the block list.

4. In the Size area, change the inches value to **0.1**, if necessary.

5. Select the **3D Geometry** tab; if necessary, change the Point Display mode to **Flatten Points to Elevation** and the elevation to **0.0**.

6. In the **Display** panel, click in the cell containing the marker layer name (V-NODE).

   This displays the Layer Selection dialog box. You can select a layer from the current drawing's layer list, from another open drawing (the Layer Source drop-list at the top left of the dialog box), or create a new layer (the New... button at the top right of the dialog box).

7. In the Layer Selection dialog box, click **New...** at the top right of the dialog box to display the Create Layer dialog box.

8. In the Create Layer dialog box, click in the Layer name cell and type **V-NODE-VEG**.

9. Click in the Color values cell, then click the ellipsis (right side of cell). In the Select Color dialog box, assign a color, and click the **OK** button to exit.

10. Click the **OK** button of the Create Layer dialog box.

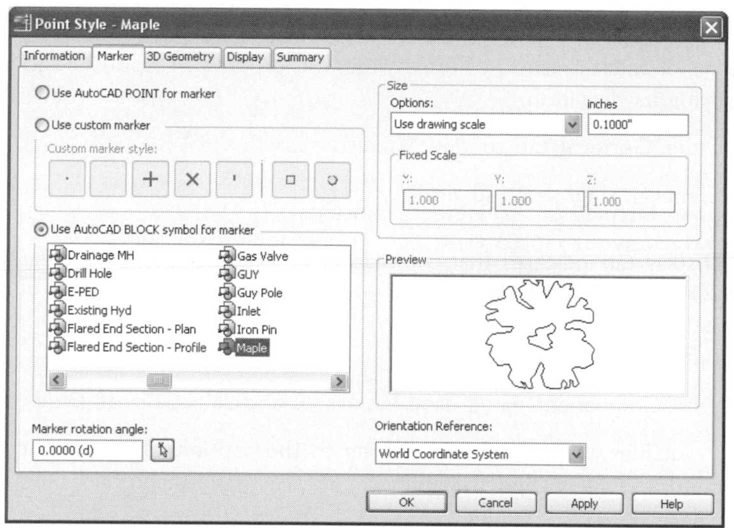

Figure 2.25

11. In the Layer Selection dialog box, locate and select the **V-NODE-VEG** layer from the layer list, and click the **OK** button to assign the selected layer to the marker.
12. Repeat Steps 7 through 11 to create a layer for the point label (**V-NODE-VEG-LBL**), but select the layer name (**V-NODE-TEXT**) to the right of the label layer entry.
13. Click the **OK** button to create the Maple point style and to exit the dialog box.

    Repeat Steps 1 through 13 and use the information in Table 2.1 to create an Oak point style. You do not need to create the **V-NODE-VEG** and **V-NODE-VEG-LBL** layers; just select them from the layer list. This style has the Point Display mode set to **Flatten Points to Elevation** and an elevation of **0.0**.

Table 2.1

Style Name	Marker Block	Display Marker	Display Label
Maple	Maple	V-NODE-VEG	V-NODE-VEG-LBL
Oak	Oak	V-NODE-VEG	V-NODE-VEG-LBL

REVIEW THE POINT#-ELEVATION-DESCRIPTION POINT LABEL STYLE

All of the points for this exercise use the Standard Point Label style.

1. In Settings, expand the Point branch until you view the list of point label styles.
2. From the list of styles, select **Point#-Elevation-Description**, press the right mouse button, and from the shortcut menu select **Edit...**.

   This displays the Label Style Composer dialog box.

3. Select the **Information** tab to view its values.

   This panel sets the name of the style and the person responsible for creating or modifying its definition.

4. Select the **General** tab to view its settings.

   The General panel sets the visibility and other values for the entire label.

5. Select the **Layout** tab to view its text component settings.

   The Component name drop-down list includes all of the types of components in the label style. This label style contains definitions for point number, elevation, and description.

6. Click the Component name drop-list arrow and select **Point Elev**.

   The Point Elev component is anchored to the top left of the Feature (General section) and its text is top-left-justified to that point (attachment value of the Text section).

7. Click the Component name drop-list arrow and select **Point Number**.

   The Point Number component is anchored to the top left of the Point Elevation label (General section) and its text is bottom-left-justified to that point (attachment value of the Text section).

8. Click the Component name drop-list arrow and select **Point Description**.

   The Point Description component is anchored to the bottom left of the Point Elevation (General section) and its text is top-left-justified to that point (attachment value of the Text section).

FORMATTING LABEL TEXT

Point Number, Point Elev, and Point Description all have their own text formatting string. The values in the string tell Civil 3D how to format the label's information on the screen (see Figure 2.23).

1. With **Point Description** as the current component, in the Text section click in the content's Value cell.
2. Click on the ellipsis at the right of the value cell to view the Text Component Editor.
3. On the left side of the dialog box, click in the Value cell for Capitalization and click the drop-list arrow to view the list of capitalization options.
4. Click the **Cancel** button to exit the Text Component Editor.
5. In Layout, click the Component name drop-list and select **Point Elev**.
6. With **Point Elev** as the current component, in the Text section, click in the content's Value cell.
7. Click on the ellipsis at the right of the value cell to view the Text Component Editor for Point Elev.

8. Click in the different value cells to view the possible formats and options for an elevation label.
9. Click the Cancel buttons until exiting the Text Component Editor and the Label Style Composer dialog boxes.

TABLE STYLE

Civil 3D has table styles that list point data in table form.

1. In Settings, expand the Point branch until you view the list of table styles.
2. From the list of table styles, select **Latitude and Longitude**, press the right mouse button, and from the shortcut menu select **Edit...**.
3. Select the **Data Properties** tab and review its values.
4. Select the **Display** tab and review its values.
5. Click the **OK** button to exit the dialog box.
6. Click the **AutoCAD Save** icon to save the drawing.

Civil 3D has several settings that control visibility and how the application works with point data. These values should be set up in advance; however, Civil 3D is flexible enough to accommodate most unanticipated situations.

SUMMARY

- Point styles can assign layers for both the marker and the label.
- Point styles define a point's marker.
- Point label styles define how and what information displays next to a marker.
- Point label styles anchor one component to the feature.
- The remaining label text anchors relative to the initial label or other label text.

UNIT 2: CREATE POINTS

The majority of point creation routines are in the Create Points toolbar. A user calls this toolbar from the Points menu or from a shortcut menu of the Prospector's Points heading.

The point creation routines create points from printed lists, external ASCII files, or calculated coordinates. This unit will not review all of the toolbar commands. The remaining routines will be reviewed in later chapters in a more appropriate context.

SETTINGS FOR THE CREATE POINTS COMMAND

When using the routines of the Create Points toolbar, users need to review and set values in a toolbar data panel or a command settings dialog box (the Command List under Settings).

CREATE POINTS

The Edit Command Settings dialog box for the Create Points toolbar contains two groups of settings: Default Layer and Points Creation (see Figure 2.26). The Points Creation section determines how to reference coordinates in a local coordinate system (northing/easting), how to assign elevations and descriptions, elevation and description parameters if using description keys, and if the user wants the coordinates echoed to the command line.

When creating a point, the Prompt For Elevations and Prompt For Descriptions parameters have three options: None, Manual, or Automatic. Set an option by clicking in the value cell containing the current setting for Prompt For Elevations or Descriptions, then click on a drop-list arrow, and from a list select the option (None, Manual, or Automatic). The None option does not prompt for an elevation or a description. Manual prompts the user to type an elevation or a description. Automatic uses the current value in the Default Elevation or Default Description value cell. Initially, this value is 0 (zero) or blank. After entering an elevation or description, that value becomes the default value for either Default Elevation or Default Description.

The use of description keys and parameters is a topic of the last unit of this chapter.

Users can access these settings in three locations: Edit Feature Settings, Edit Command Settings, and in the Data panel of the Create Points toolbar.

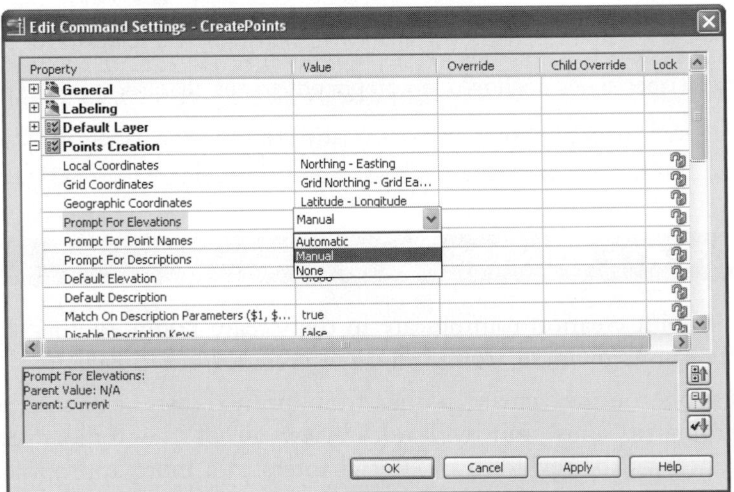

Figure 2.26

CREATE POINTS TOOLBAR

The Create Points toolbar contains seven types of point creation routines (see Figure 2.27). The first six types of point creation commands are Miscellaneous, Intersection, Alignment, Surface, Interpolation, and Slope (icons from left to right). Each type has an icon stack representing individual point-creation commands. These routines convert written data, mimic surveying techniques, interpolate between points, and create points from alignment or surface objects. The last point creation command is the rightmost icon, Import Points. This routine creates points as it reads an ASCII text file.

Figure 2.27

MISCELLANEOUS

The Miscellaneous category includes creating points manually by referencing existing points (Geodetic Direction and Distance and Resection), by referencing existing AutoCAD objects, and by Converting Softdesk point blocks into Civil 3D point objects (see Figure 2.28).

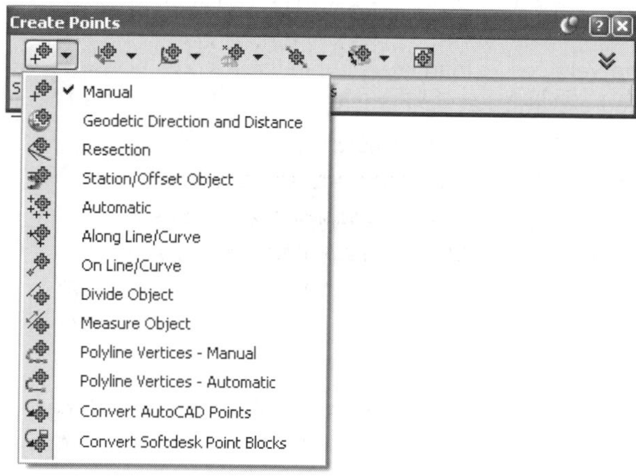

Figure 2.28

- The Manual routine places point objects at command-line entered coordinates (X,Y) or from coordinates of an object snap selection.
- The Geodetic Direction and Distance routine requires a drawing with an assigned coordinate system.
- The Resection command creates a new point whose location is derived from three known points with measured angles between them.
- The routines from Station/Offset Object to Polyline Vertices – Automatic require existing AutoCAD objects.
- The Station/Offset Object, On Line/Curve, Divide Object, and Measure Object routines work only with lines and curve segments. If selecting a polyline, the active routine will filter it out and will ask the user to select another object. All of these routines measure from a starting point on a selected object. The starting point is the endpoint nearest to its selection point. For example, if the user selects at the southern end of a line, that end is the lowest station or the starting point of the measurements.
- The Automatic routine places points at the same locations as the On Line/Curve routine. The difference between the routines is that Automatic works with a selection set of lines and curves rather than individual line and curve segments. Before the Automatic routine places points, it scans the selection set for duplicate (endpoint) coordinates and if it finds any, places one point in the drawing at the coordinates. If using the On Line/Curve routine, it places two points at the end coordinates (one point for each selected object).
- The Polyline Vertices – Manual routine uses a 2D polyline and places a point with a user-specified elevation at each vertex of the polyline. The Polyline Vertices – Automatic routine uses a 3D polyline and places a point at each vertex whose elevation is the elevation of the vertex.
- The Convert AutoCAD Points converts AutoCAD nodes into Civil 3D point objects. The routine prompts for a description for each point.
- The Convert Softdesk Point Blocks converts nodes into Civil 3D point objects.

AUTODESK CIVIL 3D CONVERT LDT POINTS

The Utilities flyout of the Points menu has a routine that converts Autodesk Land Desktop (LDT) point objects to Civil 3D point objects (see Figure 2.29).

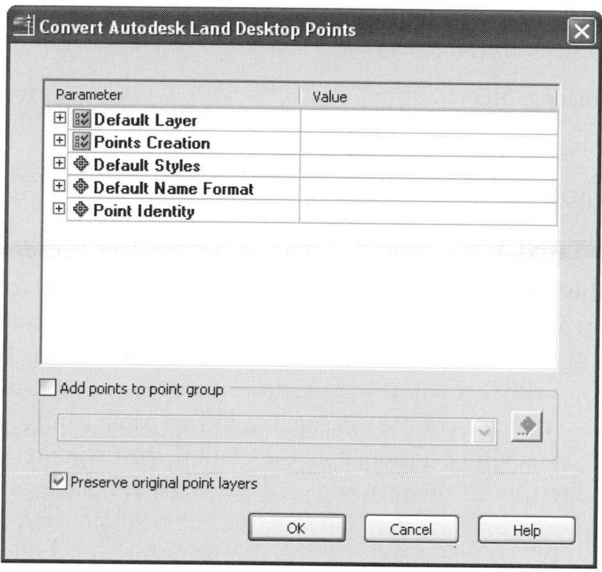

Figure 2.29

EXTENDING COMMANDS

Creating points from a printed list of northings/eastings, turned angles, bearing and distances, side shots, etc. cannot be done in the Create Points toolbar. Few if any of the Create Points routines prompt for northing/easting coordinates or understand azimuths, bearings, and distances to define coordinates. The overrides of the Transparent Command toolbar change a routine's prompting and allows the routine to understand other point placement methods. For example, if using the Manual command to type "NE" at the command line changes the routine's prompting to northing and easting. To change Manual to use a direction and distance, type 'AZ (azimuths) or 'BD (bearing). To mimic field collection techniques, use 'AD (turned angle) and 'SS (side shot). Most of the transparent commands mimic field data collection techniques. Other transparent commands reference specific Civil 3D objects (points, alignments, and profile views) and AutoCAD objects (lines, arcs, polyline, etc.).

Note: To use the transparent commands effectively, Autodesk recommends that dynamic mode be off when using them.

COORDINATES

When invoking a transparent command, the routine changes its prompting for coordinates. The Grid and Latitude transparent overrides require a coordinate zone definition to function. Set a zone in the Edit Drawing Settings dialog box. The following is a list of Coordinate transparent commands:

- Northing/easting: 'NE
- Grid northing/easting: 'GN
- Latitude/longitude: 'LL

ANGLES AND DISTANCES

The 'AD, 'DD, and the 'SS transparent commands simulate the station (pivot point), backsight, and foresight process in surveying. When using a routine with this override, users must establish a station and a backsight by selecting a line or two points. If selecting a line, the endpoint nearest to the selection point is the station (pivot point) and the farthest endpoint is the backsight point (the direction of the zero angle). When establishing a station and backsight in point mode, select two points: The first is the station (pivot point) and the second is the backsight point (the direction of the zero angle). The Angle Distance and Deflection/Distance routines use the setup once and the user must define the next station/backsight location. The Side Shot override uses the same pivot and backsight points until changing the setup. The following is a list of Angle and Distances transparent commands:

- Turned angle and distance: 'AD
- Deflection/distance: 'DD
- Side shot from point: 'SS

POINT OBJECT FILTERS

Most transparent commands have a Points option. For example, the By Turned Angle transparent command locates a setup by selecting a line or by selecting a pair of points. Any command that has a points (coordinate) mode can reference a point object or a set of northing/easting coordinates. When a routine prompts for a point, the point can be a command-line entered value or coordinates from an AutoCAD selection (usually with an object snap). If the user enters **.n** at the prompt, the prompting changes so he or she can enter northing/easting values for the point coordinates. If the user enters **.p**, the prompting changes to a point number to determine the coordinates for the point. If the user enters **.g**, the prompting switches to select a point object to determine the coordinates.

The following are code snippets to show how point object filters affect the prompting for coordinate values when using the Angle Distance override of the transparent commands.

Selecting an AutoCAD line entity for coordinate values

```
>>Select line or [Points]: (select a line entity)
```

Selecting AutoCAD coordinates for coordinate values

```
>>Select line or [Points]: P (toggles to Point mode)
>>Specify starting point or [.P/.N/.G]: (make an AutoCAD
    selection)
>>Specify ending point or [.P/.N/.G]: (make an AutoCAD selection)
```

Specifying a Civil 3D point number for coordinate values

```
>>Select line or [Points]: P (toggles to Point mode)
>>Specify starting point or [.P/.N/.G]: .P (toggles to Point
     number mode)
>>Enter point number: 240
>>Specify ending point or [.P/.N/.G]: .P (toggles to Point number
     mode)
>>Enter point number: 250
```

Entering Northings/Eastings for coordinate values

```
>>Select line or [Points]: P (toggles to Point mode)
>>Specify starting point or [.P/.N/.G]: .N (toggles to NE
     coordinates)
>>>>Enter northing <0.0000>: 5500.6395
>>>>Enter easting <0.0000>: 5892.4533
>>Specify ending point or [.P/.N/.G]: .N (toggles to NE
     coordinates)
>>>>Enter northing <0.0000>: 5830.3095
>>>>Enter easting <0.0000>: 5942.0523
```

Selecting Civil 3D point object from the display for coordinate values

```
>>Select line or [Points]: P (toggles to Points mode)
>>Specify starting point or [.P/.N/.G]: .G (toggles to select a
     point object mode)
>>
Select point object: (select a point object from the display)
>>Specify ending point or [.P/.N/.G]: .G
>>
Select point object: (select a point object from the display)
```

ANGLE FORMATS

Civil 3D uses four types of angle entries: decimal degrees, three variations of these degrees, minutes, and seconds. A routine prompting for an angle displays the current angle format. Users set the angle format in the Ambient Settings panel of Edit Drawing Settings (see Figure 2.30). The traditional method for surveyors is a decimal format for degrees, minutes, and seconds. The surveyor's convention enters the angle of 34 degrees, 52 minutes, 18 seconds as 34.5218 (dd.mmss). This convention does not allow for values greater than 59 for minutes and seconds.

DIRECTIONS

When entering directions, Civil 3D supports both the azimuth and bearing systems. The azimuth system assumes 0 degrees to be north and measures angles clockwise from 0 to 359.5959 (dd.mmss) (see Figure 2.31). The bearing method divides a circle into four 90-degree segments. The top two quadrants, the northeast (quadrant 1) and northwest (quadrant 4), assume that a vector deflects from north to the east or west. So the angle of a vector varies from 0 degrees, a line traveling due north, to 90 degrees, a line traveling due east or west. The bottom two quadrants, the southeast (quadrant 2) and southwest (quadrant 3), assume that a

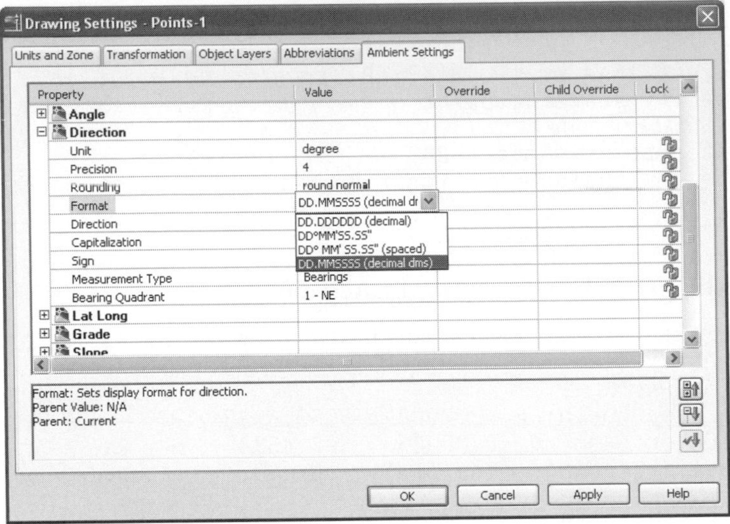

Figure 2.30

vector deflects from south to east or west. So the angle of a vector varies from 0 degrees, a line traveling due south, to 90 degrees, a line traveling due east or west.

Direction Transparent Commands

After establishing the location of a new point or starting from a known point, the two transparent commands ('BD and 'ZD) allow users to define a direction to the next point's location. Define the direction by either a bearing ('BD) or an azimuth ('ZD).

- Bearing/distance – 'BD
- Azimuth/distance – 'ZD

The following are code snippets showing how the direction transparent commands affect the prompting for directions and how a routine indicates the current angle or direction format set in the Ambient Settings panel of the Edit Drawing Settings dialog box.

Setting a point by bearing

```
Command:
Please specify a location for the new point: '_BD
>>Select starting point or [.P/.N/.G]: .P
>>Enter point number: 244
Quadrants - NE = 1, SE = 2, SW = 3, NW = 4
>>Specify quadrant (1-4): 1
Current direction unit: degree, Input: DD.MMSSSS (decimal dms)
>>Specify bearing: 57.5652
>>
>>Specify distance: 45.5234
Resuming MODELESSDISPATCH command.
Please specify a location for the new point: (6098.79 5453.6
    0.0)
Enter a point description <.>: IPF
```

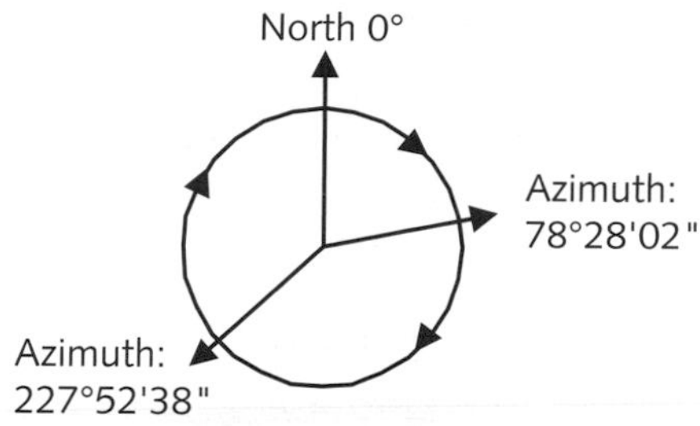

Azimuths:
Assumes North is 0°
Measures Angles Clockwise
Angles are between 0° and 360°

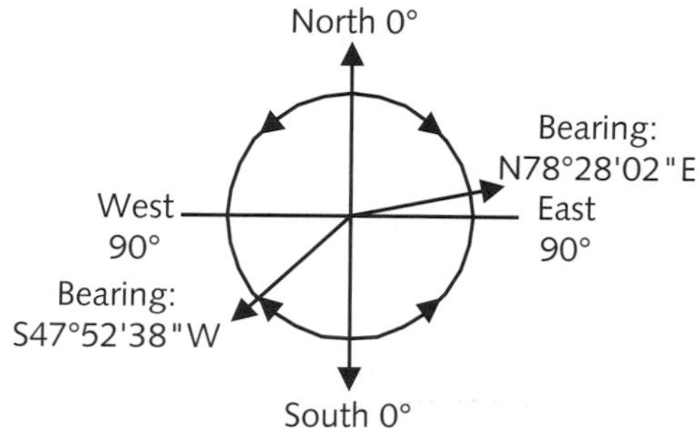

Bearing:
Assumes North or South is 0°
Angles are between 0° and 90°
East or West

Figure 2.31

```
Specify a point elevation <.>: 723.44
N: 5453.5982' E: 6098.7912'
Please specify a location for the new point:
Quadrants - NE = 1, SE = 2, SW = 3, NW = 4
>>Specify quadrant (1-4):
```

Setting a point by azimuth

```
Command:
Please specify a location for the new point: '_ZD
>>Select starting point or [.P/.N/.G]: .P
>>Enter point number: 244
Current direction unit: degree, Input: DD.MMSSSS (decimal dms)
>>Specify azimuth: 73.2644
>>Specify distance: 62.7495
Resuming MODELESSDISPATCH command.
```

```
Please specify a location for the new point: (6126.26 5439.78
    0.0)
Enter a point description <.>: IPF
Specify a point elevation <.>: 730.76
N: 5439.7831' E: 6126.2615'
Please specify a location for the new point:
Current direction unit: degree, Input: DD.MMSSSS (decimal dms)
>>Specify azimuth:
Resuming MODELESSDISPATCH command.
```

All of the angle/distance transparent commands can be used with many AutoCAD commands (lines, polyline, move, etc.). When using an AutoCAD command with a transparent Civil 3D command, simply click the appropriate transparent command after starting the AutoCAD command and follow the prompting until the AutoCAD command is complete.

ALIGNMENTS

The Alignment transparent command requires an alignment.

Station/offset: 'SO

NORTHING/EASTING, GRID, AND LONGITUDE/LATITUDE

When in a drawing with local coordinates, users can toggle a command into northing/easting mode by selecting the Northing/Easting icon or typing 'NE at the command line. When a drawing has an assigned state plane or UTM (Universal Transverse Mercator) zone, toggle on a Grid ('GN) or a Latitude/Longitude ('LL) override. The northing/easting and grid northing/grid easting overrides change the prompting to northing and then easting. Civil 3D will never allow the values to be a comma-delimited entry. When toggling on Latitude/Longitude, the command first prompts for latitude and then longitude.

POINTS

The Point transparent commands are useful when using AutoCAD commands and when there is a need to reference existing Civil 3D point objects. For example, using the line command 'PN allows the user to draw the line by point numbers and 'PO allows the user to select point objects from the display to incorporate their coordinates into the line. 'PA selects points by their names. These commands will use the elevation of the points as the elevations along the length of the line.

These transparent commands behave exactly like the .p, .n, and .g point filters. However, .p, .n, and .g can be used only when the command line lists them as options. If the dot point filters are not listed as an option, the transparent 'PN, 'NE, and 'PO should be used.

- Point number: 'PN
- Point name: 'PA
- Point object: 'PO
- Northing/easting: 'NE

CREATE POINTS: INTERSECTION

There are three types of intersection routines (see Figure 2.32). The first type uses a theoretical intersection, which includes intersections between directions and distances from known points. These are the first five commands in the list. The second intersection type assumes the existence of line and/or arc entities in the drawing (Object/Object and Perpendicular). The lines and arcs define directions and/or distances for the computations. The last type of intersection routine works only with Civil 3D alignment objects and are the last four routines on the list.

A direction can be a bearing, an azimuth, a direction of an existing line, a direction between two AutoCAD selections (including object snaps), or two point objects.

Most of the intersection routines calculate a single solution. There are, however, some routines that calculate two possible solutions. For example, the Distance/Distance routine calculates two possible solutions. The routine identifies each solution with an X in the drawing. Users have the option of selecting one or both solutions. Select an individual solution by selecting near one X with the cursor, or select all solutions by placing a point at each solution.

Many of the intersection routines allow an offset to the defined direction. The convention is a negative value as left and a positive value as right, when at the starting point facing toward the direction.

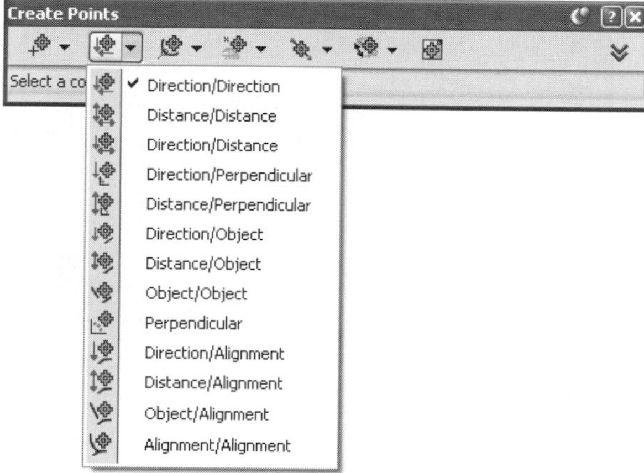

Figure 2.32

ALIGNMENT/SURFACE/INTERPOLATE/SLOPE

The routines for this group will be discussed in appropriate chapters later in this textbook.

IMPORT

The rightmost icon on the Create Points toolbar is the Import Points command. If the user receives an ASCII file containing coordinate data, the routine reads the coordinate records in the file and creates points in the drawing. The Import Points routine has several predefined formats for reading coordinate file records.

The Import Points dialog box sets the file format and file name, can assign imported points to a point group, and has advanced options for using imported data as modifiers of existing data (see Figure 2.33).

The first option with the Advanced options is modifying point elevations, Do elevation adjustment if possible. This option is similar to a datum adjustment of a point's elevation. When importing a point file, a column in the file format named +Z or –Z contains values that adjust the elevations of the points as they are placed in the drawing.

The second option transforms a point's file coordinates to new drawing coordinates based on the currently assigned drawing coordinate system. An example of this is importing points having NAD27 US Foot coordinates into a drawing using the NAD83 US Foot system.

The last option, Do coordinate data expansion if possible, forces Civil 3D to calculate longitude and latitude values for the incoming points.

Civil 3D uses the same formats and options when exporting points to ASCII files.

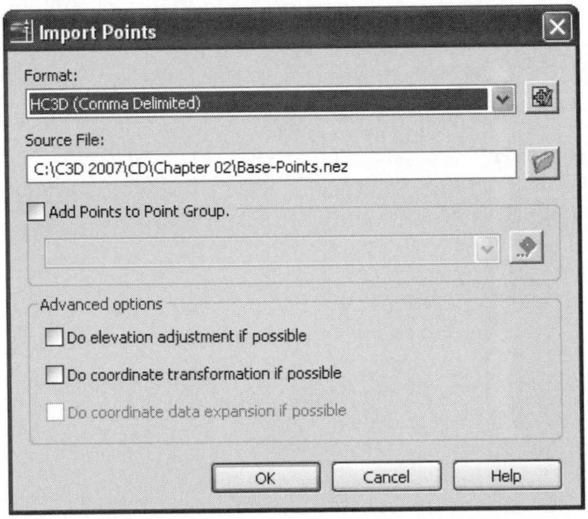

Figure 2.33

Before importing a file, users must know the structure of a line of data (or a record). Users can view an ASCII file in Notepad to determine the structure of the file. However, viewing the file does not reveal one very important piece of information. That piece of information is which fields in a record represent the northing (Y) and easting (X) coordinate values. If using the wrong format, points will appear incorrectly in the drawing. The only way to know this information is to have notes about the file's format or to talk to the person responsible for creating the file.

The following is an excerpt from a typical coordinate file:

```
1,5530.83673423,4967.36847364,723.74,PP 60
2,5635.36849278,4952.84689347,722.84,IPF
```

This file is comma-delimited and contains the following fields: point number, northing, easting, elevation, and description. Again, the only way to determine the order of data is to have some documentation identifying the fields or a conversation with the person creating the file. Civil 3D contains a format, PNEZD comma-delimited, which matches this file's structure.

The Import Options section of the previous unit discusses how Civil 3D handles point numbers and how it assigns new point numbers to duplicate points during an import session.

FORMAT MANAGER

If a user has a file with a format not on the list or wants to specify a coordinate system with a file format, he or she can create custom formats. Civil 3D's Format Manager defines file formats accommodating most any file format and coordinate system. Display Format Manager by clicking the icon to the right of the current format (see Figure 2.33).

The type and format of a file's records determines the structure of the format reader. If setting a coordinate system, the Import routine can import longitude and latitude or coordinates from a different coordinate system. When importing points from a different zone, the Import Points routine calculates new coordinates for the incoming points.

The Import routine supports files that are comma-, space-, or column-delimited. The format of the file dictates which format the Import Points routine uses. Users must exercise care when creating a column-delimited format, because in a column-delimited file, a decimal point is a part of the overall field width.

A coordinate file consists of columns and rows. A field is a column of a specific type of information, (such as northing, easting, elevation, etc.). A row is known as a record, or a unique occurrence of fields in the file. In Civil 3D, some of the field types are point number, northing, easting, elevation, description, latitude, longitude, convergence, and scale factor.

The Point File Format dialog box defines new point file formats and allows users to modify an existing point file format (see Figure 2.34). The top portion of the dialog box sets the name, extension, delimiter (if applicable), and coordinate zone (if applicable). The bottom portion of the dialog box defines the fields (columns) for a file record.

DATA SETTINGS PANEL

The Create Points toolbar has a roll-out panel containing all of the settings found in the Edit Feature Settings dialog box for point objects. This panel includes settings that affect point numbering, prompting for descriptions and elevations, and rules that affect how to

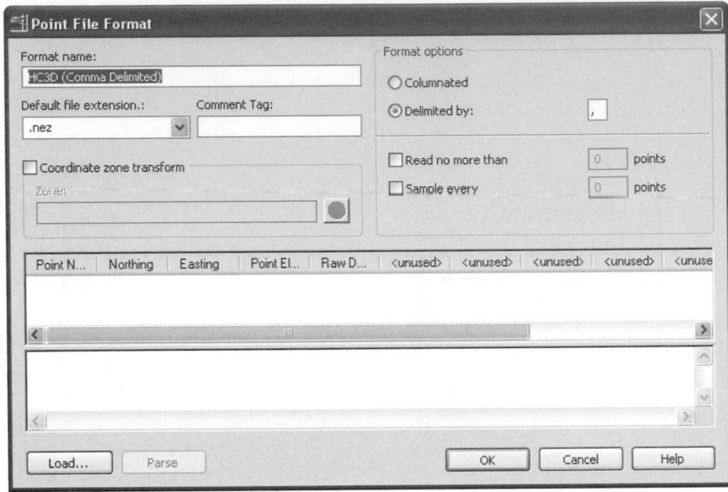

Figure 2.34

import points into the drawing (see Figure 2.35). The Next Point Number is set only in Setting's Edit Feature Settings for Point option.

Create Points

Parameter	Value
Point Identity	
Next Point Number	10052
Use Sequential Numbering	true
Point Number Offset	1
Sequence Point Numbers From	10000
If Point Numbers Are Supplied	Use
Force Names	false
If Point Numbers Already Exist	Notify
If Point Names Already Exist	Notify

Select a command from the point creation tools

Figure 2.35

LANDXML

Another method for creating points is by reading in a LandXML file. LandXML has a very specific schema for point information. The Import LandXML routine allows Civil 3D to read a file and create points in the drawing (see Figure 2.36).

When exporting points to a LandXML file, the numbers become a point name (CgPoint name). When importing a LandXML file, Civil 3D assigns a point number matching the point name as long as it is not a duplicate number. If the number is a duplicate, the settings in Point Identity determine the new point's number.

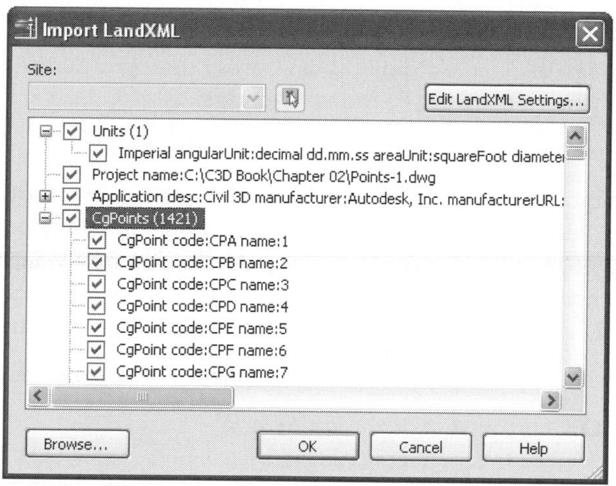

Figure 2.36

EXERCISE 2-2

When you complete this exercise, you will:

- Create an Import file format.
- Import an ASCII file.
- Import points using LandXML.
- Be familiar with setting points with routines from the Miscellaneous menu of the Create Points toolbar.
- Be familiar with setting points with routines from the Intersection menu of the Create Points toolbar.
- Be able to use transparent commands while placing points in the drawing.

EXERCISE SETUP

If you are not still in the *Points-1 drawing* file that you worked with in the previous unit, open it now. If you are just starting with this exercise, you need to open the drawing *Chapter 2 – Unit 2.dwg* to start the exercise. The *Chapter 2 – Unit 2.dwg* is in the *Chapter 2* folder of the CD that accompanies this textbook.

1. Open the *Points-1 drawing* or *Chapter 2 – Unit 2* drawing file for this exercise.

REVIEW POINT FEATURE SETTINGS

Before importing points, you need to check the values in the Point Identity section of the Point Edit Feature Settings dialog box or the Data panel of the Create Points toolbar. This section determines basic values and rules for importing points into a drawing. If the drawing already has points, these settings are critical in not destroying or modifying existing point data. These same values appear in the Create Points toolbar.

1. Click the **Settings** tab to make it current.
2. From Settings, select the **Point** heading, press the right mouse button, and from the shortcut menu select **Edit Feature Settings....**
3. Find and expand the Point Identity section of the dialog box.

   In an empty drawing most of the current settings are fine. However, if the drawing contains points, the next point number would be something other than 1. When importing points to a drawing containing points, you may want to change the Sequence Point Numbers From value. Make sure your dialog box matches the values in Figure 2.37 before exiting the Edit Feature Settings dialog box.
4. Change the value for Sequence Point Numbers From value to **10000**.
5. Click the **OK** button to exit the dialog box.

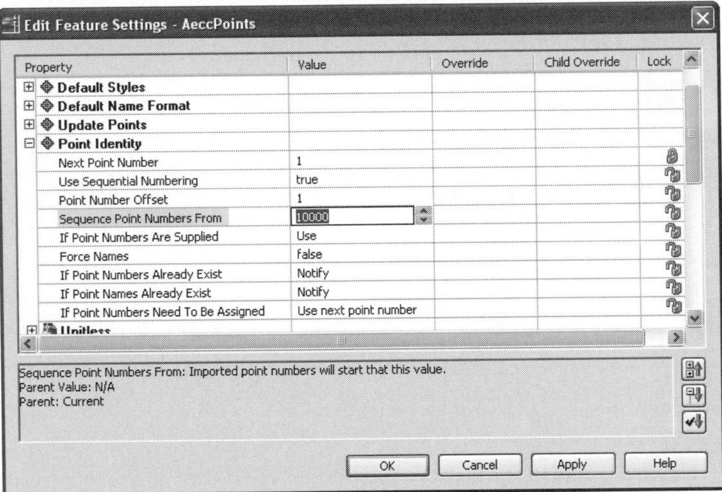

Figure 2.37

IMPORT AN ASCII FILE

The Import Points command is an icon on the Create Points toolbar.

1. Click the **Prospector** tab to make it current.
2. In Prospector, select the **Points** heading, press the right mouse button, and from the shortcut menu select **Create....**

   This displays the Create Points toolbar.
3. Click on the various drop-list arrows to view the lists of point commands under the icons.

CREATE A NEW POINT FILE FORMAT

The first step to reading the *Base-Points.nez* file is creating a new file format. The file format is point number, northing, easting, elevation, and raw description with a comma delimiter.

1. From the Create Points toolbar, select the **Import** icon at the right side.
2. In the Import Points dialog box, click the **Point Formats** icon to the right of the Format list at the top right of the dialog box.

   This displays the Point File Format dialog box containing all the current file formats.
3. In the Point File Format dialog box, click the **New...** button to display the Point File Formats – Select Format Type dialog box.
4. In the Point File Format – Select Format Type dialog box, select **User Point File** and click the **OK** button to return to the Point File Format dialog box.

   To create a new file format, enter a format name, set its extension, set its delimiter, and define a record's types of data and order. When finished setting the values, your file should look like Figure 2.38.
5. In the Point File Format dialog box at the top left, enter **HC3D (Comma Delimited)** as the name and set the extension to **.nez**.
6. In the Point File Format dialog box, in Format Options at the top right, select **Delimited by** and type in a comma (,) in the box to the right of the toggle.
7. In the Point File Format dialog box, in the middle left, click the first unused heading.

   This displays the Point File Format – Select Column Name dialog box.
8. In the Point File Format – Select Column Name dialog box, click the drop-list arrow to display the list of data types and select **Point Number** from the list.

   When selecting Point Number, the Point File Format – Select Column Name dialog box now includes an entry defining an invalid point number.
9. Click the **OK** button to assign Point Number to the first field of a file record and to return to the Point File Format dialog box.
10. Click the second column in from the left (<unused>).
11. In the Point File Format – Select Column Name dialog box, click the drop-list arrow to display the list of data types and select **Northing** from the list.

    After selecting Northing, the Point File Formats – Select Column Name dialog box now contains entries identifying invalid coordinates and the precision for northing coordinates. The precision value is important in column-formatted data, not comma- or space-delimited data. So for this exercise, you can ignore precision setting. However, if working with a columned file or with invalid indicators, these values need to be set.
12. Click the **OK** button to return to the Point File Format dialog box.

13. Repeat Steps 10 through 12 and add Easting, Point Elevation, and Raw Description to the file format. The new headings are to the right of the heading you just defined.

    After adding the three headings, your Point File Format dialog box should look like Figure 2.38.

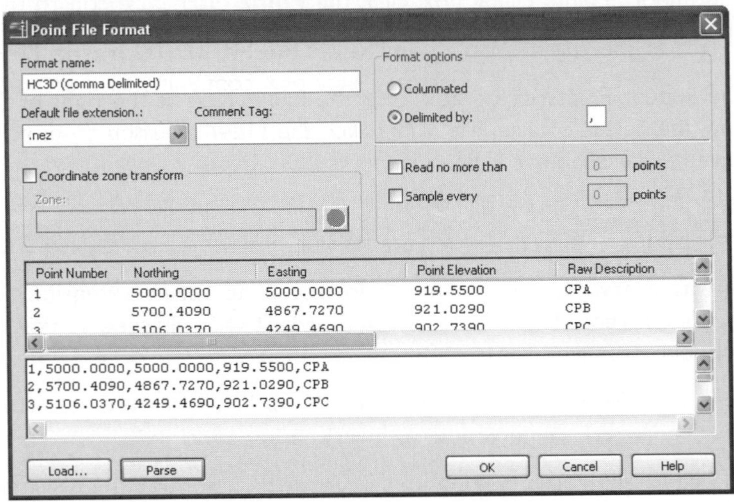

Figure 2.38

TESTING THE FORMAT

The testing file for the new format is *Base-Points.nez*. This file is in the *Civil 3D* project folder you copied from the CD to the *Civil 3D Projects* folder on your hard drive (see the exercise for Chapter 1, Unit 4) or you can use a copy of the file from the *Chapter 2* folder of the CD that accompanies this textbook. First load the file into the Point File Format dialog box and then parse the file to see if it reads the data correctly. You can change the header spacing to better see the values read by the format.

1. In the Point File Format dialog box at the bottom left, click the **Load...** button, change the Look in: folder to the local C: drive, and double-click *Civil 3D Projects* and *Civil 3D* folder (*C:\Civil 3D Projects\Civil 3D*) to look in the *Civil 3D* folder. If you didn't copy the project, locate the CD and navigate to the *Chapter 2* folder.
2. In the Select Source File dialog box at the bottom, change the Files of type to ***.nez** to list all files with that extension, select the file **Base-Points.nez**, and click the **Open** button.
3. In Point File Format dialog box, click the **Parse** button at the bottom left to read the file.

   The file reads correctly.

4. In Point File Format dialog box, click the **OK** button to exit and notice the new format is now on the format list.
5. Click the **Close** button to return to the Import Points dialog box.

IMPORTING THE FILE

1. If necessary, in the Import Points dialog box click the **Format:** drop-list arrow to the right of the current format and select the **HC3D (Comma Delimited)** format.
2. In the Import Points dialog box, click the folder icon at the right of Source File: to display the Select Source File dialog box. If the *Civil 3D* folder is not the current folder, change the Look in: folder to the local C: drive, and double-click *Civil 3D Projects* and then the *Civil 3D* folder to open it folder (*C:\Civil 3D Projects\Civil 3D*), or navigate to the *Chapter 2* folder on the CD.
3. Once in the folder, set the Files of type to ***.nez**, select the **Base-Points.nez** file, and click the **Open** button to return to the Import Points dialog box.
4. In the Import Points dialog box, do not toggle on any advanced options.

   Your Import Points dialog box should look like Figure 2.39.

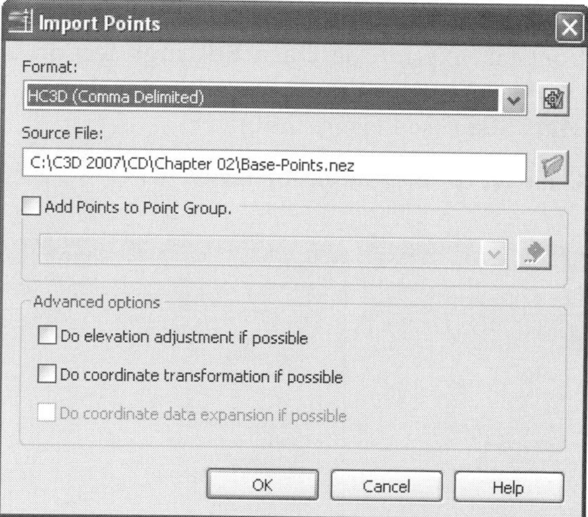

Figure 2.39

5. Finally, click the **OK** button to import the points.
6. If you cannot see the line work or points, use the Zoom Extents command to view the entire drawing.
7. In Prospector, expand the Point Groups heading.
8. Select **_All Points** from the list, press the right mouse button, and select **Properties...** from the shortcut menu.

9. Select the **Information** tab, change the Point label style to **Point#-Elevation-Description**, and click the **OK** button to exit.

10. Click the **AutoCAD Save** icon to save the drawing.

LANDXML: ELLIS.XML

Before Importing the LandXML file you need to verify and, if necessary, change the LandXML settings. These settings are in the Edit LandXML Settings dialog box of the Settings tab.

REVIEW LANDXML SETTINGS

1. Click the **Settings** tab to make it current.

2. At the top of the Settings panel, select the drawing name, press the right mouse button, and from the shortcut menu select **Edit LandXML Settings....**

3. In the LandXML Settings dialog box, click the **Import** tab, expand the Data Conversion section, and make sure **Convert Survey foot to International Foot** is set to **OFF**.

   There are two types of feet units in this world: US Survey and International.

   If exporting data to a LandXML while using the US Survey Foot, you need to change the value in the Data Settings of the Export tab.

4. Click the **Export** tab, expand the **Data Settings** section, and change the Imperial Units to **US Survey Foot** and the Angle/Direction format to **Degrees decimal dms (DDD.MMSSSS)** (see Figure 2.40).

5. Click the **OK** button to exit the dialog box.

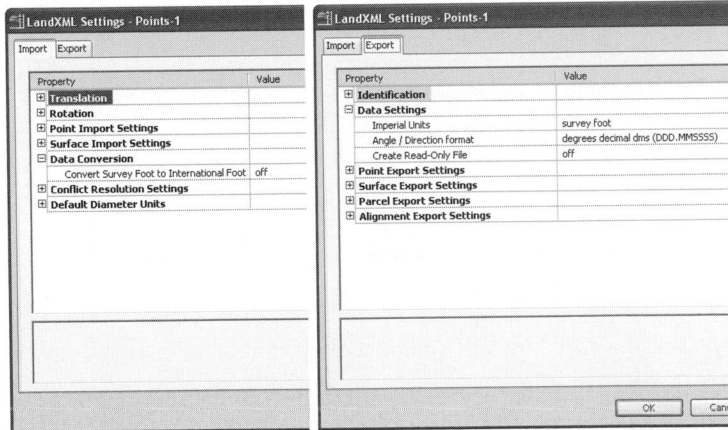

Figure 2.40

IMPORT LANDXML

1. From the File menu, the Import flyout, select **Import LandXML....**

   The Import LandXML dialog box displays.

2. If you are not in the *Civil 3D* folder, change the Look in: folder to the Local C: drive, and double-click *Civil 3D Projects* and then the *Civil 3D* folder to open it (*C:\Civil 3D Projects\Civil 3D*) or navigate to the *Chapter 2* folder of the CD that accompanies this textbook.

3. From the list, select **Ellis.xml** and click the **Open** button to read the file.

4. In the Import LandXML dialog box, toggle **OFF** the Alignments and expand CgPoints to view its contents (see Figure 2.41).

   Notice that the dialog box lists the points as point names, not point numbers.

5. Click the **OK** button to import the points and to close the dialog box.

6. Click the **AutoCAD Save** icon to save the drawing.

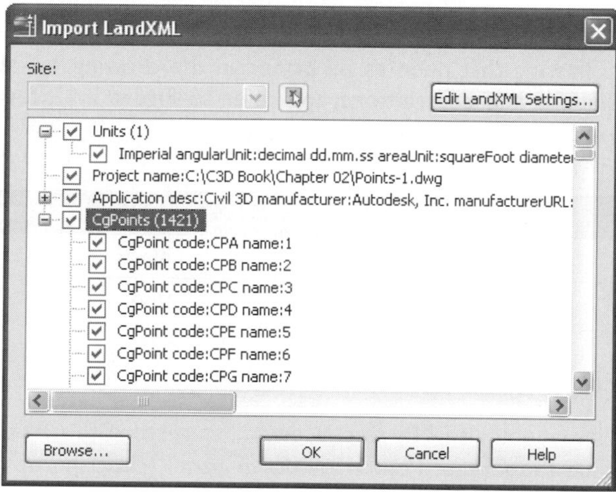

Figure 2.41

The new points are to the west of the north–south road at the western edge of the drawing's line work.

SET NEXT POINT NUMBER

The Next Point Number is the first unused point number.

1. Click the **Settings** tab to make it current.

2. In Settings, click the **Point** heading, press the right mouse button, and from the shortcut menu select **Edit Feature Settings....**

3. In Edit Feature Settings, expand the Point Identity section and check the value of the Next Point Number. If necessary, change the value of the Next Point Number to **10000**.
4. Click the **OK** button to set the value and exit the dialog box.

EDIT CREATE POINTS COMMAND SETTINGS

When setting a point, you need to set how a command assigns the elevation and description to the new points (see Figure 2.26).

1. In the Settings tab, expand the Point branch until you view a list of commands under the **Commands** heading.
2. Select **CreatePoints**, press the right mouse button, and from the shortcut menu select **Edit Command Settings....**
3. Expand the Points Creation section and review the values.
4. After reviewing, click the **OK** button to exit the dialog box.
5. Click the **Prospector** tab to make it current.

SET POINTS MANUALLY

There are a few points missing that need to be placed in the drawing. In Table 2.2, use the values to enter the point elevations and descriptions, and refer to Figure 2.42 for the points' locations.

Table 2.2

Point Number	Elevation	Description
10000	920.07	BLDG
10001	917.33	SWK

Frame Shed and Residence Points

1. From the View menu, use **Named Views...** to restore the **Frame Shed** view and, if necessary, hide the Create Points roll-down panel (click the up arrow at the right of the toolbar).
2. From the Create Points toolbar, click the first drop-list arrow at the left and select **Manual** from the list.
3. Using the AutoCAD Endpoint object snap, select the southeast corner of the shed to set point 10000. When prompted for the elevation and description, use the values from Table 2.2.
4. Using the AutoCAD Endpoint object snap, select the easterly intersection of the gravel path, east of the Framed Residence. When prompted for the elevation and description, use the values from Table 2.2.

5. Press the **Enter** key to end the command.
6. Click the **AutoCAD Save** icon to save the drawing.

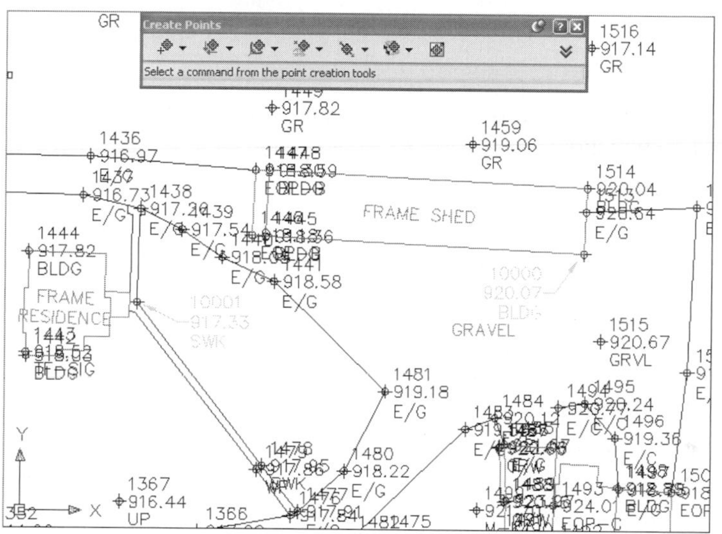

Figure 2.42

Set Points – Measure

The next series of points do not have an elevation, but they all have the same description. To do this, you need to change the prompting settings in the Data panel of the Create Points toolbar.

1. From the View menu, use **Named Views...** to restore the **Lot Line** view.
2. In the Create Point toolbar, click the roll-down arrow on the right side of the Create Points toolbar.
3. Expand the Points Creation section, set Prompt For Elevations to **None**, set Prompt For Descriptions to **Automatic**, and set Default Description to **LLPOINT** (see Figure 2.43).

   When clicking in the value cells for Prompt For Elevations and Prompt For Descriptions, the cells display a drop-list arrow. Select the arrow and select the setting from the list.
4. Click the roll-up arrow to close the Settings panel.
5. In the Create Points toolbar, click the drop-list arrow for Miscellaneous (first icon on left) and select **Divide Object** from the command list.
6. The routine prompts to select a line, arc, etc. Select the northerly line of lot 16, type **4** as the number of segments, and set the offset to **0.0**.

   The routine divides the line into four equal segments with five new points. Each point has no elevation and the description is LLPOINT.

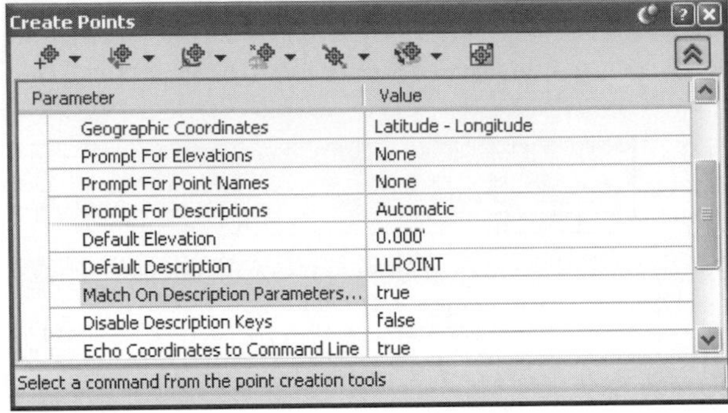

Figure 2.43

7. Press the **Enter** key to end the routine.
8. In the Create Points toolbar, click the drop-list arrow for Miscellaneous and select the **Measure Object** command.
9. The routine prompts to select a line, arc, etc. Select the southerly lot line of lot 16, press the **Enter** key twice to accept the beginning and ending stations, set the offset to **0.0**, and enter **25** as an interval.

   The routine creates seven new points with a spacing of 25 feet.
10. Press the **Enter** key to end the routine.
11. Click the **AutoCAD Save** icon to save the drawing.

    Your screen should look like Figure 2.44.

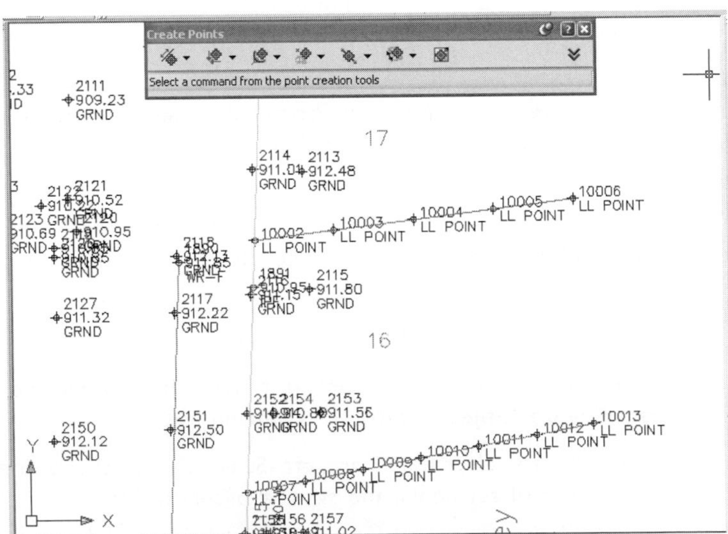

Figure 2.44

Polyline – Vertices

1. In the Create Points toolbar, click the roll-down arrow at the right of the Create Points toolbar to expand the Points Creation section, set Prompt For Elevations and Descriptions to **Automatic**, and set the Default Description to **BERM** (see Figure 2.45).

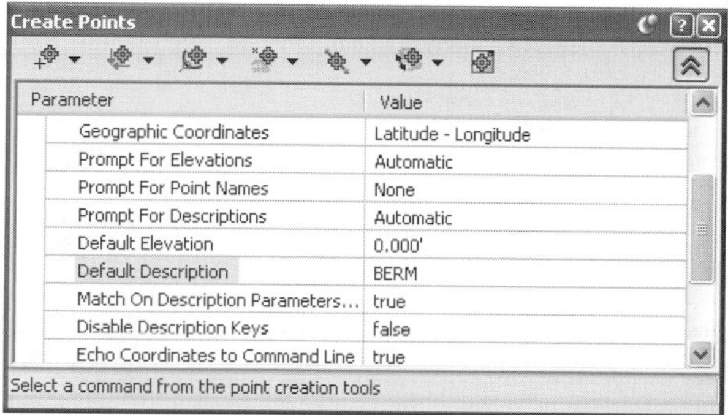

Figure 2.45

2. Click the roll-up arrow to hide the Settings panel.
3. From the View menu use the Named Views... command to restore the **Berm** view.
4. In the Create Points toolbar, click the drop-list arrow for Miscellaneous and select the **Polyline Vertices – Automatic** command from list.
5. The routine prompts for a polyline selection. Select the polyline in the center of the screen.
6. Press the **Enter** key to exit the routine.

   The berm line is a 3D polyline. The routine places a point at each vertex, assigning BERM as the description and the elevation of the vertex to each point (see Figure 2.46).
7. Click the **AutoCAD Save** icon to save the drawing.

HOUSE CORNERS BY INTERSECTION

This section of the exercise uses commands from the Intersections icon stack of the Create Points toolbar.

1. If on, toggle **OFF** AutoCAD's dynamic mode.
2. If necessary, in the Create Points toolbar, click the roll-down arrow (at the right side) to view the Settings panel.
3. Expand the Points Creation section, set Prompt For Elevations and Prompt For Descriptions to **Manual** and click the roll-up arrow to hide the Point Settings panel.
4. Click the **Settings** tab to make it current.

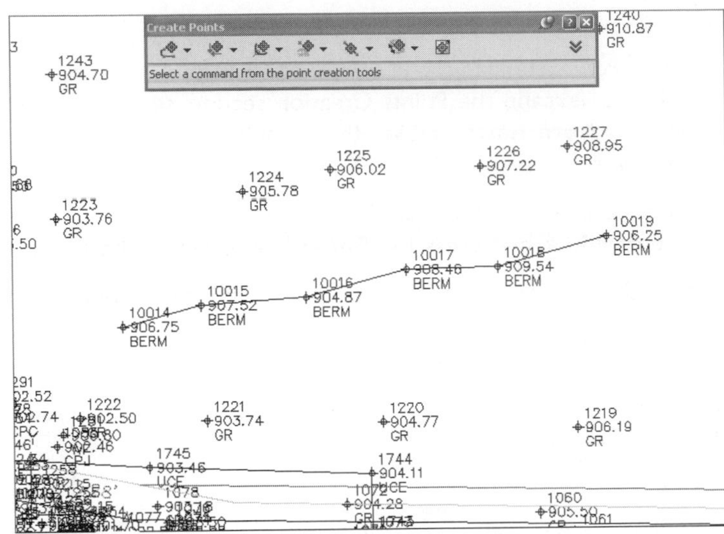

Figure 2.46

5. In Settings, select the **Point** heading, press the right mouse button, and from the shortcut menu select **Edit Feature Settings....**
6. Expand the Point Identity section, set the Next Point Number to **10050**, and click the **OK** button to exit.
7. Click the **Prospector** tab to make it current.
8. From the View menu, use Named Views... to restore the **Lot 11** view.
9. If the Transparent Commands toolbar is not on the screen, press the right mouse button over any icon and select **Transparent Commands**.

   To set this series of points, you will need to reference a point number as the beginning of the distances to the new points.

   The field crew measured the distances from two points to locate points representing the southerly corners of a house in Lot 11. From the Create Points toolbar, Intersection icons, the Distance/Distance command, create the next four points. Table 2.3 lists the points and their distances.

Table 2.3

From Point	Distance	New Point	Elevation	Description
1001	88.94	10050	920.54	BLDG
1002	113.14			
1001	116.21	10051	920.52	BLDG
1002	141.06			

10. In the Create Points toolbar, click the drop-list arrow to the right of the Intersections icon (second icon in from left) and select **Distance/Distance**.
11. In the command line, the routine prompts you to Please specify a location for the radial point location. Click the **Point Number** filter of Transparent Commands to override the current prompt. The prompt changes to Enter point number.
12. In the command line, type **1001** as the first radial point, press the **Enter** key, type the distance of **88.94**, and press the **Enter** key to define the first radial point.
13. The prompt changes for the location of the second radial point. Enter **1002** as the point number, press the **Enter** key, enter the distance of **113.14**, and press the **Enter** key to define the second radial point.
14. The routine then prompts again for a point number. Press the **Esc** key to exit the Point Number prompt and to return to the Point or All prompt.
15. Select a point near the green X at the southerly corner of the house.
16. After selecting the point, the routine prompts for the description and then the elevation. Use the values from Table 2.3 for the description and elevation of the new point.

    After entering the two values, the routine places point 10050 at the corner of the house.
17. Repeat Steps 11 through 16 to set point 10051. Remember to reselect the **Transparent Command** point filter and to stop it by pressing the **Esc** key.
18. Exit the Distance/Distance command by pressing the **Esc** key.
19. Click the **AutoCAD Save** icon and save the drawing.

HOUSE CORNERS BY SIDE SHOTS

The next set of points represents the northern corners of the current building and two additional points that define the southerly corners of a building in the next parcel to the north (Lot 12). Table 2.4 contains the setup and backsight points, the side shot angles, distances to each new point, and the point's elevations and descriptions.

The Side Shot override uses point 2341 as the setup point and point 1001 as the backsight point. Make sure AutoCAD's dynamic mode is off while entering the point values.

Table 2.4

Setup	Backsight	Angle	Distance	Description	Elevation
2341	1001	318.2034	56.55	BLDG	920.54
		294.0043	67.05	BLDG	920.53
		252.5419	128.29	BLDG	920.55
		243.5822	96.75	BLDG	920.53

1. In Create Points, click the drop-list arrow to the right of the Miscellaneous icon stack and select **Manual** from the list.
2. From the Transparent Command toolbar, select the **Side Shot** override.

   The Side Shot transparent command changes the prompting. The first prompt is now for a line or point to establish a setup and backsight for the new point observations (turned angle and distance from the setup).

   To establish points 2341 and 1001 as the setup and backsight points, type the letter **P** (for points mode) and press the **Enter** key. When in points mode, .p changes the prompts to point number and .g changes the prompts to select a point object.
3. In the command line, at the Select line or [Points]: prompt, type **P** and press the **Enter** key.
4. In the command line, at the Specify starting point or [.P/.N/.G]: prompt, type **.p** and press the **Enter** key.
5. The prompt changes to Enter Point Number; enter point **2341** and press the **Enter** key.
6. The next prompt is Specify ending point or [.P/.N/.G]:. Again type **.p** and press the **Enter** key.
7. The prompt changes to Enter Point Number; enter point **1001** and press the **Enter** key.
8. In the command line, the prompt is for an angle; enter **318.2034** for the angle and press the **Enter** key.
9. In the command line, the prompt is for a distance; enter **56.55** for the distance and press the **Enter** key.
10. The next two command prompts are for the description and elevation of the new point. Enter **BLDG** and **920.54** for their respective prompts. Press the **Enter** key to enter the description and elevation for the new point.
11. Repeat Steps 7 through 10 to create the three remaining building corner points.
12. Press the **Esc** key to exit the Side Shot entry and press the **Enter** key to end the Manual point creation.
13. In the Create Points toolbar, select the "**X**" at the right to close it.
14. Click the **AutoCAD Save** icon and save the drawing.

    This completes this exercise. After creating points you need to evaluate and possibly edit their values.

SUMMARY

- Users can create custom file formats to read additional ASCII files.
- By changing the prompting options in the Data panel of the Create Points toolbar, users change how a command prompts for its values.
- The only place to set the Next Point Number is in the Point's Edit Features dialog box.
- The Polyline Automatic routine uses the elevations of a 3D polyline for the point's elevation.
- Polyline Manual prompts for an elevation and assigns it to all of the new points.
- Any command prompting for a point can use an AutoCAD selection, point object (.p or .g), or northing/easting coordinate (.n).
- The transparent commands expand the point creation commands capabilities by using distances, angles, and directions to calculate new point coordinates.

UNIT 3: POINT ANALYSIS

The analysis and listing tools for points also function as the interface for editing their values. Prospector's preview panel and the Edit Points vista are the primary point analysis tools. When selecting Prospector's Points heading, the preview area displays all of the points currently in the drawing. In the preview area or the Edit Points vista, users can sort the point list simply by clicking on any of the column headers.

When selecting Prospector's Points heading and pressing the right mouse button, a shortcut menu displays (see Figure 2.47). Edit Points... calls the point editing vista. Edit Points... is also in the Points menu.

The shortcut menu Zoom to command changes the display to the point's location. If the current screen height is correct for viewing the next point, the user can use the Pan to command.

To create a LandXML file containing the points in the drawing, select the Point Groups heading, press the right mouse button, and select Export LandXML... from the shortcut menu.

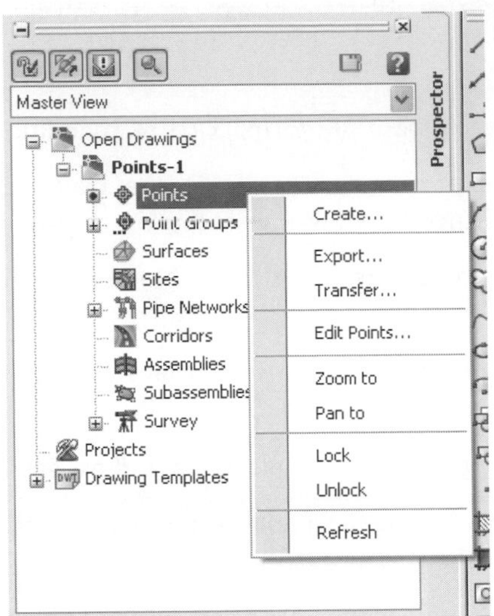

Figure 2.47

EXERCISE 2-3

When you complete this exercise, you will:

- Generate a points report.
- Be familiar with the LandXML Report Application.

GENERATING A POINT REPORT

If you are not in the *Points-1* drawing file you worked with in the previous unit, open it now. If you are just starting with this exercise, you need to open the drawing *Chapter 2 - Unit 3.dwg* to start the exercise.

1. Open the *Points-1* drawing or *Chapter 2 - Unit 3* drawing file. *Chapter 2 - Unit 3.dwg* is in the *Chapter 2* folder of the CD that accompanies this textbook.
2. Click the **Prospector** tab to make it the current panel.
3. Click the **Point Groups** heading, press the right mouse button, and from the short-cut menu select **Export LandXML....**

   This displays the Export to LandXML dialog box (see Figure 2.48).
4. If not selected, toggle **ON** the _All Points point group and click the **OK** button to exit the dialog box.
5. In the Export LandXML dialog box, locate a folder for the file and click the **Save** button.
6. Minimize Civil 3D.
7. On the Desktop, double-click the Autodesk LandXML Report Application.

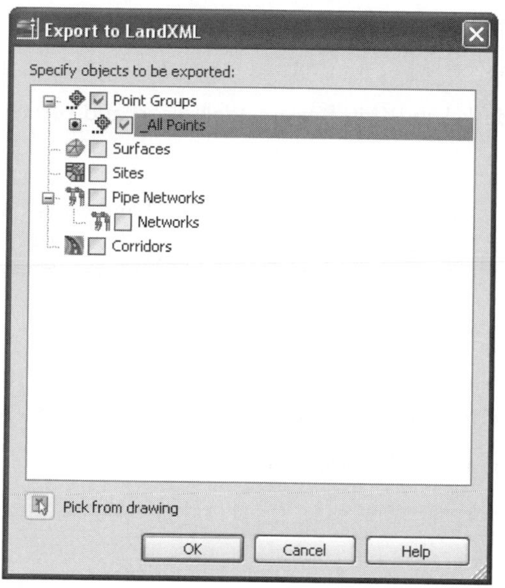

Figure 2.48

8. If the Getting Started dialog box appears, click the **OK** button to continue.

   The Report Application shows a Data panel showing the data for a report.

9. From the Report Application's File menu select **Open Data File....**
10. In the Open dialog box, browse to the folder containing the LandXML file, select the file, and click the **Open** button.
11. Click the expand tree icon to the left of CgPoints to view the point's list.

    The Forms tab contains reports for the different Civil 3D object types.

12. Click the **Forms** tab and expand Point Report Forms.
13. From the list of reports, select **Points List**.

    The Output tab previews the report based on the data and the selected form.

14. Click the **Output** tab to view the report.

    You can print or save this report.

15. Select the **Forms** tab, change the report to **Points in CSV**, and select the **Output** tab to view the new report.

    If you have Excel or a file extension association to CSV, a dialog box may appear about what to do with this file type.

16. In the File Download dialog box, select **Cancel** in the dialog box.

    The report appears in the Output panel in a cell format.

17. Exit the Autodesk LandXML Report application and return to the Civil 3D applicaion.

SHORTCUT MENU

Locating an individual point in a drawing can be difficult. The Points heading and the preview area have a shortcut menu with two point viewing commands: Zoom to and Pan to.

1. In Prospector select the **Points** heading.

   The preview area lists all of the Points in the drawing.

2. In the preview area, scroll through the list of points and select a point, press the right mouse button, and select **Zoom to** from the shortcut menu.

   The display zooms to the point's location.

3. In the preview area, scroll and select another point from the list and use **Pan to** to view the point.

4. Select a few points from the preview area and use the Pan to command from the shortcut menu to view their locations.

5. Click the **AutoCAD Save** icon and save the current drawing.

SUMMARY

- Civil 3D LandXML exports points as point groups.
- Civil 3D point reports are in the LandXML Report Application.

UNIT 4: EDIT POINTS

Mistakes are inevitable and there needs to be a way to fix them. Users edit errors in point data in Prospector's preview area or in a Point Editor vista of a panorama. Each location edits all of the point's information: number, elevation, description, location, styles, and so on.

PROSPECTOR'S PREVIEW AREA

Users can edit all of a point's values by clicking in an entry for a point (see Figure 2.49). They can also edit point numbers, point northings and eastings, point names, elevations, raw and full descriptions, and other point data.

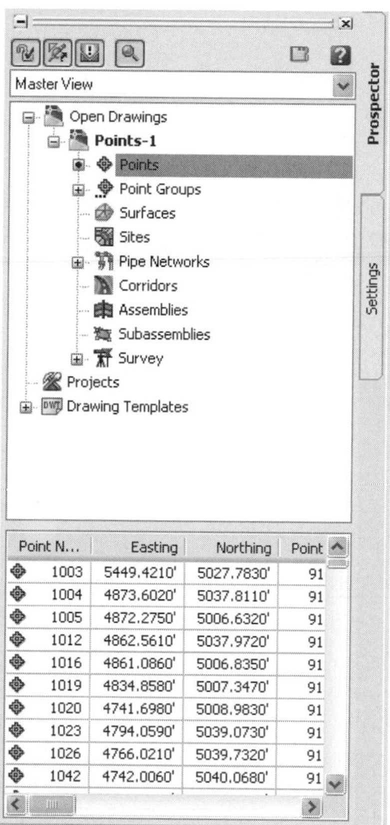

Figure 2.49

EDIT POINTS

The Edit Points command has four starting points: the Points menu, Prospector's Points heading shortcut menu, a right mouse shortcut menu after selecting one or more points on the screen, or a shortcut menu from Prospector's preview panel.

The Edit Points command displays the Point Editor vista that lists all of the same values seen in the Prospector preview panel (see Figure 2.50).

RENUMBER

The Points menu, Edit flyout, has the Renumber routine. This command assigns new point numbers by an additive factor to selected points. If a new point number is a duplicate number, the routine prompts for a solution; assign the next available point number or overwrite the existing point with the newly renumbered point.

DATUM

The Points menu, Edit flyout, has the Datum routine. The command manipulates the elevation for one or a set of selected points. Select the points only from Prospector's

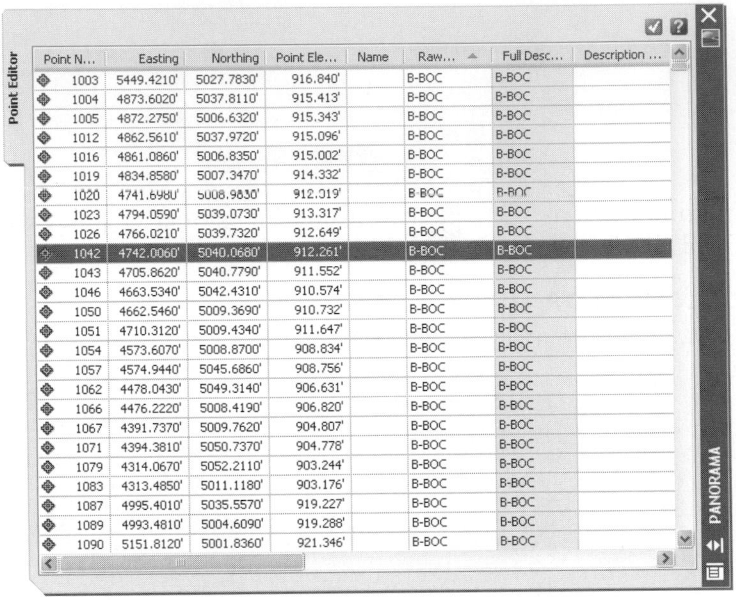

Point Editor

PANORAMA

Point N...	Easting	Northing	Point Ele...	Name	Raw...	Full Desc...	Description ...
1003	5449.4210'	5027.7830'	916.840'		B-BOC	B-BOC	
1004	4873.6020'	5037.8110'	915.413'		B-BOC	B-BOC	
1005	4872.2750'	5006.6320'	915.343'		B-BOC	B-BOC	
1012	4862.5610'	5037.9720'	915.096'		B-BOC	B-BOC	
1016	4861.0860'	5006.8350'	915.002'		B-BOC	B-BOC	
1019	4834.8580'	5007.3470'	914.332'		B-BOC	B-BOC	
1020	4741.6980'	5008.9830'	912.019'		B-BOC	B-BOC	
1023	4794.0590'	5039.0730'	913.317'		B-BOC	B-BOC	
1026	4766.0210'	5039.7320'	912.649'		B-BOC	B-BOC	
1042	4742.0060'	5040.0680'	912.261'		B-BOC	B-BOC	
1043	4705.8620'	5040.7790'	911.552'		B-BOC	B-BOC	
1046	4663.5340'	5042.4310'	910.574'		B-BOC	B-BOC	
1050	4662.5460'	5009.3690'	910.732'		B-BOC	B-BOC	
1051	4710.3120'	5009.4340'	911.647'		B-BOC	B-BOC	
1054	4573.6070'	5008.8700'	908.834'		B-BOC	B-BOC	
1057	4574.9440'	5045.6860'	908.756'		B-BOC	B-BOC	
1062	4478.0430'	5049.3140'	906.631'		B-BOC	B-BOC	
1066	4476.2220'	5008.4190'	906.820'		B-BOC	B-BOC	
1067	4391.7370'	5009.7620'	904.807'		B-BOC	B-BOC	
1071	4394.3810'	5050.7370'	904.778'		B-BOC	B-BOC	
1079	4314.0670'	5052.2110'	903.244'		B-BOC	B-BOC	
1083	4313.4850'	5011.1180'	903.176'		B-BOC	B-BOC	
1087	4995.4010'	5035.5570'	919.227'		B-BOC	B-BOC	
1089	4993.4810'	5004.6090'	919.288'		B-BOC	B-BOC	
1090	5151.8120'	5001.8360'	921.346'		B-BOC	B-BOC	

Figure 2.50

preview area or from the Point Editor vista. The change of elevation is an amount of change or a reference to an old and new elevation. When using the change in elevation option, enter a positive or negative value that is added to the elevation of the selected points (a negative value will reduce the elevations). The reference option uses the difference between two elevations to determine what value to add to the selected elevations. For example, a field crew assumes a benchmark elevation of 100 feet for a topographic survey. After researching the benchmark, the surveyor determines the true elevation of the benchmark as 435.34. The Datum elevation routine prompts for a reference elevation (the old benchmark elevation) and a new elevation and then calculates the amount to add to each selected point. The change of elevation can be a positive or negative value.

Another situation where the Datum routine comes in handy is for a blown setup elevation of a survey. If the surveyor records a set of points with an incorrect setup elevation, all the point elevations recorded during the setup are incorrect. If the surveyor discovers the error, he can use the datum routine to adjust all the affected points. Using the Datum routine adjusts all the selected points.

DELETE AND AUTOCAD ERASE

The Civil 3D's Delete command and AutoCAD's Erase command permanently erase points from the drawing.

POINT UTILITIES

The Points menu, Point Utilities flyout, provides useful tools to export, transfer, calculate geodetic coordinates, and review project point information.

EXPORT

The Export routine creates a file or Access database from the points in the drawing. The routine uses the same interface as the Import/Export routine. By default this command exports all of the points in the drawing. Limit the export to the points in a single point group.

TRANSFER

The Transfer routine converts the contents of one file format or point database (Access) to a user-specified file format (including Access).

GEODETIC CALCULATOR

The Geodetic Calculator calculates grid coordinates and latitude and longitude for points in a drawing (see Figure 2.51).

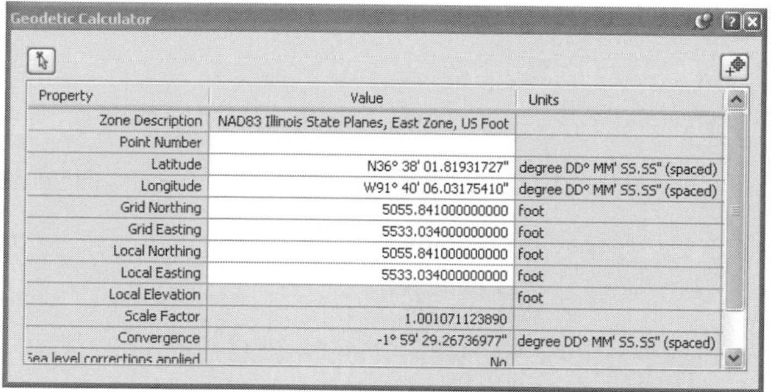

Figure 2.51

EXERCISE 2-4

When you complete this exercise, you will:

- Be able to edit points from various locations in the Civil 3D interface.
- Be able to edit points in a vista.
- Renumber selected points.
- Adjust the datum and the elevation of selected points.
- Export points to an ASCII file.

EXERCISE SETUP

This exercise continues with the drawing from the previous exercise. If you are starting with this exercise and have not done the previous exercises, you can open the *Chapter 2 - Unit 4.dwg* that is found in the *Chapter 2* folder of the CD that accompanies this textbook.

1. Open the drawing you have been using for this chapter or the *Chapter 2 - Unit 4* drawing file for this exercise.

PROSPECTOR PREVIEW - EDIT POINTS

1. Click the **Prospector** tab to make it the current panel.
2. In Prospector, click the **Points** heading to list the points in the preview area.
3. In the preview area, use the scroll bars to view the points.
4. In the preview area, click the **Point Elevation** heading to sort the point elevations.
5. In the preview area, scroll the point list and locate point **3001**.
6. In the preview area, scroll the point 3001 record until viewing its elevation (0.0). Click in the elevation cell, change the elevation to **898.76**, and press the **Enter** key to accept the new value.
7. In Prospector, click the **Points** heading, press the right mouse button, and from the shortcut menu select **Edit Points...**.

   This displays Point Editor vista.
8. In the Point Editor vista, sort the list by clicking the **Point Number** heading.
9. Click in the Elevation cell for point 10001 and change its elevation to **914.15**.
10. Click in the Elevation cell for point 10002 and change its elevation to **914.90**.
11. After editing the points, click the "**X**" at the top right of the panorama to close it.
12. In Prospector, select the **Points** heading to list the points in the preview area.
13. Scroll the point list and locate point 10003.
14. In the preview area, select the point entry, press the right mouse button, and from the shortcut menu select **Zoom to**.
15. In the drawing, select point **10003**, press the right mouse button, and from the shortcut menu select **Edit Points...**.
16. In the Point Editor, click in the elevation cell, enter the value of **916.78**, and close the vista.
17. In the preview area, locate and select point number **3002**, press the right mouse button, and from the shortcut menu select **Zoom to**.
18. In the drawing, select point 3002, press the right mouse button, and from the shortcut menu select **Edit Points...**.
19. In the Point Editor, click in the elevation cell, enter the value of **914.96**, and close the vista.
20. In the preview area, locate and select point number **3003**, press the right mouse button, and from the shortcut menu select **Zoom to**.
21. In the preview area, select point **3003**, press the right mouse button, and select **Edit Points...** from the shortcut menu.

22. In the Point Editor, click in the elevation cell, enter the value of **912.40**, and close the panorama.
23. Click the **AutoCAD Save** icon to save the drawing.

RENUMBER POINTS

There will be times when you need to renumber existing points to new numbers.

1. In the preview area, if necessary, click the **Point Number** column to sort the points.
2. In the preview area, scroll the point list until viewing point number 3001.
3. While holding down the **Shift** key, select point number 3003 to select points 3001 to 3003.
4. With the cursor over the highlighted points, press the right mouse button, and from the shortcut menu select **Renumber....**
5. In the command prompt area the command prompts for the additive factor. Enter **2000** to renumber the points to 5001, 5002, and 5003.

   The points change to the new point numbers.
6. In the preview area, scroll the point list until viewing points 5001 through 5003.
7. Click the **AutoCAD Save** icon to save the drawing.

EDIT DATUM

There will be times when points are observed with an assumed or incorrect setup elevation. All of the points using this setup can have their elevations adjusted by using the Datum routine. The Datum routine prompts the user for an old and new elevation or an amount (positive or negative) to add to the point's elevation.

1. In Prospector, select the **Points** heading to list the points in the preview area.
2. In the preview area, scroll through the point list until locating point 2025.
3. Select point 2025, press the **Ctrl** key, and select point 2024.
4. With both points highlighted, press the right mouse button, and from the shortcut menu select **Datum....**
5. In the command line, the routine prompts for an amount of change; enter **5**, and press the **Enter** key.

   This adds 5 feet of elevation to the two points.
6. In the preview area, with the two points highlighted, press the right mouse button, and select **Datum...** from the shortcut menu. You may have to reselect the two points (2024 and 2025).

7. In the command line, the routine prompts for a change in elevation; enter the letter **R** for reference, enter **100** as the reference elevation, and enter **95** as the "new" elevation.

   This subtracts 5 feet of elevation from the two points.

8. Save the drawing by clicking the **AutoCAD Save** icon.

EXPORT POINTS

There will be times when you need to export points to an ASCII file. This occurs when you need to share data with someone.

1. In Prospector, select the **Points** heading, press the right mouse button, and select **Export...** from the shortcut menu.
2. In the Export Points dialog box, select the **HC3D (Comma Delimited)** format. If you didn't make the format, use the **PNEZD (comma delimited)** format.
3. Click the folder icon at the right of Destination File, set a folder, and name the file.
4. Click the **OK** buttons to export the points and to dismiss the dialog boxes.
5. Click the **AutoCAD Save** icon to save the drawing.

This ends the exercise for editing points. The next unit covers the assignment of layers and point and label styles by using a Description Key Set.

SUMMARY

- The point editor is called from a selected point's shortcut menu, the Points menu, Prospector's preview area shortcut menu, or from Prospector's Points heading shortcut menu.
- No matter where the Edit Points... routine is called, it displays the same Point Editor vista.
- The Renumber command adds a fixed offset to the selected points.
- Select the points first before using the Datum command. If not, the routine assumes you are adjusting the elevations for all points.
- Datum adds a negative or positive amount to the elevations of selected point(s).
- A Datum routine option references an old and a new elevation to determine the value of the additive factor.
- Users can export all points or just the points in a point group when using the Export command.

UNIT 5: ANNOTATING AND ORGANIZING POINTS

A Description Key Set translates raw descriptions (codes) into drawing descriptions (full descriptions). For example, the code "IPF" means "Iron Pipe Found" as a full description. A Description Key Set is essentially a table that lists codes and their full descriptions (format column) (see Figure 2.52). Additional functions of a Description Key Set include assigning point (marker) styles (symbols if appropriate) and label styles, scaling and rotating markers, and assigning user-defined layers for point objects.

A Description Key Set does not have to assign any layers. The assigned point style can assign layers for the marker and its label style. The Description Key Set references layers in a drawing, and they should be a part of the drawing template.

EXERCISES

CODES

As previously mentioned, a raw description is also known as a code. A code is shorthand for what was observed in a survey or during a data collection session. As a result, a code assignment occurs in the field, in data collectors, in external coordinate files, or in any point-creation routine (even in Civil 3D). Many codes have a more verbose description, known as a full description. For example, a raw description of MH represents a Manhole. So, some codes need translating from their raw description value to their full description. The only mechanism in Civil 3D that translates a code into a full description is the Description Key Set.

Codes do not need to be alphabetic characters. They can be numerical, alphabetical, or a combination of both. When codes are numerical, a Description Key Set is vital in translating the codes into meaningful words.

CODE PARAMETERS

A code may contain spaces (for example, OK 6, MP 3, etc.). Civil 3D considers the space as a separator between parameters. In the example of OK 6, the first parameter is "OK" and "6" is the second parameter. Civil 3D allows up to 10 parameters in a description and identifies them with a $ (dollar sign) followed by a number. Civil 3D numbers the parameters starting with 0 (zero). Using the example of OK 6, "OK" is parameter 0 ($0) and "6" is parameter 1 ($1).

The Format (full description) column in a Description Key Set contains the translation format. When translating a code, users can rearrange a code's parameters to create a full description. With the example of OK 6, the Full Description is 6" Oak, so the Format column entry is $1" OAK. This roughly translates to use parameter $1 as the first entry and follow it with an inch sign and the word "OAK." To create a full description from a code with no translation, the Format column contains $* (use code).

Civil 3D also allows parameters to be scaling and rotating factors. With the example of the oak tree, each measured trunk size can scale the oak symbol by multiplying it with parameter ($1). A Description Key Set has Scale and Rotate Parameter columns.

POINT ANNOTATION METHODS

The first method of annotating points is by using the default point and label styles found in the Edit Feature Settings or Command Settings dialog boxes under Settings. This method assigns all points the same styles. In a previous exercise, two point styles were defined: OAK and MAPLE. These point styles define layers for the marker and its label.

When there are several different codes, a Description Key Set is very helpful when wanting to assign styles other than the default styles (see Figure 2.52).

There should be a Description Key Set entry for all the possible point codes. This may be difficult, especially if users do not have any control over the source of the raw descriptions. A text editor with a find-and-replace feature is the only way to substitute the different descriptions with the in-house values. If the work is done by a contractor, users should encourage the contractors to use the user's descriptions for point data. Even with this effort, many offices face commonly occurring naming inconsistencies.

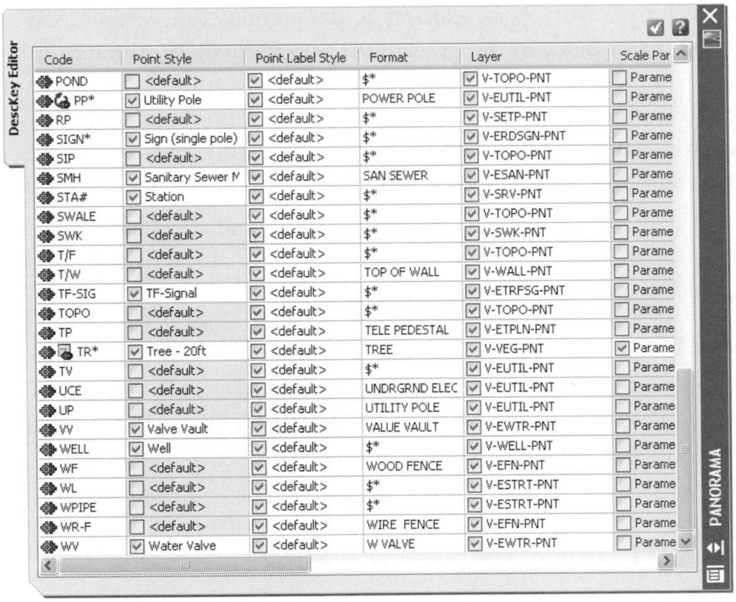

DescKey Editor

Code	Point Style	Point Label Style	Format	Layer	Scale Par
POND	<default>	<default>	$*	V-TOPO-PNT	Parame
PP*	Utility Pole	<default>	POWER POLE	V-EUTIL-PNT	Parame
RP	<default>	<default>	$*	V-SETP-PNT	Parame
SIGN*	Sign (single pole)	<default>	$*	V-ERDSGN-PNT	Parame
SIP	<default>	<default>	$*	V-TOPO-PNT	Parame
SMH	Sanitary Sewer M	<default>	SAN SEWER	V-ESAN-PNT	Parame
STA#	Station	<default>	$*	V-SRV-PNT	Parame
SWALE	<default>	<default>	$*	V-TOPO-PNT	Parame
SWK	<default>	<default>	$*	V-SWK-PNT	Parame
T/F	<default>	<default>	$*	V-TOPO-PNT	Parame
T/W	<default>	<default>	TOP OF WALL	V-WALL-PNT	Parame
TF-SIG	TF-Signal	<default>	$*	V-ETRFSG-PNT	Parame
TOPO	<default>	<default>	$*	V-TOPO-PNT	Parame
TP	<default>	<default>	TELE PEDESTAL	V-ETPLN-PNT	Parame
TR*	Tree - 20ft	<default>	TREE	V-VEG-PNT	Parame
TV	<default>	<default>	$*	V-EUTIL-PNT	Parame
UCE	<default>	<default>	UNDRGRND ELEC	V-EUTIL-PNT	Parame
UP	<default>	<default>	UTILITY POLE	V-EUTIL-PNT	Parame
VV	Valve Vault	<default>	VALUE VAULT	V-EWTR-PNT	Parame
WELL	Well	<default>	$*	V-WELL-PNT	Parame
WF	<default>	<default>	WOOD FENCE	V-EFN-PNT	Parame
WL	<default>	<default>	$*	V-ESTRT-PNT	Parame
WPIPE	<default>	<default>	$*	V-ESTRT-PNT	Parame
WR-F	<default>	<default>	WIRE FENCE	V-EFN-PNT	Parame
WV	Water Valve	<default>	W VALVE	V-EWTR-PNT	Parame

PANORAMA

Figure 2.52

When using a Description Key Set to assign layers, defining a point style with both point and label layers may not be the best strategy. As previously mentioned, the Description Key Set only assigns a layer for the marker, so how is the layer for the label assigned? There are only two places that assign a label's layer: the label style or point style. The best place to assign the label's layer is in the point style's Display panel settings. This strategy places the marker on the layer assigned by the Description Key Set and the label is on the layer defined in the point's style. Figure 2.53 shows this strategy for the code CB*. The Description Key Set assigns the marker's layer, and the point style, Catch Basin, defines the label's layer.

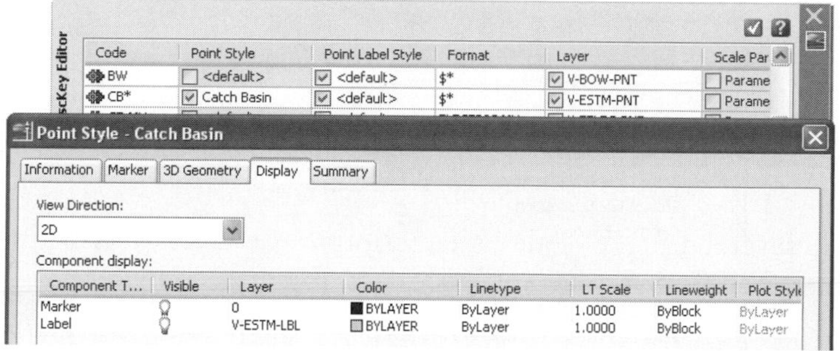

Figure 2.53

Controlling marker and label visibility occurs in three possible locations. The first place to control visibility is in Layer Properties Manager, by turning on and off the layers specified by the Description Key Set and point style. The second place to control marker and label visibility is in the Display panel of the point style. By turning off the Marker and Label component of the style, all of the markers and labels for the style are hidden. The third location for controlling point visibility is with Point Group properties.

Whatever the implemented strategy, users should be mindful that Civil 3D is in early releases, and how Description Key Sets work may change, or even morph into another process.

If deciding to use Description Key Sets, it is best to include the set as part of a template. Adding a Description Key Set after placing points in the drawing does not make the points react to the settings in the key set. If adding description keys in the middle of a project, users need to export the points and re-import them before the points follow the values in the set.

A Description Key Set is always active, and the settings in Setting's Edit Feature Settings are in control if they are active. There are additional controls in the Points Creation settings of the roll-down panel of the Create Points toolbar (see Figure 2.54). The Disable Description Keys option should always be set to false.

A second toggle, Match on Description Parameters, indicates whether the numbers in a description are for scaling and rotating the point. If the parameters contain values to scale and rotate the point object, this value needs to be set to true. This also means that spaces in a code must be correct (e.g., Man Hole 48 needs to be Manhole 48). If this is set to false, there can be spaces in the description but there can be no values to scale or rotate the point objects.

A drawing can have more than one Description Key Set. In this case, the searching order is down the key sets list. Civil 3D displays an ordered list of key sets as a property of Setting's Description Key Set heading. The Properties dialog box allows the user to manipulate the key set search order.

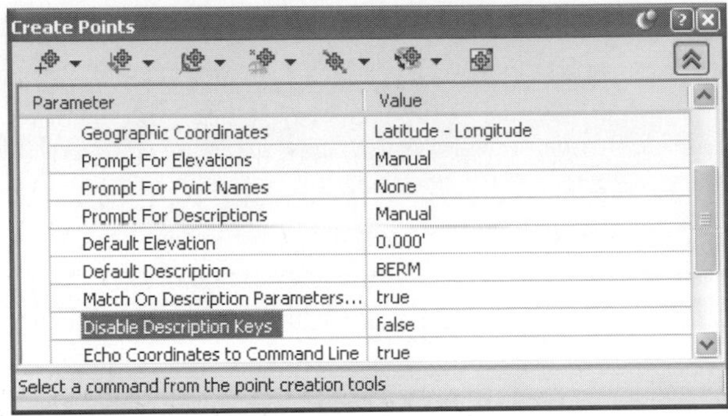

Figure 2.54

POINT GROUPS

In Civil 3D a point group organizes points and has a name (for example, existing utilities, EG surface points, and so on). A point group can assign point and/or label style overrides besides assigning a new layer for the points in the group. By having a point group assign a different layer, users can control the visibility of its members by turning on or off this one layer. This reduces the display of point groups to a single layer rather than combinations of point and label layers.

If a point is a member of more than one point group, it is the point group that is displayed last that controls the visibility of that point. A point's display can be a complex balance between styles, key sets, and point group overrides and display order.

DESCRIPTION KEY SET

A Description Key Set is a matrix that associates a point (marker or symbol), label style, layer, translation, and scaling and rotating factor to a point object. The association occurs when a code matches a code entry in the key set (see Figure 2.55). Symbols appear at the coordinates of a point because the point style assigned in the Description Key Set specifies a point style containing a symbol as a marker. For example, the Catch Basin point style uses an AutoCAD block as a symbol for the catch basin.

A point label style defines the location and amount of information about the point displaying at or near the marker's coordinates. The point (symbol) style usually is a part of the final document, while the label (point number, elevation, description, etc.) may not be.

Civil 3D installs with templates that contain some blocks. If a user wants to include his or her own symbols, the user needs to insert them as block definitions in a drawing template (or current drawing). When starting a new drawing and using this modified template, the symbols will appear in the Use AutoCAD BLOCK symbol for marker area of a point style (see Figure 2.55).

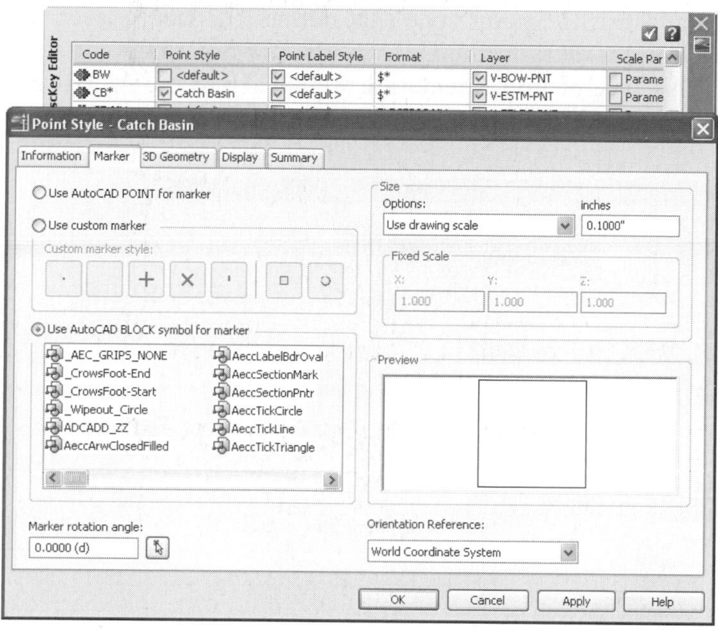

Figure 2.55

If a user makes his or her own symbols, there are some rules to follow:

- Draw all symbols the size they are to appear on a plotted sheet.
- All symbol entities should be drawn on layer 0.
- The (0,0) coordinates should be the insertion point of the block.

The first condition states the user should draw symbols their plotted size. If a manhole symbol is to be 0.2 inches in diameter when plotted, its drawing size has a diameter of 0.2 (foot-based drawing). When the point style references the manhole symbol, it scales the symbol by the drawing scale. If the drawing scale is 1 inch = 20 feet, the scaling factor for the symbol is 20.

The second condition refers to AutoCAD's handling of blocks. When inserting a block whose entities are layer 0 entities, it will assume the properties of the insertion layer. For example, when placing a shrub symbol on the vegetation layer, the block assumes the properties of the vegetation layer (color, linetype, lineweight, etc.).

The last rule is that AutoCAD assumes the inserted block's (0,0) coordinates are its insertion point.

The user can exaggerate the size of a symbol by setting a scaling factor in the scale entries of a code. See the discussion on setting the scale factors for a code that follows.

The marker and label need independent visibility. To accomplish this, it is best that each has its own layer. One possible implementation uses a point style and the Description Key Set to assign point marker and label layers. The Description Key Set assigns the layer for a point's marker and it is the point style that defines the label's layer. With the marker's layer originating from an entry in the Description Key Set, the point style needs to define only one layer. This type of implementation would define the MAPLE and OAK point styles as those in Figure 2.56.

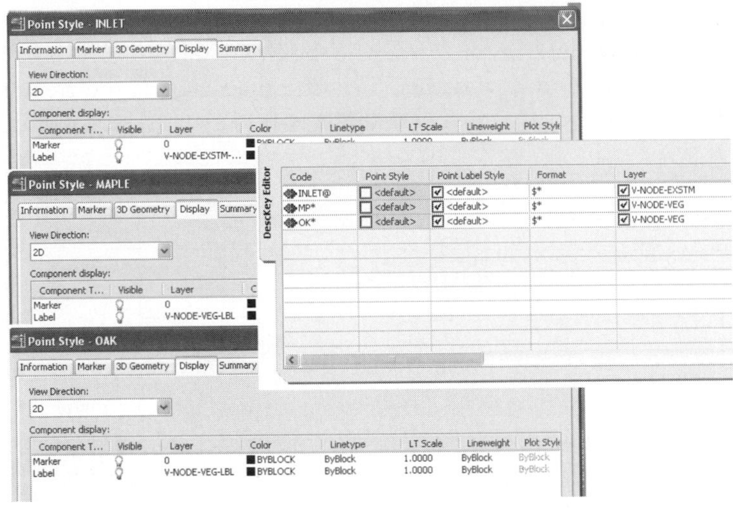

Figure 2.56

One effect of a Description Key Set is the sorting of points onto layers, which gives users the ability to control the visibility of points by turning off or freezing the marker and label layers (for example, all centerline shots on the layer V-NODE-RDCL, curb shots on the layer V-NODE-CURB, and so forth). Users can group similar point descriptions on the same layer. Implementing this strategy means all utilities are on the E-UTIL layer or all survey control is on the Control layer. This would reduce the number of layers in the drawing, but manipulating visibility affects a wider group of points.

If a code does not match an entry in the Description Key Set, the routine creating the point assigns the default styles to the point.

When using Description Key Sets, remember these three rules:

- The Points Creation toggle and Disable Description Keys option must be set to false.
- The Description Key Set must contain the correct code entries.
- A match occurs only when a code exactly matches an entry in the Description Key Set (including case sensitivity).

USING PARAMETERS

The default setting for the Match On Description Parameter option is true. This allows a raw description to contain numerical values separated by spaces to scale and rotate a marker. A Description Key Set uses the value of a specified parameter as scale factor or rotation amount.

As an example, the raw description of OK 6 has two parts: "OK" and "6." The "OK" represents the type of tree (oak) and the "6" is the diameter of the tree trunk (6 inches). As a result, OK is parameter 0 (zero) and the 6 is parameter 1. To use the parameters of the raw description in the Format entry of the Description Key Set, reference them by using $0, $1, . . . $9 in the value cell for the format. The format of $1" OAK creates the full description of 6" OAK from the raw description of OK 6. This format uses the "6," the second parameter of the raw description ($1), as the first part of the new full description.

CREATING A DESCRIPTION KEY SET

After installing Civil 3D, the content templates contain a simple Description Key Set. The expectation is that users develop their own key sets from scratch or that they migrate LDT description keys. The commands Import Data from Land Desktop... or Import LandXML... (located under the File menu, in the Import flyout) create Description Key Sets from existing LDT or other implementations. If migrating a prior implementation to Civil 3D, users still need to edit the resulting key set to add the point and label styles and, if necessary, layer assignments.

Users can drag and drop a key set from one drawing to another to add, update, or overwrite the existing set.

The Settings panel is where users create or edit an existing Description Key Set. "Description Key Set" is a heading in the Point branch. After locating the heading, select it, press the right mouse button, and select New... from the shortcut menu. After selecting New..., a Description Key Set dialog box displays (see Figure 2.57). Enter the name of the Description Key Set in this dialog box and exit.

After creating the Description Key Set, expand the Description Key Set branch to display the list of sets. Select the set name, press the right mouse button, and select Edit Keys... from the shortcut menu. This displays a DescKey Editor vista containing the current Description Key Set (see Figure 2.58). When clicking in the first cell of the first entry, press the right mouse button and select New... from the shortcut menu. This adds the code to the cell. Then fill out the remaining information for the code. To add the next code to the key set, repeat the process on the next record in the set.

Each code in the key set has 13 entries. Of the 13 entries for a code, five are for manipulating the raw description and the remaining eight are parameters for scale and rotation. The entries for the raw description are Code, Point style, Label style, Format (Full Description), and Layer. A description key can use one or all of the entries. If a key only

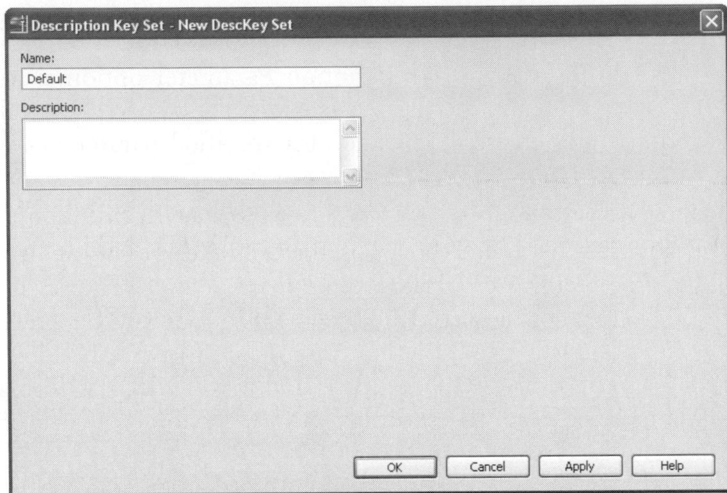

Figure 2.57

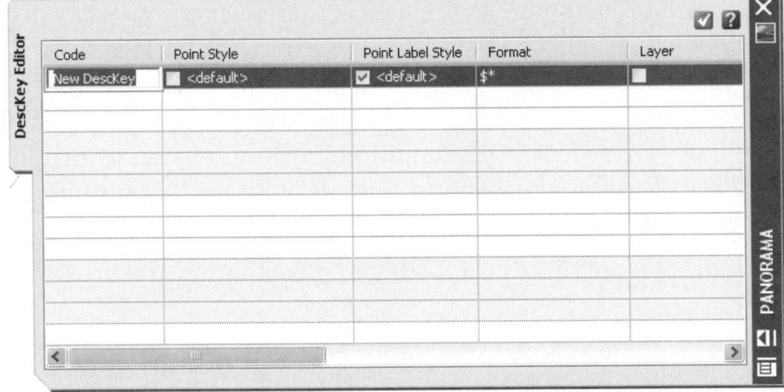

Figure 2.58

sorts points, the user needs only to identify the code, the point and label styles, and the layer name. If using parameters, the user will have to set entries in the Format, Scale, and Rotation areas.

ENTRY 1: CODE

When creating a point, the routine references the Description Key Set and compares the point's raw description to the Code entries of the key set. If matching, the routine assigns the listed point and label styles, does a translation (if set), and assigns a layer to the point object (if set).

Alphabetical and Alphanumeric Description Codes The process of literal character matching does not work well when adding numerical values to a raw description. Examples of this type of raw description are MH1, MH2, OK 3, or OK 6. Each description has a common root, manholes (MH) and oak trees (OK). What changes for each code is the added numerical value or a parameter. The field crew records manholes as a single alphanumeric value, MH1, meaning manhole 1, 2, and 3, or in the case of the Oak tree, various trunk sizes. To handle these types of codes, Civil 3D uses an asterisk (*) wildcard after the common root in the code. The asterisk wildcard solves the problem of having to match every character or number in a raw description. Using this wildcard with codes MH and OK creates the entries of MH* and OK* in the Code column of a Description Key Set. This tells the key set that only two characters need to match, and the remainder of the raw description can be ignored. The MH* entry matches MH1, MH2, etc., and the OK* entry matches the raw descriptions of OK 3 and OK 6.

Matching while using alphabetical codes is case-sensitive; that is, IP is different from Ip, iP, and ip.

In addition to the asterisk wildcard, a Description Key Set can contain the following wildcards: ? (number or character), # (single-digit only), @ (single alphabetical character only), . (any single non-alphanumerical character), [] (a list), ~ (logical not), and * (anything).

Wildcards cannot be used at the beginning of a code.

A wildcard in a code is a literal match. If a user defines a code, STA#, it matches STA0 through STA9 (which is STA with a single number following). If the range of station numbers is 1000 to 9999, the Code needs to be STA####. With the code set to STA####, the raw descriptions STA1...STA999 will not match. The range of values from STA1 to STA999 do not match because they do not have enough digits after STA to match the four digits expected by STA####. It is important to remember that the match is a literal per-character match when using any wildcard other than * (asterisk).

Examples of this type of code are as follows:

- STA# matches only STA0 through STA9.
- STA## matches only STA01 through STA99.
- STA### matches only STA001 through STA999.
- STA*—matches STA0 through STA99999999; also, it STAKE, STAR12, or anything with STA first.

A code may have multiple raw descriptions that translate to a single full description. For example, in the field, the crew differentiates between recovered iron pins (IPF) and iron rods (IRF). In the office and in the drawing the important information is that iron is found. Since IPF and IRF translate to the same full description and have a common beginning and end (letters I and F), the key set can contain a code that includes a list of possible character matches. In this example, the list of possible matching letters, P and R, are bracketed.

- I[PR]F – Matches IPF or IRF

The list wildcard allows users to include and exclude letters or numbers. To exclude values from matches, use a tilde; ~ (not).

- CP[A ... M]—Control points with only the letters A through M match.
- CP[~N ... Z]—Control points that do not include the letters N through Z match.

If the code match uses an @ (alpha character only) the following raw descriptions would match CP@:

- CPA
- CPC
- CPM

Numerical Codes Raw descriptions can be numerical. If using numerical descriptions, how would a code describe a 4-inch-diameter tree? The code references a single key set entry for that diameter or a range of diameters. For example, the key 1231 indicates a 1-inch- to 2.99-inch-diameter deciduous tree, and 1232 indicates a 3-inch- to 5.99-inch-diameter deciduous tree. There will have to be two codes, 1231 and 1232, to handle these raw descriptions.

ENTRY 2: POINT STYLE

This entry specifies the point style for a matching code (see Figure 2.59). This is how Civil 3D assigns a symbol to a matching code.

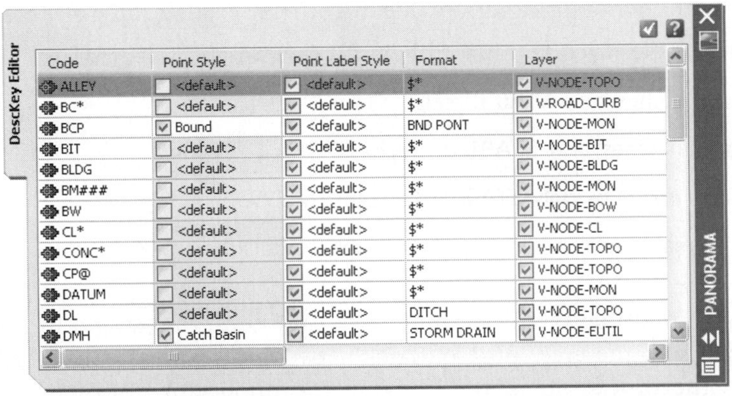

DescKey Editor

Code	Point Style	Point Label Style	Format	Layer
ALLEY	<default>	<default>	$*	V-NODE-TOPO
BC*	<default>	<default>	$*	V-ROAD-CURB
BCP	Bound	<default>	BND POINT	V-NODE-MON
BIT	<default>	<default>	$*	V-NODE-BIT
BLDG	<default>	<default>	$*	V-NODE-BLDG
BM###	<default>	<default>	$*	V-NODE-MON
BW	<default>	<default>	$*	V-NODE-BOW
CL*	<default>	<default>	$*	V-NODE-CL
CONC*	<default>	<default>	$*	V-NODE-TOPO
CP@	<default>	<default>	$*	V-NODE-TOPO
DATUM	<default>	<default>	$*	V-NODE-MON
DL	<default>	<default>	DITCH	V-NODE-TOPO
DMH	Catch Basin	<default>	STORM DRAIN	V-NODE-EUTIL

PANORAMA

Figure 2.59

ENTRY 3: POINT LABEL STYLE

This entry sets the point label style for a code in the Description Key Set.

ENTRY 4: FORMAT FULL DESCRIPTION

The next entry for a code is the format or translation of the raw description to a full description (see Figure 2.59). The purpose of the translation is to make the raw description more meaningful to the people in the office. There are three basic types of translation.

The simplest translation from raw description to full description is no translation. When a raw description does not need translation, enter $* (a dollar sign followed by an asterisk) in the Format cell. An example of this is the raw description of TOPO, the full description being the same.

Another method of translating is to completely replace the raw description to make the full description. An example of this is having the raw description of IP, and the Format value is IRON FOUND. When matching the code IP, the full description of the point is IRON FOUND.

The last translation changes the order of and/or adds information to the full description from a raw description containing parameters. An example is translating the raw description OK 6 to the full description. For this translation, the Format value is $1" OAK. This example uses the second ($1) parameter of the raw description to create a tree diameter in the full description.

ENTRY 5: LAYER

The fifth entry of a Code in a Description Key Set is a layer for the specified point object (see Figure 2.59). If this entry specifies a layer, the routine creating the point places it on this layer. If the assigned point style defines a layer for the marker and the label, there is no need to use this entry.

ENTRY 6: OPTIONAL SCALE PARAMETER

By default the point style marker scale setting defines the marker scaling. The default value uses the scale of the drawing to size the marker. This toggle adds a second scale factor to sizing the marker. The entry identifies which parameter the Set should use as a secondary scale factor (see Figure 2.60).

ENTRIES 7–10: APPLIED SCALES

These four entries allow users to scale markers based on what the symbol represents. Depending on the type of symbol, users may want to use various combinations of these toggles and values. Be careful to understand the consequences of toggling on more than one scaling factor. If all the toggles are on, the scaling of the symbol is the product of all the checked options.

The raw description of OK 6 represents a 6-inch-diameter oak tree. If there are other oak trees that vary between 2 inches and 12 inches in the point data, the user may want the markers to show the diameters of the trees as differently sized symbols. To make this happen, he or she needs to set three entries in this area. First, toggle on the Scale Parameter. Next, clicking in the cell to the right of the toggle displays a drop-list arrow.

Select the arrow to display a parameter list and set it to the correct parameter (parameter 1). The last setting is Apply to X-Y. Toggle this on to scale the symbol in two dimensions only. If the symbols are too large or too small, users can use the fixed-scale entry to adjust sizing or increase/decrease the size in the point style, marker panel.

Another scenario for symbol scaling is scaling a symbol to a specific size. For example, instead of measuring the diameter of each tree, the survey measures the diameter of tree drip lines. The desired result is a symbol correctly representing the drip-line diameter. If the raw descriptions MP 20, MP 15, and MP 7 indicate the diameter of each maple tree's drip line, the parameter is a scaling factor for sizing a block representing the canopy.

The first requirement is having an AutoCAD block whose diameter is 1 foot (unit). When the symbol is scaled by parameter, the scaled symbol represents the measured tree drip line. The user would set the scale parameter to parameter 1 ($1) and toggle on Apply to X-Y.

The last scaling option is scaling vertically, Apply to Z. This toggle is useful when a user wants to display a symbol with a vertical exaggeration. Civil 3D uses multi-view blocks that display a plan view and a second three-dimensional view symbol. For example, this could be used in a survey recording tree heights. The point's raw descriptions would be OK 15, OK 37, and OK 23. The user would toggle on the scale parameter and set the parameter to 1 ($1) and then would toggle on the Apply to Z. This scales the blocks vertically. The multi-view block would have to have a definition height of 1 to correctly show the different vertical heights.

ENTRY 11: OPTIONAL SYMBOL ROTATION

The rotation parameter allows the user to change the default rotation of a marker (see Figure 2.60). The default rotation is the orientation of the drawn symbol using AutoCAD's 0 (zero) direction.

ENTRIES 12 AND 13: FIXED SCALE AND DIRECTION OF ROTATION

The fixed rotation sets a fixed angle for the marker.

The default rotation direction is counterclockwise. When using a rotation parameter, this setting defines the direction of the rotation.

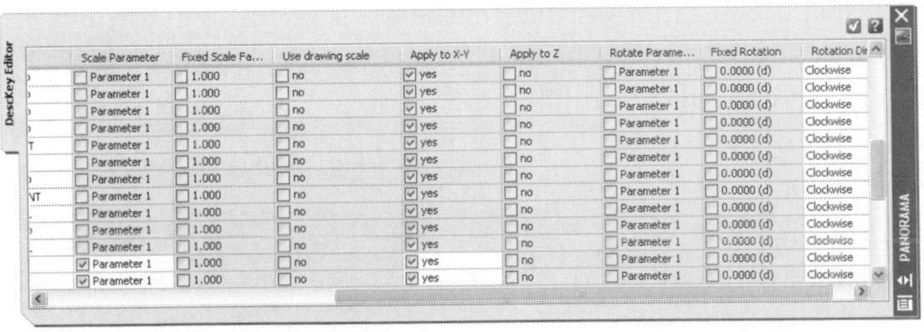

DescKey Editor

	Scale Parameter	Fixed Scale Fa...	Use drawing scale	Apply to X-Y	Apply to Z	Rotate Parame...	Fixed Rotation	Rotation Dir
	Parameter 1	1.000	no	yes	no	Parameter 1	0.0000 (d)	Clockwise
	Parameter 1	1.000	no	yes	no	Parameter 1	0.0000 (d)	Clockwise
	Parameter 1	1.000	no	yes	no	Parameter 1	0.0000 (d)	Clockwise
	Parameter 1	1.000	no	yes	no	Parameter 1	0.0000 (d)	Clockwise
T	Parameter 1	1.000	no	yes	no	Parameter 1	0.0000 (d)	Clockwise
	Parameter 1	1.000	no	yes	no	Parameter 1	0.0000 (d)	Clockwise
	Parameter 1	1.000	no	yes	no	Parameter 1	0.0000 (d)	Clockwise
NT	Parameter 1	1.000	no	yes	no	Parameter 1	0.0000 (d)	Clockwise
	Parameter 1	1.000	no	yes	no	Parameter 1	0.0000 (d)	Clockwise
	Parameter 1	1.000	no	yes	no	Parameter 1	0.0000 (d)	Clockwise
	Parameter 1	1.000	no	yes	no	Parameter 1	0.0000 (d)	Clockwise
	Parameter 1	1.000	no	yes	no	Parameter 1	0.0000 (d)	Clockwise
	Parameter 1	1.000	no	yes	no	Parameter 1	0.0000 (d)	Clockwise

PANORAMA

Figure 2.60

EXAMPLE DESCRIPTION KEY SET

The following is a sample Description Key Set listing:

Code	Point Style	Label Style	Description	Scale
BCP	Benchmark	<default>	BND POINT	
BLDG	<default>	<default>	$*	
BM####	<default>	<default>	$*	
CL*	<default>	<default>	$*	
DATUM	<default>	<default>	$*	
DL	<default>	<default>	DITCH	
DMH*	Storm drain	<default>	STORM DRAIN	
EOP	<default>	<default>	$*	
IN@	Inlet	<default>	$*	
I[PR]F	Iron pin	<default>	IRON FOUND	
LC	Iron pin set	<default>	LOT CNR	
LP*	Light	<default>	LIGHT POLE	
MD	<default>	<default>	MOUND	
MH*	Manhole	<default>	MANHOLE	
MP*	Maple	<default>	$1" MAPLE	Parameter 1
OK*	Oak	<default>	$1" OAK	Parameter 1
POND	<default>	<default>	$*	
PP	Utility pole	<default>	POWER POLE	
RP	Survey	<default>	$*	
SMH*	Sanitary Manhole	<default>	SAN SEWER	
SWALE	<default>	<default>	$*	
TOPO	Standard	<default>	$*	
WV	WV	<default>	W VALVE	

CREATING A NEW CODE

To create a new code entry, the user selects a code entry, presses the right mouse button, and selects New... from the shortcut menu. Click in each cell of the new entry to change or set the new values.

IMPORTING A DESCRIPTION KEY SET FROM LDT

Civil 3D reads the description key file from a Land Desktop project and creates a Description Key Set. The LDT description keys do not contain point and label style assignments; the user sets these manually. As Civil 3D imports the keys, it assigns default point and label styles to each entry. Default point and label styles are set in Settings, Point heading, the Edit Feature Settings dialog box. Civil 3D uses the LDT point layer assignment for the key set's layer assignment. After importing the file, the user can edit the resulting Description Key Set, assigning the correct styles, layers, and toggles.

After developing a Description Key Set, the user can transfer the Description Key Set by dragging and dropping the file to other drawings or the template.

All Civil 3D point styles have a scaling setting. The default value is Use drawing scale. This makes the points the right size for the plotting scale of the drawing. The Description Key Set also contains this setting and, if toggled on in the key set, it also multiples the marker or symbol by the scale of the drawing. This creates points that are too large. Users should toggle off Use drawing scale in the Description Key Set.

POINT GROUPS

Point groups organize points into meaningful groupings. A reason for grouping points is they have similar descriptions, functions, locations, or the user may want to isolate the points for further processing. For example, a group of points can represent surface data, lots, tree locations, or survey control of a site. The points of a group may have different descriptions (different origins of location). Some Civil 3D commands, surfaces for example, require point groups to use points as surface data.

Point groups also supply an alternative method of assigning point and label styles to its members. This uses point and label overrides, altering the way in which a point displays on the screen.

POINT GROUP PROPERTIES

Civil 3D creates point groups from commands in the Points menu or from the Prospector's Point Groups heading shortcut menu. When creating a point group, users can use four different methods of selecting points: Existing Point Groups, Raw Description Matching, options from the Include tab, and the query builder (SQL statements). At times it is easier to select a large number quickly and then remove a few from the list. To remove points from those already selected, the Point Group Properties dialog box includes an Exclude panel.

Information

The Information tab contains the name and description of a point group. The panel also sets the layer, if Civil 3D should display a tool tip when the cursor hovers over a point from the group, if the group is locked, and the current point and label styles.

Point Groups

A point group can be a combination of other point groups. For example, a user can name a point group EG Surface points, and it can be a combination of the point groups E-UTIL, GRND SHOTS, and BREAKLINES (see Figure 2.61).

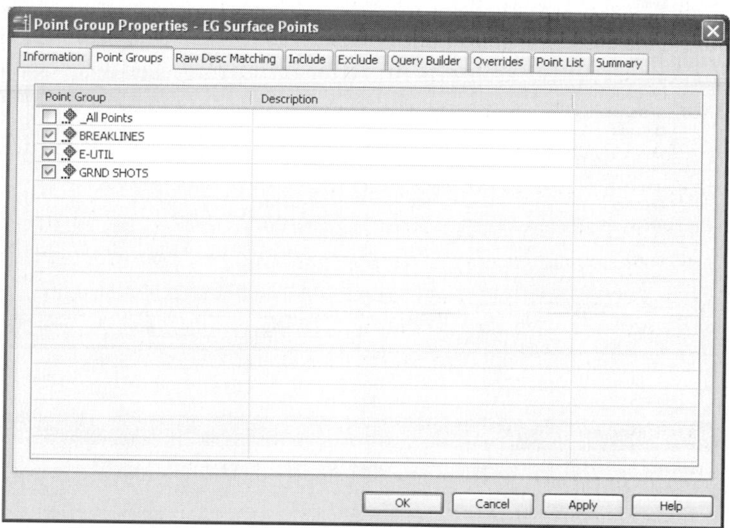

Figure 2.61

RAW DESCRIPTION MATCHING

The list of raw descriptions from which to create a group is a list of the entries from the Description Key Set(s). Users select individual codes by toggling them on to the left of their respective entries (see Figure 2.62).

INCLUDE TAB

The Include tab selects points by point number, selection set, raw or full descriptions, point names, and/or elevations (see Figure 2.63). All entries can have multiple values when separating them with a comma. See the With raw descriptions matching entry in Figure 2.63.

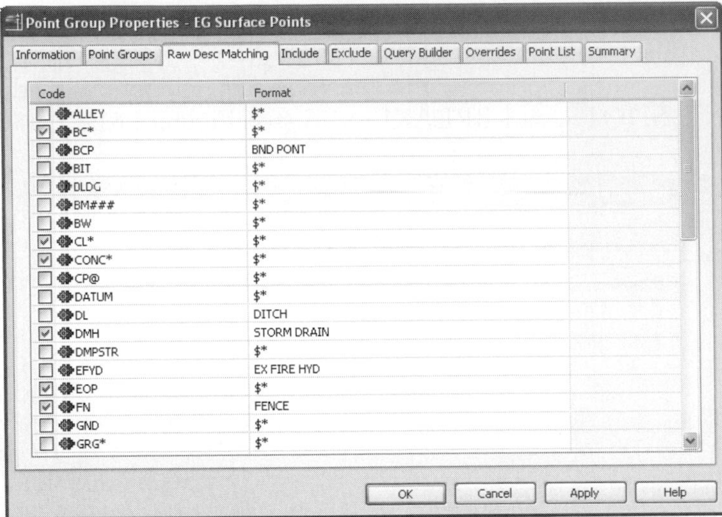

Figure 2.62

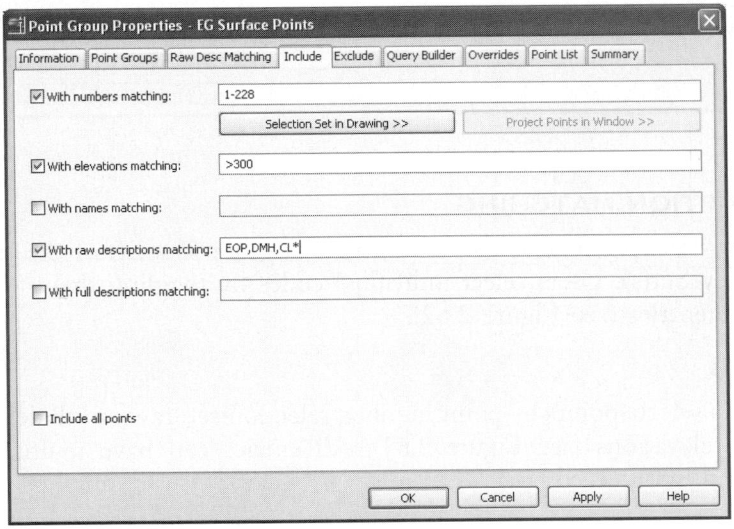

Figure 2.63

EXCLUDE

The Exclude tab is exactly like the Include tab; however, this panel removes points from the point group when they match any entry. Users can remove points by point number, selection set, raw or full descriptions, point names, and/or elevations (see Figure 2.64).

QUERY BUILDER

The Query Builder tab allows a user to select points directly from the drawing using an SQL (Standard Query Language) query. The user can select points by point number, raw

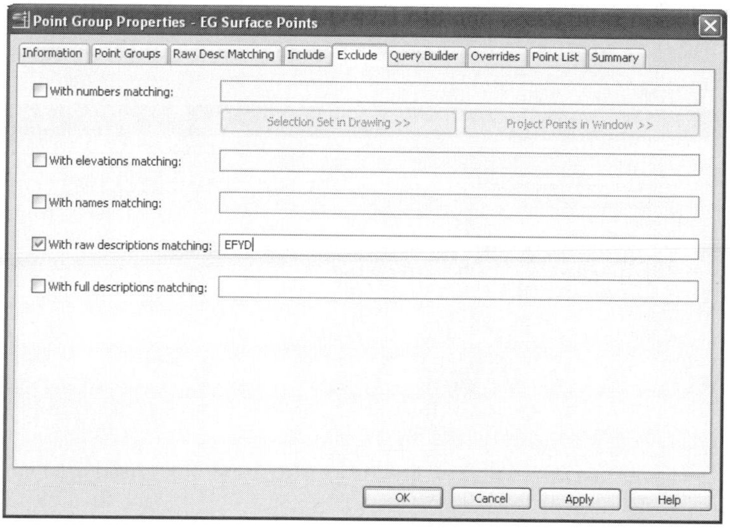

Figure 2.64

or full description, point names, elevations, and with combinations of logical AND or OR logic (see Figure 2.65).

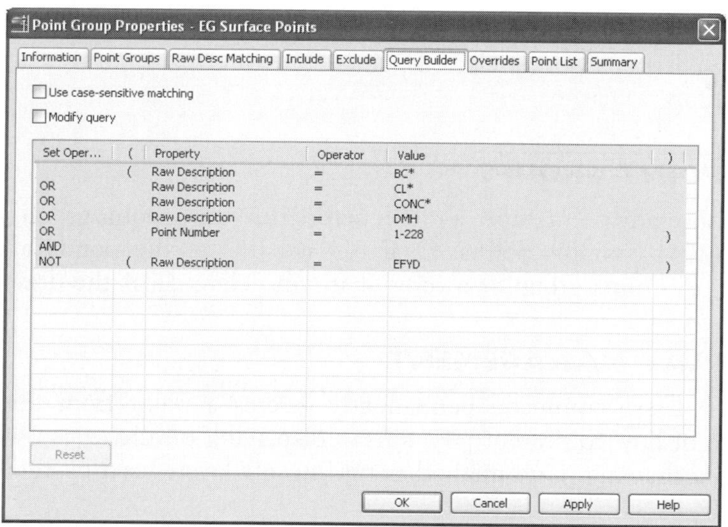

Figure 2.65

OVERRIDES

A point group can override the raw description, point elevation, and point and point label styles of the members of the group (see Figure 2.66). When defining the overrides, they also appear in the Information tab. The raw description and point elevation overrides set a new raw description and elevation for all of the points in the group. The point style and label override allow the user to assign new markers and labels to the members of the group.

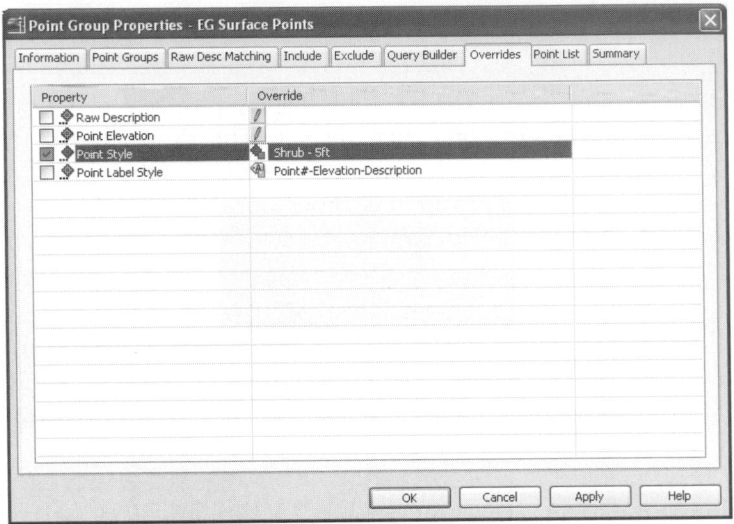

Figure 2.66

POINT LIST AND SUMMARY TABS

The Point List tab displays the selected points. The Summary tab lists all of the settings defining the group.

DELETING POINT GROUPS

Users must remove any points from a group before deleting a point group. If a user deletes a point group that contains points, Civil 3D will delete the points and point group. Deleting the _All Points group or a copy of it will delete all of the drawing points.

POINT DISPLAY MANAGEMENT

The AutoCAD Erase command permanently deletes a point from the drawing. This creates an issue of how to prevent points from displaying on the screen while still having them active in a drawing. One method is turning off layers used by the point and label styles.

There is a second method of hiding markers and labels. This method uses a None option for point and label styles. This option hides both the point marker and label. This takes advantage of the Prospector's Point Groups heading display properties and point group style overrides. When defining a point group, there is an option to assign a point and label style override for its member points. The result of assigning None as an override prevents the points in the group from showing markers and labels on the screen. Combining this override with controlling point group display order allows a user to view only the marker and labels of a single point group, to view the marker and labels that are a combination of two or more point groups, or to hide all markers and labels from showing on the screen.

POINT GROUP PROPERTIES

With all of the points stored in the drawing and without the ability to remove them, a new strategy for isolating points is a necessity. The display property of point groups helps in isolating point groups and controlling their display. The Properties of Prospector's Point Group heading sets the display order for all point groups in a drawing. Civil 3D draws the lowest point group first and follows in this order until reaching the top. The top point group is the last drawn. What then appears on a screen is all of the points in all of the point groups.

If there is a point group with no point or label styles and it is at the top of the display list, all of the points would disappear from the screen. A No Point and No Label style hides the marker and labels. The positioning of this group relative to other point groups allows the user to display a single point group or combinations of point groups on the screen while hiding other point groups from view.

If a point belongs to only one point group, it is drawn once. If a point is a member of several point groups, the point's last point group (which is the highest on the list) controls how the point appears. For example, a drawing has maple and oak trees each with their point style and point group. If a third point group, Trees, includes the Maple and Oak point groups, overriding the point style with a Generic Tree point style and placing it at the top of the display list displays all the trees as generic trees (see Figure 2.67). If the user changes the point group properties and places the Maple group above the Generic Tree group, the maple trees display their maple symbol and the oak trees display the generic tree symbol (see Figure 2.68).

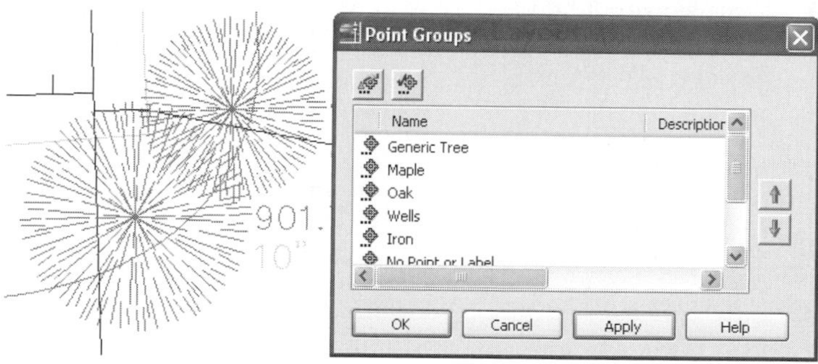

Figure 2.67

Figure 2.68

POINT TABLE STYLES

Another method of annotating points in a drawing is creating a point table. A point table can have entries for all of the points in a drawing, a selection set of points, or the points in a point group. Point tables are defined simply. The contents of the table are limited to information about the points: point numbers, point names, elevations, raw or full description, etc. (see Figure 2.69).

The blue plus sign at the lower right of the dialog box adds new columns for a table. The red X deletes a selected column or columns from the table style. By double-clicking in a column value cell, the Text Component Editor displays, allowing the user to change the text format of the information (see Figure 2.70).

Table Style - PNEZD format

Information | Data Properties | Display | Summary

Table settings
Wrap text
Maintain view orientation
Repeat title in split tables
Repeat column headers in split tables

Sort data
Sorting column: 1 Order: Ascending

Text settings
Title style: Standard Height: 0.1500"
Header style: Standard Height: 0.1000"
Data style: Standard Height: 0.1000"

Structure

	Point Table				
	Point #	Elevation	Northing	Easting	Description
Column Width	Automatic	Automatic	Automatic	Automatic	Automatic
Column Value	<[Point Numb...	<[Point Eleva...	<[Northing(U...	<[Easting(Uft...	<[Raw Descri...

OK Cancel Apply Help

Figure 2.69

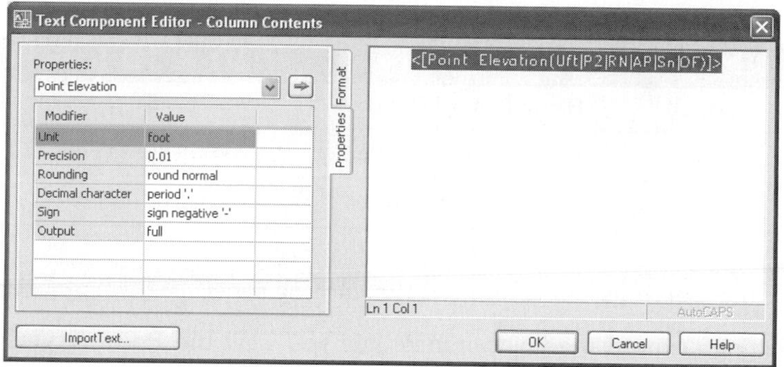

Figure 2.70

EXERCISE 2-5

When you complete this exercise, you will:

- Import a Description Key Set.
- Add entries to a Description Key Set.
- Create point groups.
- Assign point group style overrides.
- Adjust the point group display properties.
- Create a point table style.
- Create a point table.

EXERCISE SETUP

This exercise begins with a new drawing using the *Chapter 2 – Unit 5.dwt* file. The file is on the CD that accompanies this textbook and is in the *Chapter 2* folder.

1. From the File menu, select **New...**, locate the *Chapter 2 – Unit 5.dwt* template file, and click the **Open** button to start the new drawing.
2. From the File menu, select **Save As...** and save the drawing as **Points-2**.

ADD POINT STYLES

There are three missing point styles: Inlet, Maple, and Oak. This exercise creates point styles that define a layer for the point's label. The Marker component layer is 0 (zero) for all three point styles because the Description Key Set will assign the layer for the point style (see Table 2.5).

Table 2.5

Point Style	Marker	3D Geometry	2D & 3D Label Layer
Inlet	Inlet	Flatten Point to Elevation	V-ESTM-LBL
Maple	Maple	Flatten Point to Elevation	V-VEG-LBL
Oak	Oak	Flatten Point to Elevation	V-VEG-LBL

1. Click the **Settings** tab to make it the current panel.
2. In Settings, expand the Point branch until you view the **Point Styles** heading with its list of styles.
3. Select the **Point Styles** heading, press the right mouse button, and from the shortcut menu select **New....**
4. Click the **Information** tab, and for the name enter **Inlet**.
5. Click the **Marker** tab to make it the current panel.
6. Click the **Use AutoCAD BLOCK** symbol for marker, scroll through the blocks, and select **Inlet**.
7. Click the **3D Geometry** tab and change the Point Display Mode to **Flatten Points to Elevation**.
8. Click the **Display** tab to make it the current panel, leave the Marker layer 0 (zero), and change the Label layer to **V-ESTM-LBL**.
9. If necessary, change the Color and Linetype properties to **ByLayer** for both the Marker and Label entries.
10. Change the View Direction to **3D**, leave the Marker layer 0 (zero), and change the Label layer to **V-ESTM-LBL**.
11. If necessary, change the Color and Linetype properties to **ByLayer** for both the Marker and Label entries.
12. Click the **OK** button to close the dialog box.
13. Repeat Steps 3 through 12 to make the Maple and Oak Point Styles using the information in Table 2.5.
14. Click the **AutoCAD Save** icon and save the drawing.

IMPORTING A DESCRIPTION KEY SET

Civil 3D reads the description key file from a Land Desktop project and creates a Description Key Set. If you did not copy the Civil 3D project to the *Civil 3D Projects* folder, the project is on the CD that accompanies this textbook.

1. In Settings, click the **Point** heading, press the right mouse button, and from the shortcut menu select **Edit Feature Settings....**

2. Expand the Default Styles section to view the names of the default Point and Label styles.

   Each point will have Standard as its initial style.

3. Click the **OK** button to exit the dialog box.

   The Import Data from Land Desktop command imports the Description Key Set file from a LDT project.

4. From the File menu, Import flyout, select **Import Data from Land Desktop....**
5. Set the Project Directory to **Civil 3D Projects**, set the project to **Civil 3D**, and toggle **OFF** everything except for description keys (see Figure 2.71).

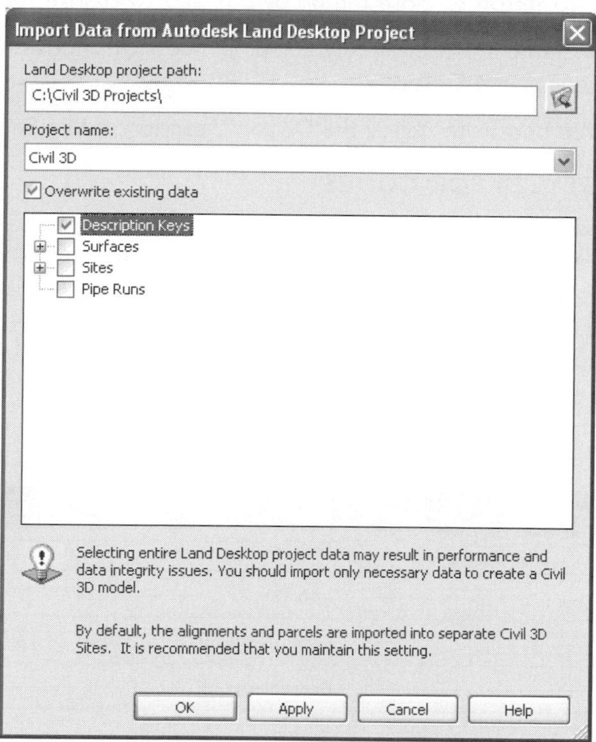

Figure 2.71

6. Click the **OK** button to migrate the description keys.

   The command displays a Description Keys Migration Completed dialog box.

7. Click the **OK** button to exit the dialog box and to complete the migration of the description keys.
8. Click the **AutoCAD Save** icon to save the drawing.

Review Imported Description Keys

1. In Settings, expand the Point branch until you view the list of Description Key Sets.

   The Import from Land Desktop command imported two description key files: Default and HNC3D-DEFAULT.

2. In Settings, in the Point branch, click the **Description Key Set** heading, press the right mouse button, and from the shortcut menu select **Properties...**.

   This shows the current search order for the key sets.

3. Click the bottom set name and click the up arrow to move the set to the top of the search order.
4. Click the **OK** button to close the dialog box and to change the search order.
5. Click the **Default** Description Key Set, press the right mouse button, and from the shortcut menu select **Delete...**.
6. Click the **Yes** button to delete the Default Description Key Set from the drawing.

CHANGE POINT STYLES FOR CODES

When importing a Description Key Set file, the Import routine assigns the default point style for all codes. However, several of the points need a style that displays a point marker symbol.

Changing the point style is a two-step process. First, toggle on the point style and then set the name of the new point style. When clicking in the point style name cell, a Select Point Style dialog box displays with a drop-down list of available point styles.

You need to change the current point styles for the codes in Table 2.6.

Table 2.6

Code	Point Style
BM###	Benchmark
BP	Iron Pin
CB*	Catch Basin
CP@	Benchmark
DATUM	Station
DMH	Storm Sewer Manhole
EFYD	Hydrant (existing)
I[PR]F	Iron Pin
LC	Iron Pin
LP*	Light Pole
MH*	Manhole

Table 2.6 (continued)

Code	Point Style
PP*	Utility Pole
SIGN*	Sign (single pole)
SMH	Sanitary Sewer Manhole
STA#	Station
TF-SIG	TF Signal
TR*	Tree - 20ft
VV	Valve Vault
WELL	Well
WV	Water Valve

1. In Settings, the Description Key Set list, select the **HNC3D-DEFAULT** key set, press the right mouse button, and from the shortcut menu select **Edit Keys....**
2. In the DescKey Editor vista, change the Use Drawing Scale toggle to **NO** for each entry in the key set.
3. In the DescKey Editor vista, the Point Style column for the BM#### entry, click **ON** the toggle box, indicating the intent of overriding the default point style.
4. In the DescKey Editor vista, the Point Style column for the BM#### entry, click <Default> to display the Select Point Style dialog box, click the drop-list arrow, select **Benchmark** from the drop list, and click the **OK** button to select the style and to close the dialog box.

   The BM#### now has Benchmark as its point style.
5. Repeat Steps 3 and 4 and assign the remaining keys in Table 2.6 to the appropriate point styles.

CREATE NEW ENTRIES IN THE DESCRIPTION KEY SET

There are three missing entries in the Description Key Set: Inlet, Maple, and Oak. The next portion of the exercise creates these keys.

The new keys use their related point styles (Inlet, Maple, and Oak). The Maple and Oak codes (MP 2 and OK 6) have a parameter that is the diameter of the tree's trunk. The format entry for these description keys use the $1 parameter and it is the first item in the full description (representing the size of the tree trunk). The resulting full description for MP 2 is 2" MAPLE and for OK 6 it is 6" OAK. The coding for the format is in Table 2.7.

The first parameter is also a multiplier of the symbol. This allows Civil 3D to show a 6-inch oak tree as a larger symbol than an oak tree that is only 2 inches in diameter.

EXERCISES

The scaling settings for the three keys are in Table 2.8. All three keys use the drawing scale to make them the correct size for the drawing. The Maple and Oak keys use parameter 1 as a secondary scaling factor. The settings in Table 2.8 make parameter 1 a scaling factor to the X-Y component of the marker. This makes a point using the MP 4 code have a larger symbol than a point with the code MP 2.

Table 2.7

Code	Point Style	Format	Layer
IN@	Inlet	$*	V-ESTM-PNT
MP*	Maple	$1" MAPLE	V-VEG-PNT
OK*	Oak	$1" OAK	V-VEG-PNT

Table 2.8

Code	Scale Parameter1	Drawing Scale	Apply to X-Y
IN@	ON	OFF	ON
MP*	ON	OFF	ON
OK*	ON	OFF	ON

INLET KEY

When creating a new entry in a Description Key Set, you have to select an existing entry in the key list. You should review all of the settings for the new key before going on to the next key.

1. In the Edit DescKey Editor vista, select the **ALLEY** entry, press the right mouse button, and from the shortcut menu select **New....**

   This makes a new key entry at the bottom of the key list.
2. Click in the code cell and enter **IN@** as the new code.
3. Click the Point Style override toggle to turn it **ON** and assign the **Inlet** point style to the entry.
4. Make sure the Format entry is **$***.
5. Click the Layer override toggle to turn it **ON** and assign the layer of **V-ESTM-PNT** to the entry.
6. If the Scale Parameter is on, toggle it **OFF**.
7. If the Use Drawing Scale is on, toggle it **OFF**.

MAPLE KEY

1. While still in the HNC3D-DEFAULT key set, select the **ALLEY** entry, press the right mouse button, and select **New...** from the shortcut menu.

   This makes a new key entry at the bottom of the key list.

2. Click in the code cell and enter **MP*** as the new code.
3. Click the Point Style override toggle to turn it **ON** and assign the **Maple** point style to the entry.
4. Change the Format to **$1" MAPLE**.
5. Click the Layer override toggle to turn it **ON** and assign the layer **V-VEG-PNT**.
6. Make sure the Scale Parameter is **ON** and is set to **Parameter 1**.
7. If the Use Drawing Scale is on, toggle it **OFF**.
8. Toggle **ON** Apply to X-Y.

OAK KEY

9. Repeat Steps 1 through 8 used that were for the Maple entry to create a new entry for the OK code and close the DescKey Editor vista by clicking the "**X**" in the upper-right corner.

IMPORTING POINTS

Everything is in place to import points using the current Description Key Set. The *Base-Points.nez* file is the ASCII file to import. The file is in the *Chapter 2* folder of the CD that accompanies this textbook.

1. From the Points menu, select **Create Points....**
2. On the right side of the toolbar click the **Import Points** icon.
3. In the Import Points dialog box, click the drop-list arrow and select the **PNEZD (comma delimited)** format.
4. To select the Source file, on the right side of the dialog box click the **Folder** icon.
5. In the Select Source File dialog box, to the right of Files of Type at the bottom, click the drop-list arrow, and select ***.nez** from the list of file extensions.
6. Navigate to the location of the *Base-Points.nez* file, click on the file name, and click the **Open** button to return to the Import Points dialog box.
7. In the Import Points dialog box do not set any Advanced Options, and click the **OK** button to import the points.
8. Click the "**X**" on the right side of the Create Points toolbar to close it.
9. Click the **Prospector** tab to make it the current panel.
10. In Prospector, expand Point Groups, select **_All Points**, press the right mouse button, and from the shortcut menu select **Properties....**
11. In _All Points Properties, change the Label Style to **Standard** and click the **OK** button to exit.
12. In Prospector, select the **Points** heading to list the points in the preview area.

EXERCISES

13. In the preview area, scroll the point list to the right until you view the Raw Description heading, and click the heading to sort the list.
14. In the preview area, scroll through the list of points until you view a WELL description, and then scroll the listing back to the left until you view the point number.
15. In the preview area, select the point number, press the right mouse button, and select **Zoom to** from the shortcut menu.
16. Click the **AutoCAD Save icon** to save the drawing.

You can zoom to any point from the preview area, the Edit Points vista, and from a preview of point group points.

CREATING POINT GROUPS

The easiest way to create a point group is by raw descriptions. The raw description entries of the Description Key Set create the selection list for point groups.

Use the following description keys listed in Table 2.9 to define three point groups.

Table 2.9

Name	Key(s)
Wells	WELL
Iron	BP, I[PR]F, LC, SIP
Site Control	BM####, CP@, STA#

1. If necessary, click the **Prospector** tab to make it the current panel.
2. In Prospector, click the **Point Groups** heading, press the right mouse button, and from the shortcut menu select **New...**.
3. Select the **Information** tab and enter **Well** as the name of the point group.
4. Select the **Raw Desc Matching** tab, scroll through the list, and toggle **ON** the WELL description.
5. At the bottom right of the dialog box, click the **Apply** button to select the points for this group.
6. Select the **Point List** tab to view the selected points.
7. Click the **OK** button to exit the Point Group Properties dialog box.
8. Repeat Steps 2 through 7 and create the Iron and Site Control group using key entries from Table 2.9.

ASSIGNING POINT AND LABEL STYLE OVERRIDES

A point group that includes all points with None for the point and label styles if it is at the top of the point group display list will hide all points in the drawing.

CREATING A NO POINT OR LABEL POINT GROUP

1. Click the **Prospector** tab to make it current.
2. In Prospector, select the **Point Groups** heading, press the right mouse button, and from the shortcut menu select **New....**
3. Click the **Information** tab to make it the current panel and enter as the name **No Point or Label**.
4. In the **Information** panel, assign **<none>** to the point and label styles.
5. Select the **Include** tab, and at the bottom of the panel toggle **ON** Include all points.
6. Select the **Overrides** tab and toggle **ON** Point Style and Point Label Style overrides.
7. Click the **OK** button to create the point group.
8. In Prospector, select the **Point Groups** heading, press the right mouse button, and from the shortcut menu select **Properties....**

   This displays the Point Group Properties dialog box containing the display order list.
9. In the Point Group Properties dialog box, move **No Point or Label** to the top of the list and click the **OK** button.

   All points disappear from the screen.
10. In Prospector, select the **Point Groups** heading, press the right mouse button, and from the shortcut menu select **Properties....**
11. In the Point Group Properties dialog box, move **Well** to the top of the list, move **No Point or Label** to the second position, and click the **OK** button.

    Only the Well points display on the screen.

    Use the Display Properties of Point Groups to manipulate the visibility of different points on the screen.
12. In Prospector, select the **Point Groups** heading, press the right mouse button, and from the shortcut menu select **Properties....**
13. Finally, in the Point Group Properties dialog box, move **_All Points** to the top and click the **OK** button.
14. Click the **AutoCAD Save** icon to save the drawing.

DEFINING A NEW POINT TABLE STYLE

Many times the documentation of a site includes a list of the benchmarks and/or survey control points. A point table is a convenient method of creating this kind documentation. In this exercise you need to identify the wells and their coordinates in a table.

The current drawing does not contain a table style that is useable for the current drawing's point data.

1. Click the **Settings** tab to make it the current panel.

2. In Settings, expand the Point branch until you view a list of styles for the **Table Styles** heading.
3. Select the **Table Styles** heading, press the right mouse button, and from the shortcut menu select **New....**
4. In the Table Style dialog box, select the **Information** tab and enter **Point NE & Description** as the name.
5. Select the **Data Properties** tab to view its settings and values.

   This panel of the dialog box defines the sorting order, text styles and heights, data options, and the structure of the table.

   The initial table definition has Point Number, Raw Description, Elevation, Northing, and Easting (see Figure 2.72).

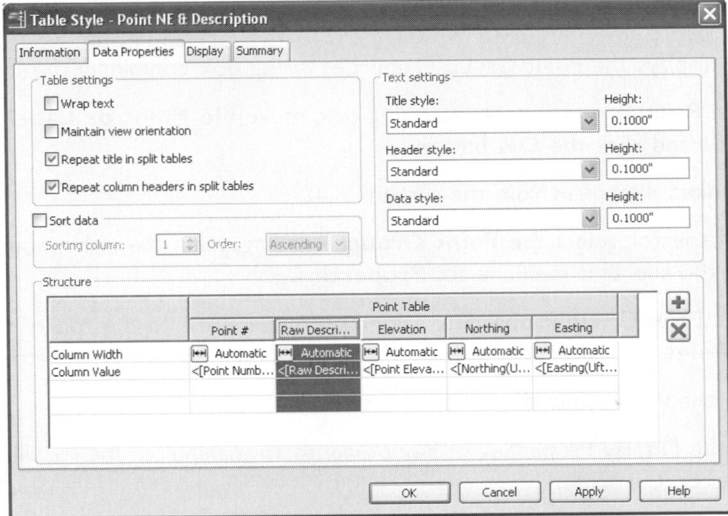

Figure 2.72

6. Click **Raw Description** to highlight the column and click the red **X** to delete the column from the table.
7. Click **Elevation** to highlight the column and click on the red **X** to delete the column from the table.
8. Click the blue **+** (plus sign) to add a new column.
9. Double-click the heading cell for the new column, and in the Text Component Editor enter **Description** and click the **OK** button to return to the Table Style dialog box.

   This creates Description as the column name.

10. For the Description column, double-click in the column value cell to display the Text Component Editor.

11. In the Text Component Editor, click the Properties drop-list arrow, select **Full Description**, click the blue arrow (in the top center of the dialog box), and click the **OK** button to return to the Table Style dialog box.

    Your dialog box should look like Figure 2.73.

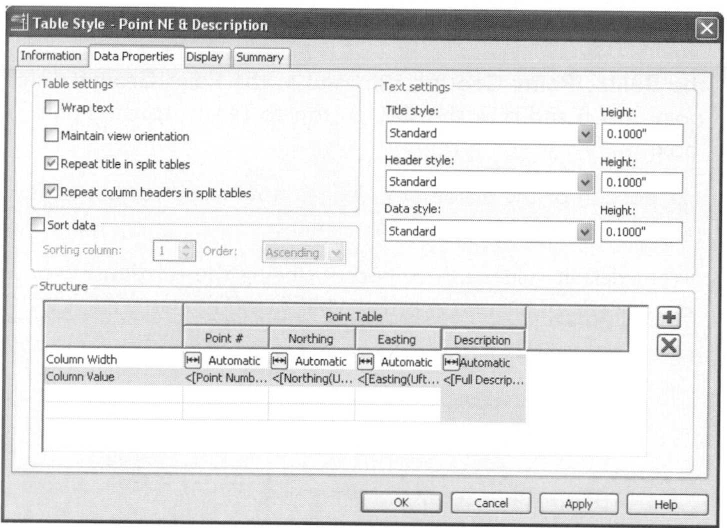

Figure 2.73

12. Click the **OK** button to create the table style.

CREATING A POINT TABLE

1. Click the **Prospector** tab to make it the current panel.
2. In Prospector, select the **Point Groups** heading, press the right mouse button, and from the shortcut menu select **Properties....**
3. In the Point Groups Properties dialog box, select the **Well** point group, move it to the top of the list, select and move the **No Point or Label** group to the second position, and click the **OK** button to dismiss the Point Group Properties dialog box.
4. Use the Pan and Zoom commands of AutoCAD to display a clear area to the east of the site.
5. From the Points menu select **Add Tables....**

   This displays the Point Table Creation dialog box (see Figure 2.74).
6. At the top of the Point Table Creation dialog box, change the Table style to **Point NE & Description**.

The center of the dialog box selects the points for the table. You can select points by Point Label Style Name, Point Groups, or by selecting a group of points from those showing on the screen. The points must be showing on the screen to be a part of a table.

7. Select the **Point Group** icon at the center left of the dialog box.

   This displays a list of point groups.

8. Click the **Select Point Groups** icon, and in the Point Groups dialog box select the **Well** point group and click the **OK** button to return to the Point Table Creation dialog box.

   The next portion of the dialog box defines how many rows a table will have before breaking it and creating a new table segment.

   The current default values are correct, including the relationship between the points in the drawing and their entry in the table, i.e., Dynamic.

9. Click the **OK** button to close the dialog box.

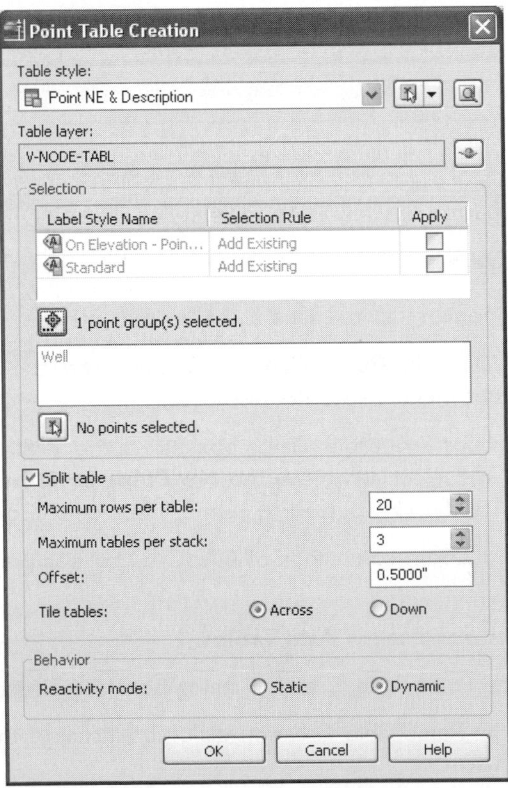

Figure 2.74

10. Select a point to locate the table in the drawing.

11. Click the **AutoCAD Save** icon to save the drawing.
12. Use the AutoCAD Zoom command to get a better view of the table.

    If you edit the data of a point in the table, the table will update to the new current data values.

DYNAMIC TABLE ENTRIES

1. In Prospector, click the **Points** heading to preview the points in the drawing.
2. In the preview area, scroll through the points list until you locate point **1473**.
3. Click in the Point Number cell for 1473 and change it to **3050**.

   A number of reactions occur. The Well point group goes out of date and displays an out-of-date icon.
4. In Prospector, Point Groups, select the **Well** entry, press the right mouse button, and from the shortcut menu select **Show Changes....**

   The Point Group Changes dialog box displays that point number 3050 is a new member, while 1473 is to be removed.
5. Click the **Close** button to exit the dialog box.
6. In Prospector, while Well is still highlighted, press the right mouse button and select **Update**.

   The Well group updates, the out-of-date icon disappears, and the new point number (3050) appears in the table.

   In Prospector, a reference (triangle) icon appears to the left of the Well point group name. This icon indicates that another object references the point group. Because there is a reference to the group, Civil 3D will not show the Delete command on the shortcut menu for this group.
7. In Prospector, click the **Well** point group, press the right mouse button, and view the commands of the shortcut menu. The command list does not include **Delete....**
8. Click the **AutoCAD Save** icon to save the drawing.
9. Exit Civil 3D.

This concludes the discussion on points. There are many more point commands in Civil 3D that were not discussed in this chapter. The point commands that create points from surface, alignment, profile, and corridor data will be discussed in later chapters.

SUMMARY

- A Description Key Set assigns a point and label style and a layer to the marker portion of a point.
- A Description Key Set can translate a point's field code (raw description) to a full description.
- With a Description Key Set assigning a layer to the marker, the point style should define the layer for the label of the point.
- An AutoCAD Erase permanently deletes a point from the drawing.
- A No Point or Label point group allows Civil 3D to hide points in a drawing.
- The properties of point groups control what point groups to display on the screen and in what combination.
- To delete a point group definition, deselect the points so it is empty, and then delete the point group definition.
- Civil 3D selects points for a point table by point label style name, by point groups, or by a selection set.
- A point table can be dynamic or static.

The next chapter examines the parcel layout, annotation, and management tools of Civil 3D. In Civil 3D, parcels start with an outer boundary, a site. A site is a container for a number of Civil 3D objects, parcels, alignments, profiles, corridors, and sections and manages the relationships between each object type.

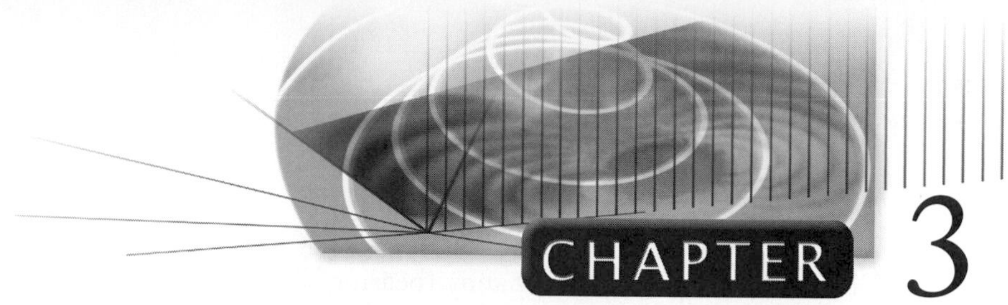

CHAPTER 3

Site and Parcels

INTRODUCTION

The division of a boundary into specific land-use blocks or resalable parcels is an exercise in practical methods and artistic flair. Civil 3D provides several tools to programmatically divide a large parcel into smaller, resalable parcels.

A boundary is the outermost polygon that contains blocks of homogenous land use. The process of subdividing blocks into small parcels is the focus of this chapter.

OBJECTIVES

This chapter focuses on the following topics:

- Organizing Parcels from the Site to the Individual Parcels
- Definition of Parcel Object Styles
- Definitions of Parcel Area, Line, and Curve Label Styles
- Development of Parcels from the CreateParcelFromObjects and CreateParcelByLayout Commands
- Parcel Sizing Tools in the Parcel Layout Tools Toolbar
- Editing of Parcel Numbers and Renaming Them
- Creation and Modification of Parcel Segments Labels
- Creation of Parcel Tables

OVERVIEW

Civil 3D introduces the concept of an overall site that is carved into smaller and smaller parcels. Both a site and its parcels are custom objects that maintain a dynamic relationship to their boundaries and roadway alignments. An example of this dynamic relationship is in the creation of two parcels from a single site. When drawing an alignment across the site, Civil 3D will divide the site into two parcels with the alignment acting as the dividing line between the new parcels. If deleting the dividing alignment, the site returns to the original site definition. Users can also divide a site or parcel into new parcels using individual parcel segments. Users can also draft parcel lines using the Parcels Layout toolbar; by using the AutoCAD line, arc, and polyline commands; or by using AutoCAD line and polyline commands with Civil 3D transparent commands (bearing/distance, direction/distance, etc.). When using AutoCAD lines, arcs, and polylines, the user must define them as parcel lines using the CreateParcelFromObjects command within Civil 3D.

A Civil 3D site is the "container" for parcel objects. Included in a site are alignments, because they, too, divide a site and/or parcels into smaller units. There can be more than one site in a drawing, and each site can have its own set of parcels. This allows the user to have different subdivision designs within the same drawing.

As mentioned in previous chapters, Civil 3D operates on the basis of styles. These styles place objects, their graphical representations (line work), and labeling on predefined layers or within the objects. It is imperative that users have as many styles defined as possible before using Civil 3D. Each Civil 3D style requires the user to change a multitude of settings before producing a "look" that a specific office might want in each drawing. In the Land Desktop product line, the "office" look was primarily a drafting exercise. However, in Civil 3D it is a blend of Civil 3D styles, text styles, and basic company blocks (frames, symbols, etc.) that produce this "look."

The process of dividing a single parcel into smaller units is an office or personal preference. Some parceling methods use circles, others use frontage measurements, others use rectangular templates, and so on. The implementer of the parceling methodology believes *his* process produces the best results for developing a parcel plan. In truth, the greatest control over developing a parcel plan is the rules and regulations governing the development of the site. Each site will have rules or covenants affecting the size and shape of the individual parcels.

UNIT 1

Civil 3D has several settings that affect the creation and annotation of sites and parcels. The focus of this unit includes the parcel styles and the basic drafting tools of Civil 3D (lines, arcs, and polylines). The display of the parcel and its annotation is a result of the currently assigned parcel styles.

UNIT 2

A site and its parcels are drafted and defined by converting existing line work or by using Civil 3D Parcel Layout tools. The CreateParcelFromObjects command converts AutoCAD line work into Civil 3D parcels or parcel segments. The CreateParcelFromObjects command requires preexisting lines, arcs, and/or polylines to create the site, parcels, or parcel segments. Civil 3D has transparent command modifiers that let users draft lines by bearing and distance, azimuth, deflection, or from Civil 3D points. The transparent commands have their own toolbar and work with the AutoCAD line, arc, and polyline commands. When using the CreateParcelByLayout toolbar to subdivide a site or its parcels, a user can only work with Civil 3D parcel segments.

UNIT 3

The third unit of this chapter focuses on the Civil 3D tools that evaluate parcels. The main tools for parcel analysis are in the Properties dialog box of a parcel, the LandXML Report application, and the Report Manager. The Parcel Properties dialog box reports the map check and inverse properties for the selected parcel. The LandXML Report application supplies additional reports (metes and bounds, surveyor's certificates, parcel areas, etc.) for one or a group of parcels from their LandXML data.

UNIT 4

The focus of the fourth unit is editing sites and parcels. The Parcel Layout Tools toolbar provides several tools that modify a site and its parcels. Users can modify a site and its parcel segments graphically by grip-moving their segments on the screen and by erasing individual parcel segments.

UNIT 5

The fifth unit of this chapter reviews the annotation of a site and parcel. There are two types of labeling: one for the site's or parcel's overall values and the other for the line and curves making up the parcel (parcel segments). The line and curve labels are either on the lines and curves of the parcel or they are an entry in a table. All parcel annotation reacts and changes to correctly display the new values of an edited site or parcel. All parcel labels are view-dependent and will rotate to read correctly in any view orientation.

UNIT 1: SITES, PARCELS, AND PARCEL STYLES

A site and its parcels are defined by one of two methods: from existing objects or by drafting them with Civil 3D parcel-sizing tools. The CreateParcelFromObjects command requires preexisting objects (lines, curves, and polylines) to create a site and/or its parcels. Civil 3D has transparent command modifiers that let users draft lines by turned angle, bearing, azimuth, deflection and distance, or with Civil 3D points. A user can start these commands by selecting one of the command's icons from the Transparent Command toolbar. Transparent commands can be used when creating curves; however, their application is limited. The CreateParcelByLayout command presents a toolbar containing parcel segment and curve tools (limited), as well as other parcel-sizing commands. The commands in the Parcel Layout Tools toolbar work only with Civil 3D site and parcel segments.

The display (layer and other object properties) and annotation (label and segments) of a site and its parcels is a result of the assigned parcel style. Parcel styles can define land use and may contain design criteria. The Parcel Labels section of the Settings menu defines different parcel labels for parcel information, styles for annotating the line and curve segments, and styles for parcel and segment tables (see Figure 3.1).

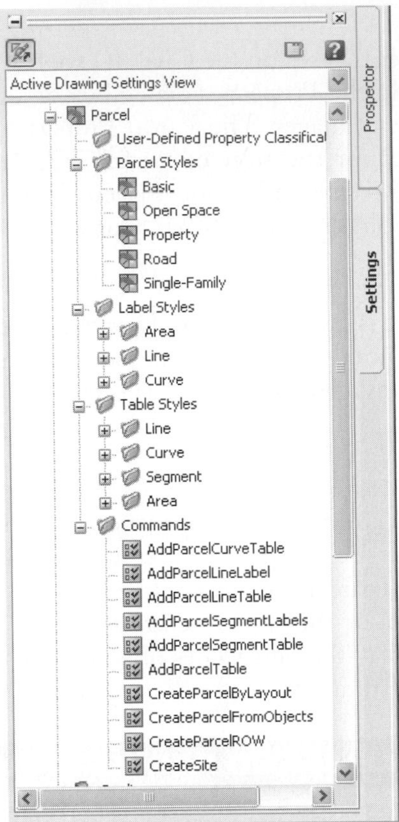

Figure 3.1

There can be more than one site in a drawing. Each site has its own set of parcel definitions (alignments, profiles, etc.). This strategy allows the user to document more than one subdivision strategy in a drawing.

PARCEL SETTINGS

The Drawing Settings, Edit Parcel Feature Settings, and Edit Command Settings affect parcel creation. Civil 3D uses three basic types of parcel labels (Parcel, Label, and Table) to create parcel objects and annotate them.

The Object Layers tab of the Drawing Settings dialog box assigns basic layers for parcel objects (see Figure 3.2). The base layers apply to parcels, their segments, and parcel tables. When assigning a modifier to these layers, using an asterisk makes the site name the value of the prefix or suffix.

The Ambient Settings panel contains several critical values that affect Civil 3D commands. These settings affect area, angle, direction, and distances for the parcel settings branch. If using the direction or angle methods to define parcel segments, the user needs to change his or her settings to use the surveyor shorthand for angles and directions: dd.mmss (Decimal dms).

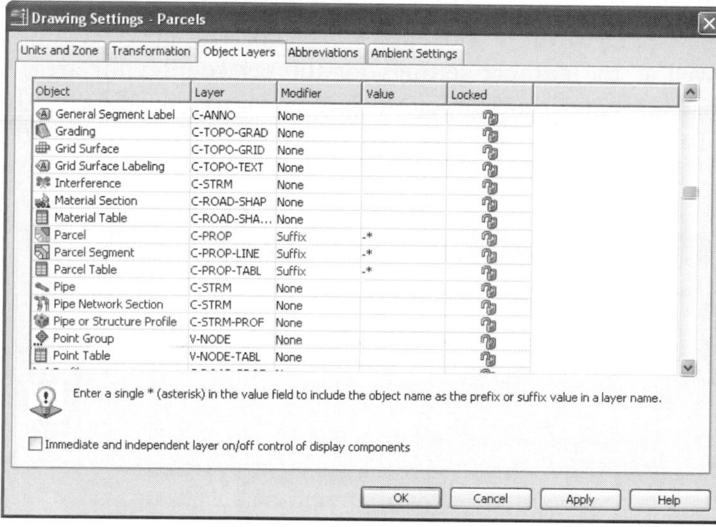

Figure 3.2

EDIT FEATURE SETTINGS

The Edit Features Settings dialog box assigns the default parcel, area label, line label, and curve label styles for new parcels (see Figure 3.3). All of the commands that create parcels use this list of styles, unless a command overrides this assignment.

Edit Feature Settings - AeccParcel

Property	Value	Override	Child Override	Lock
General				
Labeling				
Default Styles				
Parcel Style	Property			
Parcel Area Label Style	Parcel Number			
Parcel Line Label Style	Bearing over Distance			
Parcel Curve Label Style	Delta over Length and...			
Unitless				
Distance				
Dimension				
Coordinate				
Grid Coordinate				
Elevation				
Area				

OK Cancel Apply Help

Figure 3.3

COMMAND: CREATESITE

The CreateSite command creates a site. A site is critical to the functioning of Civil 3D parcels. A site is in essence the "container" for all parcels, alignments, and grading solutions (see Figure 3.4). In Prospector, under the Sites heading shortcut menu, the New... command uses the settings from this dialog box. The parcel default styles (parcel, area, line, and curve), as well as the number settings for the tag counter for area, line, and curves entries in various parcel tables are important settings of this dialog box. Users do not have to use this command if they are using the CreateParcelByLayout or CreateParcelFromObjects commands. These routines will automatically create a site.

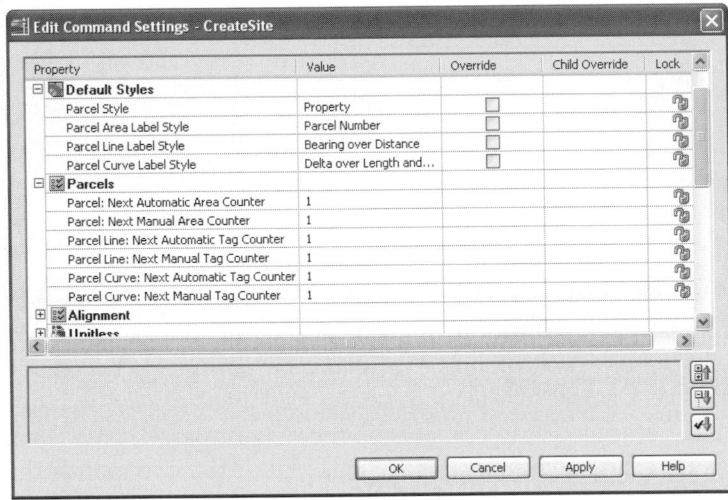

Figure 3.4

COMMAND: CREATEPARCELFROMOBJECTS

The settings in this command dialog box affect converting parcels from AutoCAD lines, curves, and polylines (see Figure 3.5). The styles in the dialog box are from values in the Edit Feature Settings dialog box. However, when defining parcels, users can change what styles this command assigns. The Convert from Entities section controls erasure of lines, curves, and polylines when converting them to parcels, and it also controls whether to label their resulting segments.

COMMAND: CREATEPARCELBYLAYOUT

This dialog box contains four sections that affect new parcel creation with the CreateParcelByLayout toolbar (see Figure 3.6). The first section, New Parcel Sizing, sets the minimum area and frontage for the newly drafted parcels. The second section, Automatic Layout, toggles on automatic parcel creation and the way in which the command distributes the remainder of the excess area: Create parcel from remainder, Place remainder in last lot, or Redistribute remainder. The third section, Manual Layout and Edit, toggles

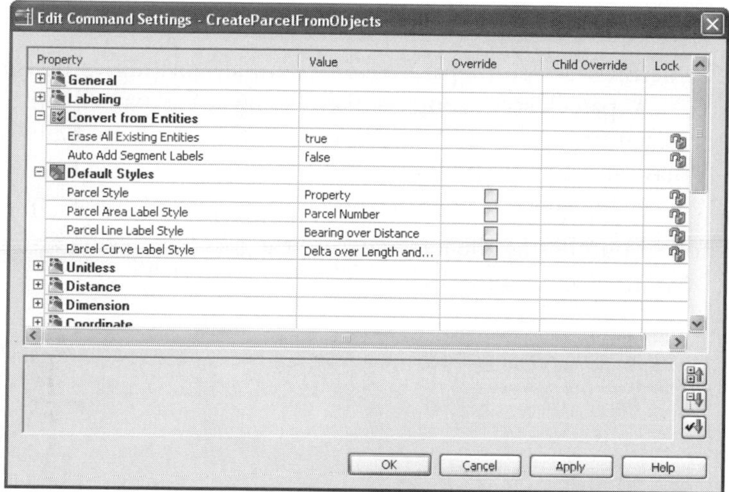

Figure 3.5

Autosnap and its increment. The sizing increment controls the amount by which users increase the size of a manually sized parcel. As users create the last parcel segment, the routine reports and controls the location of the parcel line in increments set by this setting's value. For example, if a user were to set Autosnap ON and the increment to 250 square feet, the routine would snap the location of the parcel segment by calculating each additional increment of 250 square feet over the minimum parcel size. The Slide and Swing parcel commands use these values while dynamically sizing the parcels. The last section, Convert from Entities, tells Civil 3D whether to label the new parcel object segments.

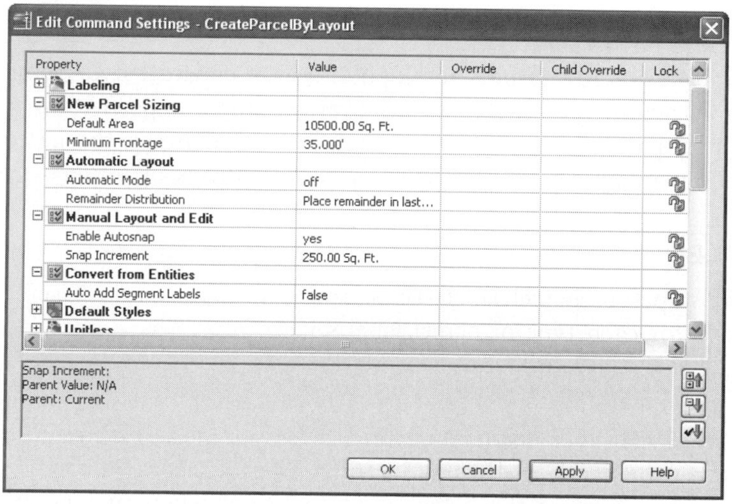

Edit Command Settings - CreateParcelByLayout
Property
Value
Override
Child Override
Lock
Labeling
New Parcel Sizing
Default Area
10500.00 Sq. Ft.
Minimum Frontage
35.000'
Automatic Layout
Automatic Mode
off
Remainder Distribution
Place remainder in last...
Manual Layout and Edit
Enable Autosnap
yes
Snap Increment
250.00 Sq. Ft.
Convert from Entities
Auto Add Segment Labels
false
Default Styles
Snap Increment:
Parent Value: N/A
Parent: Current
OK
Cancel
Apply
Help

Figure 3.6

PARCEL STYLES

Each parcel style defines a type of parcel (for example, existing, proposed, open space, single family, etc.). A parcel style controls the display of a parcel's line work and the reporting of its characteristics. Each parcel style definition contains four panels: Information, Design, Display, and Summary.

The Information panel displays the name and description of the style. The Design panel defines the method of numbering the lots, whether the lots have a fill pattern, and the scale factor of the fill pattern (see Figure 3.7).

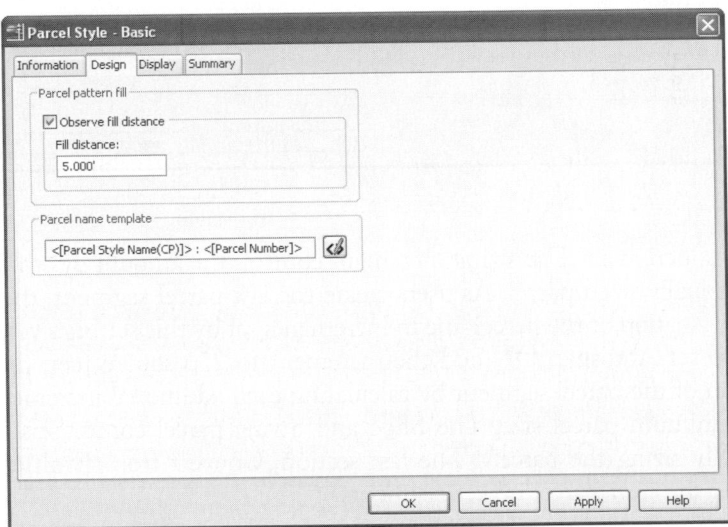

Figure 3.7

The display panel sets the layer and the fill pattern for the parcel style (see Figure 3.8). The color of the parcel segments differentiates the various parcel types within a subdivision. As with point groups, parcels have a display order. When changing the parcel display order, a user produces an image with different color combinations of parcel segments.

LABEL STYLES

There are three types of parcel label styles: area, line, and curve. Users can define any number of styles for each type of parcel label style.

The Label Style Composer interface creates or modifies these labels styles. The General tab of each style defines the text style, label style visibility, and plan readability. The Layout tab of the style defines the label contents and their attachment to the centroid of the parcel and other label elements. See Figure 3.9 for an example of a parcel area label. The Dragged State tab values set the behavior of the label when moving it from its initial position to a different location in the drawing.

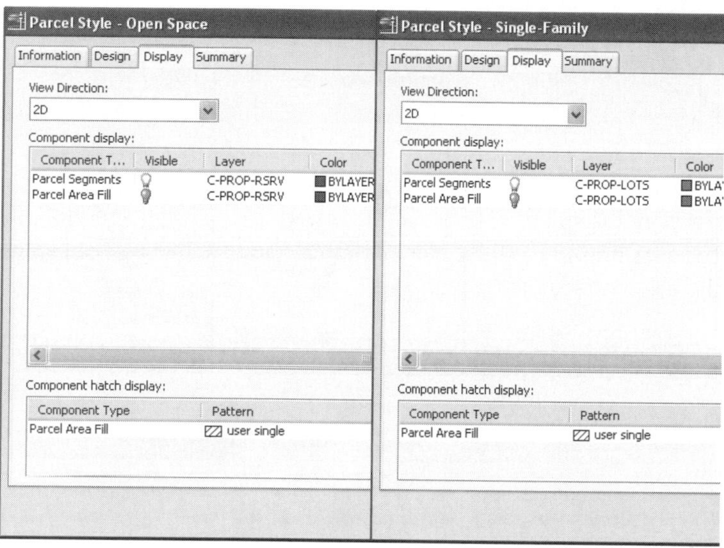

Figure 3.8

Each text component can have different text sizes. All properties within a single text component have one text size. There cannot be one border around several individual label components, only a border around a single component with several properties.

When creating a border for a text component, the user assigns color properties, masks, and gap parameters. When assigning a color to the border, the assignment overrides the color of the label layer.

In the dragged state of a label the same conditions apply to leaders and stacked text when defining their color, lineweight, and linetype properties. Users can assign a border that is different from the Layout setting or assign a color override for the label and the leader. Users can also use a block from the drawing as a part of a label (e.g., crow's feet, etc.).

AREA LABEL STYLE

An area label style can contain the name of a parcel, its description, a parcel number, its area, parameter, address, site name, and a tax ID. The three panels, General, Layout, and Dragged State, control the components, visibility, text, form, and shape of the label in its original position (layout) or at a new location dragged from its original location (dragged state) (see Figure 3.9).

LINE LABEL STYLE

A line label style can contain a table tag, bearing and/or distance, a direction arrow, and other properties. The three panels, General, Layout, and Dragged State, control the components, visibility, text, form, and shape of the label in its original position (layout) or at a new location dragged from its original location (dragged state) (see Figure 3.10).

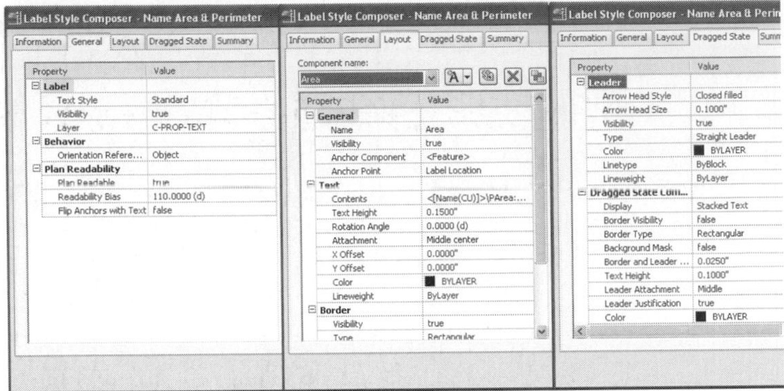

Figure 3.9

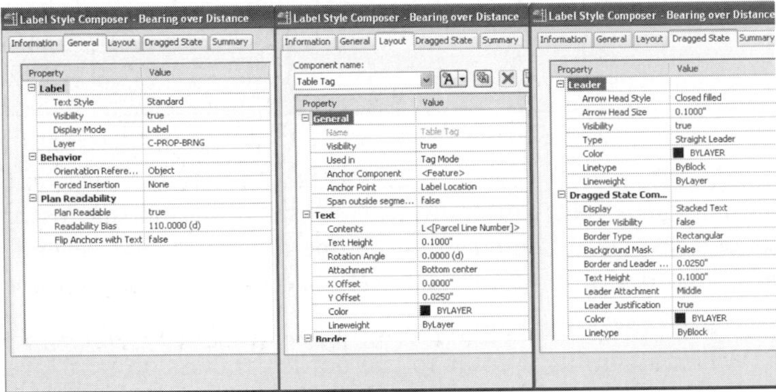

Figure 3.10

CURVE LABEL STYLE

A curve label style can contain a table tag, distance, radius, and/or other properties. The three panels, General, Layout, and Dragged State, control the visibility, text, form, and shape of the label in its original position (layout) or at a new location dragged from its original location (dragged state) (see Figure 3.11). Users can add a curve label to the drawing by selecting individual curve segments, or to all of the curve segments of a parcel by selecting the parcel label.

TABLE STYLES

The table style lists in tabular form parcel information. Civil 3D can create tables for lines, curves, line and curve segments, parcels, and areas (see Figure 3.12).

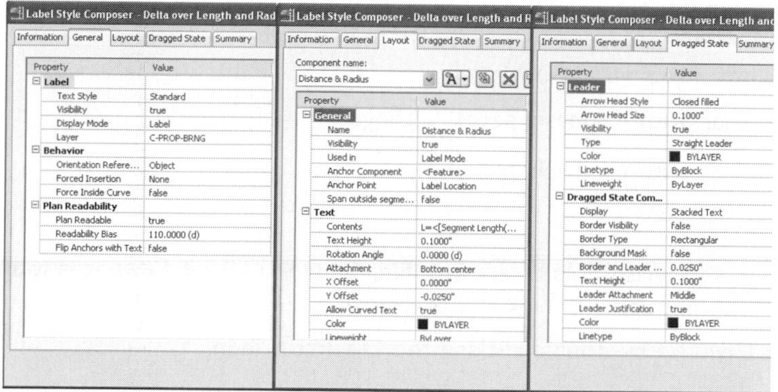

Figure 3.11

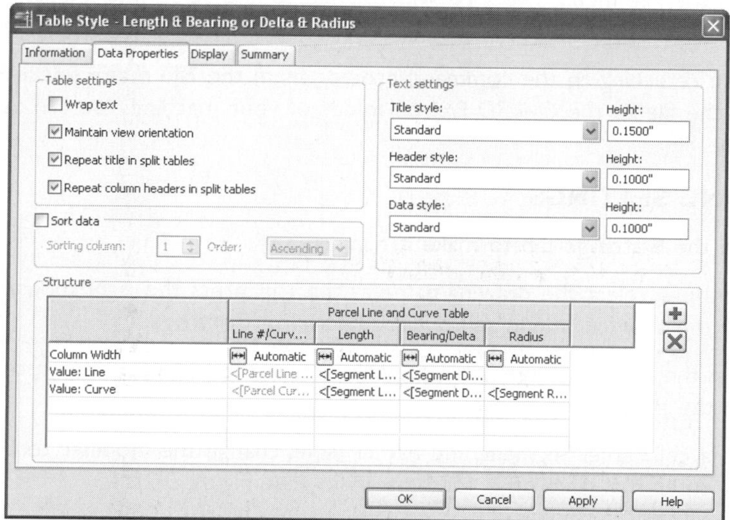

Figure 3.12

EXERCISE 3-1

When you complete this exercise, you will:

- Be familiar with the basic parcel settings.
- Be familiar with the basic parcel styles.

EXERCISE SETUP

The drawing for this exercise uses a template file, *Chapter 3 - Unit 1.dwt*. It is found in the *Chapter 3* folder on the CD that accompanies this textbook.

1. Start Civil 3D by double-clicking the **Desktop** icon.
2. Close the startup drawing and do not save it.
3. From the File menu, select **New...**, browse to and select the *Chapter 3 - Unit 1* template file from the *Chapter 3* folder of the CD that accompanies this textbook.
4. When returning to the command prompt, from the File menu select **Save As...**, and save the file to the *Civil 3D Projects* folder of your machine. You can assign any name to the file.

EDIT DRAWING SETTINGS

1. Click the **Settings** tab to make it current.
2. In Settings, select the drawing name at the top, press the right mouse button, and from the shortcut menu select **Edit Drawing Settings...**.
3. Select the **Object Layers** tab to make it current and scroll down the list until you view the Parcel layer entries.
4. For Parcel, Parcel Segment, and Parcel Table, change the modifier to **Suffix** and for each enter the value of **-*** (dash asterisk) (see Figure 3.2).
5. Select the **Ambient Settings** tab, and expand the Angle section.
6. Click in the value cell for Format and, if necessary, change it from DD.DDDD to **DD.MMSS** (decimal dms).
7. Expand the Direction section and, if necessary, set the Format to **DD.MMSS** (decimal dms).
8. Click the **OK** button to create the suffixes and to close the dialog box.

EDIT FEATURE SETTINGS

The Edit Feature Settings dialog box sets the initial parcel and label styles for all parcel routines.

1. In Settings, select the **Parcel** heading, press the right mouse button, and from the shortcut menu select **Edit Feature Settings...**.
2. Check and, if necessary, change styles to match those listed in Figure 3.3.
3. Click the **OK** button to close the dialog box.

CREATE SITE

The CreateSite command creates a site definition, and the values in its dialog box set the counters and increment values for parcel numbers and tags. If using the Create From Objects or Create By Layout routines, Civil 3D automatically creates a site and names it Site1.

1. In Settings, expand the Parcel branch until you view the Commands heading and its command list.
2. From the Commands list select **CreateSite**, press the right mouse button, and from the shortcut menu select **Edit Command Settings...**.
3. Review the settings in the dialog box and, if necessary, match them to the settings in Figure 3.4.
4. After making any needed changes, click the **OK** button to close the dialog box.

EXERCISES

CREATE FROM OBJECTS

When converting lines, curves, and polylines to parcels, Civil 3D wants to know what to do with the existing line work. The choices are to preserve or erase the selected line work. You can assign different default parcel, area, line, and curve styles unique to this command.

1. From the Commands list select **CreateParcelFromObjects**, press the right mouse button, and from the shortcut menu select **Edit Command Settings...**.
2. Review the settings in the dialog box and, if necessary, match them to the settings in Figure 3.5.
3. After making any needed changes, click the **OK** button to close the dialog box.

CREATE BY LAYOUT

The CreateParcelByLayout command presents the user a toolbar that contains tools to define, size, edit, erase, and review parcel segments or an entire parcel.

1. Select **CreateParcelByLayout** from the Commands list, press the right mouse button, and from the shortcut menu select **Edit Command Settings...**.
2. Review the settings and change them to match the settings in Figure 3.6.
3. After changing the settings, click the **OK** button to close the dialog box.

PARCEL STYLES

The parcel object style sets the layers, hatch pattern, and area label for the type of parcel.

1. In Settings, expand the Parcel branch until you view the Parcel Styles heading and its list of parcel styles.
2. From the list of styles, select the **Property** style, press the right mouse button, and from the shortcut menu select **Edit...**.
3. Select the **Information** tab to view its settings.
4. Select the **Design** tab to view its settings.

5. Select the **Display** tab to view its settings and, if necessary, match them to the settings shown in Figure 3.13.
6. After changing the settings, click the **OK** button to dismiss the dialog box.

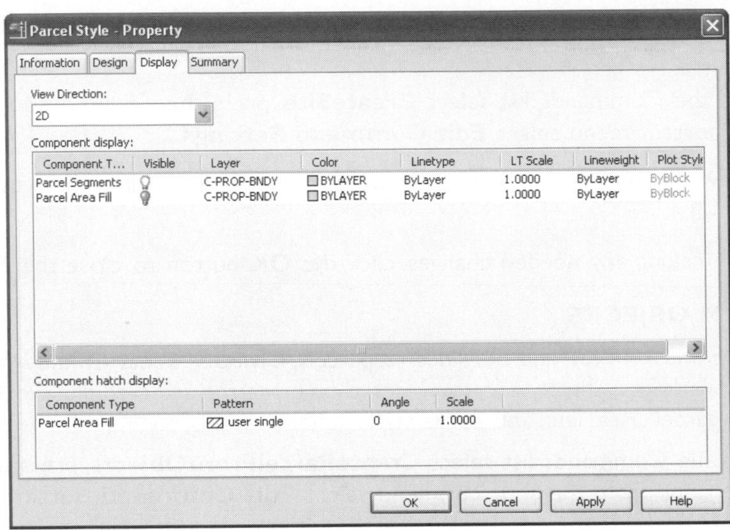

Figure 3.13

7. From the list of styles, select the **Single-Family** style, press the right mouse button, and from the shortcut menu select **Edit....**
8. Select the **Information** tab to view its settings.
9. Select the **Design** tab to view its settings.
10. Select the **Display** tab to view its settings and, if necessary, match them to the settings shown in Figure 3.14.
11. After changing the settings, click the **OK** button to close the dialog box.

LABEL STYLES

When you define a parcel, you can assign area and segment label styles to the new parcel. The segment styles include labels for lines (bearing and distance) and curves (radius, length, chord, etc.).

Area Label

The content template files of Civil 3D provide basic area labels.

1. In Settings, expand the Parcel branch until you view the list of area label styles under the Label Styles heading.
2. From the list, select **Name Area & Perimeter**, press the right mouse button, and from the shortcut menu select **Edit....**
3. Select the **Information** tab to view its settings.

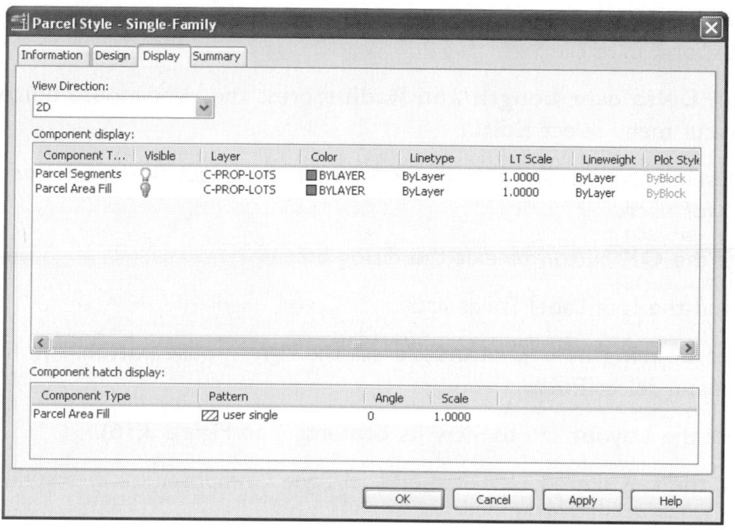

Figure 3.14

4. Select the **General** tab to view its settings.
5. Select the **Layout** tab to view its settings for the label.

   The Name Area & Perimeter label has one Component name: Area. This component has three label elements: Parcel Name, Area, and Perimeter (see Figure 3.15).

6. Click in the Text Content value cell and then click on the ellipsis to view the text format that includes parcel name, area, and perimeter.
7. After reviewing the various settings, click the **OK** buttons to close the dialog boxes, and return to the command prompt.

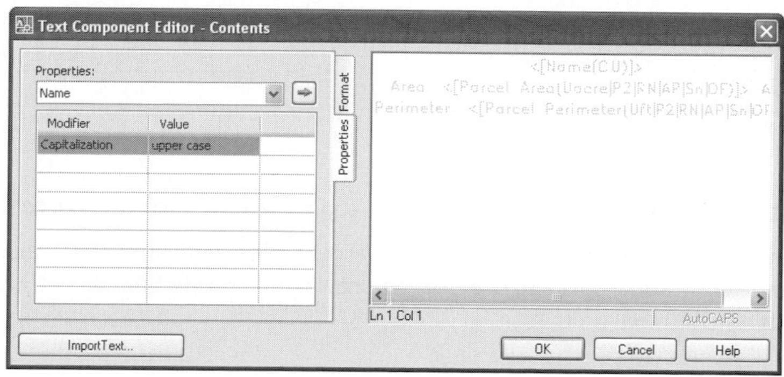

Figure 3.15

Line and Curve Label Styles

The Edit Feature Settings dialog box sets the default line and curve styles.

1. In Settings, expand the Parcel branch until you view the Label Styles heading and the Styles for Curve list.
2. Select **Delta over Length and Radius**, press the right mouse button, and from the shortcut menu select **Edit...**.
3. First select the **Layout** tab, then select the **Dragged State** tab and review the settings for each.
4. Click the **OK** button to exit the dialog box.
5. Expand the Line Label Styles list.
6. Select **Bearing over Distance**, press the right mouse button, and from the shortcut menu select **Edit...**.
7. Select the **Layout** tab to view its contents (see Figure 3.16).
8. Click the **OK** button to exit the Label Style Composer for this style.

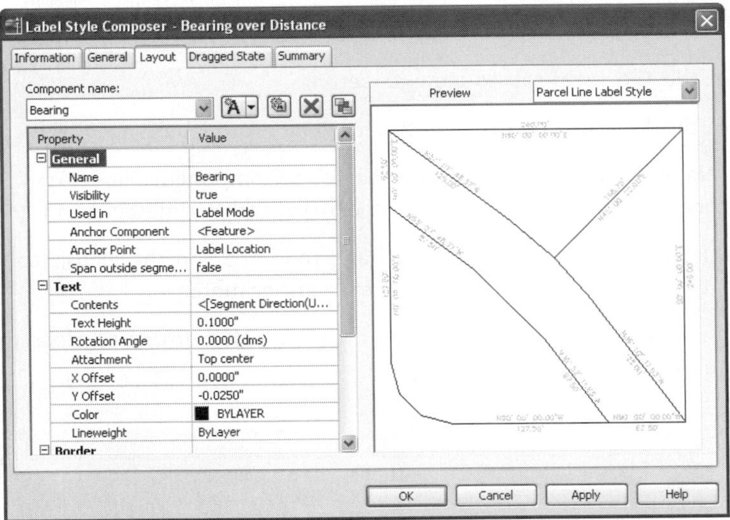

Figure 3.16

9. Click the **AutoCAD Save** icon at the top of left to save the current drawing.

This completes the drawing setup and review of parcel styles. The next unit reviews drafting and defining a site and its interior parcels. The settings reviewed in this unit directly affect the results of the commands in the next unit.

SUMMARY

- The Edit Drawing Settings dialog box sets a prefix or suffix for each parcel base layer.
- The Create Parcel commands will create a site, if it does not yet exist.
- Each command creating parcels can specify different default parcel styles.
- When creating a parcel, Civil 3D places a label near the center of the parcel.
- Users can label parcel segments as they are created, or they can be labeled afterward.
- Users can define their own parcel and segment labels (line and curve).

UNIT 2: CREATING A SITE AND PARCELS

Creating a property boundary can be a simple drafting exercise. The transparent commands of Civil 3D extend AutoCAD's line and polyline commands and allow the drafting of lines with turned angles, bearings, or azimuths and distances. The data for property lines may be a published plat of survey, a set of calculated directions and distances, or northings/eastings. Whatever the source, Civil 3D provides all of the necessary drafting tools to transcribe written values into AutoCAD entities.

Any closed polygon is a potential parcel. If the user is drafting a boundary from values on a plat of survey or a list of coordinates, the boundary probably will not close because of rounding errors in the values. When defining a parcel from an open boundary, Civil 3D creates parcel segments, but not a parcel. To close the boundary, a knowledgeable person will need to be contacted for information on values to achieve closure.

Once a boundary parcel is closed, the focus of subdivision design turns to dividing the boundary parcel into smaller parcel blocks of homogeneous land use (residential, common, open, etc.) and finally into individual parcels (single-family, commercial, etc.). Strategies for subdividing vary greatly and are at times more art than science. The subdividing process will also involve drafting roadway alignments, which divide parcels they cross. A final parcel design is a balance between the governing subdivision regulations, developer demands, and engineering constraints. In Civil 3D the process of subdividing a boundary into parcels includes drawing lines, curves, polylines, and using the parcel-sizing commands from the Parcel Layout Tools toolbar.

In Civil 3D, a site is an overall parcel boundary that acts as a "container," organizing the internal parcels and alignments. Civil 3D allows multiple sites in a drawing. A second site can be an alternative parcel design or a data set to be designed around (wetlands or other protected areas). Objects in different sites do not interact.

AUTOCAD LINE AND POLYLINE COMMANDS

The AutoCAD Line and Polyline commands create potential parcel segments (and arcs when using the Polyline command) from coordinates or object snaps. Civil 3D extends these commands by allowing them to access Civil 3D points and to make them mimic survey (angle and distance) methods. The survey angle and distance methods are similar to AutoCAD's polar toggle. For example, when in AutoCAD's Line command with polar on while dragging toward a direction, all that is required to draw the line is to enter a distance. When drafting line work and selecting the bearings and distances transparent command overrides, the prompting will change in order to ask for bearings and distances for the line segments. The transparent command overrides include bearing/distance, azimuth/distance, turned angle, deflection angle, station offset, and referencing Civil 3D points. Users can mix methods of describing the direction and distance by changing the active transparent command. See Unit 2 of Chapter 2 for a discussion on the transparent commands.

CREATE FROM OBJECTS

The Create From Objects is the Parcels menu entry that calls the CreateParcelFromObjects command. This menu selection creates a site and parcels from lines, curves, and polylines. If there is no defined site, the routine will create the site from a perimeter boundary and all of the parcels will be inside of the site boundary.

A setting in Parcel Feature Settings specifies if selected line work is erased. The Create From Objects routine uses this value as a default and, if on, will erase the selected line and arc segments, replacing them with parcel object segments. The user can disable the erasure at the bottom of the Create Parcels - From Objects dialog box.

CREATE BY LAYOUT

The Create By Layout is the Parcels menu entry that calls the CreateParcelByLayout command. The command displays the Create By Layout toolbar containing tools to create parcel lines and arc segments, size parcels (manually or automatically), and to create parcels from user-specified criteria. The main parcel-sizing routines are Slide Angle, Slide Direction, Swing Line, and Free Form Create. The routines are an icon stack, fifth in from the left. To display the individual commands within the icon stack, select the drop-list arrow. After displaying the command list, select the command to use for the next parcel segment. Each routine holds the minimum parcel size that is set in the New Parcel Sizing section of the Parcel Layout Tools toolbar.

The Slide Angle routine creates parcels by drawing the last side yard line at a specified angle to the frontage line of the parcel. The Slide Direction routine is the same as Slide Angle, but it uses a direction (bearing, azimuth, or two selected points) to draw the last side yard line. The Swing Line command draws the last side yard line from a point on the frontage to a point at the back of the lot. This side yard line can be perpendicular to the frontage, two user-specified points, a bearing, or an azimuth.

The Parcel Layout Tools toolbar has a roll-down settings panel (see Figure 3.17). The values in the panel set the minimum lot size, a minimum frontage measurement, a snap value when manually sizing a parcel, and settings that affect automatic parcel creation.

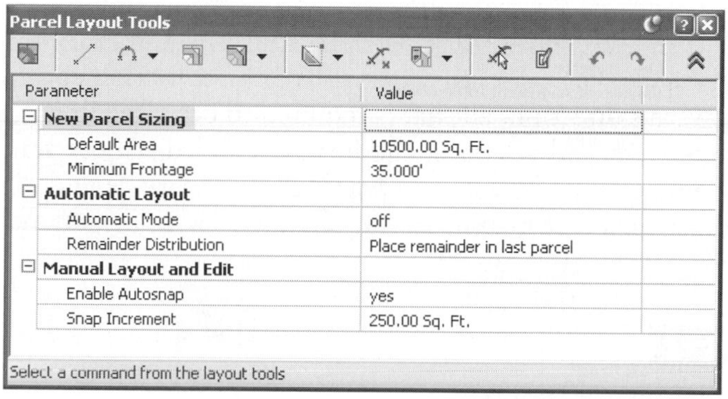

Parcel Layout Tools

Parameter	Value
New Parcel Sizing	
Default Area	10500.00 Sq. Ft.
Minimum Frontage	35.000'
Automatic Layout	
Automatic Mode	off
Remainder Distribution	Place remainder in last parcel
Manual Layout and Edit	
Enable Autosnap	yes
Snap Increment	250.00 Sq. Ft.

Select a command from the layout tools

Figure 3.17

The second and third icons draw parcel segments (line and curve). Users can use the transparent commands to define the segment's direction and length. The middle icon inserts, deletes, and breaks apart points of intersection (PIs) within parcel segments.

The Delete Sub-entity icon, located to the right of the center of the toolbar, deletes parcel segments. The icon stack to the right of Delete Sub-entity creates a single parcel from two parcels or dissolves a parcel union. When using the Parcel Union routine, the first selected parcel defines the properties of the resulting union.

The Sub-entity editor to the left of the undo arrow displays the values of a selected parcel segment. Any value that is black is editable and any value in gray cannot be changed. Select parcel segments to edit with the Sub-entity selection tool to the left of the Sub-entity editor icon.

RIGHT-OF-WAY PARCEL

The Right-of-Way (ROW) parcel is a parcel that uses an alignment to define its geometry. The remaining parameters of the ROW parcel are its width to the right and left of the alignment and how to trim intersecting parcel segments (trimmed or filleted).

A ROW parcel is not dynamically linked to alignments or other parcels. If the alignment needs to move, the user must delete and redefine the ROW parcel and its parameters.

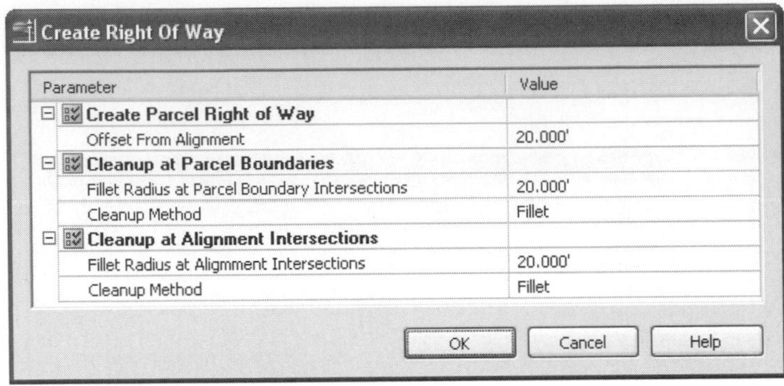

Figure 3.18

EXERCISE 3-2

When completing this exercise, you will:

- Be able to draw a boundary with the transparent commands.
- Be able to define a site and its parcels using the Create From Objects command.
- Be able to subdivide parcel blocks using the Create By Layout command.

EXERCISE DRAWING SETUP

This exercise continues with the drawing from Unit 1. If you did not do this exercise, you can open the drawing *Chapter 3 - Unit 2* found in the *Chapter 3* folder of the CD that accompanies this textbook.

1. If your drawing from the previous exercise is not open, open it now, or open the *Chapter 3 - Unit 2* drawing.
2. In Layer Properties Manager, freeze the following layers: **BLDG**, **BNDY**, **BNDYTXT**, **CONT**, **CONT2**, and **EX-ADJLOTS**.
3. Create a New layer, name the layer **BNDY-INPUT**, assign it a color, and make it the current layer.
4. Click the **OK** button to exit Layer Manager.
5. Click the **AutoCAD Save** icon to save the drawing.

DRAFTING A PROPERTY BOUNDARY

When drafting a boundary with bearings and distances, there will be a small gap from the end of the last boundary segment to the beginning of the first segment. This gap represents the closure error. Close this gap by using the Close option of the Drawing the Boundary command. If using a transparent command override, you need to press the Esc key to exit the override and to return to the Drafting command. If exiting before completing the boundary, you will have to pick up where you left off and use an endpoint object snap to draw the last small segment to the point-of-beginning (POB).

This exercise uses the AutoCAD Line command with the Northing/Easting and Bearing and Distance transparent commands. Use Figure 3.19 as a reference, or you can print a property line callout file. The callout file is a Notepad file titled *Boundary Calls.txt* that has the beginning coordinates and bearings and distances for the property boundary. You can also print a copy of the boundary from the *Property Outline.PDF* file. Both files are in the *Chapter 3* folder of the CD that accompanies this textbook.

The property boundary bearings and distances are in Table 3-1. The northing and easting coordinates, 14742.1740 and 6716.9054, represent the POB for the boundary's northwest corner. The bearings and distances follow clockwise around the boundary. Using the values from Table 3-1, the boundary almost closes, but not quite.

To correctly complete this exercise, you must have the direction set to Decimal dms (in the Ambient Settings panel). Decimal dms is a surveyor's method of entering bearings and distances. When the direction is set to Decimal dms, the first bearing in the table, N 88d 25'50" E, is entered as a quadrant 1 bearing with an angle value of 88.2550. See the discussion of transparent commands in Chapter 2, Unit 2 of this textbook.

Point 1:

Northing = 14742.1740

Easting = 6716.9054

EXERCISES

Table 3.1

Segment	From/To Point	Quadrant	Bearing	Distance
1	1 to 2	1	N 88°25'50" E	504.9805
2	2 to 3	2	S 1°34'10" E	1001.1900
3	3 to 4	1	N 87°59'17" E	333.0686
4	4 to 5	2	S 1°37'46" E	1322.0609
5	5 to 6	3	S 88° 03'56" W	1.8090
6	6 to 7	2	S 1°31'46" E	1028.4872
7	7 to 8	4	N 59°22'31" W	392.2954
8	8 to 9	4	N 1°33'54" W	817.3547
9	9 to 10	4	N 1°34'10" W	614.2499
10	10 to 11	3	S 87°56'21" W	504.9986
11	11 to 1	4	N 1°34'10" W	1712.8900

Figure 3.19 lists the starting coordinates and the quadrant, bearing, and distance for each segment.

1. Click the **Settings** tab to make it current.
2. In Settings, select the drawing name at the top, press the right mouse button, and from the shortcut menu select **Edit Drawing Settings....**
3. Select the **Ambient Settings** tab and expand the Direction section.
4. If necessary, change the **Format** to **DD.MMSS (decimal dms)**.

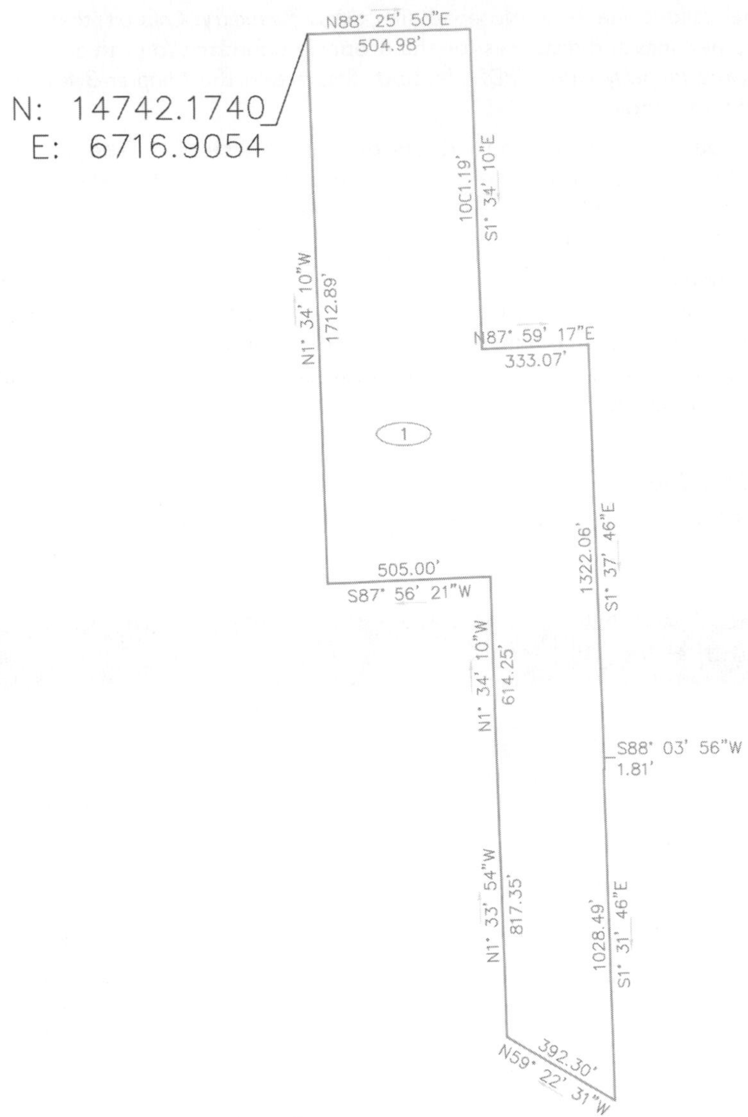

Figure 3.19

5. Click the **OK** button to set the format and to exit the dialog box.
6. Make sure the Civil 3D Transparent Commands toolbar is visible.
7. Start the AutoCAD Line command. From the Transparent Command toolbar select the **Northing/Easting** icon, and the Line command will prompt you for a starting northing coordinate.

   After entering the northing coordinate value, press the **Enter** key for the easting prompt. After entering the easting value, press the **Enter** key to enter both coordinates.

8. At the Northing prompt, enter **14742.1740** and press the **Enter** key.
9. At the Easting prompt, enter **6716.9054** and press the **Enter** key to set the starting point of the line.
10. Press the **Esc** key once to end the Northing/Easting transparent override and to return Civil 3D back to the Line command.
11. Without exiting the Line command, from the Transparent Command toolbar select the **Bearing Distance** icon and answer the prompts for quadrant, bearing, and the distance information for each remaining boundary segment.

    Civil 3D displays a tripod and crosshairs to represent the starting coordinates and direction of each segment. You can graphically define a quadrant, direction, and distance while in the override.

    You will not return to the exact starting coordinates. This is because the distance and angle values contain rounding errors even at four decimal places. If you are looking at the bearings and distances on a plat of survey, the precision is a whole second for bearings and only two decimal places for distances. This precision is less than what you have just entered for this boundary.
12. After entering the last bearing and distance, exit the bearing/distance override by pressing the **Esc** key only once.
13. At the command line, enter the letter **C** and press the **Enter** key to close the boundary by drawing the last line segment to the beginning.
14. Click the **AutoCAD Save** icon to save the drawing.

EXERCISES

CREATE FROM OBJECTS

The current line work represents the site definition.

1. Isolate the **BNDY-INPUT** layer to view the line work.
2. From the Parcels menu, select the **Create From Objects** command.
3. Select all of the lines just drawn and press the **Enter** key after selecting the lines.

    The Create Parcel - From Objects dialog box displays.
4. Match your settings to those in the dialog box (Figure 3.20) and when done, click the **OK** button to exit the dialog box and to create the site, parcel, and label.
5. Click the **Layer Previous** icon to restore previous layer display state.
6. If necessary, open **Layer Properties Manager**, toggle **ON** the C-PROP-Site1 and C-PROP-LINE-Site1 layers, and click the **OK** button to exit the Layer Properties Manager dialog box.
7. If necessary, type **Regenall** at the command prompt and press the **Enter** key to view the outside property line.

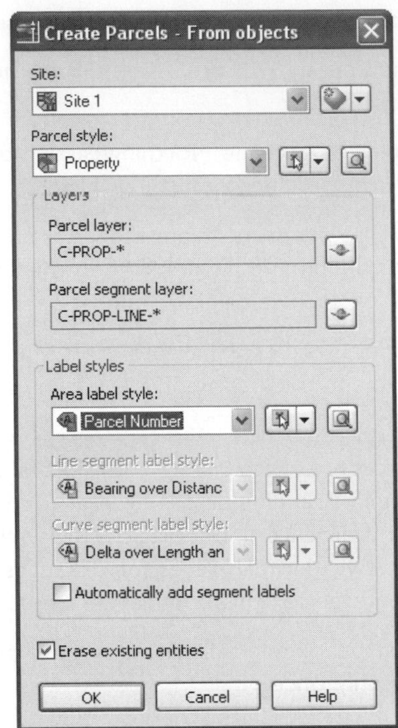

Figure 3.20

DEFINING THE RIGHT-OF-WAY AND WETLANDS PARCELS

The ROW and wetlands boundary are polylines on the PR-ROW and Wetlands layers.

1. Isolate the two layers **PR-ROW** (magenta) and **Wetlands** (rust).
2. From the Parcels menu, select **Create From Objects,** select the line work on the screen, and press the **Enter** key.
3. The Create Parcels - From Objects dialog box appears. Adjust the settings to match those in Figure 3.20.
4. After adjusting the settings, click the **OK** button to define the parcels.
5. Click the **Layer Previous** icon to restore previous layer display state.
6. If necessary, type **Regenall** at the command prompt and press the **Enter** key to view the parcels.
7. Click the **AutoCAD Save** icon and save the drawing.

CREATING BACK YARD PARCEL LINES

The previous steps create parcels with uniform types of land use. These "block" parcels will be further subdivided into individual parcels. Some of the blocks will have to be subdivided once more to create front and back yard parcel lines. These particular blocks have two frontage lines and no back yard line. The front and back yard lines make creating individual parcels much easier. You only have to create one more side yard parcel segment to create a final parcel. This will be done by selecting the polylines on the Interior Boundary layer.

1. Open the Layer Properties Manager, isolate the **Interior Boundary** layer, and click the **OK** button to exit the dialog box.
2. From the Parcels menu, select **Create From Objects**, select the line work on the screen, and press the **Enter** key.
3. The Create Parcels - From Objects dialog box displays. Adjust the settings to match those in Figure 3.20.
4. After adjusting the settings, click the **OK** button to define the parcels.
5. Click the **Layer Previous** icon to restore the previous layer ON/OFF state.
6. If necessary, type **Regenall** at the command prompt and press the **Enter** key to view the parcels.
7. Click the **Prospector** tab to make it current.
8. In Prospector, expand the Sites branch until you view the parcel list for Site 1.
9. Click the **AutoCAD Save** icon to save the drawing.

CREATING SINGLE-FAMILY PARCELS

The final step is to divide the parcel blocks into individual resalable parcels with the Parcel Layout Tools toolbar sizing tools.

1. From the **View** menu, select **Named Views...** and restore the view **Residential Parcels 1**.
2. In the Layer Properties Manager, turn on the **Area** layer and click the OK button to exit the dialog box.
3. From the Parcels menu, select **Create By Layout...**.

   This displays the Parcel Layout Tools toolbar.
4. At the right side of the toolbar, click the arrow to roll down the Settings panel.
5. If necessary, adjust your settings to match the values in Figure 3.17.
6. Click the roll-up arrow at the right of the toolbar to hide the settings.
7. Go to the fifth icon in from the left, click the drop-list arrow to right of the icon, and select **Slide Angle - Create**.

   The Create Parcels - Layout dialog box appears.
8. In the dialog box, set the parcel style to **Single-Family** and the parcel area label style to **Name Area & Perimeter** (see Figure 3.21).
9. Click the **OK** button to start creating parcels.

   The command prompts you to select a point within the lot to subdivide.

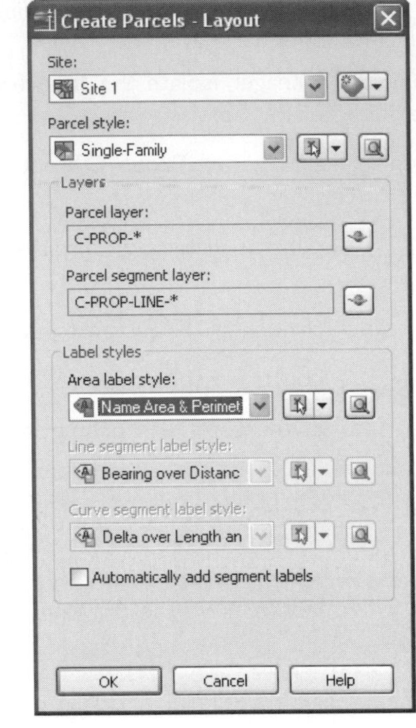

Figure 3.21

10. If Object Snap is on, toggle it **OFF**.
11. Select a point in the parcel that contains the three parcel sizes (15,350 and 2 @ 12,500).

    The next prompt is to select the beginning and ending point of the parcel frontage.

12. In the drawing, select a point at the northeast end of the frontage (use the endpoint object snap) as the beginning point and select a point at the southwestern end of the frontage (again use the endpoint object snap) to define the parcel frontage (see Figure 3.22).

    As you do this, the routine displays a jig following the frontage geometry. The jig recognizes the changing frontage geometry.

    The next prompt is for the angle turned from the frontage line.

13. At the angle from frontage prompt, enter **90** degrees as the angle.

    A lot side segment appears on the screen between the front and back yard segments with a tooltip reporting the parcel area and frontage distance.

14. Toggle **OFF** object snap (F3), move the cursor to the rightmost side of the lot near the frontage line, and the cursor will hold the side yard line at the point where the parcel matches the minimum lot size (10,500).

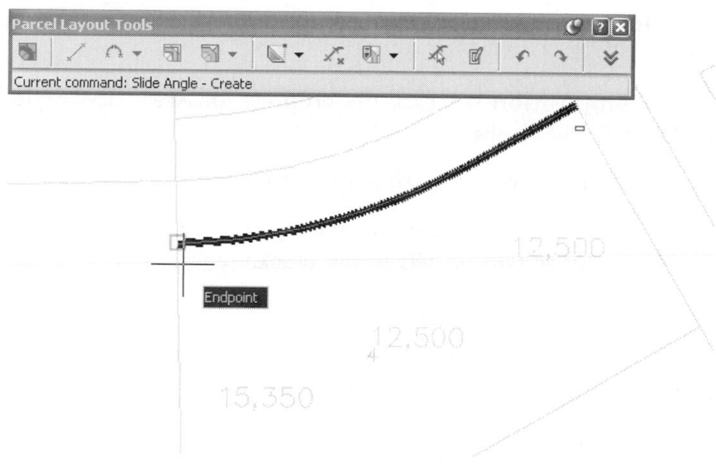

Figure 3.22

15. Slowly move the cursor to the left to move the side yard line toward the west. As you do this, the tooltip will report the increase of lot size in increments of 250 square feet as set in the toolbar.
16. When the lot size reaches 12,500 square feet, press the left mouse button to set the side yard segment to define the last side of the first parcel. A label appears in the new lot and the parcel lines change color.

    The routine prompts for another angle to turn from the front yard line.
17. Enter **90** to create a second parcel by moving the cursor toward the east until the new side yard line appears, sizing the parcel to 10,500 square feet.
18. Move the cursor slowly to the west until the new parcel is 12,500 square feet.
19. Select the point defining an adjacent 12,500-square-foot parcel.

    The routine prompts for another parcel to subdivide.
20. Press the **Esc** key to end the Parcel Sizing command.

    The last lot is the remaining area of this section of the parcel block.
21. Type **X** and press the **Enter** key to exit the Create By Layout command.
22. Click the **AutoCAD Save** icon to save the drawing.

CHANGING A PARCEL'S TYPE AND LABEL

You need to change the parcel type of the parcel at the end of the block. The Parcel property contains the Object type (located on the Information panel) and the Parcel Label style is set in the Composition tab.

1. In the drawing, select the west-end parcel's label, press the right mouse button, and from the shortcut menu select **Parcel Properties...**.

2. Select the **Information** tab, click the drop list for Object Style, and select **Single-Family**.
3. Select the **Composition** tab, click the drop list for Area Label Style, and select **Name Area & Perimeter**.
4. Click the **OK** button to change the parcel type and label.

   The remaining parcel is now a single-family parcel with an area label and segments that match the other two parcels in the block.

SLIDE DIRECTION - CREATE ROUTINE

Many times the side yard segment angle is not 90 degrees to the front yard line, but instead a bearing or perpendicular to the back yard line. The Slide Direction - Create routine is best for this situation.

1. Use the AutoCAD Zoom and Pan commands until you view the entire adjacent southerly parcel sharing the back yard line.
2. From the Parcels menu, select **Create By Layout....**
3. Go to the fifth icon in from the left, click the drop-list arrow to right of the icon, and select **Slide Direction - Create**.
4. The Create Parcels - Layout dialog box appears. Match the settings to those in Figure 3.21.
5. Click the **OK** button to start creating parcels.

   The command prompts you to select a point within the lot to subdivide.

6. Select a point in the parcel (in the southwest corner of the parcel boundary).

   The next prompt is to select the beginning and ending points of the parcel frontage.

7. In the drawing, select a point on the frontage line at its southeast end (use the endpoint object snap). This point is the beginning point of the parcel frontage. Next select a point at the northeastern end (again use the endpoint object snap) to define the end of the parcel frontage.

   As you do this, the routine displays a jig following the frontage geometry. The jig recognizes the changing frontage geometry (see Figure 3.23).

   The next prompt asks for a direction from the frontage line.

8. Select a point on the frontage using the nearest object snap and a second point perpendicular to the back yard.

   A lot side segment appears on the screen between the front and back yard segments.

9. Toggle **OFF** object snaps, move the cursor to the rightmost side of the lot, and the cursor will hold the new segment at the point the lot has the minimum lot size (10,500).

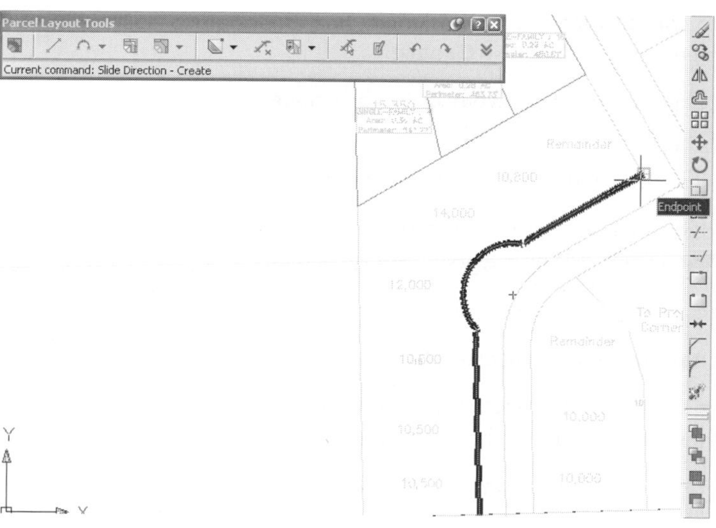

Figure 3.23

10. Slowly move the cursor to the north along the parcel frontage line. As you do this, the tooltip will report the increase of lot size in increments of 250 square feet as set in the Parcel Layout Tools toolbar.

11. The command line prompts with the minimum parcel size (10,500 square feet). Press the **Enter** key to set the parcel side yard line.

    A label appears in the new parcel and the lines change their properties.

12. The command then prompts for the next bearing and distance.

13. Repeat Step 8 to establish the direction for the next parcel, and size the parcel to 10,500 square feet.

14. Press the **Esc** key twice to close the Parcel Sizing command.

15. Type **X** and press the **Enter** key to exit the Create By Layout command.

16. Click the **AutoCAD Save** icon to save the drawing.

MIXING MANUAL AND AUTOMATIC PARCEL SIZING

The next area uses the manual method for the first parcel, and then, after switching to automatic mode, creates the remaining parcels. The last parcel is the remainder of the block.

1. From the View menu, use the Named Views... command to restore the **Residential Parcels 2** view.

2. From the Parcels menu, select **Create By Layout....**

3. At the right of the toolbar, select the roll-out arrow and check the values. If necessary, adjust your settings to match the values in Figure 3.17.

4. Click the roll-up arrow to hide the panel.

5. Go to the fifth icon in from the left, click the drop-list arrow to right of the icon, and select **Slide Angle - Create**.

   The Create Parcels - Layout dialog box appears.

6. In the dialog box, set the Parcel Style to **Single-Family** and the Parcel Area Label Style to **Name Area & Perimeter** (see Figure 3.21).
7. Click the **OK** button to exit the dialog box.
8. The routine prompts for a point in the interior of the parcel to subdivide. With the Object snaps **OFF**, select a point near 12500 (southern end of the block).

   The next prompts define the frontage.

9. Using an endpoint object snap, define the frontage by selecting the southerly and northerly endpoints of the easterly side of the block (see Figure 3.24).

   A jig appears, tracing over the frontage line and recognizing its geometry.

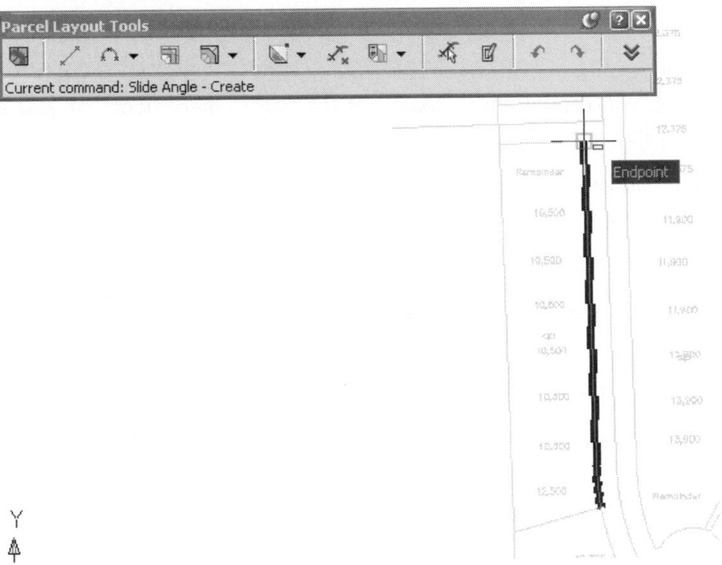

Figure 3.24

10. The next prompt is for an angle. Enter **90** and press the **Enter** key.

    A side yard segment appears in the drawing.

11. Move the cursor slowly to size the parcel to 12,500 square feet. Once you reach the size, select that point by pressing the left mouse button.
12. Press the **Esc** key twice to return to the Select from the Layout tools [or eXit] prompt.

13. On the right side of the Parcel Layout Tools toolbar, click the roll-down arrow, in Automatic Layout change Automatic Mode to **ON**, and set Remainder Distribution to **Redistribute remainder**.
14. Click the roll-up arrow to hide the panel.
15. Click the **Slide Angle - Create** icon to start creating the parcels.
16. The routine prompts you to select in the interior of the lot. Select a point just north of the last parcel you just made (by 10,500).

    Next you are prompted to define the parcel frontage.

17. Using an endpoint object snap, define the frontage by selecting the southerly (the northerly end of the parcel just made) and northerly endpoints of the easterly side of the block.
18. After defining the frontage, the command prompts for the angle; enter **90** and press the **Enter** key.
19. In the command prompt is the target size of the new parcels (10,500); press the **Enter** key to accept the target parcel size.

    The routine automatically creates the new parcels.

20. Press the **Esc** key twice to exit the Parcel Layout Tools toolbar.
21. Click the **AutoCAD Save** icon to save the drawing.

    The remaining parcel (top of the block) needs to have its type and label changed to Single-Family and Name Area & Perimeter.

22. Select the northerly parcel's label, press the right mouse button, and from the shortcut menu select **Parcel Properties...**.
23. Select the **Information** tab and the change the style to **Single-Family.**
24. Select the **Composition** tab and assign the **Name Area & Perimeter** label style, and click the **OK** button to exit.
25. Click the **AutoCAD Save** icon to save the drawing.

The next phase is reviewing the areas and boundaries of the new residential parcels.

SUMMARY

- When subdividing a parcel, one strategy is to create a set of parcel blocks (areas of uniform land use) from the overall boundary.
- Each parcel-sizing routine of the Parcel Layout Tools toolbar defines the last parcel segment (side yard).
- Each parcel style defines the layer for parcel segments.
- Users can change the parcel type in the Parcel Properties dialog box.
- When creating parcels by converting line work, users can preserve or erase them from the drawing.

UNIT 3: EVALUATING THE SITE AND PARCELS

Civil 3D provides several parcel evaluation tools. The analysis tools are in the parcel's Properties dialog box and reports from the LandXML Report application. A parcel's Properties dialog box reports map check and inverse values, and the LandXML Report application generates several reports including metes and bounds, surveyor's certificates, parcel areas, and so on.

PARCEL PROPERTIES

The Parcel Properties dialog box contains three panels: Information, Composition, and Analysis (see Figure 3.25). The Information panel contains the name and style of the parcel object. The Composition panel contains the label style, area, and perimeter of the parcel. The Analysis panel shows the parcel's Inverse or Map Check report. An Inverse report uses the precision of AutoCAD, 14 decimal places, to calculate the direction, distances, and coordinates of the parcel segments. A Map Check report calculates the perimeter of a parcel to the precision of the drawing. This reduced precision introduces distance and angle errors into the perimeter calculations. As a result, the coordinates of the starting point of the perimeter will not match the coordinates of the ending point. The distance and angle between the ending and starting points are the "closure error" of a parcel Map Check report.

Each report starts at a point, the POB, which may not be the starting point the user wants. A user can change the starting point interactively by selecting the pick icon to the right of the current POB coordinates (the top right of the Analysis panel). When selecting a new POB, Civil 3D displays a glyph at the current POB and gives the option to select a new POB, to traverse the vertices of the parcel, or to abort the redefinition process. The Inverse and Map Check reports will change to accommodate the new POB and order of parcel segments.

All Inverse and Map Check reports traverse clockwise around a lot. If the user wants to process the reports in a counterclockwise order, a toggle at the top right of the dialog box reverses the direction.

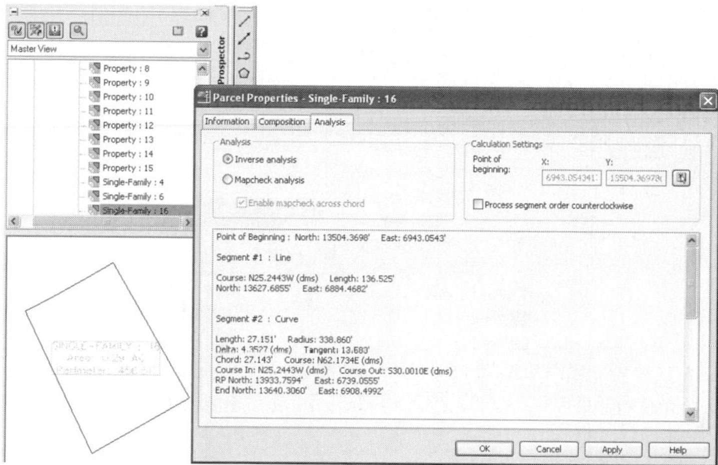

Figure 3.25

LANDXML REPORTS

The second parcel analysis method is LandXML reports. These reports depend on LandXML data generated from Prospector's parcel list. The shortcut menu of a selected parcel creates the LandXML file (see Figure 3.26). After creating the data, use the Report application to create area, inverse, metes and bounds, and surveyor-specific reports (see Figure 3.27). Users are able to modify settings within each report, or if proficient in XSL (eXtensible Style Language), users can author their own style sheets. Visit the LandXML.org Web site to learn more about this data format and the support it has in the civil industry.

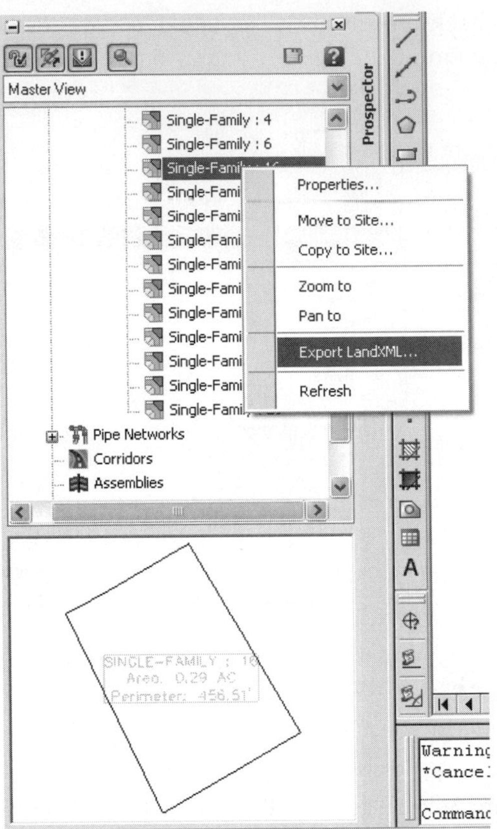
Master View
Prospector
Single-Family : 4
Single-Family : 6
Properties...
Move to Site...
Copy to Site...
Zoom to
Pan to
Export LandXML...
Refresh
Pipe Networks
Corridors
Assemblies
SINGLE-FAMILY : 16
Area: 0.29 AC
Perimeter: 456.51'

Figure 3.26

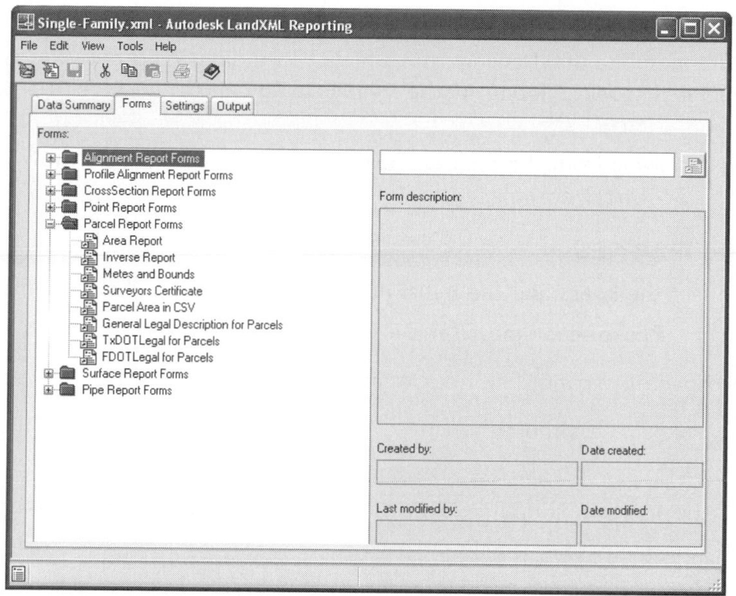

Figure 3.27

TOOLBOX/REPORT MANAGER

Rather than exporting to a LandXML file and opening the LandXML Reports application, a user can create the same reports from the Toolbox or Report Manager. In the General menu, the Toolbox and Report Manager commands add the same Toolbox tab to Civil 3D's Toolspace. This tab includes the same report forms as does the Report application.

The procedure for creating a report is to identify the correct report form in the Toolbox panel, press the right mouse button, and from the shortcut menu select Execute. An Export to LandXML dialog box appears and the user can select which items to report on in the dialog box or by selecting objects in the drawing. When clicking the OK button to export the LandXML file, the report appears in Internet Explorer, formatted by the selected form.

EXERCISE 3-3

After completing this exercise, you will:

- Be able to view site properties.
- Be able to view parcel properties.
- Be able to create parcel reports with the LandXML Report application.
- Be able to create a LandXML file containing parcel data.

EXERCISE DRAWING SETUP

This exercise continues with the drawing from Unit 2. If you did not do this exercise, you can open the drawing *Chapter 3 - Unit 3* found in the *Chapter 3* folder of the CD that accompanies this textbook.

1. If your drawing from the previous exercise is not open, open it now, or open the *Chapter 3 - Unit 3* drawing.

VIEW SITE PROPERTIES

Site properties affect the layers and the initial numbers for parcels and alignments within the site.

1. Click the **Prospector** tab to make it current.
2. In Prospector, expand the Sites branch until you view Site 1.
3. Click **Site 1**, press the right mouse button, and from the shortcut menu select **Properties....**
4. Click the **Information**, **3D Geometry**, and **Numbering** tabs to view their current values.
5. Click the **OK** button to close the dialog box.

PREVIEW PARCELS

When selecting an item, named site, or individual parcel from a list, Prospector previews an image of the item.

1. At the top of the Prospector panel, select the magnifying glass until it displays a border around the icon.
2. In Prospector, expand the Sites branch until you view the Parcels heading and the list of parcels.
3. Select the **Parcels** heading, press the right mouse button, and if not on, select **Show Preview**.

   Selecting the Parcel heading displays a list of parcels.
4. Select a parcel from the list of parcels.

   The Prospector preview area shows an image of the selected parcel.

PARCEL PROPERTIES

The properties of a parcel include its name, parcel type, parcel label, and statistics about its area and perimeter.

1. In Prospector, return to the list of parcels, select a **Single-Family** parcel from the list, press the right mouse button, and from the shortcut menu select **Zoom to**.
2. In Prospector, with the parcel entry still highlighted, press the right mouse button, and from the shortcut menu select **Properties....**
3. Click the **Information** tab to view its contents.

The parcel name is grayed out and is not editable.

4. Click the **Composition** tab to view its contents.

   The Composition panel lists the area, perimeter, and the currently assigned parcel area label style.

5. Click the **Analysis** tab to view its contents.

   This panel displays an inverse or map check analysis of the parcel. These two reports summarize the quality of the boundary definition based on the line work that defines the parcel.

6. Click **Inverse Analysis** and review its values.
7. Click **Map Check Analysis** and review its contents.
8. Click the pick new POB icon to the right of its current coordinates.

   The dialog box disappears and a glyph indicates the current position of the POB on the parcel perimeter. The default is to move to the next parcel corner.

9. Press the **Enter** key to move the POB clockwise around the boundary.
10. Type **S** to select a new POB.

    The routine returns to the dialog box with a new set of coordinates and a new analysis of the parcel starting at the point you just selected.

11. Click the **OK** button to exit the Parcel Properties dialog box.
12. Click the **AutoCAD Save** icon to save the drawing.

EXERCISES

GENERATE PARCEL REPORT

Civil 3D creates parcel reports that you can print and customize. These reports are through the Autodesk LandXML Report application.

1. With a single-family parcel highlighted, press the right mouse button, and from the shortcut menu select **Export LandXML...**.

   This displays the Export to LandXML dialog box.

2. Select a few more single-family parcels in the Export to LandXML dialog box by toggling on a check mark to the left of the parcel's name.
3. After selecting the parcels, click the **OK** button to start writing the file.
4. Place the LandXML file with the working drawings.
5. Minimize Civil 3D.
6. On the Desktop start the LandXML Report application by double-clicking its icon.
7. The Report application displays. Click the **OK** button of the Getting Started dialog box.
8. From the File menu select **Open Data File...** and open the LandXML file just created.

9. Select the **Data** tab to view its values.

   The Data panel identifies the parcels and their data.

10. Expand the tree that lists the parcels and view the geometry that is in the report.
11. Click the **Forms** tab and expand the Parcel Report Forms section.

    The Form panel lists the available reports.

12. Select the **Metes and Bounds** form and click the **Output** tab to view the report.

    The Output panel displays the report output.

13. Click the **Forms** tab and select the **Surveyor's Certificate** form. Click the **Output** tab to view the report. If more than one parcel is selected, only the first parcel selected in the Data Summary tab is output.

    Each report contains values you can adjust. These settings are found in the Settings tab of the Report application.

14. Click the **Settings** tab to view its contents.

    Each heading has values that affect precision, data, or wording in a report. By adjusting these values you can produce reports to your liking.

15. Click the "**X**" at the top right to exit the report application and to return to Civil 3D.
16. Click the **AutoCAD Save** icon to save the current drawing.

TOOLBOX REPORTS

The toolbox reports are the same as those of the Report application.

1. From the General menu, select **Toolbox**.

   A Toolbox tab appears in the Workspace.

2. In the Toolbox, expand Reports Manager until you view the Parcel report section.
3. Select the **Metes and Bounds** report, press the right mouse button, and from the shortcut menu select **Execute...**.
4. In the Export To LandXML dialog box, click the **OK** button to create the report for all of the selected parcels.
5. Scroll through the report and, when finished, select **Save As...** from the File menu to save the report as a text file.
6. Close Internet Explorer.
7. Click the **AutoCAD Save** icon to save the drawing.

This completes this unit on analyzing parcels. The next unit reviews the various tools for editing parcel geometry.

SUMMARY

- The Inverse or Map Check report starts at a POB on the perimeter of the parcel.
- Users can redefine the POB of a parcel.
- The property of a parcel includes an Inverse and Map Check reports of its perimeter.
- The LandXML Report application allows users to customize many of the values it uses for a report.
- Metes and bounds or a surveyor's certificate report uses LandXML parcel data.
- Civil 3D exports parcel geometry as a LandXML file.

UNIT 4: EDITING A SITE AND PARCELS

The Parcel Layout Tools toolbar provides tools that modify a site and its parcels. In addition to the layout tools, users can graphically adjust the parcel segments. All parcel labels will change to reflect the new parcel area and segment values.

The editing tools of the Parcel Layout Tools toolbar include deleting parcel lines and arcs; merging parcels; adding, deleting, and breaking PIs in parcel segments; and changing the values (coordinates, radii, etc.) associated with parcel lines and curves. The toolbar also includes an undo and redo for all parcel commands.

The easiest way to adjust the size of a parcel is to grip-edit the location of a parcel segment. If the parcel has a side yard segment that starts on the frontage of the parcel, when the user selects the segment, it shows a grip intersection of the segment and the frontage line. The user can slide this segment along the frontage line to adjust the area of the parcel. Users can even transfer the segment to another parcel block by simply selecting a new point on the frontage.

When finalizing a subdivision plan, the parcel numbers may not be the desired number sequence. The Renumber and Rename Parcels commands of the Parcels menu provide tools that modify the numbers and names assigned to parcels.

Renumbering parcels is a process of selecting what parcel number to start with, the increment value, and the parcels to renumber (see Figure 3.28). For example, a user may renumber parcels based on parcel blocks. A block of parcels can be renumbered by starting parcel numbers at 100 with a lot number increment of 1. A second block may start at 200 and increment the parcel numbers again by 1.

The renaming of parcels can use user-defined prefixes and number sequences, or the user can use a naming template to define the new parcel name (see Figure 3.29). If the user wants to change the naming template, Use name template in parcel style must be toggled

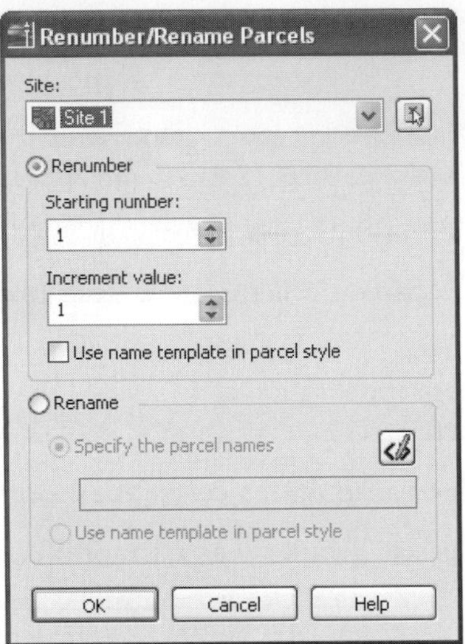

Figure 3.28

off. If this toggle is on, the user cannot change the base name of the parcel style (for example, single-family). If this is toggled off, the user can edit the value. However, from that point on he or she must manually update its value.

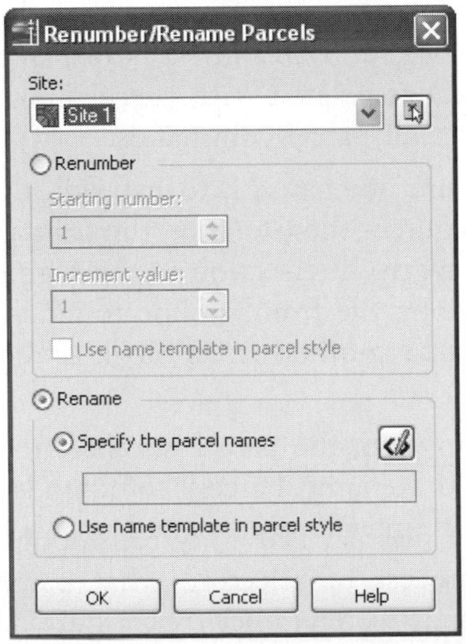

Renumber/Rename Parcels
Site:
Site 1
Renumber
Starting number:
1
Increment value:
1
Use name template in parcel style
Rename
Specify the parcel names
Use name template in parcel style
OK
Cancel
Help

Figure 3.29

When the parcel name includes a sequential number or a parcel value (address, parcel tax ID, parcel number), the user can define a renaming format in the Name Template dialog box. This dialog box can be found by selecting the icon to the right of the Specify the parcel names toggle at the bottom of the Renumber/Rename Parcels dialog box (see Figure 3.30).

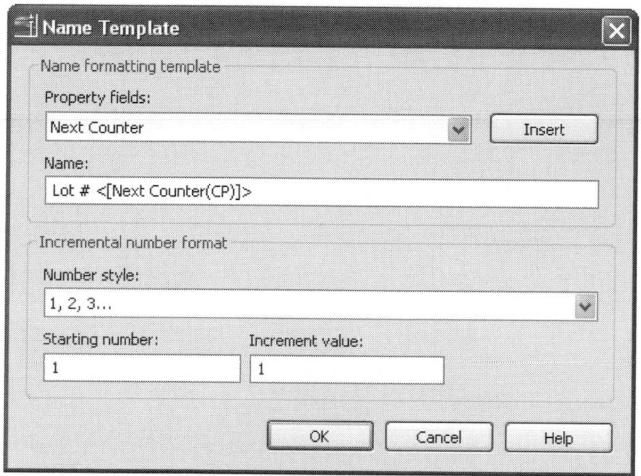

Figure 3.30

EXERCISE 3-4

When you complete this exercise, you will:

- Be able to edit parcels by manipulating parcel segments.
- Be able to edit with the tools in the Parcel Layout Tools toolbar.

EXERCISE SETUP

The exercise continues with the drawing from the previous exercise. If you are starting with this exercise, you can open the drawing *Chapter 3 - Unit 4* to start the exercise that is in the *Chapter 3* folder on the CD that accompanies this textbook.

1. If your drawing from the previous exercise is not open, open it now, or open the *Chapter 3 – Unit 4* drawing.

GRIP-EDITING PARCEL SEGMENTS

You can slide a side yard segment along the frontage line to adjust the area of the parcel. You can even transfer the segment to an adjacent parcel block by selecting a new point on the frontage (see Figure 3.31).

1. From the View menu, use Named Views... to restore the **Residential Parcels 1** view. You may want to zoom in on the parcels to see them better.
2. Click a side yard line between two adjacent parcels. A special grip appears, indicating that the side yard is "tied" to the frontage of these two lots.

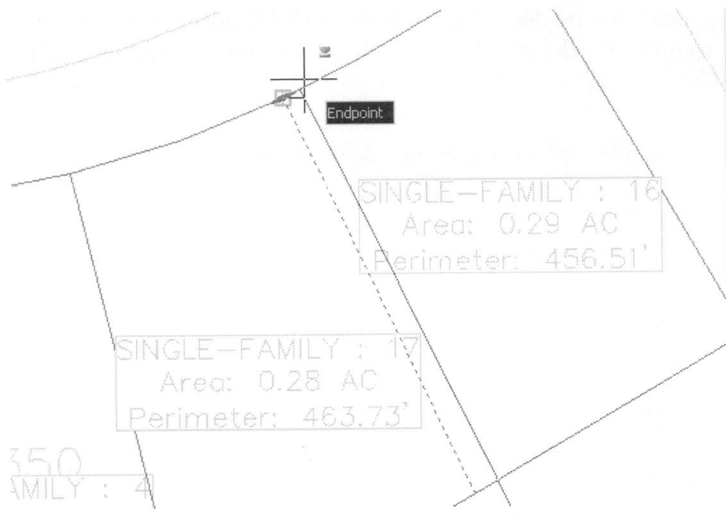

Figure 3.31

3. If on, toggle **OFF** AutoCAD Osnap and slide the segment along the frontage line.
4. Once you have the side yard line in the desired position, set its location by pressing the left mouse button, and press the **Esc** key to deselect the object.

   If dragging a parcel segment into an adjacent parcel block, the segment will divide the block into two new parcels. The original two parcels will then be merged into a single parcel. If you move the parcel segment back to where it divided the original parcels, the newly merged parcel will divide with the returned segment and the newly divided parcel block will return to a single-block parcel.

PARCEL LAYOUT TOOLS TOOLBAR

There will be times when you need to delete parcel segments and redraw them, change the segment angle used to define a parcel, or further divide existing parcels. The Parcel Layout Tools toolbar is the only location for these types of edits.

Delete Parcel Sub-Entity

1. From the Parcels Menu, select **Edit Parcel Segments...**, and select a lot line.

   This displays the Parcel Layout Tools toolbar.

2. Click the **Delete Sub-entity** icon located near the middle of the toolbar (see Figure 3.32).

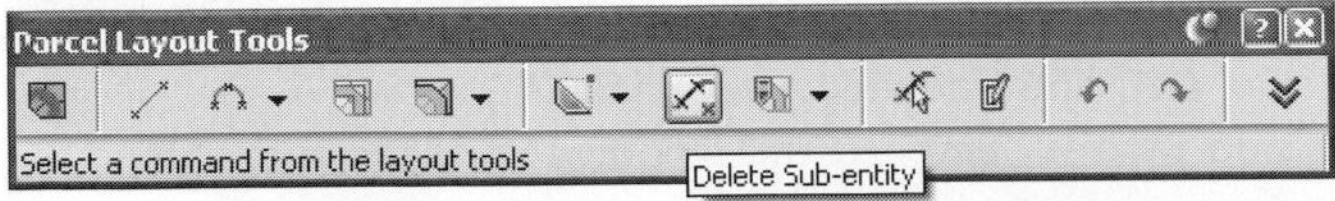

Figure 3.32

You can delete any side yard or parcel segment with this command. If you use Delete Sub-entity on a frontage, back yard, or boundary, you will destroy the parcel definition and you will have parcel segments. If deleting a side yard segment, the two parcels merge into a single parcel.

3. Delete one or both side yard segments in the current view.

Add New Parcel Segment

You can use a different method, or the same method, to redraft the side yard segment. Your choices are Slide Angle, Slide Direction, or Swing Line to complete the last segment.

1. Redraft the segment(s) with one of the previously mentioned commands from the Parcels Layout Tools toolbar. Use the Single-Family parcel style for the new lot(s).

Edit Parcel Segment Angle/Direction

There will be times when you want to change the angle that defines the side yard segment. The Slide Angle - Edit routines give you the ability to define an angle from the frontage of the lot (Angle) or to use a direction or bearing (Direction) to draft the side yard segment.

1. Click the **Slide Angle - Edit** routine from the Parcel Layout Tools toolbar.
2. If the Create Parcel From Layout dialog box appears, set the parcel type to **Single-Family**, the label to **Name Area & Perimeter**, and click the **OK** button.
3. The routine prompts you to select a parcel line to adjust. Select a side yard segment between two parcels.
4. The next prompt is for a new angle. Enter in a new angle for the side yard line.
5. The next prompt is to select a parcel to adjust. Select one of the parcels.

   The routine displays a dynamic sliding segment that uses the new angle. Once you determine the new position of the segment with the new angle, you select the point to set the side yard, and the routine redefines the parcel.
6. The next prompt is for a new area, or you can slide the new side yard direction to resize the lot.
7. Edit a few more lots in other areas of the site.
8. When you are done editing the lots, click the **AutoCAD Save** icon and save the drawing.

RENUMBERING PARCELS

After dividing the site into parcels with the sizing routines, the parcel numbers may not represent the numbering system you want for them. The Renumber/Rename Parcel command renumbers selected parcels to the numbers you want.

1. If you are not working in the area of the Residential Parcel 1 view, from the View menu, use **Named Views...** to restore the **Residential Parcels 1** view.
2. From the Parcels menu select **Renumber/Rename Parcels**.

   This displays the Renumber/Rename Parcels dialog box (see Figure 3.28).

3. Set the Starting Number to **100** and click the **OK** button to dismiss the dialog box.
4. Next select a point within the westernmost parcel and then select a point in the easternmost lot in the short block.

   This step selects the lots for renumbering.
5. Press the **Enter** key twice to end the selection and renumber the lots from west to east and from 100 to 102.

RENAMING PARCELS

Sometimes you want to use a different name than the one assigned by the style. It may be that the parcel number needs to be a Parcel Identification Number (PIN) or a more general name, such as Lot #.

1. From the Parcels menu, select **Renumber/Rename Parcels**.
2. This displays the Renumber/Rename Parcels dialog box (see Figure 3.29).
3. Toggle ON **Rename** and **Specify the Parcel Names** and click the **Edit Name Template** button (the icon to the right of the toggle).
4. In the Name Template dialog box, Name Formatting Template section, set the Property Fields to **Next Counter**, in the Name area type **Lot #**, and click the **Insert** button to put a next counter after "Lot #."
5. In the Incremental Number Format area, set the Number Style to **1, 2, 3...**, and set the Starting Number to **300**. Your settings should match Figure 3.30.
6. Click the **OK** button to exit the Name Template dialog box and return to the Renumber/Rename Parcels dialog box.
7. Click the **OK** button to close the dialog box.
8. Next select a point within the westernmost lot and then select a point in the easternmost lot in the short block.

   This step selects the lots for renumbering.
9. Press the **Enter** key twice to rename the lots.
10. Expand the Parcel branch in Prospector to view the parcel list with the new names and numbers.
11. Click the **AutoCAD Save** icon to save the drawing.

This completes the exercise on editing parcels. The last section reviews the parcel label and the annotating of parcel segments.

EXERCISES

SUMMARY

- Users can graphically edit the position of a parcel's segment.
- When graphically moving a segment, it follows the path of the object to which it is attached.
- A parcel responds to the changes of its segments (calculates a new area and perimeter).
- If deleting a boundary segment of a lot, the lines are considered open parcel segments and do not receive a label.
- When closing an open set of parcel segments, Civil 3D adds a parcel label to the object.
- Users can change the angle or direction of a parcel segment from the Parcel Layout Tools toolbar.
- The Parcel Layout Tools toolbar also has commands to add vertices to parcel segments and to add new parcel line and curve segments.

UNIT 5: PARCEL AND SEGMENT ANNOTATION

Parcel labels can be a part of the parcel-sizing process. When defining parcels, users can identify an area label and segment labels for lines and curves. If not labeling when sizing parcels, the Add Labels command has tools to create line and curve labels for defined parcels.

There are two types of parcel labels: one that labels the parcel object, and the second labels the parcel line and curve segments. The line and curve labels are either on the lines and curves or have entries in a table. All parcel annotation reacts and changes to correctly display the new values of an edited parcel or its segments.

Users can pin a label to a location in a drawing. As the view of the drawing rotates, the label will rotate around the pinned point and will remain plan readable.

PLAN READABILITY AND SCALE SENSITIVITY

By default, all parcel labels are view-dependent and will rotate to be plan readable in any view orientation (reading from left to right). Users can control the angle that changes the orientation of the label. All area and segment labels are scale dependent. The definition of a label style sets the paper size of the text and the labels size themselves by the scale of a viewport. (For example, a label with a text height defined as 0.1 is 5 units tall in a 1"=50' viewport, and 3 units tall in a 1"=30' viewport.) If each viewport has a different scale, the text size will be different to create text that is 0.1 tall when plotting the layout.

AREA LABELS

When Civil 3D creates a parcel, it places a label identifying the parcel at the parcel's centroid. An area label can be as simple as a parcel number or as complicated as the user wants to make it. Figure 3.33 shows a list of what parcel data is available for a parcel area label.

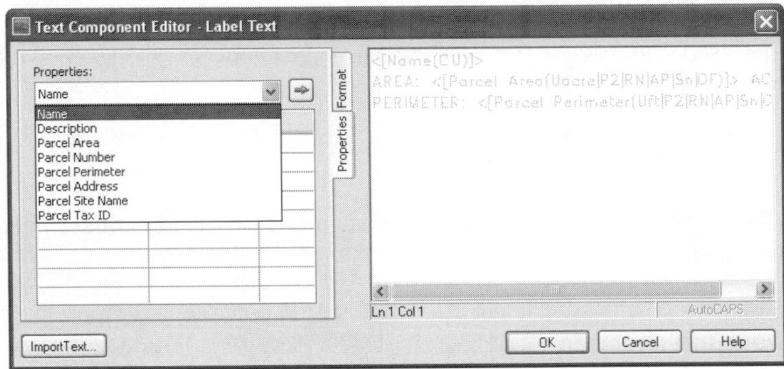

Figure 3.33

An area label can contain more that one line of text. However, when this is done, it can only have a border around each element. If a border around all of the lines of text is desired, they need to be defined as a single multi-line text entry. When using a single multiple-line entry, all text is the same height (see Figure 3.34).

Text Component Editor - Label Text

Properties:

Parcel Area

Modifier	Value
Unit	acre
Precision	0.01
Rounding	round normal
Decimal character	period '.'
Sign	sign negative '-'
Output	full

<[Name(CU)]>
AREA <[Parcel Area(Uacre|P2|RN|AP|Sn|OF)]> AC
PERIMETER <[Parcel Perimeter(Uft|P2|RN|AP|Sn|C

Ln 2 Col 6

Figure 3.34

Users can include a block around a piece of label text. A style places an ellipsis or any block in the drawing around the parcel number (see Figure 3.35).

LABELING SEGMENTS

All Civil 3D segment labels are dynamic and view-dependent. The labels rotate to be plan readable after rotating the view of parcels in the drawing. When placing labels in a viewport, they size themselves to be the correct height for the viewport scale.

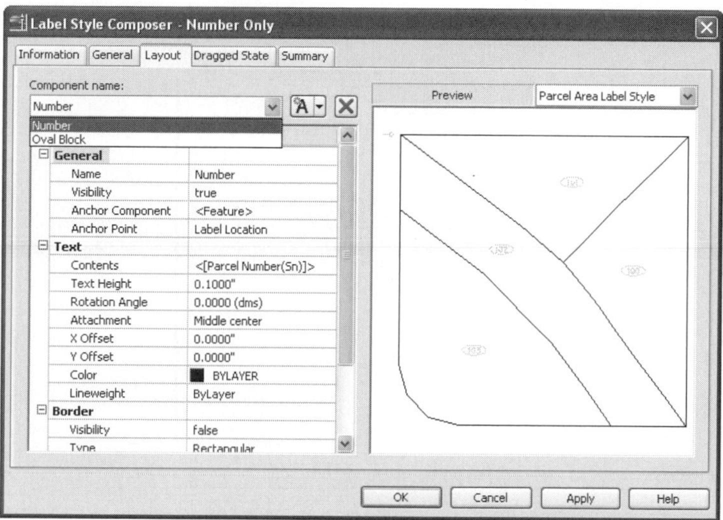

Figure 3.35

When defining parcels, users can assign labels to the new parcel segments. Or, users can label individual or all of the parcel segments with the Add Labels dialog box (see Figure 3.36). The Add Labels dialog box sets the label styles for the parcel line and curve segments.

When the labeling is simple and straightforward, the multiple segment option is quickest. However, there are times when more control over the placement of labels is needed, or the labeling needs to identify a segment that extends beyond individual lot segments (i.e. back yard lines, block parcel lines, etc.). The single segment option allows users to place each segment label at the point selected for the label.

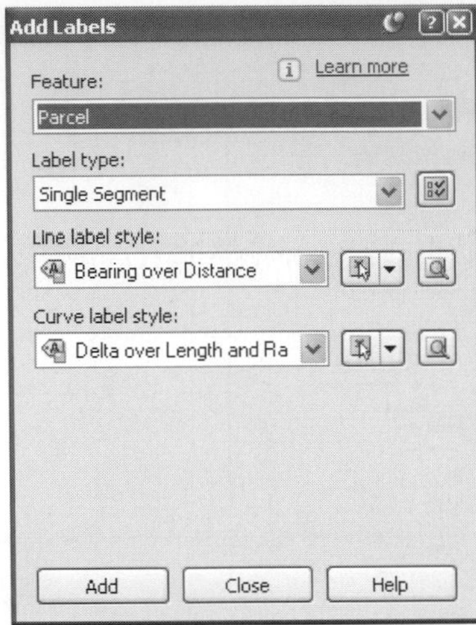

Figure 3.36

LINE AND CURVE SEGMENT STYLES

Each segment label style defines a verbose and tag style (see Figure 3.37). The verbose label style annotates specified values of the lines or curves. The amount of information for a line or curve label depends on the label definition. The tag variant of the label identifies the segment's number that corresponds to an entry in either a line or curve table. The contents of the table depend on the table definition.

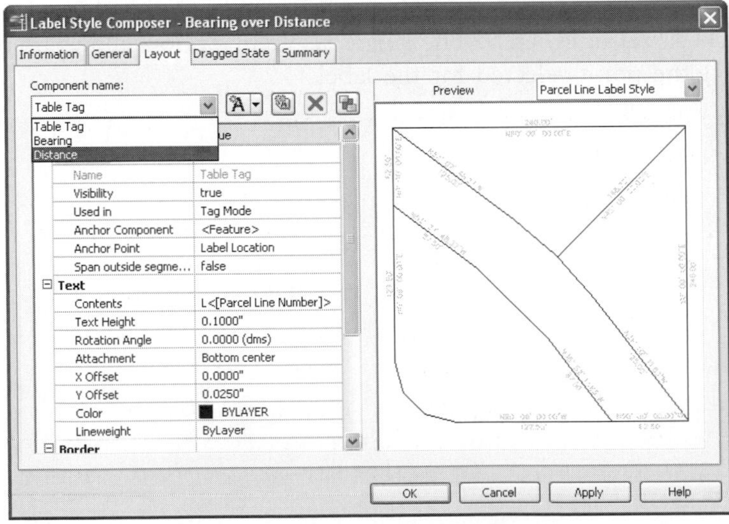

Figure 3.37

The span line label styles include crow's feet. These styles typically indicate a parcel segment line has a distance greater than any one of the lots using this line as a parcel segment. The individual parcels sharing the line will have a distance label with no bearing, because the extended segment labels the direction.

The span line label styles include a direction arrow. Users can erase the Direction Arrow Component or toggle its visibility off in the Layout panel of each style (see Figure 3.38).

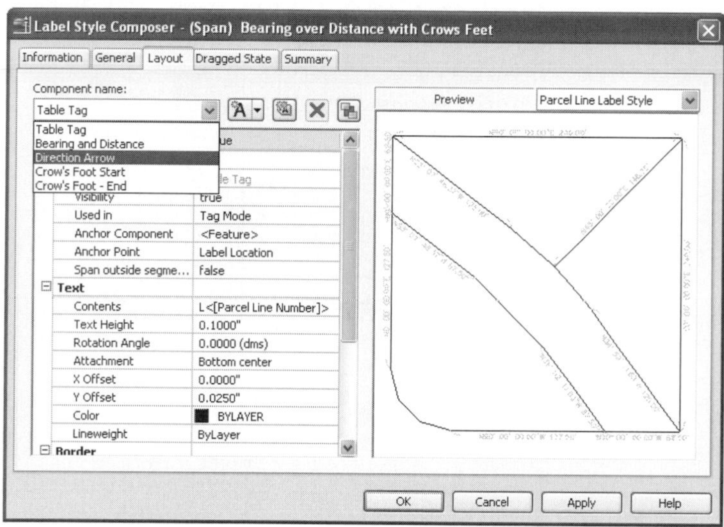

Figure 3.38

All Civil 3D segment labels attach to a parcel segment. If extending a label over several lots is desired, the segment must be a single line not broken at the intersection with the individual side yard lines.

The Plan Readability bias of the General settings panel of each style affects how the labels orientate themselves on the screen. Users may have to adjust the value to get the lefthand reading of the parcel labels.

As with plan readability, Civil 3D labels and annotation are scale-sensitive when they are viewed from a paper space viewport. When a scale is set for a viewport and the regenall command is executed, the labels and annotation text size themselves for the viewport scale. The reason the text and labels can react this way is their definition specifies a paper text size (height). When Civil 3D displays the labels and text in a viewport, they increase or decrease their text size to be correct for the set plot scale.

When selecting a label and pressing the right mouse button, a shortcut menu displays that has commands to reverse a label's direction, exchange its elements from one side of a segment to the other, and adjust the label's properties (different label style, reversed, or flipped) (see Figure 3.39).

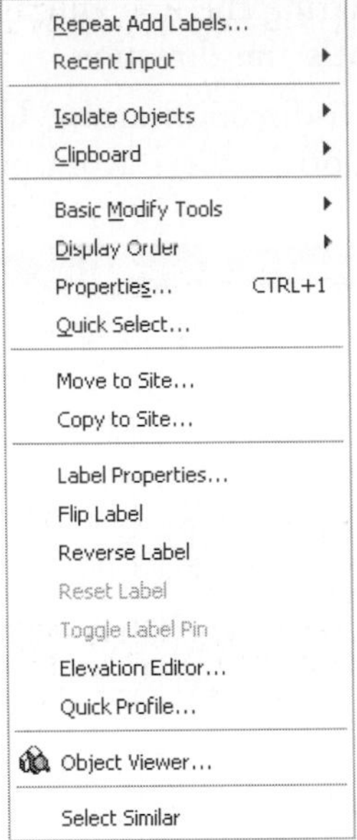

Figure 3.39

EDITING SEGMENT LABELS

There are times when the label displays the wrong direction or the positions of the label text needs to be exchanged. These situations are addressed by selecting a label and pressing the right mouse button to display a shortcut menu with segment label editing routines (see Figure 3.39).

FLIP LABEL

Selecting Flip Label causes the selected label to exchange the location of the label text. For example, if a label has the direction above and the distance below, when selecting the label, press the right mouse button, and select Flip Label. Then the direction changes to the bottom position and the distance moves to the above position.

REVERSE LABEL

Selecting Reverse Label causes the selected label to change the bearing 180 degrees. For example, if a label is N 88° 45' 54" E and the user selects Reverse Label, the label changes to S 88° 45' 54" W.

RESET LABEL

Selecting Reset Label undoes a single change made to a label. If removal of more than one change is desired, the AutoCAD Undo command should be used.

DRAGGING A LABEL

Every label has two states: original position and dragged. The original position represents the label definition found in the Layout tab of the style. When a label is dragged from its original position, it changes its display to represent the label style's Dragged State settings. The label can remain as defined or can change completely. Reset Label returns a dragged label back to its original location.

LABEL PROPERTIES

When selecting Label Properties from the shortcut menu, the user can assign a new label style, reverse the direction of the label (if a line segment), flip the label elements, and pin a label (see Figure 3.40). When Label Is Pinned is set to false, a label is free to move as the associated segment is edited. If Label Is Pinned is set to true, the label remains at its current position even if the associated object is edited. If rotating the view of a pinned label, the label rotates around the current location of the label.

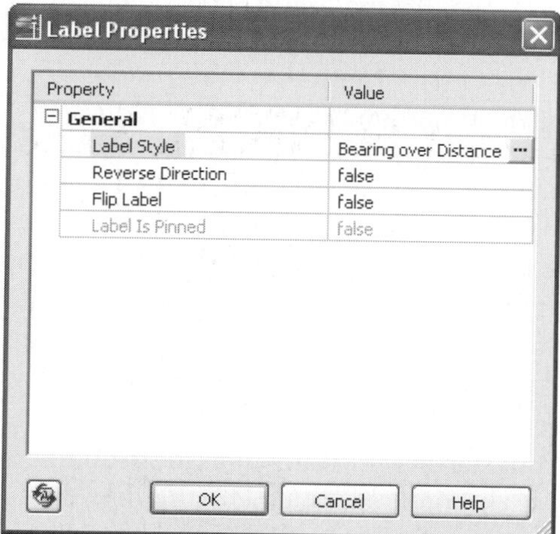

Figure 3.40

TABLE STYLES

Civil 3D defines four basic table styles for parcels: lines, curves, segments, and areas. The line and curve tables list values from parcel lines and curve segments. The entries in the table correspond to segment tags on the individual parcel segments. Each parcel segment has a unique number and a corresponding entry in a table that displays the values of the segment type. For example, an L1 tag identifies a line segment in the drawing and has a corresponding L1 entry in the table. The table record for L1 lists the bearing and distance of the tagged line segment. A segment table lists both line and curve parcel segments. The area table lists the areas for parcels.

A Table Style dialog box is broken into three sections: Table settings, Text settings, and Structure (see Figure 3.41). All of these settings and values are found on the Data Properties panel of the style.

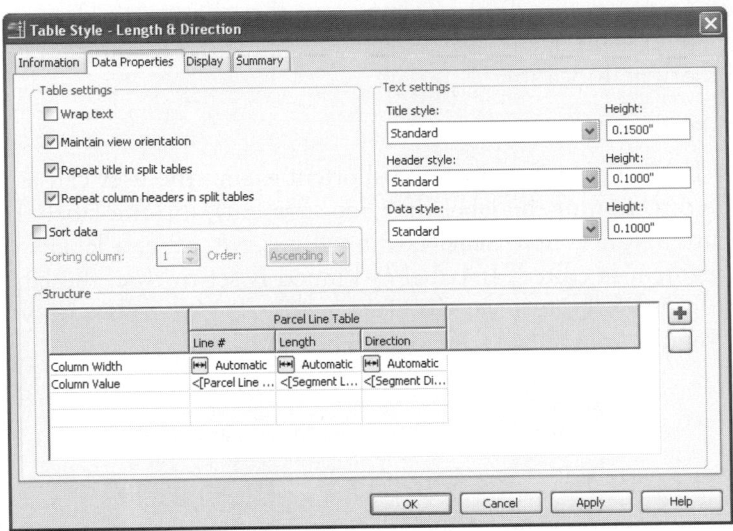

Figure 3.41

The Table settings portion of the panel contains toggles for wrapping text, maintaining plan view in rotated viewports, repeating the title and column headings in split tables, and sorting table entries. The Text settings portion of the table sets the title text style and size, column headers, and data. The Structure portion of the dialog box displays the type and order of the information in a table record. Each column of the table can contain different object information: length, direction, starting northing or easting, ending northing or easting, and tag number. The two buttons at the right side of the dialog box add or delete new columns in a table definition. Edit the contents format by double-clicking in the column value entry. This action displays the Text Component Editor, where users define what value (drop-list from Properties) appears in the column and its text format (see Figure 3.42).

CREATING A TABLE

To create a table, select the type and the information needed in the table. The type of table is a selection from the Tables flyout menu of the Parcels menu. The flyout menu lists the four types of tables: line, curve, segment (line and curve), and area.

When selecting a table type, a Table Creation dialog box displays. In the Table Creation dialog box, users select the table style, the layer for the table, the labels to include in the table, the layout of the table, and whether the table is dynamic or static (see Figure 3.43).

Select the segments by style or by manually selecting the annotation from the drawing. The Selection rule, Add Existing, indicates that only existing labeling becomes a part of the

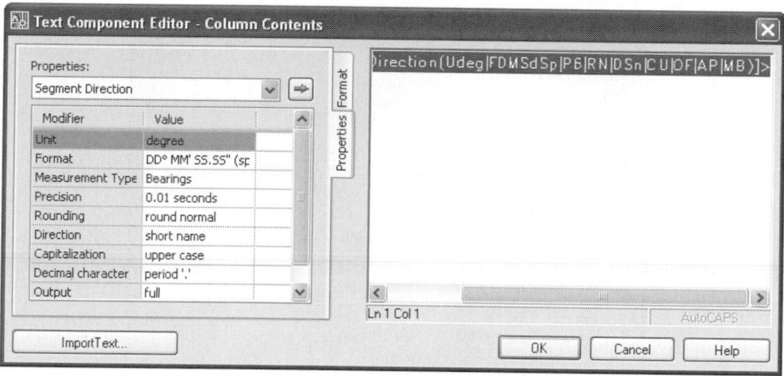

Figure 3.42

table. If a user sets the rule to Add Existing and New, the table will change to accommodate any new labeling.

Note: The table layer is the layer set in the Object Layers settings of Edit Drawing Settings.

If lengthy, at the bottom of the dialog box are settings that set a breakpoint for the next table segment.

When a label is in a table, it changes its visibility mode to Table Tag. When viewing the labels on the segment, the user will see L1... or C1.... If the user returns the labels display mode (in the General tab of the style definition) back to Label and the table is dynamic, the segments will disappear from the table and the annotation returns to verbose.

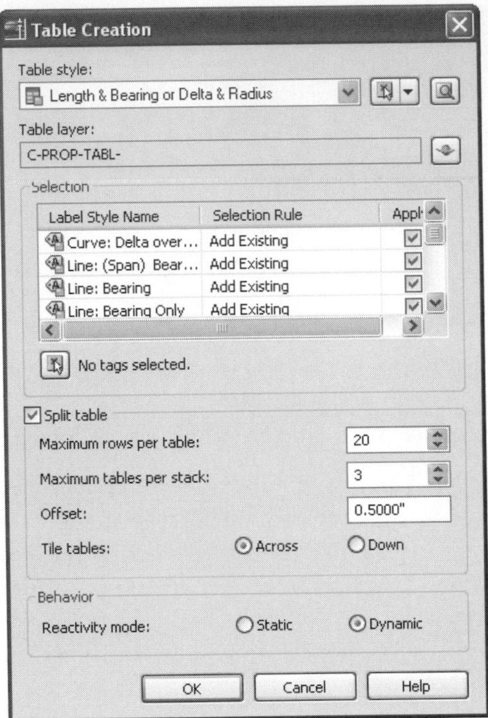

Figure 3.43

USER-DEFINED PROPERTY CLASSIFICATIONS

A user can define a parcel property that can be a part of a lot label (for example, defining an Industrial classification). The definition of a user-defined property is a two-step process. The first step is defining the classification name (see Figure 3.44).

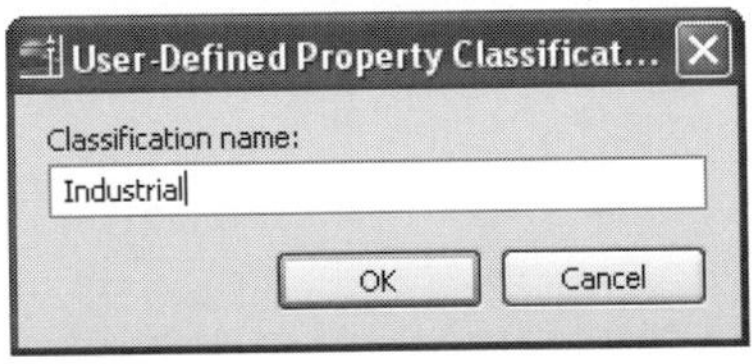

Figure 3.44

The second step is defining the properties of the classification (see Figure 3.45). In the case of Industrial, the enumerated types are Light, Moderate, and Heavy. These values can be a part of a parcel's label. With the information in the label, the values can be a part of a table or report.

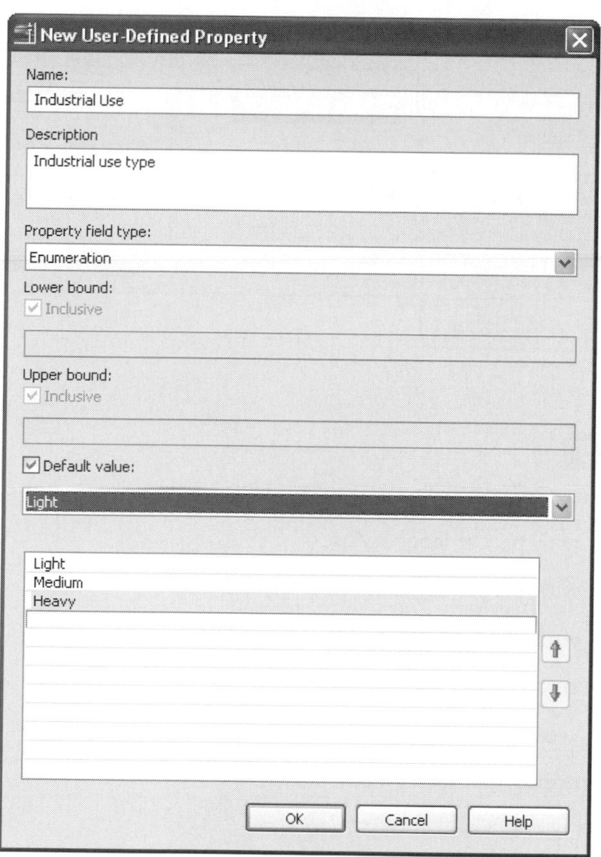

Figure 3.45

EXPRESSIONS

Each parcel label type has an expression option for building custom calculations for the label type (see Figure 3.46). Users must first define expressions, and when defining a label the expression appears as a label property.

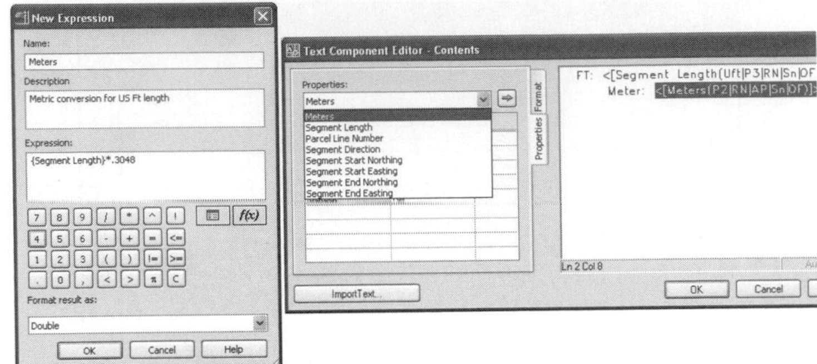

Figure 3.46

EXERCISE 3-5

When finishing this exercise, you will:

- Be familiar with parcel label styles.
- Be familiar with parcel segment label styles.
- Be familiar with parcel table styles.
- Be able to create parcel segment labels.
- Be able to create parcel area tables.
- Be able to create parcel segment tables.

EXERCISE SETUP

This exercise continues with the same drawing from the previous exercise. If starting with this exercise, open the drawing *Chapter 3 - Unit 5* that is in the *Chapter 3* folder on the CD that accompanies this textbook.

1. If your drawing from the previous exercise is not open, open it now, or open the *Chapter 3 - Unit 5* drawing.

PARCEL AREA LABELS

Labeling the area of a parcel occurs as you create the parcel.

1. Click the **Settings** tab to make it current.
2. In Settings, expand the Parcel branch until you view the list of area label styles.
3. Click **Name Area & Perimeter** area label style, press the right mouse button, and from the shortcut menu select **Edit....**
4. Click the **Layout** tab to view its contents.

   The box that encompasses the label contains the three lines of text. These text strings are not individual entries linked together like a point's label, but are three values in one text component entry.

5. Click in the **Value** cell of **Contents** in the Text section of the panel.
6. Click the ellipsis at the right side of the Value cell to view the label text in the Text Component Editor dialog box.
7. Click in the format string at the right of the dialog box to view the settings for each line of the label on the left side of the dialog box.

   All three of these lines represent one text component for the label. The box encompasses all three lines, because Civil 3D treats the three lines of text as a single block of text.

8. Click the **Cancel** button to exit the Text Component Editor.
9. Click the **Dragged State** tab to view its contents.

   If you drag this label from its original position, it becomes a stack of text with a leader. Again you can change it to remain as composed with or without a leader.
10. Click the **Cancel** button to exit the Label Style Composer.

SEGMENT LABELS

The segment label styles have varying combinations of setting values. As with area labels, each segment label has a layout and dragged-state definition.

Line Segments

The line segment labels annotate the distance, bearing, or both attributes of a line segment.

1. In Settings, the Parcel branch, expand Line Label Styles until you view a list of styles.
2. Select **Bearing over Distance**, press the right mouse button, and from the shortcut menu select **Edit....**
3. Click the **Layout** tab and view the different components of the style. You do this by clicking the drop-list arrow at the top left and then selecting a different component name from the list. This style has a tag and two components: Bearing and Distance.
4. Click the **Dragged State** tab to view its contents.
5. Click the **Cancel** button to exit the style.
6. Click **(Span) Bearing and Distance with Crow's Feet**, press the right mouse button, and from the shortcut menu select **Edit....**
7. Click the **Layout** tab and view the different components of the style. You do this by clicking the Component Name: drop-list arrow and then selecting the items from the list. This style has a tag and several components representing the bearing and distance, direction arrow, and crow's feet (start and end).

   You would use this label on segments that extend longer than just an individual line segment.
8. Click the **Dragged State** tab to view its contents.

EXERCISES

9. Click the **Cancel** button to exit the style.

Curve Segments

A curve segment label annotates the radius and length of a curve segment. A curve label can add additional curve data to a label.

1. In Settings, expand the Parcel branch until you view a list of styles for a curve.
2. Click the **Delta over Length and Radius** style, press the right mouse button, and select **Edit...** from the shortcut menu.
3. Click the **Layout** tab and change the component name to **Distance & Radius**.

   This part of the label anchors to the bottom center (Attachment value in the Text section) of a curve (this appears below the curve segment).
4. Change the component name to **Delta**.

   This part of the label anchors to the top center (Attachment value in the Text section) of a curve (this appears above the curve).
5. Click the **Dragged State** tab to view the settings for the curve label.
6. Click the **Cancel** button to exit the style.

CREATING SEGMENT LABELS

When defining parcels, you can label them as they are sized. If you are not labeling at this time, you can use the Add Labels routine to create segment labels. Add Labels has two methods of creating parcel segment labels: multiple and single segment.

Multiple Segment

1. From the Parcels menu select **Add Labels...**.
2. In the Add Labels dialog box, set the Feature to **Parcel**, the Label type to **Multiple Segment**, the Line label style to **Bearing over Distance**, and the Curve Label style to **Delta over Length and Radius** (see Figure 3.47).
3. Click the **Add** button in the dialog box.

   The routine prompts you to select the parcel labels to create the segment labels.
4. Select the parcel labels to create the segment labels.

   The routine labels the segments that define the selected parcels. There is only one label at each common side yard segment.
5. Click the **Close** button to exit the Add Labels dialog box.
6. Undo the labels.

   You may have to use the AutoCAD Erase command to remove all of the labels.

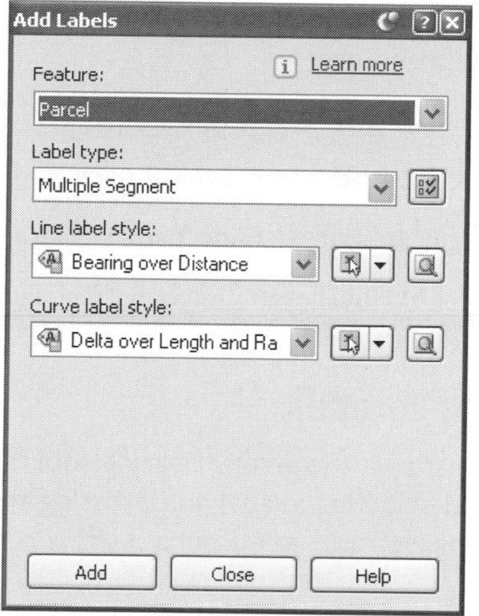

Figure 3.47

Single Segment

One problem with multiple-segment labeling is the labeling of the individual back yard segments. What may be necessary is a label spanning the back yard with a bearing and distance, as well as distances for the individual parcel back yards (see Figure 3.48). In Figure 3.48, the left side is the result of multiple segment, and the right side shows what single segments can do with a spanning label style.

When you use the single-segment label option, the point you select on the segment is the anchoring point for the label. If the label is not in the correct position, you can graphically slide the label down the line or arc to correctly locate the label.

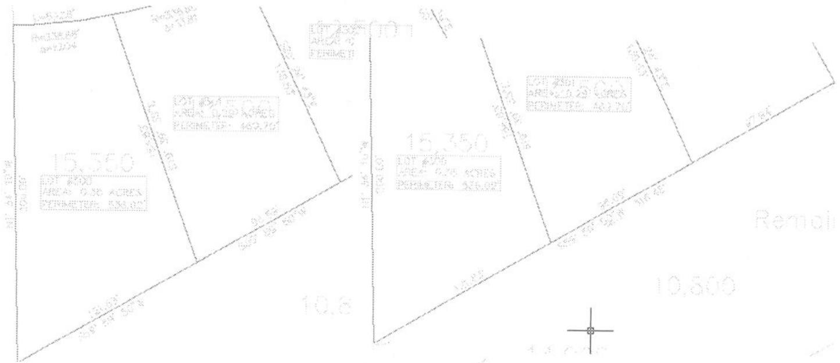

Figure 3.48

1. If you are not working in the area of the **Residential Parcel 1 view**, from the Views menu use the Named Views... command to restore this view.
2. From the Parcels menu, select **Add Labels...**.
3. In the Add Labels dialog box, set the Feature to **Parcel**, the Label type to **Single Segment**, the Line label style to **Bearing over Distance**, and the Curve label style to **Delta over Length and Radius** (see Figure 3.46).

4. Click the **Add** button and select the parcel segments to label all of their segments except the back yard segments.
5. Change the Line label style to **Distance**.
6. Click the **Add** button and label the individual back yard distances for the three parcels.
7. Click the **Close** button to exit the Add Labels dialog box.

   Your parcel labels should now look similar to Figure 3.48. You will now need the Flip Label tool to match the figure.

CHANGING THE SEGMENT LABELS

There will be times when a label shows the wrong bearing (north instead of south), or it is on the wrong side of the segment. When selecting a label and pressing the right mouse button, a shortcut menu displays with several editing options (see Figure 3.49).

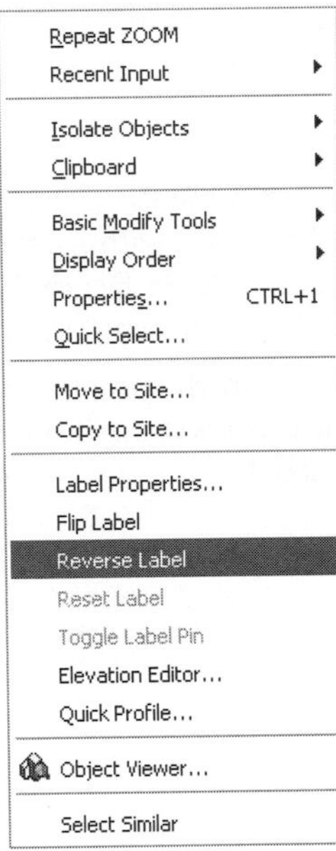

Figure 3.49

Reverse Label

If the direction of the label is incorrect, use the Reverse Label command to change the direction of the label.

1. Zoom in to view a few segment labels.
2. In the drawing, select a label, press the right mouse button, and from the shortcut menu select **Reverse Label**.

Flip Label

If a label is on the wrong side of the segment, use the Flip Label to change the label elements to opposite sides.

1. In the drawing, select a label, press the right mouse button, and from the shortcut menu select **Flip Label**.

Drag Label

If selecting and activating a label's grip, you can drag a label from its original position. Depending on the label's dragged-state settings, the label may change dramatically from its original definition. You can change the label to be right- or left-justified by selecting different locations for the dragged label.

1. In the drawing, select a label, then select its grip and move it to a new location.
2. Try different locations to see the label switch from left- to right-justified.

Reset Label

If you want to return a label to its original position and composition, use the Reset Label option.

1. In the drawing, select the dragged label, press the right mouse button, and from the shortcut menu select **Reset Label**.

Change Label

Changing a label's label style occurs in the Label Properties dialog box.

1. In the drawing, select a label, press the right mouse button, and from the shortcut menu select **Label Properties....**
2. In the Label Properties dialog box, click in the **Value** cell for Label Style. This action displays an ellipsis at the right side of the cell; click the ellipsis to display a Label Style dialog box.
3. Click the drop-list arrow and select a new segment style (see Figure 3.50).
4. Click the **OK** buttons until you exit the dialog boxes.

   When you exit the dialog boxes, the new style shows on the segment.
5. Click the **AutoCAD Save** icon to save the drawing.

PLAN READABILITY OF LABELS

All parcel labels and annotation are sensitive to rotation to maintain plan readability (reading from left to right). Whether a rotation is in model view or a layout viewport, the labels react to the rotation and read correctly on the screen.

1. Start the **Dview** command and select the parcels and line work on the screen. Use the **Twist** option and twist the view until the back yard line is nearly horizontal.

   As you rotate the view, the labels remain plan readable. No matter what rotation angle you use, the area labels and segment annotation will be plan orientated.
2. Press the **Enter** key to exit the Dview command.

Figure 3.50

3. From the View menu, select **Named Views...**, click the **New...** button, and save the current view as **Parcels Twisted**. Click the **OK** buttons until you exit the dialog boxes.
4. Use the Plan command to return to the World view.

SCALE SENSITIVITY OF LABELS

All labels and annotation are sensitive to a layout viewport's scale. The labels and annotation will react to the scale value and will correctly size themselves for that scale. You use the AutoCAD Regenall command to make the label resize.

1. Click the **Layout 1** tab and enter paper space. If the layout tabs are not displayed along the bottom of the Drawing Window, click the Layout 1 button on the status bar (the second icon after LWT).
2. Double-click the viewport to enter its model space.
3. Select **Named Views...** from the View menu and set current the **Parcels Twisted** view.
4. Double-click the paper (outside the viewport) to enter paper space.
5. Click the viewport border, press the right mouse button, and from the shortcut menu select **Properties...**.
6. In the Properties dialog box, the Misc section, set the Standard Scale to **1"=50'**.
7. Use the **Regenall** command to resize the text (if necessary).

   The label and annotation text is now correct for the viewport's scale.
8. Set the scale of the viewport to **1"=20'**.
9. Run the **Regenall** command to resize the text (if necessary).

10. Close the Properties dialog box.
11. Click the **Model Space** tab to reenter model space. If the layout tabs are not displayed along the bottom of the Drawing Window, click the **Model** button on the status bar (the first icon after LWT).

PARCEL SEGMENT TABLE

Rather than adding labels to the line work in the drawing, you can create a table listing the segments and their geometry values.

This table is created from existing labeling in the drawing. The Table routine reads the values, creates the table, and replaces the segment annotation with line or curve tags (L1, L2..., C1, C2...).

1. From the Tables flyout of the Parcels menu, select **Add Segments....**

   This displays the Table Creation dialog box. At the top the Table Style heading lists the current table style.

   The Selection area lists all of the label styles in the drawing.

2. Scroll through the list of label styles and set a check mark to the right for all style names.

   The next setting sets the type of table: static or dynamic.

3. Set the Table type to **Dynamic**.

   Your dialog box should look like Figure 3.43.

4. Click the **OK** button to dismiss the dialog box.
5. The routine prompts for the upper-lefthand corner of the table. Select a point to create the table.

   The table is created from existing labeling in the drawing. The Table routine reads the values, creates the table, and replaces the segment annotation with line or curve tags (L1, L2..., C1, C2...).

6. Zoom to the parcels to view the segment tags.
7. Click the **AutoCAD Save** icon and save the drawing.

PARCEL AREA TABLE

Many times when submitting design documents, you have to list the areas of the lots proposed in the subdivision. The listing of parcels and their areas is ideal data to place in a table.

1. From the Tables flyout of the Parcels menu, select **Add Area....**

   This displays the Table Creation dialog box.

2. In the Selection area toggle **ON** *all* of the styles.
3. Change Selection Rule to **Add Existing and New.** You do this by clicking in the cell for Add Existing, then clicking on the drop-list arrow and selecting **Add Existing and New**.

EXERCISES

4. Leave Split Table **ON**.
5. The Table type should be **Dynamic**.

   Your Create Table dialog box should look like Figure 3.51.

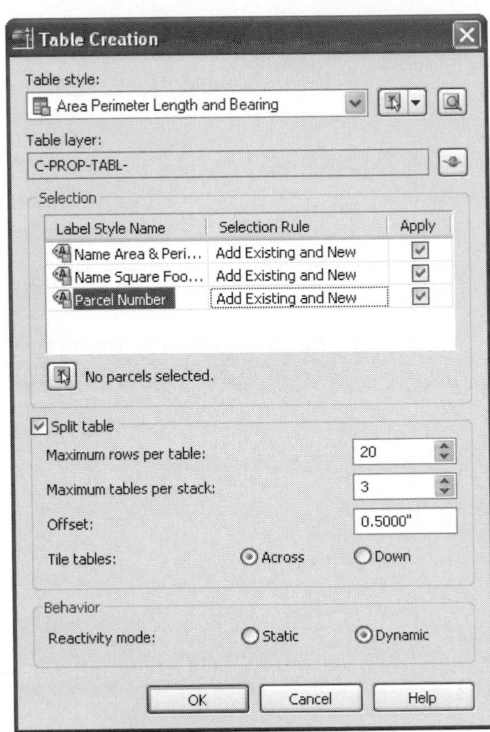

Figure 3.51

6. Click the **OK** button to close the dialog box.
7. The command prompts for an upper-lefthand corner to place the table into the drawing. Select a point in the drawing to draw the table.
8. Click the **AutoCAD Save** icon to save the drawing.

SUMMARY

- Civil 3D supports three types of labels: area, line, and curve.
- All labels placed with the multiple-segment option of Add Labels are at the midpoint of each segment.
- All labels placed with the single-segment option of Add Labels are at the point the user selects the segment to label.
- Users can flip a label's text to opposite sides of the segment, change its direction, drag it away from its original position, and reset it back to the original position and composition.
- Civil 3D creates labels and annotation that are sensitive to a rotated view. The labels and annotation will rotate to be plan readable in that view (model space or paper space).
- Civil 3D creates labels and annotation that are sensitive to a viewport's plotting scale. The labels and annotation will resize themselves to the correct size based on the current plotting scale of the viewport.
- Users can create tables for line, curve, or both types of segments.
- Users can create tables for parcel numbers, their areas, and their line and curve segments.

This concludes the chapter on parcels. The goal of any subdivision is to create marketable parcels from an overall boundary. These parcels may define homogeneous areas of land use (detention ponds, open space, wetlands, etc.) or a lot that can be sold.

Civil 3D contains all of the tools to design the individual parcels, document the subdivided land graphically, and create reports that aid in the creation metes and bounds and other necessary documents for the design solution.

The next chapter covers the creation of the Civil 3D surface. A surface is a critical element in the design process of a site.

CHAPTER 4

Surfaces

INTRODUCTION

This chapter introduces the Autodesk Civil 3D 2007 surface. A surface is a net of triangulated data, and the triangulation represents the surface's elevations and slopes. Surface data can be points from AutoCAD nodes and vector objects, Civil 3D point objects, polylines, contours, boundaries, and breaklines. Civil 3D provides tools that create, edit, evaluate, and annotate surface components and characteristics. Civil 3D surface styles allow a user to evaluate a surface as it is being built, its slopes, and its elevations as 2D and 3D objects.

OBJECTIVES

This chapter focuses on the following topics:

- Basic Settings for Surfaces
- Overview of Surface Objects and How to Label the Styles
- The Basic Surface Data Types
- How to Edit and Evaluate a Surface
- How to Analyze Slope and Elevation Characteristics
- Develop and Assign a Contour Style
- Annotate Surface Slopes and Contours
- How to Smooth a Surface
- How to Create Point Objects from Surface Elevations

OVERVIEW

A surface is a fundamental element in the civil design process. If not caught early in the review process, errors or misinterpretation of surface data may have grave consequences in every following design phase. It is important that a surface be an accurate interpretation of the data it uses.

Civil 3D displays surface components to help evaluate and edit a surface so it correctly interprets the data. Civil 3D uses the term components for all the displayable elements of a surface. The list of components includes triangle legs, points, borders, contours, elevations, slope arrows, etc. In this textbook, surface components are surface border(s), triangles, points, and boundaries. Anything that refers to surface elevations, relief, water flow (drainage), watersheds, or slopes and direction of slope are surface characteristics.

The above mentioned characteristics influence the design strategies and present a designer with several challenges. These challenges include how best to define land-use, create access to the site, manage water runoff, or develop manageable earthwork volumes. The design solutions applied to the site are in response to these challenges and the solutions may require compromises between cost and best design practices.

To create usable surfaces for any project, develop a process that produces consistently "good" surfaces (i.e. correctly interprets the data). The process should include gathering and evaluating initial data, evaluating the results of surface builds, finding and editing errors, analyzing surface characteristics, and, finally, annotating the surface.

There are seven steps to the process of developing consistent surfaces. These steps range from naming a surface to having the right contours show surface elevations. To take a surface from initial data to a finished product involves using some commands from the Surfaces menu. However, the majority of the commands are called from shortcut menus of the various Surfaces branch headings. These commands allow the user to create, edit, analyze, and annotate the components and characteristics of a surface.

The initial data for a surface can come from external data files or from object data within the drawing (points, contours, breaklines, etc.). The surface object and label styles display the components of a surface and allow the user to analyze the surface characteristics.

Civil 3D has two types of terrain surfaces; TIN (Triangular Irregular Network or Digital Terrain Model [DTM]) and Grid. A TIN connects all surface data with triangle legs and each end of a leg has a known elevation. A Grid terrain surface is a mesh structure having elevations at the intersections of the mesh. These elevations are generally interpolated values, not actual values. A DEM (Digital Elevation Model) is an example of this type of surface.

Civil 3D has two types of volume surfaces; the TIN and Grid volume surfaces. The structure of the volume surfaces is the same as a TIN or Grid terrain surface; however, the elevations of the volume surfaces are the difference in elevation between two terrain surfaces. It is from these two volume surfaces that Civil 3D calculates an earthworks volume.

SEVEN SURFACE STEPS

1. Create the surface.
2. Collect and assign initial data.
3. Add breaklines in known problem areas.
4. Evaluate resulting triangulation.
5. Edit (add, delete, modify) surface triangles, points, and/or breaklines.

6. Analyze the surface components and characteristics.
7. Annotate the surface.

The first step is naming the surface and setting its type. The command starting the process is in the shortcut menu of the Surfaces heading of Prospector (New...) or in Create Surface... in the Surfaces menu (see Figure 4.1). This selection displays the Create Surface dialog box. In the Create Surface dialog box, you assign a name, a description, and a type to the surface.

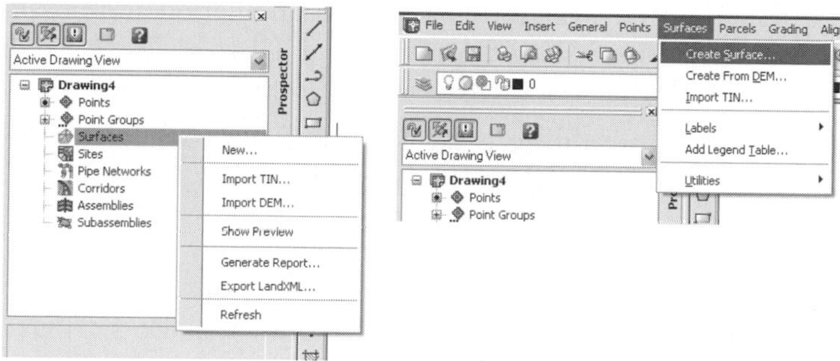

Figure 4.1

After naming a surface, the next step is to collect the initial data for the surface. Data for a surface can come from many diverse sources. These sources include points (AutoCAD objects and Civil 3D point objects), point files, contours, three-dimensional polylines, boundaries, and/or DEMs (Digital Elevation Models). No matter the data source, you need to review the data for any obvious errors and omissions before assigning it to a surface. The reason for this is that Civil 3D automatically builds a surface from a data assignment.

The third step is adding breaklines to known or obvious problem areas. If the initial data is point data, the user will need breaklines to correctly control the triangulation of linear features. Points represent three types of surface information: spot elevations, spot elevations with symbols, or spot elevations along a linear feature. When a set of points represents a linear feature, breaklines are necessary to control the triangulation along the set so the feature shows correctly in the surface. Examples of points needing breaklines are points representing edges-of-pavement, breaks in slope, stream banks, swales, ditches, and other linear features.

Breaklines can be drawn automatically from field observations or manually by you. Breaklines can be either two- or three-dimensional polylines. When using two-dimensional breaklines, the only condition is that there must be a point object at each vertex of the polyline. When using three-dimensional breaklines the user does not have to have a corresponding point object at each vertex.

The user identifies areas needing breaklines by reviewing groupings of points with similar descriptions. If there is a question about creating a breakline, there are a couple of options.

Consult with the original source of the data, ask about the area in question, and add breaklines to the data set as necessary. Or, after reviewing the surface and discussing the problem with someone, edit the surface to adjust the triangulation. Editing a surface creates the same effect as adding breaklines to the surface data. Be sure to understand the consequences of adding new breaklines before editing a surface.

How does a user decide when to add data or edit a surface to correct problems? If it takes several edits to correct a problem, adding more data or breaklines to the surface's data set is advisable. If fixing the problem involves only a few edits, edits is the preferred method.

The fourth and fifth steps review the initial surface triangulation and possibly lead to more editing or more data. These steps should use surface analysis styles that emphasize specific surface components and characteristics that can indicate problems with a surface.

After adding new data and/or editing a surface, the last two steps are analyzing surface characteristics (elevations and slopes) and creating contours.

UNIT 1

When working with Civil 3D surfaces there are a number of settings and styles that affect a surface. The first unit reviews and sets default values for these settings and styles.

UNIT 2

The second unit reviews creating a surface instance, the types of data you can use in a surface, and how Prospector manages surface data. After naming and identifying the type of surface, Civil 3D builds the surface after adding any data type to the surface definition tree.

UNIT 3

The third unit of this chapter covers the initial review and editing of a surface. Civil 3D has several options that affect what data a surface uses and how it responds to editing. The Prospector panel contains the editing tools for correcting surface errors and tweaking the triangulation to best represent the surface data. The styles assigned to a surface aid in its editing by emphasizing what facet of the surface the user is reviewing and editing. For example, a style displaying triangulation and points is best for looking for incorrectly triangulated linear features; a 3-dimensional style is for viewing bad elevations, etc.

UNIT 4

The fourth unit reviews the development and assignment of styles to analyze a surface. Examples of these style types are ones showing amounts of slope as ranged values, slope direction as down slope arrows, or elevations as ranged values. These styles display their information as two- or three-dimensional representations.

UNIT 5

The fifth unit of this chapter reviews developing a contour style and its settings for intervals, smoothing, and labeling values. After reviewing the contours, it may be necessary to smooth the surface. After creating the contours, they need annotation.

UNIT 6

The sixth unit discusses the creation of point data from a surface. These tools are found in the Create Points toolbar.

UNIT 1: SURFACE SETTINGS AND STYLES

The Edit Drawing Settings dialog box has the highest level of control over a drawing and its settings. The settings in this dialog box assign basic layer names for surface objects and labels. The Edit Feature Settings dialog box of the Surface branch controls three basic aspects of surfaces: it assigns the default object and labeling styles, sets the surface naming template, and sets the default surface type and its initial settings.

Surface styles determine how a surface displays its components and characteristics. The content templates that ship with Civil 3D provide several starter object styles for these components and characteristics. The user can create custom object styles that reflect individual work assignments and information needs.

EDIT DRAWING SETTINGS

Civil 3D manages surface visibility through AutoCAD layers and/or by styles. The Drawing Settings dialog box assigns a base layer name with an optional prefix or suffix (see Figure 4.2). By default the surface layer name is C-TOPO. By having more than one surface and not changing the default settings, all surfaces in a drawing will be on the same layer, C-TOPO. This situation gives you no control over the visibility of each surface except by assigning a style (i.e. since all surfaces use the same layer, they all show or are all hidden when setting the layer to on or off). To remedy this situation, assign a prefix or suffix to the base surface layer name. Use an asterisk (*) to assign the name of the surface to the base surface layer name. For example, creating a base layer whose name is suffixed by a surface named EXISTING. First, set the Modifier to suffix and enter the value of -*. This creates a layer in the drawing of C-TOPO-EXISTING. The dash is a separator between the base layer name and the name of the surface. This naming convention creates a layer for each named surface in the drawing.

EDIT FEATURE SETTINGS

The Surface Edit Feature Settings dialog box controls the default styles and naming template. An object style sets how a surface displays its components and characteristics (triangulation, border, breaklines, points, slopes, and elevations) and may show all, one, or any combination of these components and characteristics. One strategy for defining styles would be creating styles that show surface components and characteristics that are appropriate for an assigned task. While reviewing a newly built surface, the user's interest is the surface's triangulation, breaklines, boundaries, and points. While evaluating a surface for a preliminary design, view the elevations, relief, and slopes. With specific function styles emphasizing critical components and characteristics, the user can make better decisions during the design process.

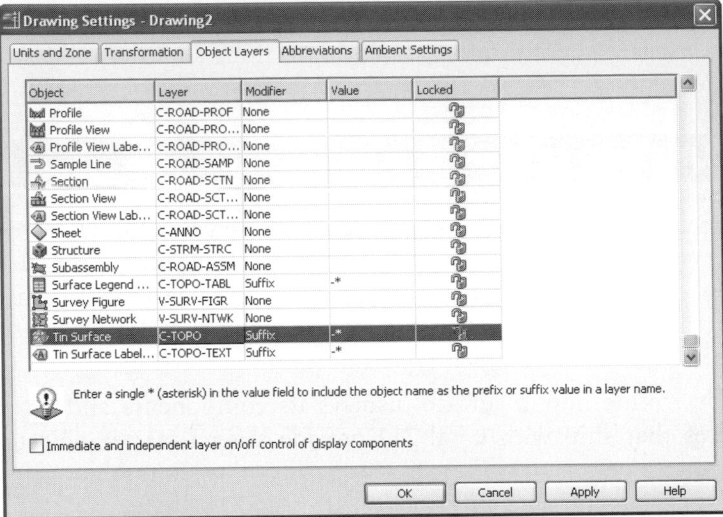

Figure 4.2

DEFAULT STYLES

The Default Styles section of the dialog box sets styles for the object and the surface annotation routines (see Figure 4.3). The label styles set includes spot elevation, slope, contours (major and minor), render material, and triangulation markers.

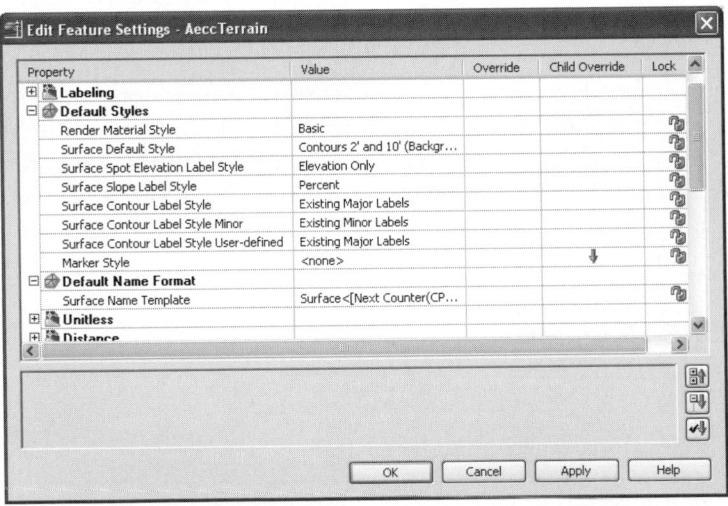

Figure 4.3

DEFAULT NAME FORMAT

This section sets the default naming method for surfaces (see Figure 4.3). The naming convention is Surface followed by a sequentially assigned number (i.e. Surface1, Surface2, etc.). You can change the default text (Surface to Terrain, for example) and/or change the method of assigning a number. You can rename a surface when you create a surface.

COMMAND: CREATESURFACE

The Surface branch of the Settings dialog box has a command branch that defines default values for all of the surface commands. This strategy allows a user to assign different and specific behaviors to each command. The CreateSurface command dialog box has two section controlling surface building: Surface Creation and Build Options. The Surface Creation section sets the default surface type, the grid X and Y spacing, and the rotation for a grid surface type (see Figure 4.4). There are four types of surfaces in Civil 3D: TIN, Grid, TIN volume, and Grid Volume.

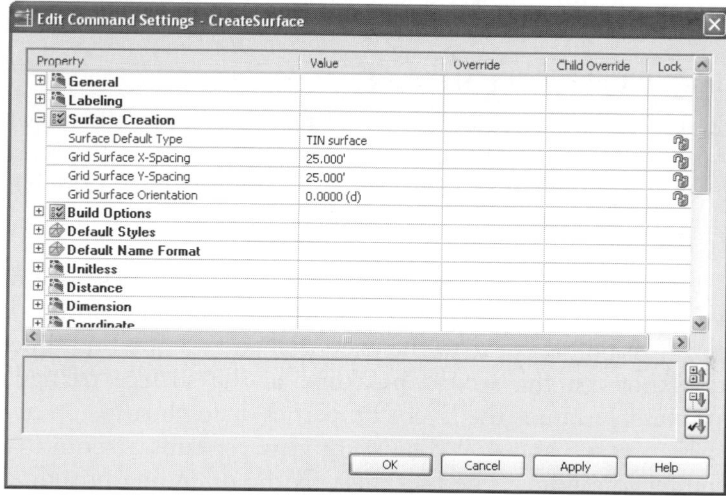

Figure 4.4

BUILD OPTIONS

The Build Options section controls how to build a surface: what data to exclude, the maximum triangle length, if proximity breaklines are to become standard breaklines, and how to handle crossing breaklines (see Figure 4.5).

Copy Deleted Dependent Objects

The Copy Deleted Dependent Objects setting specifies whether a drawing object's data is preserved in the surface definition after deleting the object from the drawing. The default setting is "no" and assumes that the original object is always present in the drawing. When using the default value (no), the original object's effect on the surface is removed from the surface triangulation when the data object is deleted from the drawing. If the user wants to preserve the effect of the original data object even after it is erased, he or she must set this value to "yes." With the value set to "yes" and the original object deleted, a copy of the object's data is added to the surface definition and preserves the effect of the data object.

Exclude Elevations Above/Below

Exclude Elevation Less Than and Exclude Elevations Greater Than set minimum and maximum elevations to exclude data from a surface. If a data item is below the minimum or above the maximum elevation, it will not be a part of the surface.

Maximum Triangle Length

The Maximum Triangle Length option has two settings. The first is a toggle to turn on or off the option. The second value sets the maximum length of a surface's triangle leg. The reason for this setting is to delete sliver triangles along the periphery of a surface. With this setting off, a surface will connect points even if there isn't data to support the triangle leg. With the setting on and a value set, a surface will remove any triangles greater than this value from the triangulation. If the user sets this value to low, it may prevent the surface form creating some interior triangles, and as a result the surface will have interior boundaries. The user can set this value any time during surface creation.

Convert Proximity Breaklines to Standard

With this toggle set to "yes," when building a surface, it will convert proximity breaklines, (2D) into standard breaklines (3D). The elevations for the converted breakline vertices are from the point elevations nearest to each vertex on the original 2D polyline.

Allow Crossing Breaklines

This toggle and setting allows a surface to have crossing breaklines. The default setting is off, or a surface does not allow crossing breaklines. When encountering crossing breaklines, a surface adds the crossing breakline to the data, triangulates one breakline, issues an error message, and does not use the second breakline in the surface triangulation. Double-clicking the error message causes the Event Properties dialog box to appear with the details of the error (see Figures 4.6 and 4.7). The event view contains a zoom to link to take the user to the offending data point. The user must fix the offending breakline before its data is included in the surface.

When the user toggles this setting on, Allow Crossing Breaklines, he or she needs to set a method to determine what elevation to use at the point of crossing. There are three methods of determining this elevation: use the first breakline elevation at intersection, use the last breakline elevation at intersection, or use average breakline elevation at intersection. If allowed crossing breaklines and their elevations are similar, the effect on the surface is minimal. However, if the elevations vary greatly, it would be better to resolve the issues with the crossing breaklines before using them in a surface.

All of the settings in the build tree can be set at any time. When the user changes a setting in the tree, the change triggers a rebuild of the surface with the new settings. For example, the user has crossing breaklines and wants to use them both as surface data. Both breaklines appear in surface definition area, but only the first one currently affects the surface triangulation. To make the second (the crossing) breakline active, toggle Allow Crossing Breaklines to on and set how to resolve the elevation at the point of intersection.

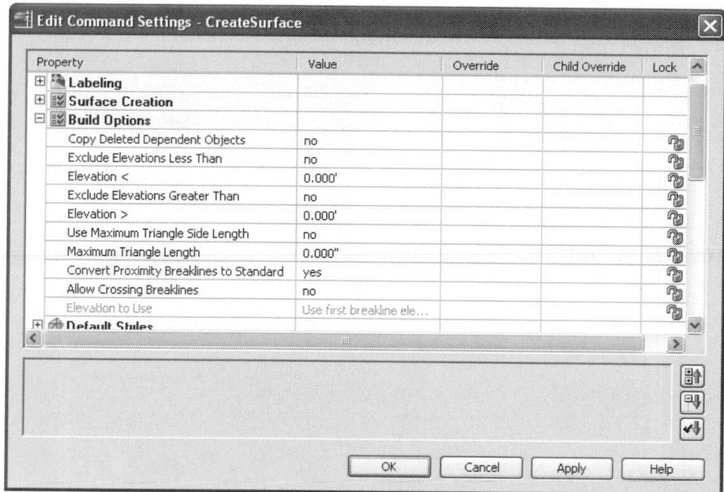

Figure 4.5

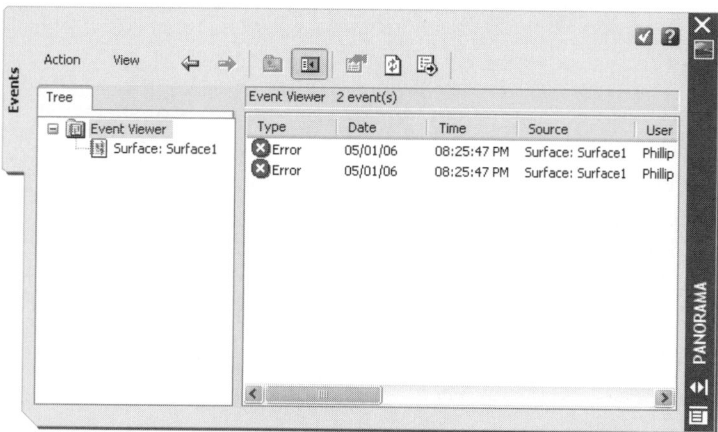

Figure 4.6

After the user sets the values and exits the Properties dialog box, the surface prompts for permission to rebuild the surface with the new settings.

Data Operations

The Data Operations section of the dialog box sets what types of surface data to include in building a surface. The intent of this section is to define what data is a part of a data set for a surface. However, this section can be a way of excluding a data type that is already a part of the surface data set. For example, the user has a surface with point groups, breaklines, and boundaries and wants to see what the surface triangulation would be like without any breakline data. He or she simply goes to this section of the Surface Properties dialog box and toggles off the use of breaklines. When the user exits the dialog box, the surface will rebuild without using any breakline data.

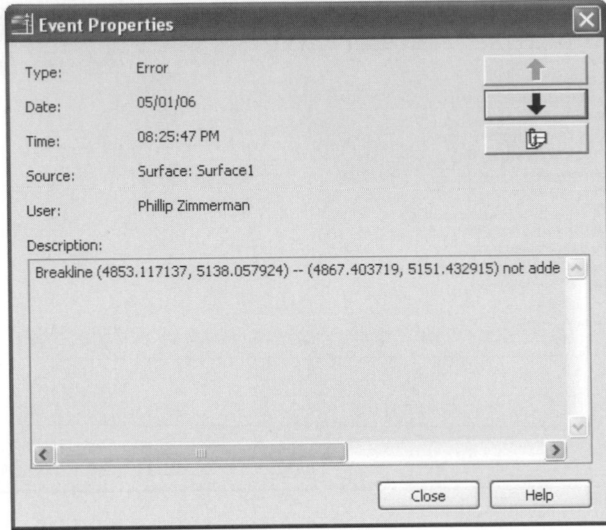

Figure 4.7

Edit Operations

The Edit Operations section of the dialog box sets what types of surface editing can be done on a surface. The intent of this section is to define what editing operations are allowed before they occur. However, the section can be a way of excluding edits types that are already a part of a surface. For example, the user has a surface with add point, delete line, and surface smoothing allowed and he or she wants to see what the surface triangulation would be like without the delete line edits. To do this, the user simply goes to this section of the Surface Properties dialog box and toggles off the use of delete line. When the user exits the dialog box, the surface will rebuild without using any delete line edits.

SURFACE STYLE

A surface object style determines how a surface displays its components and characteristics. A style showing border(s), triangulation, points, and other essential surface components is crucial to evaluating a surface under development (see Figure 4.8). After building a surface, other styles set range and display values for the analysis of surface slopes and elevations (see Figure 4.9). What a surface style displays is set by the values of the style's display panel.

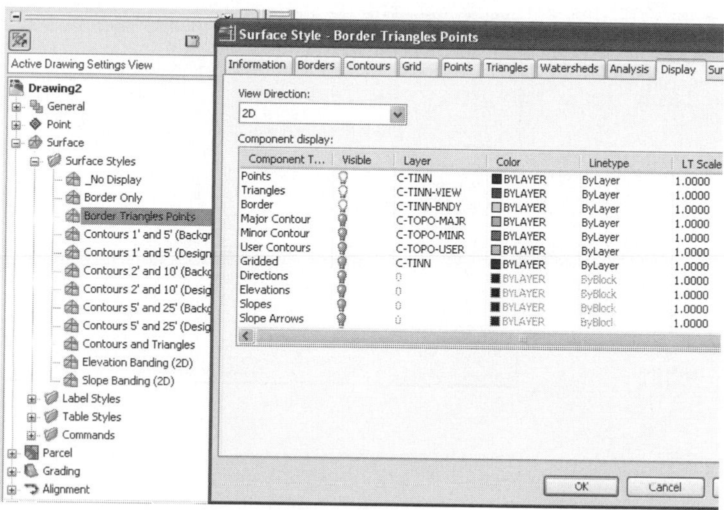

Figure 4.8

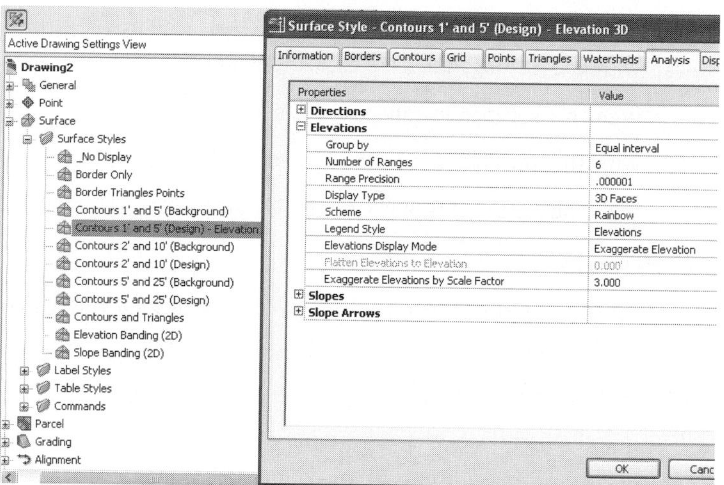

Figure 4.9

LABEL STYLES

There are four basic surface label styles: Contour, Slope, Spot Elevation, and Watershed (see Figure 4.10).

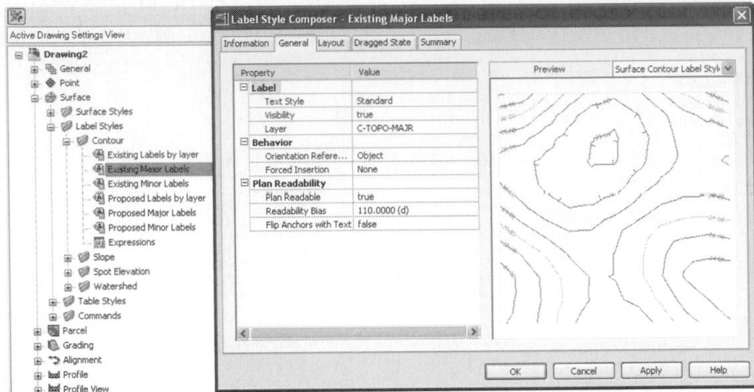

Figure 4.10

EXERCISE 4-1

When you complete this exercise, you will:

- Be familiar with the settings in Edit Drawing Settings.
- Be familiar with the settings in Edit Feature Settings.
- Be familiar with the settings in Edit Label Default Settings.
- Be familiar with the settings in Surface Object Styles.
- Be familiar with the settings in Surface Label Styles.

START A NEW DRAWING

1. If you are not in Civil 3D, start Civil 3D by double-clicking the Autodesk Civil 3D 2007 icon on the desktop.
2. Close and do not save *Drawing1*.
3. From the File menu, choose **New....**
4. Select the *Chapter 4 – Unit 1.dwt* template file from the CD that comes with this textbook (see Figure 4.11). The file is in the *Chapter 4* folder of the CD that accompanies this textbook.

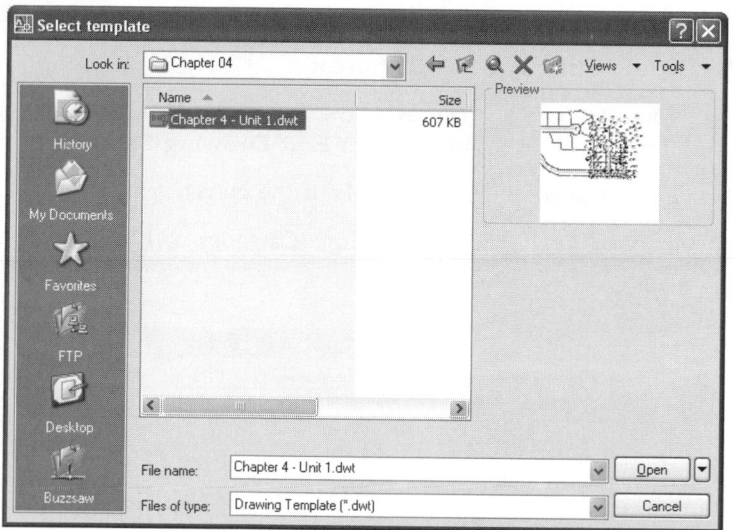

Figure 4.11

5. Using Windows Explorer create a folder, **Surface**, below the Civil 3D Projects folder.
6. From the File menu, choose **Save As...**, save the drawing to the **Surface** folder of Civil 3D Projects, and name the file **Surfaces** (see Figure 4.12).

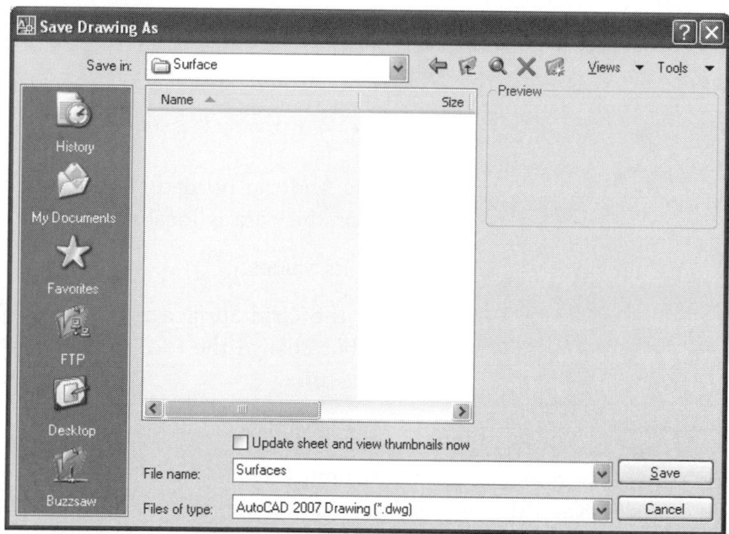

Figure 4.12

EDIT DRAWING SETTINGS DIALOG BOX: REVIEW AND SET VALUES

1. Click the **Settings** tab to make it current.
2. Click the drawing name, **Surfaces**, at the top of the tree, press the right mouse button, and from the shortcut menu select **Edit Drawing Settings....**
3. Click the **Units and Zone** tab to make it the current panel.
4. Set the scale to **1"=40'** and set the Zone Category to **No Datum, No Projection** (see Figure 4.13).

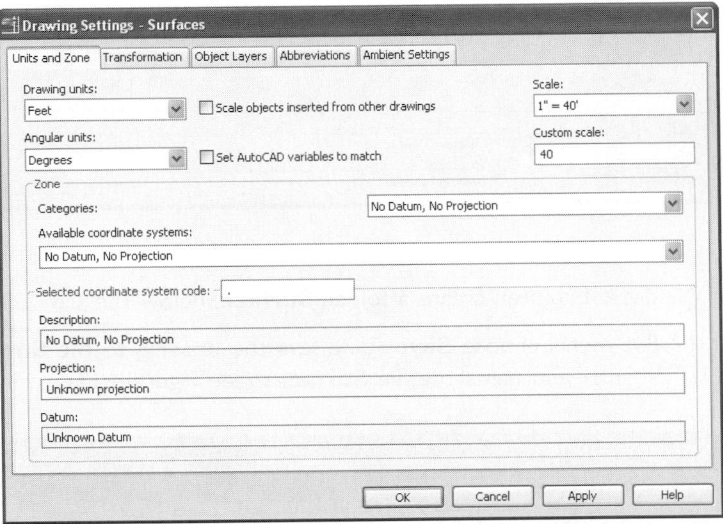

Figure 4.13

The scale setting sets the plotting scale, and the no datum and projection setting indicates that the coordinate system for the data is local coordinates.

5. Click the **Object Layers** tab to view its values.
6. Scroll down the layer list until viewing the Grid Surface and Grid Surface Labeling entries. Keep the default layer names, but change the Modifier by clicking in the cell and changing the value from None to **Suffix**.
7. Set the value of the Modifier to **-*** (a dash followed by an asterisk).
8. Continue scrolling down the layer list until viewing the TIN Surface and TIN Surface Labeling entries. Keep the default layer names, but change the Modifier by clicking in the cell and changing the value from None to **Suffix**.
9. Set the Value of the Modifier to **-*** (dash followed by an asterisk) (see Figure 4.14).

These steps set the modifier to a suffix that uses the name of the surface as part of the layer name. The dash separates the surface name from the root layer name, e.g. C-TOPO-EG.

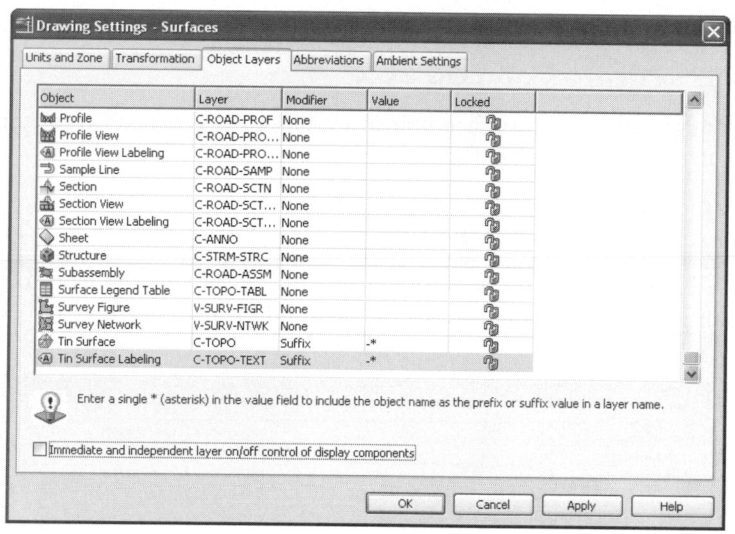

Figure 4.14

10. Click the **Ambient Settings** tab.
11. Select the expand tree icon (the plus sign to the left of the entry) next to Coordinate and Elevation to view their current settings (see Figure 4.15).

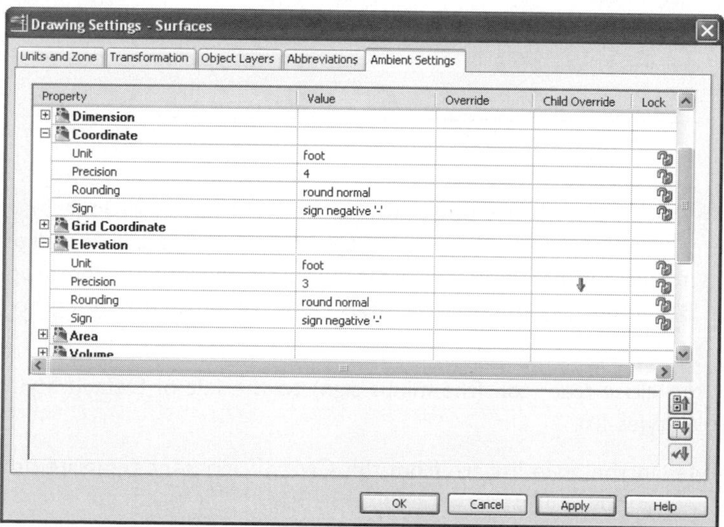

Figure 4.15

The settings for these entries control the display and reporting of coordinates and elevations when reviewing and/or editing point and surface data. The downward pointing arrow in the Child Override column indicates the precision for elevation is overridden by a lower style.

12. Click the **OK** button to close the Edit Drawing Settings dialog box.

EDIT FEATURE SETTINGS DIALOG BOX

1. In Settings, expand the surface branch by clicking the expand tree icon (plus sign) to the left of the surface heading.
2. Click the heading **Surface**, press the right mouse button, and from the shortcut menu select **Edit Feature Settings....**
3. In the Edit Feature Settings dialog box, click the expand tree icon to the left of Default Styles to view the assigned object and Label Styles (see Figure 4.16).

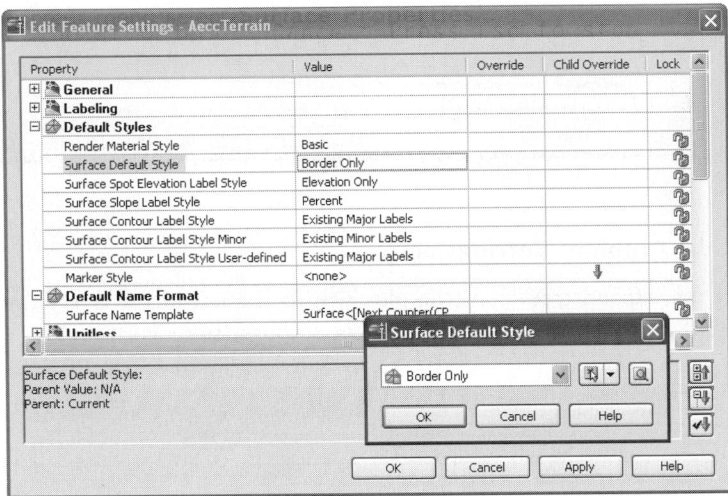

Figure 4.16

4. Change the Surface Default style to **Border Only**. This is done by clicking in the Value cell to the right of the Surface Default Style to display an ellipsis. Click the ellipsis to display the Surface Default Style dialog box. Click the drop list arrow, select **Border Only**, and click the **OK** button to exit the dialog box and assign the style.
5. Click the close tree icon (the minus sign) to the left of Default Styles to collapse the Default Styles list.

   All styles in the style list are from the Civil 3D content template. You can add, delete, or modify the styles to address your needs and standards.

6. Click the expand tree icon to the left of Default Name Format to view the naming convention for a surface.

   The default naming method is Surface1, Surface2, etc.

7. Click the collapse tree icon (the minus sign) to the left of Default Name Format to rollup the tree and click the **OK** button to exit the Edit Feature Settings dialog box.

   When creating a surface, you can change the default name.

COMMAND: CREATESURFACE

The command settings control the default values for new surfaces and set surface build parameters.

1. In Settings, expand the Surface branch until you are viewing a list of commands under the Commands heading.
2. Click on **CreateSurface**, press the right mouse button, and from the shortcut menu select **Edit Command Settings...**.
3. Click the expand tree icon (the plus sign) to the left of Surface Creation to view its values (see Figure 4.4).
4. The Default Surface Type is TIN surface.

   The grid spacing values and rotation apply only if selecting a grid surface as the default type.
5. Click the collapse tree icon to the left of Surface Creation to close the tree.
6. Expand the Build Options section to view its current settings (see Figure 4.5).

   This section contains critical surface building parameters. The settings and their values are discussed in later sections of this chapter.
7. Click the **OK** button to exit the Edit Command Settings dialog box.
8. Click the close tree icon (the minus sign) to the left of Commands to collapse the tree.

EXERCISES

SURFACE STYLE: BORDER & TRIANGLES & POINTS

A surface style for this exercise is Border & Triangles & Points. This style displays important surface components when building and reviewing surface quality.

The object style dialog box contains settings and values that affect how the surface shows its triangulation. Viewing the triangulation network allows you to decide whether additional data and/or breaklines are necessary, or if editing the surface will suffice. The Points tab defines how surface points display and the Triangles tab defines the geometry of the surface triangles. The Display tab sets the layer names, their properties, and what surface components display.

1. In the Settings, expand the Surface branch until viewing the styles under the Surface Styles heading.
2. Click the **Surface Styles heading**, press the right mouse button, and from the shortcut menu select **New...**.

Information Tab

3. Click the Information tab, enter **Border & Triangles & Points** as the style name, and give the style a short description.

4. Review the remaining values assigned to this panel.

   The information on this panel identifies the Name, Description, and the Created and Modified dates for the style (see Figure 4.17).

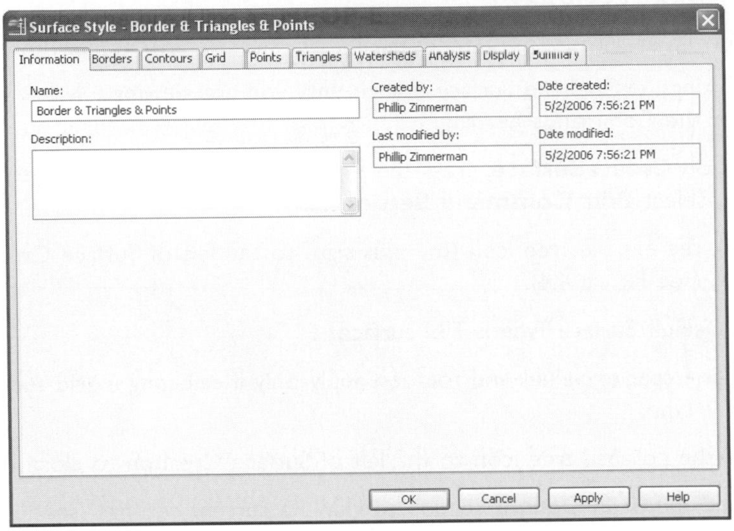

Figure 4.17

Points Tab

1. Click the **Points** tab to view the Points panel.

   When using this style, the 3D Geometry section defines how points represent surface elevations (they will be at their true Z-elevation). The Point Size section sets the size of the points (their size is set by the scale of the drawing). The Point Display section defines the shape and color of the three surface point types: Point Data (actual points from the drawing), Derived Points (smoothing points), and Non-Destructive Points (points on the surface from boundaries) (see Figure 4.18).

Triangles Tab

1. Click the **Triangles** tab to view its values.

   The value of the settings affects the elevations of the surface triangles.

2. If necessary, set the value to Use Surface Elevation (see Figure 4.19).

Display Tab

1. Click the **Display** tab to view its values.

   The Display panel sets the layers and their properties for all of the surface components and characteristics. The initial settings are for two-dimensional views (directly over the surface) of the surface border, triangles, and points. When clicking on the

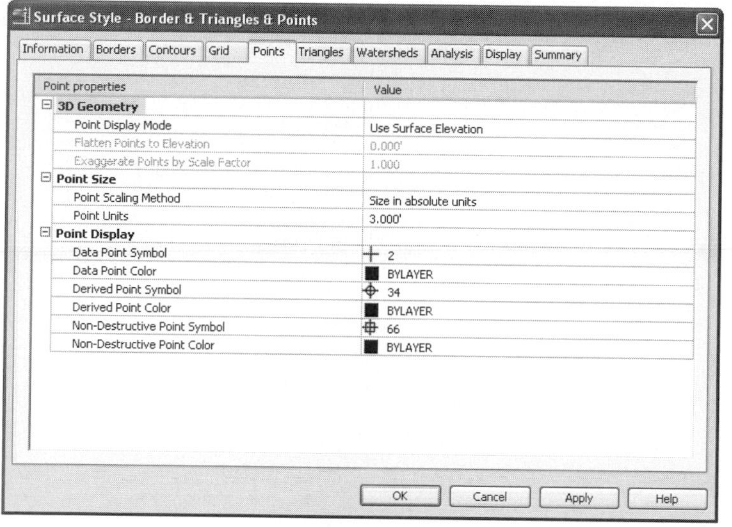

Figure 4.18

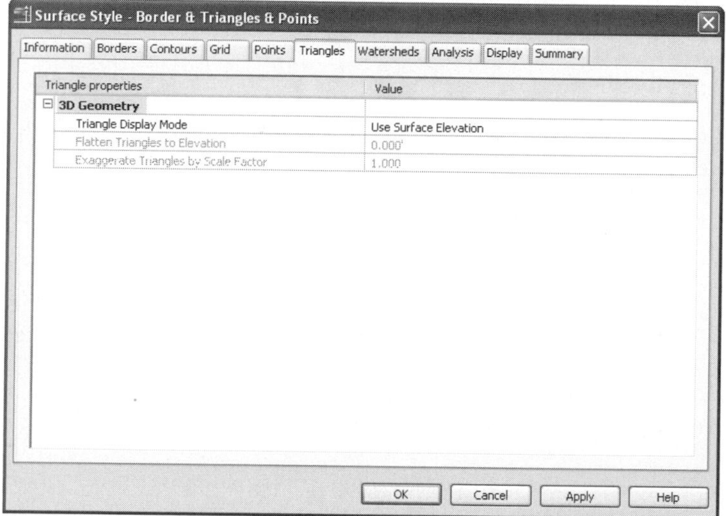

Figure 4.19

drop arrow in the View Direction list (top left), you change the settings list for three-dimensional views (see Figure 4.20).

For this surface style, the two-dimensional view displays the border, triangles, and points, and the three-dimensional view will show only the surface triangles.

9. In the Display panel, click the **View Direction** drop list arrow and change the View Direction to 3D.

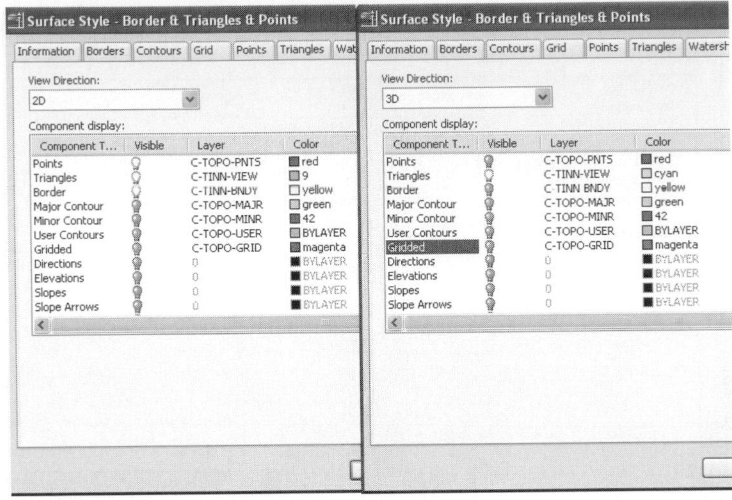

Figure 4.20

10. In the Display panel for 2D View, change the color for Triangles to color **9** and make Triangles and Points components visible.
11. Click the **OK** button to exit the Border & Triangles & Points style.
12. Click the close tree icon (the minus sign) to the left of Surface Styles to collapse the tree.
13. Click the **AutoCAD Save** icon to save the drawing.

SPOT ELEVATION LABEL STYLE - ELEVATION ONLY

The Spot Elevation label style, Elevation Only, defines a label annotating a surface's elevation. The drawing contains several additional labeling styles for contours, slopes, watersheds, and spot elevations.

1. Still in the Surface branch of Settings, click the expand tree icon to the left of Label Styles to view the list of label style types.
2. Click the expand tree icon to the left of Spot Elevation to view its label styles list.
3. Click **Elevation Only**, press the right mouse button, and from the shortcut menu select **Edit....**

   This displays the Label Style Composer dialog box for the style.

Information Tab

4. Click the **Information** tab to view its contents.

   This panel contains the Name and Description of the label style. The right side of the panel displays the creation and modified date for the style (see Figure 4.21).

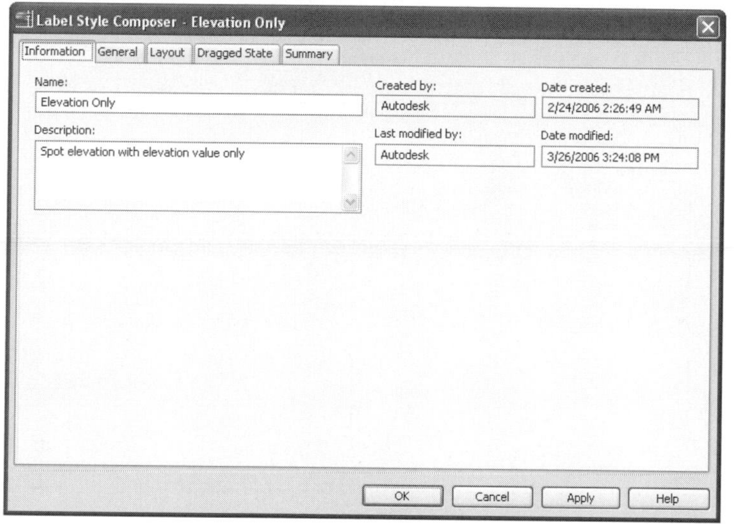

Figure 4.21

General Tab

5. Click the **General** tab to view its settings.
6. If necessary, click the drop list arrow at the upper-right of the panel and set the preview to **Surface Spot Elevation Label Style**.

   The general tab sets the overall visibility of this label style. If you change Visibility to false in the Label section of this panel, all labels using this style will not display in the drawing. The Orientation Reference sets how the label draws itself into the drawing. This setting allows you to create a label that uses the direction of the object as the 0 (zero) direction, displays horizontally in any view orientation (View), or displays its text relative to the World Coordinate System (WCS). The Plan Readability section defines how the Elevation Only label displays in any view (see Figure 4.22).

Layout Tab

7. Click the **Layout** tab to view its settings.

   The Layout panel defines how the label anchors itself in the drawing. For this label, it anchors on the <Feature> (surface) at the middle right of the selected point. The Border section defines if there is any box around the label text. The Text section defines the text height, rotation, and color of the label. The Attachment setting for the text is middle left justified to the anchor point of the feature (see Figure 4.23).

8. In the Text section, click in the Value cell to the right of Contents to display an ellipsis, and click the ellipsis to view the Text Component Editor for the value of the cell.

   The Text Component Editor shows the format string of the text labeling a surface's elevation. The right side of the dialog box displays the current format string and it is the sum of the modifiers and their values from the left side. For example, the top

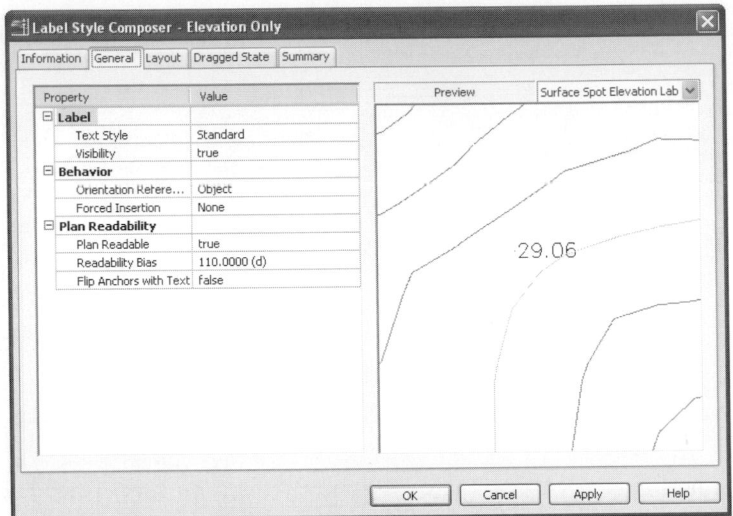

Figure 4.22

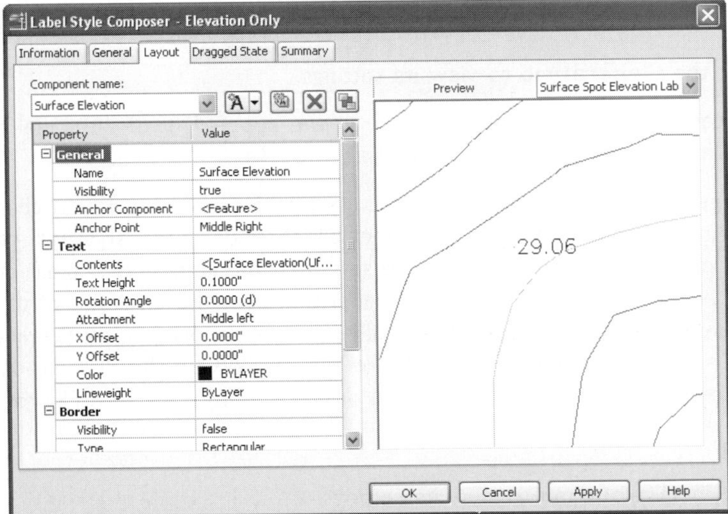

Figure 4.23

entry, Uft, indicates the units are feet and P2 sets the precision of the label to 2 decimal places. When selecting a string on the right, the left side shows each modifier and its value. You change values by clicking into a cell, clicking a drop list arrow, and selecting a new setting for the Modifier (see Figure 4.24).

9. Click the **OK** button to exit the Text Component Editor.

EXERCISES

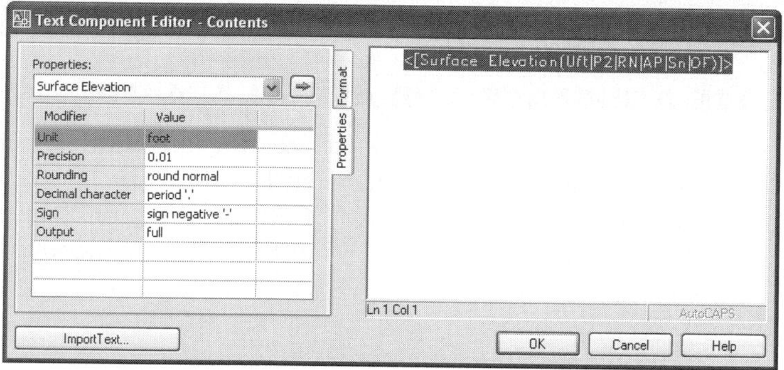

Figure 4.24

Dragged State Tab

10. Click the **Dragged State** tab to view the settings controlling the behavior of a label if it should be dragged from its initial location.

    If you drag a label from its original location, these settings define what happens to the label. In this label style, the dragged label will have a leader and the text will be stacked.

11. Click the **OK** button to close the Label Style Composer dialog box for the Elevation Only label style.

12. Click the **AutoCAD Save** icon to save the drawing.

EXERCISES

SUMMARY

- Surfaces are affected by settings in the Edit Drawing Settings and Edit Command Settings dialog boxes.
- A drawing with multiple surfaces should have prefixes or suffixes added to the base object layer name.
- Default surface names and styles are set in the Edit Feature Settings.
- The settings of a command allow a user to change the default surface styles.
- There are three style groups for surfaces: Object, Label, and Table

This completes the review and changing of values for surface settings and styles. The next step is creating the surface, adding data, and beginning the review process.

UNIT 2: SURFACE DATA

STEP 1: CREATE A SURFACE AND ASSIGN A NAME

The first step to creating a surface is to simply create and name a new surface object (to create a named surface instance). The Surfaces menu has the Create Surface... command, or select the Surfaces heading of Prospector, press the right mouse button, and select New... from the shortcut menu. Both Prospector and the Surfaces menu displays the same starting dialog box, Create Surface.

CREATE SURFACE

All of the values shown in the Create Surface dialog box are the result of the values in Edit Drawing Settings (surface layer) and Edit Feature Settings (name format and style assignments) (see Figure 4.25). The top left of the dialog box lists the default surface type. It can be changed to any one of the four surface types. The top right of the dialog box displays the surface layer (from Object Layers of Edit Drawing Settings). If the user sets a prefix or suffix for the layer in Edit Drawing Settings, the setting affects the Surface layer name after he or she enters the surface name.

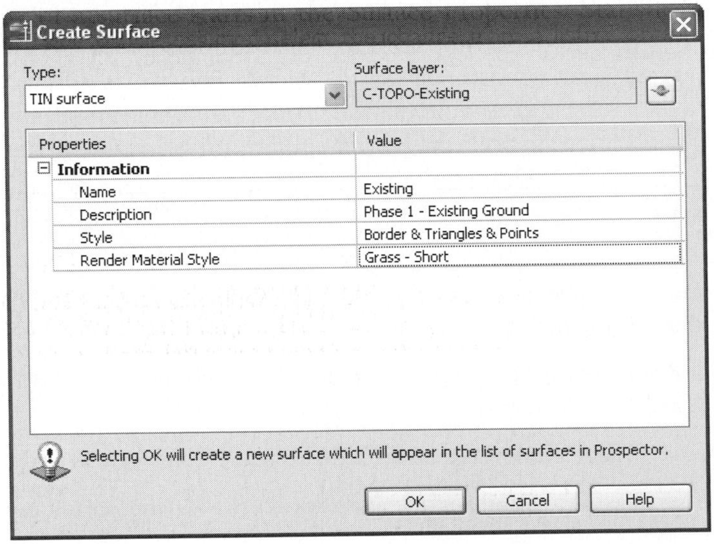

Figure 4.25

The body of the dialog box contains the name, description, and the initial object and rendering material styles. If the user wants to change the object and render styles, he or she calls a Select Style dialog box to change the current style.

When exiting the Create Surface dialog box, Civil 3D creates an entry in Prospector with the name of the surface under the Surfaces heading. The surface name is now at the head of a branch that has surface information and data. Surface information is surface masks and watersheds. The surface data, the Definition branch, lists the surface data types (points, point groups, boundaries, etc.) (see Figure 4.26).

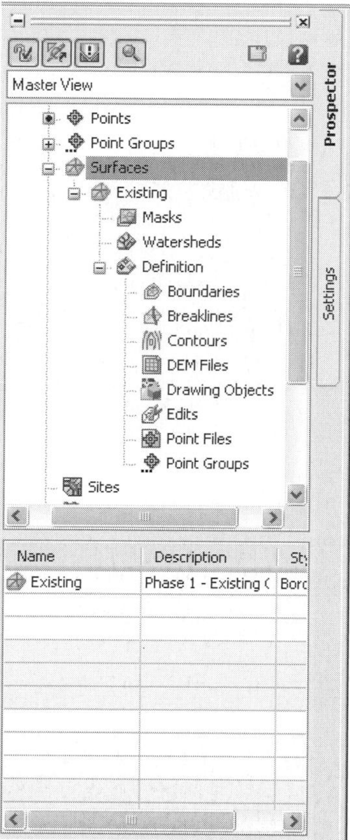

Figure 4.26

STEP 2: CREATE INITIAL DATA

The Definition branch of a surface contains a list of data types that you can add, delete, manage, and edit to create a surface's triangulation. Associated with each heading is a right mouse click that displays a shortcut menu containing tools available for that data type (see Figure 4.27).

BOUNDARIES

A surface can have three types of boundaries: Outer, Hide, and Show (see Figure 4.27). When a boundary crosses triangulation, there are two options on how to handle the crossed triangulation. The first option produces new triangles from the enclosed data to the boundary (i.e. Non-destructive breakline option). This new triangulation represents as best as possible the elevations of the original surface where the border intersects the triangle legs. If the boundary contains curves, there is a mid-ordinate setting that increases the

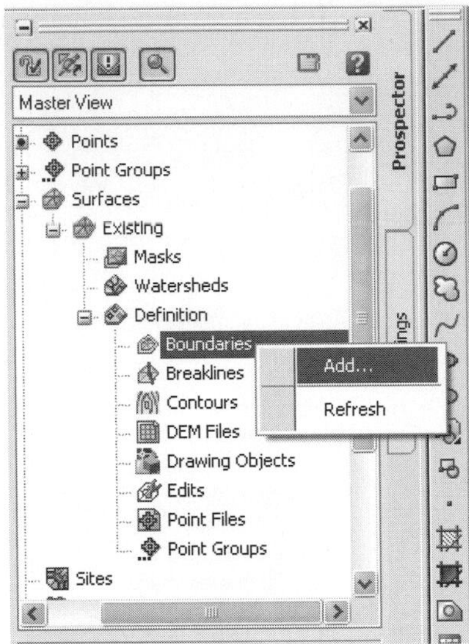

Figure 4.27

number of the new triangles along the boundary's curves. The second option is stopping the triangulation at the data. This option does not create triangles between the data and the boundary (see Figure 4.28).

Outer

An outer boundary forces a surface to define a new surface border that is smaller than the original border. The main use of an outer boundary is to limit the surface to a specific area of data (i.e. to exclude data outside of the closed polyline) (see Figure 4.28).

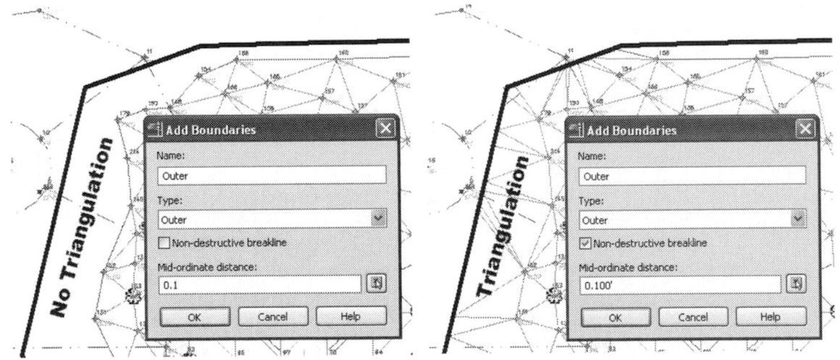

Figure 4.28

Hide

A hide boundary suppresses any surface triangulation within its boundary. The boundary does not need points in its interior to hide the triangulation (see Figure 4.29).

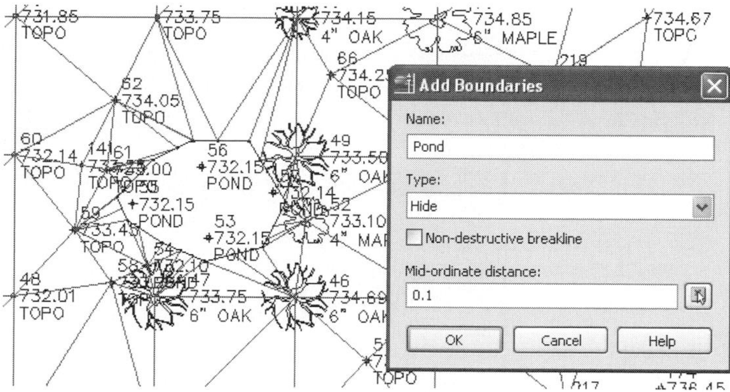

Figure 4.29

Show

A show boundary displays any surface triangulation within its boundary. The boundary does not need points in its interior to work. The main function of a show boundary is to display any triangulation within a hide boundary.

CONTOURS

Contours provide a convenient way of developing a surface. Since contours represent known elevations, the user needs only to draw them and then add them to the surface data. Surfaces using contour data do not have the same types of triangulation problems as with surfaces from points. Each contour line acts like a breakline and forces the triangulation between contours.

Caution: If you receive a contour drawing file that has been outside your control, you should thoroughly check the drawing and all the objects within it before using it as data.

Crossing Contours

A contour data set should not include crossing contours. Even though this occurs in reality, a Civil 3D surface cannot accommodate this situation. When assigning contour data that includes crossing contours, Civil 3D adds both contour lines to the data set, triangulates the surface based upon the first contour, issues an error message about the crossing contours, and ignores the data of the second contour. An error message in the Event View identifies the point of intersection of the two contours and it is up to the user to edit the contours so they correctly represent their elevations (see Figure 4.30). The contour editing may be as simple as grip editing the 3Dpolyline/contour data on the screen.

If there are several crossing contours, change the default build option from "do not allow crossing breaklines" to "allow crossing breaklines." If crossing breaklines are allowed, the surface needs to know which rule to use to include the data. The rules are as follows: to use the elevation of the first breakline (contour), use the elevation of the last breakline (contour), or to use the average elevation of the crossing breaklines (contours) at the point of intersection.

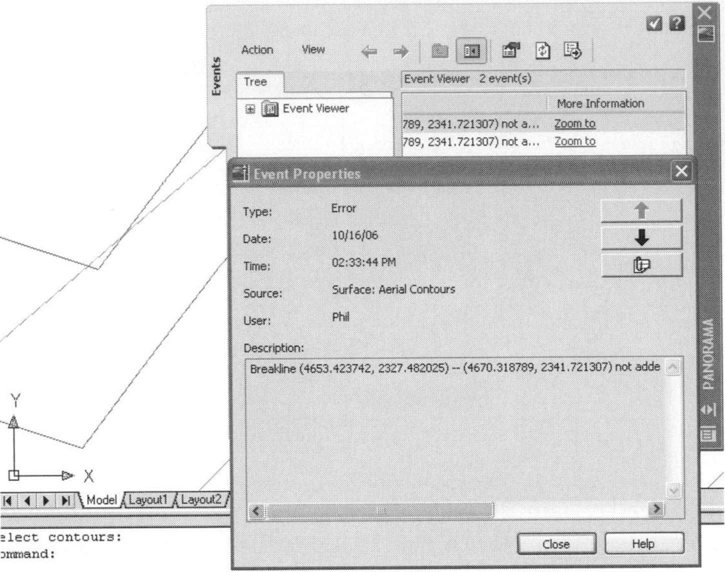

Figure 4.30

Weeding Factors

When adding contour data to a surface, the Add Contour Data dialog box contains settings that affect the amount of data to add to the surface (Weeding and Supplementing factors). Weeding analyzes distances and turning angles between straight contour segments. The Weeding factors reduce the amount data from contours by removing redundant vertices along an almost straight contour. Weeding does not exclude contour vertices that represent significant changes in direction. Weeding factors are always lower than the Supplementing factors (see Figure 4.31).

Weeding removes potentially redundant contour (polyline) vertices from the surface data. When reading the Weeding factors, you should read a Boolean "AND" between the two values (distance "AND" angle). To remove data (weeding), the vertices must be less than both weeding factors.

The first decision is distance. When evaluating the distance, it is the overall distance between three adjacent vertices that is evaluated. If the overall distance is less than the distance factor, the analysis moves to the second factor, angle (see Figure 4.32).

The angle factor is a right or left turning angle (deflection) made by the three points of the contour (see Figure 4.32). If the angle factor is 4 degrees, this limits the insignificant turning angle to 4 degrees or less to the right or left of the line from vertex 1 to 2. If the angle is less than 4 degrees, the second vertex is considered redundant and not included. If the angle is greater than 4 degrees, the second vertex is kept as data.

If the distance between three adjacent vertices is greater than the distance weeding factor, the routine moves on to find the next set of three vertices under the Weeding's distance factor.

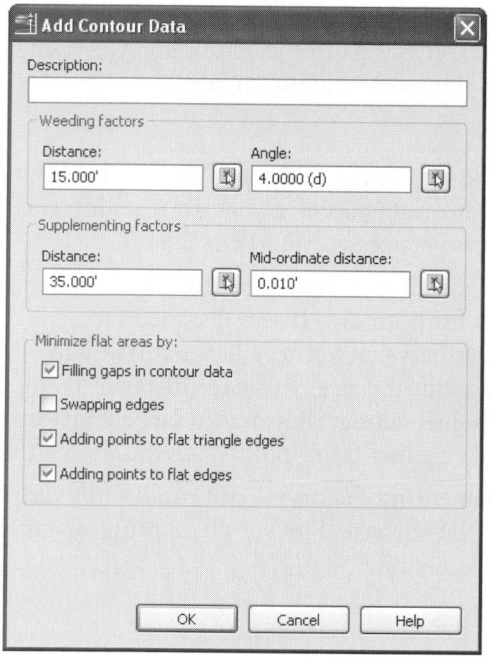

Add Contour Data

Description:

Weeding factors

Distance: 15.000'

Angle: 4.0000 (d)

Supplementing factors

Distance: 35.000'

Mid-ordinate distance: 0.010'

Minimize flat areas by:

☑ Filling gaps in contour data

☐ Swapping edges

☑ Adding points to flat triangle edges

☑ Adding points to flat edges

OK Cancel Help

Figure 4.31

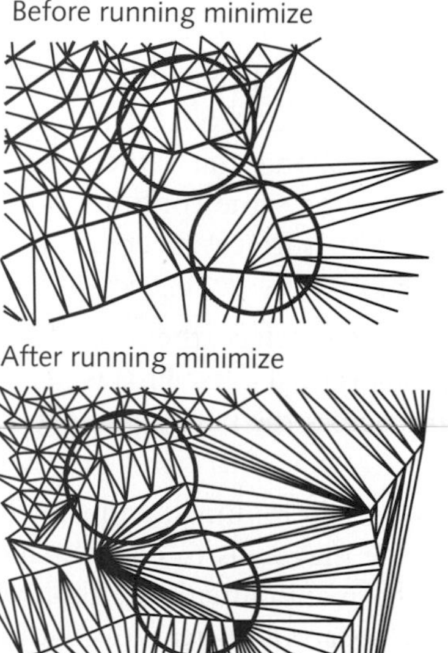

Figure 4.32

The weeding process does not change the objects in the drawing, but determines which vertices become data. How much data can be removed before encountering problems? Some experts suggest that a 50 percent reduction of vertices still produces a viable surface.

The number of vertices weeded is not the issue. What vertices remain after the weeding process is important. The only way to evaluate the effects of weeding is to review the resulting triangulation of the weeded contour data.

Supplementing factors

Supplementing factors add new data vertices to line and curve contour segments. If Add Contour Data does not encounter a vertex before reaching the distance or mid-ordinate distance value, it adds new vertices to the data.

The distance factor is the distance along a straight segment, and the mid-ordinate distance is the distance a chord is from an arc. If the distance factor is 20.0, any straight contour segment more than 20 feet between vertices has a vertex added along the contour as new data. If the segment is an arc, the mid-ordinate distance factor adds vertices to the data representing calculated points along the arc. A large mid-ordinate value creates fewer points and a smaller value creates more points for the arc in the surface data.

Like Weeding, the supplementing factors do not modify the original polylines or contours; they only create data for the surface. The supplementing distance factor can not be lower that the Weeding distance factor.

CONTOUR DATA ISSUES

There are two problems with contour data: bays and peninsulas and the lack of high and low elevations.

Bays and Peninsulas

Contours may switch back and as a result create peninsulas and bays (see Figure 4.33). With these loops in the contour data, a surface will switch from triangulating between different contours to triangulating along the same contour. This happens because the nearest triangulation point is along the same contour and not to the next higher or lower contour, or it is impossible for the surface to create a triangle between two different contours in the bay or peninsula.

When applying a contour style to the surface, the re-contouring will have contours that cut off the bays and peninsulas in the original data. This is because the contour follows the first triangle leg that makes the jump across the bay or peninsula. The result of this is a flat spot on the surface.

The net result of these errors is usually minimal, and they may even cancel each other out. Civil 3D attempts to optimize the diagonals to control this problem with the surface edit routine Minimize Flat Faces. In many situations, this routine corrects the triangle leg problems; however, this routine does not correct all of the flat spots (see Figure 4.33).

In Civil 3D 2007 a new algorithm for contour data solves the bay and peninsula problem. The algorithm looks at the surface elevation trend and adds point data that is slightly higher or lower than the contour. This allows a surface to correctly represent contour data and to be able to faithfully re-create the original contours (see Figure 4.34). This does create larger surface files, but is well worth the overhead.

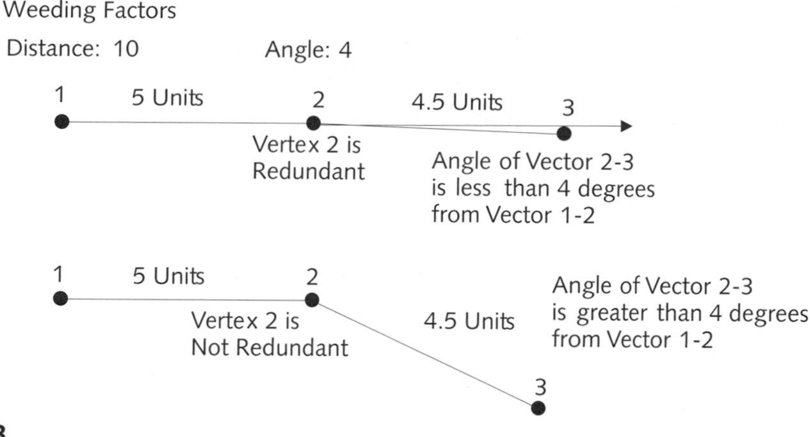

Figure 4.33

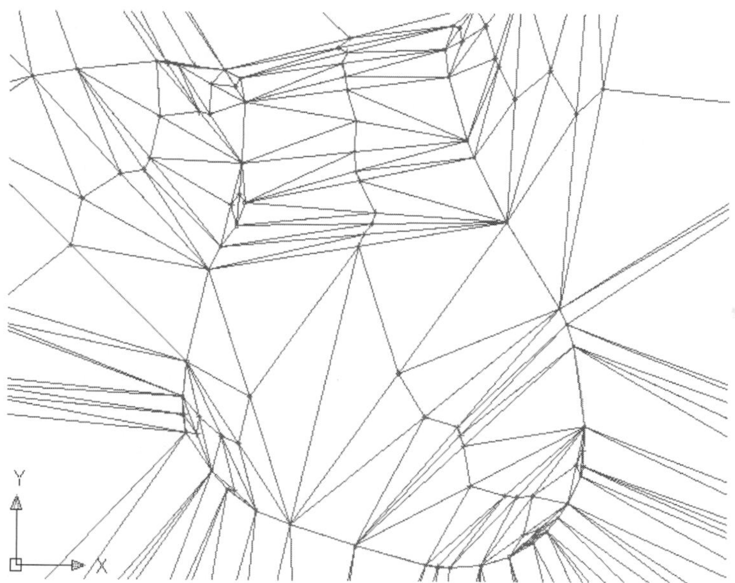

Figure 4.34

The settings on the lower portion of Figure 4.31 represent the correct values for creating optimal surface contour data.

High and Low Points

The second issue, the lack of high and low elevations, is creating point data representing these missing values. Most contour drawings from aerial firms contain text or blocks inserted at the elevation they label. In addition to text and blocks, point data can come from the following AutoCAD objects: nodes, 3D lines, and polymesh objects. Including these objects as additional surface data solves the problem of contour data not having high and low elevations.

DEM FILES

A DEM (Digital Elevation Model) is a USGS standard surface data file (see Figure 4.35). The file structure is a grid with elevations assigned to each grid intersection. The area of the DEM can be quite large and dwarf a project site. Also, the size of the cells makes a DEM good for showing a large area surface, but not very useful for showing elevations in a small site.

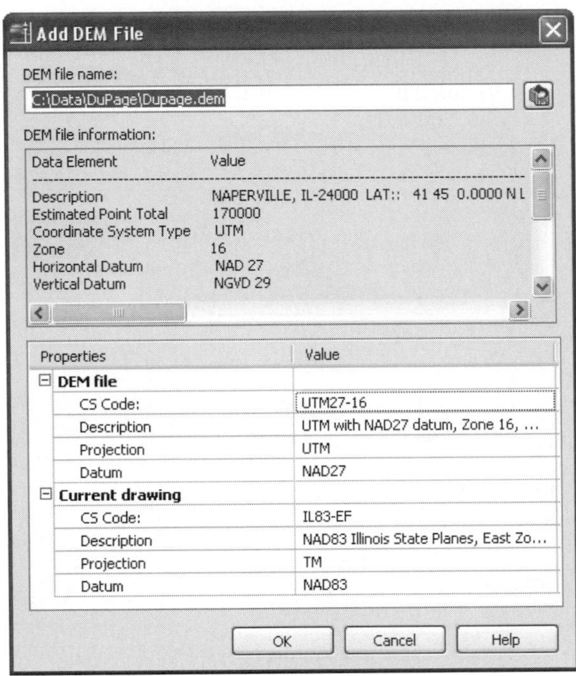

Figure 4.35

POINT DATA

Initial surface data can be from points. If the point data is in the drawing, Civil 3D requires a point group containing the points. External point data and AutoCAD entities can also be used as surface point data.

Drawing Objects

The last source for surface point data is AutoCAD objects (see Figure 4.36). When selecting objects with an elevation, this routine extracts the X, Y, and Z coordinates of each object and uses them as surface point data. The routine uses AutoCAD nodes, 3D lines, blocks, text, 3D faces, and polyface mesh objects. The point data created using this routine is independent of point file and point group data. The routine does not report or generate any connections between the endpoints and vertices of the selected objects. The exception to this is using 3D Faces and toggling on Maintain edges from objects. This preserves the edges of the 3D Faces in the surface.

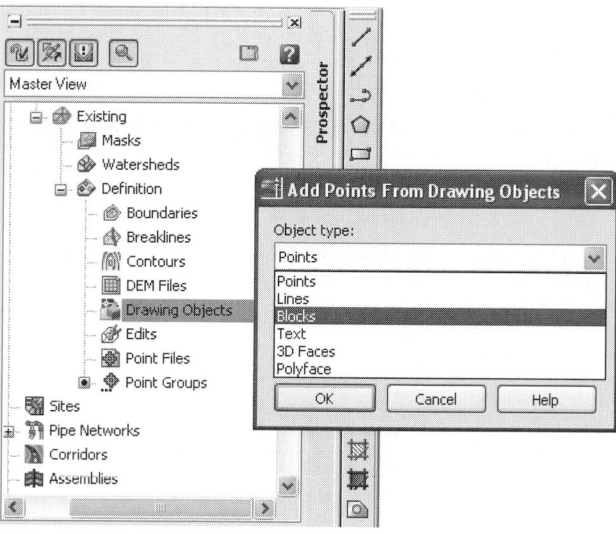

Figure 4.36

Point Files

A second source of point data is an external coordinate file. The point data imported by this process is not available to the drawing's point pool. The routine uses the same formats as the Import/Export command: the file can be ASCII text or an Access database. If the user wants the coordinate data to be an active part of the drawing, he or she can import the points using the Import/Export routine of the Create Points toolbar.

Point Groups

A point group can contain all or a selected number of points (see Figure 4.37). For information about creating and managing point groups, see the discussion on point groups in Chapter 2 of this textbook. Point groups allow the exclusion of points by number, elevation, description, or selection. For example, the point data contains elevations representing the tops of fire hydrants. These points could be anywhere from 1 to 3 feet above the surrounding ground elevations. If these points are included in the point data, they will create a hill in the surface at each location. Clearly, they should be excluded from the surface data set and removed from the point group that represents point data for the surface.

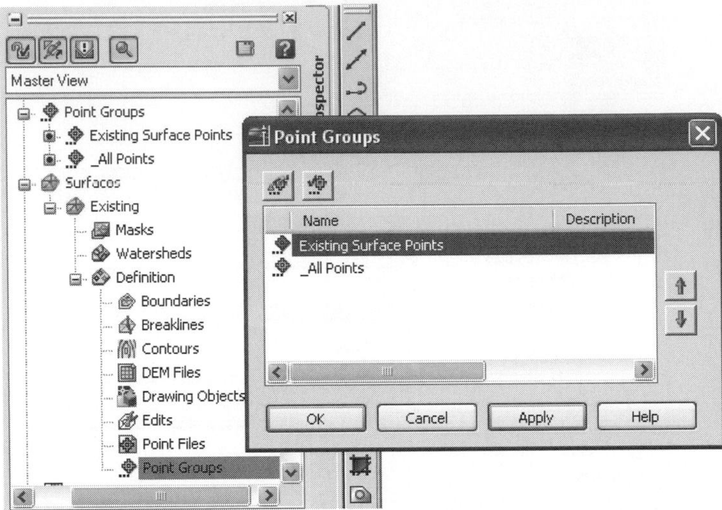

Figure 4.37

STEP 3: ADD BREAKLINES TO KNOWN PROBLEM AREAS

A surface contains many points that represent linear features. These features include manmade (edges-of-pavement, curbs, walls, etc.) and natural (swales, ditches, water edges, break in slope, etc.) features. Because of how Civil 3D builds a surface's triangulation, the triangulation will not correctly resolve these features. If the user wants to preserve the linear relationship between the points, he or she must create breaklines connecting the related points. Breaklines represent and preserve the linear trend of related surface points.

The user can draw polylines (2D or 3D) representing the linear connection between the points before adding them as breaklines. 2D or 3D breaklines can not have curves. The user can use feature lines as breakline data to support curves.

BREAKLINES

There are five types of breaklines, four of which control triangulation: Standard, Proximity, From File, and Wall (see Figure 4.38). Standard and wall breaklines are three-dimensional objects, and Proximity breaklines are two-dimensional objects. Even though the breaklines (Standard and Proximity) are different object types (2D and 3D), they produce the same effect on a surface. A wall breakline defines a face of a wall.

The fifth type of breakline, a Non-destructive breakline, fractures existing triangulation along its length. A Non-destructive breakline can be a two or three-dimensional object. Generally, Non-destructive breaklines are surface editing tools or are helpful when defining rendering areas.

On any surface, there are features that extend across several points; for example, swales, edges-of-pavement, backs of curbs, berms, etc. The elevation of each point along these features is related to the one before and after it. The only way to guarantee that the TIN triangulation correctly represents this relationship is to define a breakline that links the

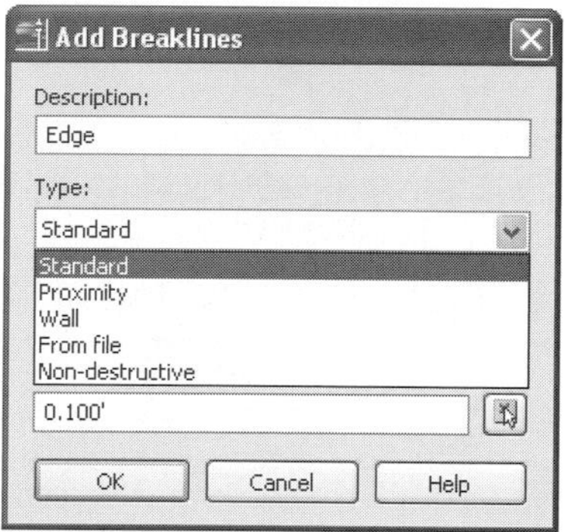

Figure 4.38

points. A breakline instructs a surface to place triangle legs linking the connected points and not to allow any triangle legs to pass across the breakline linking the points.

The misrepresentation of linear features occurs consistently in point data because of the algorithm used in creating surface triangulation. Point data must be dense enough along a linear feature to both describe and delineate it from all the other intervening points. In many surveys, this is not the case. The most difficult task for a field crew is to view their survey data as triangulated point data. They see their data as points within the context of reality. For them, it is hard to understand why a point 100 feet away is not a part of the roadway edge in a triangulated surface. In the field they see the roadway edge and they do not see the roadway edge as a series of points in a triangulated web. But an edge-of-pavement point and a similarly described point separated by 100 feet or more will lose their linear connection when creating the surface triangles. When the roadway edge points are mixed with several surrounding observed points not having the same description (trees, signs, road shoulders, manholes, etc.), the nearest neighbor to one roadway edge point is now a point of a different description (see Figure 4.39). The points surrounding the two edge-of-pavement shots (point numbers 17 and 18) may form triangles that cross between the two pavement shots and as a result negate the linear and elevation relationship between the pavement points. This same problem occurs in situations where data is too sparse to support a correct interpretation (see Figure 4.40). The situation shown in Figure 4.40 does not have enough data to correctly resolve the diagonal connection problem (between 719 or 721). Without more data or a breakline, the surface creates a diagonal that is always the shortest cross corner distance.

The solution to the above problems is to create additional control over the triangulation process. The added control preserves the linear and elevation trends between points or correctly controls the triangulation where data is sparse. Civil 3D uses breaklines for these situations. Breaklines imply a link between points and prevent the surface from triangulating across the line linking the points. The result is the correct interpretation of the linear relationships between points.

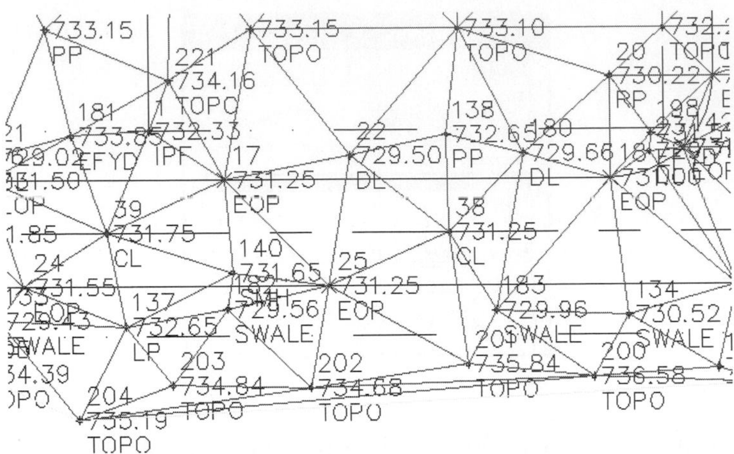

Figure 4.39

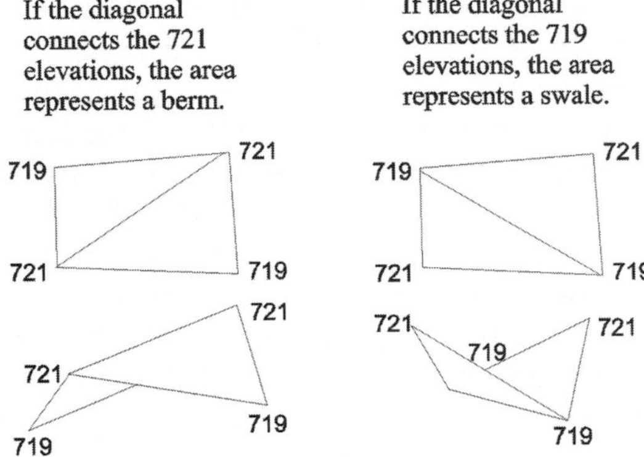

Figure 4.40

Drafting Breaklines

The Add Breakline routine expects existing 2D or 3D polylines or feature lines to define breaklines. So, the user must draw them before running the Add Breakline command. If the user is drawing a two-dimensional polyline, he or she needs to select near the point marker or use the point number ('PN) or point object ('PO) transparent command. After drafting the 2D polyline, the user can define it as a Proximity breakline.

If the user is drawing a three-dimensional polyline, he or she can use the AutoCAD 3Dpoly command. The problem with this command is it is unaware of Civil 3D point elevations that the user may want to use to create the polyline. To assign elevations to the three-dimensional polyline from Civil 3D points, either use point number ('PN) or select point object ('PO) transparent commands. By using an AutoCAD command with a Civil 3D transparent command, the elevations of the points become the vertex elevations of the resulting AutoCAD object.

The point filters are in the Transparent Commands toolbar. This toolbar allows the user to invoke any of the overrides when needed. When wanting to exit a transparent command, the user presses the Esc or Enter key, and return to the current AutoCAD command prompting.

Feature Line Arcs

The major issue with 2D and 3D polylines is that they cannot contain an arc. If the user wants to have breaklines with arcs, he or she needs to create a Feature Line object. When defining a feature line as a Standard breakline, there is an option to process its curves with a mid-ordinate value. The mid-ordinate value allows the user to control how closely the triangulation follows the arc. The smaller the mid-ordinate value, the closer the triangulation is to the original arc.

Allow Crossing Breaklines

The default surface build setting does not allow crossing breaklines. If the user is defining breaklines that cross other breaklines, the add breakline routine will add the new breakline to the surface data set, issue an error message about the crossing breakline, and does not include the breakline in the surface triangulation.

The user can allow crossing breaklines in a surface by changing the toggle in the Build section of the Definition panel of the surface's properties dialog box or in the settings of the CreateSurface command (see Figures 4.5 and 4.41). If the user sets Allow Crossing Breaklines to "Yes," he or she needs to tell the surface or command how to resolve the potential difference in elevation where the breaklines cross (intersect). Civil 3D gives three possible solutions to this condition. The first option resolves the issue by using the elevation of the first breakline at the intersection point. The second option resolves the issue by using the elevation of the last breakline at the intersection point. The third option resolves the elevation by taking the average of the elevations at the point of crossing (intersection).

Standard Breaklines

A standard breakline is a three-dimensional polyline. Since this polyline already contains elevations, it is not necessary to associate point objects with the polyline's vertices.

Standard Breaklines with Curves

The only three-dimensional drafting routine available in Civil 3D (3DPoly) only draws line segments. The best representation of an arc is a line with chord segments around it.

The best method of representing an arc is to use feature lines. Feature lines (from the Grading menu or the Feature Lines toolbar) are three-dimensional and can contain arc segments. When defining a feature line as a standard breakline, use the mid-ordinate value to better triangulate the arc's elevations. The lower the user sets the mid-ordinate value, the more data points along the path of the arc, resulting in surface triangulation that better represents the arc in the surface.

The Feature Lines toolbar Draw Feature line command draws feature lines with line and curve segments. While drafting the feature line, assign elevations to each vertex by typing in the elevation or using a point filter ('PO object or 'PN point number) from the

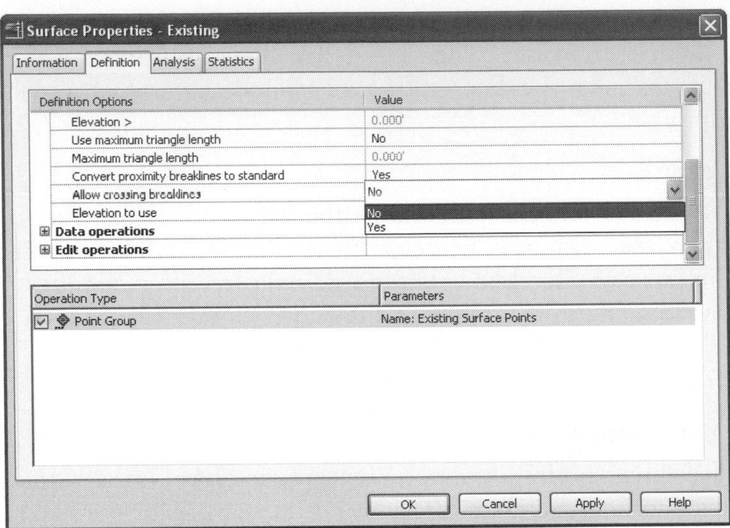

Figure 4.41

Transparent commands toolbar to use the elevation of a point as the elevation of the feature line vertex.

Proximity Breaklines

The Proximity breakline is a two-dimensional polyline. Since a proximity breakline is two-dimensional, it must have a point object at or near each polyline vertex. The point objects provide the proximity breakline with the necessary elevations and the polyline provides the line across which no triangle leg can pass. Proximity breaklines can not have arc segments.

Wall Breakline

A Wall breakline defines an (almost) sheer face from a three-dimensional polyline. The preexisting three-dimensional polyline is either the top or bottom of the wall. After describing the breakline in the Add breakline dialog box, select the polyline in the drawing. After selecting the polyline, specify the offset side by selecting a point to one side or the other of the initial polyline vertex.

The user should take care when selecting the offset side. He or she may think it has been done correctly; however, inspecting the headwall after completing the routine, may show that it is not correct. The user then needs to delete the headwall from the breakline list and redefine it.

After identifying the wall and the offset side, the routine prompts for a method of setting the offset side's elevations. The first method is all. This method prompts for a single value to add (positive or negative amount) to each vertex of the polyline to calculate the offset's elevation. The second method is individual. This method visits each vertex and prompts for the offset's elevation. At each vertex, specify an absolute elevation or a delta (a change in elevation).

The introduction of a wall into the surface will greatly change the slopes reported on the statistics panel of surface properties. Before adding the wall breakline, review the current minimum, maximum, and average slopes for the surface.

Non-destructive Breakline

A Non-destructive breakline fractures existing triangles along its length while still preserving the elevations found at the intersection of the breakline and the surface triangulation. The most common situation for this type of breakline is when deleting triangles to one side of the breakline. The breakline gives the remaining surface triangles a straight or clean boundary.

From File

Rather than drawing a breakline, create a file that defines any of the four breakline types. The file has to have the extension of .flt and contain the type of breakline, its name, coordinates, and elevations of a standard breakline. See the breakline file entry in Civil 3D's Help file for more information on this type of breakline.

FEATURE LINES

Although feature lines are in the Grading menu, they are powerful tools when defining breaklines or developing data for a surface design. The Feature Lines toolbar contains all of the tools that create or edit feature lines, and by using Quick Profile from the toolbar can evaluate a surface underdevelopment (see Figure 4.42). The Quick Profile tool does not require a feature line, but it is an essential tool when reviewing a surface under development.

Figure 4.42

This unit's discussion does not cover all of the icons of the Feature Lines toolbar. Instead, this unit reviews creating feature lines, using point filters from the Transparent Commands toolbar, editing feature line elevations, and using the Quick Profile to evaluate a surface. Unit 7 of this chapter covers the remaining commands of the toolbar.

Creating Figures

The first icon at the left of the Feature Lines toolbar draws a feature line. After selecting the draw feature line icon, a Create Feature Lines dialog box appears (see Figure 4.43). In this dialog box, assign a site, a feature line style, and a layer for the new feature line.

The user can accept or define a specific layer for the new feature line. After clicking the layer icon at the middle right of the dialog box, an Object Layer dialog box displays allowing the user to define a modifier and its value, or a new layer for the feature line (see Figure 4.44).

Figure 4.43

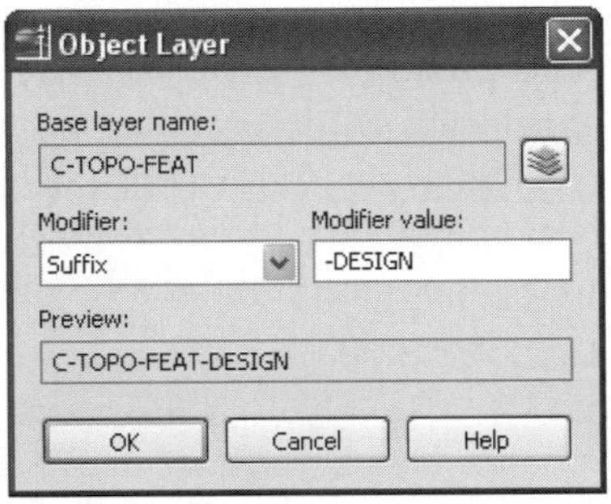

Figure 4.44

After selecting OK to exit the Create Feature Lines dialog box, the routine prompts for a feature line starting point. If the user selects a point at this prompt, the routine uses the current AutoCAD elevation (0.00) and prompts the user to accept or to edit its value. If the user isselecting the first point using a point number ('PN) or point object ('PO) filter, he or she can assign the elevation of the point to the feature line's vertex. After selecting the first point, he or she can draft a straight or arc segment. If the user is referencing point objects with the point filters, the user can set the elevation difference to 0.00 and select the remaining points. After creating the feature line, the user needs to edit its elevations to match the points selected to create the feature line.

Elevation Editor

After drafting the feature line, edit the feature line's elevations. To edit a feature line's elevations, select the line, press the right mouse button, and select Elevation Editor from the shortcut menu. This displays the Grading Elevation Editor in a vista of the panorama (see Figure 4.45). The user edits the elevations by clicking into the elevation cell for a vertex and changing the value. When done editing the elevations, a click on the "X" of the Panorama mast will hide it.

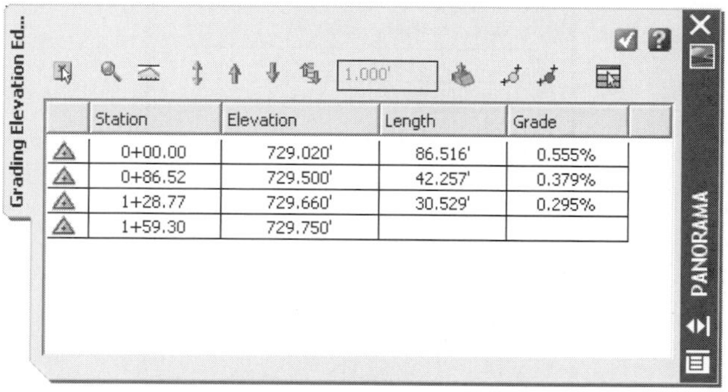

Station	Elevation	Length	Grade
0+00.00	729.020'	86.516'	0.555%
0+86.52	729.500'	42.257'	0.379%
1+28.77	729.660'	30.529'	0.295%
1+59.30	729.750'		

Figure 4.45

QUICK PROFILE

A quick profile is an indispensable tool when evaluating surface triangulation. It is difficult to create a mental image of the surface by simply looking at the triangulation. A quick profile allows the user to view a profile of a surface from selected drawing objects (lines, arcs, parcel lines, polylines, etc.). The user can move the objects that create the profile, and the profile and its profile view will update to show the new elevation along the line. The preferred object would be the AutoCAD polyline.

The Quick Profile command is on the Feature Lines toolbar (fifth icon in from the left). A quick profile is persistent only for the current drawing session. If exiting Civil 3D, the quick profile is discarded and the user will have to define it again when reentering the drawing editor.

When selecting the Quick Profile icon, select the object representing the quick section. After selecting the object, a Create Quick Profiles dialog box displays (see Figure 4.46). At the top of the dialog box, you can select all or any combination of drawing surfaces to be in the quick profile. The middle of the dialog box sets the profile view style. After setting the values for the quick profile and exiting the dialog box, the routine prompts for an insertion point and creates the quick profile (see Figure 4.47).

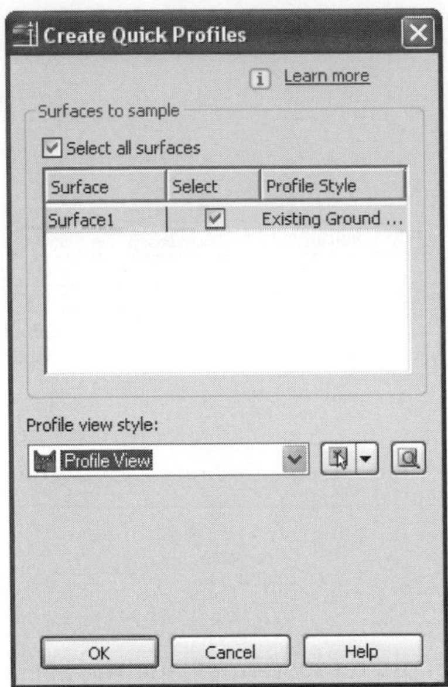

Figure 4.46

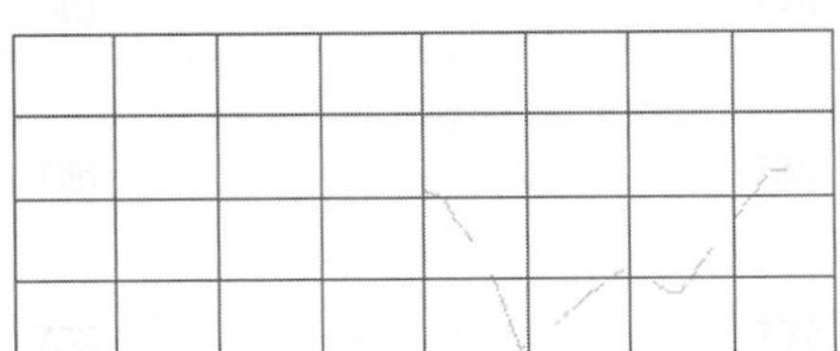

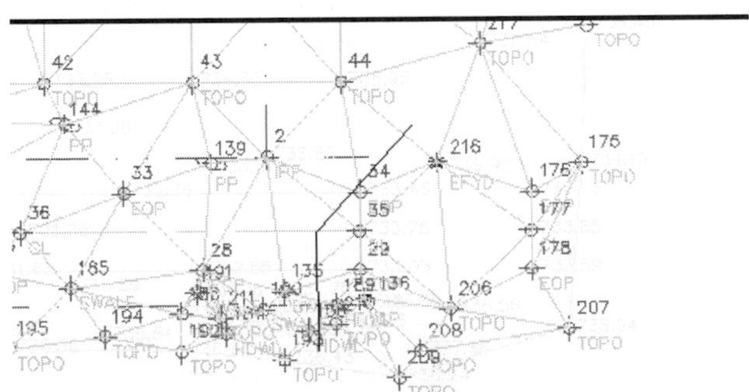

Figure 4.47

EXERCISE 4–2

When you complete this exercise, you will:

- Be able to create a surface.
- Assign surface data.
- Change assigned surface styles.
- Review the initial surface triangulation.
- Add breakline data.
- Be familiar with the Surface Properties dialog box.
- Be Familiar with the Surface Statistics panel of the Surface Properties dialog box.

EXERCISE SETUP

This exercise continues with the Surfaces drawing used in the previous unit's exercise. If you did not complete the previous exercise, you can open the *Chapter 4 - Unit 2* drawing that is in the *Chapter 4* folder of the CD that accompanies this textbook.

1. Open the Surfaces drawing from the previous exercise or open the *Chapter 4 – Unit 2.dwg* from the CD that accompanies this textbook.
2. You may need to zoom to the extents of the drawing if you do not see its contents.

CREATING A SURFACE

The first step is to create a new surface. Creating a new surface is done either from the Surfaces menu or from the shortcut menu of Surfaces in Prospector (See Figure 4.48).

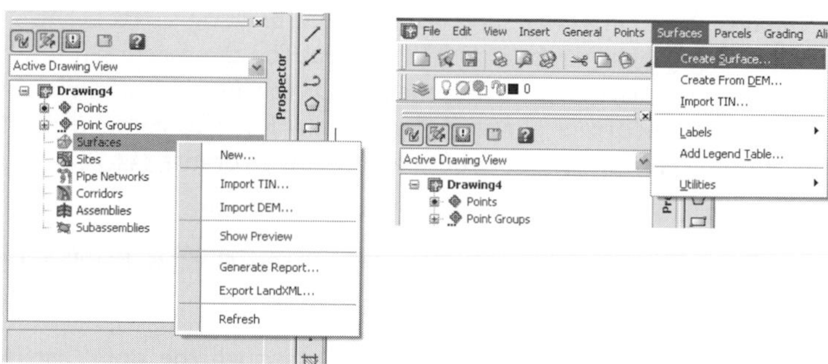

Figure 4.48

1. From the Surfaces menu select **Create Surface...** or in Prospector, click the Surfaces heading, press the right mouse button, and select **New...**. Either of these actions displays the Create Surface dialog box (see Figure 4.25).
2. In the Create Surface dialog box, enter the name of the surface, **Existing**, add the description of **Phase I – Existing Ground**, and make sure the surface type is set to TIN surface.

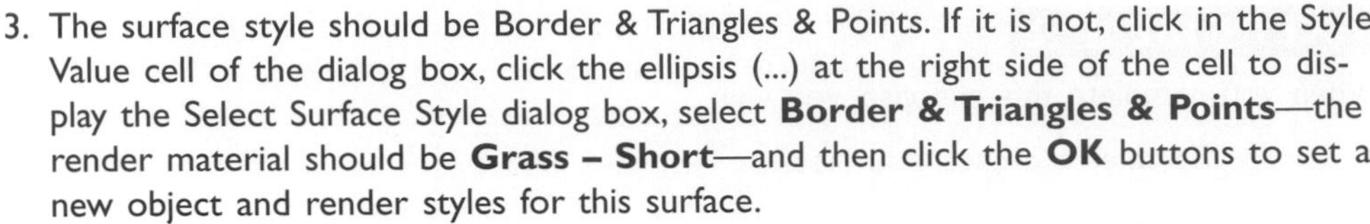

3. The surface style should be Border & Triangles & Points. If it is not, click in the Style Value cell of the dialog box, click the ellipsis (...) at the right side of the cell to display the Select Surface Style dialog box, select **Border & Triangles & Points**—the render material should be **Grass – Short**—and then click the **OK** buttons to set a new object and render styles for this surface.
4. Click the **OK** button and close the Create Surface dialog box.

   The Surfaces branch updates to show the new surface.
5. Click the expand branch icon to the left of the Surfaces heading to view the Existing instance.
6. Click the expand tree icon to the left of Existing to view the location of the surface masks, watersheds, and the definition heading.
7. Click the expand tree icon to the left of Definition to view the data tree for the surface. You may have to use the scroll bar to view all of the entries.

   Prospector lists all of the data types that can be a part of a surface in the Definition branch. When assigning type to a surface, Prospector will indicate the assignment with an icon.

CREATING POINT GROUPS

The next step is organizing the point data. This exercise creates three point groups. The first group contains the surface data points. The second point group represents points for possible surface breaklines. The last group hides all of the point markers and labels. This group allows you to view selected point groups when reviewing surface triangulation.

Existing Ground Point Group

This point group contains the points that are a data for the Existing surface. The group contains all of the points in the drawing (point numbers 1–228), except for the existing fire hydrant points. The survey crew locates fire hydrants by observing the top of the fire hydrant. By observing hydrants this way, the crew creates an incorrect elevation for each hydrant. To remove these false elevations from the surface, this point group excludes the existing fire hydrants (EFYD).

1. If Prospector is not current, click the **Prospector** tab.
2. Click the expand tree icon to the left of the Point Groups heading to view the current point group list.
3. Click the **_All Points** point group and scroll through the point list in the preview area (lower panel) of Prospector.

   The group contains point numbers 1–228, and existing fire hydrants are point numbers 149, 181, 198, and 216.
4. In Prospector, click the **Point Groups** heading, press the right mouse button, and from the shortcut menu select **New....**
5. Click on the **Information** tab and enter the name of the point group, **Existing Ground Points**.

6. Check the middle of the panel to make sure that Basic and Point#-Elevation-Description are the Point and Point Label styles.
7. Click on the **Include** tab, toggle on **With Numbers Matching**, and set the point numbers to the range of **1–228** (see Figure 4.49).

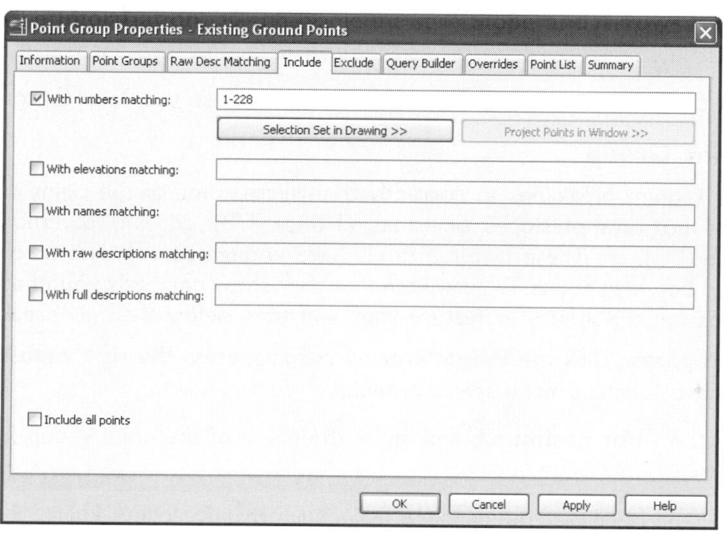

Figure 4.49

8. Click the **Exclude** tab, toggle on **With Raw Descriptions Matching**, and enter **EFYD** (see Figure 4.50).

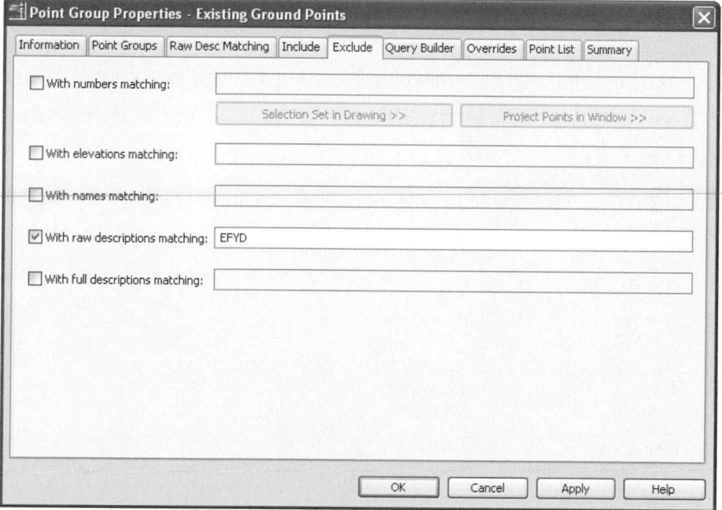

Figure 4.50

9. Click the **OK** button to create the Existing Ground Points point group.

10. Select the **Existing Ground Points** point group, press the right mouse button, and from the shortcut menu select **Edit Points...**.

    This displays a Point Editor vista containing all of the points in the point group.

11. Review the member points in the vista. Click the **Raw Description** heading to sort the list and scroll to view the absence of existing fire hydrants (EFYD).

12. Click the "**X**" on the mast of the Panorama to close the Point Editor vista.

Breakline Point Group

The surface may require breaklines to correctly triangulate some of the point data. These points include tops and bottoms of slopes or banks (T/S or T/B), edges-of-pavement (EOP), swales (SWALE), ditches (DL), etc. These features rarely have enough survey data to clearly show their linear relationship between points. To guarantee that they show correctly, you need to isolate these points, draft some initial polylines or feature lines, and then define them as breaklines.

1. In Prospector, click the **Point Groups** heading, press the right mouse button, and from the shortcut menu select **New...**.

2. Click the **Information** tab and enter the name of the point group, **Breakline Points**.

3. Check the middle of the panel to make sure that Basic and Point#-Elevation-Description Point and Point Label styles are set.

4. Click the **Raw Desc Matching** tab and toggle on **CL***, **DL**, **EOP**, **HDWL**, and **SWALE** raw descriptions (see Figure 4.51).

5. Click the **OK** button to create the Breakline Points point group.

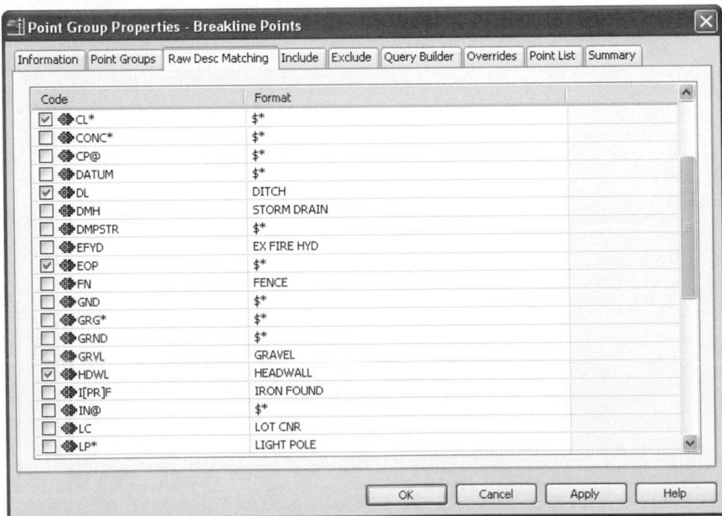

Figure 4.51

No Show Point Group

There are times when you need to view only a small number of points. You can turn off the appropriate layers or create a point group that hides the points you don't want to see. In the current drawing, you will define a point group that does not have a point or label style. When placing this point group at the top of the point group display list, all of the points will disappear from the screen. The display order list is a property of Prospector's Point Groups heading. When changing the order of the point groups in the display list, you control the visibility of points on the screen.

The No Show point group is a copy of the _All Points point group with None as the point and label style.

18. In Prospector, click the **Point Groups** heading, press the right mouse button, and from the shortcut menu select **New....** Click the Information tab and enter the name of the point group, **No Show**.
19. Click in the middle of the panel and change the Point style to **<none>** and change Point Label style to **<none>**.
20. Click the **Include** tab and toggle on **Include All Points** (at the bottom left of the panel).
21. Click the **Overrides** tab and toggle on the Point and Label Style overrides.
22. Click the **OK** button to create the No Show point group.

EXERCISES

ASSIGNING SURFACE DATA

Assigning data to a surface is simply selecting the type of data from the surface definition branch, pressing the right mouse button, and from the shortcut menu selecting **Add....**

1. In Prospector, expand the Existing surface branch until you are viewing the Definition heading and its branch.
2. In the Definition branch, click the **Point Groups** heading, press the right mouse button, and from the shortcut menu select **Add....**
3. Select **Existing Ground Points** from the point group list and click the **OK** button to close the dialog box and add the points to the surface (see Figure 4.52).

   Civil 3D immediately builds the surface from the newly assigned data.

REVIEW INITIAL SURFACE

1. Zoom and pan around the site to view the surface triangulation.
2. In Prospector, click the **Point Groups** heading, press the right mouse button, and from the shortcut menu select **Properties....**
3. In the Point Groups dialog box, select **Breakline Points**, and move the point group to the top of the list by clicking the up arrow on the right side of the dialog box.
4. Click the **No Show** point group and using the up or down arrow icon on the right, move it to the second position on the list (see Figure 4.53).

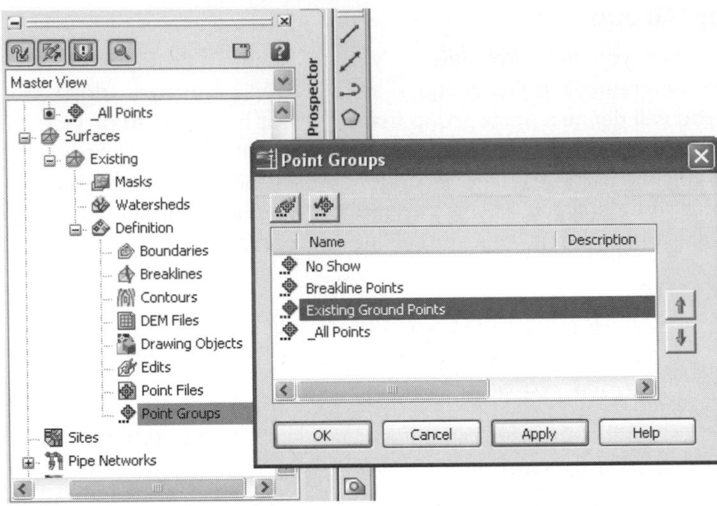

Figure 4.52

5. Click the **OK** button to close the dialog box and to change the display order for point groups. You may need to **Regenall** the drawing to see the display update.

   Changing the display order may cause the Existing surface to go out-of-date.

6. In Prospector, click the **Existing** surface name, press the right mouse button, and from the shortcut menu select **Rebuild – Automatic**.

7. Open the Layer Properties Manager dialog box, create a new layer **Existing-Breakline**, and make it the current layer.

8. Click the **AutoCAD Save** icon to save the drawing.

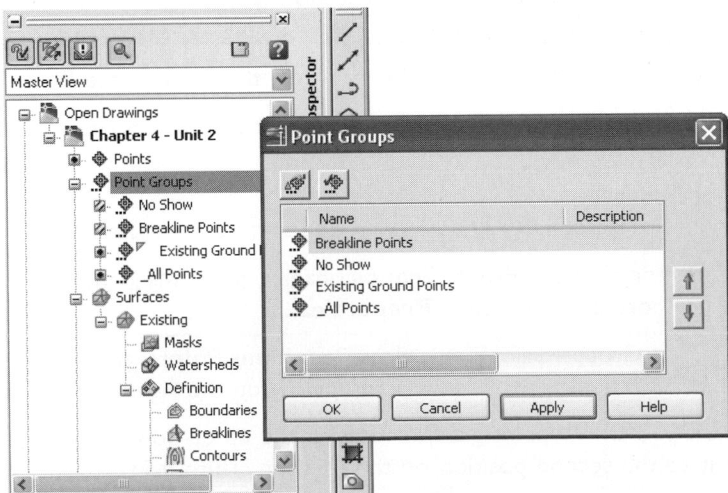

Figure 4.53

The current display order of the point groups hides all of the points in the drawing except for the last drawn (top of the list) point group (Breakline Points).

9. Use the AutoCAD **Zoom** command to view the southern portion of the site and review the triangulation for the EOP, DITCH, CL, and SWALE points.

   The initial surface triangulation does not correctly connect the points representing the north and south edges-of-pavement, the centerline, ditch, or swale (south of the roadway). These points need breaklines to correctly triangulate their elevation trend.

DITCH BREAKLINE

The ditch breakline consists of points 21, 22, 180, and 23 and is located just to the north of the western side of the road (see Figure 4.54).

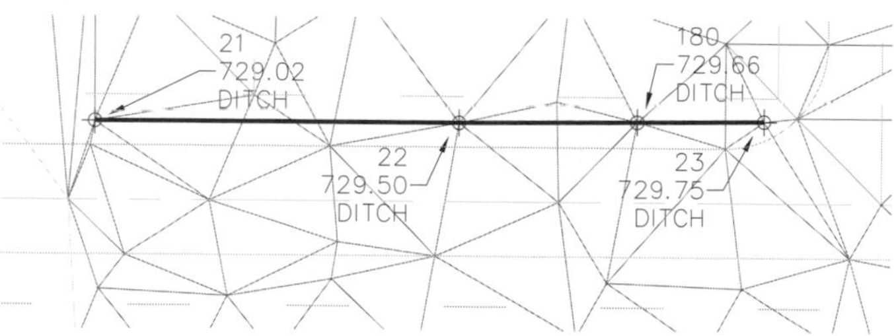

Figure 4.54

1. Zoom into the area shown in Figure 4.54.
2. Check to see if the Transparent Commands toolbar is showing on the screen. If not, toggle it **on** so you can use the toolbar.
3. Start the AutoCAD **3Dpoly** command, click either the point number or point object transparent command, and enter or select points **21, 22, 180, and 23**.

   The following are the responses from the 3Dpoly command using the point number transparent command.

```
Command: 3dpoly
Specify start point of polyline: '_PN
>>Enter point number: 21
Resuming 3DPOLY command.
Specify start point of polyline: (4962.18 5043.75 729.02)
Specify endpoint of line or [Undo]:
>>Enter point number: 22
Resuming 3DPOLY command.
Specify endpoint of line or [Undo]: (5048.7 5043.75 729.5)
Specify endpoint of line or [Undo]:
>>Enter point number: 180
Resuming 3DPOLY command.
Specify endpoint of line or [Undo]: (5090.95 5044.15 729.66)
Specify endpoint of line or [Close/Undo]:
>>Enter point number: 23
Resuming 3DPOLY command.
```

```
Specify endpoint of line or [Close/Undo]: (5121.48 5044.44
    729.75)
Specify endpoint of line or [Close/Undo]:
>>Enter point number: *Cancel* <Press the ESC key to exit
    transparent command>
Specify endpoint of line or [Close/Undo]:
Resuming 3DPOLY command.
Specify endpoint of line or [Close/Undo]: <Press the ENTER key
    to exit>
Command:
```

The command echoes back the coordinates and elevations of the selected points.

4. After drafting the 3D polyline, press the **Esc** key to exit the transparent command, and press the **Enter** key to exit the AutoCAD 3Dpoly command.
5. In the Definition branch for the Existing surface, click the **Breaklines** heading, press the right mouse button, and from the shortcut menu select **Add....**
6. In the Add Breaklines dialog box, enter **Ditch** as the description of the breakline, set the type to **Standard**, and click the **OK** button to start defining the breakline (see Figure 4.55).
7. Select the three-dimensional polyline representing the ditch and press the **Enter** key.
8. If Civil 3D issues a warning about duplicate points in the surface, close the warning panorama and view the new triangulation.

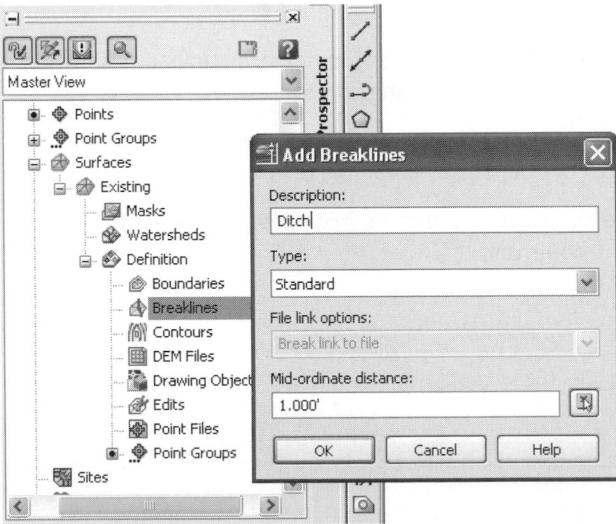

Figure 4.55

After adding the breakline, the surface updates its definition list and then rebuilds the triangulation to correctly resolve the ditch as a part of the surface.

9. In the Existing surface Definition branch, click the expand tree icon to the left of the Breaklines heading to view the Ditch entry.
10. Click the **Ditch** entry in the list, press the right mouse button, and from the shortcut menu select **Properties....**
11. Review the coordinate and elevation values of the Ditch breakline.
12. Close the **Breakline Properties** vista by selecting the "**X**" at the top of the Panorama's mast.

SWALE BREAKLINE

The second breakline is the swale just south of the roadway. The swale starts near the eastern end and extends all the way to the western edge of the surface data. The points representing the swale are point numbers (east to west) 190, 135, 191, 185, 184, 134, 183, 182, and 133 (see Figure 4.56).

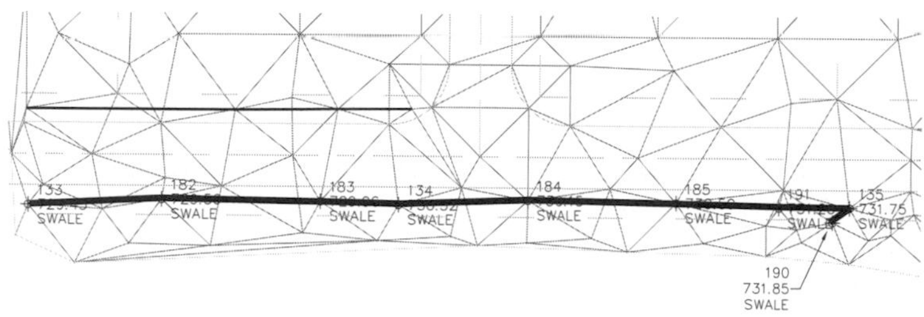

Figure 4.56

1. Zoom into the area shown in Figure 4.-51 so you can see the swale points.
2. Start the AutoCAD 3Dpoly command, click either the point number or the point object transparent command and enter or select points **190, 135, 191, 185, 184, 134, 183, 182, and 133**.
3. When you are done drawing the polyline, press the **Esc** key to exit the transparent command, and press the **Enter** key to exit the AutoCAD 3Dpoly command.
4. In the Existing Definition branch, click **Breaklines**, press the right mouse button, and from the shortcut menu select **Add....**
5. In the Add Breaklines dialog box, enter **Swale** as the description of the breakline, set the type to **Standard**, and click the **OK** button to start defining the breakline.
6. Select the three-dimensional polyline representing the Swale and press the **Enter** key.
7. If Civil 3D issues a warning about duplicate points in the surface, close the Event Viewer to view the new triangulation.

   The surface updates the triangulation to correct resolve the swale as a part of the surface.

8. If necessary, in the Existing surface Definition area of Prospector, click the expand tree icon to the left of the Breaklines heading and notice the Swale breakline is on the breakline list.
9. Click the **Swale** entry, press the right mouse button, and from the shortcut menu select **Properties....**
10. Review the coordinate and elevation values of the Swale breakline.
11. Close the **Breakline Properties** vista by selecting the "**X**" at the top of the Panorama's mast.
12. Click the **AutoCAD Save** icon to save the drawing.

The review of the triangulation does not necessarily give you a good idea of what the resulting surface looks like. One additional tool to view the effects of the breaklines and edits on a surface is the Quick Profile command of the Feature Lines toolbar.

EVALUATE SURFACE ELEVATIONS BY PROFILE

The review of a surface includes looking at the triangulation along the road at the southern end of the site. Few, if any, triangles correctly connected the ditch at the north edge of the road and the swale along the south edge-of-pavement. As a result of this initial review, two breaklines were added to the surface. Even with these two breaklines, the triangulation representing the roadway is wrong. One triangle even connects a ditch point (180) north of the road to a swale point (183) south of the road (see Figure 4.57).

It is one thing to see the triangulation, but another to understand what the triangles mean as elevations or relief on the surface. Quick Profile displays elevations along an object as a profile. This portion of the exercise creates a Quick Profile view to review the surface and the effects of adding breaklines to a surface.

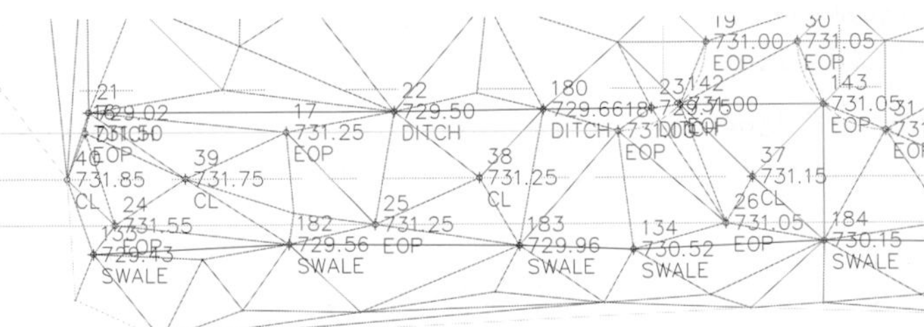

Figure 4.57

1. From the Viewports flyout menu of the View menu, select **2 Viewports** and split the screen **Horizontally**.
2. Make the top viewport the current viewport by clicking in it, and pan the site to the left until you have a clear screen.
3. Click in the lower viewport to make it the current viewport and use the AutoCAD Pan and Zoom commands to view the western half of the roadway (see Figure 4.58).

4. In the lower viewport, start the AutoCAD **Pline** command and draw a pline from south of the swale (near point number 133) to a point north of the ditch (past point 21) at the western end of the surface (see Figure 4.58).
5. Select the **Quick Profile** icon from the Feature Lines toolbar, select the polyline you just drew, and the Create Quick Profiles dialog box displays. (You can also select the polyline, right click, and from the menu select Quick Profile...).
6. Click the **OK** button to accept the defaults and to continue to drawing the profile.
7. Click in the top viewport to make it the current viewport and select a point in the lower-left of the upper viewpoint to select the profile view origin.
8. A Quick Profile appears in the upper viewport.
9. Civil 3D causes the Event Viewer to display to remind you that the profile is only temporary.
10. Click the "**X**" to close the Event Viewer.
11. Click the **AutoCAD Save** icon to save the drawing.

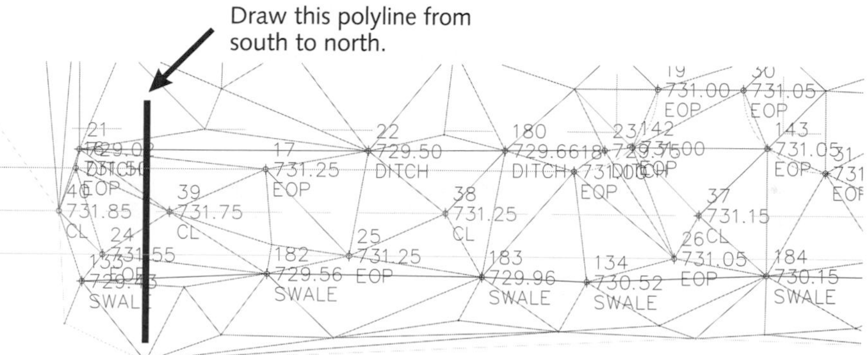
Draw this polyline from south to north.

Figure 4.58

Every time you relocate the profile segment, the quick profile updates with the new elevations. What you should see in the profile is a ditch or swale with a poorly defined road section. Because of the method of triangulation, points representing linear features rarely triangulate correctly. You must use breaklines to make these linear features (that spread over several points) appear correctly in the surface triangulation.

12. Relocate the profile segment to view various locations along the road.
13. Finally, relocate the profile segment back to its original position and press the **Esc** key to deselect the segment.
14. In the Viewports flyout menu of the View menu, select the **Named Viewports...** command.

15. Click the **New Viewports** tab and enter the name of **A and P**.
16. Click the **OK** button to exit the dialog box.
17. With the lower viewport active from the Viewports flyout menu of the View menu, select **1 Viewport**.

    This resets the drawing a single viewport.

18. Click the **AutoCAD Save** icon to save the drawing.
19. Use the AutoCAD Zoom and Pan command to view the roadway from its western end to just east of the intersection.

ROADWAY BREAKLINES

Next is adding four roadway breaklines to the surface data. Again, the Add Breakline routine expects existing objects.

The first two breaklines will be from feature lines, and the last two simply repeat the steps you used to create and define the ditch and swale breaklines. The reason for using feature lines is the north edges-of-pavement (NEOP-W and NEOP-E) have curve return segments. When defining breaklines from feature lines containing arcs, you add data to the surface by processing the curves with a Mid-ordiante value.

You will need to edit the elevations for the NEOP-W and NEOP-E feature lines.

Table 4.1 contains the breakline names, their point numbers, and their breaklines type.

Table 4.1

Name	Point Numbers	Type of Breakline
NEOP-W	16, 17, 18, 142, and 19	Standard from Feature Line object
NEOP-E	176, 34, 33, 31, 143, and 34	Standard from Feature Line object
CL	177, 35, 36, 37, 38, 39, and 40	Standard
SEOP	24, 25, 26, 27, 28, 29, and 178	Standard

1. Select the Draw Feature Line icon from the Feature Lines toolbar.

   This displays the Create Feature Lines dialog box.

2. Set the values of your dialog box to match those of Figure 4.59 and click the **OK** button.
3. From the Transparent commands toolbar, select the Point Object override ('PO), and identify the points listed in Table 1 for the NEOP-W breakline. Make sure you draw an arc from point 18 to 142 and from 142 to 19. You will want to zoom in to better view the points. There needs to be a vertex at each point on the arc. Press the **Esc** key to toggle off the transparent command and press the **Enter** key to exit the AutoCAD command.

   When selecting the first point object, the routine echoes back the elevation of point 16.

EXERCISES

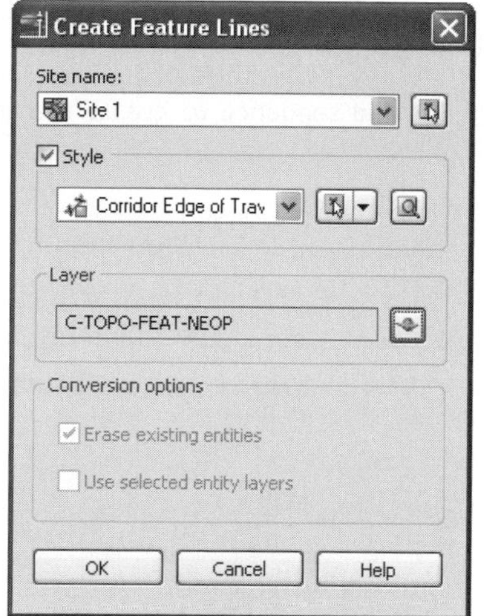

Figure 4.59

4. Press the **Enter** key to accept the elevation.

   When selecting the second point, the prompt may not be for elevation.

5. You need to press the **Esc** key to stop the override, enter the letter "**e**", press the **Enter** key to use the elevation from the point, and then reset the point object override by selecting **'PO** from the Transparent Commands toolbar.

   The following segment shows the prompting for the start of the command.

```
Command:
Specify start point: '_PO Start override
>>
Select point object: Select Point 16
Resuming DRAWFEATURELINE command.
Specify start point: (4961.15 5038.0 731.5)
Specify elevation or [Surface] <731.50>: Press ENTER
Specify the next point or [Arc]:
>>
Select point object:
Resuming DRAWFEATURELINE command.
Specify the next point or [Arc]: (5018.14 5038.0 731.25)
Distance 56.991', Grade -0.44, Slope -227.96:1, Elevation 731.
    250'
Specify grade or [SLope/Elevation/Difference/SUrface] <0.00>: e
    (To set feature line to Elevation)
Specify elevation or [Grade/SLope/Difference/SUrface] <731.250>:
(press the ENTER key)
Specify the next point or [Arc/Length/Undo]:
>>
Select point object: Select Point 17
```

6. After selecting and adding point 18, press the **Esc** key to exit the Point Object override, set the Arc option, and toggle on the Point Object filter from the Transparent

Commands toolbar. You are now ready to select points 142 and 19 to create an arc and to use their elevation on the feature line.

The following is the command sequence to create the arc segment of the feature line.

```
Select point object:
Resuming DRAWFEATURELINE command.
Specify the next point or [Arc/Length/Undo]: (5112.0 5038.0
    731.0) Press the Enter key
Distance 93.857', Grade -0.27, Slope -375.43:1, Elevation 731.
    000'
Specify elevation or [Grade/SLope/Difference/SUrface] <731.000>:
Specify the next point or [Arc/Length/Close/Undo]:
>>
Select point object: *Cancel* Press Esc to stop override
>>
Specify the next point or [Arc/Length/Close/Undo]:
Resuming DRAWFEATURELINE command.
Specify the next point or [Arc/Length/Close/Undo]: a (for Arc)
Specify arc end point or [Radius/Secondpnt/Line/Close/Undo]: '_PO
>>
Select point object: Select Point 142
Resuming DRAWFEATURELINE command.
Specify arc end point or [Radius/Secondpnt/Line/Close/Undo]:
    (5129.68 5045.32
731.0)
Distance 19.635', Grade 0.00, Slope Horizontal, Elevation
    731.000'
Specify elevation or [Grade/SLope/Difference/SUrface] <731.000>:
(Press the ENTER key)
Specify arc end point or [Radius/Secondpnt/Line/Close/Undo]:
>>
Select point object: Select 19
Resuming DRAWFEATURELINE command.
Specify arc end point or [Radius/Secondpnt/Line/Close/Undo]:
    (5137.0 5063.0
731.0)
Distance 19.635', Grade 0.00, Slope Horizontal, Elevation 731.
    000'
Specify elevation or [Grade/SLope/Difference/SUrface] <731.000>:
(Press the ENTER key)
Specify arc end point or [Radius/Secondpnt/Line/Close/Undo]:
>>
Select point object: *Cancel* (Press Esc to cancel)
>>
Specify arc end point or [Radius/Secondpnt/Line/Close/Undo]:
Resuming DRAWFEATURELINE command.
Specify arc end point or [Radius/Secondpnt/Line/Close/Undo]:
    Press the Enter key
Command:
```

7. Use the AutoCAD Pan command to view the eastern portion of the north edge-of-pavement.

EXERCISES

8. Repeat Steps 1–3 and draft the NEOP-E feature line. Use the same layer, C-TOPO-FEAT-NEOP. Start the feature line at point 176. Make sure you draw an arc from point 31 to 143 and from 143 to 30. There needs to be a vertex at each point on the arc segment.

EDIT THE ELEVATIONS OF THE FEATURE LINES

1. Select the NEOP-E feature line, press the right mouse button, and from the shortcut menu select **Elevation Editor....**
2. Compare the elevations of the feature to those in Table 4.2, and if necessary correct their elevations.

Table 4.2

Point	Elevation
176	733.75
34	733.45
33	732.35
31	731.05
143	731.05
30	731.05

3. Click on the "**X**" to close the editor vista.
4. Repeat Step 1 and select the NEOP-W feature line. If necessary, correct the elevations using those listed in Table 4.3.

Table 4.3

Point	Elevation
16	731.50
17	731.25
18	731.00
142	731.00
19	731.00

EXERCISES

ADD FEATURE LINES AS BREAKLINES

1. In the Existing surface Definition branch, click the **Breaklines** heading, press the right mouse button, and from the shortcut menu select **Add....**

2. In the Add Breaklines dialog box, name the breakline **NEOP-W,** set the type to **Standard**, set the Mid-ordinate value to **0.1**, click the **OK** button to exit, and select the western-most feature line.

3. Repeat Steps 1–2 for the eastern feature line using the same values found in Step 2, but describe the breakline as **NEOP-E**.

4. If Civil 3D issues a warning about duplicate points in the surface, close the Event Viewer and view the new triangulation.

   Each time you add a new breakline, the surface rebuilds and the triangulation contains additional data around the curve return arc segments.

EXERCISES

Draw and Define the CL and SEOP Breaklines

5. Draw the CL and SEOP with the AutoCAD 3Dpoly command using the Point Object transparent command to reference the points and their elevations.

6. After drafting the 3Dpolylines, define them as standard breaklines, and assign each one an appropriate name.

VIEW ROADWAY QUICK PROFILES

1. Select the **Named Viewports...** command from the Viewports flyout menu of the View menu.

2. Click the **Named Viewports** tab, select the **A and P** named viewport, and click the **OK** button to exit the dialog box.

3. Use the AutoCAD Zoom and Pan commands in the lower viewport to view the roadway (lower viewport).

4. Use the AutoCAD Zoom and Pan command in the upper viewport to view the Quick Profile. If it is not there, pan the screen until you are viewing an empty area of the drawing.

5. Select **Quick Profile** from the Feature Lines toolbar, select the quick profile polyline you drew earlier, click the **OK** button to close the Create Quick Profile dialog box, and select an insertion point for the profile in the upper viewport.

6. After reviewing the road, move the quick profile polyline's southern end to a point just south of point 188 and its northern end just south of the lot line north of the roadway.

   The headwall is just south of the road between points 186–189.

7. If necessary, make the lower viewport active. In the Viewports flyout menu of the View menu select the **1 Viewport** command to return the screen to a single viewport.

8. In Prospector, click the **Points** heading to preview the points.

9. In Prospector's preview area, click point **188**, press the right mouse button, and select **Zoom To**.

10. Click the **AutoCAD Save** icon to save the drawing.

WALL BREAKLINE

The last breakline is a wall breakline. It starts as a 3Dpolyline representing the top or bottom of the wall. The offset side for this breakline is north towards the road.

The introduction of a wall into the surface will greatly change the slopes reported on the statistics panel of surface properties. Before you add the wall breakline, review the current minimum, maximum, and average slopes for the Existing surface.

1. In the drawing, click the surface border or a triangle, press the right mouse button, and from the shortcut menu select **Surface Properties....**
2. Click the **Statistics** tab and then click the expand tree icon for the Extended (statistics) tree for the surface. You screen should look similar to Figure 4.60. The average slope is around 6 percent, the lowest slope is zero, and the highest is just over 200 percent. The average and maximum slope will change greatly after defining the wall breakline.

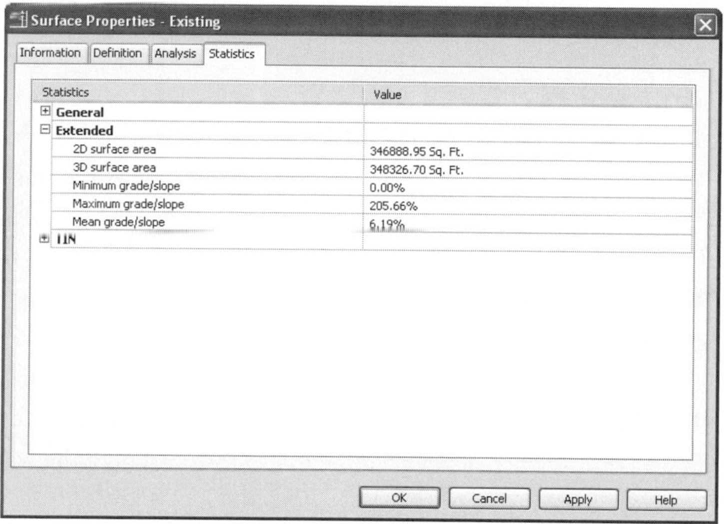

Figure 4.60

3. Click the **OK** button to exit the Surface Properties dialog box.
4. Make a new point group, **Headwall**, based on the raw description of **HDWL**. Use the Raw Desc Matching tab to identify the headwall points, and click the **OK** button to exit the dialog box.
5. After defining the Headwall point group, go into the Point Groups properties dialog box and make **Headwall** the top point group followed by the No Show point group (see Figure 4.61).
6. Use the AutoCAD **3Dpoly** routine with the **Point Object** override from the Transparent Commands toolbar and draw a 3D polyline between points 186, 187,188, and 189.

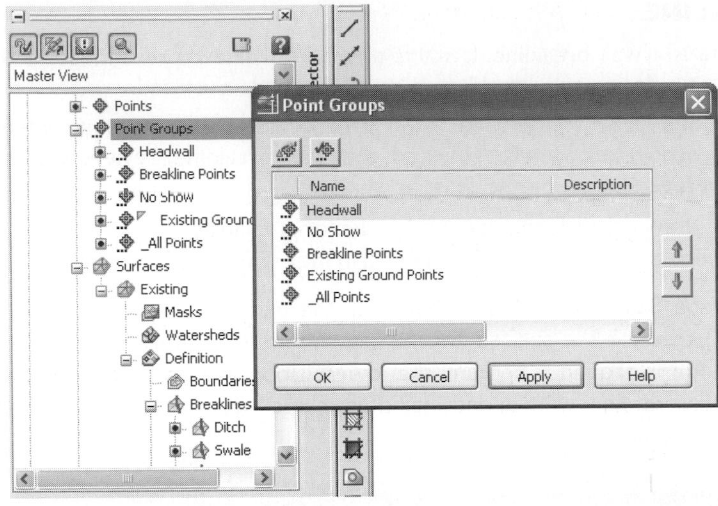

Figure 4.61

7. In Prospector, in the Definition branch of the Existing surface, click **Breaklines**, press the right mouse button, and from the shortcut menu select **Add....**
8. In the Add Breaklines dialog box, enter the description of **Headwall**, change the Type of breakline to **Wall**, and click the **OK** button to exit the dialog box.
9. Select the headwall 3Dpolyline, press the **Enter** key, and select the offset side by selecting a point to the northeast of point 188.

   The routine responds by prompting for the method of setting the elevations for the offset side.
10. Enter **I** for Individual and press the **Enter** key to visit each vertex of the headwall.

The survey crew measured a distance down from the top to the bottom of the wall. Because of this method, the offset elevation for each point is a difference in elevation.

11. At the Specify elevation prompt, enter a **D** for Delta and press the **Enter** key.
12. The prompt changes to Enter elevation difference for offset point at .... Use the elevation difference entries in Table 4.4 for the difference values.

Table 4.4

Point Number	Difference in Elevation
186	-2.5
187	-3.5
188	-3.5
189	-2.5

After entering the last value, the routine exits and rebuilds the surface.

13. Click any surface triangle or its border, press the right mouse button, and from the shortcut menu select **Surface Properties....**
14. Click the **Statistics** tab and then the expand tree icon to the left of the Extended section of the panel. Review the new settings. The average for slopes is now nearly 225 and the maximum slope is just over 350000.

    This is a considerable change and the influence of the wall on surface slopes has to be taken into consideration when reviewing the surface slope.

15. Click the **OK** button to exit the Surface Properties dialog box.
16. From the Viewports flyout menu of the View menu, use the Named Viewports... command to restore the Named Viewport, **A and P**.
17. Use the AutoCAD Zoom and Pan commands to place the profile in the upper viewport and the alignment and headwall in the lower viewport.
18. Move the quick profile view polyline to view the newly defined headwall on the surface.
19. Make the bottom viewport the current viewport.
20. From Viewports flyout menu of the View menu, use the 1 Viewport command to restore a single viewport.
21. Close the Feature Lines and, if necessary, the Transparent Commands toolbars.
22. Use the Layer Properties Manager to set layer 0 (zero) as the current layer and freeze the **Existing-Breakline**, **RDS**, **LOT**, and **CL** layers to focus on the surface and its triangulation.
23. Use the Zoom Extents command to view all of the surface and its triangles.
24. In Prospector, click **Point Groups** heading, press the right mouse button, and from the shortcut menu select **Properties....**
25. Click **No Show**, move the group to the top of the list, and click the **OK** button to exit and hide the points.
26. Click the **AutoCAD Save** icon to save the drawing.

SUMMARY

- You create a surface instance by naming, describing, and assigning a style.
- Each time you add data to a surface, the surface can rebuild itself automatically.
- Point groups are an effective way of organizing and controlling point display.
- Linear surface features using point data will need breaklines to preserve the feature on the surface.
- A breakline from a feature line is the only way to represent curvilinear surface features.
- A Quick Profile is an effective way to evaluate a surface under construction.

This exercise and unit starts the process of reviewing and modifying a surface's data. Civil 3D has other routines to review a surface's properties and several routines that edit a surface's data.

UNIT 3: SURFACE REVIEW AND EDITING

While building the initial surface, you need to evaluate the effects of data and breaklines. When evaluating each addition, the user may discover missing data, bad data, the need for additional breaklines and/or boundaries, incorrect triangles, and sliver triangles at the periphery—all indicative that a surface needs data tweaking and editing. The process of tweaking a surface is an iterative process involving reviewing the surface, adding or editing data, reviewing the effects of the changes, and again adding to or editing the surface. It may be several passes before the surface is correct and ready for analysis and annotation.

STEP 4: EVALUATE RESULTING TRIANGULATION

SURFACE PROPERTIES

The initial review of a surface should be the Surface Properties dialog box. This dialog box contains statistics about the surface, data and editing options, editing history, and many other settings and values.

INFORMATION TAB

The Information tab displays the surface name, its object style, assigned rendering material, if the surface is locked, and if the surface should display a tools tip.

DEFINITION TAB

This tab has two parts: Definition Options and Operation Type. Definition Options has three groups of information: Build, Data operations, and Edit operations (see Figure 4.62).

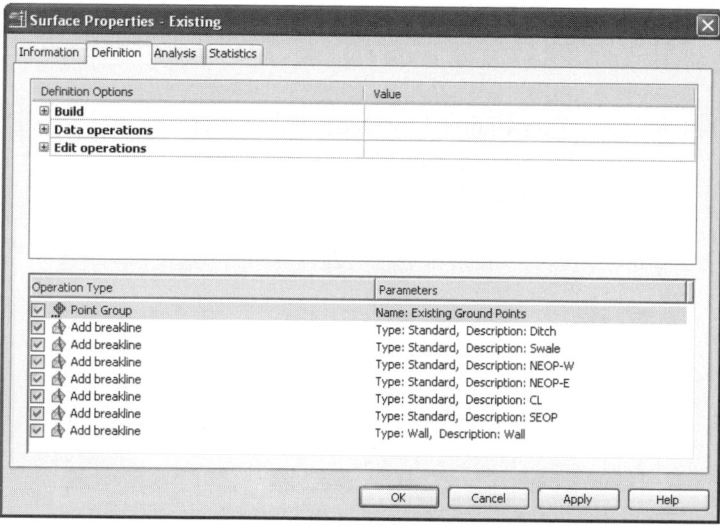

Figure 4.62

Definition Options and Operation Type

This panel maintains settings affecting the building, data use, allowed edits, and the history of the surface. See the discussion in Unit 1 of this chapter.

The Definition area sets what type of data a surface uses, removes all of a particular type of data from a surface (turning off Use Breaklines removes all of the breaklines from the surface), and sets what editing operations can affect a surface (see Figure 4.63).

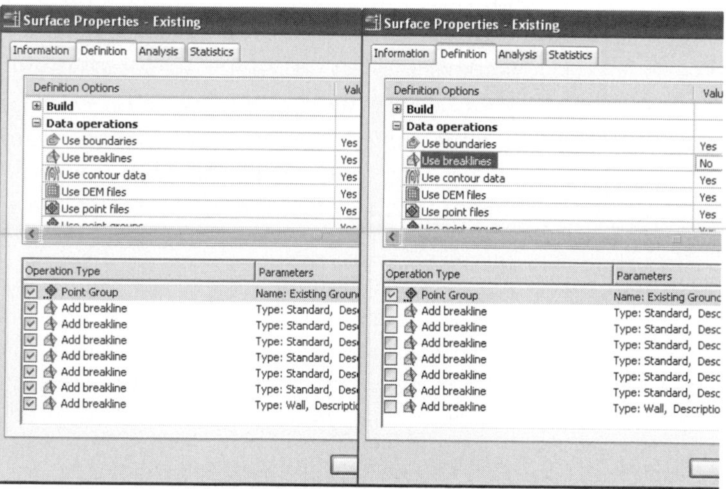

Figure 4.63

When changing Use breaklines to No in the Data operations, all breaklines present in the Operation Type (the bottom portion of the panel) are unchecked and will be removed from the surface data when approving the change.

Civil 3D then issues a warning dialog box asking if the user wants to rebuild the surface using the current settings (see Figure 4.64). If the user answers Yes, the surface rebuilds with the current settings (rebuilds the surface without breaklines).

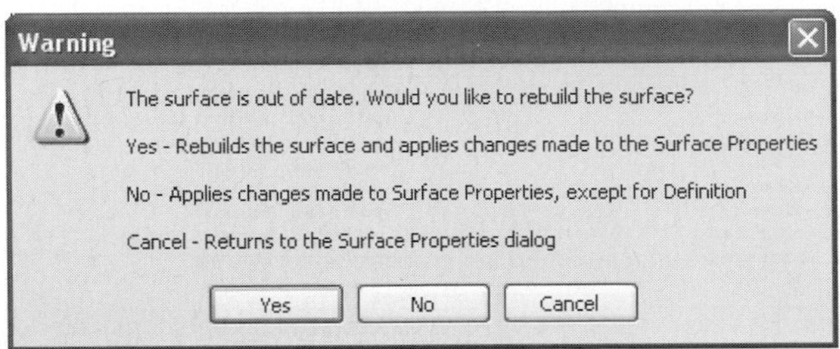

Figure 4.64

STATISTICS TAB

The initial review of a surface starts in the Surface Properties' Statistics tab. This simple review can identify the presence of bad surface elevations or coordinates in the surface data. The panel displays three grouped reports on surface properties: General, Extended, and TIN. The General listing contains information about the revision number, the number of points in the surface, and the minimum, maximum, and average surface elevation (see Figure 4.65).

The Extended report area contains information about the two and three-dimensional area of the surface, and the minimum, maximum, and average slopes across the surface. The section displays slope as a grade. A grade is the rise and fall of elevation expressed as a ratio. So, a 3:1 slope is a 33 percent grade and a 4:-1 slope is a -25 percent grade.

When creating styles that range slope values, it is important to be aware of any surface feature distorting the surface statistics (e.g. a headwall distorting the surface slope range). By turning off the headwall breakline in the Operation Type area of the Definition panel of a surface, you could view the surface slopes without the extreme influence of the wall.

The TIN report section lists the TIN area, the minimum and maximum triangle sizes, and triangle leg lengths (see Figure 4.66). Excessively long triangle legs occur around the periphery of a surface or in areas with sparse data. A review of the surface triangulation determines whether these triangles need deletion, that the points creating large triangles need to be excluded from the surface, or that there is a need to add or edit data to create a better surface. The user can prevent or control long triangle legs by setting the Maximum triangle length in the Build portion of the Definitions tab of the Surface Properties dialog box (see Figure 4.67).

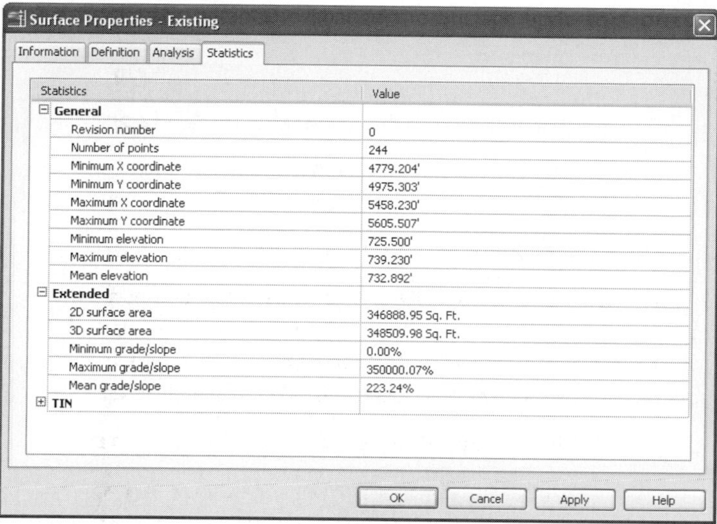

Statistics	Value
General	
Revision number	0
Number of points	244
Minimum X coordinate	4779.204'
Minimum Y coordinate	4975.303'
Maximum X coordinate	5458.230'
Maximum Y coordinate	5605.507'
Minimum elevation	725.500'
Maximum elevation	739.230'
Mean elevation	732.892'
Extended	
2D surface area	346888.95 Sq. Ft.
3D surface area	348509.98 Sq. Ft.
Minimum grade/slope	0.00%
Maximum grade/slope	350000.07%
Mean grade/slope	223.24%
TIN	

Figure 4.65

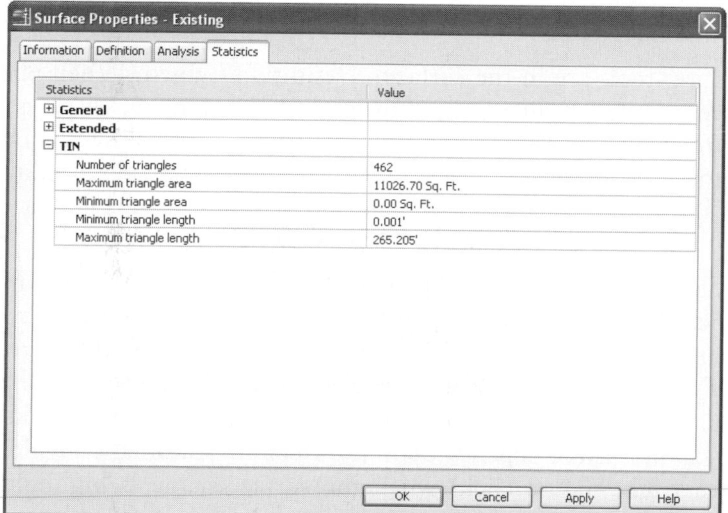

Statistics	Value
General	
Extended	
TIN	
Number of triangles	462
Maximum triangle area	11026.70 Sq. Ft.
Minimum triangle area	0.00 Sq. Ft.
Minimum triangle length	0.001'
Maximum triangle length	265.205'

Figure 4.66

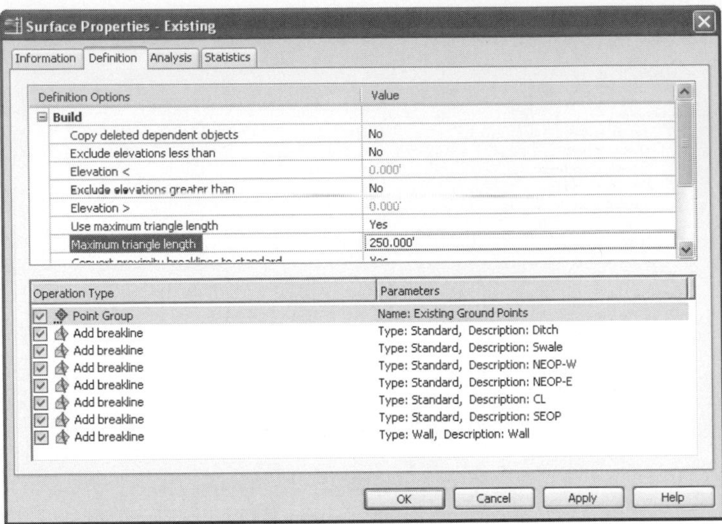

Figure 4.67

STEP 5: EDIT SURFACE

The surface editing tools are in the surface's Definition branch. When clicking the Edits heading and pressing the right mouse button, a shortcut menu displays containing the various editing routines (see Figure 4.68). The shortcut menu is divided into two types of editing tools: lines and points, and overall surface tools. The overall editing tools at the bottom of the shortcut menu raise/lower the surface, minimize Flat Faces, and smooth surface data.

The user can delete any surface edit. When he or she clicks the Edits heading in a surface's Definition branch, Prospector previews an edits list. The edits are listed from first (at top) to last (at bottom) (see Figure 4.69). Or, the user can delete an edit by removing it from the Operation Type list of the Definition panel of the Surface Properties dialog box. When deleting an edit, the surface resolves how the deleted edit changes the surface triangulation. The rebuilding of the surface depends upon the Rebuild-Automatic toggle, which is set in a shortcut menu (right mouse click) when the surface name is highlighted (see Figure 4.70). If off, the edit is removed from the list, but the surface does not update the triangulation. What appears is an out-of-date icon to the left of the name of the surface. When a surface has the out-of-date icon, to update the surface, click the surface name, press the right mouse button, and select build from the shortcut menu. If Rebuild-Automatic is on when the user is deleting the edit from the Preview list, Prospector removes the edit and automatically rebuilds the surface with the new mix of data and edits.

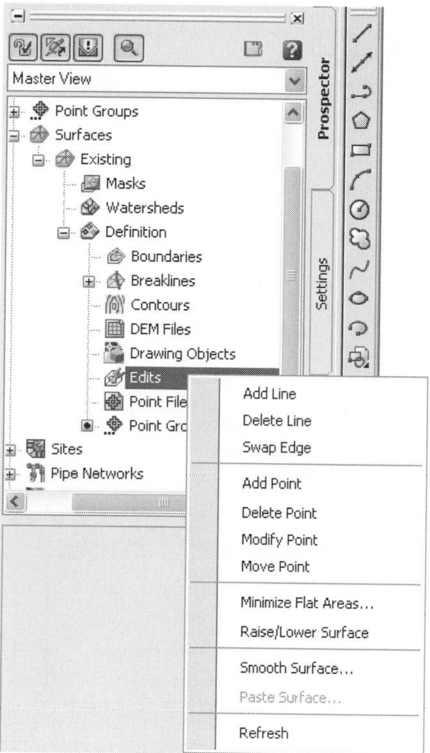

Figure 4.68

If the user prefers not to delete an edit, but would rather evaluate its impact or view the results of different combinations of edits, he or she can toggle on or off individual edits in the Operation Types area of the Definition panel of the Surface Properties dialog box (see Figure 4.71). After evaluating the impact of an edit or combinations of edits, the user can return to the Surface Properties dialog box to restore the edits.

LINE EDITS

This section of surface edits modifies the triangle legs of a surface (see Figure 4.68). The tools in this section delete, add, or swap the position of interior diagonal legs.

The Delete Line routine removes a triangle leg from the surface triangulation. Deleting an interior triangle creates a hole in the surface. After deleting a leg in the interior of a surface, Civil 3D places a surface border around the hole. If you delete a delete line edit from the edit list (from the preview area at the bottom of Prospector panel) or toggle off the edit in the Surface Properties dialog box, Civil 3D removes the border and correctly retriangulates the surface.

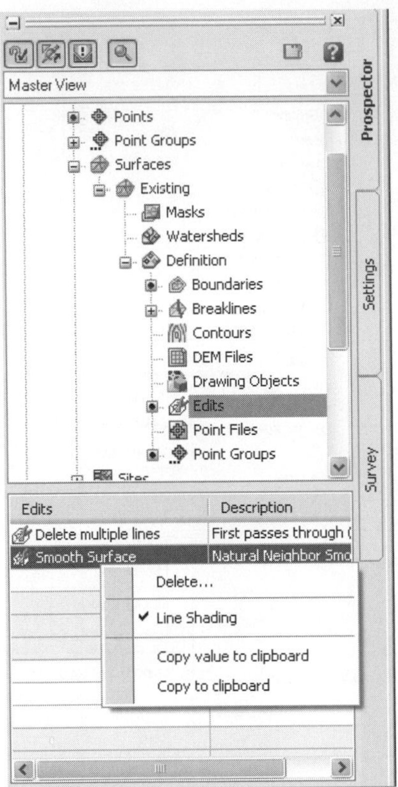

Figure 4.69

The Add line routine creates a new triangle leg at the periphery of a surface or forces a new diagonal leg in the interior of a surface. The first situation is where the surface did not create a triangle from the data and the user wants to add it to the triangulation. Beware of doing this around the edges of the surface. The reason the triangle was not made is based on surface settings and it was determined that there is no real data supporting such a triangle. For example, the triangle leg is greater than the set maximum triangle leg length.

A second use of the add line routine changes the interior diagonal solution. In the interior of a surface, a diagonal leg connects the two nearest points of a group of four neighboring points. The initial triangulation is based upon distances and groupings. If it is not the correct solution, the user can add the new leg to the surface to change the diagonal. If the line crosses an existing line, the surface deletes the original line and replaces it with the one the user drew. By using the Add Line routine this way, the user is duplicating the function of the Swap Edge routine.

The Swap Edge routine is a simpler method of changing an interior diagonal. The Swap Line routine changes the diagonal line within the group of four points.

If these edits represent an effort to correctly triangulate a linear feature, the user should use a breakline to correctly resolve the feature.

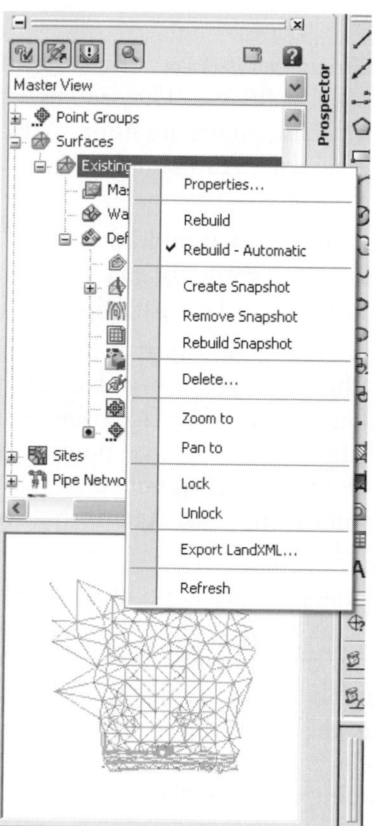

Figure 4.70

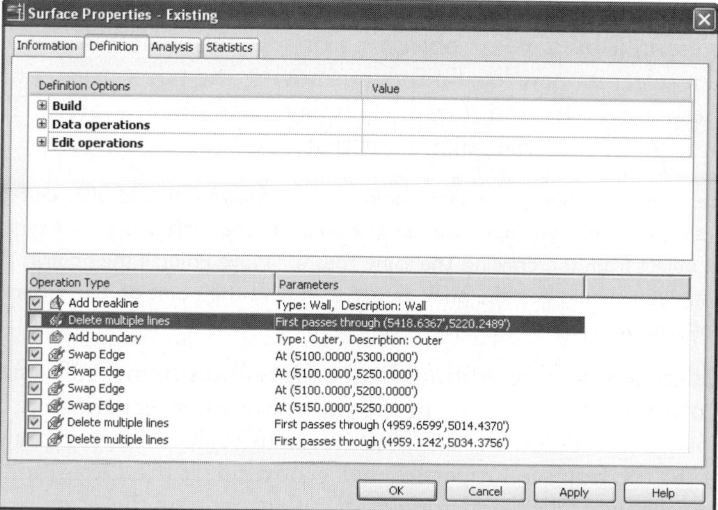

Figure 4.71

POINT EDITS

The Add Points routine places new points in a surface (see Figure 4.68). These points are not in any point group or external point file. They represent an interpolated value located within the interior of a surface. These points are not a part of the initial data, nor are they deemed necessary to be additional data. An example of this would be adding high and low elevations to contour data.

The best practice for adding new points would be as new members of an existing point group or as members of a new point group assigned to a surface. This way the user can easily trace their additions and manage their participation in surface data.

When using the Delete Points routine, be aware of how the point is a part of the surface. If the point is a point object that is a part of a surface point group, select near the point and the routine removes the point from the surface triangulation. The point object remains on the screen and a part of the original surface point group. Prospector's preview area shows only the coordinates of the deleted point. Use the 'PN (transparent override for point number) to select the point to remove. Again, the Prospector preview area displays only the coordinates of the pick.

The Modify Point routine changes the elevation of a point. If the point represents a point object, edit its elevation with Edit Points... and the surface will recalculate its triangulation based upon the edit. When deciding to edit the elevation of a point in the surface even though it is a point object, simply click near the point's location and modify its elevation. After modifying the elevation, the point object remains a member of the point group and does not display a modified elevation. The Edit Point routine only modifies the elevation of the triangulation at that point on the surface.

The Move Point routine changes the location of a point in the surface triangulation. If the point is a point object, use Edit Points... to edit its location and the surface will recalculate the triangulation based upon the edit. If the user decides to edit the location of a point in the surface represented by a point object, he or she can simply select near the point's location and then select its new location. After moving the point, the point object remains a member of the point group and does not display a modified location. The routine only modifies the location of the triangulation at that point on the surface.

Note: When modifying a surface point's elevation, you should also edit the point object's elevation. If you don't do this, the point on the surface is out-of-sync with the point object from the drawing. The same goes for Move Point: if the point is also in the drawing, you should edit the point rather than have the point on the surface become out-of-sync with the point object in the drawing.

The user can delete any surface editing. In the Definition branch of a surface, select the Edits heading to display the edits list in the preview area, select the edit, and delete it. If the user does not want to delete the edit, but wants to evaluate it or a group of other edits, he or she can go to the Surface Properties dialog box, select the Definitions tab and in the Operation Type area, and click and toggle off or on each of the edits.

To select multiple entries, select the first entry, hold down the Shift key, and select the last entry (this selects an entire block of entries). If the user wants to select several individual entries, he or she can select them while holding down the Ctrl key (see Figure 4.67).

SMOOTH SURFACE

When deciding to smooth the surface, choose a method and the amount of smoothing. It maybe that smoothing is merely a cosmetic edit needed for document production. However, the surface may have sparse data and using one of the smoothing algorithms will help in the creation of a more realistic surface. The decision to smooth should wait until after creating and reviewing the initial contours.

Civil 3D introduces surface smoothing using to two new strategies; Natural Neighbor Interpolation (NNI) and Kriging (see Figure 4.72). Both methods add data to a surface by interpolating new data for areas with large relief changes or sparse data. The interpolated data creates better transitions between elevations and more data between sparse data points. The result of each of these interpolation processes is a "smoother" surface. The NNI method operates only inside of a surface boundary. Kriging can operate within a surface boundary or extend a surface's boundary through its interpolation of data. Both methods require a smoothing region. Define a smoothing region by one of three methods: select an existing closed polyline, a parcel, or draw a polygon, select a surface boundary, or select an existing rectangle.

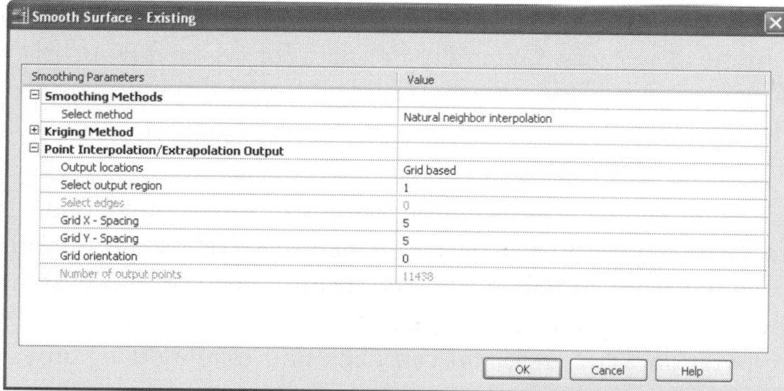

Figure 4.72

Natural Neighbor Interpolation (NNI)

The Natural Neighbor Interpolation (NNI) method uses a NNI algorithm to estimate the elevation of an arbitrary point from a set of points with known elevations (point objects, breaklines, etc.). The routine places the resulting interpolated points into the surface by one of four methods: grid, centroids, random points, or as edge midpoints.

Grid The Grid smoothing option creates new surface point data from the elevations of existing elevations within the smoothing region. The grid smoothing option requires the user to define the area of interpolation, the grid spacing, and rotation of the newly interpolated points. The result of the interpolation is new surface point data at the grid spacing and rotation within the smoothing boundary.

Centroids The Centroids option creates new surface point data from the elevations found at the center of existing triangles. The routine requires a defined area to calculate the new data. After calculating these data points, the build surface routine creates a new triangulation between the existing and new interpolation points.

Random Points The Random Points option creates new surface point data from the elevations found within a region. The first step is to define the number of new points and the second step is to define the area for interpolation. After calculating these data points, the build surface routine creates a new triangulation between the existing and new interpolation points.

Edge Points The Edge Points method creates new surface data from elevation trends found within a region. The new points are slightly higher or lower based upon the surface elevation trend. This method allows a contour data surface to correctly interpolate the elevations of bays and peninsulas in the contour data.

Kriging

Kriging is a much more speculative method for interpolating points on a surface. You can use Kriging to supplement sparse data and/or to expand the boundary of a surface. Examples of using the Kriging method are developing pollution plumes, water flow plumes, or subterranean surfaces. Generally the data for water flow or pollution plumes is sparse and the Kriging interpolation method is ideal for modifying a surface that has sparse data. An understanding of the mathematics and applications of Kriging are necessary to correctly use this method of interpolation.

The user can delete any surface smoothing. In the Definition branch of a surface, select the Edits heading to display the edits list in the preview area, select the smoothing, and delete it. If the user does not want to delete the smoothing, but wants to evaluate it or a group of other edits, he or she can go to the Surface Properties dialog box, select the Definitions tab and in the Operation Type area, and can click and toggle off or on each of the edits.

EXERCISE 4-3

When you complete this exercise, you will:

- Be familiar with the Edit Surface menu.
- Know settings and their affects in the Definition tab of Surface Properties.
- Be able to delete and control the affects of surface edits.
- Review the effects of surface edits.
- Be familiar with Nearest Neighbor Interpolation (NNI) smoothing.

This unit's exercise continues with the Surfaces drawing of Unit 2. If you did not do the exercise for Unit 2, you can open the *Chapter 4 - Unit 3* drawing that is in the *Chapter 4* folder of the CD that accompanies this textbook.

EXERCISE SETUP

1. Open the *Surfaces* or *Chapter 4 – Unit 3* drawing.

SURFACE ELEVATIONS

The simplest review of a surface is to use a surface tool tip to show elevations. The Information tab in the Surface Properties dialog box contains the tool tip toggle.

1. In the drawing, click the border of the Existing surface, press the right mouse button, and from the shortcut menu select **Surface Properties...**.
2. Click the **Information** tab to view the panel and notice the Show Tool Tips toggle at the lower-left of the dialog box. Make sure this toggle is **ON**.
3. Click the **OK** button to close the dialog box and move the cursor to different locations around the surface, letting the cursor sit for a moment. Then allow the tool tip to show the elevation at that point.

REVIEWING SLOPES AND BREAKLINE DATA

1. In the drawing, click the border of the Existing surface, press the right mouse button, and from the shortcut menu select **Surface Properties...**.
2. Click the **Statistics** tab to view its panel.
3. Review the values of the General statistics tree by clicking the expand tree icon to the left of the General heading. Note the minimum, maximum, and that the average elevation is about midway between the lowest and highest elevations.
4. Click the close tree icon to roll up the General data tree.
5. Click the expand tree icon for the Extended statistics tree.

   As reviewed in the last unit, the maximum grade (350,000+) is the result of the wall breakline. This single surface feature changes the average grade from around 6 percent to more than 220 percent. If using this slope variation for 8 ranges, each range would cover 43,000 percent (350,000/8). The first range would cover 99.9 percent of the site. When creating ranges for surface data, it is important to be aware of any surface feature that may distort the values in the surface statistics (the headwall distorting the surface slope range). By turning off the headwall breakline in the Operation Type section, you could review the surface slopes without the extreme influence of the wall.

6. Click the **Definition** tab to view the Definition and Operation type panel.
7. Click the expand tree icon to the left of Data Operations, change Use Breaklines to **No**, click the **OK** button to exit the dialog box, and answer **Yes** to rebuild the surface with all of the breakline data removed from the surface data.

8. Use the Zoom and Pan commands to view the surface triangulation around the roadway areas at the south of the surface.

   The surface has returned to a state when you only had point data as a part of the surface.

9. Select a triangle or the border of the surface, press the right mouse button, and from the shortcut menu select **Surface Properties...**.

10. Go to the Definition tab, and change Use Breaklines to **Yes** in the Data Operations tree.

    This restores all of the breaklines to the surface data set.

REMOVING THE WALL BREAKLINE

To remove the impact of a single operation, your attention moves to the lower part of the Definition panel. It is here that one can toggle on or off the effect of one or more operations in any combination. By removing the wall breakline temporarily from the surface, you can view the Extended surface statistics without the influence of the wall. After reviewing the statistics without the wall, you can reenter the Surface Properties dialog box and toggle on the wall again. This way you can always return to a point in the surface history to view slope ranges without the distortion of wall breaklines.

1. In the Operation Type, uncheck the toggle for the Wall breakline.

2. Click the **OK** button to exit the Surface Properties dialog box and click the **YES** button to rebuild the surface with the new data that does not include the wall fault.

3. If the Event Viewer appears containing warnings about duplicate points, click the "**X**" to close it and view the new triangulation.

   The triangles at the headwall have changed.

4. In the drawing, click the border of the Existing surface, press the right mouse button, and from the shortcut menu select **Surface Properties...**.

5. Click the **Statistics** tab to make it current.

6. Click the expand tree icon for the Extended statistics.

7. The slope maximum and average return to their pre-wall values (average around 6 percent and a maximum of just over 200 percent).

8. Click the **Definition** tab to view the Definition and Operation type panel, toggle **on** the Wall fault, click the **OK** button to close the dialog box, and then click the **YES** button to rebuild the surface.

REVIEWING TRIANGULATION SETTINGS

1. In the drawing, select a triangle or the border of the surface, press the right mouse button, and from the shortcut menu select **Surface Properties...**.

2. Click the **Statistics** tab to view its panel.

3. Click the expand tree icon for the TIN statistics tree.

   This area of the dialog box reports the minimum and maximum triangle lengths. There are some very small and large triangles in the surface. The question becomes, does the distance of 260 or more feet for a triangle leg represent a valid surface triangle? While building a surface, Civil 3D does try to eliminate some extraneous triangles around the perimeter. However, if you do not specify a maximum distance, a surface may have long legs around the periphery that do not have supporting data.

   While the Maximum triangle length setting controls peripheral triangle lengths, it may not give you all the control you want. After using a few values and viewing the results, you may decide it is better to edit (delete lines) or to use one or more boundaries to control the triangulation.

4. Click the **Definition** tab and click the expand icon for Build in the Definition Options.
5. Change Use Maximum Triangle Length to **YES**, set the length to **100**, click the **OK** button to close the dialog box, and then click the **YES** button to rebuild the surface.

   The surface border and the peripheral triangles change, reflecting the new settings. While this setting removes triangles at the northern end, it does not remove others from the northwestern portion of the surface. It would seem that deleting some triangle and/or defining a border would be a better solution than using triangle lengths.

6. In the drawing, click the border of the Existing surface, press the right mouse button, and from the shortcut menu select **Surface Properties...**.
7. Click the **Definition** tab and click the expand icon for Build in the Definition Options.
8. Change Use Maximum Triangle Length to **NO**, click the **OK** button to close the dialog box, and then click the **YES** button to rebuild the surface.

DELETING TRIANGLES

1. Restore the central view and review the triangulation at the east side of the surface.

   Delete the border and one additional triangle leg, because there is no data supporting them (see Figure 4.73).

2. If necessary, in the Surfaces heading of Prospector, expand the Existing surface branch until you are viewing the Definition branch and its headings.
3. In the Definitions branch for the Existing surface, click the **Edits** heading, press the right mouse button, and from the list select **Delete Line** (see Figure 4.68).
4. In the drawing, select the two lines shown in Figure 4.73. After selecting the lines, press the right mouse button to exit the command.

   The routine removes the lines and redraws a new surface boundary.

EXERCISES

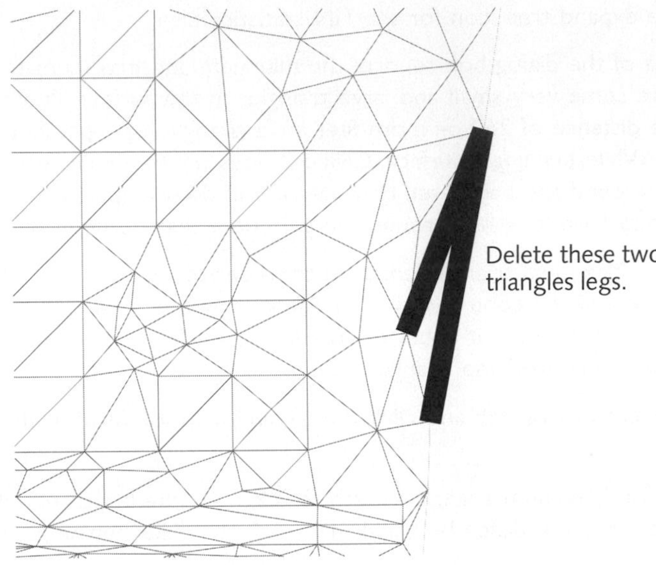
Delete these two
triangles legs.

Figure 4.73

Prospector updates and now shows the edit by changing the icon to the left of the Edits heading. When selecting the Edits heading, the preview area displays the edits for the surface (see Figure 4.74). The edit also appears in the Operation Type area of the Definition panel of Surface Properties.

5. In the Definition heading of Prospector, click the **Edits** heading and view the edit list in the preview area.
6. In the drawing, click the surface border, press the right mouse button, and from the shortcut menu select **Surface Properties....**
7. Click the **Definition** tab to view its contents.

   Notice that Delete multiple lines is an entry in the Operation Type list at the bottom of the panel. It is here that you have the option of toggling on or off the effect of the deletion or deleting it from the list.

8. Click the **Delete multiple lines** entry in the Operations Type list, press the right mouse button, and view the shortcut menu commands. Remove From Definition... is one of the commands.
9. Press the **Esc** key to close the shortcut menu.
10. Click the **OK** button to exit the dialog box.
11. In Prospector's preview area, click the **Delete multiple lines** entry (see Figure 4.74), press the right mouse button, from the shortcut menu select **Delete...** to delete the edit, and click the **OK** button in the Remove From Definition dialog box.

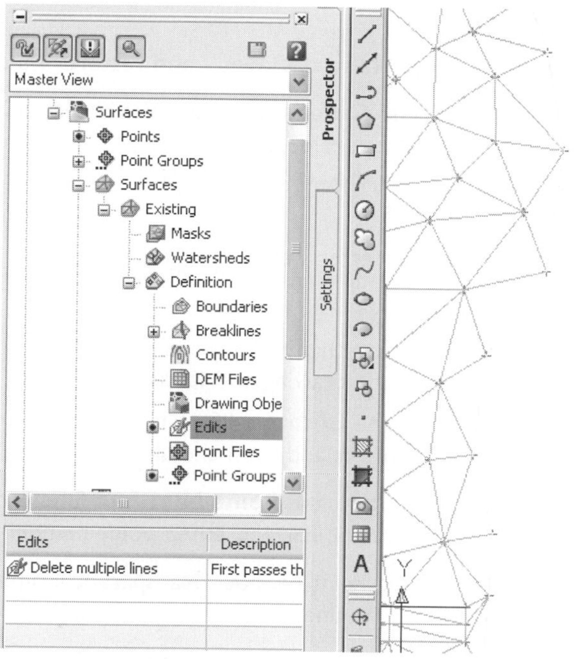

Figure 4.74

The Remove From Definition dialog box gives you the option of not deleting the edit from the definition list.

12. In the Definition heading of Prospector, click the **Edits** heading, press the right mouse button, select the **Delete Line** routine, and delete the two triangles identified in Figure 4.73.

SWAP EDGE

The Swap Edge routine simply changes a diagonal to the second solution. You use this routine to "tweak" the triangulation in areas not needing additional data or breaklines. Like Delete Lines, Prospector displays the coordinates of the swapped edges in the preview area.

1. In the Definition tree for the Existing surface, click the Edits heading, press the right mouse button, and from the shortcut menu select **Swap Edge**.
2. Click two or three diagonals to change their position in the surface and press the **Enter** key to exit the routine.
3. In Prospector's preview area, review the listings for Swap Edge.
4. In Prospector's preview area, delete the Swap Edge entries by selecting them, pressing the right mouse button, selecting **Delete...** from the shortcut menu, and clicking the **OK** button to delete the edits and rebuild the surface.

SURFACE POINT EDITING

1. From the View menu, select **Named Views...** and restore the Northeast view.
2. In Prospector, click the **Point Groups** heading, press the right mouse button, from the shortcut menu, select **Properties...**, and move the Breakline Points group to the top position and No Show to the second position to view the swale points in the northeastern section of the site.
3. Use the Zoom command to better view points 107, 111, and 112.

Delete Point

If you want to delete a point from a surface that is also point in the drawing, the point is removed from the surface but remains a part of the drawing and its assigned point group.

4. In the Definition branch for the Existing surface in Prospector, click the **Edits** heading, press the right mouse button, and from the shortcut menu select **Delete Point**.
5. When prompted for a point, click the **Point Number** filter icon (or type in **'PN**) of the Transparent Commands toolbar, and enter point number **110**.
6. Press the **Esc** key to exit the Point Number Transparent Command, and press the **Enter** key twice to exit the routine.

   When deleting a point from a surface, the surface rebuilds without point 110. Point 110 remains in the drawing and in its point group(s). Prospector's preview area lists only the coordinates of the deleted point, not its point number.

   You use the Add Point edit routine to add a new point(s) to a surface or to return point 110 to the surface. You could delete the deletion from Prospector's preview list of edits or go to the Surface Properties dialog box and toggle off the deletion.

7. In the Definition branch for the Existing surface of Prospector, click the **Edits** heading.
8. In Prospector's preview area, click the **Delete Point** entry, press the right mouse button, from the shortcut menu select **Delete...**, and click the **OK** button in the Remove From Definition dialog box. This restores point 110 to the surface.
9. In Prospector, click the **Point Groups** heading, press the right mouse button, from the shortcut menu select **Properties...**, move the No Show group to the top position, and click the **OK** button to exit the dialog box.
10. Use the Zoom Extents command to view the entire site.

BOUNDARIES

Civil 3D has three boundary types; Outer, Hide, and Show. The surface currently under review has several points in the west and northwest that seem to be peripheral to the site we are concentrating on. These points could be deleted from the surface. However, an outer boundary limits the surface to the points inside of its boundary.

1. Use Layer Properties Manager to create a new layer: **Existing-Boundary**. Make the layer the current layer, assign it a color, and click the **OK** button to create the layer and exit the dialog box.
2. Start the AutoCAD Polyline routine and draw a closed boundary similar to the one in Figure 4.75.

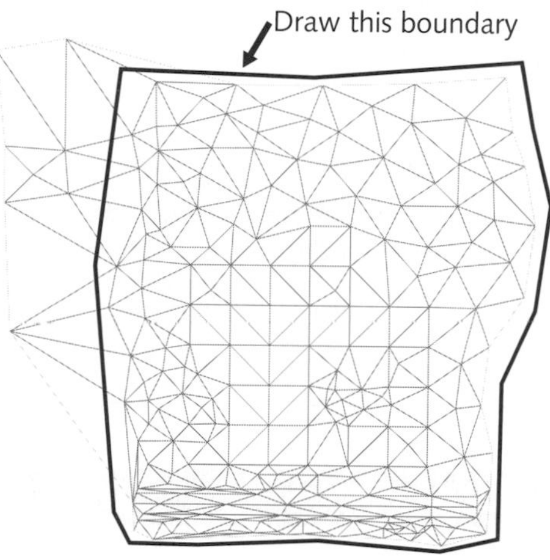

Figure 4.75

3. In the Definition tree for the Existing surface in Prospector, click the **Boundaries** heading, press the right mouse button, and from the shortcut menu select **Add...**.
4. In the Add Boundaries dialog box, name the boundary **Main Site**, set its type to **Outer**, do not toggle on Non-destructive breakline, click the **OK** button, and select the boundary polyline.
5. Open Layer Properties Manager, make layer 0 (zero) the current layer, freeze the Existing-Boundary and V-Node layers, and click the **OK** button to exit the dialog box.
6. Click the **AutoCAD Save** icon to save the drawing.

The Non-destructive breakline toggle controls how triangles will interact with the boundary. If using Non-destructive breaklines, the surface preserves the elevations along the boundary by building new triangles between the data and the boundary. If not using Non-destructive breaklines, the triangulation stops at the data within the boundary.

The surface contains your edits and changes and is now ready for analysis. However, it is difficult to decide whether to do any surface smoothing. The decision about smoothing should happen after viewing the initial contours.

The next unit reviews the task of understanding the distribution of surface slopes and elevations.

SUMMARY

- The review and editing of a surface is an iterative process.
- To rebuild a surface each time you edit it, set Rebuild – Automatic.
- Delete peripheral triangles with the Delete Line edit.
- If you delete triangles in the interior of a surface, Civil 3D places a border around the area without triangles.
- A boundary as surface data is an effective method of excluding unwanted data.
- A boundary as surface data is an effective method of controlling unwanted peripheral triangles.
- Instead of deleting triangles in the interior of a surface, consider the option of using a hide boundary.

UNIT 4: SURFACE ANALYSIS

The next step in the surface process is analyzing the surface slopes and elevations. The analysis of a surface is the function of specific use surface styles. Depending on the analysis (slope or elevation) and the distribution of data, each style may use a different algorithm to calculate ranges and to display the results of the analysis.

STEP 6: ANALYZE THE SURFACE CHARACTERISTICS

Because surfaces represent the start and end of a project, it is critical that they represent the data correctly. The surface properties dialog box contains a Statistics tab that reports specific information about that surface's slopes and elevations.

However, the best understanding of slopes and elevations is by visualizing them in two- or three-dimension views. If a surface contains bad data, each view will show the bad data as spikes or wells in the surface model. By creating analytical surface styles that emphasize surface characteristics, one can visualize a surface in a more understandable way. The surface analysis styles focus mainly on TIN structure, relief, elevation, and slope.

SLOPE AND ELEVATION

After styles focusing on triangulation, the next two surface characteristics needing visual review are slope and elevation. Slope is the amount of elevation change over a distance. There are two methods of expressing a slope: as a slope or as a grade. A slope of 3:1 represents a rise of 1 foot over an offset distance of 3 feet. A slope of 4:-1 represents a fall of 1 foot over an offset distance of 4 feet. A grade is the rise and fall of elevation expressed as a ratio. So, a 3:1 slope is a 33 percent grade and a 4:-1 slope is a -25 percent grade.

A slope surface style groups surface triangles by ranges of slope amount (i.e. 0–2 percent, 2–4 percent, etc.). Each range has a color and all triangles in the range display that color. For example, the range of 0 to 2 percent has red triangles and the range of 2 to 4 percent has yellow triangles.

The slope arrow is another useful slope analysis tool. A slope arrow displays not only the amount of slope (shows the color of its range group), but also shows the direction of down slope. When combined, a slope analysis style displays triangles and slope arrows together.

Slope arrows provide more information than just magnitude of and direction of down slope. When slope arrows point towards each other, they represent a valley or swale in the surface (see Figures 4.76). If the slope arrows point away from each other, they represent a high spot on the surface (see Figure 4.77). Slope arrows also show the direction of water flow across a surface.

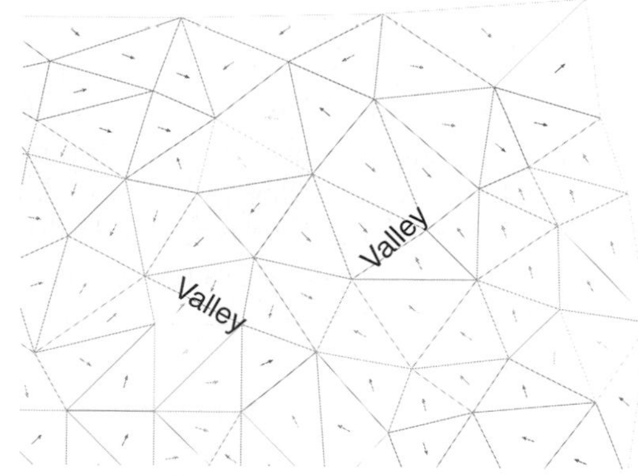

Figure 4.76

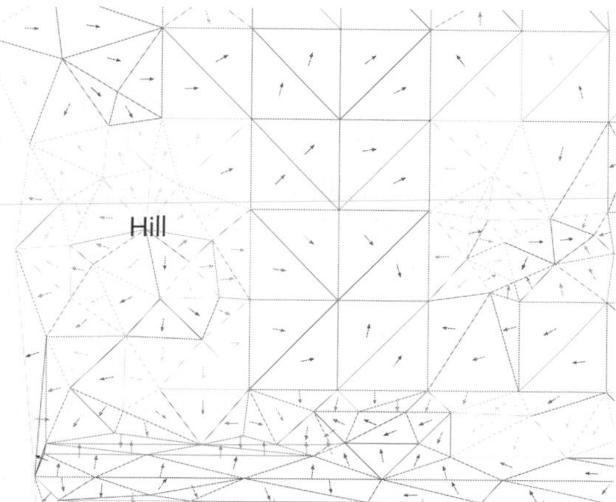

Figure 4.77

The last surface review styles analyze surface elevations (relief) grouped by amounts of elevation. Like slope styles, the elevation styles assign colors to each range of elevation and the triangles show that color on the screen. When viewing an elevation style in 3D, the user can view the relief of the surface model.

RANGING METHODS

Civil 3D provides three methods of grouping or ranging slope and elevation data: equal interval, quantile, and standard deviation.

The surface in this chapter has a wall that occupies a small surface area, but greatly affects the surface slope statistics. The site is not just this wall but has more slopes at lesser grades. The wall slopes are so extreme they bias or skew the data to the higher end values. This was apparent when comparing the slopes before and after adding the wall to the surface. The average slope jumped from 6.2 to 225 percent and the maximum slope jumped from 210 to 350,000 percent.

Equal Interval

The equal interval method divides the overall range difference (maximum minus minimum) by a user specified number of ranges. This method often over-generalizes the data. For example, the exercise surface has grades varying from 0.0 to 350,000 percent. Using 10 ranging groups, each range would represent a slope range of 35,000 percent (i.e. 0–35,000, 70,000–105,000, etc.). When evaluating surface slopes using this interval, 0–35,000 for the first range, the range would contain 99 percent of the surface. A skewed data distribution makes the Equal Interval ranging process useless.

The equal interval method is excellent for data that is not skewed. Again, the exercise surface has a range of elevations from 725 to 739 with an average of 732. The average is almost halfway between the minimum and maximum. The equal interval method will work well with the elevation data.

Quantile

The second method of ranging is quantile. This method divides the data so that each range contains an equal number of members for each interval. This methodology focuses on the more frequently occurring values and deemphasizes the few high or low values. With this method, the few high values are put in a range containing lower values to create a range that has the same number of members as the ranges for the more frequently occurring lower slopes. For example, using the slopes from the exercise surface, this method makes the last slope range contain slopes from 20 to 350,000. This range has the same number of triangles as does the range from 0 to 1.5 percent, but allows the user to correctly review the distribution of surface slopes. Because of how the method organizes data, it is used best with skewed distributions.

Standard Deviation

The last method is standard deviation. This method calculates and divides the data based on how far data values differ from the arithmetic mean. This method is most effective when the values approximate a normal distribution (bell-shaped curve). The standard deviation method is often used to highlight how far above or below a specific value is in relation to the mean value. So the best use of this method is to show areas of highest and lowest values.

WATERSHEDS

The analysis of the surface also includes the development of surface watersheds. This analysis results in a map representing areas of water collection and discharge. Civil 3D creates depressions, single and multiple discharge points, and hatches each of the watersheds with a classification hatch (single discharge, multiple discharge, depression, etc.).

ASSIGNING AND PROCESSING ANALYSIS STYLES

To view the affects of analysis styles, complete two steps: assign the style to a surface in the Information tab and process the parameters of the style in the Analysis tab of the Surface Properties dialog box (see Figure 4.78).

The Analysis panel shows the default values from the assigned style. When viewing the settings in the panel, the user can change the number of ranges and set the type of Legend before processing the analysis. After processing the analysis, the user can change the minimum and maximum range value. When changing the range values, change them from the highest range to the lowest (range 8, then 7, etc.). After changing the minimums and maximums, click on the apply button to process the data with the new range settings.

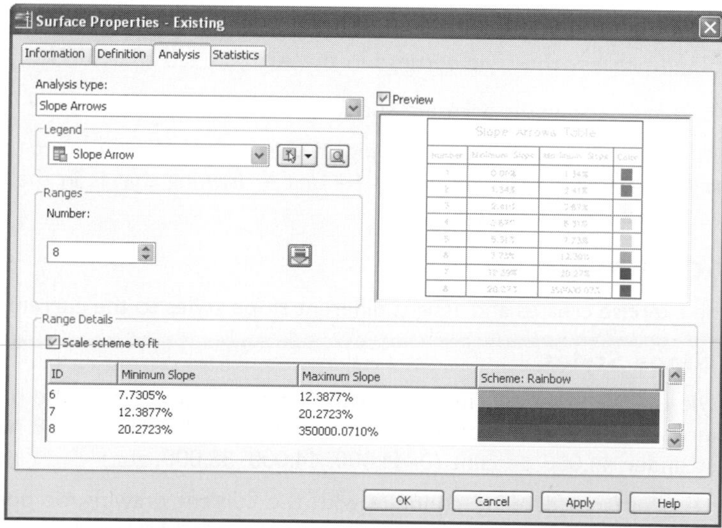

Figure 4.78

WATER DROP ANALYSIS

The slopes of a surface determine and show a general map of the flow of water across a surface. However, the path any one drop takes can be calculated based on the surface slopes. The Water Drop utility of the Surface menu allows the user to visualize the path any drop of water on a surface will take to a collection or discharge point (see Figure 4.79).

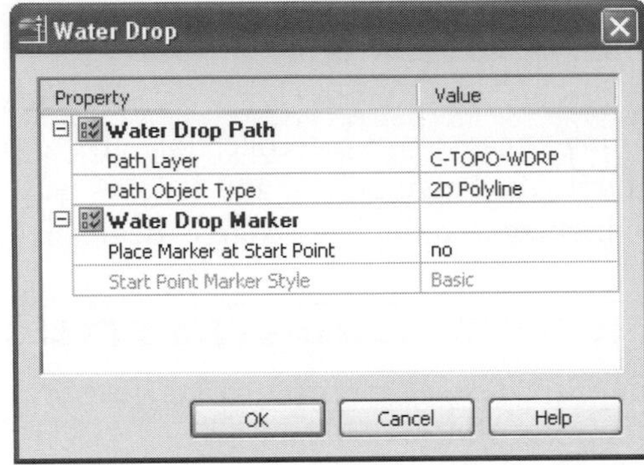

Figure 4.79

EXERCISE 4-4

When you complete this exercise, you will:

- Define surface styles for slopes and elevations.
- Define surface styles using Equal Interval, Quantile, and Standard Deviation parameters.
- Define a watershed surface style.
- Assign and process the style settings in the Analysis tab of a surface.
- Create water drop paths across a surface.

This exercise continues with the Surfaces drawing of this chapter. If you want to start from this point in the exercise, you can open the *Chapter 4 - Unit 4* drawing that is in the *Chapter 4* folder of the CD that accompanies this textbook.

SLOPE SURFACE STYLES

This portion of the exercise creates and assigns different slope styles to the current Existing surface.

Creating the Slope Styles

The first slope style is Equal Interval. This style divides the overall range by the number of ranges. In the current exercise, the range of slopes is just over 350,000. If a style uses 8 ranges, each range will represent just under 44,000 percent (0–44,000, 44,000–88,000, etc.)

1. Open the Surfaces drawing, continue with the current drawing, or open the *Chapter 4 – Unit 4* drawing.
2. Click the Settings tab to make it current.
3. In Settings, expand the Surface branch until you are viewing the list of surface styles.

Slope Surface Style: Equal Interval

4. Click the **Surface Styles** heading, press the right mouse button, and from the shortcut menu select **New...**.

   This presents the New Surface Style dialog box.

5. Click the **Information** tab and Name the style **Slope and Arrows – Equal Interval**, give the style the Description of **Style shows eight ranges of slope triangles and arrows**, and click the **Apply** button.
6. Click the **Analysis** tab to view its panel.
7. Expand the Slopes section and match your settings to those in Figure 4.80.
8. Expand the Slope Arrows section and match your settings to those in Figure 4.81.
9. Click the **Display** tab to view its settings panel.
10. Scroll down to the bottom of the list and turn **ON Border**, **Slopes**, and **Slope Arrows** components (see Figure 4.82).
11. Click the **OK** button to create and exit the dialog box.

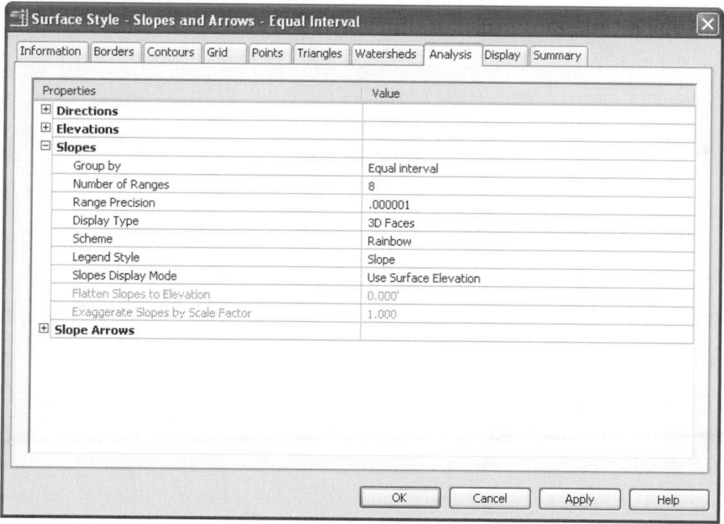

Figure 4.80

12. Create the remaining two styles, **Quantile** and **Standard Deviation**, by selecting the **Slope and Arrows – Equal Interval** style from the list of styles, pressing the right mouse button, from the shortcut menu selecting **Copy...**, and renaming the style with the adjusted parameters.

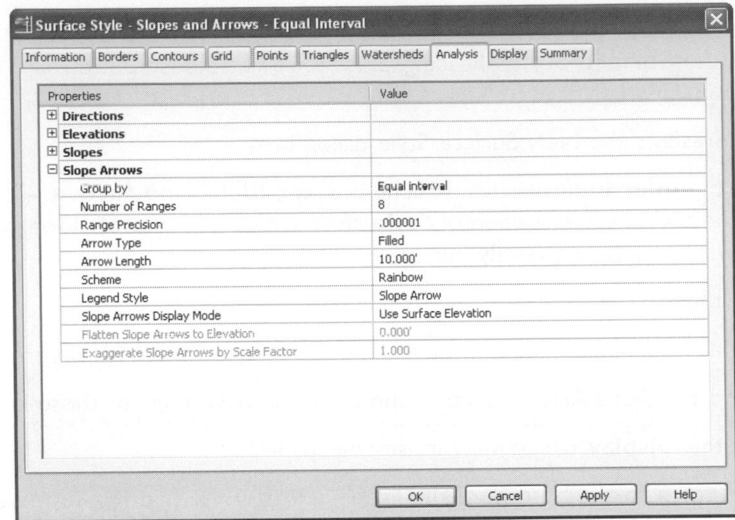

Figure 4.81

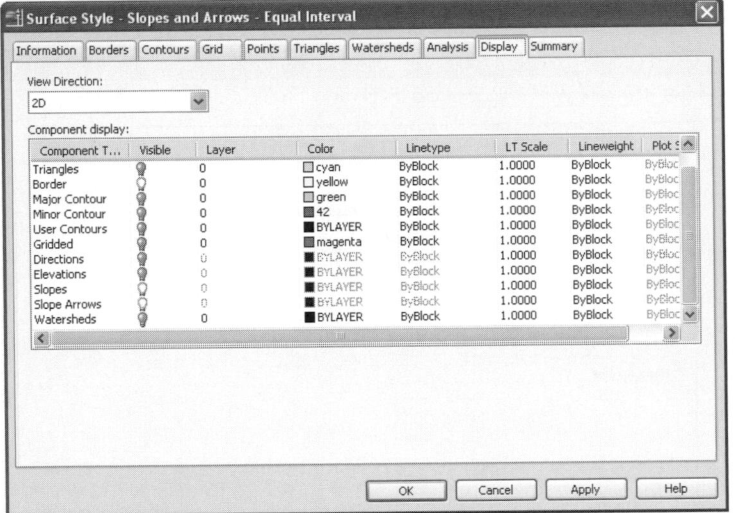

Figure 4.82

Use the following values to create the new styles:

Slope Quantile Surface Style

Information tab:

Name: **Slope and Arrows – Quantile**

Description: **Style shows eight ranges of slope triangles and arrows.**

Analysis tab:

Slope and Slope Arrows: Same settings as Figure 4.80 and 4.81, except set the Group by to **Quantile**.

Display tab:

Turn **on** Slopes and Slope Arrows (see Figure 4.82).

Slope Surface Style – Standard Deviation:

Information tab:

Name: **Slope and Arrows – Standard Deviation**

Description: **Style shows eight ranges of slope triangles and arrows**.

Analysis tab:

Slope and Slope Arrows: Same settings as Figure 4.80 and 4.81, except set the Group by to **Standard Deviation**.

Display tab:

Turn **on** Slopes and Slope Arrows (see Figure 4.82).

CREATING THE ELEVATION STYLES

This portion of the exercise continues in the Surface Styles area of the Settings panel.

Elevation Surface Style: Equal Interval

1. Click **Surface Styles**, press the right mouse button, and from the shortcut menu select **New...**.

   This presents the Surface Style - New Surface Style dialog box.

2. Click the Information tab and Name the style **Elevations – Equal Interval,** and give the style the Description of **Style shows six ranges of elevation**.
3. Click the **Analysis** tab to view its panel.
4. Click the expand icon next to the Elevations section and match your settings to the ones in Figure 4.83.
5. Click the **Display** tab to view its settings panel, click **off** all of the components, and click **ON** Elevations component (see Figure 4.84).
6. Click the **OK** button to create the style and to exit the dialog box.
7. Create the remaining two styles, Quantile and Standard Deviation, by selecting the **Elevations – Equal Interval** style from the list of styles, pressing the right mouse button, from the shortcut menu selecting **Copy...**, and renaming the style with the adjusted parameters.

Use the following values to create the styles:

Elevation Surface Style: Quantile

Information tab:

Name: **Elevations – Quantile**

EXERCISES

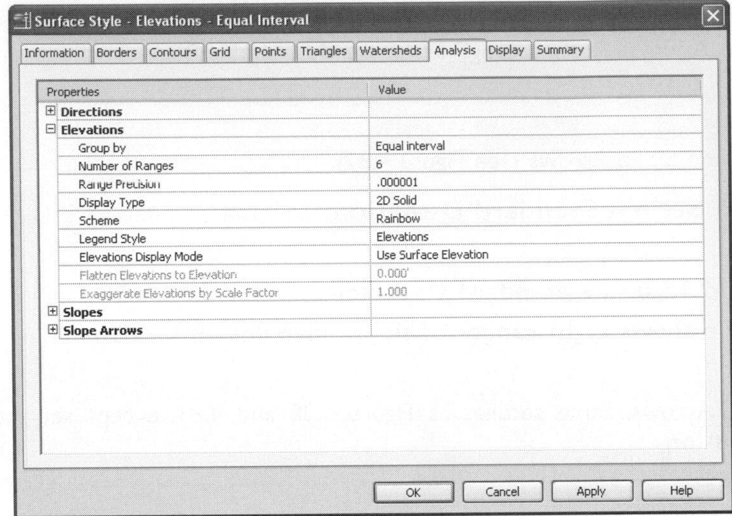

Figure 4.83

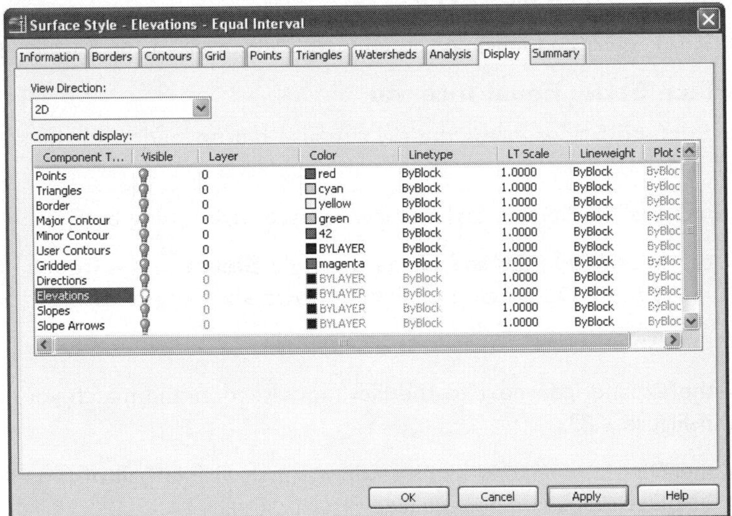

Figure 4.84

Description: **Style shows six ranges of elevation.**

Analysis tab:

Set the same settings as Figure 4.83, except set the Group by to **Quantile.**

Display tab:

Match the settings found in Figure 4.84.

EXERCISES

Elevation: Standard Deviation

Information tab:

Name: **Elevations – Standard Deviation**

Description: **Style shows six ranges of elevation.**

Analysis tab:

Set the same settings as Figure 4.83, except set the Group by to **Standard Deviation.**

Display tab:

Match the settings found in Figure 4.84.

ANALYZING SURFACE SLOPES

The steps to analyze a surface are as follows: first, assign the style, and then process the analysis in the Analysis panel of Surface Properties.

1. Click the **Prospector** tab to make it current.
2. In Prospector, expand the Surfaces branch until viewing the surface list, click the **Existing** surface, press the right mouse button, and from the shortcut menu select **Properties....**
3. Click the **Information** tab and change the Surface Style to **Slopes and Arrows – Equal Interval**.
4. Click the **Analysis** tab, change the Analysis type to **Slopes**, check the range number (it should be **8**), and click the **Run Analysis** button (the down arrow to the right of Range number) to produce the range details for slopes.
5. Change the Analysis type to **Slope Arrows**, check the range number (it should be **8**), and click the **Run Analysis** button (the down arrow to the right of Range number) to produce the range details for arrows.
6. Click the **OK** button to view the surface with colorized triangles and arrows.

   All of the triangles are red, except for those around the wall. This type of analysis style with a skewed distribution is not of much use.
7. Repeat Steps 2–6 and assign and process the **Slope and Arrows – Standard Deviation** for both Slopes and Slope Arrows. Before viewing the new triangles and arrows, review the range details to see if they better represent the slopes on the surface.

   As with the Slopes and Arrows – Equal Interval style, the Slopes and Arrows - Standard Deviation style shows that all but a few of the triangles are red (the first range group). Again, this type of style does not work well with skewed distributions.
8. Repeat Steps 2–6 and assign and process the **Slopes and Arrows – Quantile** for both Slopes and Slope Arrows. Before you view the new triangles and arrows, review the range details to see if they better represent the slopes on the surface.

This surface style handles the skewed distributions better than the Equal Interval or Standard Deviation surface styles.

The range details for this style calculate range breaks that fall between integer slope numbers (i.e. 2, 4, etc.). You can edit the range breaks after processing the analysis and set each range to a specific minimum and maximum value. The style recalculates the range membership using the new range values.

MODIFYING SLOPE RANGE VALUES

1. In the drawing, click the surface border or triangle, press the right mouse button, and from the shortcut menu select **Surface Properties...**.
2. Click the **Analysis** tab.

   Start the editing of the ranges at the eighth range and work up to the first range.
3. Set the analysis to **Slopes** and in the Range Details (lower) section of the dialog box change the ranges to the values listed in Table 4.5.

Table 4.5

Range ID:	Minimum Slope:	Maximum Slope:
1	0.0	2.0
2	2.0	4.0
3	4.0	6.0
4	6.0	8.0
5	8.0	10.0
6	10.0	100.0
7	100.00	210.00
8	210.00	351000

4. After resetting the ranges for slopes, click the **Apply** button to recalculate the range membership.
5. Change the analysis to **Slope Arrows** and in the Range Details (lower) section of the dialog box change the ranges to the same values you set for Slopes in Step 3.

   Start the editing of the ranges at the eighth range and work down to the first range.
6. After resetting the ranges for slope arrows, click the **Apply** button to recalculate the range membership.
7. Click the **OK** button to exit the dialog box, and the surface displays the new data grouping.

CREATE A SLOPE RANGE LEGEND

1. Select **Add Legend Table...** from the Surfaces menu to create a legend for slopes. If necessary, click anywhere on the surface to select it, enter **S** for the legend type, press the **Enter** key, type in **D** for a dynamic table, press the **Enter** key, and select a point to the right of the surface to place a legend in the menu.
2. Erase the slope legend from the drawing.

ANALYZING SURFACE ELEVATIONS

The steps to analyzing a surface are as follows: first, assign the style, and then process the analysis in the Analysis panel of Surface Properties.

1. In the Surfaces branch of Prospector, click the Existing surface, press the right mouse button, and from the shortcut menu select **Surface Properties....**
2. Click the **Information** tab and change the Surface Style to **Elevations – Equal Interval**.
3. Click the **Analysis** tab, change the Analysis type to **Elevations**, check the range number (it should be **6**), and click the **Run Analysis** button (the down arrow to the right of Range number) to produce the range details for elevations (see Figure 4.85).
4. Click the **OK** button to view the surface with colorized triangles.

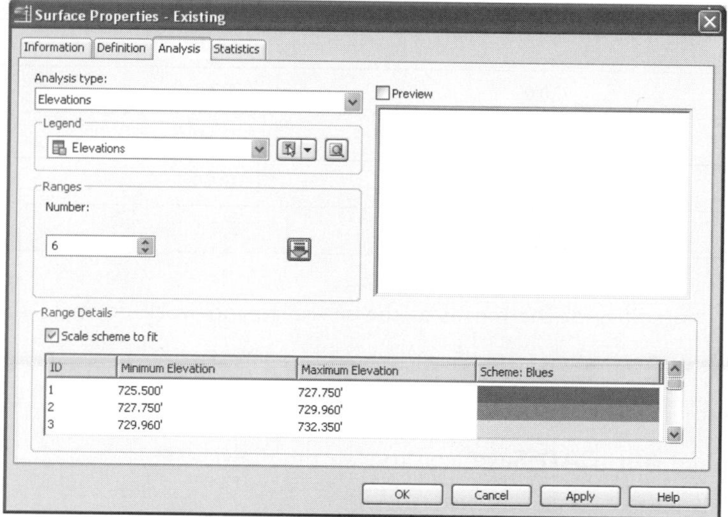

Figure 4.85

The Elevations – Equal Interval surface style correctly displays the surface elevations.

5. Repeat Steps 1–4 and assign and process the **Elevations – Quantile**.

   Before you view the new surface display, review the range details to see if they "better" represent the elevations on of the surface.

The surface elevation map is slightly different from the previous one, but it still displays a good map of the surface elevations. The Quantile map is probably the least accurate of the elevation maps.

6. Repeat Steps 1–4 and assign and process the **Elevations – Standard Deviation**.

Before you view the new surface display, review the range details to see if they "better" represent the elevations on the surface.

The surface elevation map is slightly different from the previous two maps, but it still displays a good map of the surface elevations.

All three methods correctly represent the elevation information, but in slightly different range values. It would be better if the ranges were at specific whole number elevations (i.e. 725, 728, etc.).

MODIFYING ELEVATION RANGE VALUES

1. Reassign the **Elevation – Equal Interval** surface style and process its settings in the Analysis tab of Surface Properties.
2. Edit the minimum and/or maximum range values from the first to range to the elevations in Table 4.6.

Table 4.6

Range ID:	Minimum Elevation:	Maximum Elevation:
1	725.5	728.0
2	728.0	730.0
3	730.0	732.0
4	732.0	734.0
5	734.0	736.0
6	736.0	739.5

Start the editing of the ranges at the six range and work down to the first range.

3. Click the **Apply** button to recalculate the ranges.
4. Click the **OK** button to exit and display a new elevation map.
5. Click the **AutoCAD Save** icon to save the drawing.

All of these styles emphasize and create a 2D map of surface elevations.

CREATING A 3D ELEVATION SURFACE STYLE

The new elevation style creates a 3D view of the surface with user defined elevation ranges.

1. Click the **Setting** tab to view its panel.
2. If necessary, expand the Surfaces branch to view the Surface Styles list.

3. Select the **Elevation – Equal Interval** style, press the right mouse button, and from the shortcut menu select **Copy....**

   This displays a copy of the style and allows you to edit its values.

4. Click the **Information** tab and Change the name of the style to **Elevations – Equal Interval – 3D**.
5. Click the **Analysis** tab and expand the Elevations section.
6. In the Elevations section of the Analysis tab, change the following values (see Figure 4.86):

Display Type: **3D Faces**

Number of Ranges: **6**

Scheme: **Rainbow**

Elevations Display Mode: **Exaggerate Elevation**

Exaggerate Elevation by Scale Factor: **3**

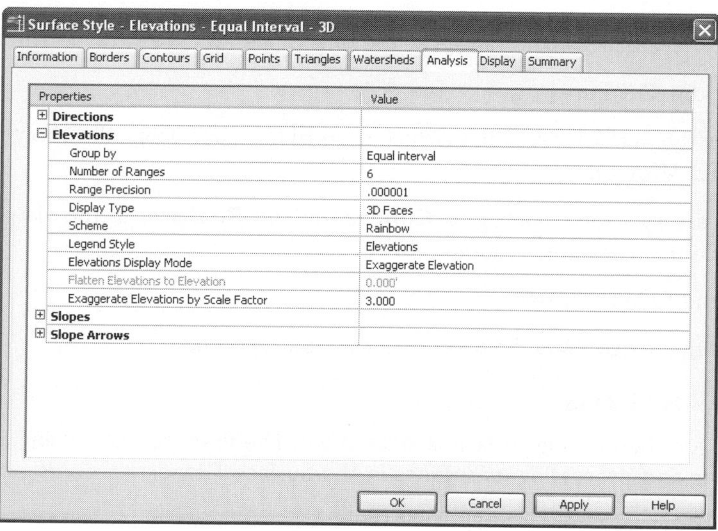

Figure 4.86

7. Click the **Display** tab, change the View Direction: to **3D**, toggle **off** all of the components, and toggle **on** only **Elevations**.
8. Click the **OK** button to exit the Elevations – Equal Interval – 3D Surface Style dialog box.
9. In the drawing, click any portion of the surface, press the right mouse button, and from the shortcut menu select **Surface Properties....**
10. Click the **Information** tab and change the Surface style to **Elevations – Equal Interval – 3D**.

EXERCISES

11. Click the **Analysis** tab, set the analysis type to Elevations, make sure the range number is **6**, click the **Run Analysis** button (the down arrow to the right of Range number), and review the elevation range values in the Range Details area of the panel.

12. Reset the ranges to the values in Table 4.7.

Table 4.7

Range ID:	Minimum Elevation:	Maximum Elevation:
1	725.5	728.0
2	728.0	730.0
3	730.0	732.0
4	732.0	734.0
5	734.0	736.0
6	736.0	739.5

Start the editing of the ranges at the end range and work down the list.

13. Click the **OK** button to exit the dialog box.

14. Select the surface, press the right mouse button, and from the shortcut menu select **Object Viewer....**

15. If necessary, toggle shading to **Conceptual or Realistic** in the viewer and view the surface from several locations.

16. Exit the Object Viewer.

17. Click the **AutoCAD Save** icon to save the drawing.

EXERCISES

WATERSHED ANALYSIS

The last analysis for the Existing surface is watersheds. The first step is defining a watershed style and then assigning and processing its settings in the Surface Properties dialog box.

1. Click **Settings** tab to make it current.
2. In Settings, expand the Surface branch until you are viewing the list of Surface Styles.

Watershed Surface Style

1. Click the **Surface Styles** heading, press the right mouse button, and from the shortcut menu select **New....**

   This presents the Surface Style - New Surface Style dialog box.

2. Click the **Information** tab, name the style **Watersheds**, and give the style the description of **Style calculates watersheds**.
3. Click the **Watersheds** tab to view its panel of settings.

6. Click the expand icon next to each of the Watershed Properties.
7. In the Surface section, change the Surface Watershed Label Style to **Watershed**.
8. Click the **Display** tab to view its settings panel and scroll down to the bottom of the list and turn **on** only **Watersheds** and turn **off** all of the remaining components.
9. Click the **OK** button to create the style and to exit the dialog box.
10. Click any portion of the surface, press the right mouse button, and select **Surface Properties...** to view the Surface Properties dialog box.
11. Click the **Information** tab and set the Surface Style to **Watersheds**.
12. Click the **Analysis** tab and change your settings to match the settings in Figure 4.87.
13. Click the **OK** button to exit the Surface Properties dialog box and view the watersheds for the Existing surface.
14. Click the **AutoCAD Save** icon and save the drawing.

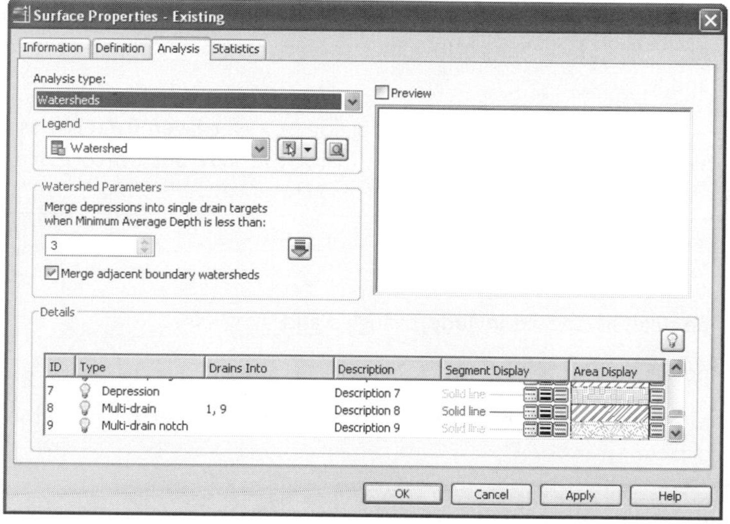

Figure 4.87

Civil 3D identifies, hatches, and labels the surface watersheds. Each watershed has a label noting the ID, type, and area.

15. Click the **Prospector** tab to make it current.
16. In Prospector, in Surfaces, expand the branch for the Existing surface.
17. Click the **Watersheds** heading and a list of watersheds for the Existing surface appear in the preview area.

18. Select a watershed from the preview area, press the right mouse button, and select **Zoom to** from the shortcut menu to view that watershed.
19. Select a couple more watersheds and use the Zoom to command to view the location in the drawing.

WATER DROP ANALYSIS

The Water Drop utility draws the path of a drop of water across a surface. This path may be critical to the review of existing conditions and design reviews.

1. Use the Zoom Extents command to view the entire site.
2. In the drawing, select an edge of the surface, press the right mouse button, and from the shortcut menu select **Surface Properties...**.
3. Click the **Information** tab, set the Surface Style to **Slope and Arrows – Quantile**, and click the **OK** button to exit the dialog box.
4. In the Surfaces menu, from the Utility flyout, select the **Water Drop...** command.
5. Click the **OK** button to accept the values and to exit the dialog box.
6. Select points on the surface in different watersheds to view the water trails as the water travels from higher to lower elevations, and press the **Enter** key to exit the command.
7. Open the Layer Properties Manager dialog box, set layer **0** (zero) as the current layer, **freeze** the C-TOPO-WDRP, and click the **OK** button to exit the dialog box.

SUMMARY

- Slope analysis should include triangles and arrows.
- Slope analysis represents the flow of water across a surface.
- Water Drop analysis draws the path of individual water drops as they travel over a surface.
- The Equal Interval and Standard Deviation range methods work best with data that is normally distributed.
- The Quantile range method works best with data that is skewed.

If the current condition of the surface components and characteristics are acceptable, then it is time to create the contours representing the surface elevations.

UNIT 5: SURFACE ANNOTATION

Surface annotation consists of contours and spot slope and/or elevation labels. In Civil 3D, contours are a displayed characteristic of a surface object (hills, valleys, depressions, etc.). A contour is a line that has a constant elevation along its entire length. Thus, a contour connects points that have the same elevation on the surface. There is a dynamic link between a surface and its contours. If a surface changes, it recalculates its triangulation and redraws the contours in response to the changes. How contours display on the screen is under the direct control of a surface object style.

STEP 7: ANNOTATE THE SURFACE

After completing the editing and analysis of a surface, the next step is to annotate it with contours and assign labels. Also, the user can add spot slope and elevation labels at critical points on the surface.

CONTOUR SURFACE STYLE

The display of contours is from the assignment of a surface style containing the appropriate settings. A contour style uses three panels of the Surface Style dialog box: Information, Contours, and Display.

INFORMATION TAB

The Information panel sets the name and description of the contour style.

CONTOURS TAB

The 3D Geometry section of the Contours panel contains settings affecting the display contours. This group of settings controls the 3D display mode of the contours. A contour can display itself to the actual elevation, flatten to a single elevation, or rise to an exaggerated elevation to show relief. The default is to display contours at the actual elevation. When flattening the contours, specify the flatten elevation. When exaggerating contours, set an exaggeration factor (see Figure 4.88).

CONTOUR INTERVAL

This section sets the contour interval for major and minor contours. The major interval is a multiple of the minor (usually a factor of 5). For example, if the minor interval is 0.5 the major is 2.5; if the minor is 1 the major is 5; if the minor is 2 the major is 10, etc. When setting a minor interval, the major setting responds with a value five times the value of the minor interval. Always check these settings to make sure they are correct for your surface (see Figure 4.89).

CONTOUR RANGES

This section defines range groupings for contours. Creating contour ranges allows a user to assign different colors to contours. For example, the contours for the elevation range of

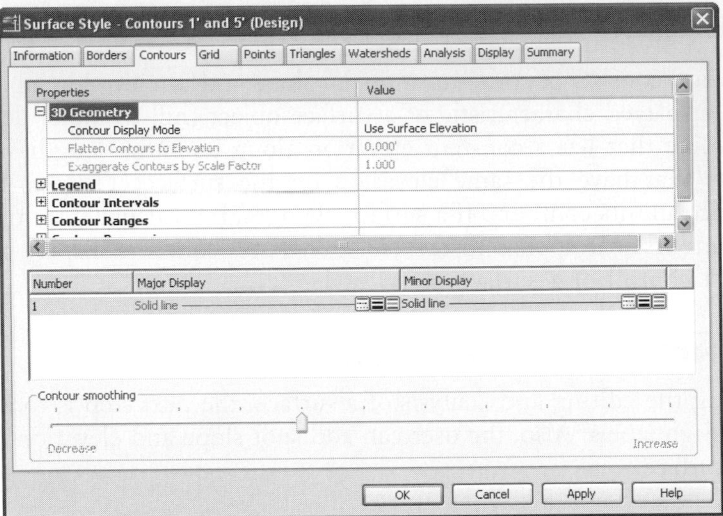

Figure 4.88

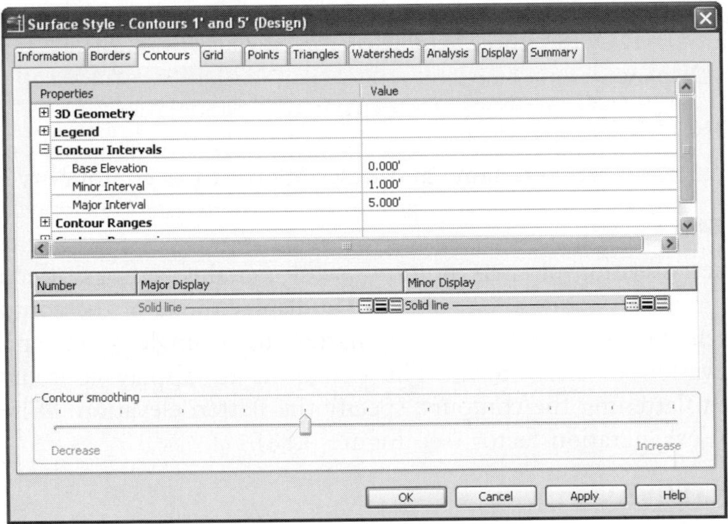

Figure 4.89

700–730 are blue, for the elevation range of 730–760 they are red, etc. These settings are related to an analysis of surface elevations and toggling on User Contours in the Display tab.

DEPRESSION CONTOURS

The contour depressions section toggles depression contour display, sets the tick interval (frequency), and size of the tick (see Figure 4.90).

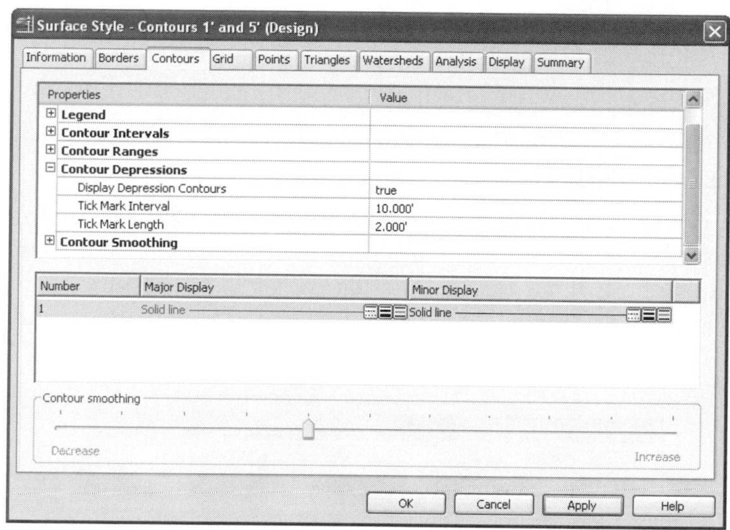

Figure 4.90

CONTOUR SMOOTHING

The Contour Smoothing option does not add or modify the surface data. There are two smoothing options: add vertices and spline. The add vertices option adds vertices to the contour, giving it a smoother look, but preserving as best as possible the actual location of the elevations. When this option is active, the bottom of the dialog box contains a slider that sets the amount of smoothing (see Figure 4.90). The spline option creates the greatest contour smoothness by passing a splined polyline through the points of the contour, but this option also creates the greatest number of crossing contours.

CONTOUR LABELS

The Add Contour Labels command annotates contours. The command calls the Create Contour Labels toolbar. The icons on the toolbar set the layer for the contour labels, the types contour labels (group, single, multiple interior), and if labeling only major or minor contours (see Figure 4.91).

Figure 4.91

CONTOUR LABEL LAYER

The first icon on the left of the Create Contour Labels toolbar calls the Select Layer dialog box. The Defpoints layer (default) represents the layer for a line defining the location of the label text. When creating the labels, the routine places the labels along a line at intersections with contours. The user can freeze or turn off the Defpoints layer and have only the labels visible. The Defpoints layer is a non-plotting layer. If this layer is on in a drawing, the layer does not plot. When erasing a contour labeling line, the user also erases the labels.

METHOD OF LABELING

There are three methods of labeling contours (see Figure 4.92). The first method is Single Label Line. This option creates a contour label at selected point on a contour. The second method is Group Label Line. This method places labels at the intersection of a line and contours. The last method, Label from Objects, labels only lines and polylines that intersect a drawn line.

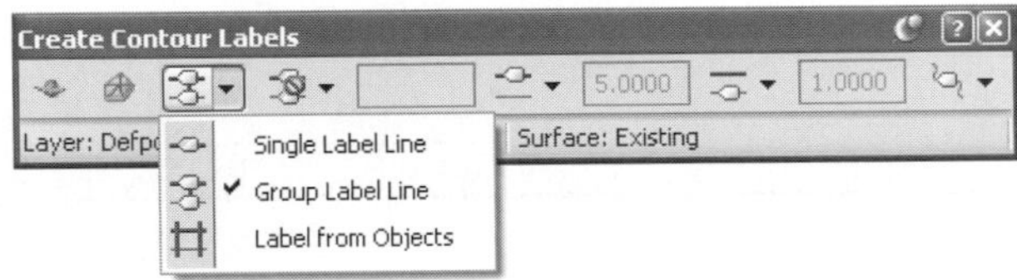

Figure 4.92

MULTIPLE LABELS

The fourth icon in from the left sets whether there are multiple labels along the length of a contour (see Figure 4.93). When selecting this option, set the distance between the labels in the box to the right of the icon. A red circle with a slash indicates that the labels appear only when the label line crosses a contour.

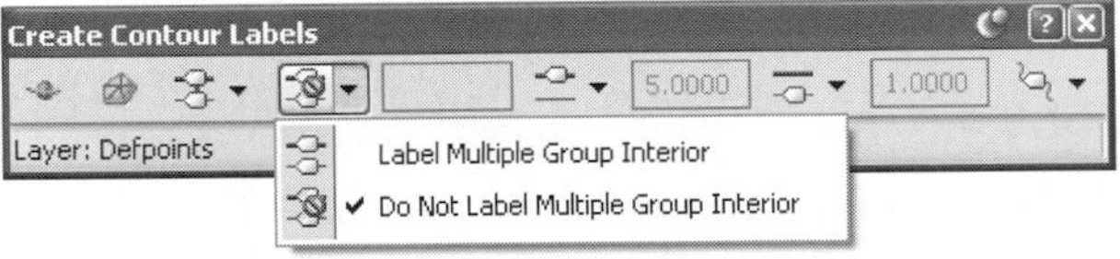

Figure 4.93

MAJOR, MINOR, AND USER DEFINED

The last three icons set what style to use for major, minor, or user define labels (see Figure 4.94). When selecting contour type to label, a toolbar displays asking the user to select a style. A red circle with a slash indicates that the contour type will not be labeled.

SURFACE SMOOTHING

Surface smoothing is a surface edit. This smoothing option interpolates better surface data for a selected region. This option is best for areas with great relief over short distances or having sparse data. After reviewing surface contours, the user may decide to use smoothing to create more pleasing contours. Review the discussion of smoothing in Unit 3 of this chapter.

Figure 4.94

SPOT SLOPE LABELS

A Spot Slope label annotates the slope at the selected point on a surface. This label documents critical surface slopes. The best method of placing these labels is with a single pick on a surface. When labeling the slope between two points, be careful not to select two points having several triangles between them. When selecting two points, the routine uses the difference in elevation between the two points to calculate a slope, not the changes that may occur between them. Place these labels from the Add Labels dialog box (see Figure 4.95).

SPOT ELEVATION LABELS

A Spot Elevation label annotates the elevation at the selected point on a surface. This label is important to documenting critical elevations on a surface. Place these labels using the Add Labels dialog box (see Figure 4.95).

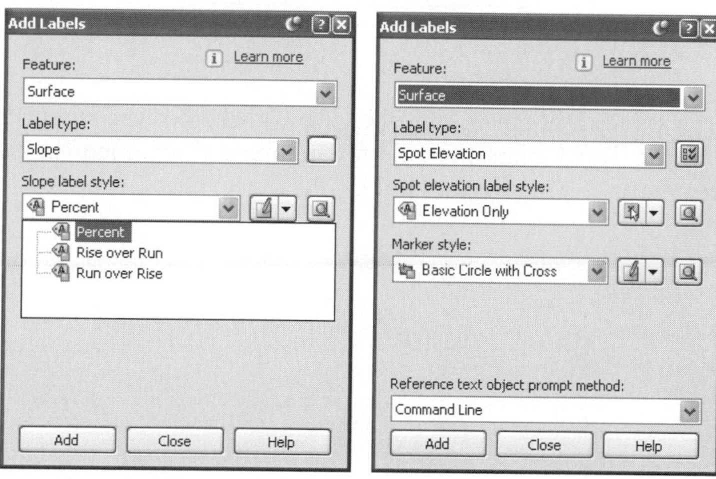

Figure 4.95

EXPRESSIONS

An expression is a user-defined label property that is specific to each label type (e.g. contour, slope, spot elevations, and watershed). An example of this type of label is a label annotating a gutter and curb top-of-face elevation. The elevation of the gutter is from a surface and an expression calculates a top-of-face elevation.

The expression dialog box presents a calculator-like interface that has access to object properties for the type of label and mathematic functions. Once defined, the user assigns a name and description to the expression and when exiting, the named expression becomes a property for the label group (see Figure 4.96).

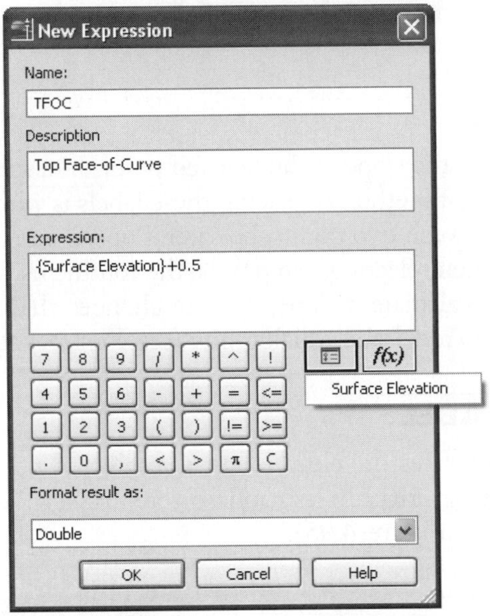

Figure 4.96

After defining the expression, create the label. Any label from the label type can now use the expression. In the Text Component Editor, combine the named expression and other properties to create a label (see Figure 4.97).

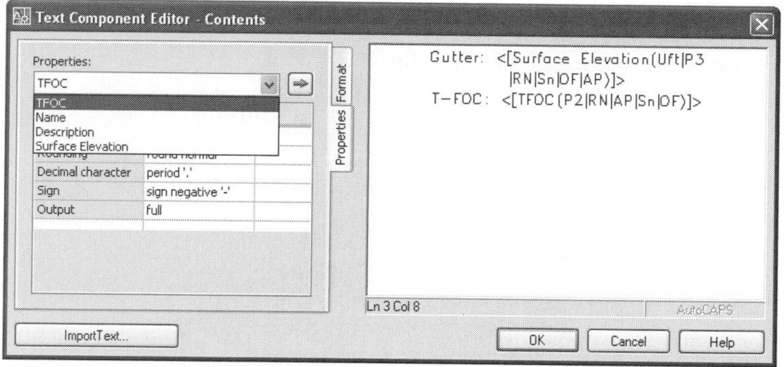

Figure 4.97

EXERCISE 4-5

When you complete this exercise, you will:

- Modify a contour style.
- Create contour labels.
- Use Nearest Neighbor smoothing on a surface.
- Create spot slope labels.
- Create spot elevation labels.

This exercise continues with the Surfaces drawing of this chapter. If you want to do only this exercise, you can load the *Chapter 4 - Unit 5* drawing that is in the *Chapter 4* folder of the CD that accompanies this textbook.

CONTOUR SURFACE STYLE

The Contours 1' and 5' (Design) is a content style from a Civil 3D template.

1. If you are not in the *Surfaces* drawing, open the drawing or open the *Chapter 4 – Unit 5* drawing.
2. Click the **Settings** tab to make it current.
3. In Settings, expand the Surface branch until you are viewing the list of Surface Styles.
4. Select the **Contours 1' and 5'** (Design) style, press the right mouse button, and from the shortcut menu select **Edit...**.
5. Click the **Contours** tab and review the setting for 3D Geometry, Contour Display Mode, and set Contour Depressions, Display Depression Contours to **True**.
6. Click the **OK** button to exit the dialog box.
7. In Settings, expand the Surface branch until viewing a list of Label styles for Contour.
8. Click the **Existing Major Labels** style, press the right mouse button, and select **Edit...** to view the Label Style Composer dialog box for the style.

EXERCISES

9. Click the **General** tab to display its settings.

   The General tab controls the visibility and plan readability for the style.

10. Click the **Dragged State** tab to display its settings.

    The Dragged State tab contains settings that control what happens to a label should it be dragged from its original location.

11. Click the **Layout** tab to view the label settings.

    The Layout tab contains settings that define and format the contour label. The General section of the panel defines how the label attaches to a contour, the Text section defines how the text is formatted and located relative to its attachment to the contour, and the Border section defines a box surrounding the label if used.

12. If necessary, expand the Text section to display its values.

13. Click in the **Value** cell for Contents, then click the ellipsis to the right to display the Text Component Editor – Contents dialog box to view the label formatting (see Figure 4.98).

14. To change the value of any modifier, highlight the format on the right side of the dialog box, on the left side click once in the Value cell, then on the drop list arrow at the right side of the cell, and select the change from a list of options.

15. Click the **Cancel** buttons to exit the dialog boxes and to return to the command prompt.

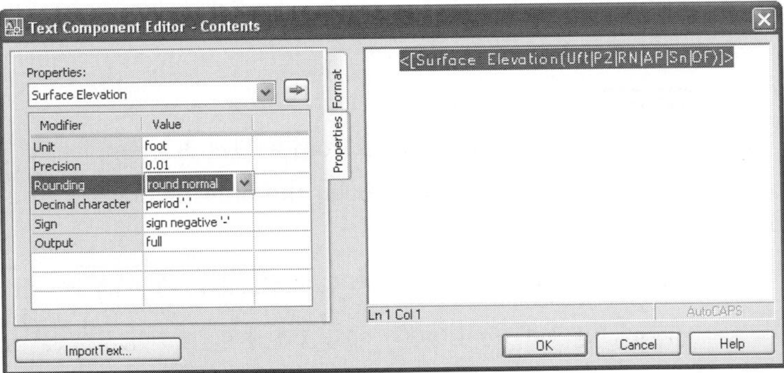

Figure 4.98

CREATING AND LABELING CONTOURS

To create the contours for the surface, assign a style that displays them.

1. Click the **Prospector** tab to display its panel.
2. In Prospector, expand the Surfaces branch until viewing a list of surfaces, select **Existing**, press the right mouse button, and from the shortcut menu select **Properties...**.

3. Click the **Information** tab and change the Surface Style to **Contours 1' and 5' (Design)**.
4. Click the **OK** button to exit the dialog box.

   By assigning this style, the surface displays its contours.
5. From the Labels flyout in the Surfaces menu, select **Add Contour Labels...** to display the Create Contour Labels layout toolbar.
6. In the toolbar, click the drop list arrow to the right of the Group Label Line icon and set it to **Group Label Line**.

   The fourth icon in from the left controls multiple labels on a contour. If the icon has a red circle with a slash, a label will be put where the label line crosses a contour. If this is set to Label Group Interior, you need to set the distance between labels on contour in the box to the right of the icon.
7. Set the icon to **Do Not Label Multiple Group Interior**.
8. Next set the icons to label **Majors** and **Minors**. If prompted for a style, select the **Existing Major Labels** for the major contour and **Existing Minor Labels** for the minor.

   The routine prompts you for a starting point of a contour label line.
9. Select and start an end point on the surface to create a few label lines.

   This places the contour labels on the contours where the line intersects the labels.
10. Type in "**X**" to exit the contour label routine.
11. Use a grip edit to relocate the label line, repositioning the labels.
12. Change the label parameters and label additional contours.
13. Erase the label lines to remove the contour labels from the drawing.

EXERCISES

SURFACE SMOOTHING

The Existing surface displays several chevron contours. The reason for these contours is the lack of data. The problem with surface object smoothing or splining is potentially creating crossing contours. Civil 3D offers another option, Edit Surface Smoothing. This routine creates interpolated surface data and as a result produces smoother contours.

1. In the drawing, click anywhere on the surface, press the right mouse button, from the shortcut menu select **Surface Properties...**, and click the **Information** tab.
2. In the Information tab, set the current surface style to **Border & Triangles & Points** and click the **OK** button to exit the dialog box.
3. From the View menu, select **Named Views...**, set the **Central** view current, and click the **OK** button to exit the dialog box.
4. If necessary, click the **Prospector** tab.

5. In Prospector, expand the Existing surface branch until viewing the Definition data type list.
6. Click **Edits**, press the right mouse button, and from the shortcut menu select **Smooth Surface....**
7. Set the smoothing method to **Nearest Neighbor Interpolation**, **Grid based**, and with an X and Y spacing of **5**.
8. Click in the Value cell of Select Output Region, click the ellipsis at the right side of the value cell, type in the letter **E** for the **rectangle option,** and select a region that covers the central portion of the surface.

   After selecting the region you return to the Smooth Surface dialog box.
9. Click the **OK** button to exit the Smooth Surface dialog box.

   The effect of the command is to create new grid triangulation in the center of the site. This allows for a better interpolation of surface elevations (see Figure 4.99).

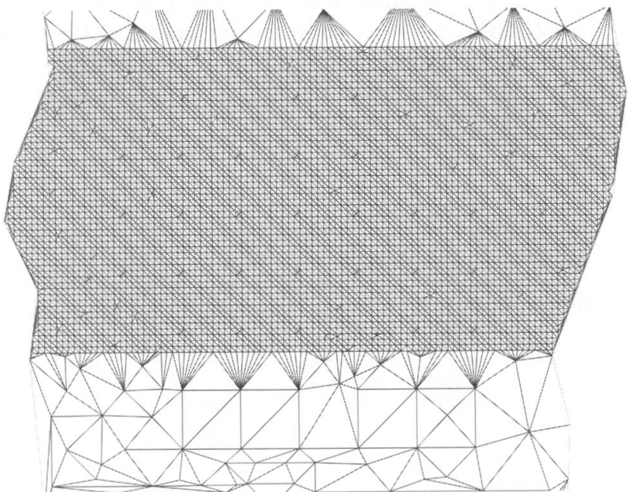

Figure 4.99

10. In the drawing, click anywhere on the surface, press the right mouse button, and from the shortcut menu select **Surface Properties....**
11. In the Information tab, set the current surface style to **Contours 1' and 5' (Design)** and click the **OK** button to exit the dialog box.
12. Click the **AutoCAD Save** icon to save the drawing.

The new contours are considerably smoother than the original surface contours (see Figure 4.100).

In the Definition tab of the Surface Properties dialog box, the smoothing Edit may not show on the Operation Type list. When the list gets long, the Operation Type area does not display scrollbars. You must select an operation from the list and press the down arrow key to view the hidden edits. You can toggle off the effects of the smoothing in the dialog box and return the surface to its previous state.

Figure 4.100

SPOT SLOPE LABELS

Slope labels are one choice from the Add Labels dialog box. While triangles and slope arrows could be part of the documentation, there is a need to annotate critical slopes on an existing or design surface.

1. In the drawing, select anywhere on the surface, press the right mouse button, and from the shortcut menu select **Surface Properties...**.
2. Click the **Definition** tab and in Operations Type toggle **off** surface smoothing.
3. Click the **OK** button to exit the Surface Properties dialog box and click the **Yes** button to rebuild the surface.
4. If Civil 3D issues a warning about duplicate points, close the Event Viewer.

   The surface returns to the previous state.
5. Click the **Settings** tab and expand the Surface branch until you are viewing the list of Slope label styles.
6. Select the style **Percent**, press the right mouse button, and from the shortcut menu select **Edit...**.
7. Click the **Layout** tab to view its settings.
8. Change the Component Name to **Surface Slope**.
9. In the Text section, click in the Contents Value cell (an ellipsis appears) and click on the ellipsis to view the Text Component Editor.
10. The Properties portion of the dialog box contains the format of the label (see Figure 4.101).

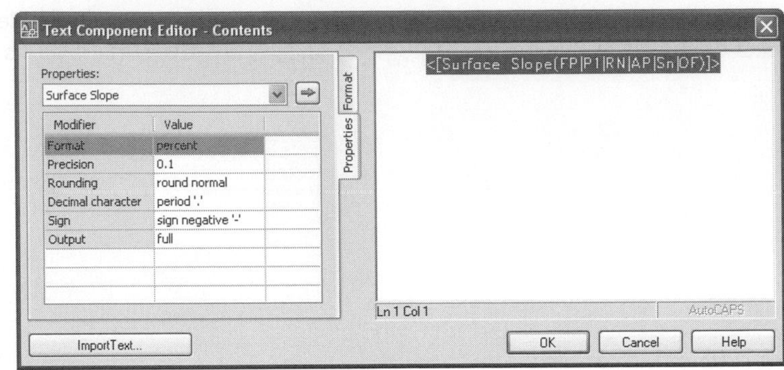

Figure 4.101

11. Click the **OK** button to exit the Text Component Editor.
12. Click the **OK** button until you exit the Percent style.
13. From the Labels flyout in the Surfaces menu, select **Add Surface Labels....**

    The Add Labels dialog box displays.
14. Set the Feature to **Surface**, the Type to **Slope**, and the Style to **Percent**.
15. Click the **Add** button to start labeling.
16. If the routine prompts for a surface, select the border or contour that represents the existing surface.
17. The next prompt asks for one or two points to create a label. Select the one point method by pressing the **Enter** key.
18. Select four or five points on the screen to create the labels.
19. When you are done, click the **Close** button to close the Add Labels dialog box.

The labels annotate the percentage of slope (negative or positive) and show the direction of down slope by orientating the text and arrow towards the down slope direction.

SPOT ELEVATION LABELS

1. Click the **Settings** tab, and expand Surface branch until you are viewing the list of Spot Elevation label styles.
2. Click **EL:100.00**, press the right mouse button, and from the shortcut menu select **Edit....**
3. Click the **Layout** tab to view the settings for the label.
4. In the Text section of the panel, click in the Contents Value cell (an ellipsis appears), and click on the ellipsis to view the Text Component Editor.
5. In the Properties for Surface Elevation are the values defining the format of the label (see Figure 4.102).

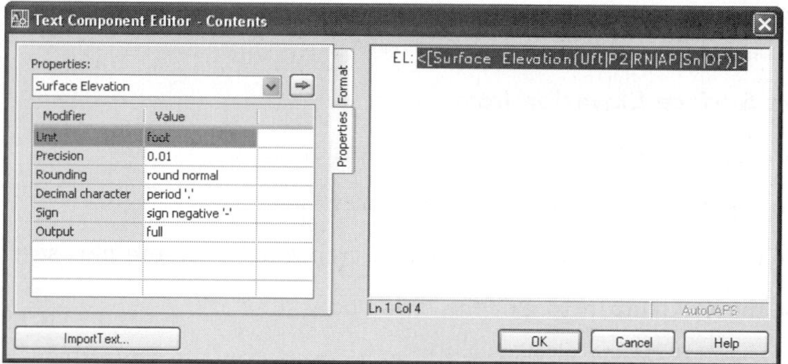

Figure 4.102

6. Click the **OK** button to exit the Text Component Editor.
7. Click the **OK** button until you exit the dialog box.
8. Select **Add Surface Labels...** from the Labels flyout of the Surfaces menu.

   This displays the Add Labels dialog box.
9. Make sure the Feature is **Surface**, the Type is **Spot Elevation**, the Spot Elevation Style is **EL: 100.00**, and the Point Style is **Basic Circle with Cross**.
10. Click the **Add** button to start labeling.
11. If the routine prompts for a surface, select the border or contour that represents the Existing surface.
12. The next prompt asks you to select a point to create a label. Select four or five points on the screen to create the labels.
13. When you are finished, click the **Close** button to close the Add Labels dialog box.
14. Erase the labeling from the drawing.
15. Click the **AutoCAD Save** icon to save the drawing.

CREATING AN EXPRESSION

An expression uses an object property specific to the type of label. You are going to create a spot elevation label that has a surface elevation and adds 0.5 feet to the elevation.

1. Click the **Settings** tab to make it current.
2. In Settings, expand the Surface branch until viewing a list of Spot Elevations styles.
3. Select **Expressions**, press the right mouse button, and from the shortcut menu select **New....**
4. For the Name, enter **TFOC** and for the description enter **Top Face-of-Curb** (see Figure 4.96).

5. Click the **Insert Property** button (to the right of the ! (exclamation point)) to display the list of properties.
6. Select **Surface Elevation** from the list.

   This adds surface elevation to the expression area.
7. Click the plus sign to add it to the expression.
8. Click the **0** (zero), a point (**.**), and **5** to create 0.5 after the plus sign.
9. Click the **OK** button to exit the dialog box.

   The Expression entry in Settings has an assigned icon and the preview area lists TFOC.

ADDING AN EXPRESSION TO A LABEL

The expression is now a property to which you choose to add to a label component.

1. Select the **EL:100.00** style, press the right mouse button, and from the shortcut menu select **Copy....**
2. Select the **Information** tab and change the name to **Gutter and TFOC**. Enter **Gutter and Top of Face-of-Curb** for the description.
3. Select the **Layout** tab to make it current.
4. Click in the Contents Value cell and click the ellipsis to display the Text Component Editor.
5. Click in the text EL: on the right side of the dialog box and change it to **Gutter:**.
6. Place the cursor after the format string of Gutter (after the ">" greater than sign) and press the **Enter** key to make a new line.
7. Type in **F-TOC:** (leave a space after the colon).
8. On the left side, click the drop list arrow for Properties and select **TFOC** from the list.
9. For the modifier precision, click in the Value cell, then click on the drop list arrow, and select **0.01** as the precision for the TFOC value.
10. Click the blue arrow at the top center to place the format into the label.
11. Click the **OK** buttons until you exit and return to the command prompt.
12. Click the **AutoCAD Save** icon to save the drawing.
13. From the Labels flyout of the Surfaces menu, select **Add Surface Labels....**
14. In the Add Labels dialog box, change the Label Type to **Spot Elevation**, change the Spot elevation label style to **Gutter and TFOC**, and change the Marker style to **Basic Circle with Cross**.
15. Click the **Add** button and add a few labels to the drawing.
16. In the Add Label dialog box, click the **Close** button.

17. Erase the labels from the drawing.
18. Click the **AutoCAD Save** icon to save the drawing.

SUMMARY

- Contour styles consist of an interval for the major and minor contours.
- You can define depression contours for a surface.
- Spot, slope, and elevation labels are important labeling styles.
- Surface smoothing is effective in areas of sparse data.

Once you have a satisfactory surface, you can create points whose elevations are the elevations of the surface. These points can be data for a survey crew or a starting point for a second surface.

UNIT 6: SURFACE POINT AND EXPORT TOOLS

The Create Points toolbar contains routines that create points from surface elevations. The export LandXML command exports surfaces so other applications can use the surface.

POINTS FROM SURFACES

The Surface icon stack of the Create Points toolbar has routines creating new points whose elevations are from a selected surface. Create Points of the Points menu or from the shortcut menu of Prospector's Points heading displays this toolbar. The Surface point creation routines are Random Points, On Grid, Along a Polyline/Contour, and Polyline/Contour Vertices (see Figure 4.103).

The Along a Polyline/Contour and (at) Polyline/Contour vertices routines create points whose elevations are at distances along the length of or at the vertices of a polyline or contour. These routines are useful when developing a grading design and allow the user to develop points that represent the intersection of the grading and existing ground.

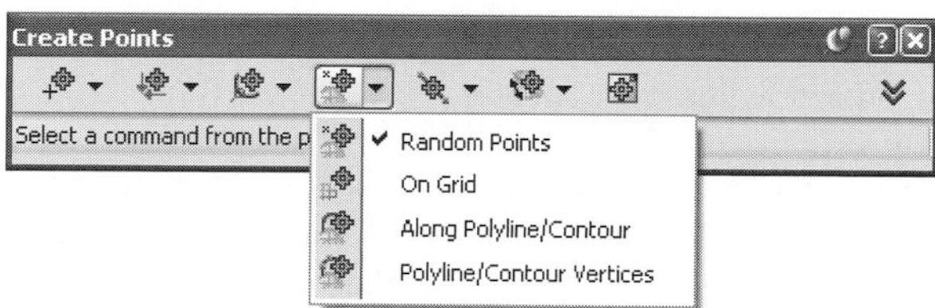

Figure 4.103

LANDXML

One method of transferring a surface to other applications, including Land Desktop, is exporting a Civil 3D surface as a LandXML file. The export LandXML command is in the Export flyout of the File menu, or from a shortcut menu of a selected surface name (see Figure 4.104).

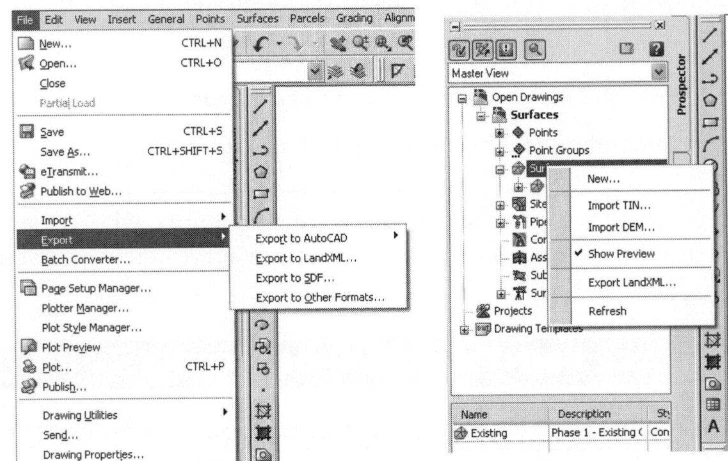

Figure 4.104

The LandXML export process prompts for a method of identifying the surface. The user can select a surface from the drawing or from a filtering dialog box (see Figure 4.105).

LANDXML SETTINGS

Before exporting any LandXML data, check the export unit settings and make sure they represent the correct units. By default, the Imperial units are set to International feet and should be set to U.S. Foot (see Figure 4.106).

EXERCISE 4-6

When you complete this exercise, you will:

- Place points on a grid.
- Place points on a polyline or contour.
- Export a LandXML file.

This exercise continues with the Surfaces drawing of this chapter. If you are starting with this exercise, you can open the *Chapter 4 - Unit 6* drawing that is in the *Chapter 4* folder on the CD that accompanies this textbook.

POINT CREATION AND POINT IDENTITY SETTINGS

The first part of this exercise sets the environment for creating points from a surface.

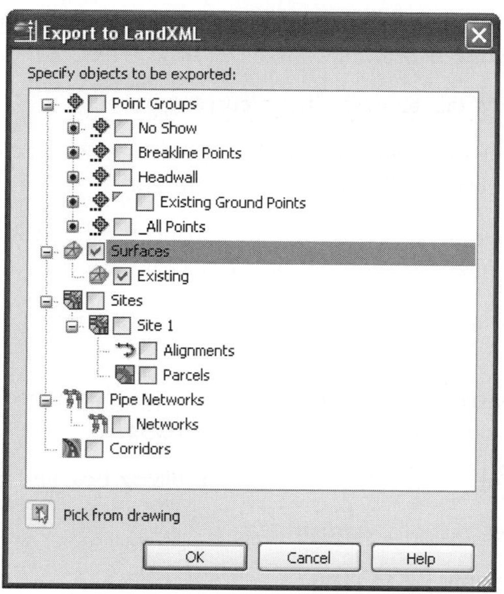

Figure 4.105

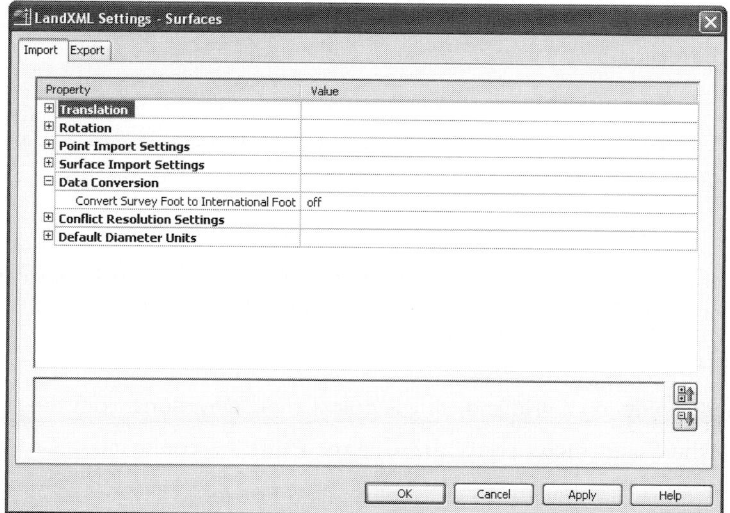

Figure 4.106

EXERCISES

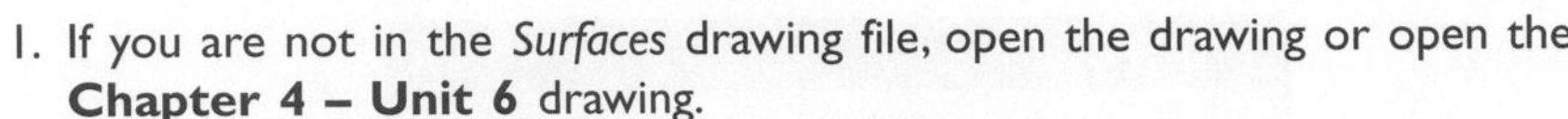

1. If you are not in the *Surfaces* drawing file, open the drawing or open the **Chapter 4 – Unit 6** drawing.
2. Click the Settings tab to make it the current panel.
3. In Settings, select the Point heading, press the right mouse button, and from the shortcut menu select **Edit Feature Settings....**
4. In the Edit Feature Settings dialog box, expand the Point Identity section, set the Next Point Number to **400**, and click the **OK** button to exit the dialog box.
5. In Settings, expand the Point branch until you are viewing the Commands heading and its list.
6. Select the **CreatePoints** command, press the right mouse button, and from the shortcut menu select **Edit Command Settings....**
7. Expand the Points Creation section of the dialog box and set the following values:

   Prompt For Descriptions: **Automatic**

   Default Description: **EXISTING**
8. Click the **OK** button to exit the Edit Command Settings dialog box.
9. In Layer Properties Manager, thaw the **V-NODE** layer, and click the **OK** button to exit the dialog box.
10. Click the **Prospector** tab to view its panel.
11. Click the **Point Groups** heading, press the right mouse button, and from the shortcut menu select **Properties....**
12. Move the **_All Points** point group to the top of the list.
13. If Civil 3D issues a warning about duplicate points in the surface, close the Event Viewer Panorama throughout this exercise.

POINTS ON GRID

This routine places points on a grid and assigns points their elevations from the surface elevations.

1. From the Points menu select the **Create Points...** command.

   This displays the Create Points toolbar (see Figure 4.103).
2. Click the drop list arrow to the right of the Surface icon stack and select the **On Grid** routine.
3. The routine prompts you to select a surface object and select a component of the surface to identify it.
4. Next, the routine asks for the base point of the grid. Select a point in the lower-left of the surface.
5. Then press the **Enter** key to accept a zero rotation and enter the grid spacing of **30** for X and Y.

6. Finally, select a point at the upper-right of the surface interior.

   The routine displays a box representing the grid area. Within the grid area is another box showing the grid spacing (lower-left corner).

7. The command prompt asks if you want to change the definition of the grid. Answer **NO** by pressing the **Enter** key to move to the next prompt.

   The routine uses the automatic description of Existing for the points and draws the points on the screen.

8. The routine prompts for another surface object. Press the **Enter** key to exit the command and return to the command prompt.

9. Click the **Points** heading in Prospector to preview the points.

   The new points display in the preview area and start with point number 400.

10. In Prospector, select the Point Groups heading, press the right mouse button, and from the shortcut menu select **New....**

11. In the Point Group Properties dialog box, click the **Information** tab and name the point group **Points from Existing**.

12. Click the **Include** tab, toggle on **With Numbers Matching**, and enter in the point numbers from **400–800** (or whatever ending point number is appropriate for the points in your drawing).

13. Click the **OK** button to exit the Point Group Properties dialog box.

14. Click the **Point Groups** heading, press the right mouse button, and from the shortcut menu select **Properties....**

15. Click the **_All Points** point group and move it to the **top** of the list.

16. Click the **OK** button to exit the dialog box.

POINTS ON A POLYLINE/CONTOUR

1. From the View menu select the **Named Views...** command and restore the **Central** view.

2. Use the Zoom Window command and zoom in on the surface.

3. In the Create Points toolbar, click the roll-down arrow, expand the Points Creation section of the panel, change the Default Description to **CONTOUR**, and click the roll-up arrow to hide the settings panel of the Create Points toolbar.

4. Click the drop list arrow to the right of the Surface icon stack and select the **Along Polyline/Contour** routine.

5. Click the surface border to identify the surface object, accept or set the distance to **10** feet, and select a contour to set the points on.

6. Press the **Enter** key to exit the routine.

7. In Prospector, select the Point Groups heading, press the right mouse button, and from the shortcut menu select **New...**.
8. Click the **Information** tab and assign the name **From Contour**.
9. Click the **Include** tab, toggle on **With Raw Description Matching**, enter in **CONTOUR**, and click the **OK** button to create the From Contour point group.
10. In Prospector, select the Point Groups heading, press the right mouse button, and from the shortcut menu select **Properties...**.
11. Click the **No Show** point group and move it to the second position on the list.
12. Click the **OK** button to exit the dialog box.

    This displays only the contour points on the screen.
13. You may need to click **Point Groups**, press the right mouse button, and from the shortcut menu select **Update**.
14. Click **Point Groups**, press the right mouse button, and from the shortcut menu select **Properties...**.
15. Click the **No Show** point group and move it to the **top** of the list.

    This hides all of the points in the drawing.
16. Click the **OK** button to exit the Point Group Properties dialog box.
17. Click the "**X**" to close the Create Points toolbar.
18. In Prospector, select the **From Contour** point group, press the right mouse button, and from the shortcut menu select **Delete Points...**. Click the **OK** button to answer **yes** to the Are you sure dialog box.
19. In Prospector, select the **From Contour** point group, press the right mouse button, and from the shortcut menu select **Delete...**. Answer **Yes** to the Are you sure dialog box.
20. In Prospector, select the **Points From Existing** point group, press the right mouse button, and from the shortcut menu select **Delete Points...**. Click the **OK** button to answer **Yes** to the Are you sure dialog box.
21. In Prospector, select the **Points From Existing** point group, press the right mouse button, and from the shortcut menu select **Delete...**. Answer **Yes** to the Are you sure dialog box.
22. Click the **AutoCAD Save** icon and save the drawing.

CHECK LANDXML EXPORT SETTINGS

1. Click the **Settings** tab to make it current.
2. Select the drawing name, press the right mouse button, and from the shortcut menu select **Edit LandXML Settings...**.

3. Click the **Export** tab and expand the **Data Settings** section of the panel.
4. Change the Imperial Units value to **survey foot**.
5. Click the **OK** button to exit the dialog box.

EXPORTING A SURFACE AND POINT GROUPS

1. Click the **Prospector** tab to make it current.
2. In Prospector, from the Surfaces branch, select the Existing surface, press the right mouse button, and from the shortcut menu select **Export LandXML....**

   The Existing surface should be toggled on.
3. Toggle on the **Existing Ground** and **Breakline Points** point groups to include their points in the LandXML file.
4. Click the **OK** button to continue the exporting process.
5. In the Export LandXML dialog box locate a folder, name the file, and click the **Save** button to export the data.

This ends the chapter on Civil 3D surfaces. The next step is to design a second surface and calculate an earthworks volume between the existing ground and the new design surface. The next chapter focuses on surface design tools and the process of calculating earthwork volumes.

SUMMARY

- You can create points whose elevation is from a surface randomly or on a grid.
- You can place points whose elevations are from a surface on a polyline at an interval.
- The default Imperial Unit for exporting data to a LandXML file is International feet. This should be set to survey feet.

Alignments

INTRODUCTION

This and the next three chapters focus on the design and documentation of a roadway. This chapter concentrates on the plan view of the design process. The next three chapters focus on profile and section views of a design and the roadway model (the corridor). As with all Civil 3D objects, styles affect the display of object components and, when done with the design, other styles annotate it with appropriate labels and notes.

OBJECTIVES

This chapter focuses on the following topics:

- Settings that Affect Designing Civil 3D Alignments
- Alignment Styles Affecting the Look and Design Criteria of an Alignment
- Modification of the Stationing and the Addition of Station Equations to an Alignment
- Alignment Reports
- Alignment Design Parameters
- Right-of-Way (ROW) Creation from an Alignment
- Point Creation from an Alignment
- Parcel Division with an Alignment

OVERVIEW

Each menu—Alignments, Profiles, Corridors, and Sections—contains tools to create, analyze, edit, and annotate a roadway model. The process of designing a roadway is a sequence of steps that use two-dimensional elements and views; alignments (plan view); profiles (side view); and sections (view to the left and right) to create a three-dimensional model of a road (a corridor). In Civil 3D a site and its parcels are related to alignments. Alignments can subdivide a site and its parcel(s), and because of this, alignments, profiles, and sections are a part of the Site branch.

Civil 3D dynamically links each of these elements, and any changes made to one of these design elements cascades from where the change is made to the corridor and its sections. The progression of a change follows the Alignments branch of Sites.

Changes travel from the first element in the dependency list to the last (i.e., from alignment to profile, to the corridor, and finally to section views). The dependency chain includes surfaces. A surface provides elevations to profile views and section views. If something should change in a surface, the dependent roadway elements adjust their values to accommodate the surface change.

The Civil 3D roadway design process produces a design model (the corridor) and a set of documentation. Civil 3D uses design views, plans, profiles, and sections to document these two-dimensional aspects of a three-dimensional model. The roadway design process reduces the three-dimensional aspects of the roadway model to three two-dimensional components (see Figure 5.1). This is a traditional design methodology; it means projecting the 3D model (the corridor) onto planes showing that portion of the road design as 2D. This same process occurs when designing a roadway on paper. The paper is the 2D plane on which the designer represents the three respective views of the roadway design model. It isn't until Civil 3D creates a corridor that a model of a road design appears. The road model documentation includes roadway earthwork volume estimations and roadway surface volumes.

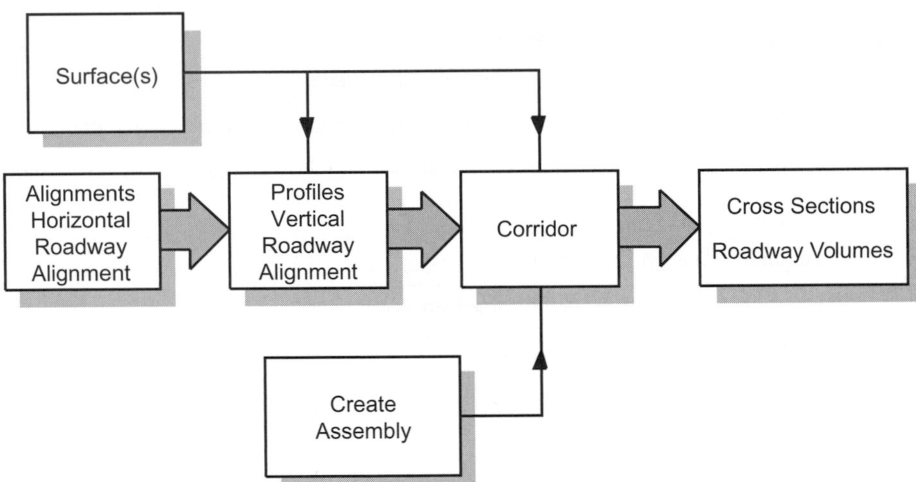

Figure 5.1

The horizontal design tools for a roadway are in the Alignments menu (see Figure 5.2). The vertical tools are in the Profiles menu. The model and cross-sectional views of the design are in the Corridors and Sections menus. When developing a vertical frame of reference (profile and cross-section views), the routines sample elevation data from the surfaces along or to either side of a roadway centerline.

The design process is not overly flexible. Users cannot design a horizontal roadway centerline and then jump to creating an assembly and a corridor. The path of an assembly along a centerline also depends on a vertical alignment developed in a profile view. What occurs at any point along the road design process depends on what occurred before the current location in the process.

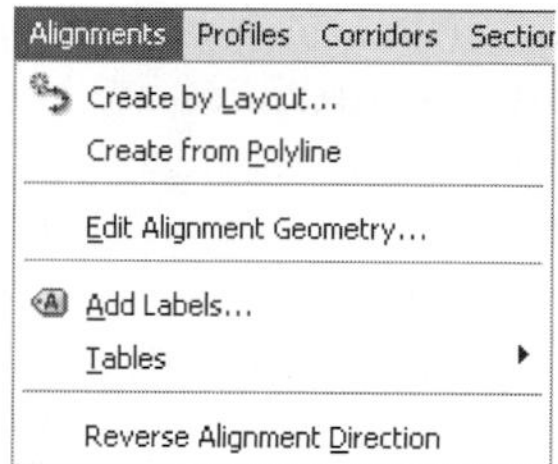

Figure 5.2

What provides the most benefit to the analysis and editing of the road model are the relationships among the design elements. If a user wants to try an alignment in a new position and view its new profile, all he or she needs to do is to grip-edit the alignment, and the profile will redraft itself to the new length, resample the new elevations, and redraw itself within an adjusted profile view.

The goals of a roadway design will vary from project to project, but generally the goals include not moving more soil material than necessary to create the road, and to use the elevations from the road design to start the overall grading of a site. Rarely is this process a single pass through the design steps; more likely, it is an iterative process. All the initial values of the design elements give the designer a starting point from which to optimize the final model. To modify a design or an element in the design means returning to a point in the design process and changing the values or elements found at that point. After making the change, Civil 3D ripples the change through the model by working the change downward from the point of change toward roadway section views.

UNIT 1

The object styles used by horizontal alignments are the topic of the first unit of this chapter. The alignment object styles control basic design parameters and visibility of alignment components and characteristics. As with other styles in Civil 3D, these styles can be saved to an office template, and when a user selects the template for a new drawing, the styles with the template become a part of the drawing.

It is difficult to separate the labeling styles from the alignment object styles. Civil 3D will automatically annotate an alignment as a user designs it in the drawing. Users may find that they will develop styles that apply when designing an alignment, and other styles to document its final design values.

UNIT 2

A centerline alignment in Civil 3D is a single object composed of tangents, curves, and possibly spirals. There are four methods of creating an alignment. The first is by importing a LandXML file. The second is by reading alignment definitions directly from a Land Desktop (LDT) project data structure. The LandXML file and LDT project can both contain one or more alignments, and when importing them the user can select which alignments to import. The third method simply converts a polyline into an alignment. The last method allows users to create an alignment with the tools in the Alignment Layout

Tools toolbar. This toolbar allows users to draft alignment tangent and cure elements. The drafting and defining of the roadway centerlines are the focus of Unit 2.

UNIT 3

Unit 3 covers the analysis of a centerline's design values. Most engineers design by calculations or design criteria. Lines and arcs on the Civil 3D screen do not tell the designer the whole story of their values. The design numbers are found in the horizontal editor or in the LandXML reports of Civil 3D.

UNIT 4

There are many times when the design changes or station equations are placed on the centerline. When a user edits an alignment, the Alignment Editor vista displays all the essential data the designer needs to make a design change. An alternative editing method is to graphically relocate alignment elements. An alignment is a collection of fixed, floating, and/or free elements. When an alignment uses these segment types, it has a predictable behavior when graphically manipulated. The editing of alignments and the addition of station equations are the subjects of the fourth unit.

UNIT 5

The annotation of the centerline is the topic of Unit 5. There are two basic annotation styles for documenting an alignment. The first style is labeling parallel to and offset from the roadway. The second uses labels perpendicular to and on the centerline. Civil 3D provides an extremely flexible system for annotating alignments.

UNIT 6

The last unit reviews routines that create new objects that reference the centerline. The types of objects that reference an alignment are station and offset labels, points, and parcels (ROW).

UNIT 1: ROADWAY STYLES

The appearance of an alignment in a Civil 3D drawing depends on drawing settings and object and label styles.

DRAWING SETTINGS

The Object Layers panel of the Drawing Settings dialog box sets the base layer name for an alignment object (see Figure 5.3). The layer can have a prefix or suffix whose value is the name of the alignment. This is done by setting the type of modifier and entering an asterisk (*) to include the name of the associated alignment. The Object Layers panel allows users to lock this naming convention for all lower styles.

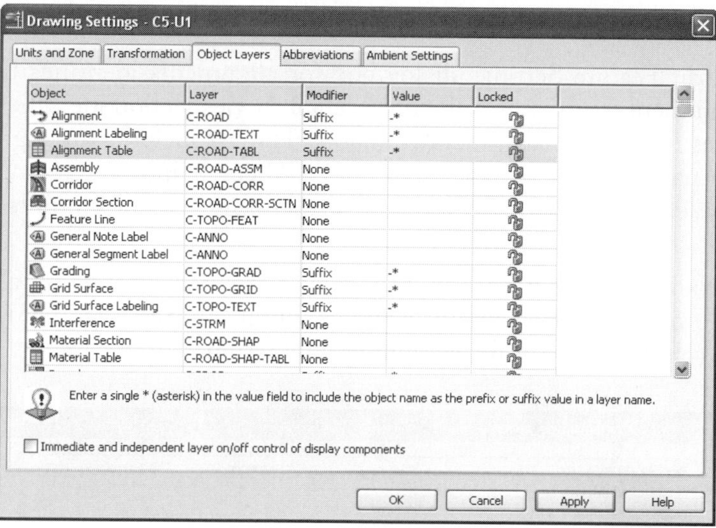

Figure 5.3

The Abbreviations panel sets critical alignment geometry points coding (see Figure 5.4). The code's values vary greatly from authority to authority, and Civil 3D allows users to enter a set of codes specific for the location. If a user is working with clients who use different codes, the user could define different drawing templates that address each client's code values. This section also sets basic labeling values for alignment label styles.

Drawing Settings - C5-U1

Units and Zone | Transformation | Object Layers | Abbreviations | Ambient Settings

Property	Value
General Text	
Alignment Geometry Point Text	
Spiral-Spiral Intersect	SS
Reverse Spiral Intersect	SPI
Alignment End	EP
Alignment Beginning	BP
Tangent-Tangent Intersect	PI
Tangent-Curve Intersect	PC
Curve-Tangent Intersect	PT
Compound Curve-Curve Intersect	PCC
Reverse Curve-Curve Intersect	PRC
Spiral-Tangent Intersect	ST
Tangent-Spiral Intersect	TS
Curve-Spiral Intersect	CS
Spiral-Curve Intersect	SC
Alignment Geometry Point Entity Data	
Alignment Beginning Point	BP: STA=<[Station Value(Uft|FS|P2|RN|Sn|...
Alignment End Point	EP: STA=<[Station Value(Uft|FS|P2|RN|Sn|...
Line Beginning	LB: L=<[Length(Uft|P3|RN|AP|Sn|OF)]> DI...

OK | Cancel | Apply | Help

Figure 5.4

EDIT FEATURE SETTINGS

The Setting's Edit Feature Settings dialog box for alignments identifies the default label styles for an alignment (see Figure 5.5). The Default Styles section identifies the object and label style names for specific alignment elements. The list includes an alignment label set. A label set is a named group of label styles (e.g., Major Minor & Geometry Point). See Figure 5-6. The effect of a set is to label all of the basic geometry points along the path of an alignment by using a single style.

The Default Naming Format section sets how Civil 3D creates an alignment name (alignment followed by a sequentially assigned number). See Figure 5-5. The Station Indexing value sets the interval for stationing annotation.

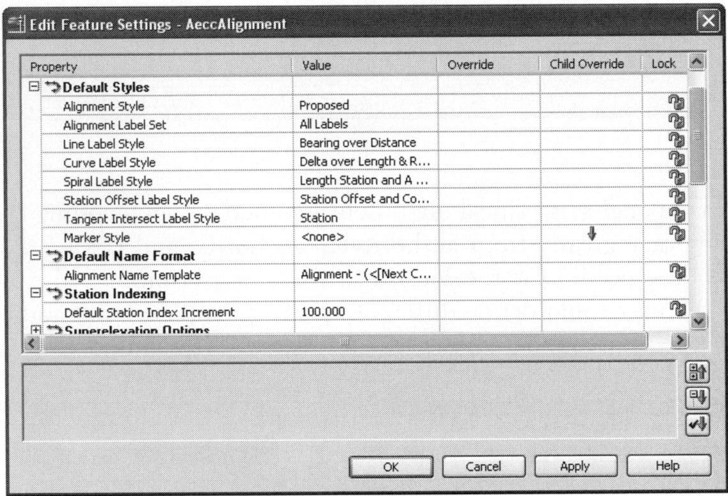

Figure 5.5

The Superelevation Options section sets the default values for any superelevating road design. This is a topic for a later chapter.

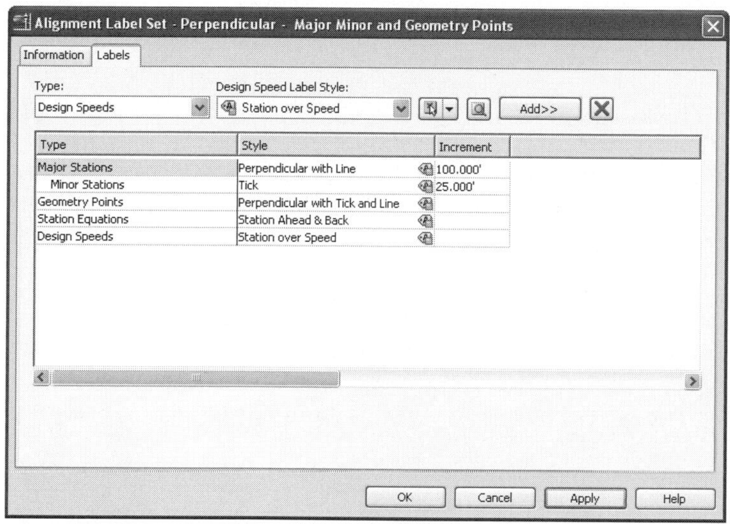

Figure 5.6

ALIGNMENT OBJECT STYLE

The alignment object style controls basic design parameters and visibility of alignment components and characteristics. As with other styles in Civil 3D, these styles can be saved to an office template, and when a user selects the template for a new drawing, the styles in the template become a part of the drawing.

PROPOSED ALIGNMENT STYLE

The Proposed style defines the behavior and display of alignment segments and components.

The Information tab sets the name of the style, description, and modification date of the style.

The Design tab's Enable radius snap toggle enables a radius snapping value when graphically editing an alignment. When editing an alignment with this toggle on, the resulting curves will have a radius that is divisible by the Radius snap value (see Figure 5.7).

The Markers tab sets the styles for markers identifying critical points along an alignment's path. The marker styles are from the Multipurpose Style branch of Settings's General heading. The bottom of the panel defines the alignment direction arrow (see Figure 5.8).

The Display tab contains a layer list for each component of an alignment (see Figure 5.9).

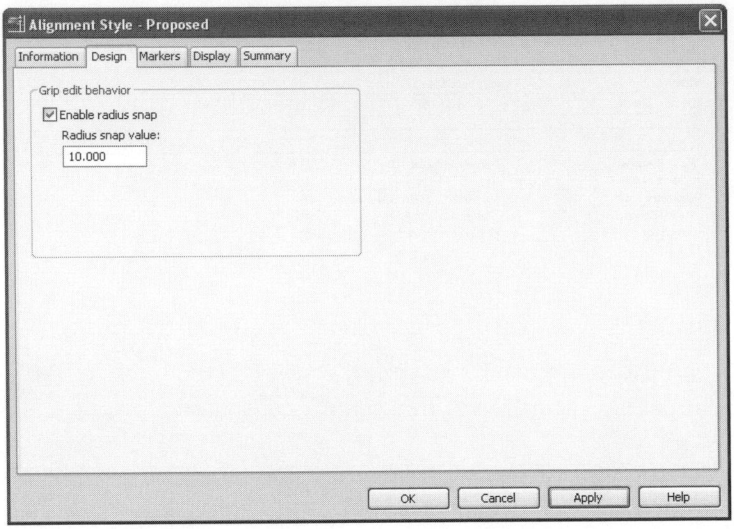

Figure 5.7

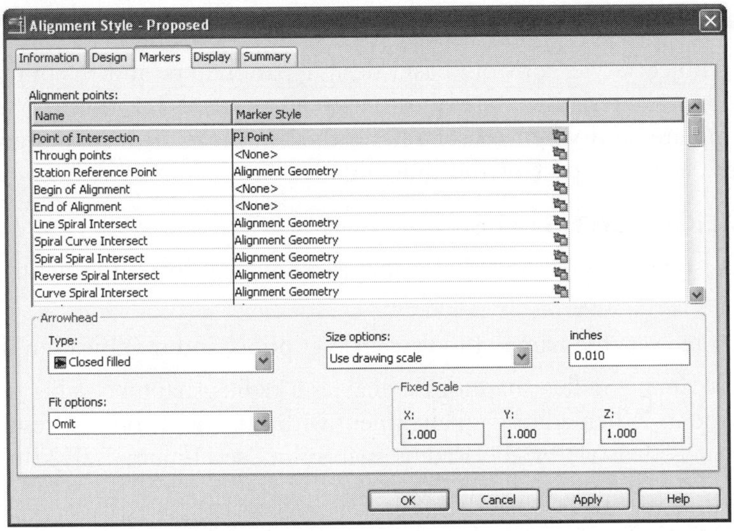

Figure 5.8

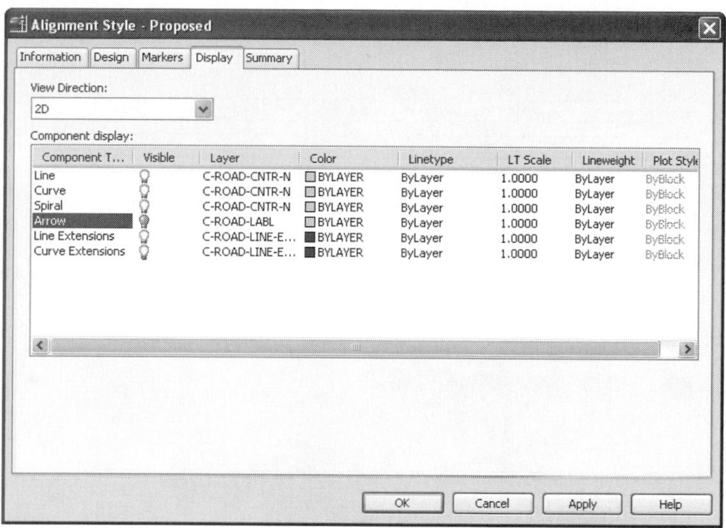

Figure 5.9

ALIGNMENT LABEL STYLES

When creating an alignment, the Create Alignment dialog box contains a setting identifying a labeling style. Users can have different styles for differing phases of design. One style can be for the design phase and another can be for final document submission. As with other styles in Civil 3D, these styles can be saved to a template, and when a user is selecting the template for a new drawing, the styles in the template become a part of the new drawing.

If a user wants more than just a single label on an alignment (major and minor stations, ticks, etc.) he or she needs to have a label set. A label set is an alias for a style containing several different label styles. Assigning the set to an alignment applies all of the included styles.

MAJOR MINOR AND GEOMETRY POINTS LABEL SET

The Major Minor and Geometry Points Label Set labels major and minor stations with text and ticks perpendicular to centerline (see Figure 5.10). The Labels tab of the Label Set dialog box identifies the styles that are a part of the style set and their labeling increments, if appropriate.

Major Stations: Perpendicular with Tick

The Perpendicular with Tick major station label places a short tick on the alignment at the major station (1+00). Adjacent to the tick is the value of the station (see Figure 5.11).

Minor Station: Tick

The Tick style places a small line object (block) along the centerline between the major ticks (see Figure 5.12).

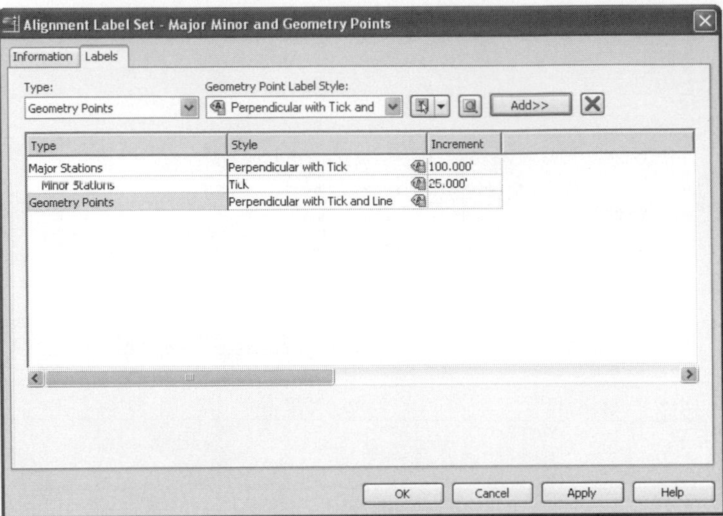

Figure 5.10

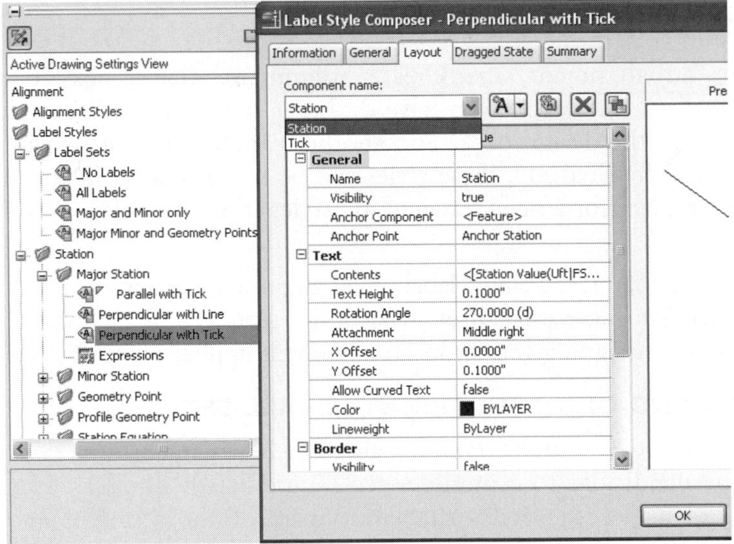

Figure 5.11

Geometry Point: Perpendicular with Tick and Line

This style places a tick, a line, and text anchored to the line to label an important point on the centerline (see Figure 5.13). This point can be a beginning or ending point of a curve or spiral, a compound curve, etc. The types of geometric points are on the Abbreviations panel of Edit Drawing Settings.

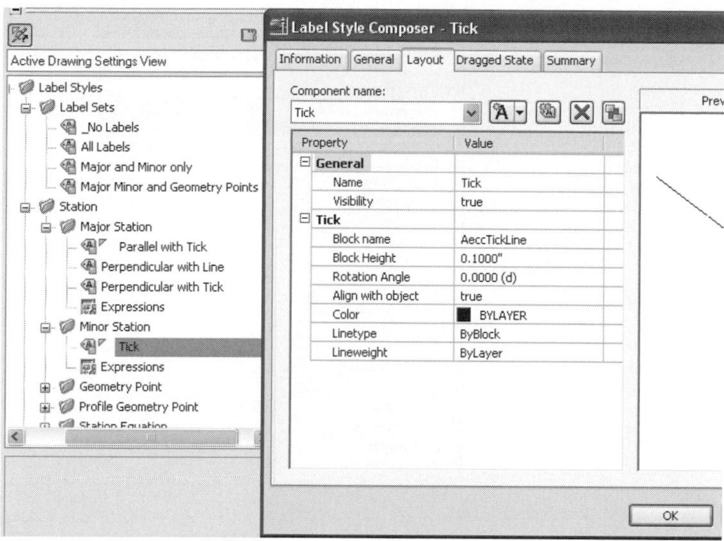

Figure 5.12

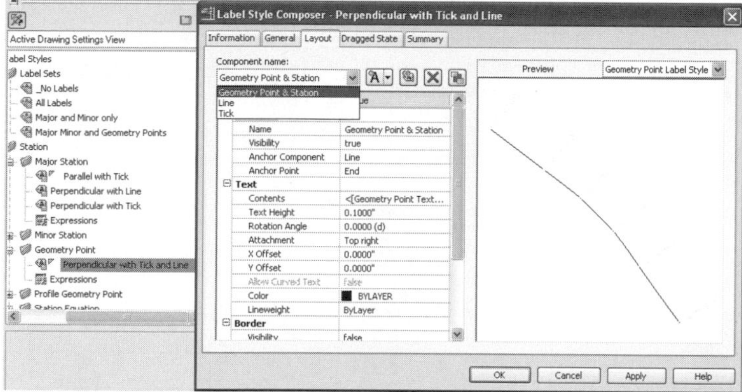

Figure 5.13

Profile Geometry Point

A Profile Geometry Point style identifies a critical point of the vertical design on an alignment (see Figure 5.14).

Station Equation

A Station Equation style defines annotation for the ahead and behind stationing of a point on an alignment (see Figure 5.15). Station equations occur when there is a change in the stationing of an alignment.

Design Speed

A Design Speed style defines annotation that labels the design speed for a portion of an alignment (see Figure 5.16).

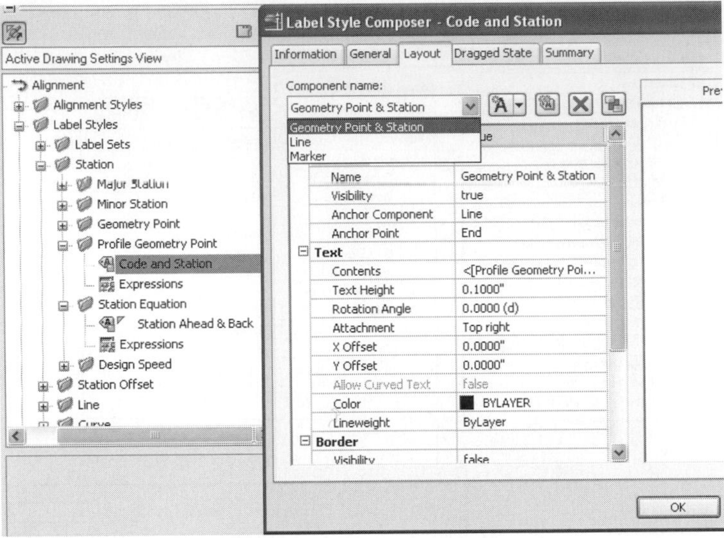

Figure 5.14

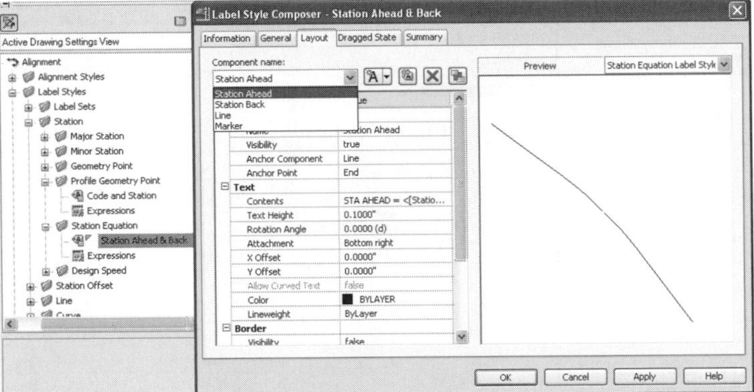

Figure 5.15

STATION OFFSET LABELS

A station offset label is a spot label containing an alignment station and the offset (perpendicular distance) from an alignment's centerline (see Figure 5.17).

LINE LABELS

A line label annotates the direction of a tangent segment of an alignment and is similar to the line segment labels of parcels. The Add Labels command adds these labels to the alignment.

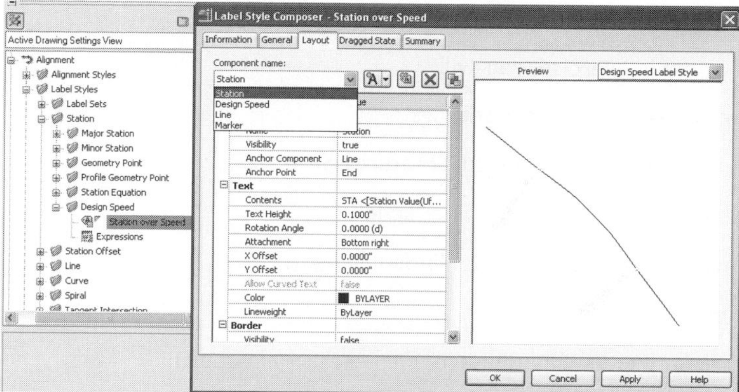

Figure 5.16

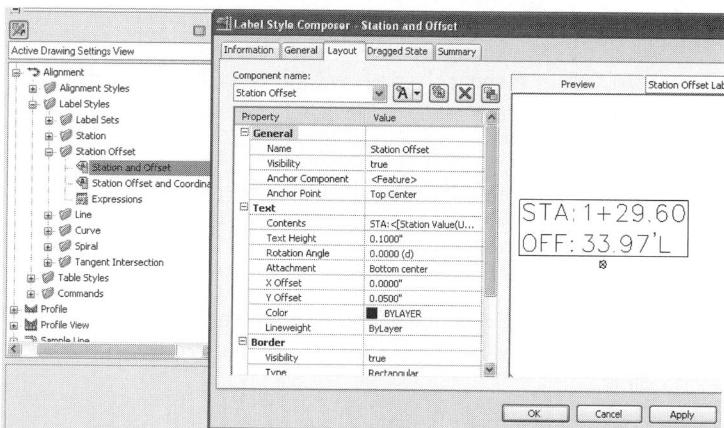

Figure 5.17

CURVE LABELS

A curve label annotates an arc segment in an alignment and is similar to the line segment labels of parcels. The Add Labels command adds these labels to the alignment.

SPIRAL LABELS

A spiral label annotates a spiral segment in an alignment and is similar to the line segment labels of parcels. The Add Labels command adds these labels to the alignment.

TANGENT INTERSECTIONS

A tangent intersections label annotates the station of an alignment intersection. The Add Labels command adds these labels to the alignment.

EXERCISE 5-1

When you complete this exercise, you will:

- Be familiar with alignment settings.
- Be familiar with Alignment Edit Features Settings.
- Be familiar with alignment object styles.
- Be familiar with the Major & Minor label set.

DRAWING SETUP

This exercise starts a new drawing using the *Chapter 5 – Unit 1.dwt* file that is in the *Chapter 5* folder of the CD that accompanies this textbook.

1. If you are not in Civil 3D, double-click the Civil 3D icon to start the application.
2. At the command prompt close the drawing and do not save it.
3. From AutoCAD's File menu, select **New...**, in the File Open dialog box, change the location to where *Chapter 5 – Unit 1.dwt* is located, select the file, and click **Open** to use the template file.
4. At the command prompt, from the AutoCAD File menu select **Save As...** and assign it the name **Alignments**.

DRAWING SETTINGS

The Edit Drawing Settings dialog box under Settings has several values affecting the alignment object. These settings include layer-naming conventions and abbreviations for labeling critical alignment geometry points.

1. Click the **Settings** tab to make it current.
2. In Settings, at the top, select the drawing name, press the right mouse button, and from the shortcut menu select **Edit Drawing Settings....**
3. Click the **Object Layers** tab to view the layer values.
4. In the Object Layers panel, set the Modifier for Alignment, Alignment Labeling, and Alignment Table to **Suffix** and enter a dash followed by an asterisk (**-***) as the value of each modifier (see Figure 5.3).
5. Click the **Abbreviations** tab to review its contents.
6. Click the **OK** button to save the changes and to close the Drawing Settings dialog box.

ALIGNMENT FEATURE SETTINGS

The Alignment Edit Feature Settings dialog box assigns several initial values and styles for routines creating and labeling alignments (see Figure 5.5).

1. In Settings, select the **Alignment** heading, press the right mouse button, and from the shortcut menu select **Edit Feature Settings....**

2. Expand the Default Styles, Default Name Format, and Station Indexing sections to view their values.
3. Click the **OK** button to close the dialog box.

ALIGNMENT OBJECT STYLES

This and the following exercises will use two basic alignment object styles, Layout and Existing. The Layout style applies different colors to various alignment elements, and the Existing style assigns the same color to all alignment elements.

1. In Settings, expand the Alignment branch until you view the list of Alignment styles.
2. From the list, select the layout style, press the right mouse button, and from the shortcut menu select **Edit....**
3. Click the **Design** tab to view its settings.
4. Click the **Markers** tab to view its settings.
5. Click the **Display** tab to view the settings for the various components of an alignment. Each component can be a different color.
6. Click the **OK** button to exit the layout style.
7. Select the existing style, press the right mouse button, and from the shortcut menu select **Edit....**
8. Click the **Design** tab to view its settings.
9. Click the **Markers** tab to view its settings.
10. Click the **Display** tab to view the settings for the different components of an alignment. Each component can be the same color.
11. Click the **OK** button to exit the existing style.

EXERCISES

ALIGNMENT LABEL SET AND LABEL STYLES

The alignment you are going to draft uses the Major Minor and Geometry Points label set. This label set contains five alignment label styles: Major Station - Perpendicular with Line, Minor Station - Tick, Geometry Point - Perpendicular with Tick & Line, Station Equations - Station Ahead & Back, and Design Speeds - Station Over Speed (see Figure 5.6). The label styles in this set are all from the Station branch of the label styles for Alignments.

1. In Settings, expand the Alignment, Label Styles branch until you view a list of styles for label sets.
2. From the list, select the **Major Minor and Geometry Points** label set, press the right mouse button, and from the shortcut menu select **Copy....**
3. Select the **Information** tab and for the name type, **Perpendicular - Major Minor and Geometry Points Label Set.**
4. Click the **Labels** tab to view its contents.

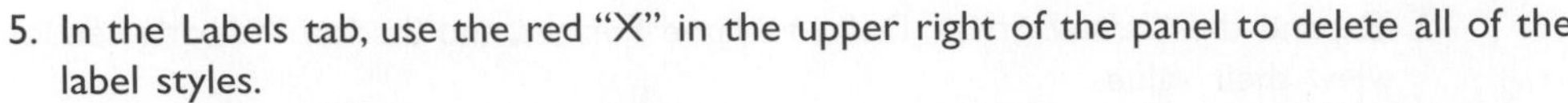

5. In the Labels tab, use the red "X" in the upper right of the panel to delete all of the label styles.
6. In the Labels tab, click the drop-list arrow to the right of Type, and select **Major Stations** from the list.
7. In the Labels tab, click the drop-list arrow to the right of Major Stations Label Style, select **Perpendicular with Line** from the list, and click the **Add>>** button to add the style to the set.
8. Repeat Steps 6 and 7, and set the label type to **Minor Stations**, the Minor Station Label Style to **Tick**, and add it to the list of label styles.
9. Repeat Steps 6 and 7, and set the label type to **Geometry Points**, the Geometry Points Label Style to **Perpendicular with Tick and Line**, and add it to the list of label styles.
10. Repeat Steps 6 and 7, and set the label type to **Station Equations**, the Station Equation Label Style to **Station Ahead & Back**, and add it to the list of label styles.
11. Repeat Steps 6 and 7, and set the label type to **Design Speeds**, the Design Speeds Label Style to **Station over Speed**, and add it to the list of label styles.
12. Click the **OK** button to create the new style.
13. Expand the Station branch until you view the list of styles for Major Station.
14. From the list of styles, select **Perpendicular with Line**, press the right mouse button, and from the shortcut menu select **Edit...**.
15. In the Label Style Composer dialog box, click the **Layout** tab and review its contents.
16. Click the drop-list arrow to the right of Component Name and view the other components of the label style.
17. Click the **OK** button to exit the dialog box.
18. Expand the Minor Station branch to view its list of styles, select the **Tick** style, press the right mouse button, and from the shortcut menu select **Edit...**, to review its settings.
19. Click the **OK** button to exit the dialog box.
20. In the same branch, expand Geometry Point to view its styles, select **Perpendicular with Tick and Line**, press the right mouse button, from the shortcut menu select **Edit...**, and review its settings.
21. Click the **OK** button to exit the dialog box.
22. Click the **AutoCAD Save** icon at the top left of the screen to save the drawing.

This completes the setup and review of the styles for the exercises in this chapter. The next step is defining and drafting the alignments.

SUMMARY

- Civil 3D assigns alignments base layer names.
- If there is more than one alignment in a drawing, use either a prefix or a suffix to place each alignment in its own layer.
- Alignment annotation appears as the user create the alignment.
- The assigned alignment label set annotates the alignment.
- Users can add or delete label styles from a label set or change the labeling in the alignment properties.
- Alignment labels immediately change when changing or adding label styles of a label set.

UNIT 2: CREATING ROADWAY CENTERLINES

In Civil 3D, a centerline is an object composed of tangents (lines), curves (arcs), and possibly spirals. There are four methods of creating an alignment. The first method imports a LandXML file and the second directly imports an alignment definition from a LDT project. Both of these methods can import one or more alignments. When there is more than one alignment, the routine displays an alignment list and the user selects the alignment(s) to import.

The third method converts a polyline into an alignment. The only drawback to this method is that the stationing corresponds to the direction of the polyline segments. However, Civil 3D has a command that reverses the direction of an alignment's stationing.

The last method drafts an alignment from the Alignment Layout Tools toolbar. The Alignment Layout Tools toolbar has routines that create tangent, curve, and spiral segments. Each segment can be fixed, floating, or free. One way of designing an alignment is sketching the controlling geometry and using the Alignement Layout Tools commands to trace alignment segments over them, or by referencing their geometry with object snaps (see Figure 5.18).

Whether creating an alignment from a polyline or from the tools in the Alignment Layout Tools toolbar, users will encounter the Create Alignment dialog box. This dialog box sets the alignment name, description, starting station, site name, alignment style, layer, and alignment label style for the new alignment (see Figure 5.19).

How the alignment appears on the screen is a result of the two associated styles: the alignment and label set styles. The alignment style controls how the line work displays and what constraints the alignment follows. The label set style annotates the alignment line work.

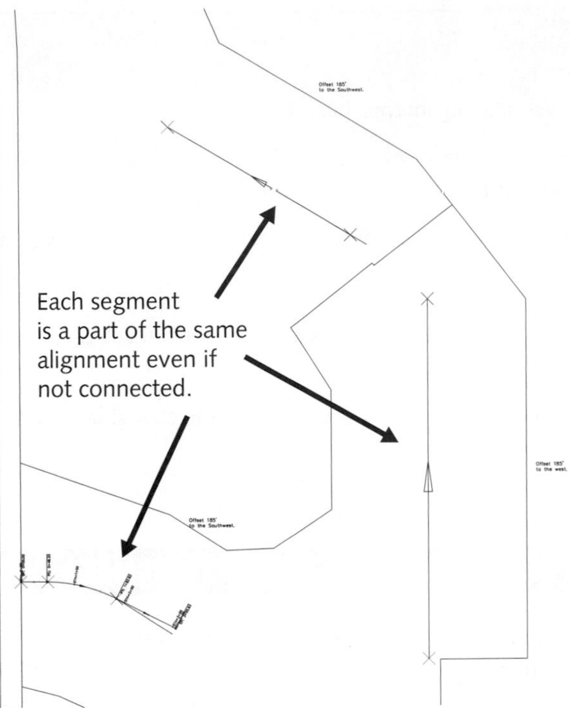
Each segment
is a part of the same
alignment even if
not connected.

Figure 5.18

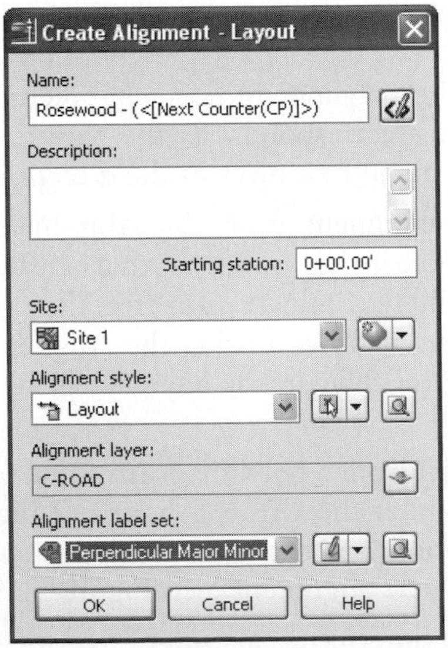

Create Alignment - Layout

Name:
Rosewood - (<[Next Counter(CP)]>)

Description:

Starting station: 0+00.00'

Site:
Site 1

Alignment style:
Layout

Alignment layer:
C-ROAD

Alignment label set:
Perpendicular Major Minor

OK | Cancel | Help

Figure 5.19

ALIGNMENTS MENU

The Alignments menu is the only place users can create an alignment (for example, centerline, horizontal centerline, and so on). The menu contains all the commands needed to create, edit (and analyze), and annotate roadway centerlines (see Figure 5.2).

CREATE FROM POLYLINE

After drawing the polyline, the Create from Polyline routine converts the polyline into an alignment object. An alignment from a polyline creates an object with fixed segments. The initial direction of the alignment follows the path of the polyline. If this is not the desired direction, the Reverse Alignment Direction command within the Alignments menu reverses the direction of a selected alignment.

CREATE FROM ALIGNMENT LAYOUT TOOLS TOOLBAR

The Alignment Layout Tools toolbar contains commands that create tangent and curve segments for centerline design (see Figure 5.20). The tangent and curve segments can be fixed, floating, or free. The behavior desired and the rules set forward by Civil 3D allow users to create an alignment that solves several constraint and relationship issues.

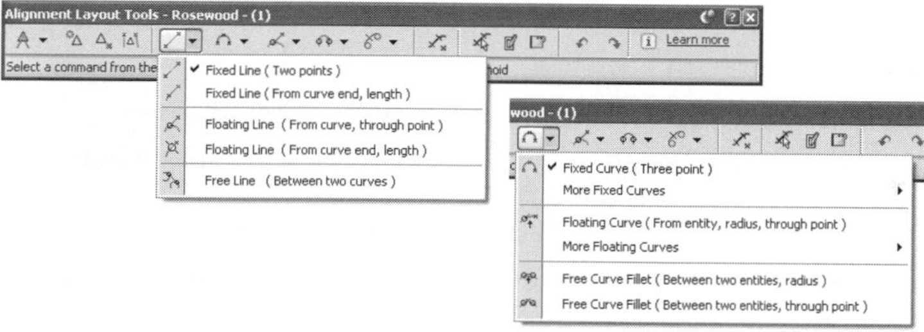

Figure 5.20

There will be times when certain properties of a segment should be held (for example, always tangent, always attached to two entities, etc.), The type of segment—fixed, floating, or free—allows users to draft a complicated alignment with implied relationships.

All of the alignment segments must have the same direction. If drafting individual segments, they must all point in a logical direction for them to be connected with other segment types. If this does not happen, you should delete the offending alignment segment and correctly redraft it.

Tangent Lines Only: Tangent with Curves

When drafting a centerline with or without curve segments, all of the tangent entities are fixed entities (two selected points) and the curves are free entities (see Figure 5.21). The radius of the curve is set by the Curve Settings selection of the first icon stack at the left side of the toolbar. When a user selects this icon, it displays a Curve Settings dialog box that includes settings for spirals as well as curves.

If a user chooses to include spirals, he or she should set their values in the Curve Settings dialog box. Civil 3D supports Bloss, Clothoid, Cubic (JP), and Sine Half-wave Diminishing Tangent.

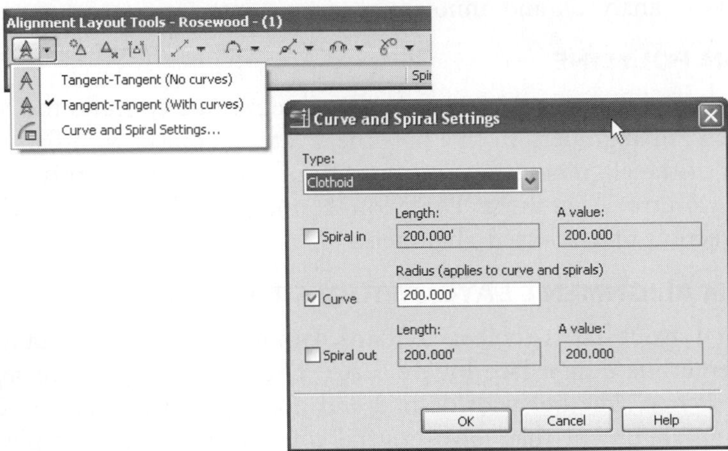

Figure 5.21

PI Management

The next three icons in from the left of the toolbar allow users to insert, delete, and break apart PIs. These commands allow users to add PIs to tangent segments where there are none.

Individual Tangent and Curve Entities

The Tangent and Curve icons provide several tools to create specific types of tangent and curve segments (see Figure 5.20). A tangent segment can reference existing points, use the transparent direction overrides (bearing, azimuth, etc.), or AutoCAD object snaps. When connecting the entities to each other or to AutoCAD objects, the user can use AutoCAD object snaps.

Civil 3D has three types of alignment segments: fixed, floating, and free.

A fixed segment:

- Is parametrically defined by specifying coordinate points or radius.
- Is editable only by adjusting the coordinates of the segment (endpoints of the line or the three points of a curve segment).
- Is not dependent on another entity for its geometry or tangency.
- Will not maintain its tangency when related entities are edited if the segment is initially created with a tangency association.

The first rule states a fixed tangent is specified by two (line) or three points (curve). The second rule implies the user can only change the coordinates of the line or curve (three points defining the curve). The Alignment Editor vista or sub-entity editor presents only the coordinates of the segment (see Figure 5.22). The editor indicates the editable values in black.

The third rule states that the entity is not specifically attached to other segments of the alignment (tangent to, from the endpoint of, etc.). The fourth rules states if the segment was drawn by being tangent to another object, editing the alignment *will not* maintain the original tangency between the objects.

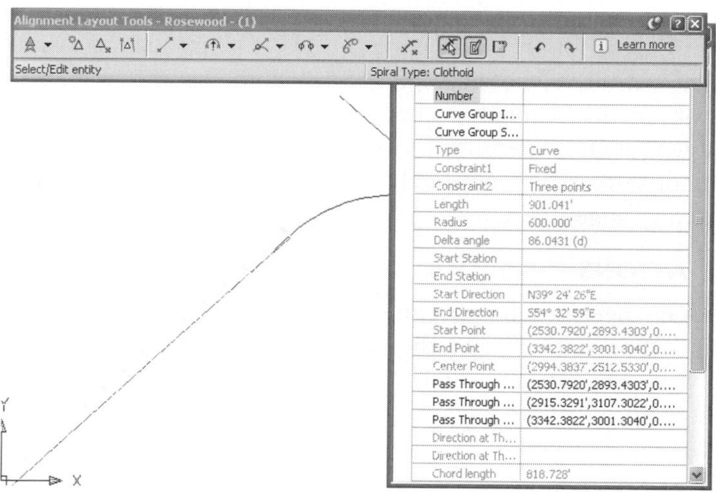

Figure 5.22

A floating segment is:

- Dependant on one other alignment segment to define its geometry.
- Always tangent to the entity to which it is attached.
- Always attached to one entity and defined by the parameters that you specify.
- Attachable only to another floating entity or a fixed entity.

The first and second rules state that a floating segment attaches to at least one other alignment segment and is tangent. The third rule states that after attaching the entity, the user must define other parameters to complete the drafting of the segment. The last rule states the segment attaches only to floating or fixed alignment segments.

A free segment:

- Always dependent on two other alignment segments for its geometry.
- Always tangent to the entity before and after
- A connector between two fixed segments, two floating entities, or one fixed and one floating entity.
- Defined by surrounding geometry and additional user-specified parameters.

The first and second rules state that a free entity requires two existing entities and is tangent to both of them. The third rule states that a free entity connects only to combinations of fixed and floating entities. The last rule states that after attaching the floating segment, the user must enter additional entity parameters.

Spiral Segments

The Alignment Layout Tools toolbar contains three icon stacks with spiral creation commands: Line and Spiral (floating), Spiral-Curve-Spiral (free or free), and Fixed and Free and Floating Compound or Reverse Sprials. A fixed spiral connects between to existing alignment entities. A floating spiral attaches to the end of an existing entity. The default method of creating a spiral is spiral, curve, and spiral (S-C-S). A user can set any one of the spiral lengths to 0 (zero). This makes the curve tangent to one of its connecting objects. The default settings for the spiral, in and out, are set in the Curve Settings dialog box of the Tangent/Tangent Curve icon at the left of the toolbar.

STATION EQUATIONS

Some alignments change stationing at one or more points along their path. This point is a station equation. The stationing of the alignment changes at this point, and from this point the stationing may increase (rarely decrease). There can be several station equations on any one alignment. Civil 3D manages station equations as an alignment property.

IMPORTING LANDXML

A LandXML file can contain alignment definitions (see Figure 5.23). It does not store points, but rather the coordinates and geometric description of the elements that make up the alignment. This allows the file to transfer design data between applications without losing any design fidelity. If the LandXML file contains more than one alignment, the user can select which alignments to import.

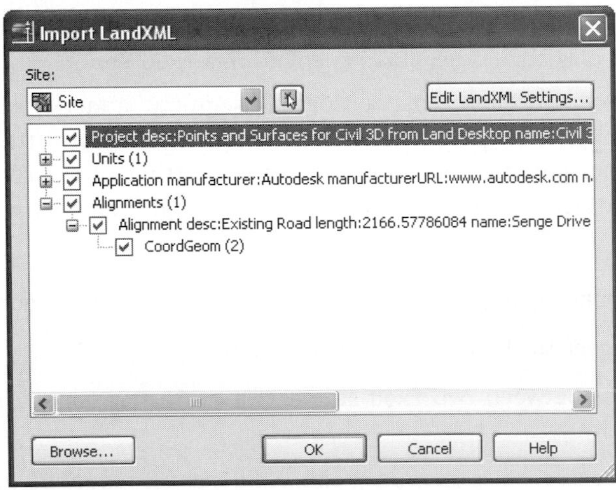

Figure 5.23

IMPORTING FROM LAND DESKTOP

Importing an alignment from a LDT project is similar to importing a LandXML file. The Import From Land Desktop command, found in the Import flyout of the File menu, displays a dialog box where the user selects a project path and a project. After the user selects the project, the dialog box displays a list of LDT project data, including alignments. Next, the user selects the alignments and, when exiting the dialog box, Civil 3D imports the alignment(s) directly from the LDT project to the drawing.

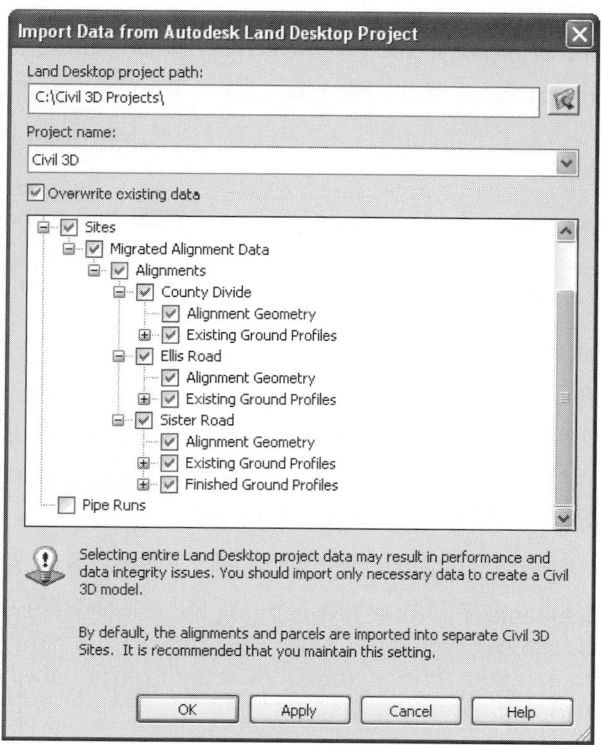

Figure 5.24

EXERCISE 5-2

When you complete this exercise, you will:

- Be able to create an alignment by importing a LandXML file.
- Be able to create an alignment from a polyline.
- Be able to create an alignment using the Alignment Layout Tools toolbar.
- Be able to use transparent commands while laying out a centerline.

EXERCISE SETUP

This exercise continues with the drawing from the last exercise. If you did not do the previous exercise, you can open the *Chapter 5 - Unit 2.dwg* file that is in the *Chapter 5* folder of the CD that accompanies this textbook.

1. Open the drawing used in the previous exercise or the *Chapter 5 - Unit 2* drawing.
2. In Layer Properties Manager, or by using the Express tools, freeze the contour and boundary layers (**3EXCONT**, **3EXCONT5**, and **Boundary**) and exit the dialog box.

IMPORT AN ALIGNMENT

A LandXML file contains alignment definitions and can be imported to create alignment objects.

1. From the Import flyout of the File menu, select **Import LandXML...**.
2. In the Import LandXML dialog box, select the **Existing Road.xml** file, and click the **Open** button.

   The file is located in the *Chapter 5* folder of the CD that accompanies this textbook.
3. In the Import LandXML dilalog box, click the **OK** button to import the Senge Drive alignment into the drawing.

CREATE FROM POLYLINE

The Create from Polyline command converts a polyline into an alignment.

1. From the View menu, select **Named Views...**, and restore **Existing Road**.
2. From the Alignments menu, select **Create from Polyline** and select the red centerline (**C-ROAD-CTLN**) near the southern end.

   Your dialog box should match Figure 5.25 after finishing Step 3.
3. The Create Alignment - From Polyline dialog box displays. In Name:, type **OMalley Phase 2**; for the Alignment style assign **Existing**; for the Alignment label set assign **Perpendicular Major Minor and Geometry Points Label Set**; and click the **OK** button to exit.

   The stationing should start at the northern end and increase toward the south.
4. If the stationing increases from the south to north, select **Reverse Alignment Direction** from the Alignments menu, select the alignment, and in the warning dialog box click the **OK** button to accept the changes and reverse the alignment stations.
5. Use the AutoCAD Zoom Extents command to view the entire site.
6. Click the **AutoCAD Save** icon at the top left of the screen to save the drawing.

OFFSETTING PRELIMINARY CENTERLINE TANGENTS

In this exercise, the engineer guiding the job wants to start the centerline from two points at the southwestern side of the parcel boundary (point numbers 1 and 2). He also wants the centerline to be about 185 feet in from the parcel boundary. There are lines on the P-CL layer that represent the bearing of the lines he wants for the new centerline (see Figure 5.26).

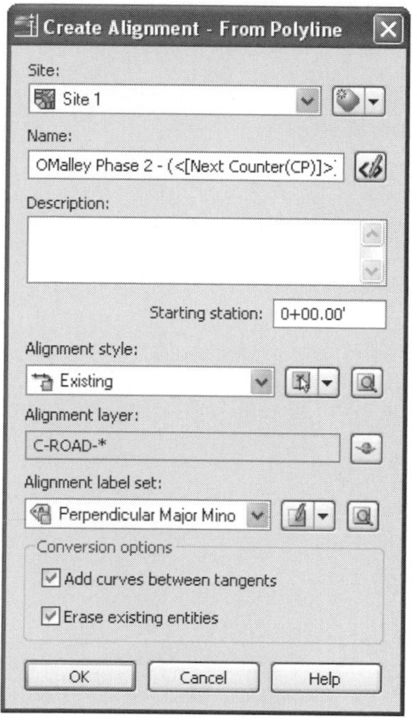

Figure 5.25

1. From the Layers II toolbar, use the **Isolate Layer** routine and select one of the entities on the **P-CL** layer.
2. Start the AutoCAD **Offset** command, set the distance to **185** feet, and offset the southerly line, southwest; the eastern line, west; and the northeastern line, southwest.
3. From the Layers toolbar, select the **Layer Previous** icon to restore the previous layer state.

DEFINE THE PARCEL BOUNDARY

The next step is defining the property boundary as a parcel.

1. From the Parcels menu, select **Create From Objects**, select the outer parcel boundary (see Figure 5.26), press the **Enter** key, and accept the defaults by clicking the **OK** button in the Create Parcels - From Objects dialog box.

DEFINE DETENTION POND PARCEL

1. From the Parcels menu, rerun the **Create From Objects** command, select the Detention Pond boundary located in the southwestern corner of the overall parcel (see Figure 5.26), press the **Enter** key, and accept the defaults by clicking the **OK** button in the Create Parcels - From Objects dialog box.
2. Click the **AutoCAD Save** icon at the top left of the screen to save the drawing.

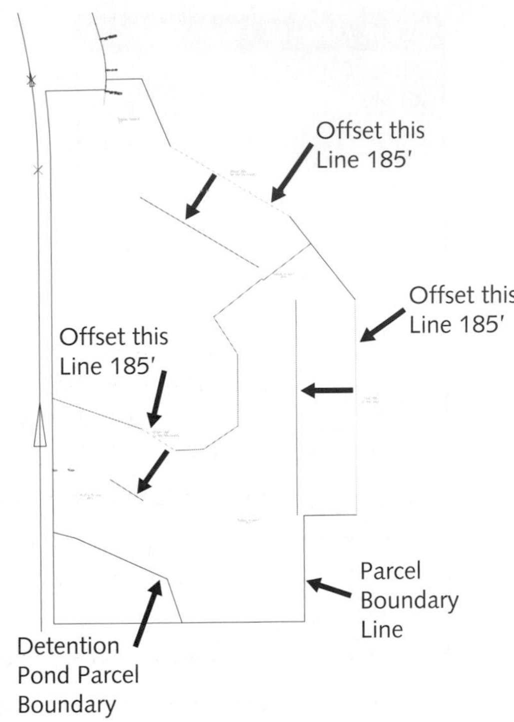

Figure 5.26

DRAFTING THE ROSEWOOD ALIGNMENT

The Rosewood alignment starts at the site's lower-left (southwest) side and winds its way around the site to a connection point with the OMalley centerline at the northern boundary. The lines previously offset are guides for fixed tangent segments. But their bearings and distances may not be held in the final design. To start drafting the alignment, you will use a Civil 3D transparent command to reference the two points, and AutoCAD object snaps to draft four fixed tangent segments from the offset lines.

1. From the View menu, use Named Views... to restore the **Proposed Starting Point** view.
2. Make sure the Transparent Commands toolbar is visible.
3. From the Alignments menu, select **Create by Layout...**.
4. In the Create Alignment - Layout dialog box, for the alignment Name type **Rosewood**, leave in the counter, set the Alignment Style to **Layout**, set the Label Set to **Perpendicular Major Minor and Geometry Point Label Set**, and click the **OK** button to start drafting the Rosewood centerline.

   Draw the first segment with the Draw Fixed Line - Two Points routine, referencing points 1 and 2 from the screen.

5. From the Alignment Layout Tools toolbar, click the **Draw Fixed Line - Two Points** icon (fifth in from the left).
6. Next, click **Point Object** ('PO) in the Transparent Commands toolbar. In the drawing, select anywhere on point 1 and then select anywhere on point 2.
7. Press the **Esc** key to return to the Alignment Layout prompt.

   After you select the two points, a fixed alignment segment appears between them with an arrow pointing from left to right (see Figure 5.27).

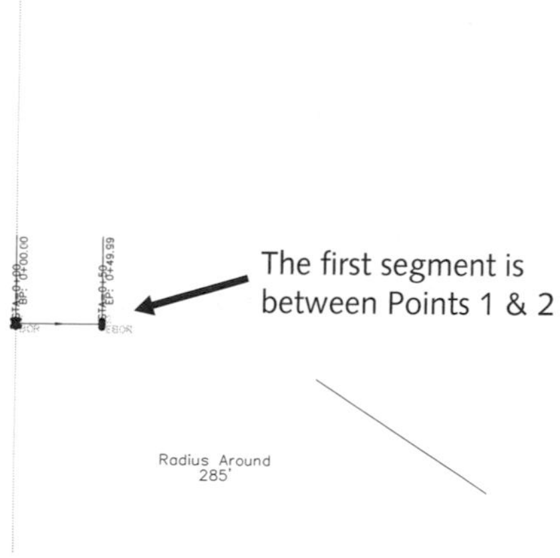

Figure 5.27

The next segment is a floating curve that attaches to the end of the segment just drawn.

8. In the Alignment Layout Tools toolbar, click the drop-list arrow for curves (sixth icon in from the left). From the More Floating Curves flyout menu, select **Floating Curve (From entity end, radius, length)**. See Figure 5.28.
9. In the drawing, select the eastern end of the alignment segment just drawn.
10. In the command line, the prompt is for arc direction; press the **Enter** key to set the direction of the arc to clockwise.
11. In the command line, the prompt is for a radius; enter **285**, and press the **Enter** key.
12. In the command line, the prompt is for arc length; define arc length by selecting a point at the eastern end of the alignment tangent and a second point near the northwestern end of the offset line segment. These two points define the length of the arc.

    This draws a blue curve segment.

    The next tangent segment is floating and is always tangent to the curve just drawn.

13. In the Alignment Layout Tools toolbar, click the drop-list arrow for tangent segments (fifth in from the left) and select **Floating Line (From curve end, length)**.

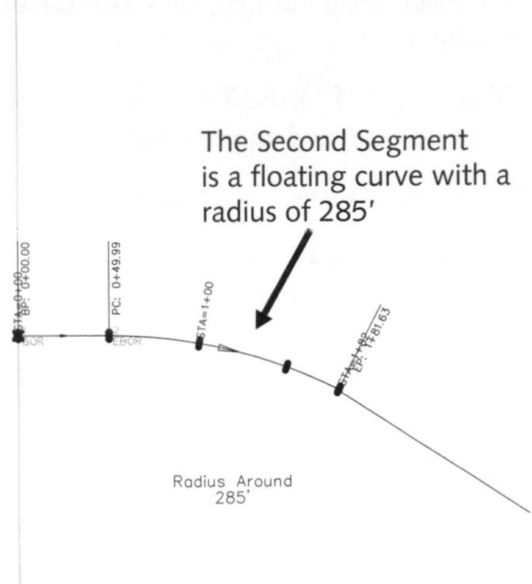

Figure 5.28

14. In the drawing, select the eastern end of the curve just drawn, then reselect the end of the curve as the starting length of the tangent (endpoint object snap), and select the eastern end of the offset line as the ending length of the line (endpoint object snap).

    A red tangent should appear off the end of the curve. The new tangent has a slightly different bearing than the offset line. This is fine for now (see Figure 5.29).

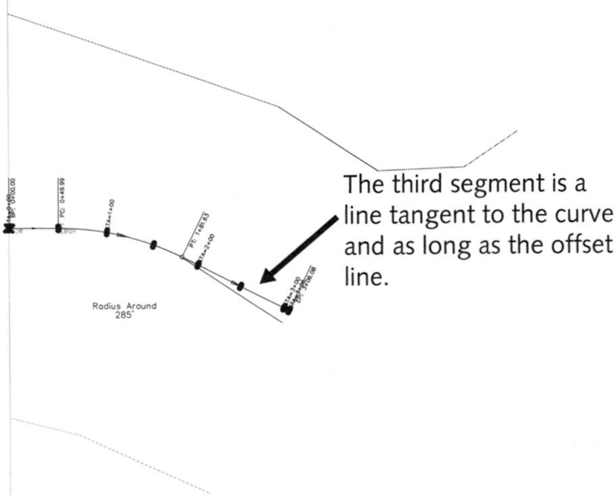

The third segment is a
line tangent to the curve
and as long as the offset
line.
Radius Around
285'

Figure 5.29

The next curve attaches to the end of the tangent just drawn, but it cannot be drawn yet; you do not have enough parameters to define the curve. The easiest method of creating the curve is first drafting the next alignment tangent segment and then creating a free curve connecting the tangents.

The next offset line (the tangent offset from the eastern parcel boundary) is the target tangent segment for the curve we are getting ready to draft. But before drafting the curve, we need to create an alignment tangent segment representing the offset of the east property line. The tangent segment should be drawn from south to north (see Figure 5.30).

15. Use the AutoCAD Zoom and Pan commands to better view the east tangent line.
16. From the Alignment Layout Tools toolbar, select **Fixed Line (Two points)** (fifth icon in from the left). In the drawings using an endpoint object snap, draw an alignment segment from the southern end of the offset property line to its northern end.

    The next step places a free curve from the end of the western tangent to a point on the tangent just drawn.

17. From the Alignment Layout Tools toolbar, select the drop-list arrow for curves, and select **Free Curve Fillet (Between two entities, radius)** from the list.
18. In the drawing, select the east end of the west tangent (the tangent before the curve) and select the tangent just drawn near its middle (the tangent after the curve).
19. In the command line, the prompt is for the curve solution angle; it is **less than 180 degrees**. Press the **Enter** key.
20. In the command line, the prompt is for a curve radius; enter **310**, and press the **Enter** key to create the curve.

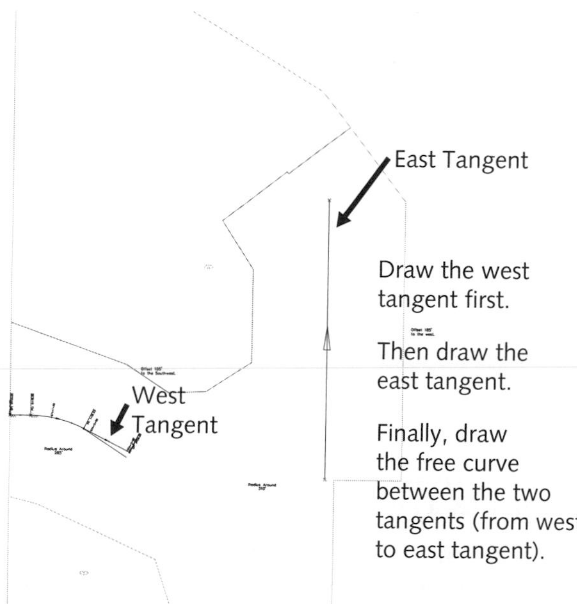

Figure 5.30

From the eastern tangent the road goes into another curve to a tangent turning northwesterly. Also, the northwestern tangent segment connects to a final tangent segment with a curve on its western end. You must draw the tangents first to draft the final curves. The next steps draw the last two tangent segments, and they must be drawn in the same direction as the current centerline (northwesterly and then northeasterly). See Figure 5.31.

21. Use the AutoCAD Zoom and Pan commands to view the last to offset property line segments and the current alignment.
22. In the Alignments Layout Tools toolbar, click the **Draw Fixed Line - Two Points** icon. In the drawing, using the endpoint object snap, draw the first tangent from the southeast to its northwestern endpoint.

    The drafting mode remains Fixed Line - Two Points.
23. In the drawing, using the endpoint object snap, draw the last tangent segment from the southwest endpoint to its northeast endpoint of the line.

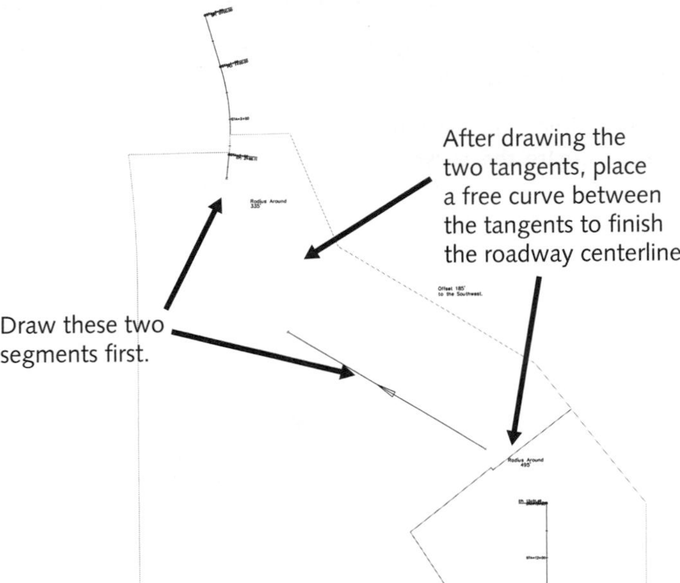
After drawing the two tangents, place a free curve between the tangents to finish the roadway centerline.
Draw these two segments first.

Figure 5.31

24. From the Alignment Layout Tools toolbar, click the curves drop-list arrow and select **Free Curve Fillet (Between two entities, radius)**.
25. In the drawing, select the northern end of the east tangent (the tangent before the curve) and then select the northwestern tangent (the tangent after the curve).
26. In the command line, the prompt is for the curve solution angle; press the **Enter** key to accept the curve being **less than 180** degrees.
27. In the command line, the prompt is for a radius; enter **495** and press the **Enter** key.

The command remains Free Curve Fillet (Between two entities, radius).

28. In the drawing, select the western end of the northwestern tangent (the tangent before the curve) and then select the southwestern end of the northeastern tangent (the tangent after the curve).
29. In the command line, the prompt is for the curve solution angle; press the **Enter** key to set it to **less than 180** degrees.
30. In the command line, the prompt is for a radius; enter **335**, and press the **Enter** key to create the final curve.

    You centerline should look like Figure 5.32.

31. Press the **Enter** key again to exit the Create Alignment By Layout command.
32. Click the **AutoCAD Save** icon to save the drawing.

EXERCISES

This completes the exercise on importing and drafting alignment segments. When using the Alignment Layout Tools toolbar, the resulting alignment is a combination of fixed, floating, and free segments (line and curve). You can also convert polylines into alignments, or import them from LandXML files or LDT projects. The next step is to review the values of the segments to see if they are optimal for the site and present a reasonable design solution.

SUMMARY

- A LandXML file can contain more than one alignment definition.
- When importing alignments from a LandXML file or a LDT project, one or more alignments can be selected.
- The Create from Polyline routine converts a polyline into an alignment.
- If any alignment has the wrong direction, the Reverse Alignment Direction command from the Alignments menu can be used to change its direction.
- When using the Alignment Layout Tools toolbar, the user can switch between drawing tangents and curves.
- Alignment segments do not need to be connected for Civil 3D to realize they are part of the same alignment.

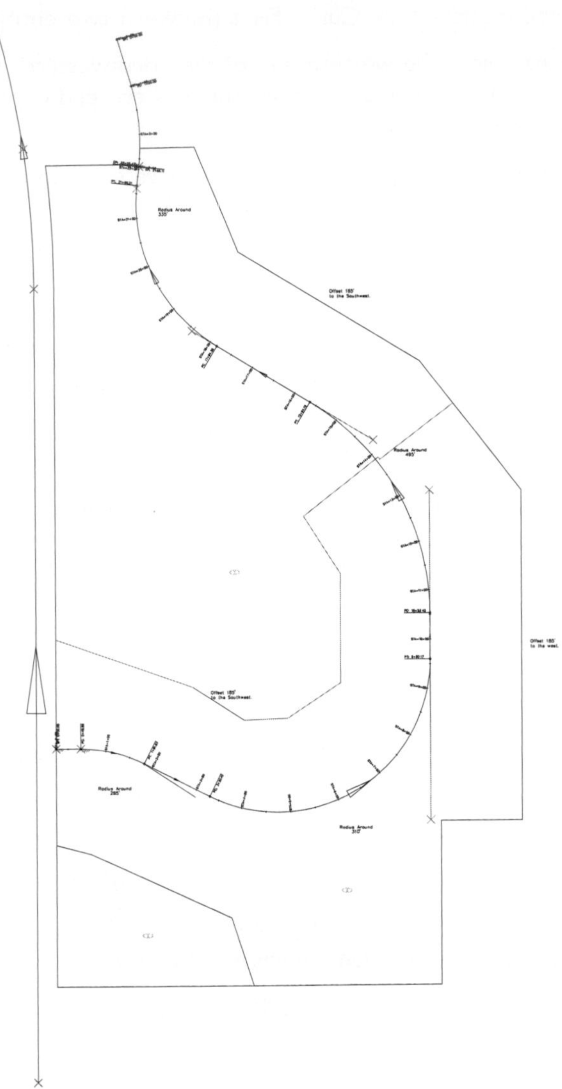

Figure 5.32

UNIT 3: CENTERLINE ANALYSIS

Most engineers design by calculations or design criteria. The alignment graphics do not tell the designer the whole story of their values. These design numbers are a part of the alignment's values displayed in the Alignment Layout Tools toolbar editors. To analyze an alignment, call the Layout Toolbar from the Alignments menu (Edit AlignmentGeometry...), or with an alignment selected from a right mouse button shortcut menu (Edit Alignment Geometry...). The editor displays the alignment's numbers in two vistas: a sub-entity (a selected segment) or as an overall alignment.

Also, Civil 3D provides several reporting routines that create reports for a thorough review of the alignment. This is done by report forms in the Report Manager or the LandXML Reporting application.

SUB-ENTITY ANALYSIS

The Sub-entity vista displays the values of a selected alignment segment. When a user uses this method, the Sub-entity Editor is displayed by clicking its icon (second icon in from the right). After displaying the Sub-entity Editor, select the Pick Sub-entity icon and select a sub-entity from the screen. When a user selects a sub-entity, its values appear in the vista (see Figure 5.33). When viewing another sub-entity's values is desired, just select another sub-entity in the alignment. When the next sub-entity is selected, its values will replace those currently in the vista. All of the values in the vista that are gray cannot be edited; only those values in black can be edited.

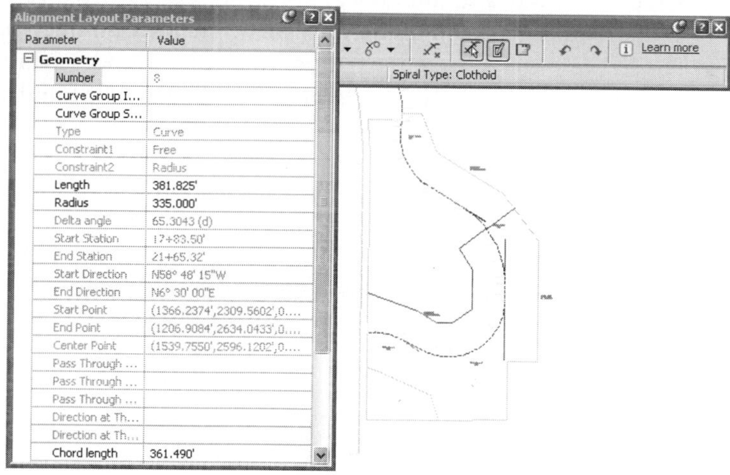

Figure 5.33

ALIGNMENT ANALYSIS

The second vista, Alignment Grid View, displays the alignment in its entirety (see Figure 5.34). The vista lists all of the critical values for each segment of the alignment. Users can edit only the values in black.

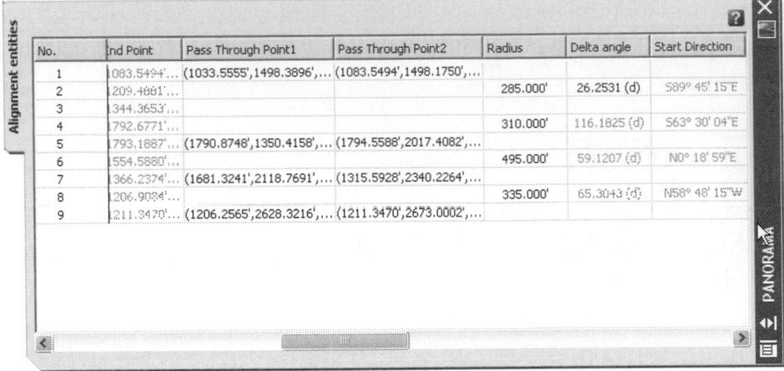

Figure 5.34

LANDXML REPORTS

Civil 3D generates reports about alignment segments. Access to these reports is through Report Manager or the Autodesk LandXML Reports application. Both applications have the same reports concerning the selected alignment(s). The Report Manager generates reports from the Toolbox of the Civil 3D Toolspace (see Figure 5.35) and the LandXML Report application is a separate application.

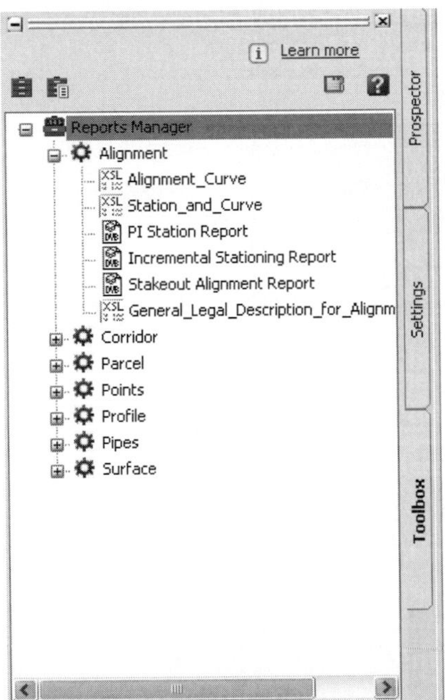

Figure 5.35

The types of alignment reports include Alignment Curve, Station and Curve, a General Legal Description for Alignments, and a Legal Description for Alignments formats for the Texas and Florida Departments of Transportation (see Figure 5.36).

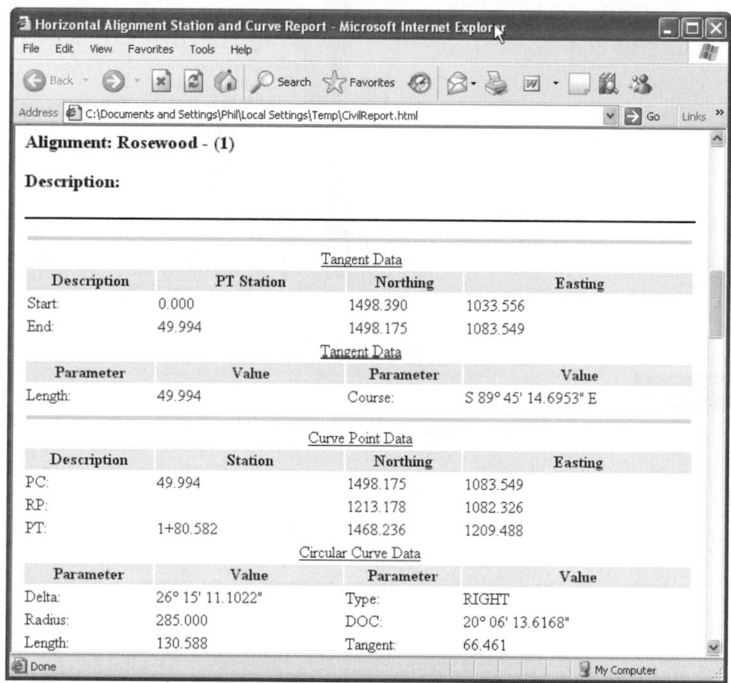

Alignment: Rosewood - (1)

Description:

Tangent Data

Description	PT Station	Northing	Easting
Start:	0.000	1498.390	1033.556
End:	49.994	1498.175	1083.549

Tangent Data

Parameter	Value	Parameter	Value
Length:	49.994	Course:	S 89° 45' 14.6953" E

Curve Point Data

Description	Station	Northing	Easting
PC:	49.994	1498.175	1083.549
RP:		1213.178	1082.326
PT:	1+80.582	1468.236	1209.488

Circular Curve Data

Parameter	Value	Parameter	Value
Delta:	26° 15' 11.1022"	Type:	RIGHT
Radius:	285.000	DOC:	20° 06' 13.6168"
Length:	130.588	Tangent:	66.461

Figure 5.36

INQUIRY TOOLS

The inquiry tools allow users to determine station and offset values, profiles, and surface elevations for selected points (see Figure 5.37).

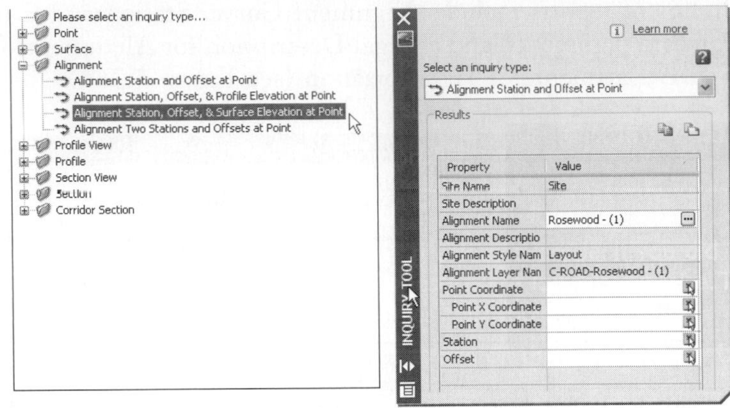

Figure 5.37

EXERCISE 5-3

When you complete this exercise, you will:

- Review sub-entity values.
- Review the segment values for an entire alignment.
- Select and view various LandXML reports about alignments.

EXERCISE SETUP

This exercise continues with the drawing used in the previous exercise. If you did not do the previous exercise, you can open the *Chapter 5 - Unit 3.dwg* file. The file is in the *Chapter 5* folder of the CD that accompanies this textbook.

1. If you are not in the drawing from the last exercises, open it now or open the *Chapter 5 - Unit 3* drawing.

SUB-ENTITY REVIEW

The first method reviews the individual segments of an alignment in a cell-based editor.

1. In the drawing, select the **Rosewood** alignment, press the right mouse button, and from the shortcut menu select **Edit Alignment Geometry....**

   This displays the Alignment Layout Tools toolbar.

2. In the Alignment Layout Tools toolbar, select the **Sub-entity Editor** icon (the second icon in from the righthand side) to display the Sub-entity Editor.

3. In the Alignment Layout Tools toolbar, click the **Pick Sub-entity** icon to the left of the Sub-entity Editor icon and select near the 335-foot radius curve at the northern end of the alignment.

   You should have something like Figure 5.33 on your screen. Again, the values in black are editable.

4. In the drawing, select other sub-entities to review their values.
5. Close the Sub-entity Editor by clicking the red "**X**".

FULL ALIGNMENT REVIEW

You can view all of the alignment's segment information in a single vista.

1. In the Alignment Layout Tools toolbar, click the rightmost icon.

   This icon calls a vista displaying all of the values for the alignment. By scrolling horizontally, you can view all of the numbers for each segment in this vista.

2. In the vista, scroll through the alignment entries.
3. Click the "**X**" to close the panorama.
4. Click the red "**X**" to close the Alignment Layout Tools toolbar.

EXERCISES

LANDXML REPORTS

You can create reports for all or a subset of alignments in the drawing. You select which alignment to generate a report on either in the initial report selection dialog box or in the Report application itself.

1. Click the **Prospector** tab if it is not current.
2. In Prospector, expand the Sites branch until you view the list of alignments.
3. Select the **Alignments** heading, press the right mouse button, and from the shortcut menu select **Export LandXML....**
4. The Export to LandXML dialog box lists all of the alignments you can export. Click the **OK** button to export the file.

   This selects all of the alignments in the drawing. You can leave all of the alignments selected or you can toggle off the alignments you don't want in a report.

5. In the File dialog box, select a location and type a name for the file, and click the **Save** button to write the data and exit the dialog box.
6. On the Desktop, double-click the **Autodesk Report** application to start it and click the **OK** button to continue to the main dialog box.
7. From the File menu of the Report application, click **Open Data File...**, browse to the location of the LandXML file just written, select it, and click the **Open** button.
8. In the Report application, click the **Data Summary** tab to make it the current tab.

   You can toggle on and off the alignments in this panel. If you leave all of the alignments selected, each alignment will be a part of a report.

9. In the Data Summary panel, leave all of the toggles **ON** and select the **Forms** tab.

   The Forms tab has several reports about alignments (see Figure 5.38).

10. Expand the Alignment section, select the **Alignment Curve** form, and click the **Output** tab to view the report.

11. In the Report application, click the **Forms** tab, select the **Station and Curve** form, and click the **Output** tab to view the report.
12. Exit the Report application and return to Civil 3D.
13. Click the **AutoCAD Save** icon to save the drawing.

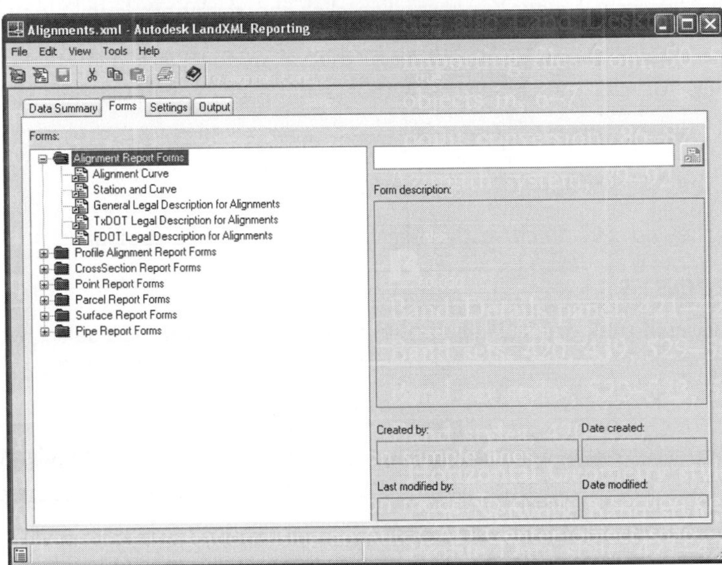

Figure 5.38

REPORTS MANAGER

1. In Civil 3D, from the General menu, select **Reports Manager....**

   The Reports Manager is another palette in the Toolbox containing the same reports as the LandXML Report Application.

2. If necessary, click the **Toolbox** tab to make it current.
3. In Toolbox, expand the Alignment section in the Reports Manager branch, select a report, press the right mouse button, and from the shortcut menu select **Execute....**
4. Review the report and then exit the report.

INQUIRY TOOLS

1. From the General menu, select **Show Inquiry Tool....**
2. In Inquiry Tool, select the drop-list arrow for Select an inquiry type:, expand the Alignment section, and select **Alignment Station and Offset at Point** from the list.
3. In the Select Alignment dialog box, select the **Rosewood – (1)** alignment, and click the **OK** button to continue.

4. In the drawing, select a point near the Rosewood alignment.

   The Inquiry Tool palette shows the selected point's coordinates, station, and offset.

5. Select a few more locations, noting the values listed in the Inquiry Tool palette.
6. In the Inquiry Tool palette, select the drop-list arrow for Select an inquiry type: and select **Alignment Two Stations and Offsets at Point** from the list.
7. In the Select Alignment dialog box, select the **Rosewood – (1)** and **Senge Drive** alignments, and click the **OK** button to continue.
8. In the drawing, select a point near the Rosewood alignment. You may need to use the select point icon in the Value cell for Point Coordinate.

   The Inquiry Tool palette shows the selected point's coordinates, station, and offset relative to the two alignments.

9. Select a few more locations, noting the values listed in the Inquiry Tool palette.
10. Click the "**X**" to close the Inquiry Tool palette.
11. Click the **AutoCAD Save** icon to save the drawing.

EXERCISES

This completes the overview of the alignment review methods. Civil 3D has an editor displaying all of the information about an alignment's segments. The LandXML Report application and the reports of the Toolbox have forms creating formal reports on alignment tangent and curve segments.

SUMMARY

- By selecting Edit Alignment Geometry and displaying a sub-entity or overall vista of the alignment segments, users can review the numbers of an alignment.
- The vistas display all of the pertinent segment geometry.
- Editable values are in black, and when you make a change, the alignment updates.
- The LandXML Report application and the Toolbox have forms creating formal alignment segment reports.

UNIT 4: EDITING A HORIZONTAL ALIGNMENT

There are times when the design changes, the design needs tweaking, or there is a need to add station equations to the centerline. Civil 3D allows users to edit an alignment by any of three methods: graphical, sub-entity, or overall. The graphical method uses selected alignment grips and as a result moves a segment to a new location (see Figure 5.39).

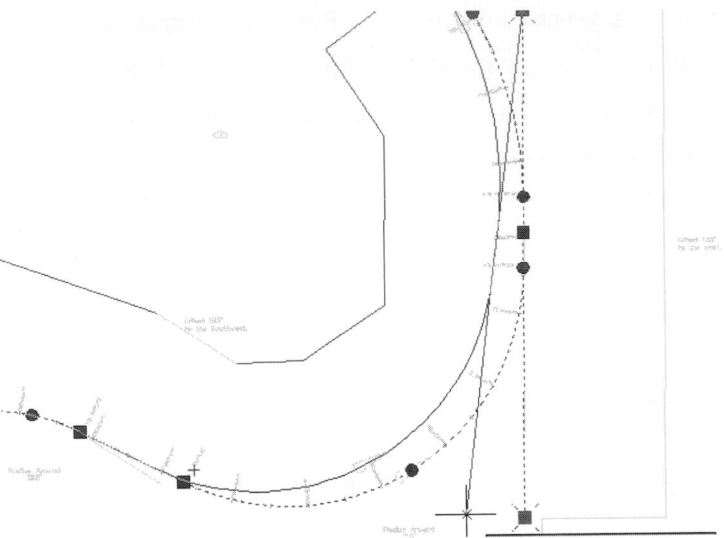

Figure 5.39

The Edit Alignment Geometry... routine of the Alignment menu displays the Alignment Layout Tools toolbar. From the Alignment Layout Tools toolbar, the Sub-entity Editor or an overall alignment editor can be displayed. The Sub-entity Editor requires the selection of an alignment segment, and after a user selects a sub-entity it displays its values in the editor. The overall editor displays a vista containing all of the alignment's values. Whether using the sub-entity or overall editor, they display the editable values in black and update the alignment after making the changes.

GRAPHICAL EDITING

When graphically editing an alignment, first select the alignment to display all of its grips and then select a segment grip and manipulate the segment's location by moving the cursor. A tangent segment has grips at its midpoint and endpoints, and a curve segment has grips at the beginning (PC) and end (PT), and midpoint. To change the alignment, click on a grip and move it to a new location. If the user attempts to locate a grip in a way that it cannot correctly resolve the alignment, Civil 3D will not allow that point to be selected.

The type of sub-entities (fixed, floating, and free) influences how the alignment "drags" and resolves the design at the location the user selects. The rules for fixed, floating, and free, and the type of entity connected before and after the active grip, all affect what the result will be after ending the edit. For example, fixed entities do not guarantee tangency, but floating and free ones do. Users need to be aware of what behavior each segment has before or during their edit. If they do not, the edit may create non-tangent segments and curves, unanticipated tangent extensions, and other unforeseen effects.

If selecting a PC or PT grip of a floating curve, the user can only lengthen the curve and cannot adjust its radius. If selecting the midpoint grip, the user must adjust the radius and change the bearing of an attached tangent. One good thing about graphically editing an alignment is that an alignment will not allow a solution that violates the minimum curve radius. And, of course, the edit cannot be undone.

When graphically editing an alignment created by the Create from Polyline command, the editing options are limited because all tangents and curves are fixed segments. Much of the original geometry will not be preserved as the alignment moves, because fixed entities do not understand tangency. When relocating the entities, they do not attempt to preserve any previous tangency, and as a result the new alignment may be incorrect.

EDIT ALIGNMENT: LAYOUT

The Alignment Layout Tools toolbar has two editors: Sub-entity and Grid (overall).

SUB-ENTITY EDITOR

When displaying the Sub-entity Editor and selecting a tangent with the sub-entity selection tool, the vista displays all the values of the selected tangent. Within the Sub-entity Editor only the northing/easting values for the ends of the tangent can be changed. When the user changes a value and presses the Enter key, the tangent changes on the screen to show the values from the panorama (see Figure 5.40).

Alignment Layout Parameters

Parameter	Value
Geometry	
Number	7
Curve Group I...	
Curve Group S...	
Type	Line
Constraint1	Fixed
Constraint2	Two points
Length	220.189'
Direction	N58° 48' 15"W
Start Station	15+63.63'
End Station	17+83.82'
Start Point	(1554.5880',2195.5103',0.0000')
End Point	(1366.2374',2309.5602',0.0000')
Pass Through ...	(1681.3241',2118.7691',0.0000')
Pass Through ...	(1315.5928',2340.2264',0.0000')

Alignment Layout Parameters

Parameter	Value
Geometry	
Number	6
Curve Group I...	
Curve Group S...	
Type	Curve
Constraint1	Free
Constraint2	Radius
Length	510.767'
Radius	495.000'
Delta angle	59.1207 (d)
Start Station	10+52.87'
End Station	15+63.63'
Start Direction	N0° 18' 59"E
End Direction	N58° 48' 15"W
Start Point	(1793.1887',1769.3518',0.0000')
End Point	(1554.5880',2195.5103',0.0000')
Center Point	(1298.1963',1772.0858',0.0000')
Pass Through ...	
Pass Through ...	
Pass Through ...	
Direction at Th...	
Direction at Th...	
Chord length	488.407'
Chord Direction	N29° 14' 38"W
Mid-Ordinate	64.431
External Tangent	280.746'
External Secant	74.073'
PI Included Angle	120.8793 (d)
Greater than 180	false
PI Station	13+33.61'

Figure 5.40

When the user selects a curve segment with the sub-entity selection tool, the Sub-entity Editor displays all the values of the selected curve. Depending on the type of curve (fixed, free, floating), users can change the curve's length, radius, chord, mid-ordinate, and the external ordinate or secant. After changing the values and pressing the Enter key, the curve changes on the screen to show the new parameters.

When selecting a spiral segment with the sub-entity selection tool, the vista displays all the values of the selected spiral. Users can edit the length, the A value of the in and out spiral segments, and the radius of the circle (see Figure 5.41).

Alignment Layout Parameters

Parameter	Value
Spiral In	
Entity	14
Curve Group I...	
Curve Group S...	
Type	Spiral-Curve-Spiral
Constraint1	Free
Constraint2	SpiIn-Radius-SpiOut
Entity Before	12
Entity After	13
Spiral type	Simple
Length	100.000'
A	111.803'
Delta angle	22.9183 (d)
Start Station	
End Station	
Start Direction	N12° 54' 43.22"W
End Direction	N35° 49' 49.15"W
Start Point	(1329.7608',2093...
End Point	(1294.9218',2186...
Incurve	Incurve
Compound	false
Radius in	Infinity'
Radius out	125.000'
Total X	98.412'
Total Y	13.182'
Short tan	33.850'
Long tan	67.234'
P	3.314'
K	49.735'
Spiral Definition	Clothoid
SPI Station	
SPI Northing	2159.3411'

Figure 5.41

Users can delete any segment from an alignment. The alignment will display a gap for each deleted sub-entity. Users can then add a new entity to replace the one deleted.

OVERALL EDITOR

The second editor displays in a vista all of the critical values for each segment of the alignment. The values in the vista are the same as those listed in the Sub-entity Editor. After selecting and editing a value in the vista and pressing the Enter key, the entity's graphics change to match its new values. All editable values are in black.

DELETING AND INSERTING SEGMENTS

Tangents, curves, and spirals cannot be deleted in the editors. They can only be deleted by using the sub-entity delete tool of the Alignment Layout Tools toolbar.

The Alignment Layout Tools toolbar has routines inserting a PI, deleting a PI, and breaking apart a PI, so additional or new alignment sub-entities can be added.

ALIGNMENT PROPERTIES

An alignment has properties that report information about the alignment and affect the resulting roadway corridor. The Alignment Properties dialog box has the ability to redefine station values, manage station equations, add design speeds, manage labeling sets, add in superelevation control, and identify profiles and profile views related to the alignment.

STATION CONTROL

The Station Control panel sets the location and starting station for an alignment (reference point). By default, the starting station is the coordinates at the starting end of the beginning segment of the alignment. In the Reference point section of the panel, users can define a new beginning station or beginning point for the alignment (selecting a new location in the drawing). See Figure 5.42.

STATION EQUATIONS

There will be times when there are points along an alignment where the stationing changes (see Figure 5.42). This happens for many reasons, such as a new road connecting to an existing one, a change of jurisdiction controlling the centerline, and so on. The station equation point signals the end of one system of stationing and the beginning of another. The stations on the other side of the equation can be totally different and can even decrease in value along the remainder of the roadway.

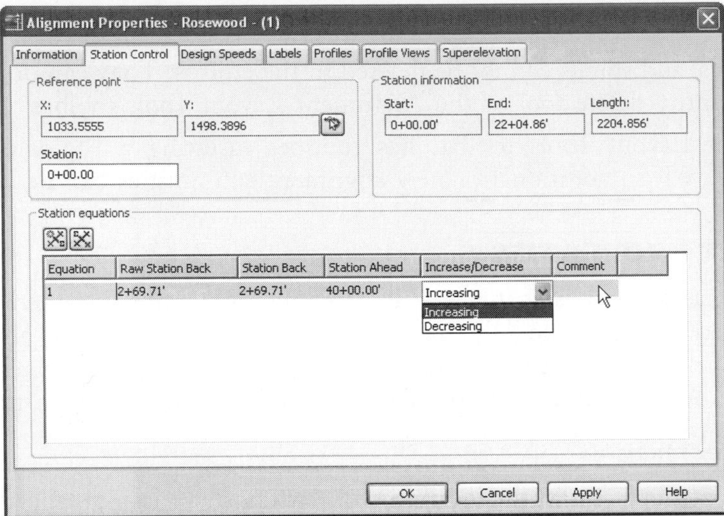

Figure 5.42

DESIGN SPEEDS

When designing a roadway with superelevations or by following certain design guidelines, users will have to enter design speeds for the roadway (see Figure 5.43). These speeds affect the roadway design criteria used when creating a design solution for superelevation.

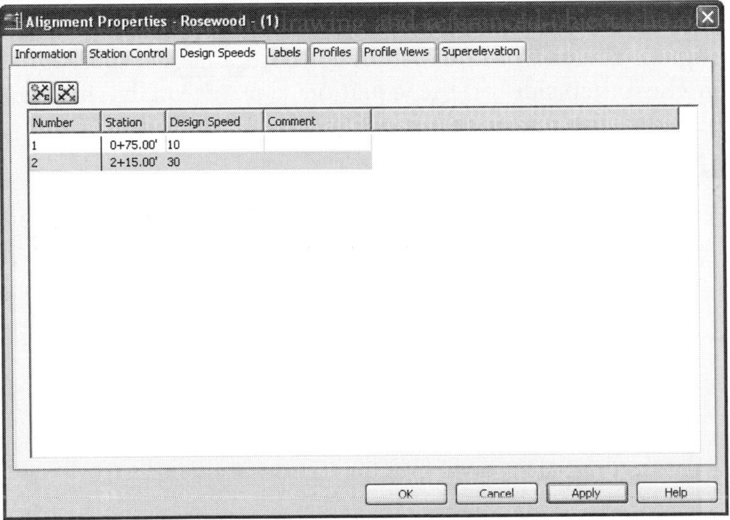

Figure 5.43

LABELS

The Labels panel lists and allows users to change the labeling styles currently annotating the alignment. This panel reflects the styles found in the label set assigned when defining the alignment (see Figure 5.6).

PROFILES AND PROFILE VIEWS

These panels list the profiles and profile views associated with the alignment. They contain data only after the user creates a profile view and profile for the named alignment.

SUPERELEVATION

This panel defines the superelevation regions within the alignment (each curve of the alignment is a region). The panel references tables containing rules governing superelevation. Each curve defines a superelevation region and each region can follow different rules, if so desired (see Figure 5.44 and Figure 5.45).

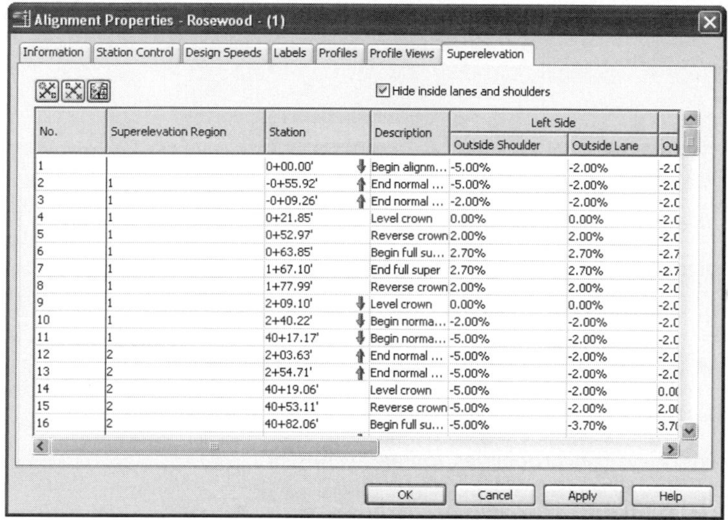

Figure 5.44

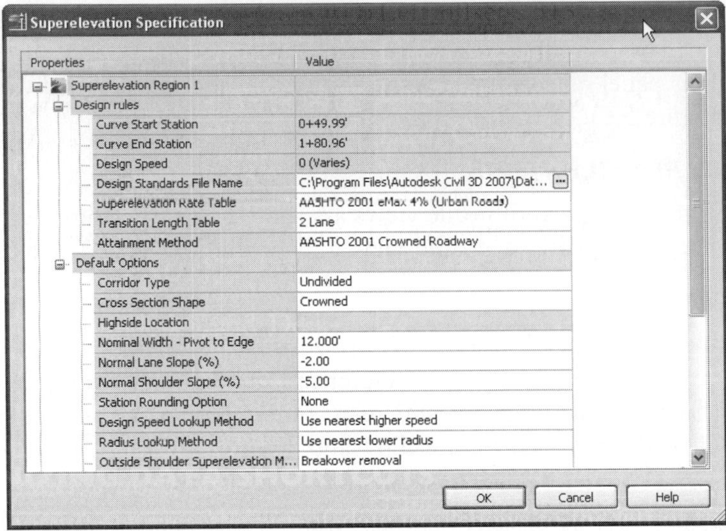

Figure 5.45

EXERCISE 5-4

When you complete this exercise, you will:

- Graphically edit an alignment.
- Edit alignment values from an overall vista.
- Edit an alignment's sub-entities.
- Review and edit an alignment from its Properties dialog box.
- Set design speeds for an alignment.
- Set superelevaiton properties for an alignment.

EXERCISE SETUP

This exercise continues with the drawing from the last exercise. If you didn't do the previous exercise, you can open the *Chapter 5 - Unit 4.dwg* that is in the *Chapter 5* folder of the CD that accompanies this textbook.

1. Open the drawing from the previous exercise or the *Chapter 5 - Unit 4* drawing.

GRAPHICAL EDITING

When a user selects any alignment segment, the entire alignment shows grips reflecting locations you can use to change the alignment. To change the alignment, click on a grip and move it to a new location. If you attempt to locate a grip in a way that it cannot correctly resolve the alignment, Civil 3D will not allow you to select that point.

The type of sub-entities (fixed, floating, and free) influences how the alignment "drags" and resolves the design at the location you select. Some entities allow you to change their bearing (fixed lines), their radius, or some other value simply by selecting the correct grip.

Adjusting Tangent Segments

There are four fixed tangents in the alignment. You can move all of the tangents by selecting the middle grip or by adjusting their directions by selecting a grip at one end of the tangent.

1. In the drawing, use the AutoCAD Zoom command to better view the eastern north/south tangent.
2. In the drawing, click the alignment to activate its grips.
3. Select the middle grip (square) of the eastern tangent segment and move the roadway east and west.

   As you move the tangent segment, all of the connecting segments change to accommodate the change to the tangent.
4. Press the **Esc** key to place the tangent back to its original location.
5. In the drawing, click the southern tangent extension grip.

   Activating this grip allows you to change the direction of the segment from its southern end while holding its northern point. As you reposition the location of the grip, all of the connecting segments adjust their values to maintain a valid alignment definition.
6. Press the **Esc** key to return the tangent to its original location.

EXERCISES

Adjusting Curve Segments

When activating a free curve grip, you can adjust the radius of the curve. Both connected tangents adjust their length to accommodate the curve changes.

The first curve is attached to the end of the first tangent segment. The curve has a floating tangent that is always attached to its other end. If selecting a PC or PT grip of the floating curve, you can only lengthen the curve; you cannot adjust its radius. If selecting the midpoint grip, you adjust the radius and change the bearing of the eastern tangent.

1. If necessary, click the 310-foot radius curve just before the east tangent segment to activate its grips.
2. In the drawing, click the middle grip and slowly move the grip toward the northwest.

   The alignment shows a solution for a short distance from the original point, but it will stop when there is no valid solution for the movement of the alignment.
3. Slowly move the grip toward the southeast.

   Both tangents lengthen (equally) to accommodate the change to the curve.
4. Press the **Esc** key to return the curve to its original location.

Lengthening a Floating Curve

The second tangent connecting to the first alignment curve is a floating tangent and is always tangent to the curve. This situation provides two ways of adjusting the curve to make the tangent's bearing more like the bearing of the line offset as a guide. The first option is to lengthen the curve and the second is to change the radius of the curve.

The method you use to solve this design issue depends on the design constraints set forth by the governing covenants. There may be criteria on lengths of tangents between curves, curve radii, and other alignment specifics that you must use to create a "correct" design for the site.

If you shorten the curve by moving the PT grip to the northwest, you make the bearing of the tangent become more northerly. If you lengthen the curve, the tangent segment will change its bearing toward the south to a point where the tangent is almost parallel to the offset guideline. As you do either motion, the attached eastern tangent and curve (east end of the tangent) react to the lengthening of the curve, preserving their tangency with the adjusted curve.

1. In the drawing, click the eastern square grip of the first curve and move it slowly toward the northwest, then slowly move it southeasterly.
2. Press the **Esc** key to return the curve to its original location.

   Changing the radius of the curve also provides an alternative solution.
3. In the drawing, click the circular grip at the midpoint of the arc and slowly move it to see it change the alignment.

   This does not provide the correct adjustment.
4. Press the **Esc** key to return the curve to its original location

   As you move the circular grip to the north, the radius increases and pushes the tangent to a more easterly bearing. If you move the circular grip to the south, the radius decreases and pushes the tangent to a more southerly bearing.
5. Click the **Settings** tab to make it current.
6. Expand the Alignment branch until you view the list of alignment styles.
7. From the list, select **Layout**, press the right mouse button, and from the shortcut menu select **Edit....**
8. In the Layout Style dialog box, select the **Design** tab, toggle **ON** Enable radius snap, set the snap value to **10**, and click the **OK** button to exit.
9. If necessary, toggle **OFF** object snaps.
10. In the drawing, reselect the alignment to activate its grips.
11. In the drawing, click the curve's midpoint grip and slowly move it first north and then south, changing the curve radius, until you reach a possible solution. Select a point to adjust the radius of the curve.
12. While the alignment is highlighted, press the right mouse button, and from the shortcut menu select **Edit Alignment Geometry....**
13. Click the **Alignment Grid View** icon (the rightmost icon of the toolbar) and review the radius for the second segment or first curve of the alignment.

    Notice that the new radius is a value divisible by 10. The alignment style has a constraint (radius snap) that is set to 10. This means when you adjust a curve for an alignment using this style, the resulting adjusted curve radii is always divisible by 10 (possible radii would be 310, 280, 250, or any value divisible by 10).

14. Close the Alignment Grid View and the Alignment Layout Tools toolbar.

    The criteria governing the alignment do not allow a radius less than 280 feet. The original radius should be preserved and the only solution is to lengthen the curve to make the tangent's bearing become more like the offset guideline.

15. Use the AutoCAD Undo command until the curve is back to its original position.
16. In the drawing, click the alignment, then click the eastern square grip and move the PT toward the southeast to a point where the tangent matches the guideline. When the tangent is at the correct location, select that point to adjust the curve.
17. Save the drawing by clicking the **AutoCAD Save** icon.

ALIGNMENT EDITORS

The next method of editing an alignment is from the Alignment Layout Tools toolbar. The Alignment Grid View shows all of the segments and their types (fixed, floating, and free). The second editor is the Sub-entity Editor, which displays a selected segment's information. Each editor indicates editable values in black.

1. In the drawing, click the alignment, press the right mouse button, and from the shortcut menu select **Edit Alignment Geometry....**
2. In the Alignment Layout Tools toolbar, click the **Alignment Grid View** icon (the rightmost icon of the toolbar) and review the segment values for the alignment.
3. In the vista for segment 2, change the radius from 285 to **280** feet and click in another cell to complete the change.

   The curve changes on the screen to accommodate the change in the editor.

4. In the panorama, click the "**X**" to close it.
5. In the Alignment Layout Tools toolbar, click the **Sub-entity Editor** icon (first icon to the left of Alignment Grid View) to display its vista.
6. In the Alignment Layout Tools toolbar, click the **Pick Sub-entity** icon. In the drawing, select the alignment curve you just edited.

   The data for the curve appears in the editor. The current radius is 280 feet and is in black print.

7. Click in the curve Radius Value cell, change the radius to **285** feet, and click in another cell to complete the change.

   The curve changes on the screen to accommodate the change in the editor.

8. Click the red "**X**" in the upper-right corner of the Sub-entity vista and in the Alignment Layout Tools toolbar to return to the command prompt.

ALIGNMENT PROPERTIES

All Civil 3D objects have properties, and the alignment is no exception. The properties of an alignment affect stationing, station equations, design speeds, labeling, superelevation, and report what profiles and profile views use the alignment's data.

1. From the View menu, the **Named Views...** command, restore **Station Equation**.
2. In the drawing, click the alignment, press the right mouse button, and from the shortcut menu select **Alignment Properties....**
3. Select the **Station Control** tab to view its contents (see Figure 5.46).

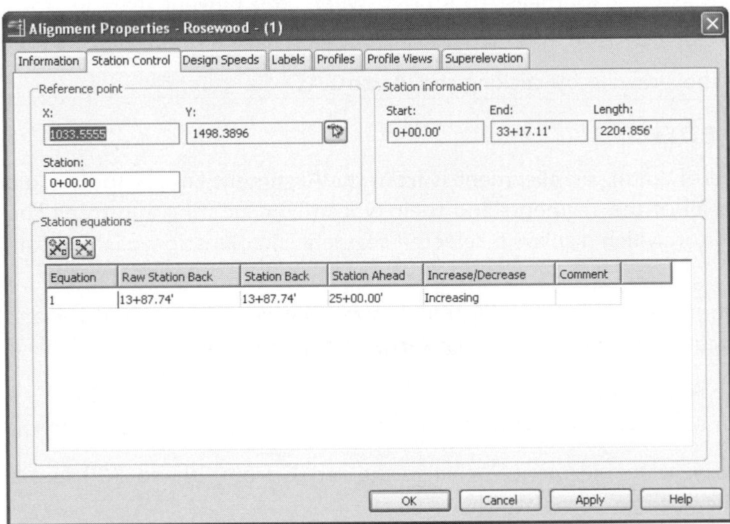

Figure 5.46

The lower portion of the dialog box is where Civil 3D manages (add and remove) station equations. Associated with an equation are the changes to the station at that point and what happens to the stationing after the equation.

4. In the dialog box, click the **Add Station Equation** icon at the middle left.

   The dialog box disappears, and in the drawing a jig attaches to the alignment reporting stations.

5. Using an Intersection object snap, select the intersection of the alignment and the Phase I line.

   You return to the Alignment Properties dialog box.

6. In the dialog box, type **2500** for station ahead, and set Increase/Decrease to **Increasing**.
7. Click the **Apply** button to modify the alignment.

The beginning station remains the same; however, the ending station changes to a higher value and the length remains the same as before.

8. Review the new ending station in the Station Control panel.

DESIGN SPEED

1. In the Alignment Properties dialog box, click the **Design Speeds** tab to view its contents.
2. Click the **Add Design Speed Station** icon (at the left of the dialog box).

   The dialog box disappears.
3. In the drawing, zoom back to see more of the site.
4. In the drawing, zoom in to the beginning of the alignment, and select a point around station 0+75.
5. After selecting the point, you return to the editor with a station value that is not exactly 0+75.
6. Click in the Station cell and change it to 0+75 (**75**) and assign it the speed of **10** (see Figure 5.43).
7. Again click the **Add Design Speed Station** icon and select a point around station 2+15. When you return to the dialog box, change the station to 2+15 (**215**) and set the speed to **30**.

   The road will use 30 for the remainder of the road length.

LABELS

1. In the Alignment Properties dialog box, click the **Labels** tab to view its contents. The labels listed in the dialog box are the same as you assigned when defining the alignment.
2. Click the **OK** button to close the Alignment Properties dialog box.
3. Click the **AutoCAD Save** icon at the top left of the screen to save the drawing.

This ends the exercise on editing an alignment. Civil 3D allows you to edit an alignment graphically, in a sub-entity or an overall editor, and in the alignment's property dialog box. The Alignment Properties dialog box is the only place you can enter data for station equations, design speeds, superelevation, and other important alignment values.

SUMMARY

- Users can graphically relocate segments of an alignment.
- When a user graphically edits an alignment, the segments react to the changes and will not allow users to select a point creating an incorrect solution.
- When a user graphically edits an alignment, the segment types (fixed, floating, and free) determine how the alignment reacts to the changes.
- When editing an alignment in the Alignment Layout Tools toolbar, users have a choice of editing segment data in an overall or a sub-entity vista.
- When a user edits data in a vista editor, only the values in black are editable.
- The Alignment Properties dialog box has panels affecting an alignment's stationing, station equations, labels sets, design speeds, profiles and profile views references, and settings for superelevations.

UNIT 5: CENTERLINE ANNOTATION

When designing or creating an alignment, Civil 3D automatically adds an annotation to the alignment from the assigned label set. A label set is a named collection of individual label styles. Users can change the label styles in a label set by changing the set's list of labels or by adding or removing label types and styles in the Labels tab of the Alignment Properties dialog box. Users can create new styles and label sets from these new styles. These labels are specifically for annotating the alignment segment geometry. All label sets and their styles are listed in the Label Sets and Stations sections under the Label Styles heading (see Figure 5.47).

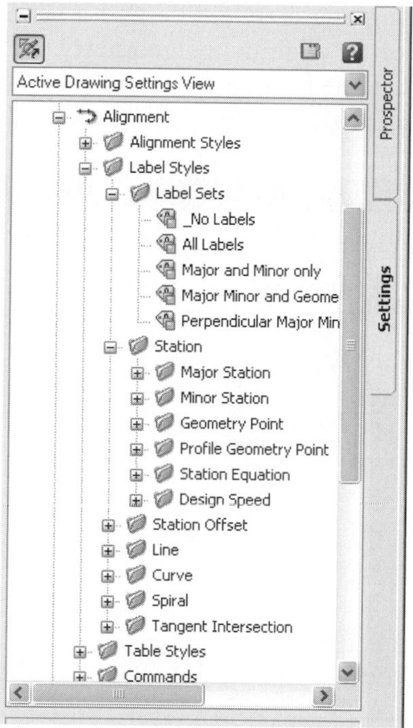

Figure 5.47

A second group of labels annotate alignments or locations near them using the Add Labels dialog box. The Alignment branch lists these labels after the Label Sets and Station headings. These labels include Station Offset, Line, Curve, Spiral, and Tangent Intersection.

All alignment labels will adjust their size when a plot scale is set in a layout viewport. All alignment labels are reactive to changes to the alignment's design and its stationing. Civil 3D provides several "starter" styles; however, users can make their own to accommodate company standards.

ALIGNMENT LABEL STYLES

The labels annotating an alignment are individual styles or style sets. A label set is a combination of individual styles grouped under a single name. When the user assigns a label set to an alignment, all of the styles annotate the alignment with their particular focus. Users can change the set or assign new styles to an alignment in the Labels tab of an alignment's Properties dialog box.

LABEL SETS

A label set contains specific purpose styles that annotate critical alignment geometry points (see Figure 5.48). The All Labels Alignment Label set annotates major and minor stations, horizontal and profile geometry points, station equations, and design speeds.

Users create a label set by assigning a style type and a style from that type to the set. A set can label all or a subset of an alignment's geometry. When adding a label style, first identify the type and the style name for that type, and then click the Add>> button to place type and style on the label set list.

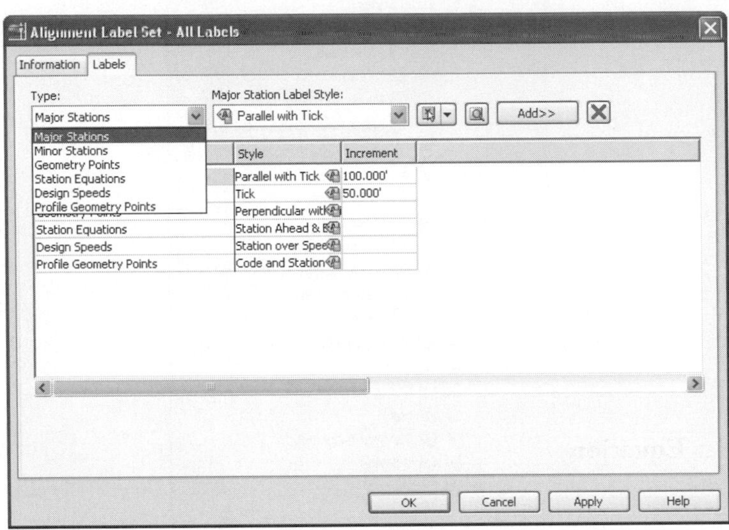

Figure 5.48

LABEL SET STYLES

The label set styles define annotation for the alignment's segments. The types of labels include Major and Minor Stations, Geometry Points, Profile Geometry Points, Station Equations, and Design Speeds. The label set styles are listed under the Stations heading of Label Styles for the Alignment branch, and these styles can be used only with a label set.

Station: Major and Minor Station

These styles define how to annotate the stations along the path of an alignment. For example, the Parallel with Tick annotates the stationing as labels offset and parallel to the roadway with a tick on the alignment. The Perpendicular with Tick style annotates the stations with labels perpendicular to the alignment and with ticks at the intervals set in the style (see Figure 5.49).

Station: Geometry Point

A Geometry Point style labels the start and end PCs and PTs of the horizontal alignment geometry.

Station: Profile Geometry Point

A Profile Geometry Point style labels the critical points (BVC, PVI, HP, and EVC) of a vertical design along the path of an alignment.

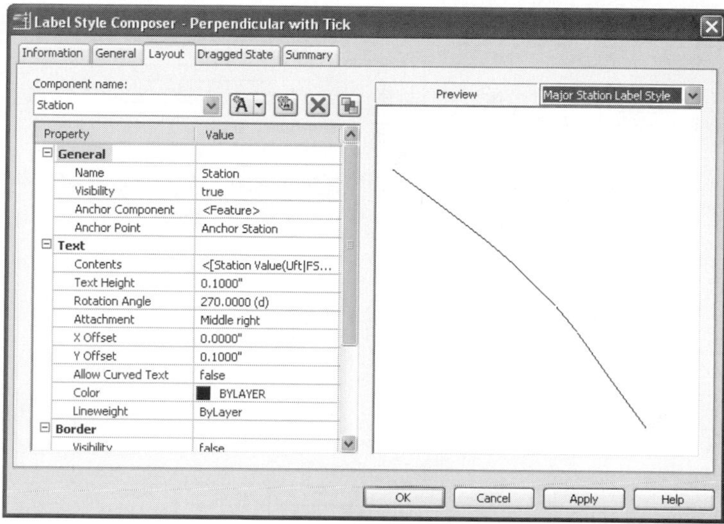

Figure 5.49

Station: Station Equation

The Station Equation style labels station equation points.

Station: Design Speed

The Design Speed styles label the design speed on the alignment at the point where the speed changes.

LABEL STYLES FOR ADD LABELS

The Add labels command places additional labels in a drawing to reference an alignment and/or its properties. Most of these labels are off of the alignment; however, the line, curve, and spiral labels appear on their respective alignment segment.

STATION OFFSET

The Station Offset labels document the station and offset of coordinates relative to an alignment. The label is complex and makes several references to the values of the associated alignment (see Figure 5.50). The label reacts to any changes to the alignment by updating the station and offset values.

This is a post-alignment creation label. Its purpose is to annotate the location of a lot corner, point, structure, etc., relative to an alignment. A fixed point station and offset label do not move if the alignment stationing changes. The label updates to show the new station value. Non-fixed station and offset labels will move to retain the original alignment station value.

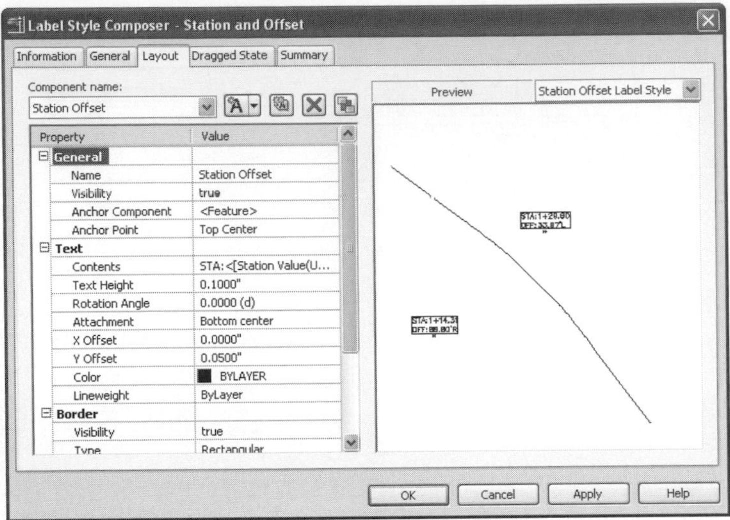

Figure 5.50

LINE, CURVE, AND SPIRAL

This group of label styles annotates the bearings and distances, curve, and spiral geometry of the various segment types of an alignment. These labels are the basis for any tables containing bearing, distances, and curve data.

TANGENT INTERSECTIONS

This label annotates the intersection of two alignment tangents, curves, or spiral-curve-spirals at a tangent intersection.

ALIGNMENT TABLE

Civil 3D allows users to create various tables representing the lines, curves, spirals, or segments within the alignment (see Figure 5.51). Each entry in a table corresponds to a label (line, curve, and spiral) on the alignment.

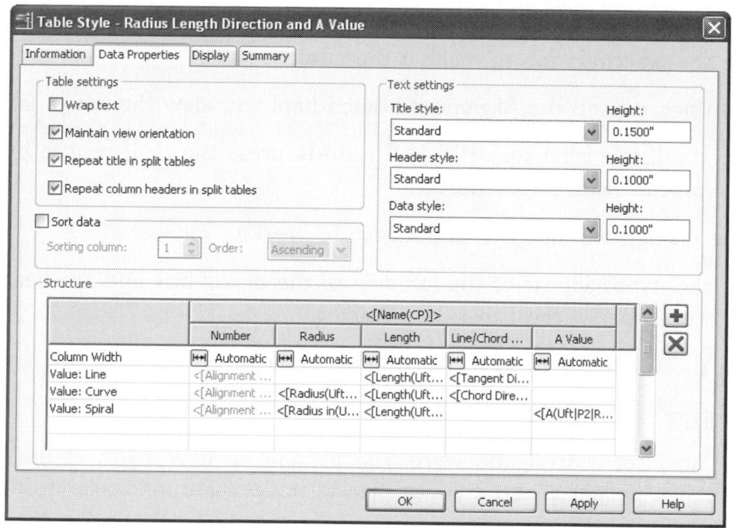

Figure 5.51

EXERCISE 5-5

When you complete this exercise, you will:

- Review alignment label sets.
- Review alignment label styles.
- Create new label styles.
- Define a new label set.
- Apply a label set to an alignment.
- Add station and offset labels.
- Add segment labels.
- Create an alignment table.

EXERCISE SETUP

This exercise continues with the drawing used in the previous exercise. If you did not complete the previous exercise, you can open the *Chapter 5 - Unit 5.dwg* that is in the *Chapter 5* folder of the CD that accompanies this textbook.

1. Open the drawing from the previous exercise or the *Chapter 5 - Unit 5* drawing.

LABEL SETS

When a user creates an alignment, the Create Alignment dialog box has an entry for a Label Set labeling the resulting alignment's geometry (see Figure 5.25).

The Label Set Labels panel lists the label types and the associated label style names.

All Labels

1. Click the **Settings** tab to make it the current panel.
2. In Settings, expand the Alignment branch until you view the list of label sets.
3. From the list of label sets, click **All Labels**, press the right mouse button, and from the shortcut menu select **Edit....**
4. In the dialog box, click the **Labels** tab to view its values.

   The label types appear at the left side of the dialog box and are headings in the Station branch below the Label Sets heading.
5. Click the **Cancel** button to exit the dialog box.

NEW LABEL SET

Creating a new label set is straightforward. The defining of new styles is the challenge of this exercise. This exercise defines a new style by placing stations perpendicular to the station on the alignment.

Major Station: Perpendicular Style

The Perpendicular style has station label text at the station and is orientated to the direction of the centerline. The text does not change when you drag it away from its original position.

1. In Settings, expand the Alignment, Label Styles, Station, Major Station branches until you view the list of styles.
2. From the Major Station list of styles, select **Perpendicular with Tick**, press the right mouse button, and from the shortcut menu select **Copy....**
3. In the dialog box, select the **Information** tab, change the Name to **Perpendicular**, and give the style a short description.
4. Click the **General** tab to view its settings.

   To make the label orientate to the path of the centerline, you cannot have the label set to plan readable.
5. In the dialog box, click the Value cell for Plan Readable and change it to **False**.
6. Select the **Layout** tab to view its contents.
7. At the top left of the panel click the drop-list arrow to the right of Component Name. Select **Tick** and delete the Tick component by selecting the red "**X**" in the upper middle portion of the dialog box.

   The current component name is now Station.
8. In the Text section, click in the Value cell for Attachment to display a drop-list arrow. Click the drop-list arrow, and from the list select **Middle Center**.
9. In the Text section, click in the Value cell for Y Offset and change the offset to **0** (zero).

   Your Layout panel should look similar to Figure 5.52.

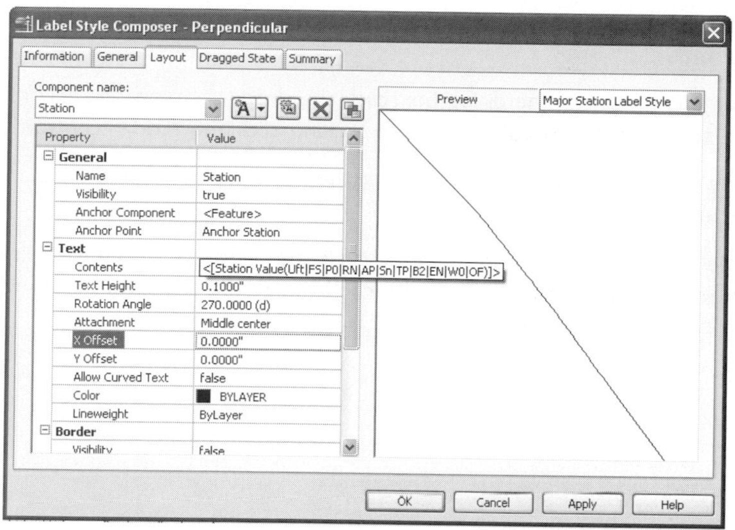

Figure 5.52

10. Select the **Dragged State** tab to review its settings.
11. In the Leader section, click in the Visibility value cell and change it to **False**.
12. In the Dragged State section, click in the Display value cell and change the value to **As Composed**.
13. Click the **OK** button to create the style.
14. Click the **AutoCAD Save** icon to save the drawing.

Perpendicular Label Set

The Perpendicular style does not belong to any label set definition. The All Labels label set contains parallel labels. The new label set is exactly like All Labels except for the Major Station label. When complete, your new style should look like Figure 5.53. Table 5.1 contains all of the label types and the styles they use.

1. From Settings, expand the Alignment, Label Styles, and Label Sets branches until you view a list of label sets.
2. From the list of label sets select **All Labels**, press the right mouse button, and from the shortcut menu select **Copy...**.
3. Select the **Information** tab and type **All Labels – Perpendicular** as the name.
4. Click the **Labels** tab to view its contents.
5. In the list of labels select the type **Major Stations** and click the red "**X**" at the upper right of the dialog box to delete both the Major and Minor Station types and their associated label styles.

6. In the dialog box at the top left, click the drop-list arrow for Type and select **Major Stations**.
7. In the top middle of the dialog box, click the drop-list arrow for Major Station Label Style, select **Perpendicular** from the list, and click the **Add>>** button at the top right.
8. In the dialog box at the top left, click the drop-list arrow for Type and select **Minor Stations**.
9. In the top middle of the dialog box, click the drop-list arrow for Minor Station Label Style, select **Tick** from the list, and click the **Add>>** button at the top right.
10. Click the **Apply** button to re-sort the list of styles.
11. Click the **OK** button to exit the dialog box.

Table 5.1

Label Type	Style
Major Stations	Perpendicular
Minor Stations	Tick
Geometry Points	Perpendicular with Tick and Line
Station Equation	Station Ahead & Back
Design Speeds	Station over Speed
Profile Geometry Points	Code and Station

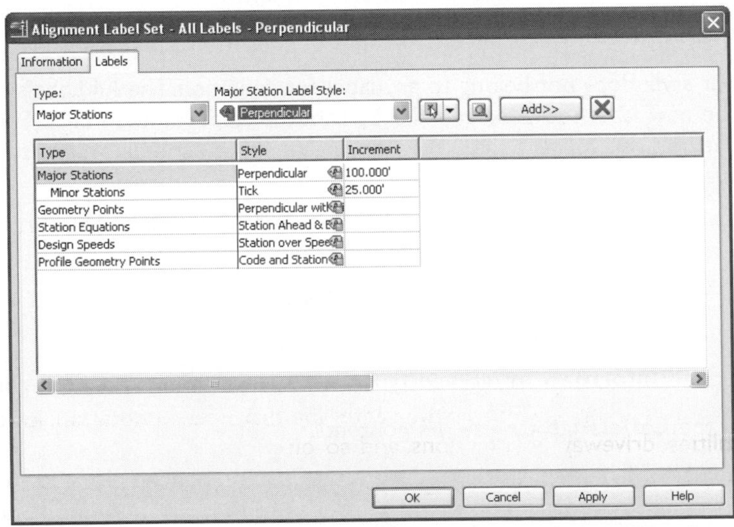

Figure 5.53

CHANGING ALIGNMENT LABEL SETS

When you change the Labels tab label list of the Alignments Properties dialog box, you change the labeling of an existing alignment. The list of styles is built by setting a label type (top left), selecting a style (center top), and adding it to the list or by importing a label set.

1. Click the **Prospector** tab to make it current.
2. Expand the Sites branch until you view Alignments and its list of defined alignments.
3. From the list select **Rosewood – (1)**, press the right mouse button, and from the shortcut menu select **Properties...**.
4. Select the **Labels** tab. Under the Type heading, select **Major Stations**, and click the red "**X**" to the right of the Add>> button to remove the label type and its style.
5. Repeat the previous step until there are no styles listed in the panel.
6. At the bottom of the Labels panel, click the **Import Label Set...** button.
7. In the Select Style Set dialog box, click the drop-list arrow. From the list select **All Labels**, and click the **OK** button to return to the Alignment Properties dialog box. When prompted for a profile, click the **Cancel** button.
8. Click the **OK** button to exit the dialog box.
9. Use the AutoCAD Zoom and Pan commands to better view the alignment labeling.

   The major stationing labels should be parallel with the alignment.
10. In the drawing, select the alignment, press the right mouse button, and from the shortcut menu select **Alignment Properties...**.
11. In the Alignment Properties dialog box, select the **Labels** tab and delete all of the labels from the list.
12. At the bottom of the Labels panel, click the **Import Label Set...** button.
13. In the Select Style Set dialog box, click the drop-list arrow. From the list select **All Labels - Perpendicular**, and click the **OK** button to return to the Alignment Properties dialog box. When prompted for a profile, click the **Cancel** button (see Figure 5.54).
14. Click the **OK** button to close the dialog box.

The labels change to indicate the major stations as text perpendicular to the roadway.

EXERCISES

ADD LABELS

There are critical points along the path of the centerline that need annotation: lot corners, existing trees, existing utilities, driveway intersections, and so on.

1. Make sure you can select icons from the Transparent Commands toolbar.
2. From the Import flyout of the File menu, select **Import XML...**.

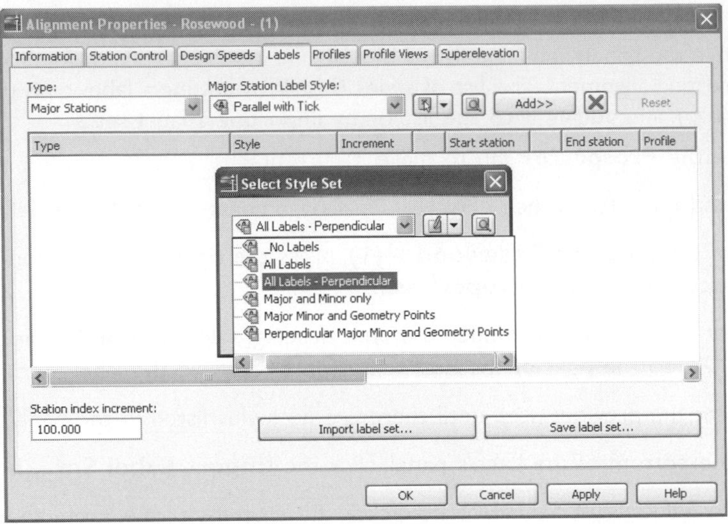

Figure 5.54

3. In the Import LandXML dialog box, browse to **Station Offset.xml**, select it, click the **Open** button, and click the **OK** button. The file is in the *Chapter 5* folder of the CD that accompanies this textbook.

   The points appear in the drawing.

4. From the View menu, select **Named Views...** and restore the view **Station Equation**.

5. From the Alignments menu select the **Add Labels...** command.

6. In Add Labels, change the Label type to **Station Offset - Fixed Point** and set the Station offset label style to **Station Offset and Coordinates**.

   Your Add Label should match Figure 5.55.

7. Click the **Add** button to start the labeling process.

   The Add Label command prompts you to select an alignment.

8. In the drawing, select the **Rosewood** alignment.

   You want to label the point's location with a station, offset, and coordinates. To be able to select the points, use the Point Object transparent command. This transparent command passes the coordinates of the selected point to the labeling routine so it can calculate the point's station and offset. The transparent command works for only one selection, so you need to reselect the transparent command for each point selection.

   The routine displays a station jig and asks you to select a point.

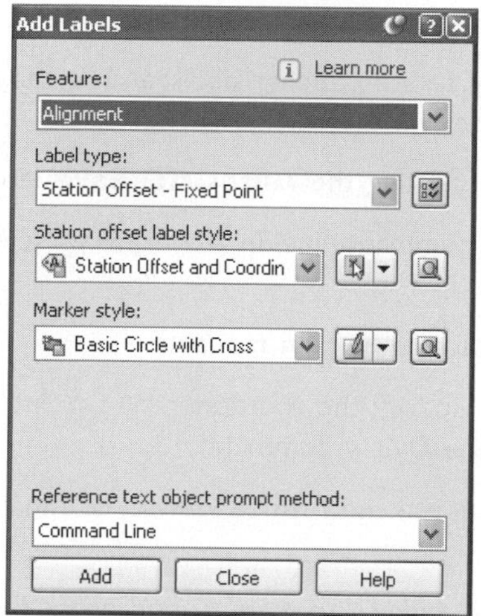

Figure 5.55

9. From the Transparent Commands toolbar, click the **Point Object** icon and in the drawing select one of the points.
10. Repeat the previous step to annotate the remaining points.
11. When done annotating the points, press the **Enter** key to end the command.

    The labels may be on top of one another.
12. In the drawing, activate the station/offset label grip by clicking the label; select the leader grip (blue diamond to the southwest of the location grip), and drag the label to a new position.

    The label changes to the Dragged State definition and has a leader pointing to the coordinates of the point object.
13. Use the AutoCAD Pan command and shift the drawing so you can see more of the alignment away from the points.
14. In the Add Labels dialog box, change the Label type to **Station Offset** and the Label style to Station Offset and Coordinates.
15. Click the **Add** button to start the labeling process.

    The Add Label command prompts you to select an alignment.
16. In the drawing, select the **Rosewood** alignment.

    A station jig appears. You locate the offset from left to right or vice versa after setting the station.
17. Using the station jig in the drawing, select a point astride the alignment to identify the station and select a second point to identify the location of the offset.

18. Repeat the previous step to create more station/offset labels.
19. Press the **Enter** key to stop adding labels, and in the Add Labels dialog box, click the **Close** button to exit.
20. Save the drawing by clicking the **AutoCAD Save** icon.
21. Type **Dview** at the command line. In the drawing, select all objects and use the Twist option to rotate the image.

    All of the labels rotate to be plan readable.
22. Move the mouse to specify the rotation, pick a point in the drawing, and press the **Enter** key to exit the Dview command.
23. From the View menu, use the Named Views... command to save the current view as **Twisted**.
24. In the command line, type **Plan**, type **C**, and press the **Enter** key twice to restore the previous display state.

LABELS AND LAYOUTS

1. In the drawing, select **Layout1**'s tab. If the layout tabs are not displayed below the drawing window, click the **Layout1** icon (second button after LWT on the status bar).
2. In the layout, double-click the viewport to enter its model space.
3. From the View menu use Named Views... to restore **Twisted**.
4. Double-click outside the viewport to return to paper space.
5. In the layout, select the viewport, press the right mouse button, and from the shortcut menu select **Properties...**.
6. In Properties, the Misc section, set Standard Scale to **1"=60'**.
7. In the command line, type **Regenall** and press the **Enter** key to resize the station offset labels.

    The text in the viewport resizes to the new plot scale and is plan readable.
8. In the layout, select the viewport and in the Misc section of Properties, change the Standard Scale to **1"=30'**.
9. In the command line, type **Regenall** and press the **Enter** key to resize the station offset labels.

    The text in the window scales to the new plot scale.
10. In the status bar, select the **Model** icon (first button after LWT) and erase the station/offset labels from the drawing.
11. Click the **AutoCAD Save** icon to save the drawing.

SEGMENT LABELS

There are times when you need to label the tangent and curve values of an alignment. You add segment labels with the Add Labels dialog box.

1. From the Alignments menu select **Add Labels....**
2. In Add Labels, change the Label type to **Multiple Segment** by clicking the drop-list arrow and selecting it from the list.
3. Click the **Add** button. In the command line the routine prompts for an alignment; in the drawing, select the alignment.

   Segment labels appear along the alignment's tangents and curves.
4. Use AutoCAD's Zoom and Pan commands to view labels.
5. Click the **Close** button of the Add Labels dialog box to close it.

ALIGNMENT TABLE

Segment labels allow you to create a table that lists the alignment's lines and curves.

1. Use the AutoCAD Zoom and Pan command to locate the site to the right of the screen.
2. From the Tables flyout of the Alignments menu, select **Add Segments....**
3. In the Alignment Table Creation dialog box, set the table to alignment and set the alignment to **Rosewood – (1)**. All of the remaining settings remain the same. Click the **OK** button to close the dialog box.

   As you exit the dialog box, the routine creates a table hanging on the cursor.
4. In the drawing, select a point to place the table in the drawing.
5. Click the **AutoCAD Save** icon to save the drawing.

This completes the review of annotating alignments. The next unit reviews commands that create new entities from an existing alignment.

SUMMARY

- When creating an alignment, the label set automatically annotates critical alignment values.
- All labels are scale and rotation-sensitive.
- All station and offset labels will update their station and offset values in response to changes to the associated alignment.
- Alignment segment labels are similar to parcel segment labels, but they must be defined in the Alignment branch of Settings.
- Users can create an alignment table listing line, curve, and spiral segment values.

UNIT 6: OBJECTS FROM ALIGNMENTS

After defining alignments, Civil 3D has routines creating new objects that reference them. Most routines create points and are in the Create Points toolbar.

POINT SETTINGS

Almost all the routines use the current point settings found in the Point Settings roll-down of the Create Points toolbar, or in the values of Edit Feature Settings.

The settings for Elevations and Descriptions in the Create Points toolbar roll-down affect how they assign descriptions and elevations. If point description and elevation are set to Manual, a routine will prompt for these values before placing each point in the drawing. If a user has descriptions and elevations set to Automatic, a routine assigns the current settings to each point without prompting.

CREATE POINTS – AUTOMATIC – OBJECT

When creating points from an alignment when the alignment's name is desired to be part of the description, users can set the Prompt for Description value to Automatic - Object. This option places the alignment's name as part of the point's description.

CREATE POINTS—ALIGNMENT

The Create Points—Alignment section of the Create Points toolbar has several point creation routines (see Figure 5.56).

STATION/OFFSET

The Station/Offset routine places points at specific stations and offsets along a selected centerline. When starting a routine from this list, it prompts for an alignment. Select the alignment in the drawing and after selecting the alignment, a station jig appears to locate

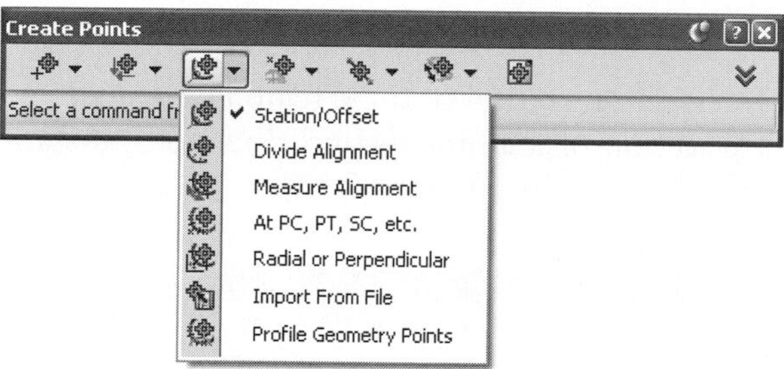

Figure 5.56

the station and offset. Instead of graphically locating a station with the station jig, users can type the station value and then select or enter an offset.

DIVIDE ALIGNMENT AND MEASURE ALIGNMENT

The Divide Alignment and Measure Alignment routines place points on the alignment. Divide places points on the alignment to reflect the number of user-specified segments (for example, 10 equal segments). Measure places points on the alignment at a user-specified distance (for example, every 20 units).

AT PC, PT, SC, ETC.

The At PC, PT, SC, etc. routine places points at critical alignment locations. The critical points are PC (point of curve beginning), PT (point of curve ending), SC (point of spiral curve intersection), RP (radial point), CC (compound curve points), and spiral curve intersections. (The same list can be found in the Abbreviations panel of Edit Drawing Settings).

RADIAL OR PERPENDICULAR

The Radial or Perpendicular routine places points on an alignment that are radial or perpendicular to points the user selects away from the alignment.

IMPORT FROM FILE

The Import from File routine reads an external file and places points in the drawing to represent the station and offset values in the external file. The file formats the routine reads are:

```
Station, Offset
Station, Offset, Elevation
Station, Offset, Rod, Hi
Station, Offset, Description
Station, Offset, Elevation, Description
Station, Offset, Rod, Hi, Description
```

The Rod and Hi are for level surveys. The Rod is the prism elevation, and the Hi is the instrument height. The file can be space- or comma-delimited.

CREATE POINTS—INTERSECTION

The Create Points—Intersection section of the Create Points toolbar contains routines that place points at the intersection of alignments and directions, distances, other alignments, and other objects in the drawing (see Figure 5.57).

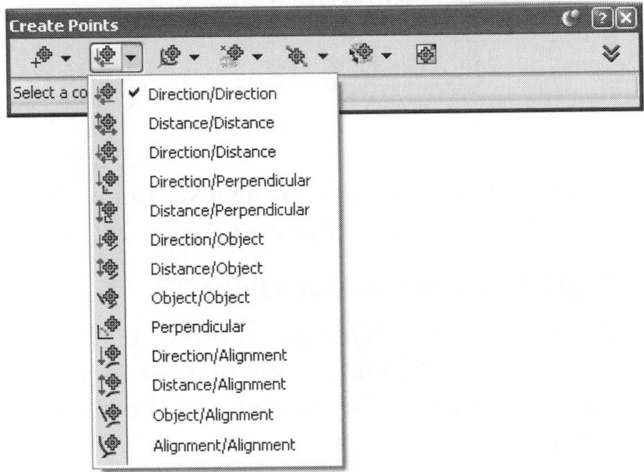

Figure 5.57

DIRECTION/ALIGNMENT

This routine places points that are at the intersection between an alignment and a direction from a known point. The intersection can be an offset from both the direction and alignment or either the direction or alignment.

DISTANCE/ALIGNMENT

The Distance/Alignment routine creates points at the intersection of a distance from a known point and an alignment. There can be an offset from the alignment.

OBJECT/ALIGNMENT

The Object/Alignment routine creates an intersection point at the intersection of an object (line, circle, or spiral) and an alignment. Both the object and the alignment can have an associated offset.

ALIGNMENT/ALIGNMENT

The Alignment/Alignment routine creates a point at the intersection of two alignments. There can be an offset from both or either alignments.

LANDXML OUTPUT

The LandXML Export routine creates a data file describing the alignment geometry (see Figure 5.58). This file allows you to transfer the alignment geometry to other applications without loss of coordinate and design fidelity and to produce reports from the Autodesk Report application.

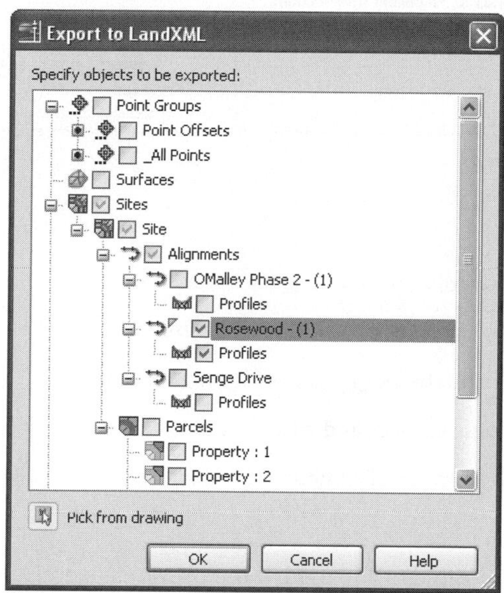

Figure 5.58

EXPORT DATA TO LAND DESKTOP

This command allows you to export an alignment's data to an existing LDT project. To call this routine the user must first start the Import Data from Land Desktop command (File menu, Import flyout) and cancel out of it. In the command line type ExportLDT-Data, and an Export Data to Autodesk Land Desktop dialog box appears. Select the appropriate project path, project, data, and select the OK button to export.

CREATE ROW (RIGHT-OF-WAY)

The Create ROW routine is in the Parcels menu and creates parcels (right and left side of the alignment) that encompass the alignment and any intersecting alignment. When running the routine, it displays a dialog box that sets fillet or chamfer values for intersecting ROW lines from intersecting alignments (see Figure 5.59).

Most of the objects in Civil 3D react to or are linked to other objects, allowing them to interact with one another. Unfortunately, the ROW parcel is not linked to its alignment and as a result will not change its geometry to match any changes in the alignment definition.

The ROW boundary can be a frontage parcel boundary.

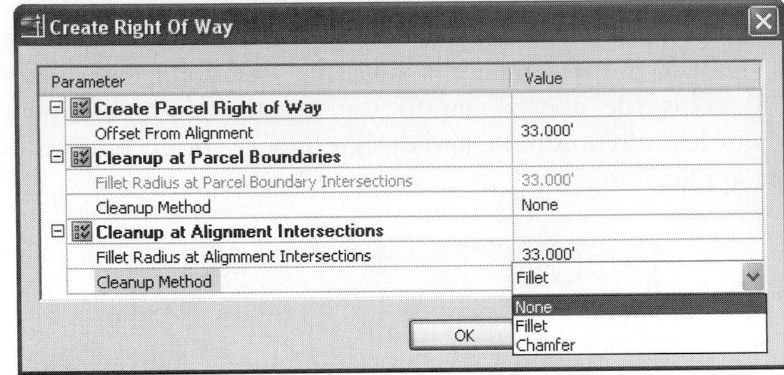

Figure 5.59

EXERCISE 5–6

When you complete this exercise, you will:

- Create points by intersection with an alignment.
- Create points from the geometry with an alignment.
- Import points with station and offset values.
- Create a ROW from an alignment.

EXERCISE SETUP

This exercise continues with the drawing used in the previous exercise. If you did not complete the previous exercise, you can open the *Chapter 5 - Unit 6.dwg* that is in the *Chapter 5* folder of the CD that accompanies this textbook.

1. Open the drawing from the previous exercise or the *Chapter 5 - Unit 6* drawing.

ALIGNMENT POINTS

Civil 3D creates points whose locations are critical points along the centerline, that reference station and offset, that measure lengths along the alignment, or that divide it into equal length segments.

Station/Offset

This is similar to labeling station and offset locations relative to an alignment.

1. From the View menu, select **Named Views...** and restore Proposed Starting Point.
2. From the Points menu select **Create Points...**.
3. In the toolbar, click the roll-down arrow to view the current point settings.
4. In the Points Creation section, use Table 5.2 to set the values for the roll-down panel.

Table 5.2

Section	Setting	Value
Points Creation	Prompt For Elevations	None
Points Creation	Prompt For Descriptions	Automatic - Object
Points Creation	Default Description	CL

Your roll-down should look like Figure 5.60.

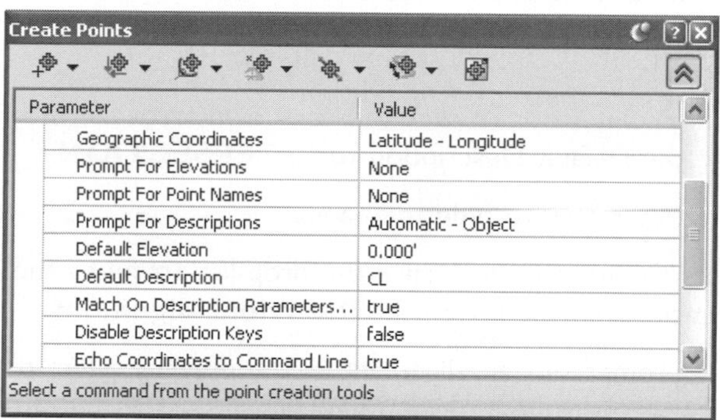

Figure 5.60

5. Click the roll-up arrow to close the point settings.
6. In the Create Points toolbar, click the drop-list arrow to the right of the Alignment icon. From the list select **At PC, PT, SC, etc.**
7. In the command line, the routine prompts for an alignment; select the **Rosewood** alignment.
8. In the command line, the routine prompts the starting and ending stations; press the **Enter** key, once for the beginning and ending station values and to exit the routine.

   The routine places points in the drawing representing curve PCs and PTs.

IMPORT FROM FILE

Civil 3D allows you to import points from a file that references an alignment.

1. From the Create Points toolbar, click the drop-list arrow to the right of the Alignment icon stack. From the list select **Import From File**.
2. In the Select File dialog box, browse to and select the **Station and Offset.txt** file that is in the *Chapter 5* folder of the CD that accompanies this textbook.
3. In the command line, the routine prompts for a file format (you may have to press the F2 key to see the list of formats). For the file format type **1** (Station, Offset), and press the **Enter** key to continue.

4. In the command line, the routine prompts for a delimiter. Type **2** (comma), and press the **Enter** key to accept the value and the invalid station/offset indicator.
5. In the command line, the routine prompts for an alignment; in the drawing select the **Rosewood** alignment.

   The points in the file reference stations 1+00 to 2+00.
6. Use the AutoCAD Zoom command to better view the points in the area of stations 1+00 and 2+00.
7. Click the **AutoCAD Save** icon to save the drawing.

EXERCISES

POINTS FROM INTERSECTIONS WITH ALIGNMENTS

1. In the Create Points toolbar, click the roll-down arrow. In the Points Creation section, change the Default Description to **INTERSECTION**.
2. Click the roll-up arrow to hide the panel.
3. In the Points Creation toolbar, click the drop-list arrow to the right of the Intersection icon stack. From the list, select **Direction/Alignment**.
4. The routine prompts for an alignment; in the drawing select the **Rosewood** alignment. Next, the routine prompts for an offset; for the offset, type **−25**.
5. In the command line, the routine prompts for a direction starting point; in the drawing, select a point near the Phase 1 line.
6. After selecting the starting point, the routine prompts for a second point to set a direction (a jig helps you define the direction). In the drawing, select a second point defining a direction that intersects the Rosewood alignment.
7. In the command line, the routine prompts for an offset; for the offset type **0** (zero). The routine places a point in the drawing to represent the intersection of offset from the alignment and the direction from the Phase 1 line.
8. In the Create Points toolbar, click the "**X**" to close the toolbar.
9. Click the **AutoCAD Save** icon at the top left of the screen.

CREATING A ROW

The Rosewood alignment divides the property into two new parcels. In each subdivision there is a buffer that extends to either side of the alignment. This buffer is the ROW parcel. Civil 3D creates a ROW from all of the defined centerlines of a drawing.

1. From the Parcels menu, select **Create ROW**.
2. In the command line, the routine prompts you to select the parcels that are adjacent to the alignment. In the drawing, select the two parcel labels on each side of the Rosewood alignment.
3. After selecting the parcel labels, the routine displays the **Create Right Of Way** dialog box.

4. In the Create Right of Way dialog box, set the Offset From Alignment value to **33** and set the Cleanup at Parcel Boundaries and Alignment Intersections to **None** (no fillet or chamfer). See Figure 5.59.
5. Click the **OK** button to create the ROW.
6. Click the **AutoCAD Save** icon to save the drawing.

This completes the exercises for alignments.

EXERCISES

SUMMARY

- Users can create points that reference the geometry or stations and offsets relative to an alignment.
- Users can create points that intersect with an alignment using other alignments, directions, or distances.
- Users can import points from a file containing station and offset values (plus additional data).
- Users can create a Right-of-Way parcel based on an alignments definition.
- A Right-of-Way parcel is not linked to an alignment and will not react to any changes in the alignment's geometry.

An alignment is the horizontal path of a roadway centerline. An alignment is a collection of segment types that understand basic rules about their connection with segments before and after the alignment. If a user graphically adjusts or manually edits segments values, the entire alignment adjusts to accommodate the changes. When the user creates alignments, they automatically include station and geometry annotation.

The next chapter focuses on the profile view of the elevations along the path of the centerline.

Profile Views and Profiles

INTRODUCTION

The second step of the road design process uses a view showing the elevations along the path of the alignment. In Autodesk Civil 3D 2007, a profile view creates a grid displaying and annotating stations and elevations. A profile is the elevations from a surface or line work representing a roadway's vertical design within a profile view. Civil 3D dynamically links the profile view (grid) and the surface profiles to the alignment. When the alignment is edited (moved, shortened, or lengthened), the profile view and the surface profiles change to show the new alignment length and its elevations. Even changing only the surface causes the surface's profile to change within the profile view.

Styles affect the format of the grid, the location and types of annotation of the grid, its profiles, and the properties of the profiles within the grid. There is a complicated web of dependencies and styles that make up the final profile view and its profiles.

OBJECTIVES

This chapter focuses on the following topics:

- Introducing and Creating a Profile View (Grid)
- Creating a Simple Profile View (Grid) Style and Modifying Existing View Styles
- Introducing and Creating a Profile
- Creating a Simple Profile Style and Modifying Existing Grid Styles

OVERVIEW

This chapter covers the second phase of the roadway design process: the profile view and its profiles. A profile view is a graph representing an alignment's stationing and elevations along the path of the roadway centerline. A profile is line work within a profile view representing surface elevations along an alignment's path or a roadway's vertical design. A profile view and its surface profiles are the backdrop for the proposed roadway vertical design.

It is in the profile view that the user may start to develop a feel for the impact of the road design on an earthworks volume result. The height of the proposed roadway above or below the existing ground elevations begins to give visual feedback as to amounts of earthwork needed to be done or what problems are needed to be resolved in order to build the road design (see Figure 6.1).

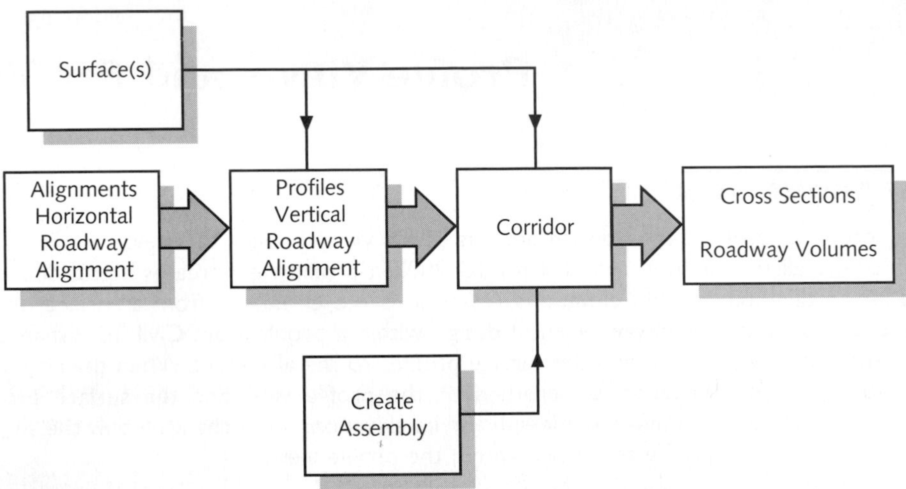

Figure 6.1

There are three steps to creating a profile and profile view. First, there must be data. Profile data can be a surface or a file containing alignment stations and their elevations. Second is creating the profile (sampling surface elevations, assigning styles, and adjusting other values). The last step is immediately creating a profile view containing the profile(s).

Profile and profile view steps:

1. Have a surface or read profile data from a file
2. Create a profile
3. Create a profile view

A profile view can contain multiple profiles (surfaces and proposed vertical designs). Each profile can have a different style allowing it to display its information uniquely in the profile view.

UNIT 1

The profile view is a graph in which Civil 3D draws profiles. A profile represents the existing ground, sub-surfaces, or one or more proposed vertical design. The styles affecting a profile view and its profiles is the focus of this first unit.

UNIT 2

The second unit of this chapter covers the steps needed to create a profile and its profile view.

UNIT 3

Within the context of a profile view and profile (the grid and an existing ground profile), a designer creates a vertical (alignment) roadway design. A vertical alignment contains tangents and vertical curves. Civil 3D supports three types of vertical curves: circular, symmetrical, and asymmetrical parabolic curves. Creating a vertical alignment is the topic of the third unit of this chapter.

UNIT 4

The analysis and editing of a vertical alignment (profile) is the focus of the fourth unit of the chapter. This unit covers the analysis of a vertical alignment with reports from the Autodesk LandXML Report Application, the Inquiry tools, the properties of a profile, and from within the vertical editor.

UNIT 5

The fifth unit of this chapter reviews the annotation of profiles and vertical alignments. The annotation is the result of the labeling styles or labeling set applied to the profile.

UNIT 1: PROFILE VIEW AND PROFILE STYLES

A profile view is a graph in which Civil 3D draws profiles. The profiles drawn within a view represent the existing ground and other surfaces along the path of an alignment and one or more proposed vertical alignments.

PROFILE VIEW

The profile view is a grid representing the stations and elevations along the path of its associated alignment. The stations are measured along the bottom of the graph and create vertical lines marking the stations in the elevation area of the profile view (see Figure 6.2). The station interval has a major and minor increment with station annotation at the major stations (minimally). All of these values are user-specified values set in the styles applied to the profile view.

Elevations are measured from a datum (lowest elevation) of the graph upward to the highest (see Figure 6.2). The interval for these lines is an even elevation (every 2 or 5 feet for example) and has secondary tick marks at minor increments with annotation at major elevations (minimally). All of these values are user-specified values set in the styles applied to the profile view.

Traditionally, a profile view is one-tenth of the horizontal scale. If you are working with a drawing that has a 1"=40' scale, the vertical scale is 1"=4' (1 inch of paper represents 4 feet of relief).

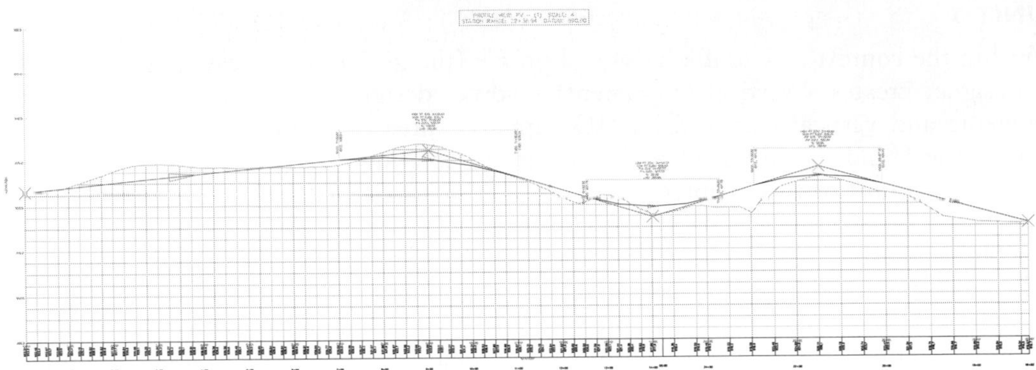

Figure 6.2

PROFILE VIEW STYLE: FULL GRID

A Profile View Style defines values affecting titles and the annotation for stations and elevations within the profile view. A Profile View Style is a multi-tabbed dialog box with each tab affecting different aspects of the view.

INFORMATION

As with all styles, there is an Information tab in the Profile View Style dialog box. This tab contains the name, description, and details on who and when the style was created.

GRAPH

The Graph tab displays the values for the title, direction of the profile, exaggeration, and grid clipping options for the profile view (see Figure 6.3). The left side of the panel sets the text style, heading, and position of the profile view title. The text icon to the right of the title text calls the Text Component Editor. Set the text and its formatting in the editor.

The upper-right portion of the panel sets the direction of the profile view (left-to-right or right-to-left). The middle portion of the panel sets the vertical exaggeration. By default all of the profile views are set to a 10 times exaggeration. A setting of 1 would mean the profile has no exaggeration.

The lower-right of the panel sets the number of additional grid lines above, below, and to the right and left of the view grid.

The middle-right portion of the panel sets the grid clipping choices. There are four possible combinations of the toggles: all on, all off, clip horizontally, and clip vertically. The effect of toggling on all of the clip toggles is seen on the right side of Figure 6.4. The left side of Figure 6.4 shows the result of having all of the toggles off.

The effect of toggling on only horizontal clipping is shown in the right side of Figure 6.5. The effect of toggling on only vertical clipping is shown on the left side of Figure 6.5.

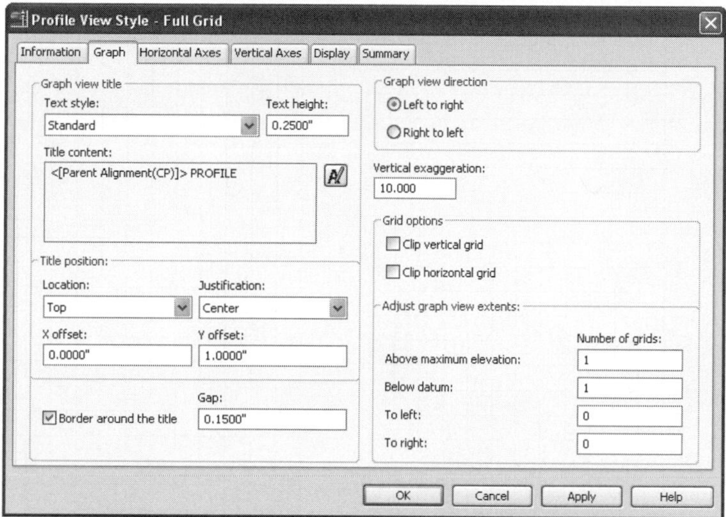

Figure 6.3

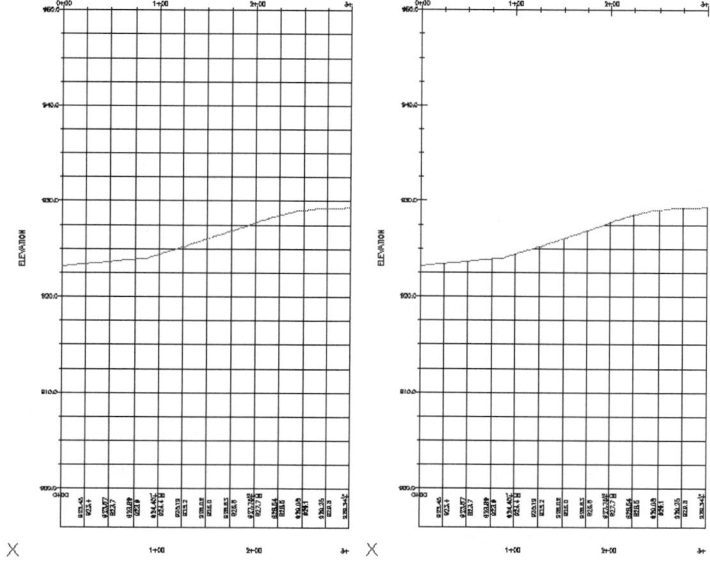

Figure 6.4

HORIZONTAL AXES

The Horizontal Axes tab defines the station text and tick intervals, text styles and size, and titles for the vertical grid components (see Figure 6.6). Station labeling can be at the top, bottom, or both sides. The Select Axis drop down-list at the top left indicates that the settings are for either the top or bottom of the vertical grid lines. The left side of the dialog box defines the axis title, its content, and text styles and size.

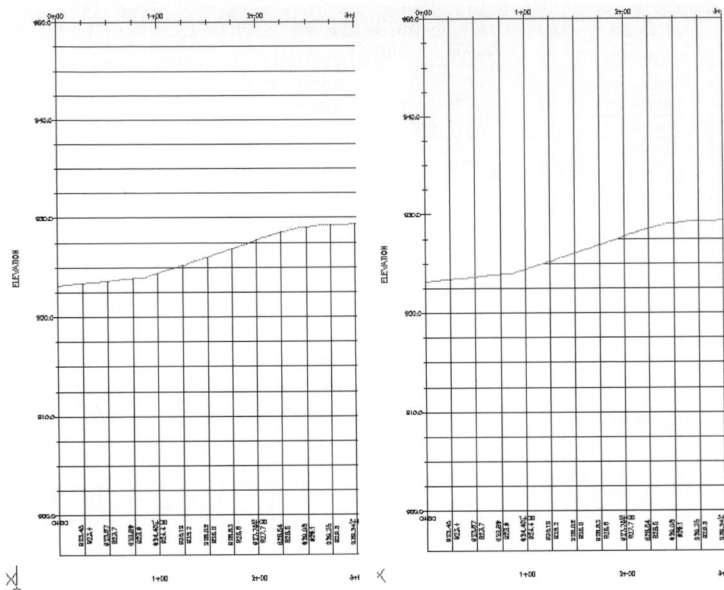

Figure 6.5

The top and middle right of the dialog box define the Major and Minor tick intervals, text style and size, X and Y offsets, content, and the format string of the label. The text icon calls the Text Component Editor to define the content and it formats for the title and major and minor tick.

The bottom right of the dialog box contains settings and format for annotating horizontal geometry points.

Even though the user defines annotation for all profile view axes, the Display tab settings of the profile view determine what annotation appears.

VERTICAL AXES

The Vertical Axes tab defines the elevation text and tick intervals, text styles and size, and titles for the vertical grid components (see Figure 6.6). Elevation labeling can be at the right, left, or on both sides. The Select Axis drop down list at the top left indicates that the settings are for either the right or left vertical axis. The left side defines the title, its content, and text styles and size.

The top and middle right define the Major and Minor tick intervals, text style and size, X and Y offsets, content, and the format string of the label. The text icon calls the Text Component Editor to define the content and it formats for the title and major and minor tick.

Even though the user defines annotation for all profile view axes, the Display tab settings of the profile view determine what annotation appears.

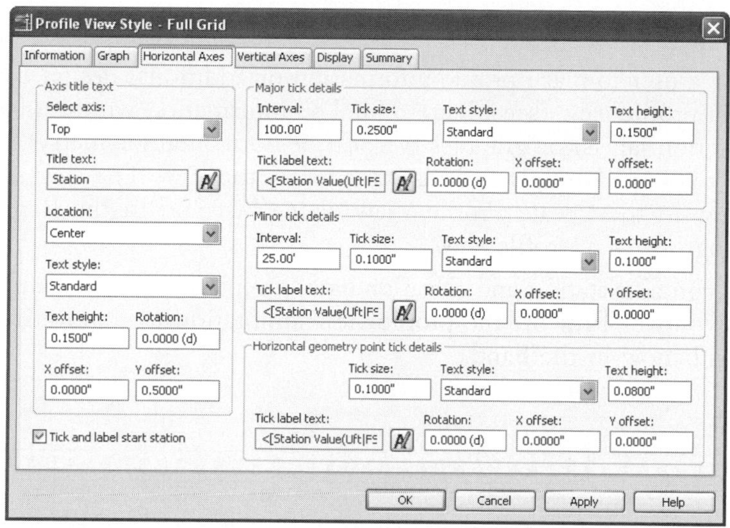

Figure 6.6

DISPLAY

The Display tab controls the visibility of all of the components defined in the Graph and Axes tabs (see Figure 6.7). While the user may define every possible tick, grid, or label, they will only show *if* they are turned on in this panel. This panel defines exactly what shows on a profile view.

Figure 6.7

BAND SETS

Band sets allow a user to place profile information or additional design data as a band at the bottom of a profile view (see Figure 6.8). The information within a band can be text (elevation or station labels) or graphics (vertical or horizontal geometry). When a profile view has more than one band, the bands stack below the view. The user assigns band sets to a profile view in the Create Profile View dialog box and in the Bands panel of the Properties dialog box of a profile view.

If the band set contains station and elevation data, it replaces the same annotation defined in a profile view style. Turn off the profile view annotation to remove it from the view; otherwise, it will show in the band.

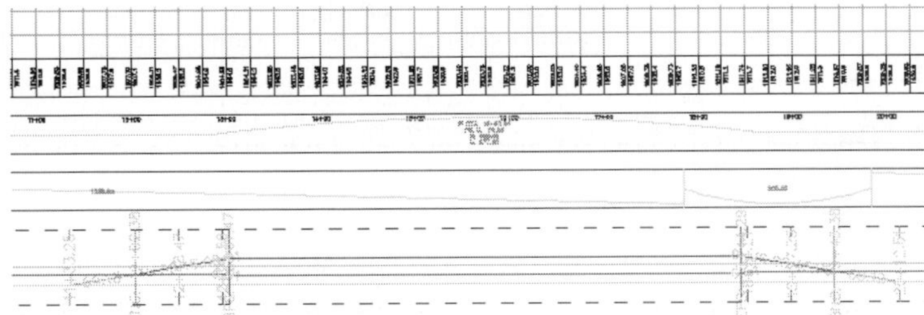

Figure 6.8

A band set name is an alias for a collection of band styles and depends on specific (band) styles to create their definition. A band set can have one or any combination of five types of band styles: Profile Data, Horizontal and Vertical Geometry, Sectional Data, Pipe Network Bands, and Superelevation Data. These five style types are themselves a collection of values from the respective object type values. For example, the Profile Data styles emphasize labels for major and minor stations, horizontal and vertical geometry points, and station equations, while the Vertical Geometry styles emphasize up and down hill tangent and sag and crest curve labeling, etc.

The Bands panel of the Profile View Properties dialog box assigns which band set to display, which profiles or alignments to reference, and which order and spacing they have below (or above) the profile view (see Figure 6.9). When creating a profile view from a surface, Civil 3D automatically assigns the first surface to Profile 1 and Profile 2. The assignment remains this way until the user reassigns a specific profile to Profile 2. Generally, this is done after drafting the proposed vertical design and assigning it to Profile 2 (see Figure 6.9).

BAND STYLES

All band style dialog boxes have four panels: Information, Band Details, Display, and Summary (see Figure 6.10). The Band Details panel defines all of the potential labels for a style and the Display panel defines what labels actually show when using the style.

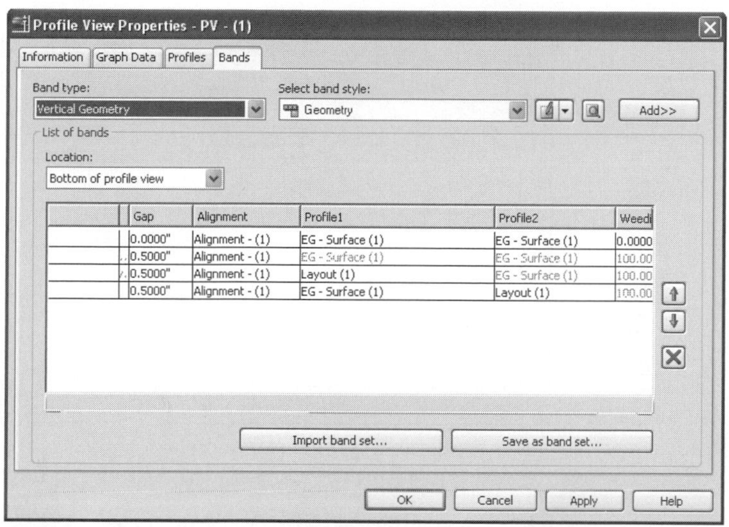

Figure 6.9

The Information and the Summary panels are the same for all styles. The Information panel names and records the creation and modification date of the style. The Summary panel reviews only the basic settings for each type of label and tick and does not review specific label values. To view the specific values of any label, review the values in the Band Details panel.

PROFILE DATA BAND STYLE

The Profile Data styles annotate profile data at major and minor alignment geometry points, and at station equations. Although the user can decide to define all types of annotation, it is the Display panel settings of the style that control what actually shows using the style.

Band Details

The Band Details panel defines the text style of the band (top left), title, its size, location (middle left), and the general layout of the band (see Figure 6.10). The layout area (bottom left) of the style includes the gap between the band and the bottom (or top) of the profile and the text box width and height containing the title of the band.

The critical part of the Band Details panel is what is not visible. What is hidden are the label definitions for each of the listed label types: At Major Station, At Minor Station, etc. When highlighting one of the label types and clicking on the Compose Label... button, the Label Style Composer dialog box appears, naming the type of label reviewed or created (see Figure 6.11).

When clicking on the drop-list arrow for Component Names, a list of all of the labels that occur for the selected label type appear. Figure 6.11 lists all of the label types for the At Major Station for the Profile Data style. Repeat selecting the label type (At Major Station, At Minor Station, etc.) by clicking on the Compose Label... button, and dropping the list of Components Names to find what labels the style defines.

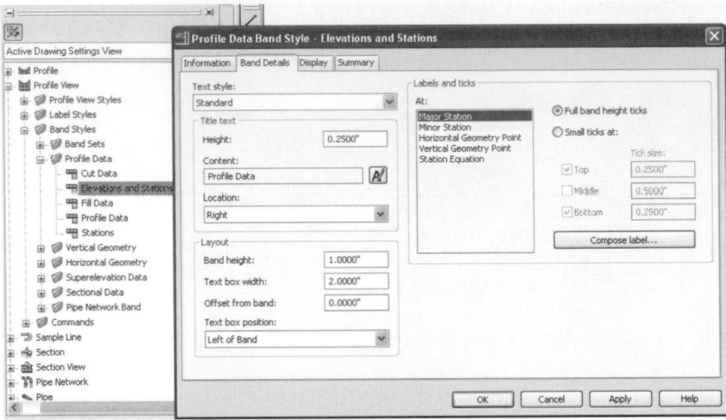

Figure 6.10

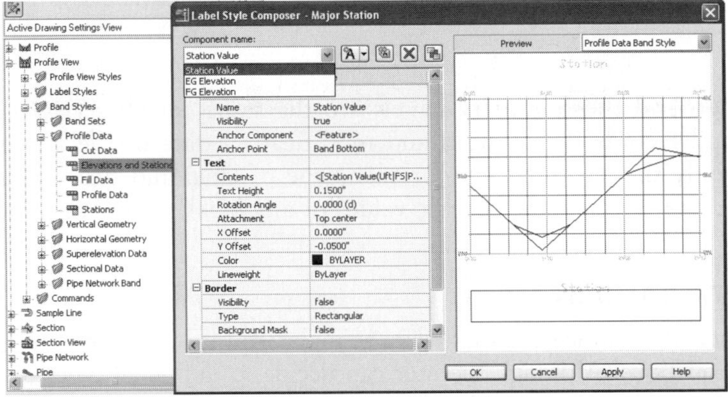

Figure 6.11

Display

The settings in the Display panel control the visibility of a band's labels, titles, ticks, and lines (see Figure 6.12).

VERTICAL GEOMETRY BAND STYLE

The Vertical Geometry styles draw and annotate the critical values of a proposed vertical design (at the bottom of a profile view) (see Figure 6.13). This style uses an assigned profile as data for the labels and sketch of the profile appearing in the band.

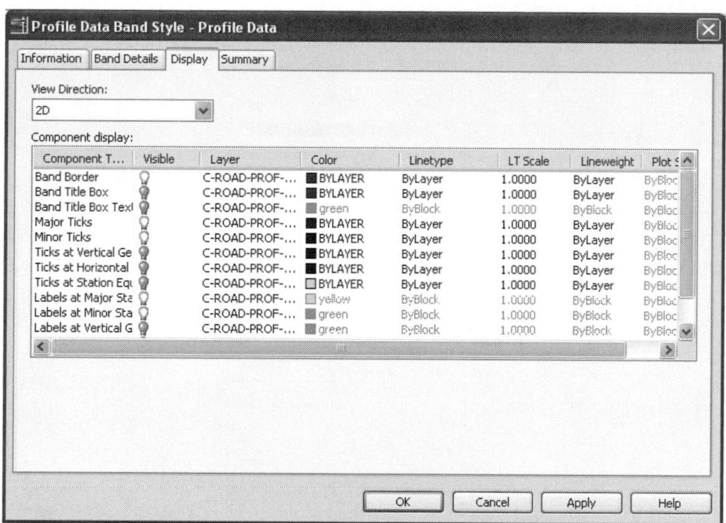

Figure 6.12

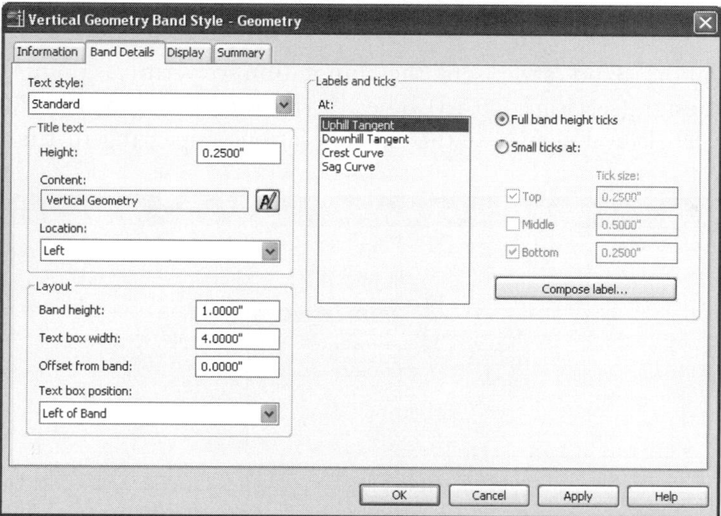

Figure 6.13

HORIZONTAL GEOMETRY BAND STYLE

The Horizontal Geometry styles draw and annotate the critical values of a horizontal alignment (at the bottom or top of a profile view) (see Figure 6.14). These styles use an assigned horizontal alignment as data for the labels and sketch of the alignment appearing in the band.

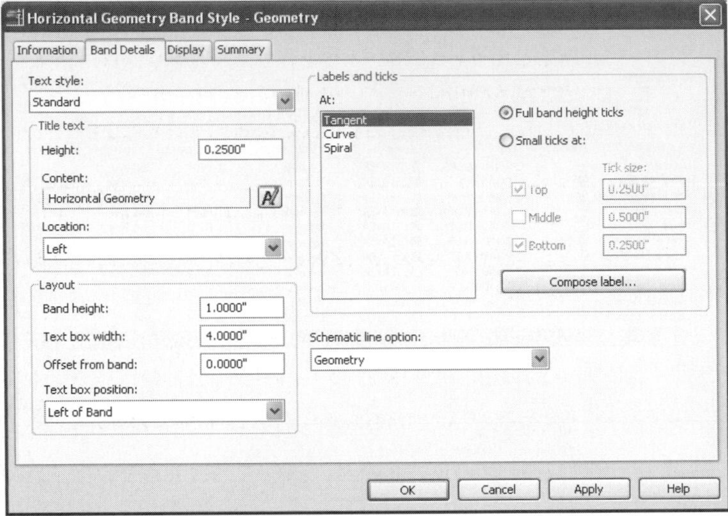

Figure 6.14

SUPERELEVATION BAND STYLE

The Superelevation Band style sets the annotation of critical points of a proposed superelevation design (see Figure 6.15). These styles use an assigned horizontal alignment as data for the labels and sketch of the superelevations appearing in the band.

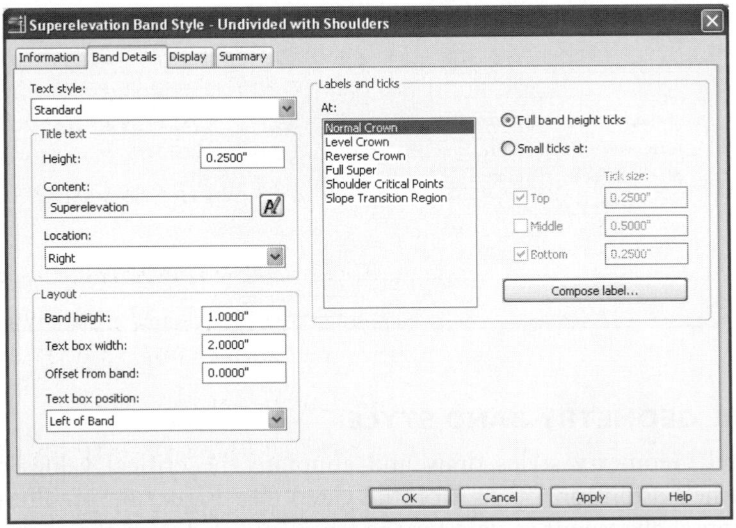

Figure 6.15

SECTIONAL DATA BAND STYLE

The Sectional Data Band style creates annotation representing the cross sections from a sample line group (see Figure 6.16). These styles use an assigned sample line group as data for the labels.

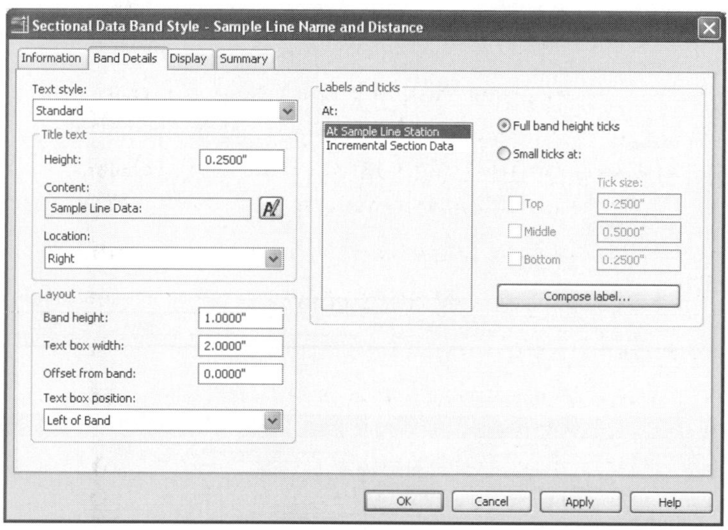

Figure 6.16

PIPE NETWORK BAND STYLE

The Pipe Network Band style creates an annotation representing an aspect of a pipe network (see Figure 6.17). These styles use an assigned pipe network as data for the labels.

PROFILE STYLES

The profile styles affect the display of the profile object in a profile view. In any implementation there will be at least three and possibly four styles: design, existing ground, plotting, and possibly a subsurface style. The purpose of having multiple styles is being able to visually differentiate profile types on the screen. Another reason is to use one style as a "design" style and another as a "plot" style. For example, the design style colorizes all of the vertical alignment components differently. A plot style assigns the profile components the correct layers for plotting submission documents. For example, Figure 6.18 shows the layer assignments for two profile styles. The left side of the figure is the Layout style and the right side shows the Existing Ground Profile (plot) style.

SURFACE PROFILE STYLE

A surface profile style defines curve smoothness (when drawing curve segments in 3D), point markers for vertical geometry, and what layer and/or properties a profile has in a profile view. Surface profile styles use only one layer (line component) and none of the settings in the Design and Markers tabs.

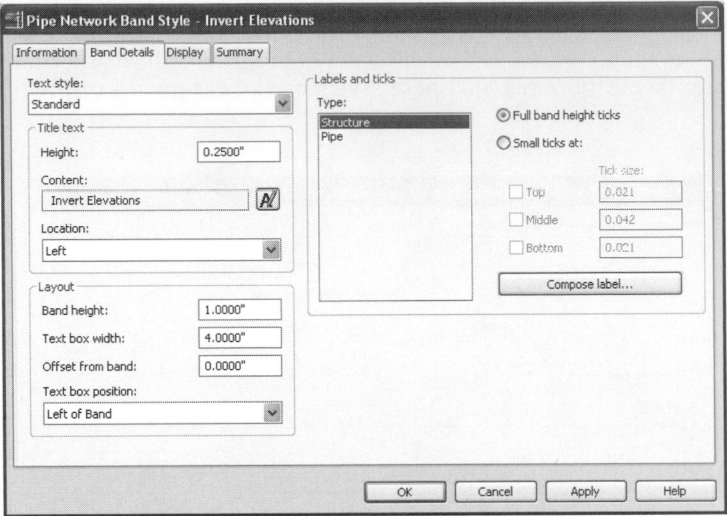

Figure 6.17

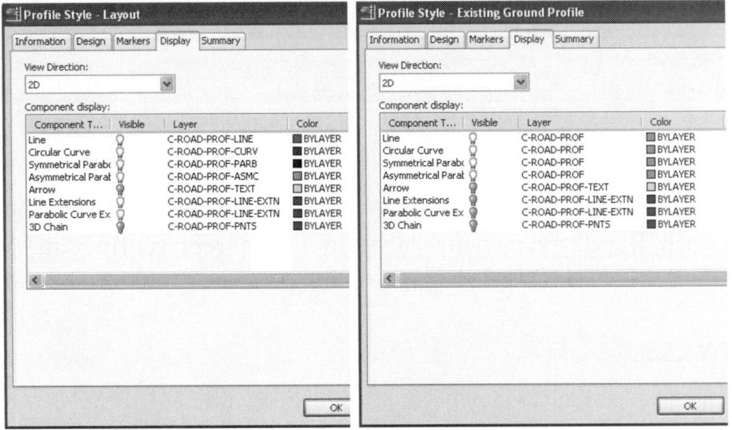

Figure 6.18

DESIGN PROFILE STYLE

A design profile style focuses on the components of a roadway vertical alignment. When the user wants to view the vertical alignment in 3D, this style type should set a 3D Chain Visualization value in the Design tab (see Figure 6.19). The lower the value is for this setting, the smoother the representation of the vertical curve. When the user wants to display markers for critical points along the vertical alignment, this style type should define values in the Markers tab (see Figure 6.20). All of the components need layer assignments, and the Display panel associates the component with a layer name (see Figure 6.18).

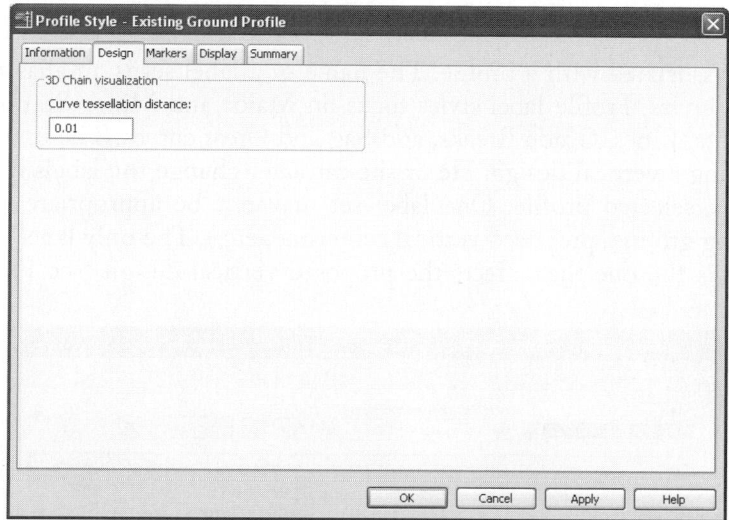

Figure 6.19

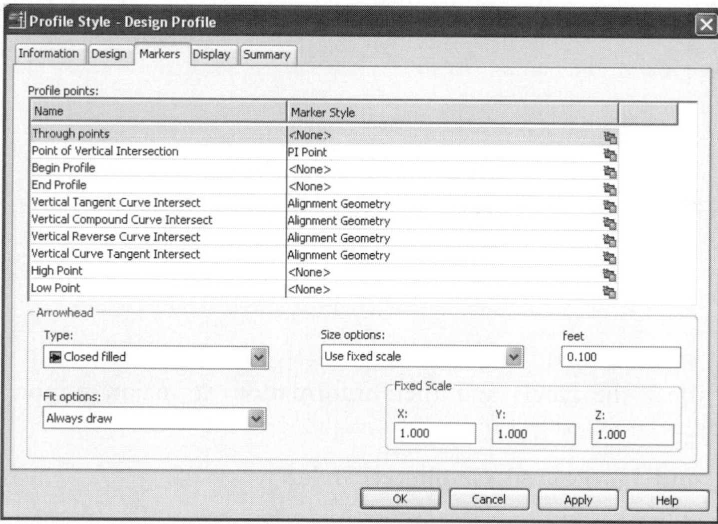

Figure 6.20

PLOT PROFILE STYLE

This style type assigns the profile components to specific layers having correct object properties for submission set plotting.

PROFILE LABEL SETS AND STYLES

Label sets are associated with a profile. The name of a label set is an alias for a collection of profile label styles. Profile label styles focus on Major and Minor Stations, Horizontal Geometry Points, Lines, Grade Breaks, and Sag and Crest curves. The user assigns a label set when drafting a vertical design. He or she can later change the labels in the Properties dialog box of a selected profile. One label set may not be appropriate for all types of profiles (existing ground, proposed vertical centerline, etc.). The only label set the user will have to define is the one that affects the proposed vertical design (see Figure 6.21).

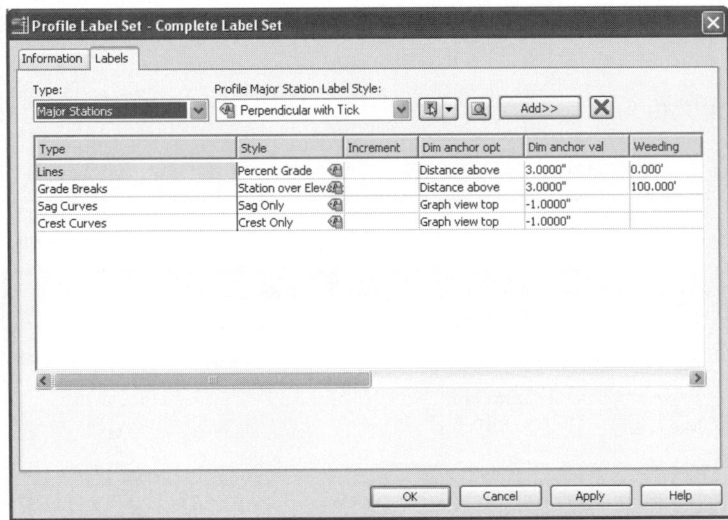

Figure 6.21

STATION

Station styles affect the labels and their information at major, minor, and horizontal geometry points.

Major, Minor, and Horizontal Geometry Styles

The Major and Minor Station styles are interval label styles. The frequency of the Major and Minor label styles is set by the stationing parameters of the profile view. The alignment geometry points define the location of the Horizontal Geometry Points labels. The type of data available for the label styles include alignment, profile, and superimposed profile data (see Figures 6.22 and 6.23).

Grade Break

A Grade Break style is primarily for labeling the beginning and ending station of a profile tangent or a break in grade point. The information available to this style includes alignment and profile data (see Figure 6.24).

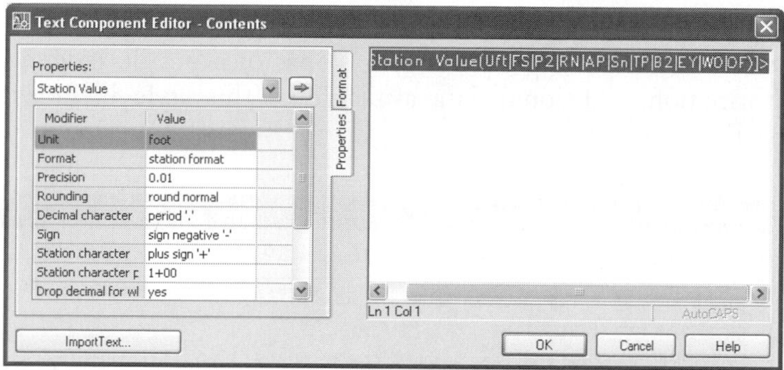

Figure 6.22

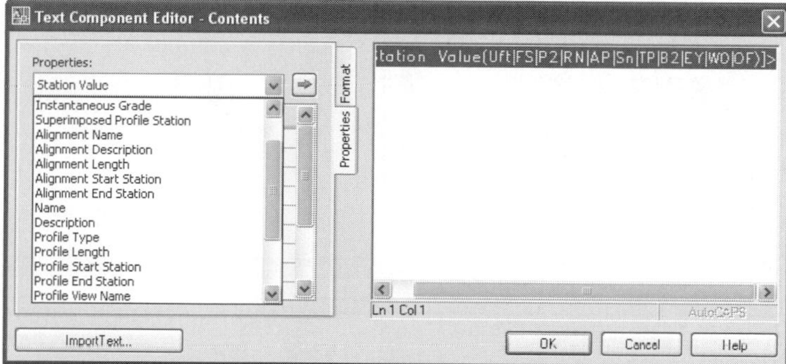

Figure 6.23

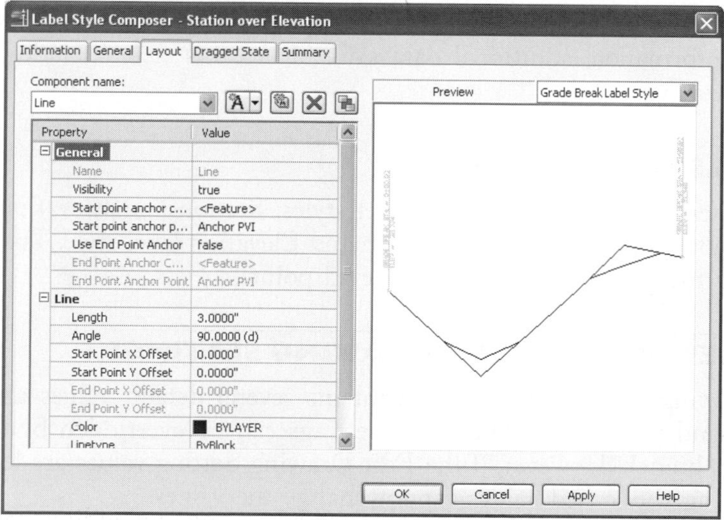

Figure 6.24

Line

A Line label style annotates the grades and/or slopes of a profile tangent. In addition to the tangent information, additional data available to this style includes alignment and profile data (see Figure 6.25).

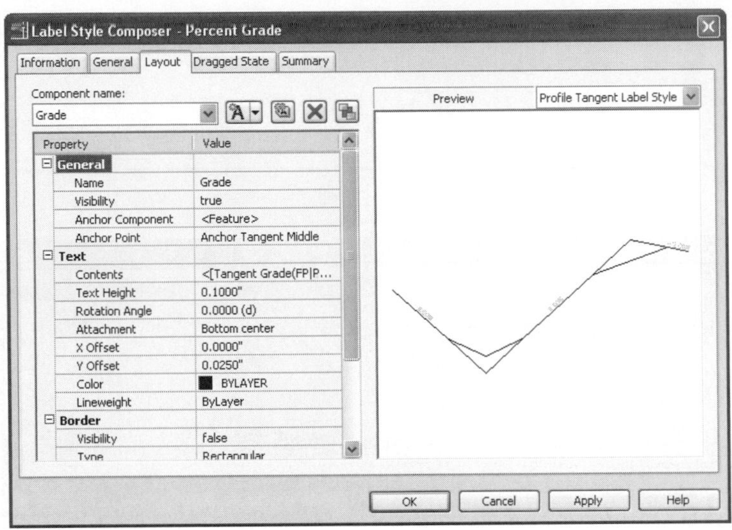

Figure 6.25

Curve

The Curve label style annotates the critical values of a profile's vertical curves. In addition to the curve information, additional data available to this style includes alignment and profile data (see Figure 6.26).

ADD LABELS LABEL STYLES

Civil 3D defines two spot profile view label styles: Station and Elevations, and Depths. Figure 6.27 shows an example of the Station and Elevation label style. These labels appear in the body of the profile view at user-selected points.

PROFILE VIEW AND PROFILE DRAWING SETTINGS

The Edit Drawing Settings dialog box contains several default values affecting profiles, profile views, and their labels. The initial settings can be overridden by any style in the Profile settings branch. However, if the Edit Drawing Setting values are locked, the lower styles referencing the locked values cannot change the values.

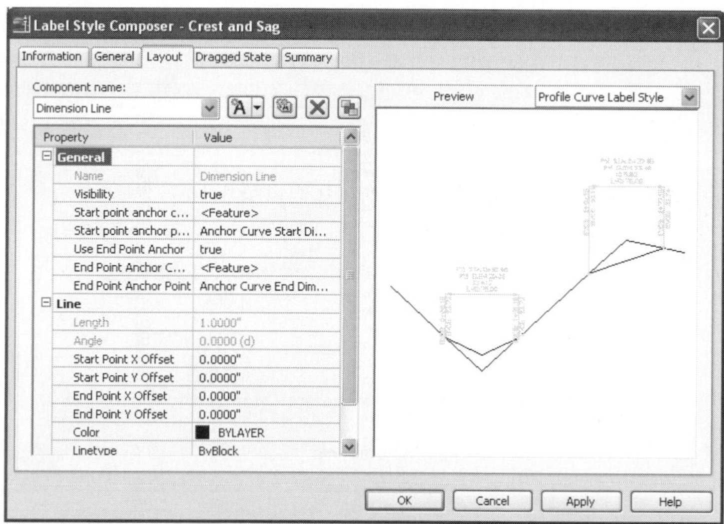

Figure 6.26

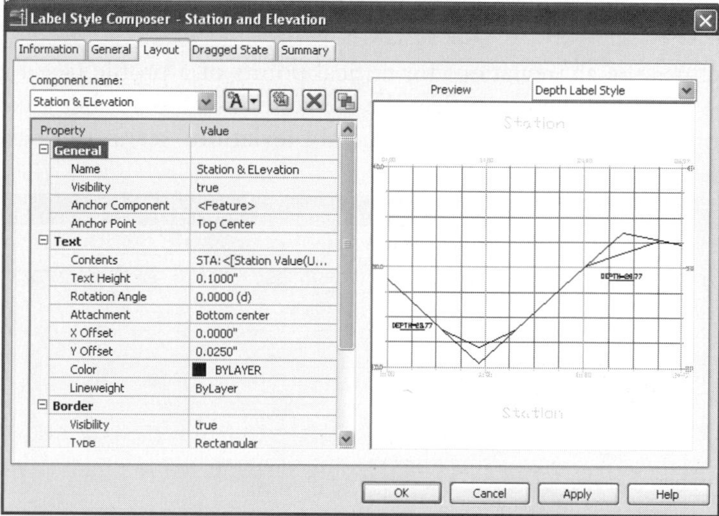

Figure 6.27

OBJECT LAYERS

The Edit Drawing Settings dialog box sets the base layer names for the various profile components. Each layer can have a modifier (prefix or suffix). The value of the modifier is the name of the referenced profile (see Figure 6.28).

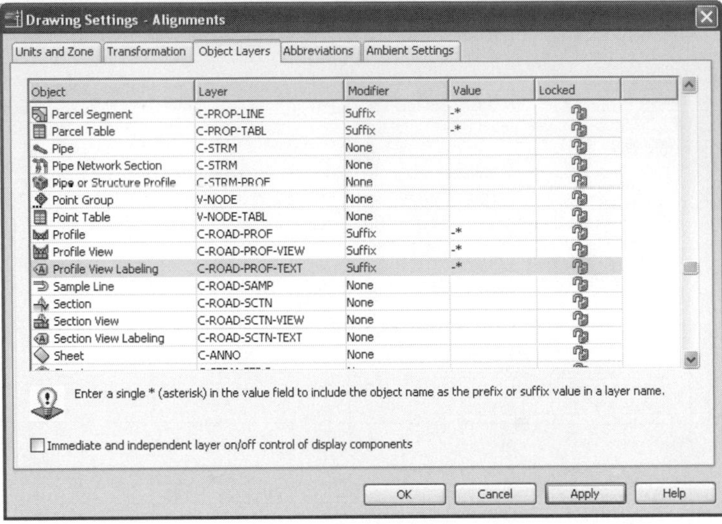

Figure 6.28

ABBREVIATIONS

This section defines the abbreviations for critical points of a profile (see Figure 6.29). The user can change the values here to reflect the conventions of his or her area. All label styles in the Profile branch of Settings will use the abbreviations set in the panel.

Drawing Settings - Alignments

Units and Zone | Transformation | Object Layers | Abbreviations | Ambient Settings

Property	Value
Profile	
Profile Start	BVP
Profile End	EVP
Point Of Vertical Intersection	PVI
Grade Break	BREAK
Vertical Tangent-Curve Intersect	BVC
Vertical Tangent-Curve Intersect Station	BVCS
Vertical Tangent-Curve Intersect Elevation	BVCE
Vertical Curve-Tangent Intersect	EVC
Vertical Curve-Tangent Intersect Station	EVCS
Vertical Curve-Tangent Intersect Elevation	EVCE
Vertical Compound Curve Intersect	VCC
Vertical Compound Curve Intersect Station	VCCS
Vertical Compound Curve Intersect Elevation	VCCE
Vertical Reverse Curve Intersect	VRC
Vertical Reverse Curve Intersect Station	VRCS
Vertical Reverse Curve Intersect Elevation	VRCE
High Point	HP
Low Point	LP

OK | Cancel | Apply | Help

Figure 6.29

PROFILE VIEW: EDIT FEATURE SETTINGS

The initial styles and band set assignments for profile views come from the values in the Edit Feature Settings dialog box (see Figure 6.30). The user can change the default styles to reflect his or her standards.

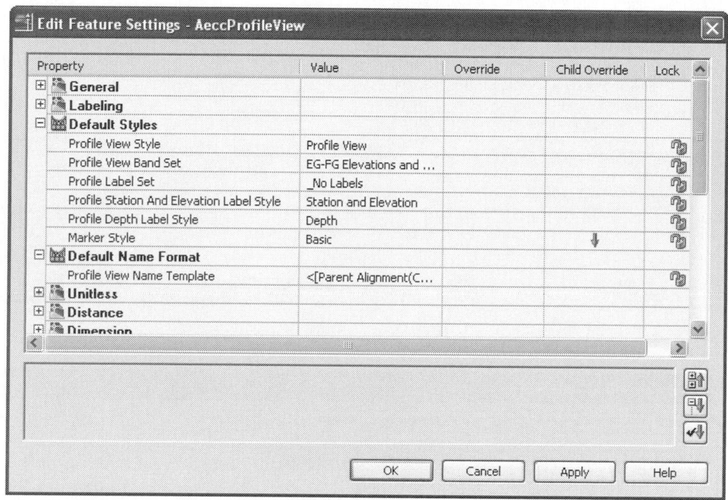

Figure 6.30

PROFILE: EDIT FEATURE SETTINGS

The initial styles and label set assignments for profiles are in the Edit Feature Settings dialog box. The styles listed in the dialog set the defaults for all profile styles (see Figure 6.31).

The Default Name section defines the naming convention for Profiles, Offset Profiles, and Superimposed Profiles.

The Profile Creation section sets the initial values for the three types of vertical curves (circular, parabolic, and asymmetrical parabolic). These values appear as defaults in the Profile Layout Tools toolbar.

Other entries in this section set the default vertical curve values. The Default Vertical Curve Type allows the specification of a type of vertical (circular, parabolic, and asymmetrical) and the Parabolic Crest and Sag settings allow the choice of the method to specify them (length or K-Value). Other values in the section are eye height, stopping height, etc. for vertical curve calculations.

PROFILE VIEW COMMAND SETTINGS

The settings for the commands creating profile views and their labels are in the Commands branch of Profile View settings (see Figure 6.32). The styles are the same as those in the Edit Feature Settings dialog box.

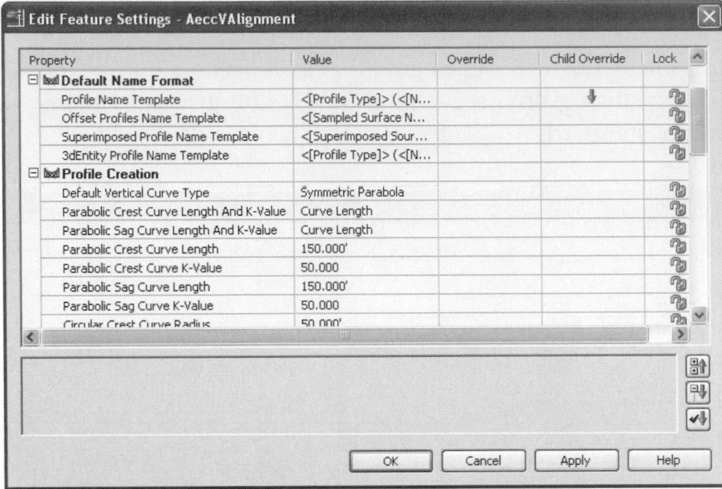

Figure 6.31

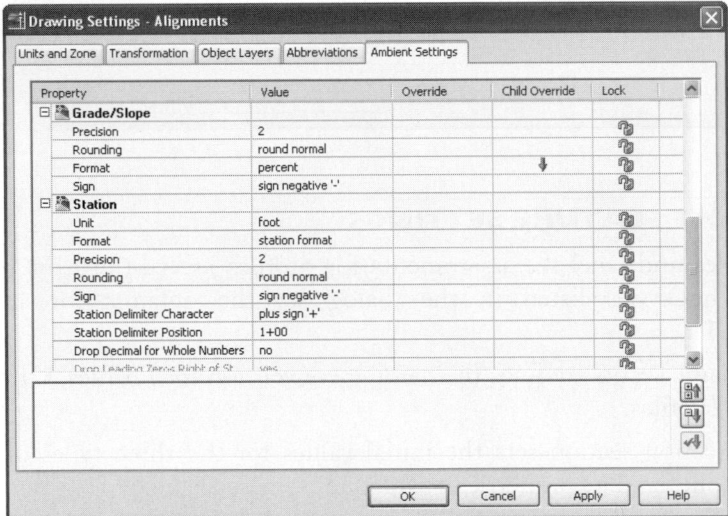

Figure 6.32

PROFILE COMMAND SETTINGS

Civil 3D has default values for each command that creates a profile. The styles assigned by a command are set in the Edit Feature Settings dialog box. Any command creating a profile object can assign its own set of styles.

The settings of the CreateProfileFromSurface command affect profile creation. The Default Styles section sets the profile style for the initial profile in a profile view. The Default Name Format sets the initial values for profile and profile view names. The default naming method is a sequential counter after the type of profile: Profile View 1, Surface 1, etc. (see Figure 6.33). In the figure, the command overrides the Profile Name Template.

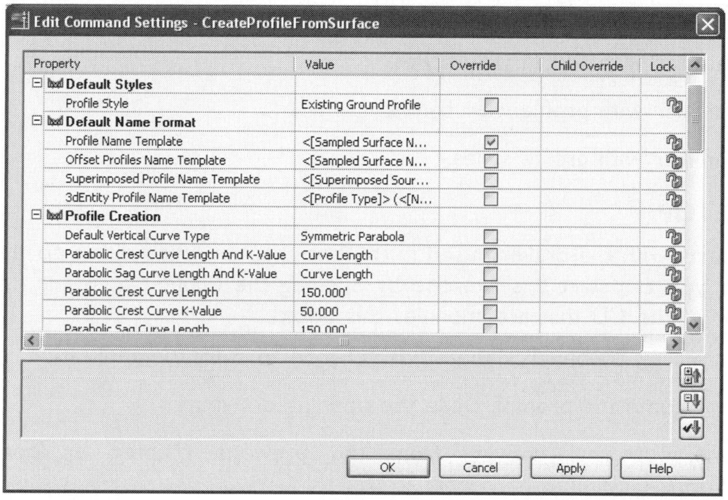

Figure 6.33

The Profile Creation section sets the default length, or K Values, for parabolic, asymmetrical, and circular vertical curves (see Figure 6.34). The lower portion of the section contains the values for passing, stopping, and sight distances for calculating vertical curves. The values of this section are set in the Edit Feature Settings dialog box of the Profile heading of Settings. Several of the values are locked and can be changed only in the Edit Feature Settings dialog box.

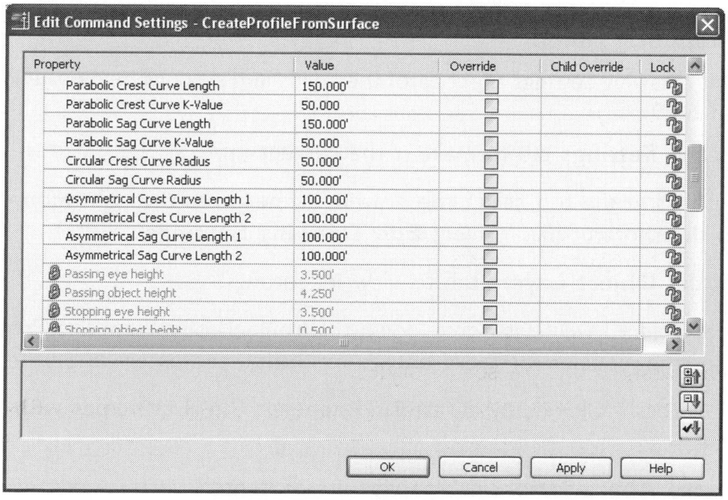

Figure 6.34

EXERCISE 6-1

When you complete this exercise, you will:

- Be familiar with profile drawing settings.
- Be familiar with profile view styles.
- Be familiar with profile view bands.
- Be familiar with profile styles.

EXERCISES

EXERCISE SETUP

This exercise starts with the drawing used in the last chapter. If you didn't do the exercises from the last chapter, you can start this exercise by opening the *Chapter 6 – Unit 1.dwg* found in the *Chapter 6* folder of the CD that accompanies this book.

1. If necessary, double-click the **Civil 3D** icon to start the application.
2. At the command prompt, close the opening drawing.
3. Open your final drawing from the previous chapter or open the drawing *Chapter 6 – Unit 1*.
4. Click the **Prospector** tab to make it current.
5. In Prospector, expand the Point Groups branch until you are viewing the list of point groups.
6. Select the **Point Offsets** point group, right mouse click, and from the shortcut menu select **Delete Points....**
7. In Layer Properties Manager, freeze the V-NODE layer, and click the **OK** button to exit.
8. Click the **AutoCAD Save** icon to save the drawing.

EDIT DRAWING SETTINGS

The values of Edit Drawing Settings affect all of the styles and settings below the Point heading (see Figures 6-28 and 6-29).

1. Click the **Settings** tab to make it the current tab.
2. In Settings at the top, select the drawing name, press the right mouse button, and from the shortcut menu select **Edit Drawing Settings....**
3. Click the **Object Layers** tab to view its settings.
4. In the Object Layers tab, scroll down to Profile, Profile View, and Profile View Labeling, set the modifier for these entries to **Suffix**, and enter **-*** for the value.

   The –* (dash followed by an asterisk) appends the layer names with the name of the profile.
5. Click the **Abbreviations** tab to view its contents.

6. Close the Alignment and Superelevation sections to view the profiles entries.
7. Review the profile abbreviation sections.

   If these abbreviations are not representative of your area, changing their values here changes all profile styles using abbreviations (see Figure 6.29).
8. Click the **Ambient Settings** tab to view its contents.
9. Expand the Grade/Slope and Station sections to view their format values.

   This section sets the drawing's grade and slope conventions (see Figure 6.35). The current setting for a grade is two decimal places and uses a negative sign for down slope.

   The Station section defines a station format. These values are the initial settings for any alignment stationing label. Again, changing any of these values here affects all station formats.

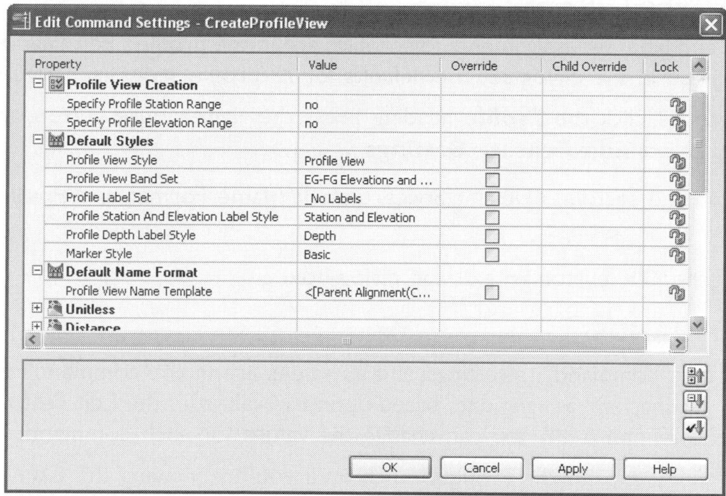

Figure 6.35

10. Click the **OK** button to close the dialog box.

PROFILE VIEW: EDIT FEATURE SETTINGS

The initial styles and band set assignments for profile views come from the values in the Edit Feature Settings dialog box (see Figure 6.30).

1. In Settings, click the **Profile View** heading, press the right mouse button, and from the shortcut menu select **Edit Feature Settings....**
2. In the Edit Feature Settings dialog box, expand the Default Styles and Default Name Format sections and review their values.
3. Click the **OK** button to exit the dialog box.

PROFILE VIEW: COMMANDS

The settings for the commands creating profile views and their labels are in the Commands branch of Profile View settings (see Figure 6.32). The styles are the same as those in the Edit Feature Settings dialog box.

1. In Settings, expand the Profile View branch until you are viewing the Commands heading and its list.
2. From the list under the Commands heading, click **CreateProfileView**, press the right mouse button, and from the shortcut menu select **Edit Command Settings....**
3. In the dialog box, expand the Profile View Creation, Default Styles, and Default Name Format section to review their values.
4. Click the **OK** button to exit the dialog box.

PROFILE: EDIT FEATURE SETTINGS

The initial styles and label set assignments for profiles are in the profile's Edit Feature Settings dialog box. The styles listed in the dialog set the defaults for all profile styles (see Figure 6.31).

1. In Settings, click the **Profile** heading, press the right mouse button, and from the menu select **Edit Feature Settings....**
2. Expand the sections (Default Styles, Default Name Format, and Profile Creation) and review their settings.
3. Click the **OK** button to exit the dialog box.

PROFILE: COMMANDS

The styles used by a command, its settings, and its values are in the command's dialog box. These dialog boxes list settings and assignments based upon the values in the Edit Feature Settings dialog box (see Figures 6-33 and 6-34). You can change the values for each command.

1. In Settings, expand the Profile branch until you are viewing the Commands list.
2. From the list, click **CreateProfileFromSurface**, press the right mouse button, and from the shortcut menu select **Edit Command Settings....**
3. Expand the sections (Default Styles, Default Name Format, and Profile Creation) and review their settings.
4. Click the **OK** button to exit the dialog box.

PROFILE VIEW: FULL GRID

Civil 3D draws a profile within a grid (a profile view). A profile view has vertical lines representing horizontal alignment stations and horizontal lines representing elevations along the alignment. The Profile View Style defines the grid and its annotation (see Figures 6-3, 6-6, and 6-7).

1. In Settings, expand the Profile View branch until you are viewing the list of Profile View Styles.

2. From the list, click **Full Grid**, press the right mouse button, and from the shortcut menu select **Edit...**.
3. Click the **Graph** tab to review its settings.

   This panel controls the title, the title's location (left half), the direction of the profile (upper right), the vertical exaggeration, if the grid is clipped, and if there are any extra vertical or horizontal grid segments.
4. Click the **Horizontal Axes** tab to view its settings.

   This tab controls the annotation for the stationing (top and bottom).
5. Click the **Text Component Editor** icon to the right side of Tick Label Text.

   This displays the Text Component Editor where the content of the label is formatted.
6. Click the **OK** button to exit the Text Component Editor.
7. Click the **Vertical Axes** tab to view its settings.

   This tab controls the annotation for elevations (right and left).
8. Click the **Display** tab to view its settings.
9. Click the **OK** button to exit the dialog box.

PROFILE VIEW: BAND SET

Civil 3D allows bands of information above or below a profile view. A band can contain profile data, horizontal and vertical geometry, superelevation, section sample line data, and pipe network values.

1. In Settings, expand the Profile View branch until you are viewing the Band Styles heading and the styles list.
2. In the Band Styles branch, expand Band Sets; from the list of styles, select **Profile Data with Geometry and Superelevation**, press the right mouse button, and from the shortcut menu select **Edit...**.
3. In the dialog box, click the **Bands** tab to view the band types and their label styles.
4. Note each style referenced in the dialog box.
5. Click the **OK** button to exit the Band Set - Profile Data with Geometry and Superelevation dialog box.

PROFILE: EXISTING GROUND

The Existing Ground profile style affects how the elevations along an alignment display within the Profile View. This style assigns each component a layer, a color, and other properties (see Figure 6.18). The Layout and the Design Profile styles are variations of this style. Each style assigns slightly different color schemes, but overall they are the same.

1. In Settings, expand the Profile branch until you are viewing the list of Profile Styles.
2. From the list, select **Existing Ground Profile**, press the right mouse button, and from the shortcut menu select **Edit...**.

3. In the dialog box, click the **Display** tab to view its contents.
4. Click the **OK** button to exit the dialog box.

PROFILE: LABEL SETS

The function of a profile label set is to be an alias for a collection of styles.

1. In Settings, expand the Profile and then the Label Styles branches until you are viewing the list of Label Sets.
2. From the list, click **Complete Label Set**, press the right mouse button, and from the shortcut menu select **Edit....**
3. Click the **Labels** tab to view the label types list and their associated label styles.

   You identify the label type and style at the top of the dialog box.
4. Click the **OK** button to exit the dialog box.

PROFILE LABEL SET STYLES

It is the label set label styles that annotate a profile in a profile view. These styles label values within the profile view, not at its edges like the profile view band sets.

Major and Minor Station

These styles place a label at the major and minor stations along the path of a profile (see Figure 6.22 and Figure 6.23).

1. In Settings, expand the Profile branch until you are viewing the list of style types under Label Styles.
2. In the Label Styles branch, expand the Station and Major Station branches to view their list of styles.
3. Select **Perpendicular with Tick**, press the right mouse button, and from the shortcut menu select **Edit....**
4. In the dialog box, click the **Layout** tab to view its contents.
5. In the Text section, Contents, click in the **Value** cell to display the editing ellipsis.
6. Click the ellipsis to view the content format string in the Text Component Editor.
7. Click the **Properties** drop-list arrow to view all of the alignment and profile properties you can include in this type of label.
8. Click the **OK** buttons until you exit the dialog boxes and return to the command line.

Horizontal Geometry Point

This style places a label on the profile at a critical alignment point station.

1. In the Station branch, expand the Horizontal Geometry Point branch until you are viewing its style list.

EXERCISES

2. From the list, select the **Horizontal Geometry Station**, press the right mouse button, and from the shortcut menu select **Edit...**.
3. In the dialog box, click the **Layout** tab to view its contents.
4. In the Layout panel, click the drop-list arrow of Component Names at the top of the dialog box and review the definition of the label components.
5. Reset the Component Name to **Horizontal Geometry**.
6. In Contents in the Text section, click in the Value cell to display the editing ellipsis.
7. Click the ellipsis to view the content format string in the Text Component Editor.
8. Click the **OK** buttons until exiting the dialog boxes and returning to the command line.

Line

This style places a label on the tangent lines of a profile using a percent format (see Figure 6.25).

1. Expand the Line Styles branch until you are viewing the list of styles.
2. From the list, select **Percent Grade**, press the right mouse button, and from the shortcut menu select **Edit...**.
3. In the dialog box, click the **Layout** tab to view its contents.
4. In Contents in the Text section, click in the Value cell to display the editing ellipsis.
5. Click the ellipsis to view the Text Component Editor.
6. Review the format of the tangent label.
7. Click the **Properties** drop-list arrow to view all of the alignment and profile properties you can include in this type of label.
8. Click the **OK** buttons until you exit the dialog boxes and return to the command line.

Curve

This style places a multi-lined label on the curves of a vertical alignment. This label is the most complex of the profile labels. Its components range from arrow blocks to formatted text (see Figure 6.26).

1. In Label Styles, expand the Curve branch until you are viewing a list of styles.
2. From the list of styles, select **Crest & Sag**, press the right mouse button, and from the shortcut menu select **Edit...**.
3. Click the **Layout** tab to view its contents.

   The Crest and Sag label is a complex label annotating the beginning and end of, the statistics of, and the length of an alignment's vertical curves.

4. In the Layout panel, click the drop-list arrow of Component Names at the top of the dialog box and review the definition of the label components.
5. Reset the Component Name to **PVI Sta and Elev**.
6. In Contents of the Text section, click the **Value** cell to display the editing ellipsis.
7. Click the ellipsis to view the Text Component Editor.
8. Review the format of the curve label.
9. Click the **Properties** drop-list arrow to view all of the alignment and profile properties you can include in this type of label.
10. Click the **OK** buttons until you exit the dialog boxes and return to the command line.

EXERCISES

STATION AND ELEVATION

The Station and Label command adds a spot elevation to a profile at the location you select within the profile (see Figure 6.27). You can drag the label from its original location. However, the dragged label may be different than the original label.

1. In Settings, expand the Profile View branch until you are viewing the list of labels under Label Styles.
2. In the list of label styles, expand the Station Elevation branch to view its list of styles.
3. Select **Station and Elevation** from the list, press the right mouse button, and from the shortcut menu select **Edit...**.
4. In the dialog box, click the **Layout** tab to view its contents.
5. At the top right, change the preview to **Station Elevation Label Style**.
6. In Contents of the Text section, click the **Value** cell to display the editing ellipsis.
7. Click the ellipsis to view the Text Component Editor.
8. Click the **OK** buttons until you exit the dialog boxes and return to the command line.

DEPTH

The Depth Label Style places a label between user-selected points. The resulting label is the slope or grade between the selected points.

1. If necessary, expand the Profile View branch until viewing the list of Label Styles.
2. Expand the Depth branch to view its styles.
3. From the list of styles, select **Depth**, press the right mouse button, and from the shortcut menu select **Edit...**.
4. In the dialog box, click the **Layout** tab to view its contents.
5. At the top right of the panel, change the Preview to **Depth Label Style**.

6. In the Layout panel, click the drop-list arrow of Component Names at the top of the dialog box and review the definition of the label components.
7. Reset the Component Name to **Depth**.
8. In Contents of the Text section, click in the **Value** cell to display the editing ellipsis.
9. Click the ellipsis to view the Text Component Editor.
10. Click the **OK** buttons until exiting the dialog boxes and returning to the command line.

    This ends the review of the styles that affect Profile Views and Profiles. The next unit covers the creation of a profile grid and the existing ground profile.

SUMMARY

- Edit Drawing Settings sets initial values used by all styles and commands in the Profile View or Profile settings branch.
- Command Settings can change the default styles and values set in Edit Feature Settings.
- Even though defining annotation for all axes of a profile grid, the display settings control what is visible on the grid.
- The profile styles define the layers and colors for profile components.
- A profile label set places the labels in a profile view and the profile view band set places its data in a band at the top or bottom of a profile view.
- The Station, Elevation, and Depth Label Styles are labels you manually add to a Profile with the Add Labels command.

UNIT 2: CREATING A PROFILE AND ITS VIEW

Creating a profile with a profile view is a simple two-step process: sample elevations along the alignment and create a profile view. These two steps can be executed as one command sequence or as two separate steps.

SURFACE DATA

The first step is determining the elevations along the path of an alignment. To do this, Civil 3D samples a surface or reads a data file containing station and elevation entries. This step associates surface or file elevations to the alignment stationing. The easiest method of creating a profile is by sampling elevations from a surface. If there are multiple surfaces, sample one or all of them.

CREATE PROFILE FROM SURFACE

The Create Profile from Surface dialog box displays initial values needed to sample a surface (see Figure 6.36). The top portion of the dialog box sets the alignment association (left side) and the surface(s) to sample (right side). The middle left of the dialog box sets the beginning and ending sampling stations. These stations represent the beginning and ending stations of the alignment. The user can use these station values or set his or her own values. The Sample offset toggle at the middle right allows the user to add offset sampling to the right and/or left of the currently selected alignment. Each offset appears as a separate entry in the Profile list with its assigned offset value. The Add button on the right side uses the settings in the upper part of the dialog box to create an entry in the Profile list (the bottom of the dialog box).

The Profile list at the lower portion of the dialog box displays a band of information for each profile sample. If there are multiple entries, create multiple profiles. Each band shows the surface name, type, update mode, style of the profile, stations, and elevation minimum and maximums. Each profile can have a different profile style and settings.

At the bottom of the dialog box are two important buttons: Remove and Draw in profile view. The Remove button allows the user to delete unwanted entries from the Profile list. The Draw in profile view button calls the Create View command to execute the second step, creating a profile view (grid) for the listed profile(s). If the user clicks the OK button, the routine creates the profiles. Creating a view is a separate step.

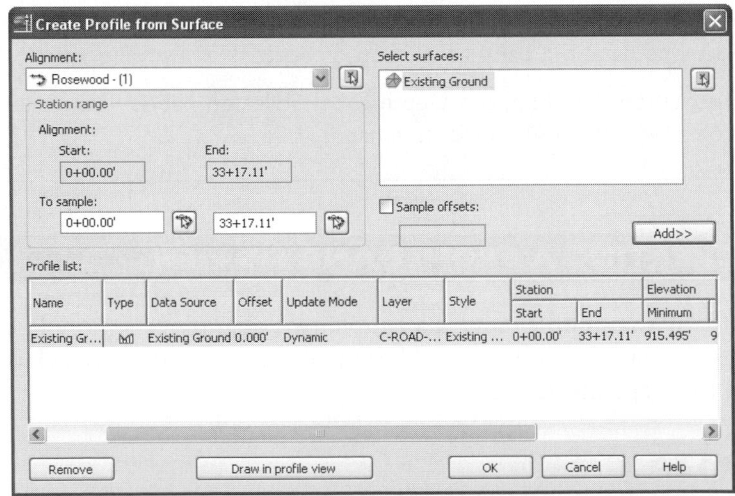

Figure 6.36

CREATE PROFILE VIEW

The second step is creating a profile view (a context grid) with the profile inside. The profile view horizontal axes, marked by vertical lines, represent stations and the vertical axes, represented by horizontal lines, represent the sampled surface elevations.

The Create Profile View dialog box contains the initial settings for the grid the user will create (see Figure 6.37). The top left of the dialog box sets the name and description of the view. The upper right sets the current alignment and the layer for the view. The layer can be a prefix or suffix that is the name of the profile.

The left middle portion of the dialog box sets the stationing for the view. The default values are the beginning and ending stations of the alignment. There is a toggle to set user-specified stations. After toggling on user stationing, enter new station values or select points off of the screen representing the new stations. The middle right portion of the dialog box sets the elevation range of the profile view. The default values are the minimum and maximum elevations of the profile(s). The user can set a specific elevation range by toggling on Specify height and entering in new minimum and maximum values.

Below the station and elevation area are the profile view style and annotation band set. The user can set, edit, and create new styles from this area. Be familiar with the different profile grid and band styles before changing these values.

The list at the bottom of the dialog box contains the profile(s) to draw in the grid. The user can change the assigned styles, clip the profile view grid with the profile, toggle on or off which profiles to draw, and change additional settings for each profile in the list.

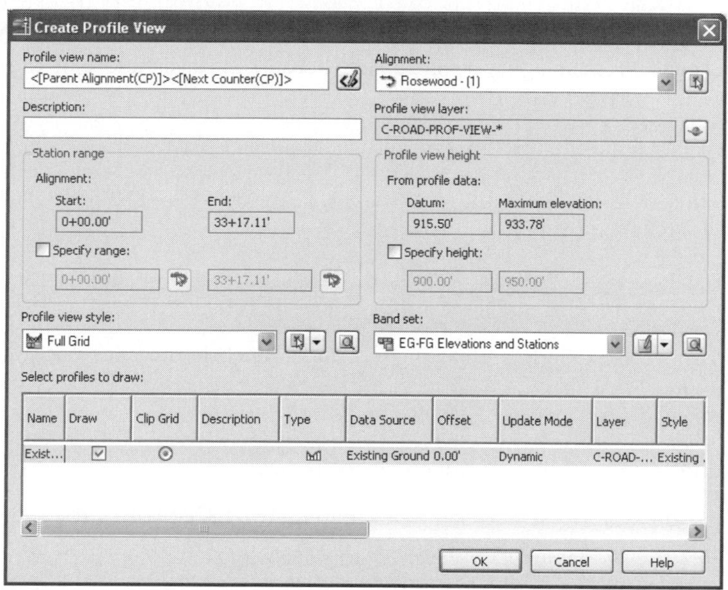

Figure 6.37

EXERCISE 6-2

When you complete this exercise, you will:

- Be able to create a profile.
- Be able to create a profile view.
- Be able to change styles assigned to the profile view.
- Be able to modify profile view annotation.

EXERCISE SETUP

This exercise continues with the previous unit's drawing. If you have not completed the previous exercise, you can open the drawing, *Chapter 6 – Unit 2.dwg* and start the exercise. The file is in the *Chapter 6* folder on the CD that accompanies this book.

1. If you are not in the drawing, open the drawing from the previous exercise or open the *Chapter 6 – Unit 2* drawing.

CREATE A SURFACE

First there needs to be a surface. If you haven't created a surface before, you should review Chapter 4 - Surfaces.

1. In the layer Properties Manager, thaw the **3EXCONT** and **3EXCONT5** layers, turn off all of the remaining layers including the current layer, and exit the Layer Properties Manager.
2. Click the **Prospector** tab to make it current.
3. In Prospector, click the **Surfaces** heading, press the right mouse button, and from the shortcut menu select **New....**
4. In the Create Surface dialog box, enter **Existing Ground** for the name, make sure it is a **TIN** surface, and click the **OK** button to make the surface.
5. In Prospector, expand the Existing Ground branch in Surfaces until you are viewing its Definition branch and its data list.
6. In the Definition list for Existing Ground, click **Contours**, press the right mouse button, and from the shortcut menu select **Add....**
7. In the Add Contour Data dialog box enter **Aero Contours** for the description and set the weeding and supplementing values as shown in Figure 6.38.
8. Click the **OK** button and in the drawing select the contours on the screen.
9. Click the **Layer Previous** icon to restore the last layer settings.
10. In the command line, enter **Regenall** and press the **Enter** key.
11. Reopen the Layer Properties Manager, freeze the layers **3EXCONT**and **3EXCONT5**, thaw the layer **Boundary**, and click the **OK** button.

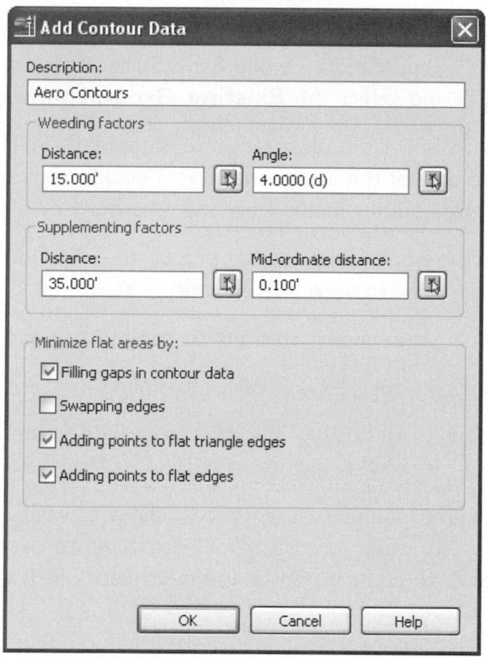

Figure 6.38

12. In the Definition branch of the Existing Ground surface, select **Boundaries**, press the right mouse button, and select **Add...** from the shortcut menu.
13. In the Add Boundaries dialog box, enter **Outer** for the name, set the type to **Outer**, and click the **OK** button.
14. In the drawing, select the polyline boundary to add it to the surface.
15. Click the **Layer Previous** icon to restore the last layer settings.
16. In the command line, enter **Regenall** and press the **Enter** key.
17. Reopen the Layer Properties Manager, freeze the layers **Boundary** and **C-TOPO-Existing Ground**, and click the **OK** button to exit.

    The C-TOPO-Existing Ground layer contains the surface you just made.
18. Use the AutoCAD Zoom and Pan command to place the site on the left side of the drawing.
19. Click the **AutoCAD Save** icon to save the drawing.

CREATE A PROFILE

After defining a surface, the next step is sampling the surface along the alignment to create a profile. If you have more than one surface, you can identify each surface to sample.

1. From the Profiles menu, select **Create From Surface...**.

There are three alignments in the drawing,

2. In the upper left of the Create Profile from Surface dialog box, set the alignment to **Rosewood – (1)** and select the **Existing Ground** surface in the upper right.

   The next decision is what station range to sample. By default, sampling is from the beginning to the end of the alignment. If you decide to sample only a portion of the alignment, you either enter in the new stations here or select them from the screen.

   The second decision is whether to sample to the right or left of the alignment (offsets). You may want to view the elevations at the edge of the travelway, ditch, etc.

3. Leave the beginning and ending stations to sample the entire length of the alignment.
4. If it is not already off, toggle **OFF** offset sampling.
5. In the middle right of the dialog box, click the **Add>>** button to place the current alignment and surface values into the Profile List.

   The Create Profile from Surface dialog box defines a profile. You could exit the box and at a later time use the Create Profile View command to draw the profile within a view. A second option is to continue on to the next step, which creates the profile view.

EXERCISES

CREATE A PROFILE VIEW

As mentioned above, the next step is to define a profile view. The Create Profile View command reads the sampled data and presents them as default station and elevation values for the profile view.

1. At the bottom of the Create Profile from Surface dialog box, click the **Draw in Profile View** button.

   This displays the Create Profile View dialog box.

2. Review your values in the Create Profile View dialog box and adjust them if needed. They should be similar to those in Figure 6.37.
3. Click the **OK** button to close the dialog box, and in the drawing select a point to the right of the site to draft the profile.
4. An Event View opens, noting that you need to set a profile2 for the band labeling to be correct; close the panorama.
5. Click the **AutoCAD Save** icon to save the drawing.
6. Zoom in to better view your profile.

CHANGE PROFILE VIEW

The profile view style defines the grid on the screen. When changing the profile view style, the profile changes reflecting the settings of the assigned style.

1. In the drawing, click anywhere on the profile view, press the right mouse button, and from the shortcut menu select **Profile View Properties...**.
2. In the Profile View Properties dialog box, select the **Information** tab, then click the **Object Style** drop-list arrow, and from the list select **Major Grids**.

3. Click the **OK** button to exit the Profile View Properties dialog box.
4. Repeat the previous thee steps, but reset the Profile View Style to **Full Grid**.

MODIFY A PROFILE VIEW STYLE

Modifying a view style changes how it displays on the screen.

1. Click the **Settings** tab to make it current.
2. In Settings, expand the Profile View branch until you are viewing the list of Profile View Styles.
3. From the list, select **Full Grid**, press the right mouse button, and from the shortcut menu select **Edit....**
4. In the Profile View Style dialog box, click the **Graph** tab to view its contents.
5. In the Grid Options area toggle **ON** both **Clip Horizontal Grid** and **Clip Vertical Grid**.
6. Click the **OK** button to view the changes to the profile view.
7. In the drawing, click anywhere on the profile view, press the right mouse button, and from the shortcut menu select **Edit Profile View Style....**
8. In the Profile View Style dialog box, click the **Graph** tab and in the Grid Options area toggle **OFF** both **Clip Horizontal Grid** and **Clip Vertical Grid**.
9. Just above the Grid Options area, change the Vertical Exaggeration factor to **5**.
10. Click the **OK** button to view the changes to the Profile View.
11. In the drawing, click anywhere on the profile view, press the right mouse button, and from the shortcut menu select **Edit Profile View Style....**
12. Select the Graph tab, change the Vertical Exaggeration factor for the profile back to **10**, toggle **ON** both **Clip Horizontal Grid** and **Clip Vertical Grid**, and click the **OK** button to exit.
13. Click the **AutoCAD Save** icon to save the drawing.

EXERCISES

This completes the process of creating a profile and its view.

The next unit describes the process of creating a proposed vertical alignment for the Rosewood roadway.

SUMMARY

- Users can sample any number of surfaces in the Create Profile from Surface command.
- Users control what profiles appear in a profile view in either the Create Profile or Create Profile View dialog box.
- A profile view grid can extend to the right and left of an alignment.
- A profile view can extend above and below the minimum and maximum elevation.
- A profile style defines color, linetype, etc. properties for profiles in a profile view.
- A profile view's annotation is around the perimeter of the grid (top, bottom, right, and left side).
- A profile's annotation is within the body of the profile view.

UNIT 3: DESIGNING A PROPOSED PROFILE

Within the context of the profile view and the existing ground profile, a designer creates a vertical road design. This design goes by many names, including vertical alignment, finished ground, etc. A vertical alignment contains proposed tangents (lines) and vertical curve segments.

Vertical curves allow a vehicle to transition from one grade to the next. There are two types of vertical curves: crest (at the top of two intersecting tangents) or sag (at the bottom of two intersecting slopes). There are three curve types for vertical curves: circular, asymmetrical (parabolic), and parabolic. Create these curves by specifying a curve length or a K Value. A K Value represents the horizontal distance along which a 1 percent change in grade occurs on the vertical curve and is a measure of abruptness.

CREATE BY LAYOUT

The Create by Layout ... command in the Profiles menu is the only command that creates a vertical alignment's tangent and curve segments (see Figure 6.39). Minimally, the user can draft the tangent (line) segments and later add the vertical curves to the design. You also have the option of defining the vertical curves as you draft the tangent's lines.

By clicking on the drop-list arrow to the right of the leftmost icon, the user can select a drafting mode, draw tangents only, draw tangents with vertical curves, or set the values of each type of vertical curve. The user can also change the default curve length or K Value (the ratio of the change of grade to the length of vertical curve). The default type, length, and design vertical curve values are a combination of the settings from the Edit Feature Settings dialog box for the profile or the command settings for Create by Layout.

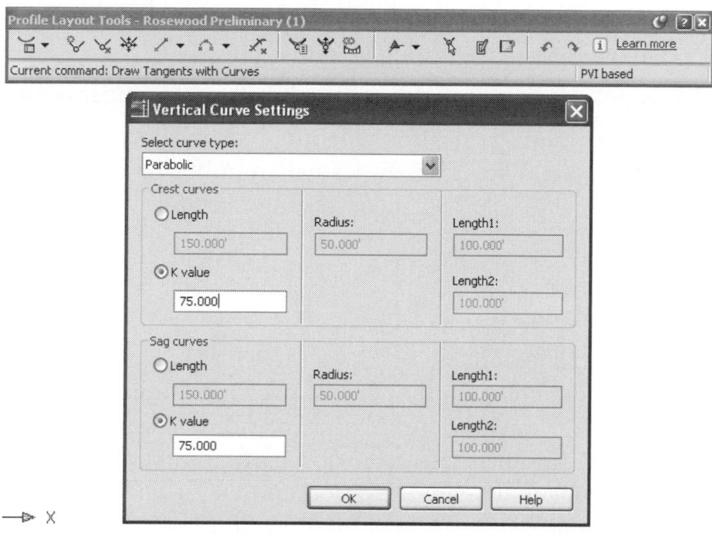

Figure 6.39

The next three icons to the right insert, delete, or move tangent PVIs (Points of Vertical Intersection).

TANGENTS

The Tangent icon stack creates three types of tangents: fixed, floating, and free. Draft a fixed tangent by selecting two points in the profile view. A floating tangent attaches to an existing tangent or curve and the user has to define a pass-through point. A free tangent is drawn between two existing symmetrical parabolas.

CURVES

The Curve icon stack drafts fixed, floating, and free vertical curves.

Fixed

There are four variations of fixed vertical curves: three points, selecting the end of an existing segment and identifying the pass through point, selecting two points and entering in the tangent grade at the start of the vertical curve, and selecting two points and entering in the tangent grade at the end of the vertical curve.

Floating

There are three variations of floating vertical curves: selecting a existing segment to attach to, identifying a pass-through point and a K value and selecting a existing segment to attach to, and identifying a pass-through point and a grade.

Free

There are four types of free vertical curves between two tangent segments: specifying a K value or a minimum radius for the vertical curve, specifying a parabolic, asymmetrical parabolic, or circular vertical curve by identifying a PVI and one additional parameter (curve length, K value, and/or pass-through point).

The remaining icons provide tools to edit and analyze profile tangents and vertical curves.

The rules governing maximum grades for tangents and what type and length of vertical curve may constraint the final solution to a small number of choices. The editing of a vertical design is tweaking a basic design or setting specific grades, elevations, or curve types and lengths.

TRANSPARENT COMMANDS

There are three transparent commands that affect vertical tangent drafting in a profile view. The first is Station and Elevation. This override prompts for a station and its elevation for the vertical segment. The second is Profile Grade Station. This override prompts for a grade, freezes the cursor at the specified grade and allows the user to move the cursor along the stations to an ending station and elevation using the fixed grade. The third override is Profile Grade Length. This method simply prompts for a grade, a starting point, and the length to travel at that grade.

PROFILE2 ASSIGNMENT

When creating a profile, an event viewer displays, reminding the user a second profile has to be assigned to the profile view. The second profile in the profile view is traditionally the proposed roadway vertical design and its elevations that appear in the profile view annotation (the profile data band). When creating a profile view, Civil 3D assigns the first profile to both the Profile1 and Profile2 (see Figure 6.9). This means that when viewing the elevations in the profile view band, both elevations are for the same profile. Also, the elevations appear as elevations with 1 or 2 decimal places of precision.

The assignment of Profile2 is a manual step and the user must identify which profile is Profile2 in the Profile View Properties dialog box. When using the Create By Layout command Civil 3D does *not* assign the resulting profile as Profile2.

EXERCISE 6-3

When you complete this exercise, you will:

- Be familiar with the Create By Layout Command Settings.
- Create a profile from the Layout toolbar.

EXERCISE SETUP

This exercise continues with the previous unit's drawing. If you have not completed the previous exercise, you can open the drawing, *Chapter 6 – Unit 3.dwg* and start the exercise. The file is in the *Chapter 6* folder on the CD that accompanies this book.

1. If you are not in a drawing, open the drawing from the previous exercise or open the *Chapter 6 – Unit 3* drawing.

EDIT FEATURE SETTINGS

Some of the settings affecting vertical curves are in Profile Edit Feature Settings dialog box.

1. Click the **Settings** tab to make it current.
2. In Settings, click the **Profile** heading, press the right mouse button, and from the shortcut menu select **Edit Feature Settings....**
3. Expand the Profile Creation section of the dialog box.

   The first three entries set the default type of vertical curve. The next two settings identify the design criteria for crest and sag vertical curves.

4. Set the Default Vertical Curve Type to **Symmetric Parabola**.
5. Set the criteria to **K Value**.

   Your dialog box should look like Figure 6.40.

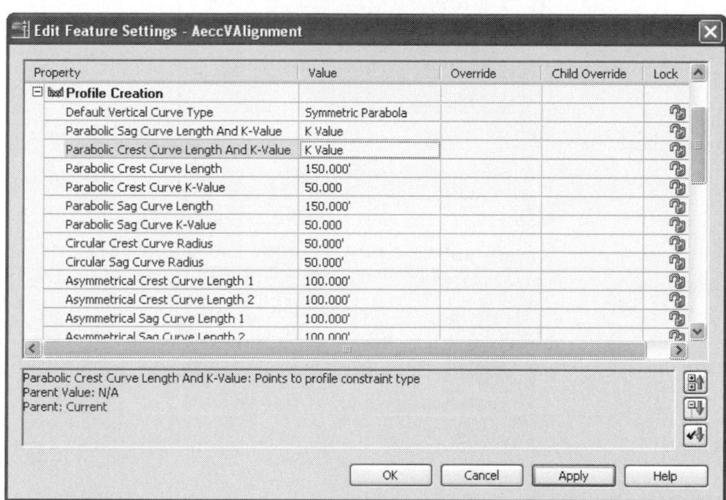

Figure 6.40

6. Click the **OK** button to set the values and exit the dialog box.

DRAFTING A PROPOSED CENTERLINE

Creating a vertical design is not always a simple-single pass process. The vertical design may have grade, curve length, K Value, or distance between restrictions affecting the vertical segments. In addition to these design issues, you may have to design a road that balances or meets some targeted earthworks amount. All of these issues combine to make the process of vertical design an exercise in design compromise.

1. Use the AutoCAD Zoom command to view the profile view.
2. From the View menu, use the **Named Views...** command to create a new view, **Profile**.

3. Click the **AutoCAD Save** icon to save the drawing.
4. Make sure that the Transparent Commands toolbar is visible.
5. From the Profile menu, select **Create By Layout...** and in the drawing select the Rosewood profile view.

   This displays the Create Profile – Draw New dialog box.

6. In the dialog box, for the name replace <[Profile Type]> with **Rosewood Preliminary**, give the profile a short description, set the Profile Style to **Layout**, and the set Label Set to **Complete Label Set**.

   Your dialog box should look similar to Figure 6.41.

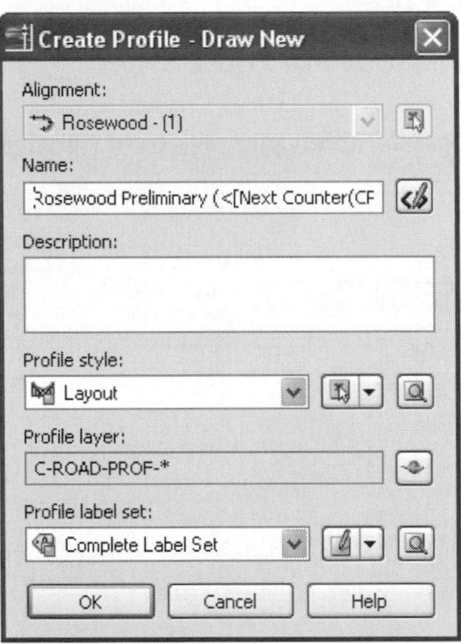

Figure 6.41

7. Click the **OK** button to close the Create Profile - Draw New dialog box.

   The Profile Layout tools toolbar appears and lists the name of the current vertical alignment.

   The Rosewood vertical design restrictions are the following:

   No tangent grades over 6% or under 1%.

   No tangent can be less than 200 feet.

   No vertical curves less than 350.

Drafting the Tangent Segments with Vertical Curves

The drafting strategy will be to first draw tangent segments with vertical curves. Table 6.1 has the From and To Stations and grades between them. The first and last points on the vertical alignment are the intersection of the existing ground and the vertical grid line. Make sure you use an object snap (endpoint or intersection) to select these points.

1. On the Profile Layout Tools toolbar, click the drop-list arrow to the right of the first icon on the left and from the list, select **Curve Settings....**
2. In the Vertical Curve Settings dialog box, if necessary, click the **Select curve type** drop-list arrow at the top and change the curve type to **Parabolic**.
3. Set the K Value to **75** for crest and sag curves.
4. Click the **OK** button to close the dialog box.

   Your screen should look like Figure 6.39.
5. In the Profile Layout Tools toolbar, click the drop-list arrow to the right of the left-most icon and select **Draw Tangents with Curves** from the list.

   The first prompt is for a starting point.
6. Select a point at the intersection (use object snap) of existing ground and the **0+00** station of the profile.

   You may have to zoom in to the beginning of the profile to select this point and zoom back after selecting the beginning point.
7. In the command line, the routine prompts for an ending point, but from the Transparent Commands toolbar, select the **Profile Grade Station** icon.
8. In the command line, the prompt changes to Select a Profile View; click the **profile view**.
9. In the command line, the prompt is for a Grade (positive up and negative down); enter **1** as the grade to use for this segment, and press the **Enter** key.

   The crosshairs now show a vertical alignment line frozen at the grade of 1%, a vertical line showing the station, and the tool tip at the intersection of the two lines listing the station.
10. In the drawing, select a point near station 900 or enter **900** at the command prompt. If entering a value at the command line, you must press the **Enter** key to use the value.
11. Use the AutoCAD transparent Zoom and Pan commands to move Station 9+00 to the left side of the screen.
12. In the command line, the routine prompts for the next Grade; set the grade to -3% (**-3**), drag the cursor to around station 13+75, and select that point or enter **1375** (and press the **Enter** key).
13. Use the transparent Pan command to move the new PVI to the left side of the screen.
14. At the Grade prompt, change the grade to 3% (**3**), and select a point near 28+50 or enter **2850**.

15. Press the **Esc** key to end the Transparent Command. The prompt changes to Specify End Point:. Select the intersection (use object snap) of the existing ground and the last vertical grid line of the profile.

Table 6.1

From Station	To Station	Grade
Intersection	9+00	1%
9+00	13+75	-3%
13+75	28+50	3%
28+50	Intersection	No specific grade

With the current vertical curve values, the last vertical cannot be resolved. Your profile should look similar to Figure 6.42.

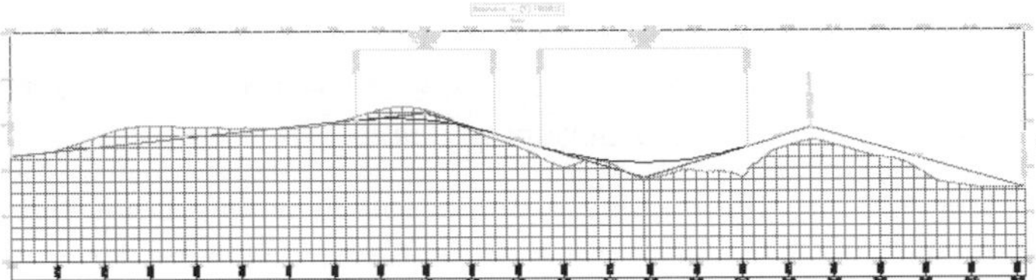

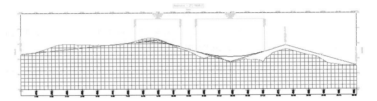

Figure 6.42

ASSIGNING THE PROFILE TO PROFILE2

The assignment of the new profile to the Profile View Band Set's Profile2 occurs in the Band Set panel of the Profile View Properties dialog box.

1. Use the AutoCAD Zoom command and zoom in on the beginning of the profile so you can view the intersection of the left and bottom axis and its major and minor station annotation for elevations.

   Currently, both band elevations represent the existing ground to one or two decimal places. This is from an automatic assignment Civil 3D makes when creating a profile view.

2. In the drawing, click the **Profile View**, press the right mouse button, and from the shortcut menu select **Profile View Properties....**
3. Click the **Bands** tab to view its contents.
4. In the Bands panel, scroll the data completely to the right to view the Profile1 and Profile2 assignments.
5. Currently both are Existing Ground - Surface.

6. In the cell under Profile2, click twice to display a drop-list arrow; click the drop-list arrow, and from the list select the **Rosewood Preliminary (1)** (see Figure 6.43).
7. Click the **OK** button to exit the dialog box.
8. Click the **AutoCAD Save** icon to save the drawing.

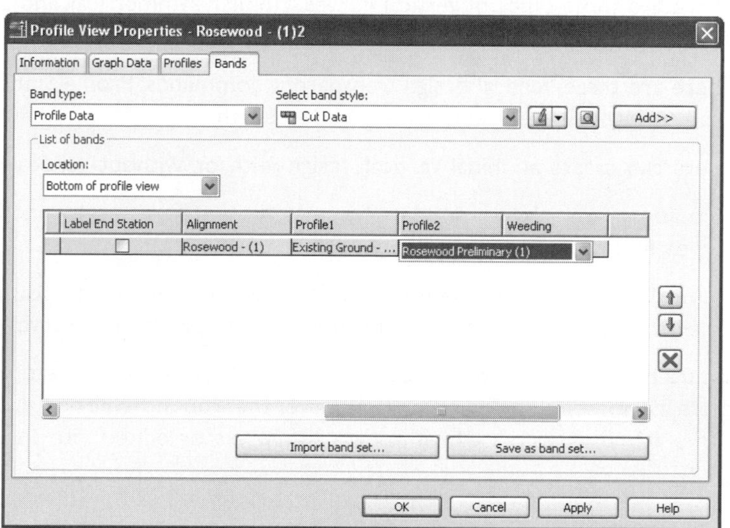

Figure 6.43

This ends the exercise for drafting vertical alignment tangents and vertical curves. The next unit focuses on the evaluation and editing of the newly created vertical alignment.

SUMMARY

- The values in the profile's Edit Feature Settings dialog box set the type, criteria, and calculation values for vertical curves.
- There are three types of vertical curves: circular, symmetrical, and asymmetrical parabolic curves.
- There are three vertical design transparent commands: Profile Station Elevation, Profile Grade Station, and Profile Grade Length.
- Users can create an initial vertical design with or without vertical curves.
- If creating an initial design with only tangent segments, users can add the vertical curves in a second pass or as an edit session.
- If entering in values that create vertical curves that overlap, the routine will issue an error message and start the prompting for new vertical curve values.
- If a user is drafting an initial vertical design with tangents and vertical curves and there is no solution for a vertical curve, the routine will not draw a vertical curve. It must be created with new parameters or edited into the design.

UNIT 4: ANALYZING AND EDITING A VERTICAL DESIGN

The analysis and editing of a vertical alignment is this unit's focus. The analysis of the vertical alignment data uses LandXML data files for reports or information editor information from the Profile Layout Tools toolbar. You can graphically edit a profile or edit its values in Profile Layout Tools vistas.

PROFILE REPORTS

The only reports for a profile are those of the Autodesk LandXML Report Application or the Inquiry Toolspace (Report Manager). To create a report, create a LandXML file from a Prospector shortcut menu and then open the Autodesk Report Application (see Figure 6.44). A LandXML file contains all of the necessary geometry to document a vertical design. When creating a report from the Toolbox panel in Reports Manager in Toolspace, select the report type, right mouse click and execute the report (see Figure 6.45).

The Autodesk LandXML Report Application has to open the LandXML data file. After loading the data file, the next step is selecting a form (report) from a form list and selecting the Output tab to view the report (see Figures 6.46 and 6.47).

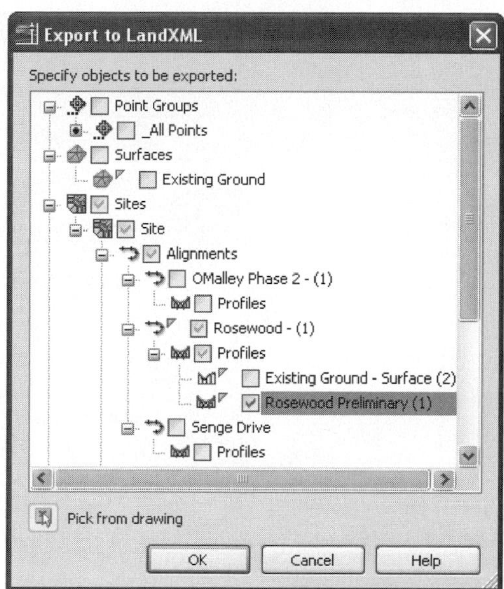

Figure 6.44

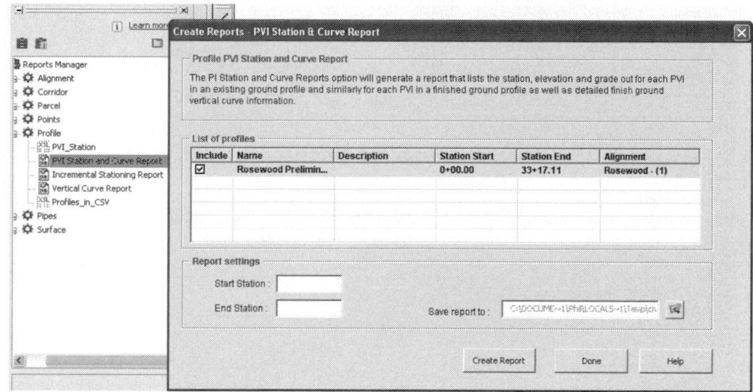

Figure 6.45

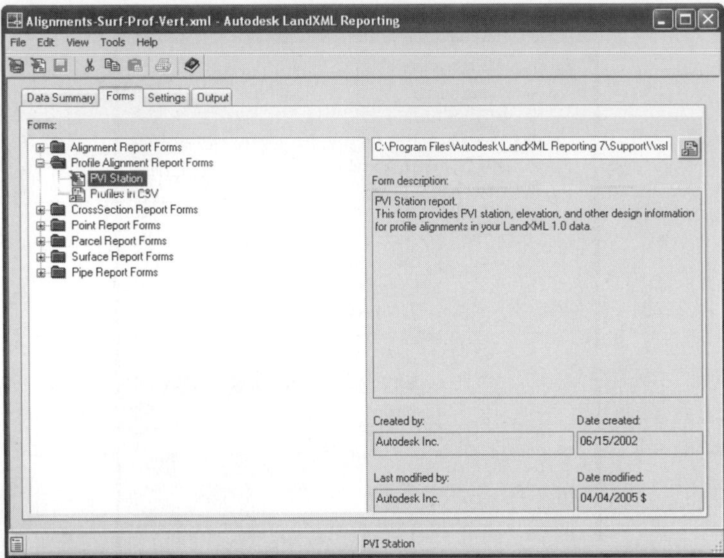

Figure 6.46

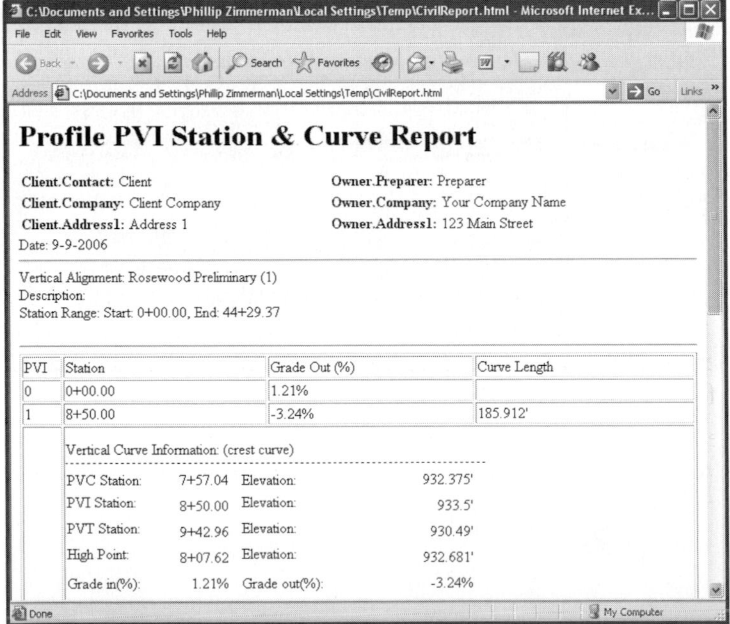

Figure 6.47

GRAPHICAL EDITING

The first profile editing method graphically manipulates the PVIs and segments. When the user selects a profile, the profile displays grips allowing the user to edit a vertical curve's length and/or its location or change the location and elevation of a PVI by holding one grade and changing the second tangent's grade. When manipulating the PVI's grip, all of the connected segments change to accommodate its new location (see Figure 6.48).

Each grip has a specific editing function. Round grips are vertical curve grips and appear at the ends and midpoint of the vertical curve. These grips allow the user to adjust curve parameters. The arrow grips appear at the end of each tangent into a PVI. By selecting one grip, the user holds the tangent grade of the active grip and changes the intersection point and grade of the unselected tangent (i.e., he or she moves the PVI based on the grade in or out). The last grip, the red triangle, is the PVI. When moving this grip, adjust the grades of each tangent and potentially change either the K value or the length of the associated vertical curve.

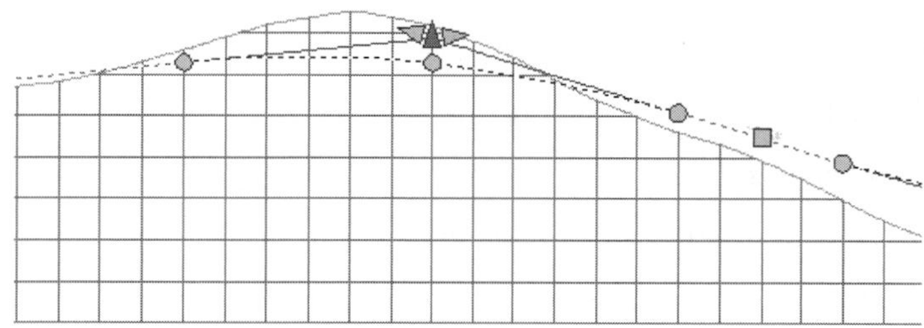

Figure 6.48

PROFILE LAYOUT TOOLS EDITORS

The Profile Layout Tools toolbar provides the second method of reviewing and editing profile tangent and vertical curve values. The toolbar has two types of vista editors: one displays all of the vertical alignment's values (the Profile Entities vista), and the Profile Layout Parameters vista displays the values for a selected vertical segment. Figure 6.49 shows the Profile Layout Parameters vista.

When the user makes any adjustments in either editor, all segments recalculate and redisplay their new values and update the drawing's profile. Each editor indicates values the user can change with black lettering.

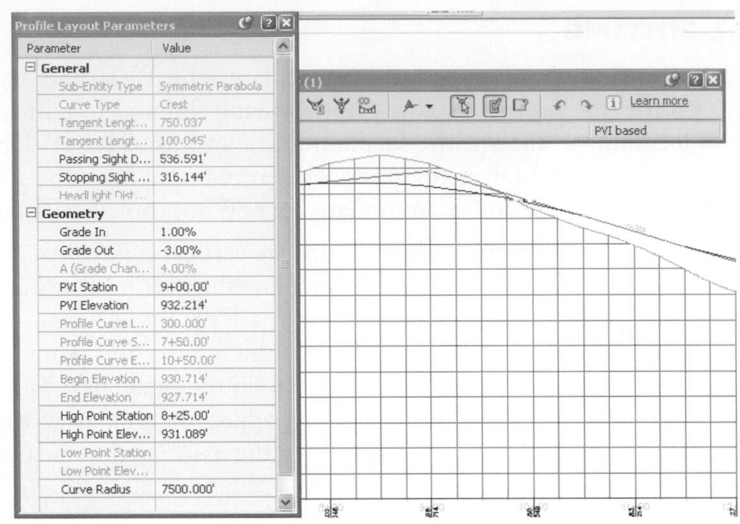

Figure 6.49

EXERCISE 6-4

When you complete this exercise, you will:

- Be familiar with producing a report on a selected profile.
- Be able to review a report on a vertical alignment.
- Be able to graphically edit an alignment.
- Be able to edit the vertical alignment in an overall or subentity view.

EXERCISE SETUP

This exercise continues with the drawing used in the previous unit's exercise. If you did not do the previous exercise, you can open *Chapter 6 - Unit 4.dwg,* which is in the *Chapter 6* folder of the CD that accompanies this book.

1. Open the drawing from the previous exercise or the *Chapter 6 – Unit 4* drawing.

REVIEWING A PROFILE

The first report reviews a profile's tangents and curves.

LandXML Report

The Report manager application provides a PVI Station and Curve report.

1. From the General menu, select **Toolbox** and expand the Profile section to view the list of reports.
2. From the list of reports, select **PVI Station and Curve Report** from the list, press the right mouse button, and select **Execute...** from the shortcut menu.

3. In the Create Reports dialog box, click the **Create Report...** button to create the report.

   The report appears.

4. Review the report, close the report, and click the **Done** button.

GRAPHICALLY EDITING A PROPOSED PROFILE

You can graphically edit a vertical alignment.

1. Use the AutoCAD Zoom to better view the first PVI.
2. Select the **Rosewood Preliminary (1)** alignment to display its grips.
3. Adjust the vertical curve by selecting the round grips and relocating them.
4. Adjust a PVI by selecting one tangent arrow and moving its location.
5. Adjust a PVI by selecting the other tangent arrow and moving its location.
6. Adjust a PVI by selecting it and moving it to a new location.

SUBENTITY VISTA

When making any adjustments in the vista editors, all of associated numbers change to accommodate the edit.

1. In the drawing, select the **Rosewood Preliminary (1)** profile, press the right mouse button, and from the shortcut menu select **Edit Profile Geometry...**.

   The Profile Layout Tools toolbar appears.

2. Click the **Profile Layout Parameters** grid icon on the right side of the toolbar to display the subentity grid.
3. In the drawing, use the AutoCAD Zoom and/or pan command until you are viewing the first PVI.
4. From the toolbar, click the **Select PVI** icon, and select a point near the PVI.
5. Change the PVI Station back to **8+50** and change the PVI Elevation to **933.50**.
6. Use the AutoCAD Zoom and/or pan command until you are viewing the second PVI.
7. In the drawing, click the **Select the PVI** icon from the toolbar and select a point near the PVI.
8. Change the location of the second PVI Station to **13+75**, raise its Elevation to **916.50**, and change the radius to **3000**.
9. Use the AutoCAD Zoom and/or pan until you are viewing the last PVI.
10. In the Profile Layout Tools toolbar, click drop-list arrow to the right of the Add Curve icon, and from the More Free Vertical Curves flyout, select **Free Vertical Parabolic (PVI based)**.

11. In the command line, the prompt is to select a point by a PVI; select the PVI with no vertical curve.
12. In the command line, the prompt is for a Curve Length; for the curve length enter **100** and press the **Enter** key.
13. From the Profile Layout Tools toolbar, select the **Select PVI** icon, and select a point near the PVI.
14. In the Profile Layout Parameters dialog box, change the location of the PVI to **28+50** and set its Elevation to **926.5**.
15. In the Profile Layout Tools toolbar, click the drop-list arrow to the right of the PVI Based icon and select **Entity Based** to change the Selection Operation.
16. In the profile view, select a point on a couple of tangents to view their values.
17. Click the red "**X**" of the Profile Layout Parameters vista to close it.

OVERALL VISTA

The second editor is the overall editor. It shows the same values as the previous editor, except in a linear format.

1. In the Profile Layout Tools toolbar, click the **Profile Grid View** icon on the right side of the toolbar to display the editor and review the alignment values.
2. Close the dialog box and close the Profile Layout Tools toolbar.
3. Click the **AutoCAD Save** icon to save the drawing.

   Your profile and profile view should now look similar to Figure 6.50.

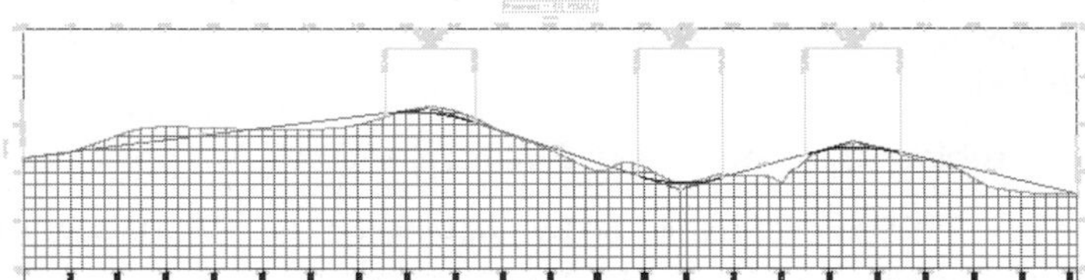

Figure 6.50

This completes the exercise on evaluating and editing a vertical alignment. The two methods of editing are by graphically manipulating the alignment's grips or by editing the numbers of the vertical design in the profile editors.

SUMMARY

- The toolbox has reports on the station, elevation, grades, and length of vertical curves in a profile.
- When graphically editing a vertical curve, users manipulate grips linked to specific geometric PVI and vertical curve points.
- When graphically editing a profile, round grips affect vertical curves.
- When graphically editing a profile, arrow grips affect the location of a PVI by holding one and changing the second grade.
- When graphically editing a profile, selecting the triangular PVI grip will change its location by changing the in and out grades, the PVI elevation, and/or the PVI station.

UNIT 5: PROFILE ANNOTATION

This unit reviews the annotation of profiles (vertical alignments). There are two type of annotation: profile view and profile. This annotation appears when the user creates the profile view and a profile. This type of annotation is the subject of earlier units of this chapter.

SPOT PROFILE LABELS

A second type of profile annotation is labels that appear within the profile view representing values of a profile. The Add Profile View Labels command of the Profiles menu creates two types of labels: Station and Elevation, and Depth. The Station and Elevation label annotates its namesake at a location selected in a profile view. After identifying the profile view, a jig appears connecting to the stationing along the bottom of the profile to the cursor. After it is selected, the station is frozen and a second jig helps identify the elevation.

The Depth label annotates the distance between two points selected within the profile view (see Figure 6.51). If the user selects a lower point rather than a higher point, he or she gets a positive grade. If the user selects a higher point and then a lower one, he or she gets a negative grade.

Figure 6.51

EXERCISE 6-5

When you complete this exercise, you will:

- Be able to apply a different profile style and label set to the profile.

EXERCISE SETUP

This exercise continues with the previous unit's drawing. If you didn't complete the previous exercise, you can open the *Chapter 6 – Unit 5.dwg* and continue with the exercise. The file is in the *Chapter 6* folder on the CD that accompanies this book.

1. Open the drawing from the last exercise to continue or open the *Chapter 6 Unit 5* drawing.

SPOT PROFILE ELEVATIONS

The Add Labels dialog box creates two types of spot profile labels: a Station and Elevation label and a Depth label. A Depth label annotates the grade/slope between two selected profile view points.

Station and Elevations

1. From the Profiles menu, select **Add Profile View Labels....**

   The Add Labels dialog box displays.

2. In the Add Labels dialog box, change the Label Type to **Station Elevation** and click the **Add** button.

3. In the command line, the prompt is Select a Profile View; in the drawing, select the **Rosewood Profile View**.

4. In the command line, the routine prompts for a Station; using the station jig, select a point in the profile view or enter in a Station value and press the **Enter** key.

The station jig freezes and a second jig appears, prompting elevation. You can enter an elevation at the command line.

5. In the command line, the routine prompts for an elevation; select a point in the profile view or enter an elevation and press the **Enter** key.
6. Place two more labels in the profile view.
7. Press the **Enter** key to end the command.
8. Use the AutoCAD Zoom command to better view the labels.

Depth

1. In the Add Labels dialog box, change the Label Type to **Depth** and click the **Add** button to start creating the labels.
2. In the command line, the prompt is Select a Profile View; in the drawing, select the **Rosewood Profile View**.
3. In the command line, the prompt is Select First Point; in the profile view, select a point (a jig appears connecting the cursor to the selected point).
4. In the command line, the routine prompts for a second point; in the profile view, select a second point.

   The routine draws a line between the two points you selected, labels the line with a distance, and exits to the command line. You must select the Add button to add more depth labels.

5. Define a couple more Depth labels, and zoom in to view the new labels.
6. Click the **Close** button to close the Add Labels dialog box.
7. Click the **AutoCAD Save** icon and save the drawing.

   This completes the review of the profile labels.

SUMMARY

- The Station and Elevation label creates a label in the profile view.
- The Depth label annotates the distance between select points in a profile view.

This ends the chapter on profiles and profile views. The next chapter focuses on the assembly, its subassemblies, and the roadway model, the corridor.

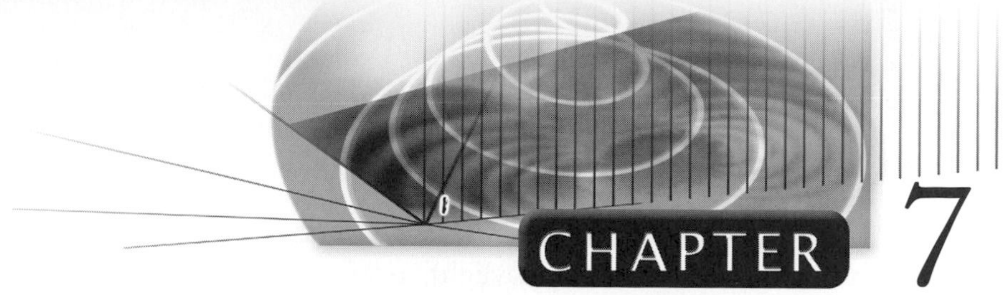

Assemblies and Corridors

INTRODUCTION

In Civil 3D, the process of developing a roadway design is the same as any other application. What is new and beneficial in Civil 3D is the interaction of the design model elements that create the roadway. Chapters 5 and 6 covered the first two steps of road design, the horizontal and vertical alignments. Chapter 7 covers the third step of the road design process, the assembly. An assembly is similar to an LDT template, but with greatly enhanced capabilities. Assemblies contain subassemblies with parametric controls and "intelligence" behind them, making them more than simple templates.

A corridor is the roadway model resulting from the combination of a horizontal and vertical alignment, a surface, and an assembly.

OBJECTIVES

This chapter focuses on the following topics:

- The Corridor Modeling Catalogs (Imperial) in Civil 3D
- Subassemblies, Their Behaviors, and Parameters
- The Creation of an Assembly and the Modification of Subassemblies
- The Creation of a Simple Corridor
- The Creation of a Corridor from a Surface, from Alignments, and from Assemblies
- Use of the Simple Corridor Command Versus the Corridor Command

OVERVIEW

After designing the roadway horizontally and vertically, the next step is defining a roadway cross section, or an assembly. In other civil applications, this is known as defining a template. However, Civil 3D assemblies are more flexible and better represent road design situations. These enhanced capabilities come from the inclusion of VBA scripting to control the assembly. The scripts and subassembly parameters define the rules that affect how an assembly solves a roadway design. This makes an assembly more powerful and useful to the civil designer.

An assembly is a vertical line representing the attachment location of the horizontal (baseline) and vertical alignment (see Figure 7.1). Subassemblies attach to the assembly to make a cross section of the roadway. Each subassembly has a set function (curb, slope to daylight, pavement, etc.), attaches to an assembly or another subassembly at a connection point, and has design constraints or parameters governing its behavior. These connection points provide the ability to connect locations for subassemblies outward from a more centrally located subassembly.

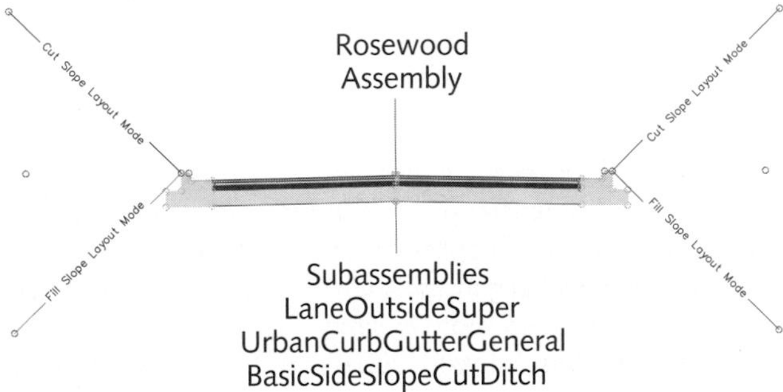

Figure 7.1

Each subassembly has right and left behavior set by its parameters list. These parameters allow the subassembly to react to varying conditions of the design. In Figure 7.2, the right side shows the simple grading slope parameters attached to the back of a curb. By using a more complex slope subassembly, the parameters can automate a solution for a complex design. A list of complex parameters is represented on the left side of Figure 7.2.

No selection

Parameters	
Version	R2006
Side	Left
Omit Daylight Link	Include Daylight link
Cut Test Point Link	3
Cut 1 Width	0.000'
Cut 1 Slope	0.000%
Cut 2 Width	0.000'
Cut 2 Slope	0.000%
Cut 3 Width	0.000'
Cut 3 Slope	0.000%
Cut 4 Width	0.000'
Cut 4 Slope	0.000%
Cut 5 Width	0.000'
Cut 5 Slope	0.000%
Cut 6 Width	0.000'
Cut 6 Slope	0.000%
Cut 7 Width	0.000'
Cut 7 Slope	0.000%
Cut 8 Width	0.000'
Cut 8 Slope	0.000%
Flat Cut Slope	16.667%
Flat Cut Max Height	5.000'
Medium Cut Slope	25.000%
Medium Cut Max He...	10.000'
Steep Cut Slope	50.000%

Design | Feature Data | Display | Extended Data | PROPERTIES

No selection

Description	
True Color	ByLayer
Layer	0
Linetype	ByLayer
Linetype scale	1.0000
Plot style	ByColor
Lineweight	ByLayer
Show Tooltips	
Default Loop In Lay...	Last
Default Loop Offset...	-10.0000
Macro Name	C3DStockSubassemblyScripts...
Macro Project	C:\Program Files\Autodesk Ci...
Use embedded proj...	No
ADVANCED	
Parameters	
Version	R2006
Side	Left
Omit Daylight Link	Include Daylight link
Cut Slope	50.000%
Fill Slope	25.000%
Foreslope Slope	25.000%
Foreslope Width	4.000'
Bottom Width	2.000'
Backslope Slope	25.000%
Backslope Width	4.000'

Design | Feature Data | Display | Extended Data | PROPERTIES

Figure 7.2

After creating one or more assemblies, the next step is to create a corridor. A corridor object combines the horizontal and vertical alignments, the assembly, and even a surface into a roadway model. A model of the roadway provides several new opportunities to review and create new data from a resulting road design. Users can slice the model in any diagonal and see a real section of the roadway. Users can also create feature lines (base grading objects) from the model. These feature lines become starting points to grade the areas surrounding the roadway. Users can use the model to begin tying a piping network to the roadway.

A corridor can create a surface, and the surface can have any surface object style. Civil 3D creates surfaces from links and corridor feature lines. A contour style displays a corridor's contours within the model. A corridor surface appears in Prospector's Surfaces list. The subassembly shapes allow users to calculate volumes for roadway materials, and a corridor surface is a comparison surface for general roadway earthworks.

The goal of the road design will vary from project to project, but generally the goal is not to move more spoil material than necessary. Editing the design may take a user to the very beginning of the design process: the horizontal alignment. Or, the modifications can occur in the proposed profile or with changes to the assembly. Wherever the editing takes place, Civil 3D moves the changes forward from that point to the corridor. The roadway design elements have a dynamic relationship.

UNIT 1

The first unit focuses on the styles and settings affecting the corridor. Each subassembly that defines an assembly has associated styles. These styles affect the subassembly connection points, fill patterns, outlines, and labeling data of critical points on the subassembly.

UNIT 2

The next step is creating the assembly. An assembly is built from subassemblies representing elements of a roadway cross section (pavement, curbs, walks, links, etc.). Each of these subassemblies has a right and left component as well as a list of constraints on their shape and corridor modeling behavior.

UNIT 3

The third unit focuses on the creation of a corridor (Roadway Model). Once all of the roadway elements are defined, the next step is to create a corridor. There are two ways in which to create a corridor: the Create Simple... command and the Create Corridor... command. The simple corridor makes certain assumptions about the model relative to the elements that are its makeup. If these assumptions are incorrect, users can edit the model with the Corridor dialog box. The Corridor dialog box is the second method of creating a model. This dialog box allows the user to control complex station and assembly assignments, surfaces, and other aspects to create a more complex roadway mode.

UNIT 4

Once a corridor is established, the next step is to evaluate and possibly edit some of its values. This is done in the View/Edit Corridor sections.

UNIT 5

This unit explores the data a corridor can generate from its model. The data can include feature lines for grading, points, offset baselines for transitioning, volumes, or other useful roadway design elements.

UNIT 1: CATALOGS, PALETTES, AND STYLES

The concept of a catalog comes from the Autodesk Architectural Desktop world and is used in Civil 3D to organize design elements of an assembly.

CORRIDOR MODELING CATALOG (IMPERIAL)

There are six Imperial Corridor Modeling catalogs: Channel and Retaining Wall Subassembly, Generic Subassembly, Getting Started Subassembly, Rehab Subassembly, Subdivision Roads Subassembly, and Transportation Road Design Subassembly (see Figure 7.3). Some of the catalogs are more than one page. Each catalog has similar design elements, but each has specific functionality or constraints that set it a part. Users can add to or customize the existing roadway elements.

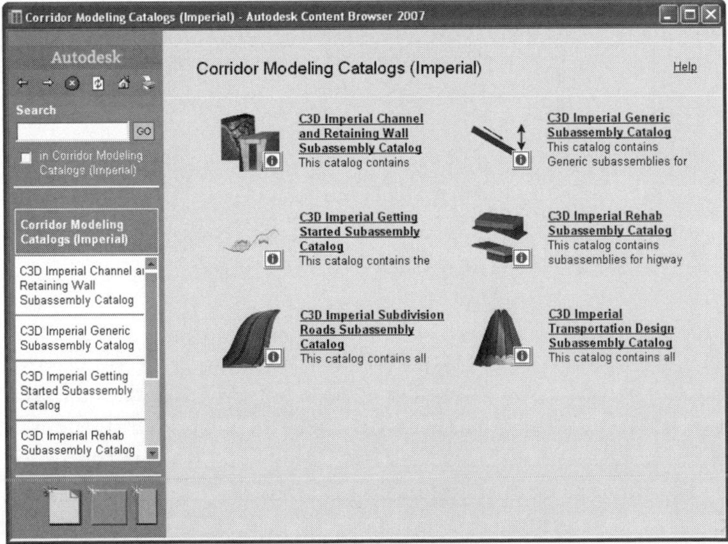

Figure 7.3

CIVIL 3D IMPERIAL PALETTE

The Civil 3D tool palette displays the subassemblies in groups based on function (see Figure 7.4). Each tab represents a portion of the Roadway Catalog (basic or complex pavement, sidewalks, sod, shoulders, rehabilitation strategies, and links between subassemblies). All of the catalog subassemblies appear in the Civil 3D Roadway palette.

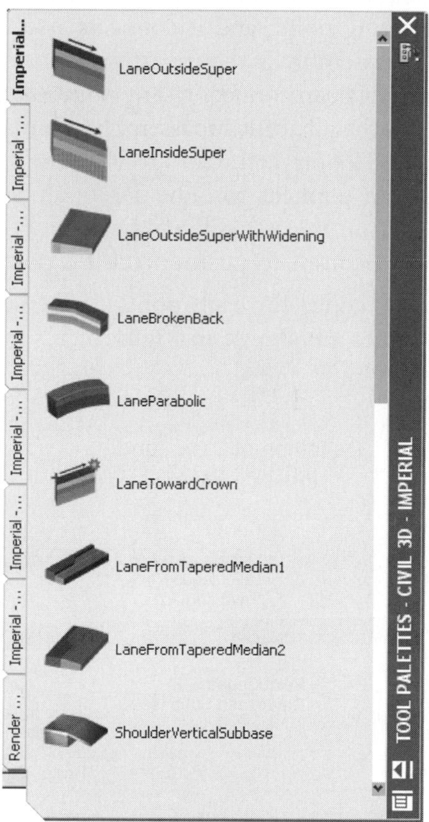

Figure 7.4

A subassembly is a combination of points, links, and shapes. Civil 3D uses point codes, links, and shapes to label data, to integrate the roadway with grading solutions, to use as data for slope staking, to use as data for construction staking, and for corridor visualization. Codes, links, and shapes are skeletal elements that cause subassemblies to react to parameters, and they require an understanding of roadway design needs and VBA coding. No simple task.

Civil 3D assigns each vertex a point code, and between each point a link code. It also defines the closed polygon as a subassembly shape.

Each point code, link, and shape has a style. A point code style defines the marker and/or color for the point on the subassembly. These are the Marker Styles in the Multipurpose

Styles branch. The link styles (lines between point codes) define the layer and color of the link in a subassembly. The link styles are the Link Styles in the Multipurpose Styles branch. The shape styles define the shape's fill color and its outline. These styles are the Shape Styles in the Multipurpose Styles branch.

SUBASSEMBLY POINTS (CODES)

Subassemblies contain a series of points that have specific functions. The first point on the subassembly (P1) is its connecting point, and it connects a subassembly to a more centrally located subassembly. If the subassembly is the innermost subassembly, this point attaches to the assembly point. A subassembly can connect to any point on an adjacent subassembly. The attachment of P1 to a point on the adjacent subassembly is a matter of selecting the "correct" point. Users may have to use the Zoom command to view the specific points on the adjacent subassembly that the point can connect to. The P1 point switches to the right side of the subassembly when working on the left side of the assembly. For example, the curb or a sidewalk subassembly will mirror itself to position the P1 point to the correct side.

Civil 3D defines unique point codes for each point of the subassembly. In the case of a pavement subassembly, the codes are shown in Figure 7.5. Civil 3D Help lists these codes under Subassembly Point Codes.

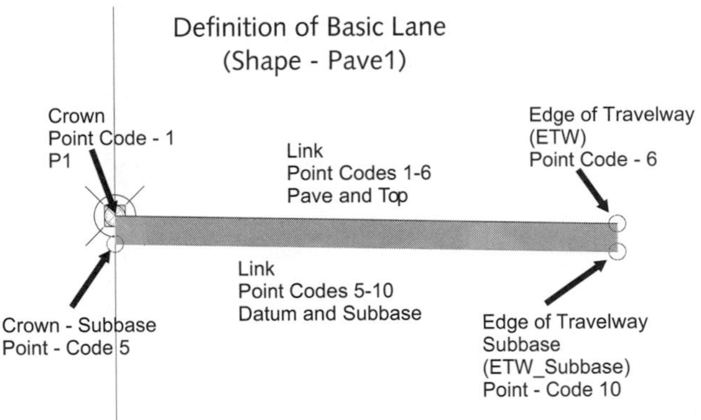

Figure 7.5

Point codes provide data for section labels (offset and elevation) and feature lines from a corridor model.

SUBASSEMBLY LINKS

Between each vertex (point code) on the subassembly is a link. This link creates part of the "shell" of a subassembly shape (see Figure 7.5). Civil 3D defines a link type for each segment of the shell. All subassemblies have a top-of-subassembly, a datum line, or a top or bottom of a shape link.

Links provide slope and distance data between the subassembly points. A label uses the link's values to determine a pavement cross slope, daylight slope, surface data, and so on.

SUBASSEMBLY SHAPES

The points (codes) and the links create a shape. In Figure 7.5, Pave1 is a shape. Its other name is "Basic Lane."

The data that a shape provides a label is the name of the shape or an area for a material volume calculation.

EDIT DRAWING SETTINGS

The assembly object has a base layer, C-ROAD-ASSM. If there is more than one assembly, it is best to add a suffix or prefix to the base layer name so each assembly has its own layer.

The base layers for a corridor object and its sections are C-ROAD-CORR and C-ROAD-SCTN-CORR. If there is more than one corridor, it is best to add a suffix or prefix to the base layer name so each corridor has its own layer.

CORRIDOR CONTROL

The Settings Corridor branch contains Command Settings that affect the creation of corridors and assign the styles it uses when creating the corridor. Most of the commands apply to situations after building the corridor and use the corridor as a source of data for new objects. For example, some of the commands create a feature line from the back of a sidewalk point of an assembly, create an alignment from point codes, or create coordinate geometry points from the point codes of the subassemblies.

CREATESIMPLECORRIDOR

The CreateSimpleCorridor command creates a corridor that has few if any regions (see Figure 7.6). A region is a portion of an alignment using a different assembly. Examples of these situations are intersecting roadways, cul-de-sacs, and knuckles.

The Assembly Insertion Defaults section defines the assembly (section) frequency along tangents, curves, and spirals. This section also defines whether the corridor includes critical geometry points from horizontal, superelevation, and profile elements or parameters.

The Default Styles section sets the styles for each element of a corridor. All of these styles are references to existing object styles for contour, alignments, code sets, slope patterns, and so on (see Figure 7.7).

The Default Name Format section defines the initial names for the corridor elements.

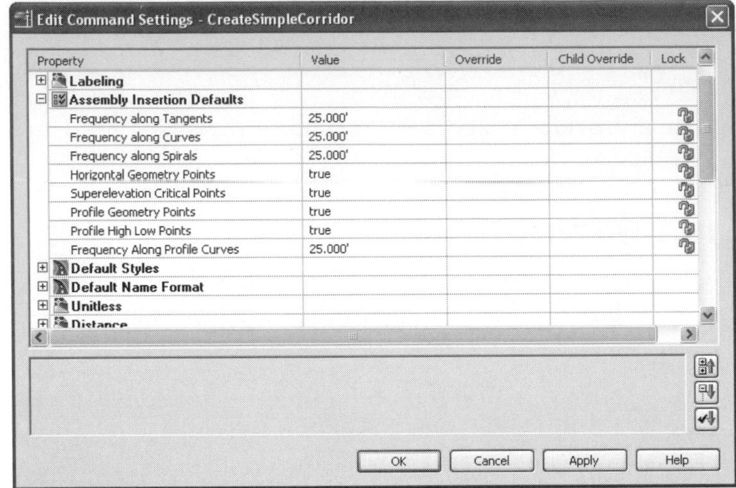

Figure 7.6

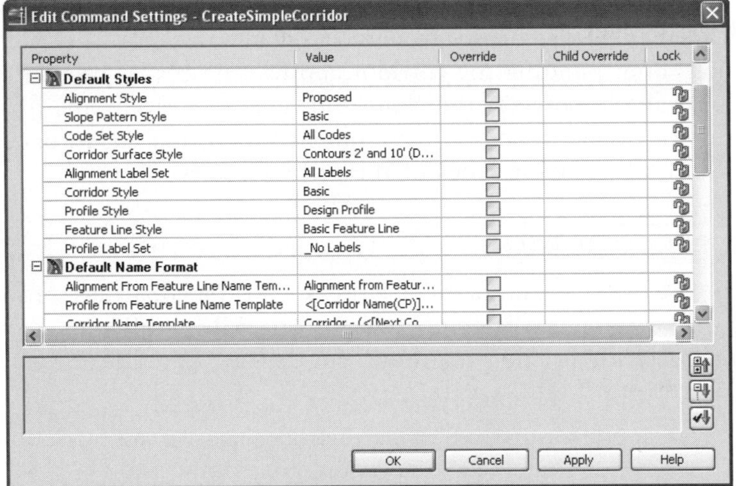

Figure 7.7

MARKER AND LINK LABEL SETS AND STYLES

The label set and styles for points and links apply more to annotation of sampled cross sections than the corridor, and they are the subject of a later unit of this chapter.

EXERCISE 7-1

When you complete this exercise you will be:

- Familiar with the Marker, Link, and Shape styles of Prospector's Settings.
- Familiar with the settings for the CreateSimpleCorridor command.

EXERCISE SETUP

This exercise starts with a drawing that contains a surface, alignment, and profile. You can use the drawing from the previous chapter or use the *Chapter 7 - Unit 1.dwg* found in the *Chapter 7* folder of the CD that accompanies this textbook.

1. If you are not in Civil 3D, double-click on the icon to start the application.
2. When you come to the command prompt, close the open drawing and do not save it.
3. Open the drawing from the previous chapter or open the *Chapter 7- Unit 1* drawing.

EDIT DRAWING SETTINGS

The base layers for this exercise are C-ROAD-ASSM (for the assembly), C-ROAD-CORR (for the corridor), and C-ROAD-SCTN-CORR (for the corridor sections). If there is more than one corridor or assembly in a drawing, it is best to add a suffix or prefix to the layer names so each object has its own layer.

1. Click the **Settings** tab to make it current.
2. In Settings, click the drawing name at the top, press the right mouse button, and from the shortcut menu select **Edit Drawing Settings....**
3. Click the **Object Layers** tab to view its contents.
4. In the Object Layers tab, change the Modifier for Assembly, Corridor, and Corridor Section to **Suffix**, and change the Value to **-*** (a dash followed by an asterisk).
5. Click the **OK** button to set the values and exit the dialog box.

POINTS STYLES

Points, links, and shapes have styles that define their color, layers, and markers. Rarely do you need to interact with these styles, but if you should create a custom subassembly, you may have to go to this area.

1. In Settings, expand the General and the Multipurpose Styles branch until you view the Marker Styles list. From the list select **Crown**, press the right mouse button, and from the shortcut menu select **Edit....**
2. In the Crown Marker Style, click the **Marker** tab to view it settings.

   The settings in this panel define what marker displays for a roadway crown. The Display tab defines what color, layer, and if the marker displays.

3. Click the **Display** tab to view its settings.
4. Click the **OK** button to exit the dialog box.
5. In Settings, collapse the Marker Styles branch.

LINK STYLES

A link style defines the color a link is in the drawing.

1. In Settings, General, Multipurpose Styles, expand the Link Styles branch until you view a list of styles. From the list select **Pave1**, press the right mouse button, and from the shortcut menu select **Edit....**

2. Click the **Display** tab to view its contents.

   The style assigns a layer and its properties to the Pave1 link.

3. Click the **OK** button to exit the dialog box.
4. In Settings, collapse the Link Styles branch.

SHAPE STYLES

A shape style defines what colors, outline, and fill pattern a shape will show in the drawing.

1. In Settings, Multipurpose Styles, expand the Shape Styles branch until you view a list of styles. From the list select **Pave1**, press the right mouse button, and from the shortcut menu select **Edit...**.
2. In the Shape Style dialog box, click the **Display** tab to view its contents.

   The style assigns a shape border, fill layer, color properties, and hatch pattern. You can change the fill pattern and its scale at the bottom of the dialog box.

3. Click the **OK** button to exit the dialog box.
4. In Settings, collapse the Multipurpose Styles branch.

COMMAND SETTINGS

The Edit Command Settings dialog box under CreateSimpleCorridor sets several default object styles and, most importantly, sets how Civil 3D creates a corridor.

1. In Settings, expand the Corridor branch until you view the Commands list.
2. From the list, select **CreateSimpleCorridor**, press the right mouse button, and from the shortcut menu select **Edit Command Settings...**.
3. Expand the Assembly Insertion Defaults section to view its contents.

   These values set the section interval for the corridor (every 25 feet), what critical geometry points to include (horizontal, vertical, and superelevation), and how often to sample a vertical curve.

4. Close the Assembly Insertion Defaults section and expand the Default Styles section.

   All of the styles in this section are existing object styles from other object types in Settings (Alignment, Multipurpose, Surface, and Profile Style groups).

5. Close the Default Styles section and expand the Default Name Format section.

   The Default Name Format section sets a counter for naming the corridor models and feature lines.

6. Click the **OK** button to close the dialog box.

THE IMPERIAL CATALOGS

The content for the Civil 3D Imperial and Metric palettes are generated from the Roadway Catalogs.

1. From the General menu, select **Catalog...**.

This displays the Imperial and Metric Catalog.

2. In the catalog, select the **Corridor Modeling Catalogs (Imperial)** icon to view its contents.
3. Select and review the contents of each catalog.
4. Close Corridor Catalogs.
5. Click the **AutoCAD Save** icon to save the drawing.

THE CIVIL 3D - IMPERIAL PALETTE

Civil 3D has a multi-tabbed palette containing subassemblies that address several road design issues. The different tabs represent basic categories of subassemblies from the Civil 3D Roadway Catalogs.

1. From the General menu select **Tool Palettes Window**.
2. Click each tab of the Civil 3D - Imperial palette to view each collection of subassemblies.

   To use one of the subassemblies you need to have an assembly present in the drawing.

This completes this exercise reviewing the settings, styles, catalogs, and palettes involved with a corridor. The next unit creates an assembly that uses several subassemblies from the Civil 3D - Imperial palette.

SUMMARY

- Point codes are critical subassembly vertices.
- Subassembly links and shapes use point codes as their endpoints or vertices defining their shape.
- Point code labels create station, offset, and elevation annotation.
- Links define the edge of a shape, provide data for slope/grade labeling, and surface data.
- Shapes provide the name and an area for material volumes.
- The CreateSimpleCorridor settings affect how a corridor model is made and the styles it uses.

UNIT 2: ASSEMBLIES AND SUBASSEMBLIES

The assembly is the anchor for a roadway cross section to create a road model (corridor). The assembly is the center of the section and all subassemblies attach to and move outward from it. Subassemblies represent discreet elements of a cross section (for example, pavement, curbs, and shoulders). As mentioned, subassemblies are a collection of points and links. Each point and link allows Civil 3D to create labels, surfaces, grading data, points, and so on for each assembly.

The horizontal and vertical alignments pass through the assembly's eyelet. As these alignments move, they pull the eyelet, moving the assembly to the right, left, up, or down. The corridor uses the subassembly shapes and location of the assembly to create a corridor model of the roadway.

Each subassembly has a set of markers around its shape. Each marker has a point code and a link that connects it to the next point on the shape. Each link also has a name (e.g., top, datum, etc.).

To create or customize subassemblies, users must be familiar with VBA scripting. The Help file documents what steps are necessary to create a custom subassembly.

SUBASSEMBLY PROPERTIES

A subassembly attaches to the right and left of the assembly. After attaching a subassembly to an assembly, the subassembly should be relabeled to a more meaningful name. The reason for this is that Civil 3D only assigns a subassembly name and number to each used subassembly. When an assembly becomes complicated, the time taken to name subassemblies pays dividends when reviewing an assembly or when assigning alignments and profiles in the Create Corridor dialog box (see Figure 7.8).

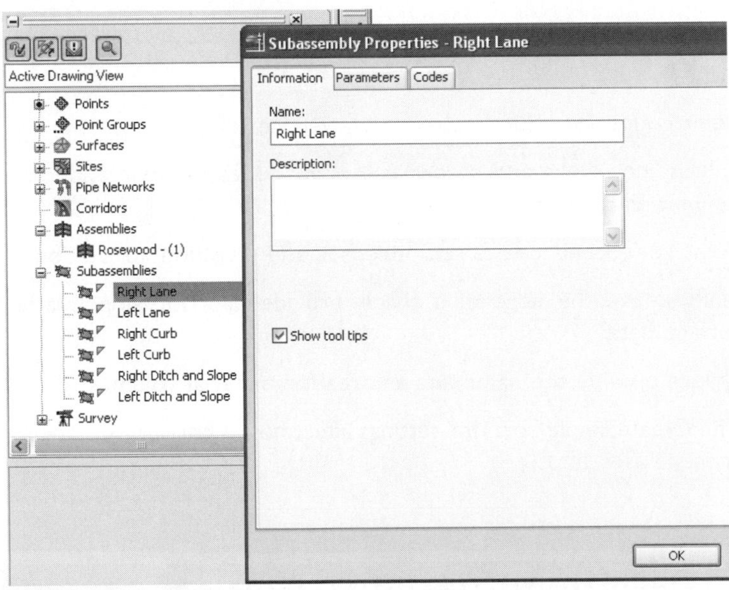

Figure 7.8

Each subassembly's Parameters panel contains the subassembly values and allows users to change them (see Figure 7.9). The Value Name and the Default Input Value depend on the selected subassembly.

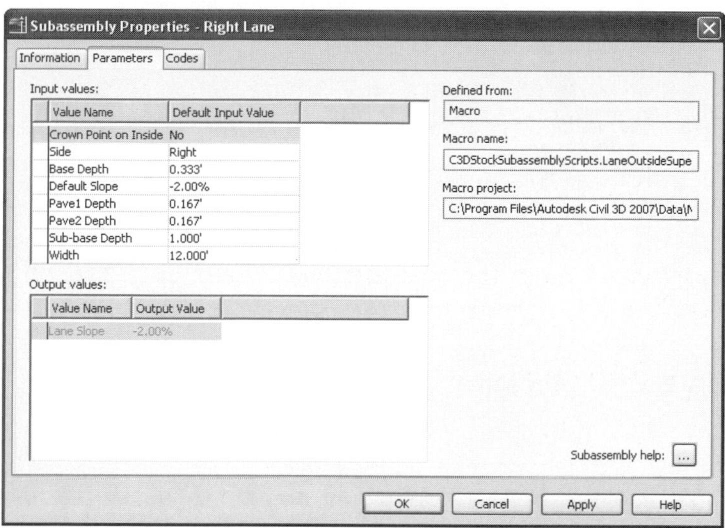

Figure 7.9

ASSEMBLY PROPERTIES

The properties of an assembly show its construction and dependencies. These relationships are especially important in a complex assembly. This is where naming subassemblies pays off. Users also should rename portions of the assembly to more meaningful names (see Figure 7.10).

Users can adjust the parameters of each subassembly by selecting the subassembly in the left panel and editing its displayed values on the right side of the dialog box.

The assembly and its subassemblies appear in tree form. The main branches are at the right and left, and below each heading are the attached subassemblies. When selecting a subassembly from the list, a user can view its values on the right side.

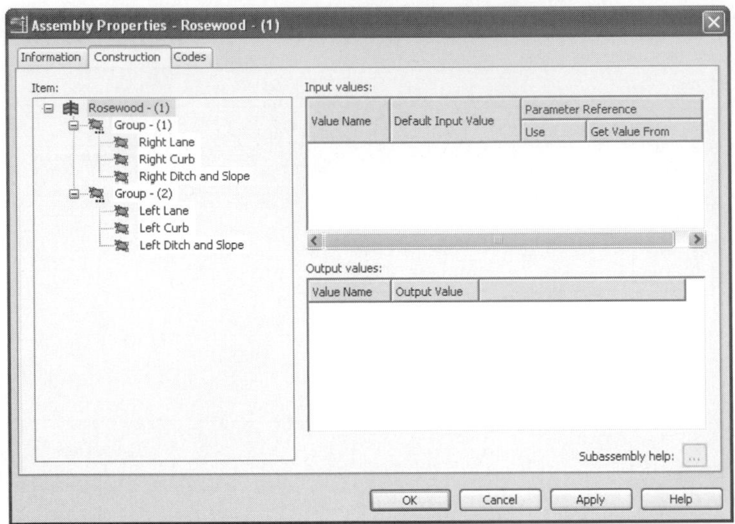

Figure 7.10

EXERCISE 7–2

After you have completed this exercise, you will be:

- Able to create an assembly.
- Familiar with various subassemblies.
- Able to select and attach to an assembly and a subassembly.
- Able to edit the properties of a subassembly.

EXERCISE SETUP

This exercise continues with the drawing from the previous exercise. If you did not complete the previous exercise, you can open the *Chapter 7 - Unit 2.dwg*. The file is in the *Chapter 7* folder of the CD that accompanies this textbook.

1. Open the drawing from the previous exercise, or the *Chapter 7 - Unit 2* drawing.

CREATE THE ASSEMBLY

Creating an assembly requires a selection from the Corridors menu, naming it, and placing it in the drawing. After placing the assembly in the drawing, the next step is to add the appropriate subassemblies to make the road section.

1. From the Corridors menu, select **Create Assembly....**
2. In the Create Assembly dialog box, replace "Assembly" with **Rosewood** as the name of the assembly, leave the Assembly Style as **Basic** and the Code Set Style to **All Codes**, and click the **OK** button to close the dialog box (see Figure 7.11).
3. In the command line, the routine prompts for the location of the assembly; select a point just to the left of the lower-left corner of the Rosewood Profile View.

4. If necessary, use the AutoCAD Zoom command to better view the assembly.
5. The assembly is a vertical line with a connection symbol at its midpoint.

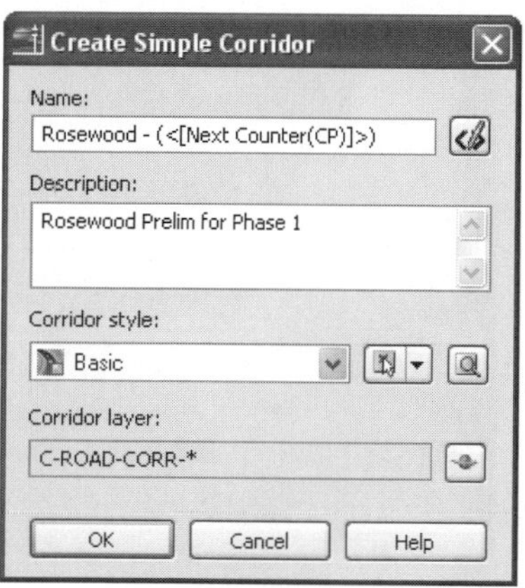

Figure 7.11

ADD SUBASSEMBLIES

The subassemblies create the roadway section. The Civil 3D - Imperial palette contains the subassemblies necessary to make the roadway section.

Travel Lanes

The travel lanes for this assembly are 12 feet wide, have four materials of various thickness, and have a cross slope of –2 percent. All of these parameters are properties you adjust before attaching the subassembly to the assembly, or to an adjacent subassembly (see Figure 7.12).

Most subassemblies have a right and left property. When attaching a subassembly to an assembly or an inner subassembly, you need to make sure you have set the correct side parameter.

1. From the General menu select the Civil 3D **Tool Palettes** window.
2. Click the **Imperial - Roadway** tab to display its subassemblies.
3. From the Imperial - Roadway palette, select the **LaneOutsideSuper** subassembly; the selection displays the Properties palette (see Figure 7.12).
4. In Properties, set the Side to **Right**, change the Pave1 and Pave2 depths to **0.167**, and in the drawing select the assembly line to attach the subassembly.

   The right lane subassembly attaches to the assembly.
5. In the Properties palette, change the Side to **Left**, and in the drawing select the assembly to attach the subassembly to the left side.
6. Press the **Enter** key twice to end the routine.
7. If necessary, select the **Prospector** tab to make it current.

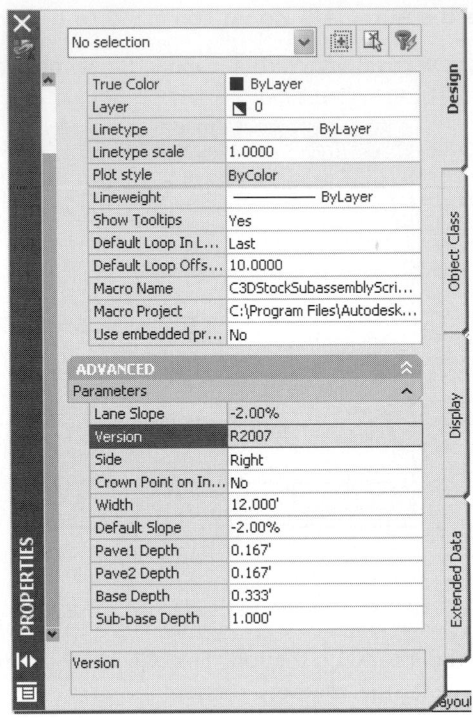

Figure 7.12

8. In Prospector, expand the Subassemblies branch, select the first subassembly from the list, press the right mouse button, and from the shortcut menu select **Properties...**.
9. In the Subassembly Properties dialog box, click the **Information** tab, and change the Name to **Right Lane** (see Figure 7.8).
10. Click the **Parameters** tab to view its contents.

    It is very helpful to take a little extra time to identify the subassemblies with descriptive names. When building complex assemblies, the names easily identify the correct subassembly.

11. Click the **OK** button to change the name of the subassembly.
12. Click the left lane, press the right mouse button, and from the shortcut menu select **Properties...**.
13. In the Subassembly Properties dialog box, click the **Information** tab, and change the Name to **Left Lane**.
14. Click the **Parameters** tab to view its contents.
15. Click the **OK** button to change the name of the subassembly.

Curbs

1. On the tool palette select **Imperial - Structures** tab and from the palette and click **UrbanCurbGutterGeneral**.
2. In the displayed Properties palette, review the Side property for this curb, and change the Side to **Left**. In the drawing, pan to the left side of the assembly, and attach the subassembly to the left side of the travelway, at the upper outside left attachment point (see Figure 7.13).

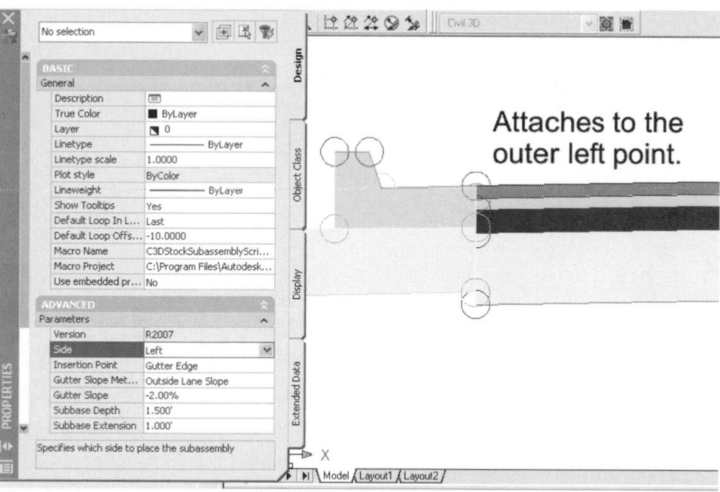

Figure 7.13

The Curb subassembly attaches to the red ringlet at the top of pavement on the left side.

3. In Properties, change Side to **Right**. In the drawing, pan to the right side of the assembly and attach the subassembly to the red ringlet at the top outside travelway.
4. Press the **Enter** key twice to end the routine.
5. Click the right curb, press the right mouse button, and from the shortcut menu select **Subassembly Properties....**
6. In the Subassembly Properties dialog box, click the **Information** tab, and change the Name to **Right Curb**.
7. Click the **Parameters** tab to view its contents.
8. Click the **OK** button to change the name of the subassembly.
9. Click the left curb, press the right mouse button, and from the shortcut menu select **Subassembly Properties....**
10. Click the **Parameters** tab to view its contents.

11. In the subassembly Properties dialog box, click the **Information** tab, and change the Name to **Left Curb**.
12. Click the **OK** button to change the name of the subassembly.

    The Subassemblies should now appear in the tree and your subassembly should look like Figure 7.14.

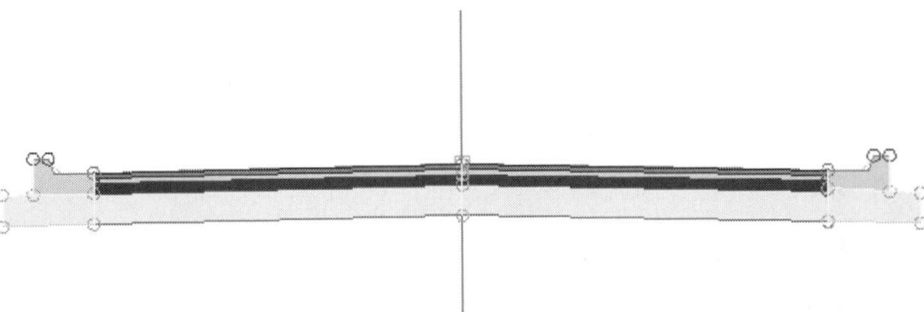

Figure 7.14

Ditch and Daylight Slopes

The last subassembly, BasicSideSlopeCutDitch, has Ditch and Daylight Slope parameters. When attaching it to the outside back edge of the curb, it shows as a sideways "V." When processed as part of a corridor, it creates the expected sections.

1. In the Civil 3D - Imperial tool palette, click the **Imperial - Basic** tab, and select **BasicSideSlopeCutDitch** from the palette.
2. In the Properties palette, review its settings, change the side to **Right**, and in the drawing attach the subassembly to the right outside top back of curb.
3. In the Properties palette, change the side to **Left**, and in the drawing attach the subassembly to the left side of the top back of curve.
4. Press the **Enter** key twice to exit the command.

   Your subassembly should look like Figure 7.15.

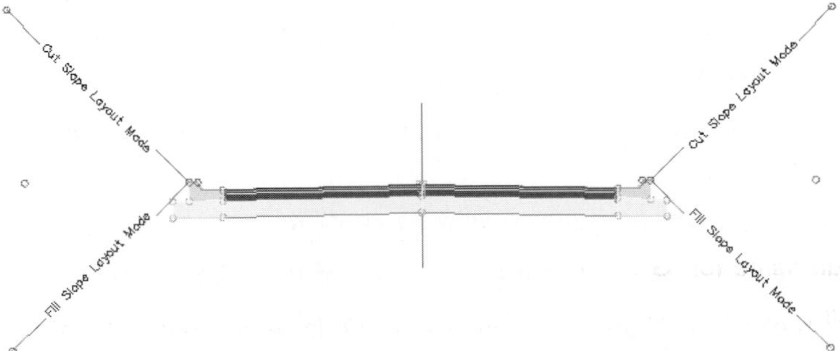

Figure 7.15

5. Click the right ditch and slope subassembly, press the right mouse button, and from the shortcut menu select **Subassembly Properties...**.
6. Click the **Information** tab and change the name to **Right Ditch and Slope**.
7. Click the **Parameters** tab to view its contents.

   You can adjust any of the parameters in this panel.
8. Click the **OK** button to change the name of the subassembly.
9. Click the left ditch and slope subassembly, press the right mouse button, and from the shortcut menu select **Subassembly Properties...**.
10. Click the **Information** tab and change the name to **Left Ditch and Slope**.
11. Click the **Parameters** tab to view its contents.

    You can adjust any of the parameters in this panel.
12. Click the **OK** button to change the name of the subassembly.
13. Click the "**X**" to close the Civil 3D Subassembly Tool Palette.
14. Click the **AutoCAD Save** icon to save the drawing.

EXERCISES

SUBASSEMBLY PROPERTIES

The vertical depth of the curb's flange and back-of-curb depth do not match the depth of the travelway. You can adjust a subassembly's parameters in the Properties palette.

1. If necessary, select the **Prospector** tab to make it current.
2. Expand the Subassemblies branch until you view the subassembly list.
3. From the list, select **Right Curb**, press the right mouse button, and from the shortcut menu select **Properties...**.
4. In the Subassembly Properties dialog box, select the **Parameters** tab, and in the lower right select the ellipsis for Subassembly Help.

   In Help, the A and G links represent the flange and back of curb depth. These two values need to change to match the new depth of the pavement.
5. Close Help and return to the Subassembly Properties dialog box.
6. In the Input Values list, locate the value for Dimension A; change the default input value of A to **8.0**.
7. Scroll down the Input Values list, locate the value for Dimension G; change the default value for G to **14.0**.
8. Scroll down the Input Values list, locate the value for Subbase Depth; change the default value for Subbase Depth to **1.666** (1.67).
9. Click the **OK** button to exit the dialog box.
10. Repeat Steps 3 through 9 and update the values for the Left Curb (see Figure 7.16).

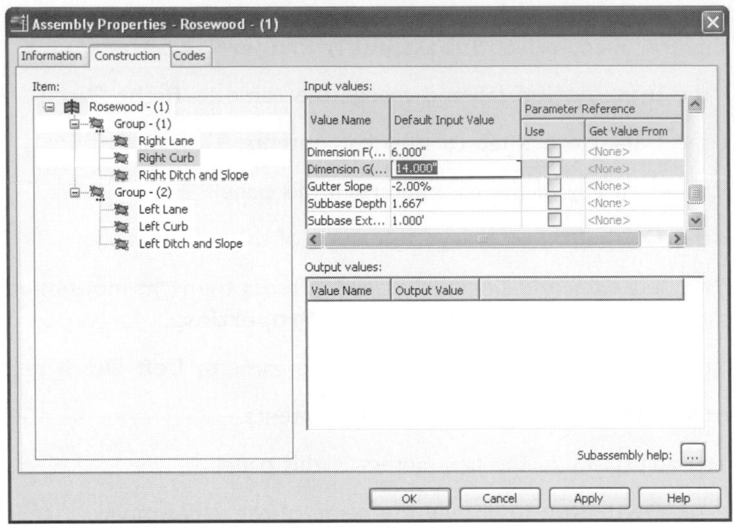

Figure 7.16

11. Click the **AutoCAD Save** icon to save the drawing.

ASSEMBLY PROPERTIES

The Construction tab of the Assembly Properties dialog box lists the assembly's components. When selecting a subassembly on the left, its parameters appear on the right.

1. Expand the Assemblies tree in the Prospector panel.
2. From the list, click **Rosewood - (1)**, press the right mouse button, and from the shortcut menu select **Properties....**
3. In the Properties dialog box, click the **Construction** tab to view its contents.
4. Click **Group - (1)**, press the right mouse button, and select **Rename** from the shortcut menu. Rename the group **Right Side**, and click the **OK** button to exit the dialog box.
5. Click **Group - (2)**, press the right mouse button, and from the shortcut menu select **Rename**. Rename the group **Left Side**, and click the **OK** button to exit the dialog box (see Figure 7.17).
6. Click the **AutoCAD Save** icon to save the drawing.

This ends the review of defining a simple assembly. Next is creating a corridor from the horizontal and vertical alignment data, and the assembly.

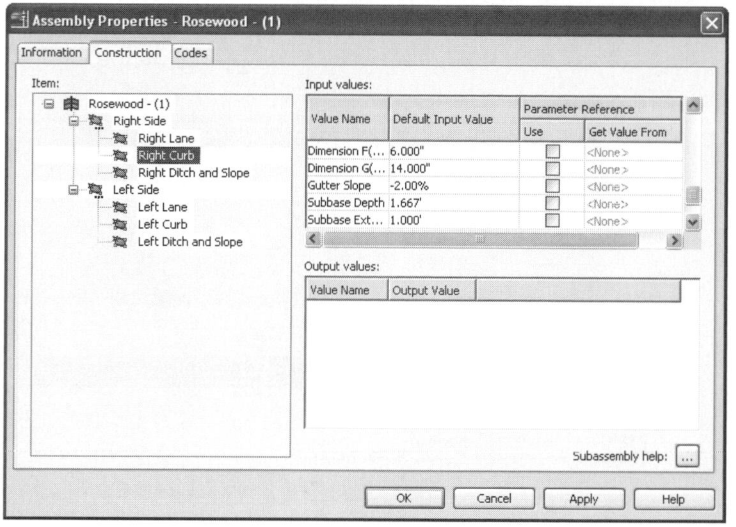

Figure 7.17

SUMMARY

- The first step in making a road section is creating the assembly object.
- After creating the assembly object, attach subassemblies from the assembly in an outward direction.
- Each subassembly has a Right or Left property. Set this before attaching the subassembly to an assembly or a more central subassembly.
- Each subassembly has design parameters. Set them before attaching the subassembly to an assembly or a more central subassembly.
- Each subassembly has a Properties dialog box. Users can adjust all of the parameters in this dialog box after attaching the subassembly to an assembly.
- Each assembly has a Properties dialog box. Users can adjust all of the parameters of all subassemblies in this dialog box.

UNIT 3: CREATING A SIMPLE CORRIDOR

The final step in making a corridor (roadway model) is using the CreateSimpleCorridor command from the Corridors menu. A corridor combines all of the settings and parameters of each roadway element and produces a model. When using the Simple Corridor command, a Target Mapping dialog box displays, prompting the user to review the corridor elements (alignment(s), profile(s), and assemblies). With most simple corridors, the only necessary assignment is the name of the daylight surface (see Figure 7.18). The assignments are made in the Logical Names Mapping dialog box.

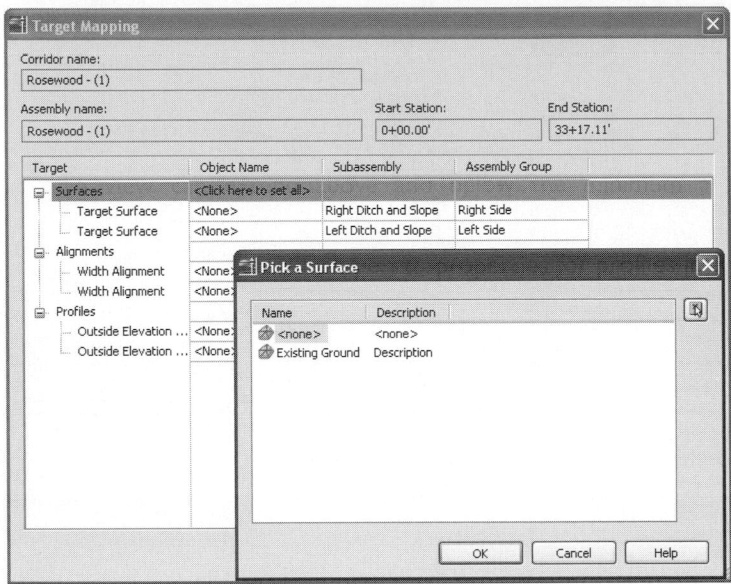

Figure 7.18

When viewing the model with Object Viewer or 3D Orbit, Civil 3D displays the corridor sections as assemblies with strings connecting each subassembly point to the next or previous section (see Figure 7.19).

It is evident from Figure 7.19 that the points (crown, gutter, etc.) act as eyelets (the point codes of the subassembly) through which threads are strung (feature lines). These threads are in fact data for many new Civil 3D objects, including feature lines, design annotation, points, and surfaces.

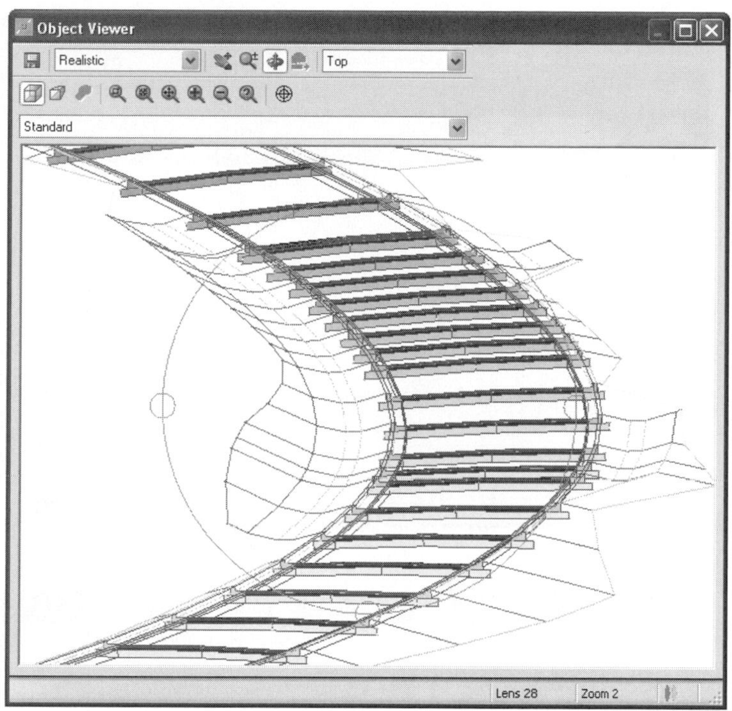

Figure 7.19

CORRIDOR PROPERTIES

Each corridor has properties, as well as a Properties dialog box displaying its values.

PARAMETERS

The Parameters tab contains the regions of a corridor (see Figure 7.20). There may be times when a user wants to use different assemblies at various locations along a corridor. This is where the user defines roadway design regions to accommodate those decisions. Regions represent alignment transitions, mergers, or portions of a cul-de-sac. Users assign alignments, profiles, and surfaces to a region after creating a new region.

CODES

The Codes panel lists all of the points, links, and shapes in the corridor. The entries reflect the codes found in the subassemblies of the assembly (see Figure 7.21).

FEATURE LINES

Feature lines are strings or threads passing through the point codes on the template. Their names are the same as the points they pass through (see Figure 7.22). At the bottom of the panel, two settings affect complex sections by using a point code multiple times. The Branching Inward and Outward settings affect the merging of these lines. Merging inward means the feature lines, if not present in the next section, should merge to the next interior

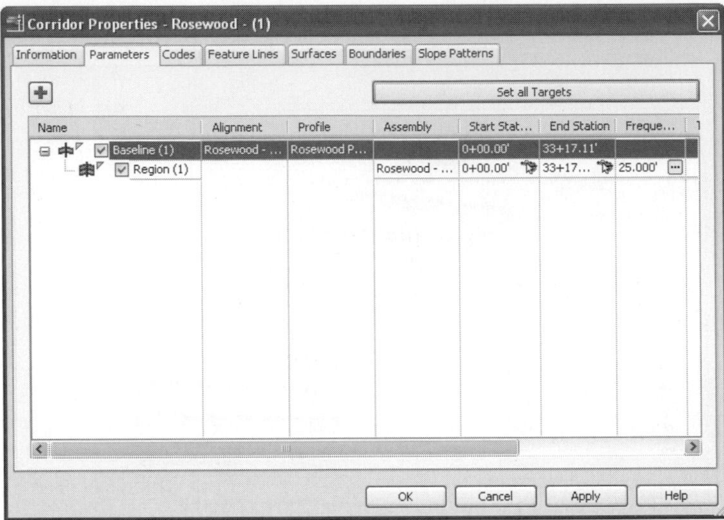

Figure 7.20

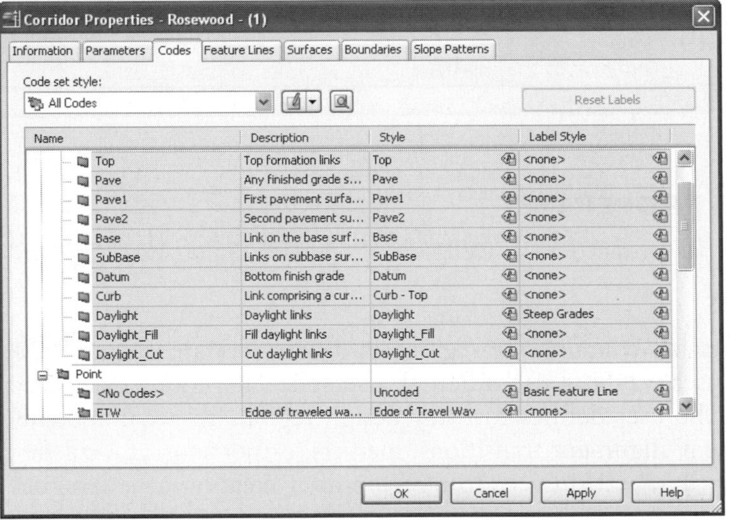

Figure 7.21

point of the same description. The definition of outward is the opposite of inward: If more points are present in the next section, they connect to the outside points of the same name. The second condition, Connect extra points, is a toggle. This tells the feature lines to join between sections when there are varying numbers of the same point code between two sequential sections.

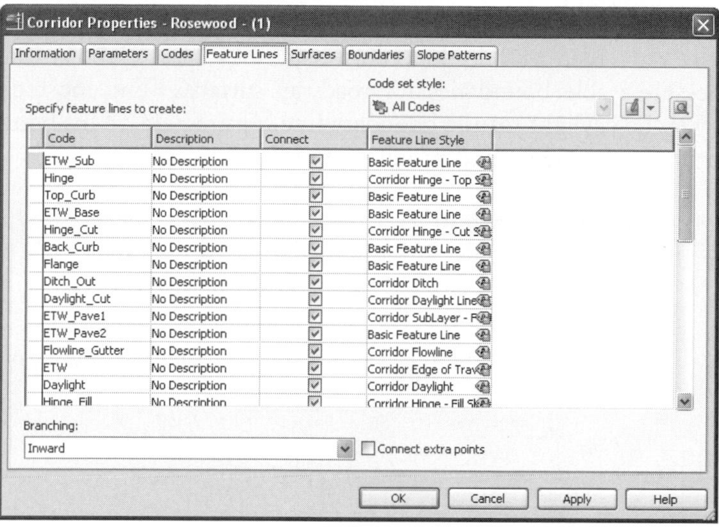

Figure 7.22

SURFACES

A corridor model contains data representing a corridor surface (see Figure 7.23). The surface data can come from any of the section links or feature lines. For example, the corridor top link represents the top surface of a road. A surface representing the top of the road is to use the top link as its data source. This surface would represent the final road product. Another important corridor surface is the datum surface, which represents the limit of cut and fill. When the datum and the existing ground surfaces are compared, the mass earthworks quantities for a road design are given.

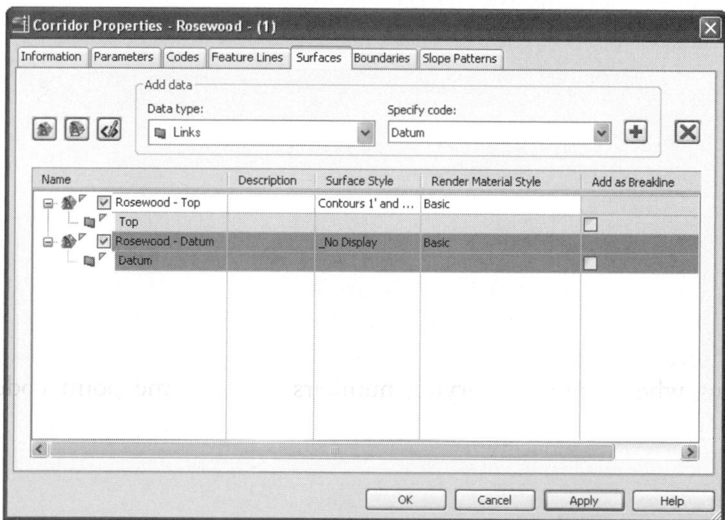

Figure 7.23

BOUNDARIES

As with all surface data, there is a need to control spurious triangulation around its periphery. The feature lines are viable boundaries for roadway surfaces. The most common roadway surface boundary is the daylight feature boundary (see Figure 7.24). This boundary focuses the triangulation within its boundary and produces a clean corridor surface.

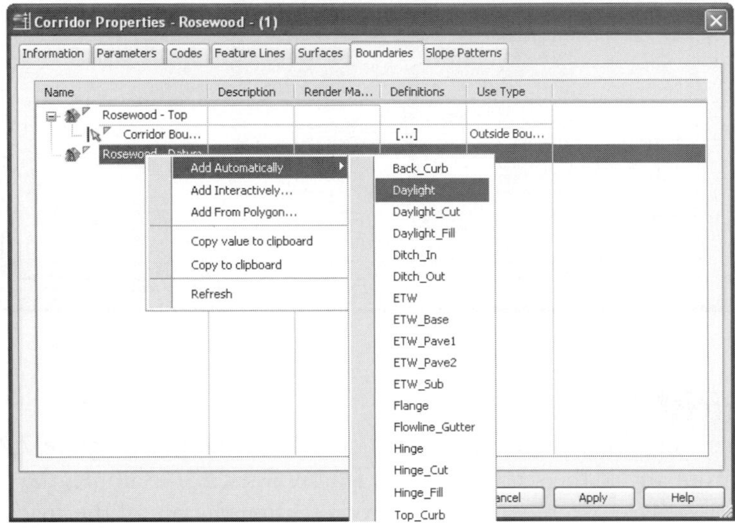

Figure 7.24

SLOPE PATTERNS

The Slope Patterns panel indicates the type of slope along a corridor's path or region (see Figure 7.25). The Multipurpose Styles branch of Settings defines the patterns.

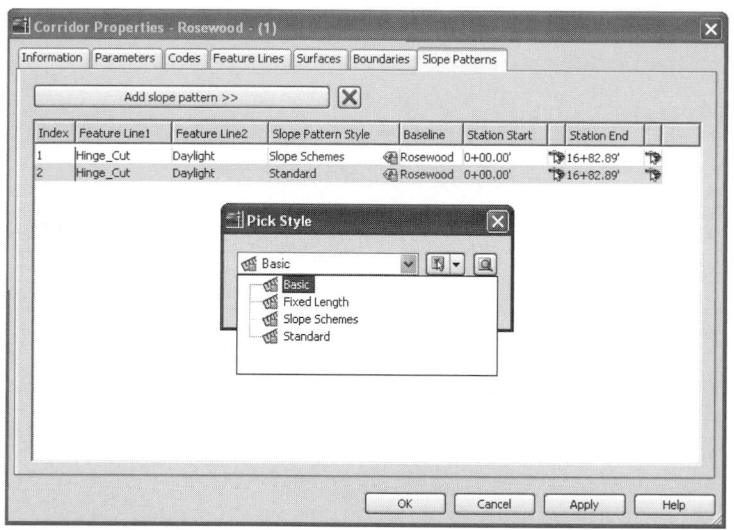

Figure 7.25

EXERCISE 7-3

After completing this exercise, you will be:

- Able to create a simple corridor.
- Familiar with corridor properties.

EXERCISE SETUP

This exercise continues with the drawing from the last exercise. If you did not complete the previous exercise, you can open the *Chapter 7 - Unit 3.dwg* found in the *Chapter 7* folder of the CD that accompanies this textbook.

1. Open the drawing file from the last exercise or open the *Chapter 7 - Unit 3* drawing.

CREATE SIMPLE CORRIDOR

1. From the Corridors menu, select **Create Simple Corridor....**
2. In the Create Simple Corridor dialog box, change "Corridor" to **Rosewood**. Leave in the counter, give the corridor a short description, and click the **OK** button to begin identifying the corridor components.

   The command line prompts for an alignment. You can select the alignment from the drawing or you can select it from the list by pressing the right mouse button.
3. Press the **Enter** key. From the dialog box, select the **Rosewood - (1)** alignment, and click the **OK** button.

   The command line prompts for a profile. You can select the profile from the drawing or you can select it from the list by pressing the right mouse button.

4. Press the **Enter** key. From the dialog box, select the **Rosewood Preliminary - (1)** profile and click the **OK** button.

   The command line prompts for an assembly. You can select the assembly from the drawing or you can select it from the list by pressing the right mouse button.

5. In the drawing, select the **Rosewood - (1)** assembly.

6. The Target Mapping dialog box will display. Click the Object Name cell with the Value **<Click here to set all>**. This will display the Pick a Surface dialog box, where **Existing Ground** should be listed. Select the surface, and click the **OK** button to specify the surface.

7. Click the **OK** button to begin building the corridor.

   The program builds the corridor and displays it on the screen as a mesh.

8. Click the **AutoCAD Save** icon to save the drawing.

EXERCISES

VIEWING THE CORRIDOR

The corridor is a roadway 3D model and is viewable in the Object Viewer or with 3D Orbit.

1. Use the AutoCAD Zoom and Pan commands to view the new corridor in the site.

2. In the drawing, select any segment representing the corridor, press the right mouse button, and from the shortcut menu select **Object Viewer...**.

3. In Object Viewer, begin viewing by clicking and holding the left mouse button down in the ring at 6 o'clock, and slowly moving the cursor toward the center of the circle. This tilts the roadway toward your point of view.

4. Just before viewing the surface edge-on, release the left mouse button.

5. Repeat the same process, but click in the ring at 3 o'clock. This rotates the roadway east and west.

6. After viewing the corridor, select the "**X**" in the upper right to close the Object Viewer.

CORRIDOR PROPERTIES

Corridor properties range from parameters affecting the alignments, profiles, surfaces, boundaries, slope patterns, and feature lines.

1. In the drawing, select any corridor segment, press the right mouse button, and from the shortcut menu select **Corridor Properties...**.

2. In the Corridor Properties dialog box, click the **Parameters** tab to view its contents.

   The Parameters panel lists the alignment, profile, and assembly assignments for the corridor. You change the name of any of these elements by clicking the named element and selecting a new alignment, profile, etc. from a listing dialog box.

3. Click the **Set All Targets** button at the top right of the panel.

This panel reports the target surface (existing ground) for the Right and Left Ditch and Slope subassembly (daylight surface). This assignment was originally set when creating the simple corridor.

4. Click the **OK** button to exit the Target Mapping dialog box and to return to the Corridor Properties dialog box.
5. Select the **Codes** tab to view its contents.

   The Codes panel lists all of the links, points, and shapes that are a part of the corridor.

6. Click the **Feature Lines** tab to view its contents.

   Feature lines are threads passing through the point codes of an assembly. Each code has a name and a specific function. A feature line is a 3D object representing the grades along the path of the road. You can import feature lines and use them to design a surface or use as grading objects.

7. Click the **Surfaces** tab to view its contents.

   This panel allows you to create surfaces from feature lines and boundary data.

8. Click the **Boundaries** tab to view its contents.

   As mentioned, a surface may need a boundary to control spurious triangulation. This is where you select the boundary for a corridor surface.

9. Click the **Slope Patterns** tab to view its contents.

   If you want to include a slope pattern in the corridor object, you define it here. You identify where in the corridor the pattern appears and what pattern to use.

10. Click the **Cancel** button to exit the Corridor Properties dialog box.

FEATURE LINES

The All Codes style assigns each feature line a style.

1. From the View menu use the Named Views... command and restore the view to **Proposed Starting Point**.
2. Use the AutoCAD Zoom and Pan commands to better view corridor in cut (see Figure 7.26).
3. In the drawing, select any segment of the corridor, press the right mouse button, and from the shortcut menu select **Corridor Properties...**.
4. In the Corridor Properties dialog box, click the **Feature Lines** tab to view its contents.

   This panel lists all of the feature lines currently in the corridor. All of them have a name, are visible, and have styles.

5. Click the **OK** button to exit the Corridor Properties dialog box.

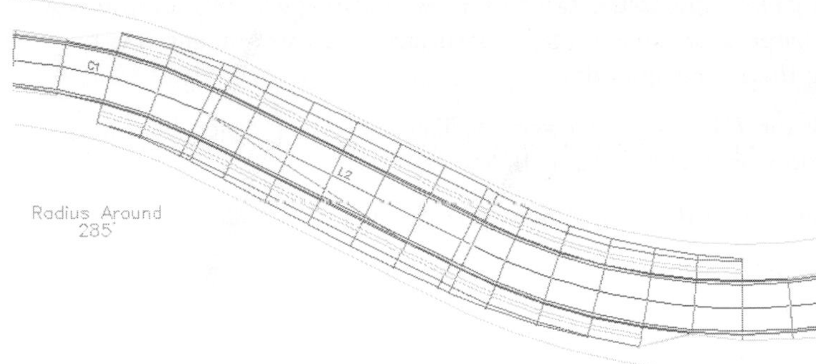

Figure 7.26

6. Click the **AutoCAD Save** icon to save the drawing.

This ends the exercise on creating and viewing the properties of a corridor. Once you have a corridor, you can create new objects from its data. The next unit covers the review and editing of a corridor's values.

SUMMARY

- Creating a simple corridor is a one-step process.
- The alignments, profiles, assemblies, and surfaces all contain data and constraints, and the Create Corridor command blends these elements together to create a corridor.
- A corridor has an extensive list of properties.
- The Corridor Properties dialog box also creates corridor surfaces, boundaries for surfaces, and can add slopes markings to areas within the corridor.

UNIT 4: CORRIDOR REVIEW AND EDIT

The Corridor Properties dialog box displays values describing the overall character of the roadway model. However, reviewing and possibly editing some of the individual sections of a corridor may be necessary. This is done in the View/Edit Corridor command of the Corridor menu.

The View/Edit Corridor routine displays a toolbar that controls viewed stations, sets section overrides, extends the section edit to a range of stations, and contains a roll-down menu that allows users to edit all of the section parameter values (see Figure 7.27). The toolbar displays the station nearest the selection point or the first corridor section. The middle of the toolbar identifies the current section and has controls affecting which section displays. The left side of the toolbar identifies the name of the corridor, and the center identifies the current section on display. The controls on the right control how extensive the changes are for the corridor: this station only, a range of stations, or the entire corridor. Other tools, located to the right of the station control, allow users to add or delete points, links, or shapes to the section.

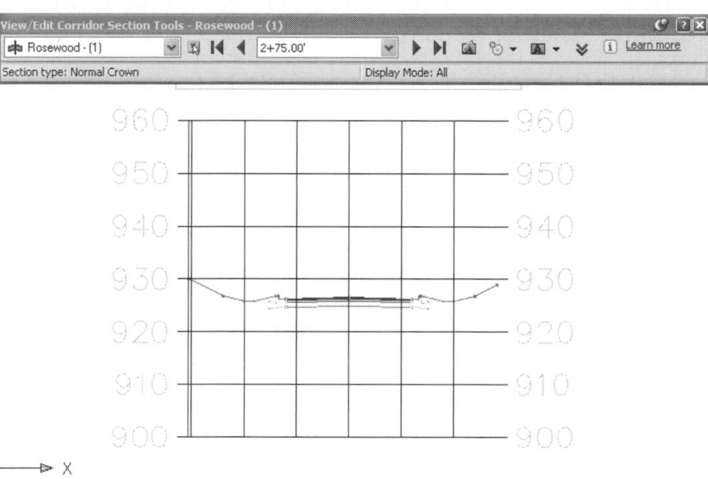

Figure 7.27

The View/Edit command is not interactive, but when making a change to a parameter, the section reacts to the changes. Users can apply changes to the current section or extend it over a range of stations.

EXERCISE 7-4

After you complete this exercise, you will be:

- Able to use the View/Edit Section Editor.
- Familiar with editing a section.

EXERCISE SETUP

This exercise continues with the drawing used in the previous exercise. If you did not complete the previous exercise you can open *Chapter 7 - Unit 4.dwg*. The file is in the *Chapter 7* folder of the CD that accompanies this textbook.

1. Open the drawing from the previous exercise or the *Chapter 7 - Unit 4* drawing.

VIEWING CORRIDOR SECTIONS

Viewing sections is a way of reviewing the road design and its solution based on the conditions and parameters of the design elements.

1. From the Corridors menu, select **View/Edit Corridor Section**.
2. In the drawing, select any corridor segment.

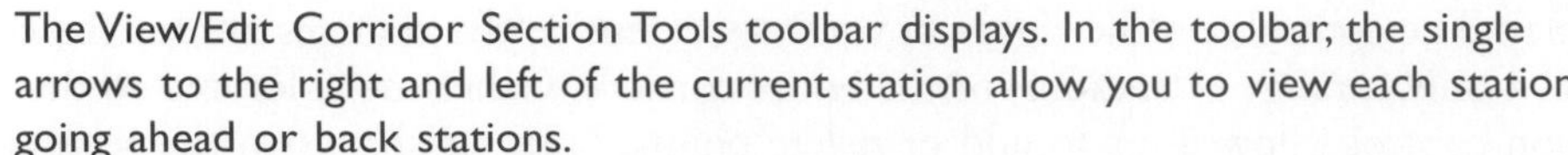

The View/Edit Corridor Section Tools toolbar displays. In the toolbar, the single arrows to the right and left of the current station allow you to view each station going ahead or back stations.

3. On the View/Edit Corridor Section Tools toolbar, click the **right** single arrow a few times to move to higher stations.
4. On the View/Edit Corridor Section Tools toolbar, click the **left** single arrow a few times to move lower station values.
5. On the View/Edit Corridor Section Tools toolbar, click the barred-arrow icon to the right of the current section to view the last station of the corridor.
6. On the View/Edit Corridor Section Tools toolbar, click the barred-arrow icon to the left of the current section to view the first station of the corridor.

EDITING SECTIONS

There are times when it is necessary to edit parameters of a section to successfully meet design constraints. Each subassembly of the section displays all of its parameters in the roll-down panel of the toolbar. You can change any value in this panel and apply the change to the current section or a range of sections.

1. In the View/Edit Corridor Section Tools toolbar, click the station list drop-down arrow and select station **9+25**.
2. With station 9+25 as the current station, at the right of the View/Edit Corridor Section Tools toolbar, click the roll-down arrow to view the parameters for the current section.
3. In the Parameter panel for the section, locate the Default Slope entry in the Right Lane section. Change the grade by clicking in Value cell and typing **6**.

   This changes the right lane slope to a 6 percent upslope. The entry in the panel places a check in the override box and the section view responds by raising the right pavement segment.
4. In the View/Edit Corridor Section Tools toolbar, click the roll-up arrow on the right side of the toolbar to close the panel.
5. In the View/Edit Corridor Section Tools toolbar, click the drop-list arrow to the left of the last icon on the right and select **Apply to a station range**.
6. In the Apply to a Range of Stations dialog box, set the range from **9+25** to **13+50** and click the **OK** button to apply the change to the station range.
7. In the View/Edit Corridor Section Tools toolbar, click the single arrow to the right of the current station and view the changed value until you get to station 14+00.
8. To remove the change and restore the value back to the original value, click the station drop-list arrow in the View/Edit Corridor Section Tools toolbar. Select station

9+25 from the list, and click the arrow to unroll the parameters panel. In the Right Lane Default Slope entry, uncheck the override (the value returns to –2%), and roll up the panel.

9. In the View/Edit Corridor Section Tools toolbar, click the **Apply to a station range** icon, apply the change to the full station range, and click the **OK** button to exit the Apply to a Range of Stations dialog box.
10. Verify the change by viewing stations ahead of the current station.
11. Close the View/Edit Corridor Section Tools toolbar.

This is ends the reviewing and editing of corridor values. The View/Edit Section editor tweaks values for any parameter of a corridor section. A section reacts to the change, and changes can apply to the current station or a station range.

SUMMARY

- The View/Edit Corridor Section editor displays each section with a panel listing all of the current section's parameters.
- You can change any parameter in the Data panel for a section.
- When changing a value in the Data panel, the section responds to show the change on the screen and in the corridor.
- You can limit a change to the current section or apply it to a range of stations.

UNIT 5: OBJECTS FROM A CORRIDOR

Feature lines appearing between corridor sections are integral parts of the corridor's definition. These lines can become additional alignments, profiles, polylines, grading feature lines, and surface data.

To create an object from a corridor feature line, use a command from the Utilities flyout menu under Corridors. Each command produces a different object type (see Figure 7.28).

Each feature line of a corridor displays a tooltip identifying its name. The tooltip is an effective way to determine which feature line to export (see Figure 7.29).

If there is more than one feature line where the user selects the corridor, the routine displays a Select a Feature Line dialog box. Select the desired feature line from the list (see Figure 7.30).

CORRIDOR FEATURE LINE AS POLYLINE

This routine exports a 3D polyline whose elevations are the corridor elevations of the exported feature line. The routine places the resulting polyline on the current layer.

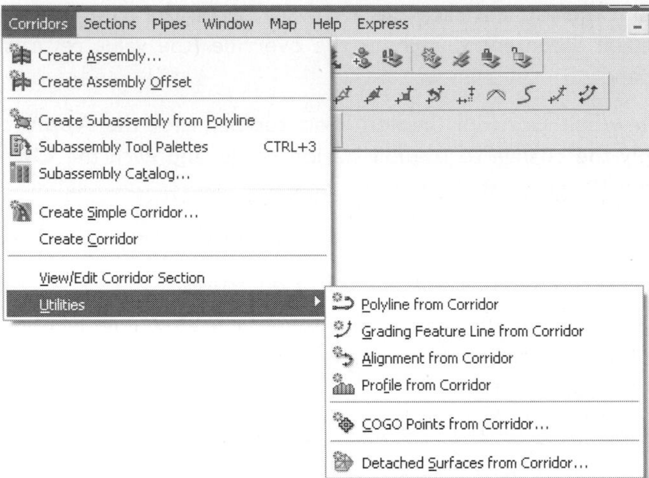

Figure 7.28

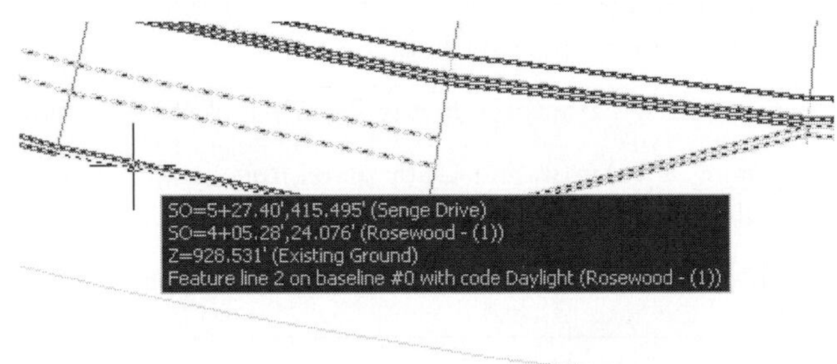

Figure 7.29

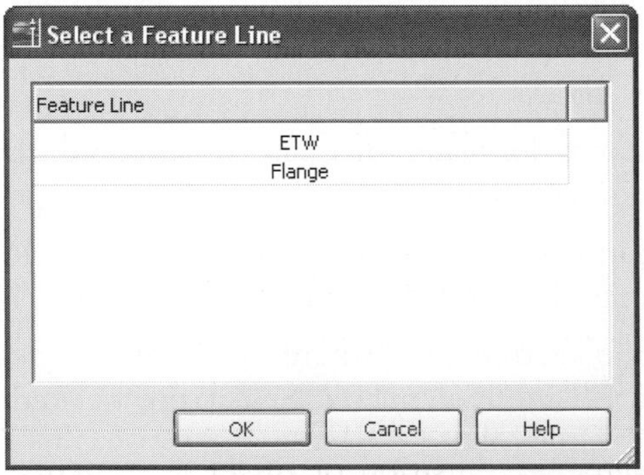

Select a Feature Line

Feature Line
ETW
Flange

OK Cancel Help

Figure 7.30

CORRIDOR FEATURE LINE AS GRADING FEATURE LINE

This routine exports a feature line object whose elevations are the corridor elevations of the exported corridor feature line. The routine places the resulting feature line on the feature line layer.

CORRIDOR FEATURE LINE AS ALIGNMENT

This routine exports an alignment object whose path is the same as the exported corridor feature line. The routine displays the Create Alignment - From Polyline dialog box and names the alignment using the feature line's name (see Figure 7.31). Users can create a profile for the new alignment at the same time.

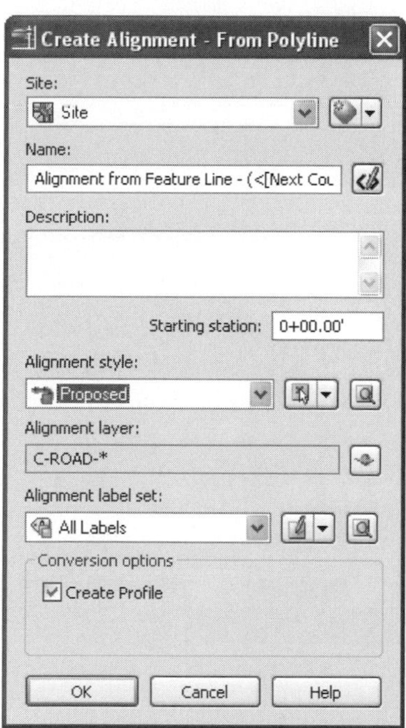

Figure 7.31

CORRIDOR FEATURE LINE AS PROFILE

This routine creates a profile object whose path and elevations are the same as the exported corridor feature line. The routine displays the Create Profile - Draw New dialog box (see Figure 7.32). A profile created this way presents the user with a starting point for developing a vertical design for transition or offset alignments.

CORRIDOR GEOMETRY POINTS AS COGO POINTS

This routine exports points whose elevations are the elevations of the selected corridor feature line(s). The points are at each corridor section. After selecting a corridor, the Export COGO Points dialog box appears (see Figure 7.33).This dialog box lists *all* of the

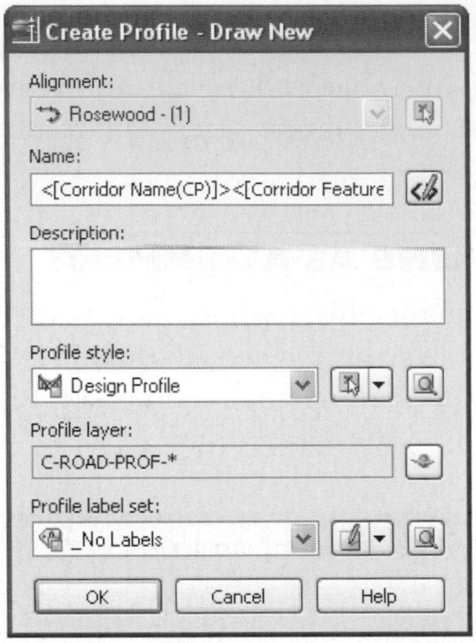

Create Profile - Draw New

Alignment:

Rosewood - (1)

Name:

<[Corridor Name(CP)]><[Corridor Feature

Description:

Profile style:

Design Profile

Profile layer:

C-ROAD-PROF-*

Profile label set:

_No Labels

OK Cancel Help

Figure 7.32

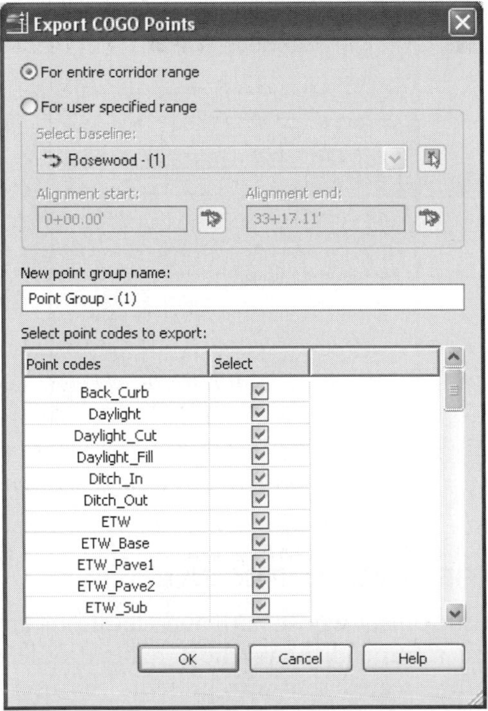

Figure 7.33

corridor feature lines and creates points only for those toggled on when exiting. The routine can create a point group from the exported points. Users can export points for the entire corridor's length, or for a range of stations.

CORRIDOR SURFACES

Creating a surface from a corridor is a three-step process: naming the surface, identifying its data, and defining a boundary to control spurious triangles. A corridor surface is a part of the Prospector surface list, and any changes to the corridor model changes the surface any objects referencing its data.

The Surfaces tab of the Corridor Properties dialog box defines the surface (name and the data used in creating the surface). See Figure 7.34.

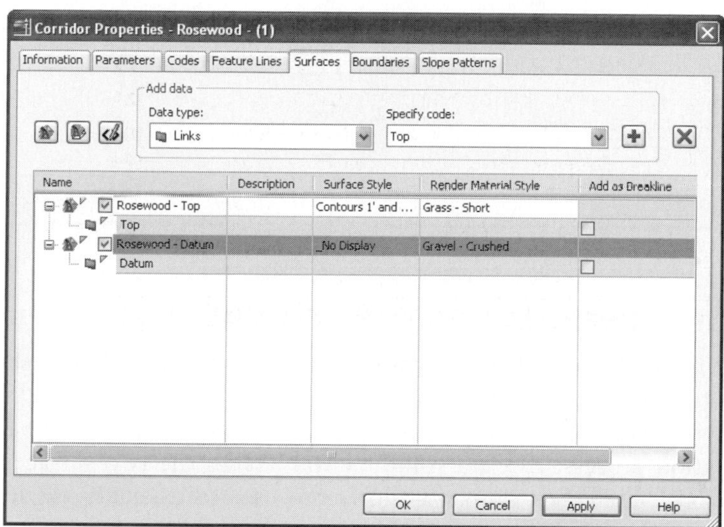

Figure 7.34

The Boundaries tab has a corresponding entry for each surface. The boundary acts to limit the triangulation to the data between the two outermost feature lines (see Figure 7.35).

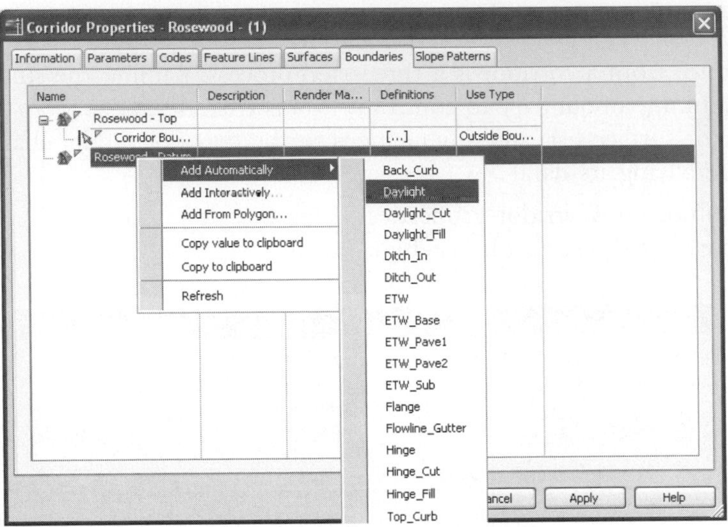

Figure 7.35

CALCULATING OVERALL EARTHWORK VOLUMES

Civil 3D has three tools to calculate roadway earthwork volumes. The first tool is a simple comparison between two surfaces. The Volumes command of the Utilities menu of Surfaces displays a vista that contains a composite volume calculator (see Figure 7.36). To start the calculation process, create a volume entry, assign the two surfaces, and click in a Cut or Fill output cell. This command allows the user to calculate an up-to-the-minute volume without having to define any surfaces.

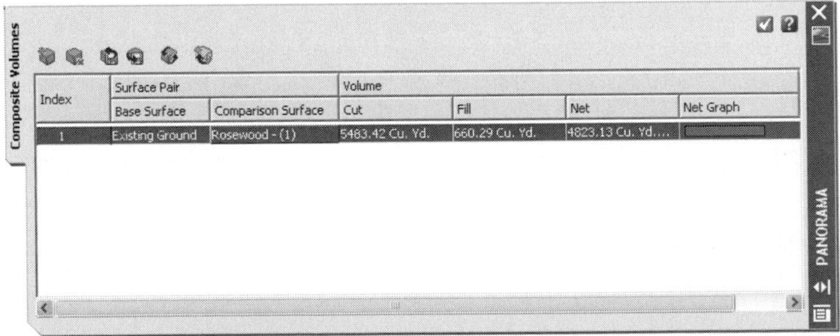

Figure 7.36

Users can define a volume surface with the Create Surface... command using the existing ground and one of the corridor surfaces.

The last method of calculating a volume is as a Quantity Takeoff report from the Sections menu. This method is a part of the next chapter.

EXERCISE 7-5

After you complete this exercise, you will be:

- Able to create 3D polylines from the corridor.
- Able to create feature lines from a corridor.
- Able to create points from a corridor.
- Able to build surfaces from a corridor.
- Able to add a boundary to corridor surfaces.
- Able to add a slope pattern between two feature lines.

EXERCISE SETUP

This exercise continues with the drawing from the previous exercise. If you did not complete the previous exercise, you can open the *Chapter 7 - Unit 5.dwg* that is in the *Chapter 7* folder of the CD that accompanies this textbook.

1. Open the drawing from the previous exercise or the *Chapter 7 - Unit 5* drawing.

POLYLINE

You export a 3D polyline by selecting a corridor feature line.

1. Open the **Layer Properties Manager**, make a new layer, and name the layer **3D poly**. Make it the current layer, assign it a color, and click the **OK** button to exit the dialog box.
2. From the Views menu, use the Named Views... command and restore the **Proposed Starting Point** view.

   The yellow line, the Daylight feature line, represents the intersection of a slope out of the ditch to an intersection with the Existing Ground surface.
3. From the Corridors menu, Utilities flyout, select **Polyline From Corridor**.
4. In the drawing select the **Daylight Line** to the north of the corridor (yellow line) and press the **Enter** key to exit the routine to create the polyline.

   A new object appears on the screen.
5. Click the new object, press the right mouse button, and from the shortcut menu select **Properties....**

   The new object is a 3D polyline with vertices that have elevations taken from the daylight feature line of the corridor.
6. Click the "**X**" to exit the Properties dialog box.
7. Erase the 3D polyline you just created.

FEATURE LINES

Features Lines are similar to 3D polylines. They have varying elevations at each vertex, but feature lines are custom Civil 3D grading objects. The Feature Line export routine works exactly like the Export Polyline command.

1. Select **Grading Feature Line From Corridor** from the Utilities flyout menu of the Corridors menu.
2. Select the **Daylight Line** to the north of the corridor (yellow line) and press the **Enter** key to exit the routine and create the feature line.

   A new object appears on the screen on the predefined layer.
3. Click the new object, press the right mouse button, and from the shortcut menu select **Properties....**

   The new object lists as a feature line. The feature line is on the C-TOPO-FEAT. All feature lines that you create from the corridor will be on this layer.
4. Click the "**X**" to exit the Properties palette.
5. Erase the feature line you just created.

POINTS

Critical to placing a design in the field are points representing corridor elevations. Also, the point numbers should be offset from other existing points. There should be a point group that identifies these points as design rather than existing condition points.

1. Click the **Settings** tab to make the panel current.
2. In Settings, click the **Point** heading, press the right mouse button, and from the shortcut menu select **Edit Feature Settings....**
3. Expand the Point Identity section, change the Next Point Number to **10000**, and click the **OK** button to exit the dialog box.
4. From the Corridors menu, the Utilities flyout, select **COGO Points From Corridor....**
5. In the command line, the routine prompts for a corridor. In the drawing, select a segment of the **Rosewood – (1)** corridor.
6. In the Export COGO Points dialog box, set the station range to the entire length of the corridor, name the point group **Design Points**, and use the following descriptions to select points to export.

- Point Codes to Export:
- Back_Curb (Back of Curb)
- Daylight
- ETW
- Flowline_Gutter

7. Click the **OK** button to create the points and their point group.

   The points appear on the screen and are a part of the Design Points point group.

8. Click the **Prospector** tab to make it the current panel and expand the Point Groups branch.

9. In the point groups list, select the **Design Points** point group, press the right mouse button, and select from the shortcut menu **Edit Points....**

10. After reviewing the points in the vista, click the "**X**" to hide the Point Editor vista.

11. In the point groups list, select the **Design Points** point group, press the right mouse button, and from the shortcut menu select **Delete Points....**

12. Click the **OK** button in the Are you sure? dialog box.

    This deletes all of the points from the point group.

13. With the Design Points point group still highlighted, press the right mouse button, and from the shortcut menu select **Delete....**

    This deletes the Design Points point group.

SLOPE PATTERNS

When preparing for a submission or a presentation, you may want to show slope patterns along the corridor's path.

1. The View menu, the Named Views... command, restore the **Proposed Starting Point** view.

2. If necessary, use the AutoCAD Pan and Zoom commands until you view the corridor cut area (stations 1+25 to 2+50).

3. Use the AutoCAD Zoom command to zoom in to the northern section of the corridor; you can identify the Ditch_Out and Daylight_Cut lines around station 2+50. Use the tooltip to identify the location of the feature lines.

4. In the drawing, select any segment on the corridor, press the right mouse button, and from the shortcut menu select **Corridor Properties....**

5. Click the **Slope Patterns** tab and view its contents.

6. Click the **Add Slope Pattern >>** button at the top of the dialog box.

7. In the command line, the routine prompts for the first feature line. Select the **Ditch_Out** line just north of the Ditch_In feature line, on the north side of the corridor.

8. In the command line, the routine prompts for the second feature line. In the drawing, select the **Daylight_Cut** line on the northern side of the corridor.

   The routine returns to the Corridor Properties dialog box, listing the first and second feature lines.

9. In the dialog box, click the **Slope Pattern Style** icon in the Slope Pattern Style column and select a new style, **Slope Schemes**.
10. Again click the **Add Slope Pattern >>** button at the top of the dialog box.

    You may have to pan to the southern side of the corridor view the next two lines to select.
11. In the drawing on the corridor's southern side, select the same two lines, **Ditch_ Out** line and **Ditch_Cut** line.

    The routine returns to the Corridor Properties dialog box, listing the first and second feature lines.
12. For the new entry, click the **Slope Pattern Style** icon and select a new style, **Slope Schemes**.
13. Click the **OK** button to exit the dialog box and to assign the slope pattern to cut hinge area of the roadway.
14. Use the AutoCAD Zoom and Pan commands to view the pattern. The pattern is now on both sides of the corridor for its entire length.
15. Click the **AutoCAD Save** icon to save the drawing.

EXERCISES

CORRIDOR SURFACES

The calculation of mass earthworks for the corridor is a comparison of the elevations between the existing conditions and the corridor datum.

1. In the drawing, select any corridor segment, press the right mouse button, and from the shortcut menu select **Corridor Properties...**.
2. Click the **Surfaces** tab to view its contents.

   This tab contains the names of the surfaces and the data they use to create their triangulation.
3. In the dialog box, at the top left, click the icon **Create a Corridor Surface**.

   An entry appears in the panel.
4. Click in the Name column for the first entry and change the Surface Name to **Rosewood - Top**.
5. In the dialog box top center, in the Add Data area in the top center of the dialog box, set the Data Type to **Links**. Set the Specify Code to **Top**, and click the **+** (plus sign) to add the data type to the surface.
6. Assign the Surface Style as **Contours 1' and 5' (Design)**.
7. Assign the Render Material Style as **Grass - Short** to the surface (see Figure 7.34).
8. At the top left of the dialog box, click the icon **Create a Corridor Surface** to make a second surface.

9. Click in the Name column for the second surface and change the Surface Name to **Rosewood - Datum**.
10. In the center of the dialog box, in the Add Data area in the top center of the dialog box, set the Data Type to **Links**. Set the Specify Code to **Datum**, and click the **+** (plus sign) to add the data type to the surface.
11. Assign the Surface Style as **_No Display**.
12. Assign the Render Material Style as **Gravel - Crushed** to the surface (see Figure 7.34).
13. Click the **Boundaries** tab to view its contents.
14. Click **Rosewood - Top**, press the right mouse button, and in the Add Automatically flyout menu select **Daylight** from the list of boundaries. Make sure the boundary type is set to **Outside** (see Figure 7.35).
15. Repeat the previous step and add the same boundary to the **Rosewood - Datum** surface.
16. Click the **OK** button to create the surfaces and to exit the dialog box.
17. Place your cursor over the corridor and review the elevations and station of the corridor and its surfaces.
18. Click the **AutoCAD Save** icon to save the drawing.

    Any changes made to the corridor automatically become a part of the corridor surfaces.

CALCULATE AN EARTHWORKS VOLUME

1. In the Surfaces menu, Utilities flyout, select the command **Volumes....**
2. In the upper left of the Composite Volumes vista, click the **Create New Volume Entry** icon.

   Clicking in the Base and Comparison Surface cells displays a drop list of surfaces to chose from.
3. Change the first entry to the Base Surface to **Existing Ground**.
4. Change the second entry to the Comparison Surface to **Rosewood - Datum**.
5. Click in the Cut cell to make the routine calculate a volume.
6. Click the "**X**" at the top right to hide the panorama.

This ends the chapter on simple corridors. The corridor is a dynamic model of the roadway and reacts to any change to the objects that are its data. The next step is to document critical sections of this model, cross sections.

SUMMARY

- A simple corridor combines the alignment, profile, and assembly data and parameters to make a corridor (roadway model).
- A corridor is dynamic and changes if any one of its dependent objects (alignment, profile, or assembly) changes.
- The Corridor Properties dialog box can assign feature line styles to each feature line in the corridor.
- The Corridor Properties dialog box can assign slope patterns to portions of a corridor.
- The Corridor Properties dialog box can create surfaces, data for the surfaces, and boundaries for the surfaces.
- All Corridor surfaces appear as surface data in the Volumes... command of the Utilities flyout of the Surfaces menu.

Cross Sections and Volumes

INTRODUCTION

This chapter is the final chapter that discusses the basics of roadway design in Civil 3D. Chapters 5, 6, and 7 cover the creation of the horizontal alignment, profiles, and the corridor. This chapter covers the creation of cross-section documentation from a roadway model. The cross-section documentation combines the roadway assembly and the existing ground conditions to the right and left of the roadway centerline.

OBJECTIVES

This chapter focuses on the following topics:

- Settings for Roadway Cross Sections
- Sample Group Lines Creation for Cross Sections
- Roadway Volumes Review
- Importation of Design Cross Sections
- Annotation of the Design Results
- Creation of Section Sheets
- Creation of Plan and Profile Sheets

OVERVIEW

The design itself is not the end of the design process. While the design solves engineering issues, it must be reviewed, documented, and must have volumes calculated (see Figure 8.1). The alignment view, profile view, and annotation document, the horizontal and side view aspects of the design. One more view is needed to complete the design documentation: the section view. This view looks at the design starting at the lowest station, and from a distance measured offset from left to right of the centerline. A left offset is always a negative value, and the right is always a positive value. Some firms do not show stations as negative or positive numbers, but only as an offset distance.

The offset's overall left-to-right distance is known as swath width. A swath width is generally wider than the Right-of-Way (ROW). While the ROW distance may

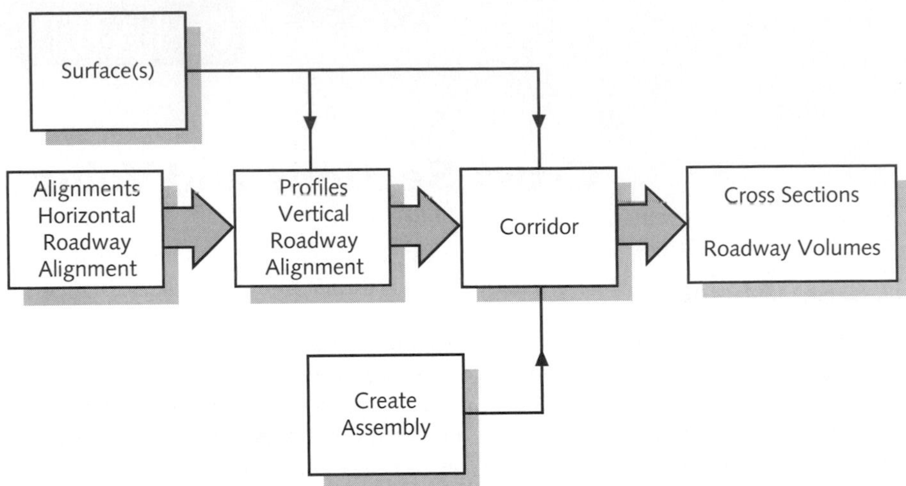

Figure 8.1

vary along the path of a road design, the swath width will stay the same. What values to use to create a swath width is generally a decision based on office standards.

Civil 3D samples a corridor with sample lines. A corridor can have as many groups of sample lines as necessary to document the corridor. Each group of sample lines can produce different section views. By applying different styles and having different sampled data, users can produce completely different looking sections.

Once the section lines are in place, it is difficult to add new data to them. If adding new data to a corridor—for example, a pipe network—it is best to add the network first and then create the sample lines. If the pipe network comes after the sample lines, the easiest way to add the network to the sections is to create a new set of sample lines and a new series of section views that include the pipe network.

After creating the sample lines, the next step is to create section views. Users create views individually, as a page(s) of a section, or for all of the sections. A section view is a grid with basic annotation. Much like profile views, a section view can have annotation bands. Even though users may define annotation for all sides of the grid and extra bands of annotation, it is the Display panel settings that control what is visible on the screen.

The process of creating a section view also creates a section. The Existing Ground is an example of a section that can be a part of a section view. The second section in a section view is the road section. The road section goes by many different names: FG, Finished Ground, etc. This book will use FG as the identifier of the roadway section. This section represents the assembly and its grading solution.

The section annotation comes from two different style groups: Section and Multipurpose (of the General branch). The section styles annotate the offsets and elevations, grade breaks, etc. of a surface. These types of labels are not frequently used in section views. The Multipurpose styles of the General branch use the corridor model to create grades/slopes, offsets, and elevations labels from the points, links, and shapes of the roadway assembly. These styles create the typical labels seen in a sections view.

UNIT 1

The settings and styles used by the document phase are the focus of the first unit of this section. There are three types of styles that affect Civil 3D sections: sample lines, section views, and sections. Each have their own settings found in the Edit Feature Settings for each type of object. These styles control layer names (if using a layer template), sampling, annotation (spot and grade), and page layout.

UNIT 2

The focus of the second unit is the first step in working with cross sections. This step is creating a sample line group that defines how to sample the corridor.

UNIT 3

After sampling the corridor, Civil 3D creates section views containing sections. Unit 3 covers the importing of section views with sections and their initial annotation into a drawing. This unit also discusses the properties of the section view.

UNIT 4

Unit 4 reviews the Multipurpose styles and how they annotate the assembly within the section view.

UNIT 5

This unit focuses on quantity takeoffs for earthwork calculations and material estimates.

UNIT 1: CROSS-SECTION SETTINGS AND STYLES

There is a myriad of settings and styles for cross sections and section views. Sections and section view styles are the most complex and demanding of a user's attention.

EDIT DRAWING SETTINGS

As with all Civil 3D objects, the values within the Edit Drawing Settings dialog box affect basic object layers and initial values (see Figure 8.2). If a project contains more than one grouping of sections and sample lines, it is best to assign either a prefix or suffix to the base object layer name. This allows the program to place the sample lines, sections, and section views on different layers.

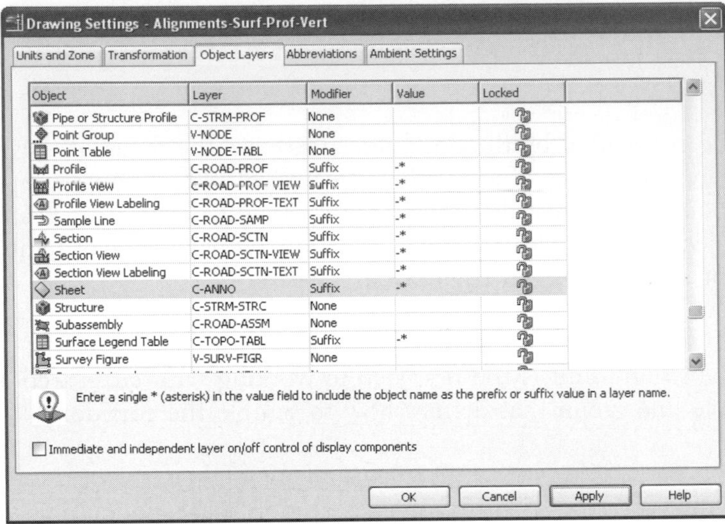

Figure 8.2

EDIT FEATURE SETTINGS

The Edit Feature Settings dialog box assigns initial values to several sample line, section view, and section elements. The values in Edit Feature Settings affect all the lower commands and styles in the Sample Line, Section View, and Section branches.

SAMPLE LINE

The Sample Line Edit Feature Settings dialog box includes assigning object and label styles and defining their naming convention (see Figure 8.3). Civil 3D uses sample line groups to extract corridor information and this is the basis of annotation for section and section views.

SECTION VIEW

The Section View Edit Feature Settings dialog box sets the object styles for the sections and section view, their label styles, plotting style, and the default styles for the Add Labels command (see Figure 8.4). Bands are information placed at the top and/or bottom of a section view.

The Section Label Set defines what labels appear on the section in the section view. The Group Plot Style defines how the sections appear in the drawing. Whether working with band or label sets, their names are an alias for a collection of styles. Each of the styles within the set or label affect some aspect of the section. The last entry in Edit Feature Settings defines the naming convention for the section views.

SECTION

The Section Edit Feature Settings dialog box defines initial styles for a section and their naming convention (see Figure 8.5).

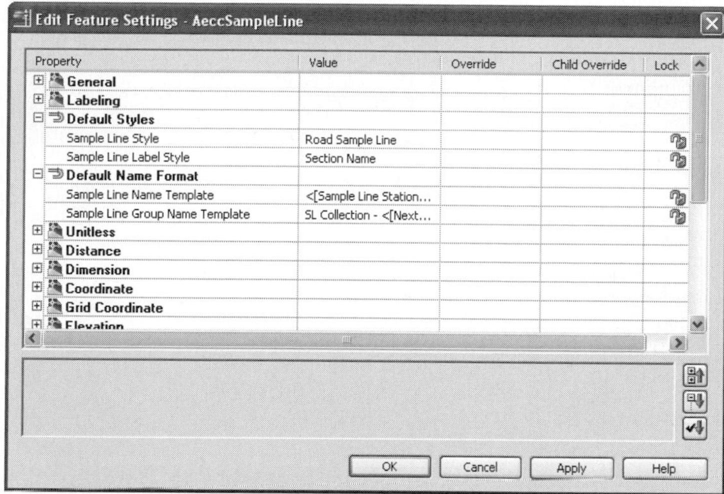

Figure 8.3

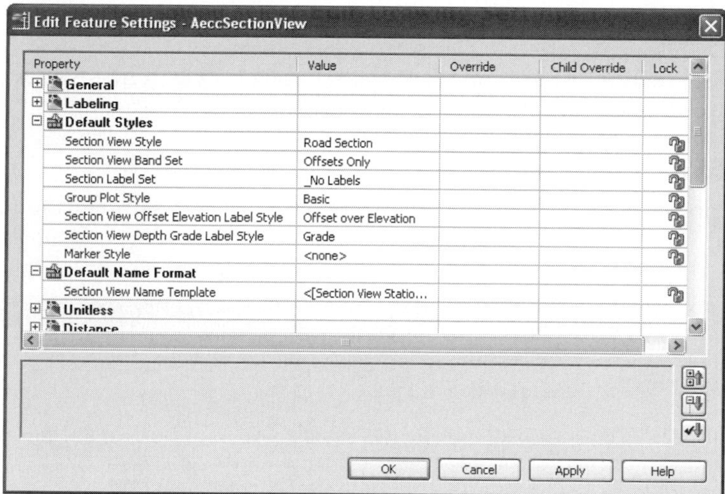

Figure 8.4

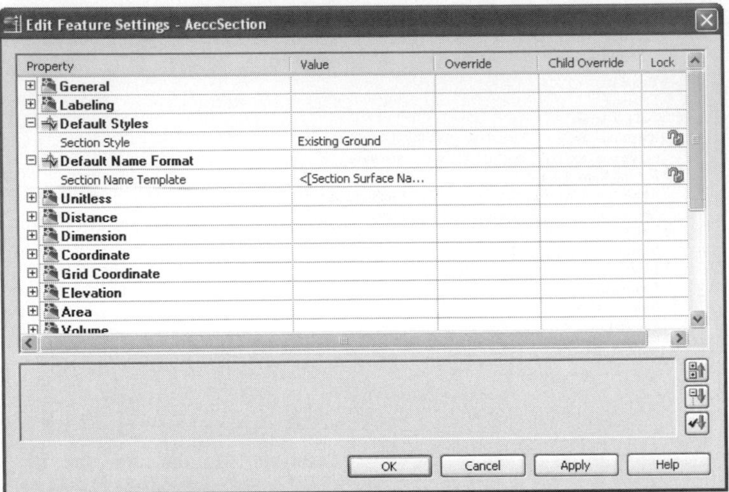

Figure 8.5

OBJECT STYLES

Each sample line, section, or section view has an object style. The style defines the shape, layer names, and display properties of its components.

SAMPLE LINES

A sample line object style defines the layer properties of the sample line (see Figure 8.6).

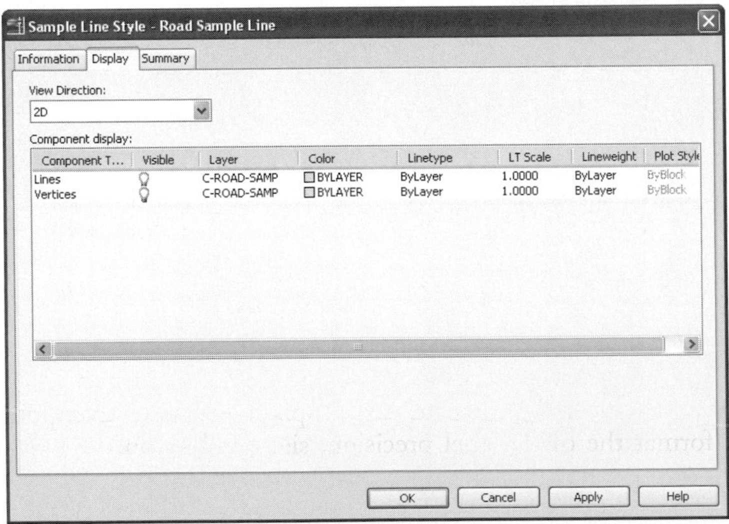

Figure 8.6

SECTION VIEW

A section view object style defines the grid, title, basic annotation, and display properties of the view.

Graph Panel

The Graph panel controls the title of the section and basic grid parameters (exaggeration and number of extra grid lines). See Figure 8.7. The values on the left side define the title, its size, location, text style, justification, and if it has a border.

The values on the right side set the vertical scale, if the grid is clipped and in what combination, and the number of extra grid lines around the body of the section.

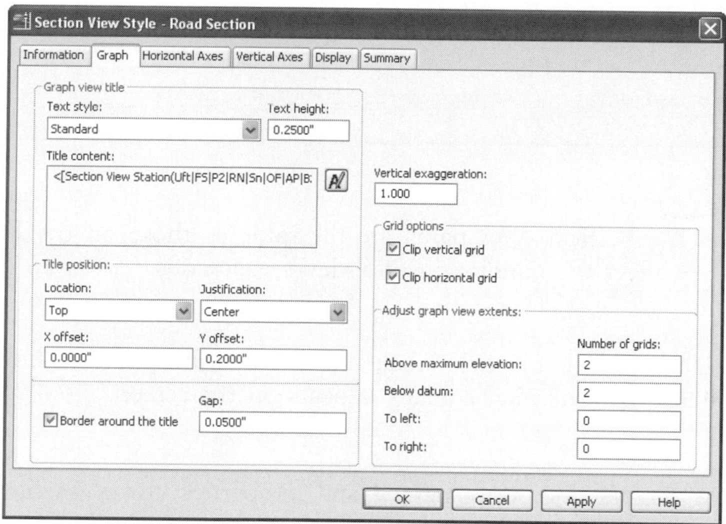

Figure 8.7

Horizontal Axes Panel

The Horizontal Axes panel defines the station annotation of a section view (see Figure 8.8). A section view uses vertical lines to demark and annotate horizontal section offsets.

The left side of the panel defines offset distance annotation for the top and bottom of the grid. The control at the top left of the panel sets the side that the settings affect (the top or bottom).

The right side of the panel defines annotation for the section top or bottom. The settings on this side of the panel define the major and minor tick intervals, their size, style, height, and other attributes. Selecting the text icon displays the Text Component Editor and allows users to format the offset label precision, sign, and so on.

Vertical Axes Panel

The Vertical Axes panel defines the elevation annotation of a section view. A section view uses horizontal lines to demark and annotate section elevations.

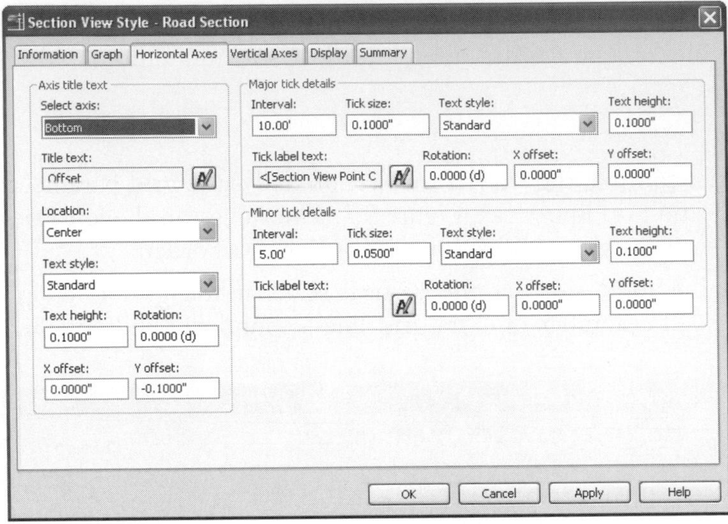

Figure 8.8

The controls on the Vertical Axes panel are the same as those on the Horizontal Axes panel, except they affect the annotation of a view's elevations.

Display Panel

Even though each view style has tick, label, and title definitions for each axis, the settings in the Display panel control what actually appears on the screen.

SECTION

A section style defines the layer names and properties for a section's components (see Figure 8.9).

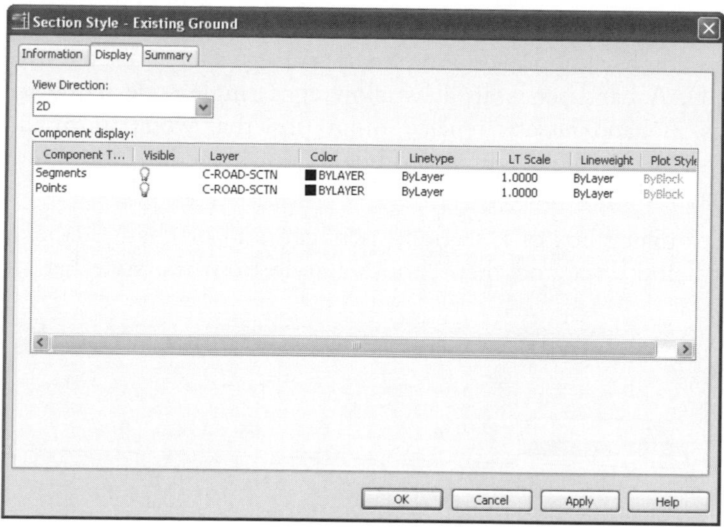

Figure 8.9

LABEL STYLES

Civil 3D has label styles for sample lines, section views, and sections.

SAMPLE LINES

A sample line label style defines the sample line, its look, and any notations (generally a station value). See Figure 8.10.

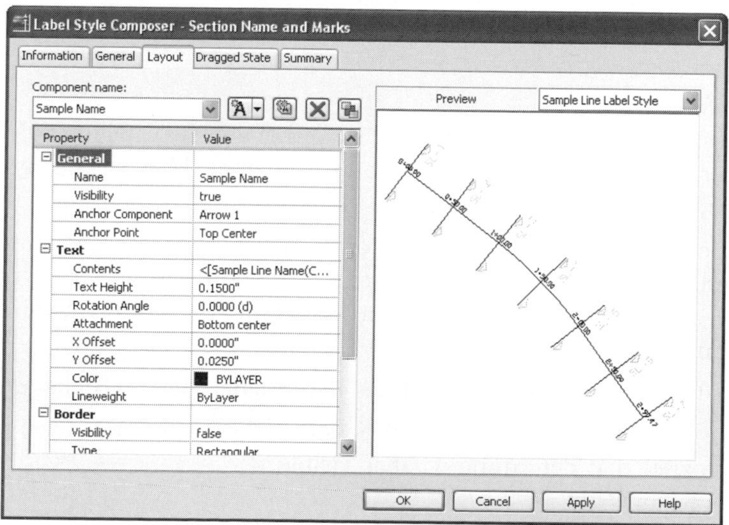

Figure 8.10

SECTION VIEW BAND SET

The section view can have station annotation appearing as a band set above or below a view (see Figure 8.11). A band set is an alias allowing a single style name to include several individual styles. A band set can replace annotation that would normally appear on the periphery of a section. By changing the band set of a section, the annotation around a section view may radically change.

In the Properties dialog box of a section, users can add or change label styles. Click the drop-list arrow below Type and then choose a style from the style list.

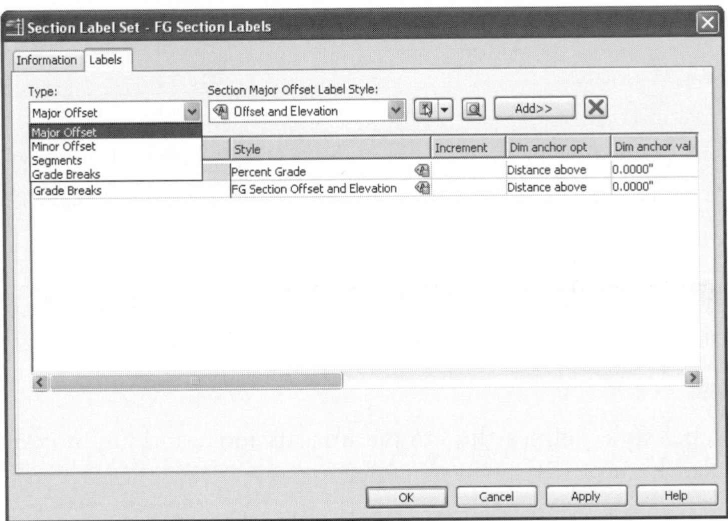

Figure 8.11

Band Set Styles

Each band set style uses the same interface to define the styles. The style's Band Details panel sets values for the title on the left side of the panel. These values include text style, height, location, etc. (see Figure 8.12).

The right side of the panel defines a label for each type listed at the center of the dialog box. To view, create, or modify a label's values, select the label type from the list and click the Compose Label... button. This calls the Label Style Composer, which displays the label's definition (see Figure 8.13).

The upper-right section of the panel defines if the label has any ticks and their size.

Even though it is here that users define all possible label styles and ticks, it is the settings of the Display panel that determine what is visible.

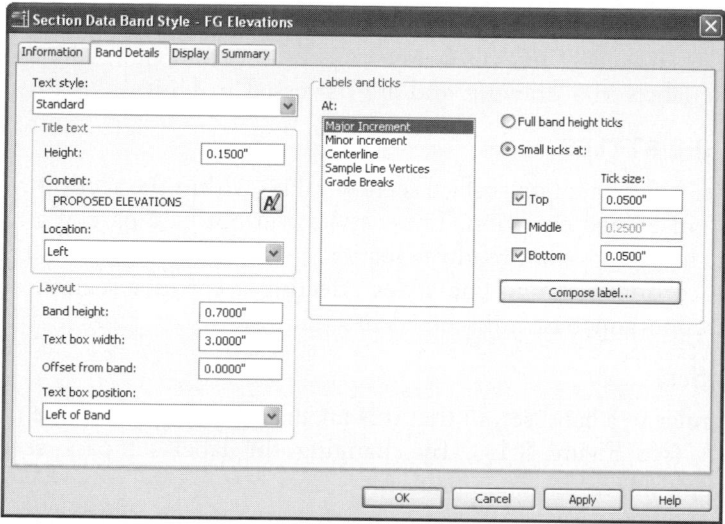

Figure 8.12

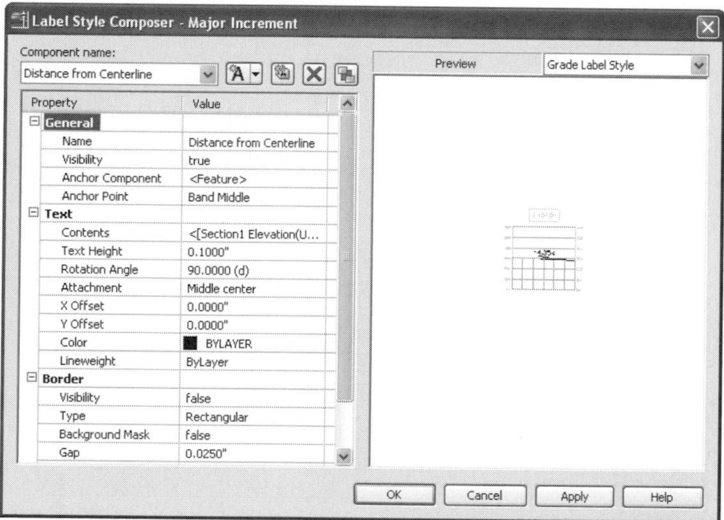

Figure 8.13

Section Data The major offsets band style defines annotation appearing in a band at major offsets from a centerline. This label can include distance from centerline, offset from centerline, and elevations from Section1 or Section2.

Section Segment This style annotates the length of section segments, the beginning and ending offsets, and the surface name and its elevations.

SECTION VIEW SPOT LABEL STYLES

These styles label selected points within the body of a section view. The Add Labels dialog box places these labels in a drawing and are discussed in Unit 4.

SECTION LABEL STYLES

The section label styles annotate surface section values within the section view. These labels annotate grade, offset, and elevation. These styles will not be a part of a roadway section's annotation. The most important roadway section annotation comes from the code set styles assigned to the corridor section. The styles affecting a corridor section are a part of the discussion of section annotation discussed in Unit 4.

Section Label Sets

A label set is similar to a band set, in that it is an alias allowing one name to include several individual styles (see Figure 8.14). By changing the label set of a section, users may completely change the section's annotation.

Label sets use a combination of following label styles: Major and Minor Offsets, Grade Breaks, and Segments.

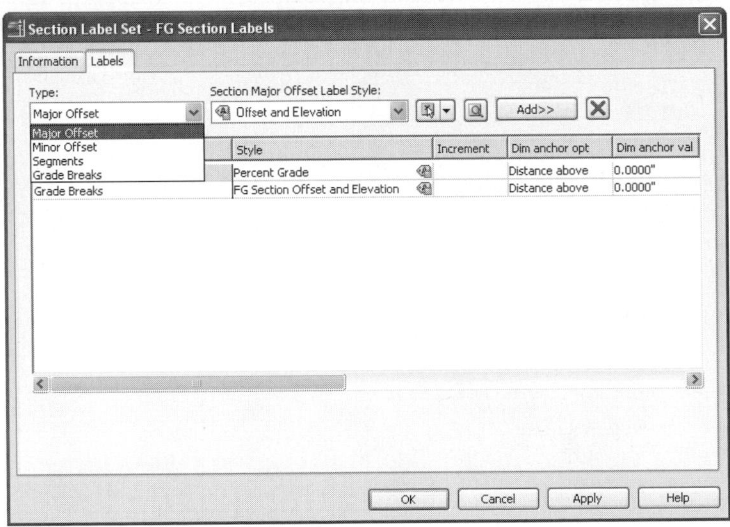

Figure 8.14

Major and Minor Offsets The Major and Minor Offset styles label the offset and elevation of a section at a major and minor interval defined by the section view.

Grade Break The Grade Break styles label the offset of a grade break (change of grade) of a section.

Segment The Segment styles label the length and grade of each segment of a section.

COMMAND SETTINGS

The sample lines, section view, and sections all have their own command set values. The values determine the default styles and other initial values each command uses.

SAMPLE LINES – CREATE SAMPLE LINES

The settings of the Create Sample Lines command affect how the command samples the corridor (see Figure 8.15). The Default Swath Widths section settings define the maximum left and right offset of a sample line. The Sampling Increments section values set whether the sampling is incremental and the sampling interval for tangents, curves, and spirals segments. The Additional Sample Controls section settings determine if there are additional sample lines at the beginning and ending station of the sampling range, at horizontal geometry points, at critical stations for superelevation, and at the beginning and ending stations of an alignment.

In the Miscellaneous section there is an important setting: Lock to Station. This setting, when true, tells the sample lines to update if the alignment changes.

In the Default Styles section is the sample line and its label styles.

The Default Name section defines the naming format for the section lines.

After creating a sample line group, Prospector displays all of the sample lines as a list under the sampled alignment. This list includes the surface names, corridors, and pipe networks that are a part of the section's data.

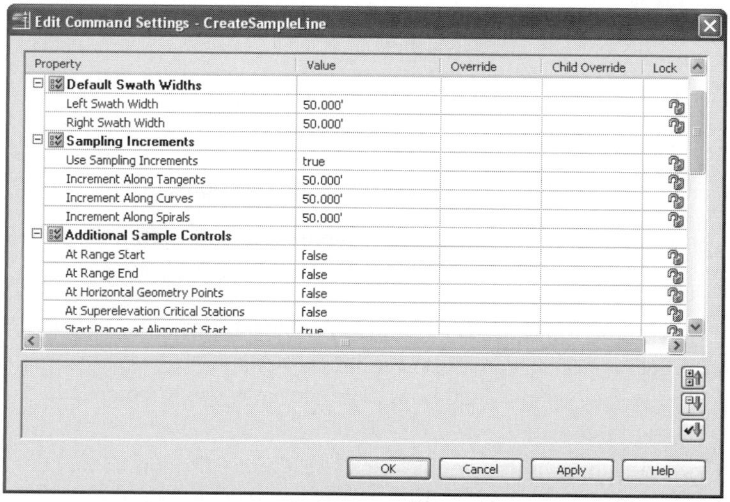

Edit Command Settings - CreateSampleLine

Property	Value	Override	Child Override	Lock
Default Swath Widths				
Left Swath Width	50.000'			
Right Swath Width	50.000'			
Sampling Increments				
Use Sampling Increments	true			
Increment Along Tangents	50.000'			
Increment Along Curves	50.000'			
Increment Along Spirals	50.000'			
Additional Sample Controls				
At Range Start	false			
At Range End	false			
At Horizontal Geometry Points	false			
At Superelevation Critical Stations	false			

OK Cancel Apply Help

Figure 8.15

SECTION VIEW – CREATE SECTION VIEW

Whether the command is Create Section View or Create Multiple Section Views, the settings are the same (see Figure 8.16). The initial command settings come from the values of a section view's Edit Feature Settings dialog box. The values of the Default Styles section assign styles for the section view, band sets, section labels styles, section view group plot, and default styles of the Add Labels dialog box. The Default Name Format section defines the naming convention for section views.

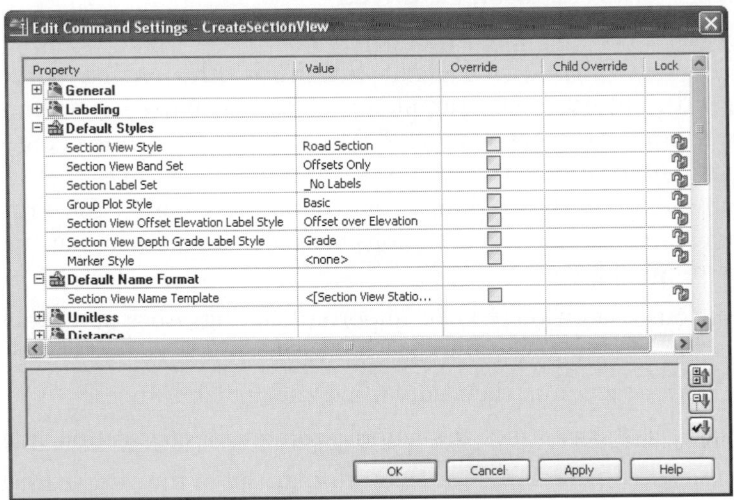

Figure 8.16

EXERCISE 8-1

When you complete this exercise, you will be:

- Familiar with Edit Drawing Settings.
- Familiar with Edit Feature Settings.
- Familiar with label styles for sample lines and sections.
- Familiar with command settings for sample lines and sections.

EXERCISE SETUP

This exercise starts with the drawing from the previous chapter. The drawing contains an alignment, profile, assembly, and corridor. If you didn't complete Chapter 7's exercises, you can start this exercise by opening *the Chapter 8 - Unit 1.dwg*. The file is in the *Chapter 8* folder of the CD that accompanies this textbook.

1. If you are not in Civil 3D, double-click the **Civil 3D** icon to start the application.
2. When you come to the command prompt, close the open drawing and do not save it.
3. Open the drawing from the previous chapter or open the drawing *Chapter 8 - Unit 1*.

EDIT DRAWING SETTINGS

The values in the Edit Drawing Settings dialog box affect all of the settings and styles in the Settings panel.

1. Click the **Settings** tab to make it the current view.
2. In Settings, at the top, click the name of the drawing, press the right mouse button, and from the shortcut menu select **Edit Drawing Settings....**
3. Select the **Object Layers** tab to view its contents.

   If there are multiple instances of sample line and section objects, their layers entries should have a modifier and a value (see Figure 8.2).
4. In the Object Layers panel, change the modifier for Sample Line, Section, Section View, Section View Labeling, and Sheet to **Suffix**. For the value, type (**-***) (a dash followed by an asterisk).
5. Click the **OK** button to exit the dialog box.

EXERCISES

EDIT FEATURE SETTINGS

Each object has an Edit Feature Settings dialog box containing values that assign styles and default naming values.

Sample Line

The values of Sample Line's Edit Feature Settings dialog box set default styles and name formats (see Figure 8.3).

1. In Settings, select the **Sample Line** heading, press the right mouse button, and from the shortcut menu select **Edit Feature Settings....**
2. In the Edit Feature Settings dialog box, expand the Default Styles and Default Name Format sections to view their values.
3. Click the **OK** button to close the dialog box.

Section View

The values of Section View's Edit Feature Settings dialog box set default view styles, its section styles, plotting styles, label styles, and default name formats (see Figure 8.4).

1. In Settings, select the **Section View** heading, press the right mouse button, and from the shortcut menu select **Edit Feature Settings....**
2. In the Edit Feature Settings dialog box, expand the Default Styles and Default Name Format sections to view their values.
3. Click the **OK** button to close the dialog box.

Section

The values of Section's Edit Feature Settings dialog box set the default style and naming format for a section (see Figure 8.5).

1. In Settings, select the **Section** heading, press the right mouse button, and from the shortcut menu select **Edit Feature Settings....**

2. In the Edit Feature Settings dialog box, expand the Default Styles and Default Name Format sections to view their values.
3. Click the **OK** button to close the dialog box.

OBJECT AND LABEL STYLES

Sample lines, section views, and sections styles specifically label their critical values. Section view band sets and section label sets are aliases for a collection of specific purpose styles.

Sample Line

The object style for a sample line sets the layers it uses in a drawing. The label sample line styles affect its appearance and labeling.

1. In Settings, expand the Sample Line, Label Styles branches until you view the list of label styles under the Sample Line heading.
2. From the list of styles, select **Section Name and Marks**, press the right mouse button, and from the shortcut menu select **Edit...**.
3. Click the **Layout** tab to view the contents of the panel.
4. In the dialog box, click the Component Name drop-list arrow to view a list of label components.
5. Review the settings for each component.
6. Set the Component Name to **Sample Name**; in the Text section, click in the Contents Value cell to display an ellipsis, and click the ellipsis to display the Text Component Editor.
7. Review the format string for this label.
8. Click the **Cancel** buttons to return to the command line.

Section View

A Section view is a grid that encloses a section's station and elevation values. The view styles provide basic station and elevation annotation.

1. In Settings, expand the Section View branch until you view a list of styles under the Section View Styles heading.
2. From the list, select **Road Section**, press the right mouse button, and from the shortcut menu select **Edit...**.
3. Select the **Graph**, **Horizontal Axes,** and **Vertical Axes** tabs to view their contents.
4. Click the **OK** button to exit the Road Section view style.

Band Sets and Band Set Styles

A band set is an alias for a group of styles that appear below or above the section grid (see Figure 8.11).

1. In Settings, expand the Section View and Band Styles branches until you view a list of band sets under the Band Sets heading.
2. From the list, select **Major Stations Offsets and Elevations**, press the right mouse button, and from the shortcut menu select **Edit...**.
3. Click the **Bands** tab to view its label list.
4. In the dialog box, click the drop-list arrow to the right of Band Type and from the list select **Section Data**.
5. At the top middle of the panel, click the drop-list arrow to the right of Select Band Style to display a list of styles.
6. In the dialog box, click the drop-list arrow to the right of Band Type and from the list select **Section Segment**.
7. At the top middle of the panel, click the drop-list arrow to the right of Select Band Style to display a list of styles.
8. Click the **Cancel** button to close the dialog box.

   The band set styles are below the Band Set heading. The Section Data and Section Segment headings list their style below each heading.

1. In Settings, expand the Section View, Band Styles branch until you view a list of styles for the Section Data heading.
2. From the list of styles under Section Data, select **Offsets**, press the right mouse button, and from the shortcut menu select **Edit...**.
3. Click the **Band Details** tab to view its contents.

   The left side of this panel defines the title of the band. The Labels and Ticks area in the center lists all of the locations of available labels. The Compose Label... button displays the Label Style Composer dialog box.

4. In the top middle of the Band Details panel, from the At: area select **Centerline** and click the **Compose Label...** button to view the label definition.
5. In the Label Style Composer dialog box, in the Text section, click in the Contents Value cell. An ellipsis appears; click the ellipse to view the format string for this label.
6. In the Text Component Editor, at the left side of the dialog box, click the drop-list arrow to the right of Properties to view a list of properties this label type uses.
7. Click the **Cancel** buttons to return to the command line.

Section Segment

This label type annotates the lengths and grades of section segments in a band.

1. In Settings, expand the Section View, Band Styles branch until you view a list of styles for the Section Segment heading.
2. From the list of styles for Section Segment, select **Segment Length**, press the right mouse button, and from the shortcut menu select **Edit...**.
3. Click the **Band Details** tab to view its contents.

   The left side of this panel defines the title of the band. The Labels and Ticks area in the center lists all of the locations of available labels. The Compose Label... button displays the Label Style Composer dialog box.
4. In the top middle of the Band Details panel, in the At: area, select **Segment Labels** and click the **Compose Label...** button.
5. In the Label Style Composer dialog box, in the Text section, click in the Contents Value cell. An ellipsis appears; click the ellipsis.

   This label annotates segment lengths from a surface in a band.
6. In the Text Component Editor, at the top left of the dialog box, click the drop-list arrow to the right of Properties to view a list of properties for this label.
7. Click the **Cancel** buttons to return to the command line.

Section Labels

Section labels annotate surface elevations and stations within a section view.

Section Label Set

A section label set is an alias containing one or more styles from the Major and Minor Offsets, Grade Breaks, and Segments style types. A set also specifies a location for the label and a weeding factor. A weeding factor prevents labels from overwriting.

1. In Settings, expand the Section, Label Styles branches until you view a list of styles under the Label Sets heading.
2. From the list of styles, select **FG Sections Labels**, press the right mouse button, and from the shortcut menu select **Edit...**.
3. Click the **Labels** tab to view the labels assigned to this set.
4. At the top left of the dialog box, click the drop-list arrow to the right of Type: to view a list of style types.
5. From the list of types, select **Segments**.
6. In the top middle of the dialog box, click the drop-list arrow to the right of Section Segment Label Style: to view styles for this section type.
7. Click the **Cancel** button to return to the command line.

Major and Minor Offset

Major and Minor Offset label styles annotate offsets and elevations at major or minor station intervals. The section view defines the intervals for both major and minor styles.

1. In Settings, expand the Section branch until you view a list of styles for Major and Minor Offset branches.
2. From the list of styles for Major Offset, select **Offset and Elevation**, press the right mouse button, and from the shortcut menu select **Edit....**
3. Click the **Layout** tab to view its contents.
4. In the dialog box, in the Text section, click in Contents Value cell. An ellipsis appears; click the ellipsis to view the format string for this label.

   The label annotates the offset and the elevation at the major station interval.
5. Click the **Cancel** buttons to return to the command line.

Grade Break

The Grade Break style labels a break in grade for the assigned section, EG or FG. The labeling appears in the section view.

1. In Settings, expand the Section branch until you view a list of styles for the Grade Break branch.
2. From the list of styles, select the **FG Section Offset and Elevation**, press the right mouse button, and from the shortcut menu select **Edit....**
3. Click the **Layout** tab to view its contents.
4. Set the Component Name to **txtOffset**. In the Text section, click in the Contents Value cell. An ellipsis appears; click the ellipsis to view the format string for this label.
5. Click the **Cancel** buttons to return to the command line.

Segment

This label type annotates the length and cross slope surface section segments.

1. In Settings, expand the Section branch until you view a list of styles for the Segment branch.
2. From the list of styles, select **Percent Grade**, press the right mouse button, and from the shortcut menu select **Edit....**
3. Click the **Layout** tab to view its contents.
4. In the dialog box, in the Text section, click in the Contents Value cell. An ellipsis appears; click the ellipsis to view the format string for this label.
5. Click the **Cancel** buttons to return to the command line.

COMMANDS

Each command to create sample lines, section views, and sections has default settings.

Create Sample Lines

This command samples the corridor at an interval, at specific stations, or at a station range.

1. In Settings, expand the Sample Line branch until you view a list of commands under the Commands heading.
2. From the list, select the **CreateSampleLine** command, press the right mouse button, and from the shortcut menu select **Edit Command Settings...**.
3. Expand the section Default Swath Widths.

   A swath width is the distance sampled to the left and right of the roadway.
4. Expand the Sampling Increments section.

   This section sets the initial values for the frequency of sampling the corridor.
5. Expand the Additional Sample Controls section.

   This section sets the other critical points the sampling should include.
6. Expand the Miscellaneous section.

   If Lock To Station is true, the sections resample if any change occurs to the alignment geometry and properties.
7. Expand the Default Name Format and Default Styles sections.

   These styles are from the Sample Line Feature Settings and you can change their values.
8. Click the **Cancel** button to return to the command line.

Create Section View

The menu picks Create Section View (single section) and Create Multiple Section View (multiple sections) commands use the same settings.

1. In Settings, expand the Section View branch until you view a list of commands under the Commands heading.
2. From the list, select **CreateSectionView**, press the right mouse button, and from the shortcut menu select **Edit Command Settings...**.
3. Expand both the Default Styles and Default Name Format sections and review their values.

   The Default Styles area lists the style the command will apply to in any new section view. The Default Name Format specifies the station and the number of the view.
4. Click the **Cancel** button to return to the command line.
5. Click the **AutoCAD Save** icon to save the drawing.

This ends the exercise of reviewing the objects, styles, and commands that create sample lines, section views, and sections. The next unit reviews the process of creating the sample lines for the sections.

SUMMARY

- Civil 3D's most complex object is the section view.
- A section is a sampled object (surface, corridor section, and/or pipe network) along and perpendicular to the alignment's path.
- If designing pipe networks, design them before using the Create Sample Line command.
- A section view is a grid with sections inside of it.
- A section view has annotation around all four axes plus bands that can appear at the top and the bottom.
- When creating a section view, the command assigns label styles and/or sets to the section.

UNIT 2: CREATE SAMPLE LINES

Sample lines are the link between the corridor and contents of a section and section view. It is best to define all of the content for a corridor before creating the sample lines. For example, if creating sample lines and then adding a piping network, it is best to define another sample line group that samples the new data.

CREATE SAMPLE LINES

The process starts by identifying which roadway elements to sample (see Figure 8.17). The Create Sample Line Group dialog box reports the sample line group name format (upper left), the current sample line style, label style, and layer (upper right), and the current alignment in the middle left.

The bottom of the dialog box indicates all of the elements available for sampling along the path of the selected alignment. There are four types of data: surface(s), corridors, corridor surface(s), and pipe network(s). Each has a unique icon to indicate its data type. The section sampling defaults identifies the source, if it is sampled, the style it uses in the section view, its layer, and its updating mode.

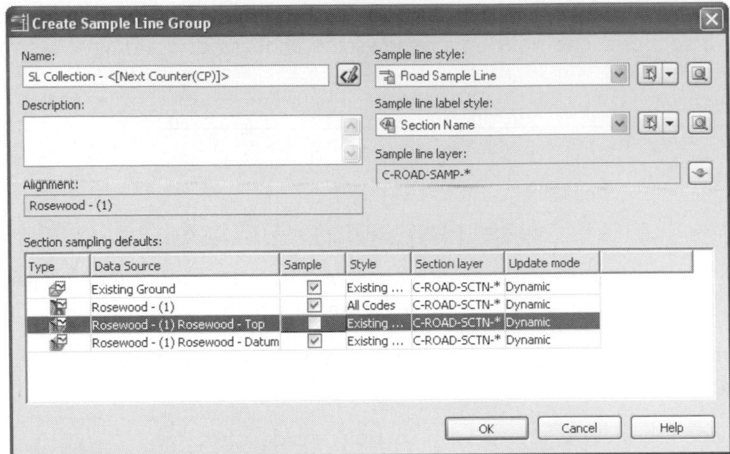

Figure 8.17

After identifying what elements to sample, the routine displays the Sample Line Tools toolbar. The Sample Line Tools toolbar contains routines that create sample lines, view their content, and delete individual sample lines (see Figure 8.18). The methods of defining sample lines are by stations, by selecting points in the drawing, by selecting existing polylines, by a station range, or from the corridor sections. When defining sample lines, users may use a combination of methods (e.g., by station range and by selecting points along the centerline of the corridor). The toolbar always defaults to defining by stations, even after the user creates sample lines with one of the methods.

The toolbar reports the current sample line group (SLG) at the center of the toolbar, reports the current method of defining a sample line, the current alignment, and the layer for the sample lines. The methods of defining the sample stations are in a drop list on the right side of the toolbar.

The toolbar allows users to edit the SLG, delete existing groups, create new groups, and select a sample line that changes the current group to the one represented by the selected sample line.

Figure 8.18

When using the By stations method, users select a point to identify the station, and the routine prompts for the offset to sample. If the user is defining sample lines by selecting points in the drawing, the selection determines the station and offset. When using the By station range... or From corridor stations methods, the routine displays the Create Sample Lines – By Station Range dialog box (see Figure 8.19). This dialog box sets right and left swath width and the sampling increments.

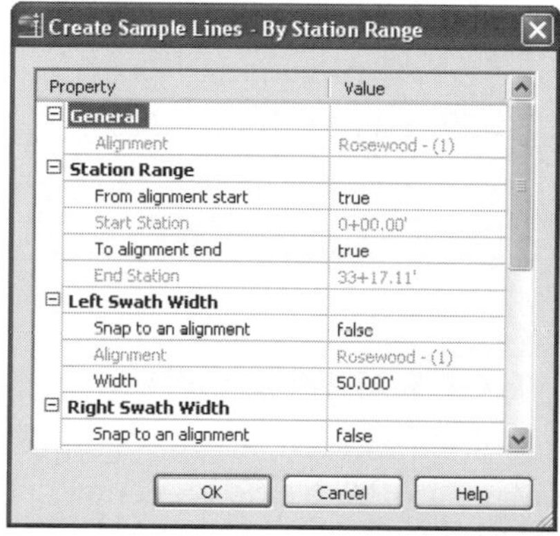

Figure 8.19

If creating sample lines that duplicate existing lines, Civil 3D will issue a warning dialog box with options for resolving the duplication of sample lines (see Figure 8.20).

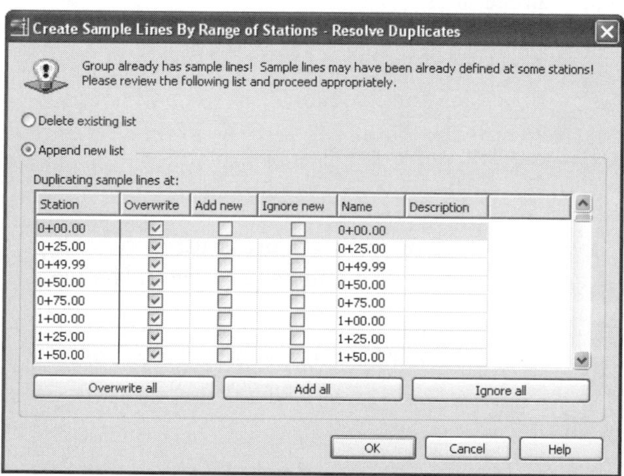

Figure 8.20

All Civil 3D layout toolbars include a cell-based viewer to allow the user to review the properties of the section (see Figure 8.21). At the bottom of the dialog box are buttons to display the previous or next vertex's information (Center, Left 1..., and Right 1...).

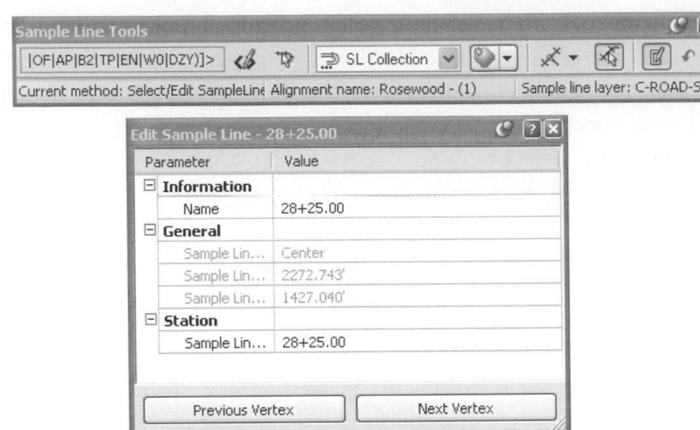

Figure 8.21

EDITING AND REVIEWING SAMPLE LINES

The review and editing of sample lines occur in the Sample Line Tools toolbar. There is an Edit Sample Lines... command in the Sections menu that calls the Sample Line Tools toolbar and presents the Edit Sample Line dialog box.

EXERCISE 8-2

When you have completed this exercise, you will be:

- Able to create sample lines.
- Able to create sample lines by different methods.
- Familiar with the grid view of a section.
- Able to view sample line properties.

DRAWING SETUP

This exercise continues with the drawing from the previous exercise. If you did not complete the previous exercise, you can open the drawing *Chapter 8 - Unit 2*. The file is in the *Chapter 8* folder of the CD that accompanies this textbook.

1. Open the drawing from the previous exercise or the *Chapter 8 - Unit 2* drawing.

CREATE SAMPLE LINES

The Create Sample Lines command displays a Create Sample Line Group dialog box. This dialog box identifies what elements to include in a sample line. After identifying the elements to sample, the routine displays the Sample Line Tools toolbar. The Sample Line methods icon stack lists the methods for establishing sample lines.

1. From the Sections menu, select **Create Sample Lines...**.
2. In the command line, the routine prompts for an alignment.

You can select the alignment from the screen or you can select it from a list of alignments by pressing the **Enter** key.

3. Press the **Enter** key to display the alignment list and select **Rosewood - (1)** from the list.
4. Toggle **OFF** the Rosewood - Datum entry, and click the **OK** button to exit the dialog box.

   The Sample Line Tools toolbar appears with a jig attached to the alignment. The jig's tooltip reports the current alignment station.
5. Move your cursor around the corridor to view the station reporting. If necessary, zoom or pan to better view the corridor.
6. In the Sample Line Tools toolbar, to the right of the Sample Line Group name, click the second icon stack drop-list arrow and select from the list **From corridor stations**.
7. Click the **OK** button to create the SLG.

   The routine makes the sample lines, and it resets the mode to By station.
8. If you need to zoom in to better view the section lines, do so now.
9. Without changing the sampling mode, select a couple stations from the screen and, when prompted, enter a swath width of **50** for both offsets.

   These stations are added to the end of the sample line list.

REVIEW SAMPLE LINE DATA

1. In the Sample Line Tools toolbar, click the Sample **Line Entity View** icon on the right side of the toolbar.

   This displays a cell-based editor.
2. In the Sample Line Tools toolbar, click the **Select/Edit Sample Line** icon to the left of the Sample Line Entity View icon and in the drawing, select a sample line.

   The sample data appears in the viewer. You can change the station number or view the data for the various section vertices (center, left, and right).
3. Click the **Next Vertex** or **Previous Vertex** buttons to view the section line information.
4. Click the red "**X**" to close the viewer.
5. Press the **Enter** key to end the Create Sample Lines command.
6. Click the **AutoCAD Save** icon and save the drawing.

SAMPLE LINE GROUP PROPERTIES

Each SLG has properties. Sample Line Properties include sample line data, what has been sampled, and what section views use the section's data.

1. Click **Prospector** tab to make it the current panel.
2. In Prospector, expand the Sites branch until you view the list of samples line for the SL Collection -1 for Rosewood - (1).
3. Select the **SL Collection - 1** heading, press the right mouse button, and from the shortcut menu select **Properties...**.
4. Click the **Sample Lines** tab view its information.
5. Click the **Sections** tab to view its contents.

   The top portion of the Sections panel displays all corridor elements along the path of the alignment (see Figure 8.22). You may have to expand the columns to be able to read all of the pertinent information. You can change the combination of data and resample the new data combination by selecting the Resample button at the top right, or you can change the assigned style to each section element.

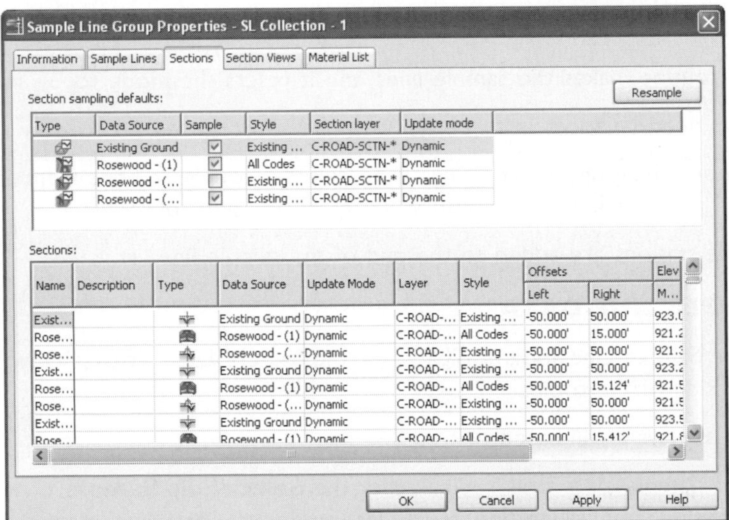

Figure 8.22

The bottom portion of the panel indicates what sections are available for each sample line, the data source, its assigned layer, its update mode, the statistics of the section (offset, elevation, and station), and most importantly, the style the section will use in the section view.

When clicking on any of the headers at the bottom of the Sections panel, the entries sort for the selected heading. You can change the object style for each sample line. If you want to change a range of sample lines to another style, select the first entry for that group, hold down the Shift key, and select the end entry.

6. Click the **OK** button to exit the dialog box.

   There are currently no section views in the drawing.

SAMPLE LINE PROPERTIES

The properties of a sample line are not trivial. Each sample line has a tree structure that categorizes the section's information (see Figure 8.23).

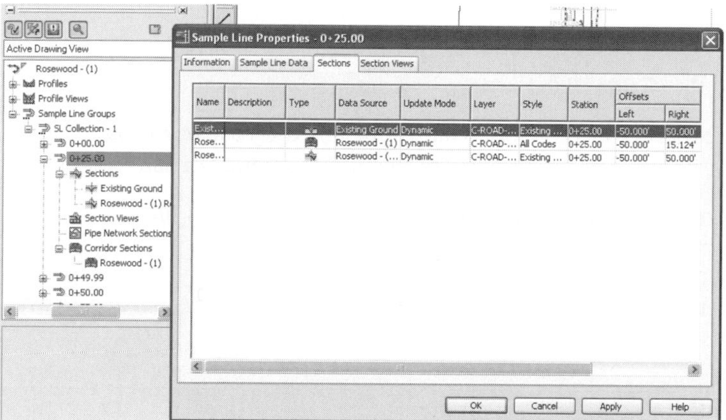

Figure 8.23

1. In Prospector, expand the Sites, Sample Line Collection - 1 branches until you view the sample line list.
2. Scroll through the list. From the list, select **10+00**, press the right mouse button, and from the shortcut menu select **Properties...**.

   This displays an overall review of 10+00.
3. Click the **Sample Line Data** tab to view its contents.

   This reviews the location, label style, alignment, and whether the sample line is locked to a station.
4. Click the **Sections** tab to view its contents.

   This panel lists what sections are parts of 10+00. You can change the layer, the update mode, or the styles the sections are using.
5. Click the **Cancel** button to exit the dialog box.
6. Expand the Station 10+00 branch to view the section list under the Sections heading.
7. From the list of sections, select the **Datum** surface section, press the right mouse button, and from the shortcut menu select **Properties...**.

   The Properties dialog box appears, displaying information about the section, including the section style. You can change the style for the section in this dialog box.
8. Click the **Cancel** button to exit the dialog box.

9. In the same Sections branch, expand the Corridor Sections branch. Select the **Corridor** section, press the right mouse button, and from the shortcut menu select **Properties...**.
10. Click the **Information** tab to view its contents.
11. Click the **Section Data** tab to view its contents.
12. Click the **Codes** tab to view its contents.

    The contents of the Codes tab represent the roadway cross section annotating. The point, link, and shape entries are from the assembly's subassemblies.
13. Click the **Cancel** button to return to the command line.
14. Click the **AutoCAD Save** icon to save the drawing.

EXERCISES

This ends the section on creating sample lines and viewing their properties. The next step is creating the section views.

SUMMARY

- Sample lines sample what exists along an alignment.
- Users should have all of the data present along the alignment before creating the sample lines.
- If adding data to a corridor, it is best to redefine or create new sample lines to include the data in the sample line data.
- Users can define sample lines by station range, by corridor sections, by selecting multiple points, by existing polylines, and by points that determine a station and request a left and right offset.
- A sample line's properties list individual vertices of a section, the station of the sample line, and what section views use the sample lines data.
- A sample line contains sections data and what styles they use in a section view.
- A corridor section uses styles from the General, Multipurpose style list.

UNIT 3: CREATING ROAD SECTIONS

The last step in documenting a roadway design is creating the corridor sections. Civil 3D imports sections one at a time, all at once (as an array), or as pages of organized sections. When importing multiple sections, a group style organizes the sections.

The station and elevation data is part of a section view object. This means users can move section views anywhere in the drawing and still produce valid offset and station labels.

CREATE VIEW

The Create View dialog box contains the current settings for a sample line and its sections. In Figure 8.24, the top portion of the dialog box reviews the current section view name format, its description, and its layer. The current sample line and its station area are located at the middle left of the dialog box. Clicking the drop-list arrow to the right of the sample line name displays all of the sample lines available for section views. The elevations in the sample line data and its offsets are on the middle right of the dialog box.

The view and band set style for the section is in the middle part of the dialog box. At the bottom is a list of sections available for the current sample line. Even though all of the sections are here, a user may not want to draw them all in one section. When exiting the routine, it draws the section views with the checked section(s) on the screen.

The Sections to Draw area contains all of the assigned styles and statistics of each section. When a user scrolls to the Style column, there is one last opportunity to override the current section style with another. This is where a user can block a section from being a part of a section view.

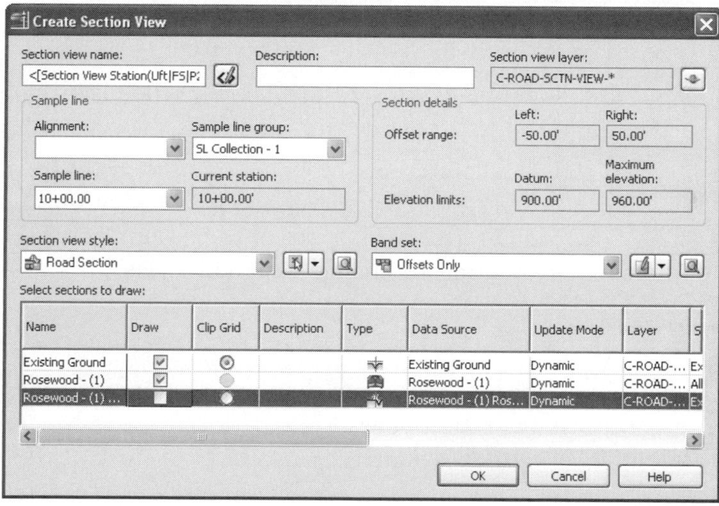

Figure 8.24

CREATE MULTIPLE VIEWS

The Create Multiple Views command plots all or a range of sections (by station). When starting this command, the Create Multiple Section Views dialog box appears (see Figure 8.25). This dialog box is similar to the Create Section View in that the top of the dialog box displays the summary statistics about the alignment and the current SLG. The User specified toggle at the middle right sets a range of stations for plotting.

The middle of the dialog box shows the Section view style (left) and a Band set assignment (right), and the center shows the method of plotting the multiple section views. Users can

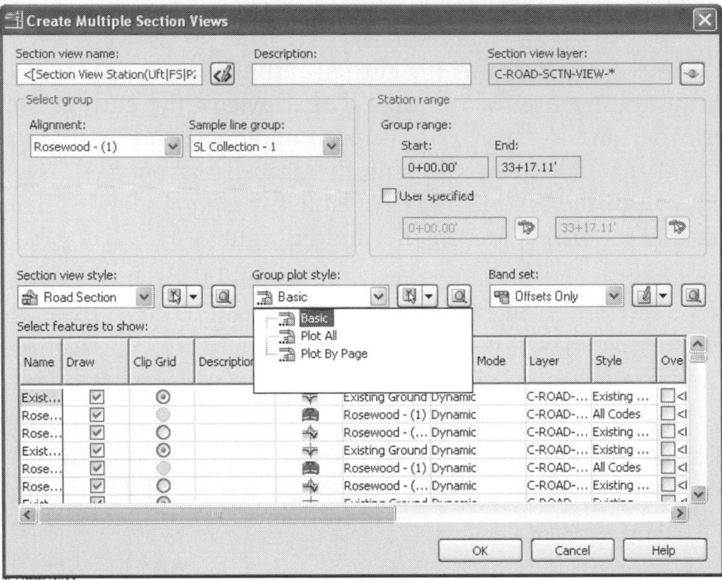

Figure 8.25

organize section views by Plot By Page or by Plot All. The Plot By Page style references a page layout style. The Plot All style defines an array of section views.

The bottom of the dialog box lists all of the sample lines and their sections. Users may not want to plot sections of a certain type. Sorting the list by clicking on a heading (section type) groups similar type sections. After sorting, select the group and turn off the plotting. The same can be done when wanting to change assigned styles.

GROUP PLOT STYLES

There are times when a user needs to plot more than one section at a time. The group plot styles define sheets or arrays of section views.

PLOT BY PAGE

A page style defines a sheet for sections.

Array Panel

The Array panel defines how to distribute the sections in the drawing, the number of sections (per row/column), and the spacing between rows/columns (top right) (see Figure 8.26). The spacing distance (on the left) is the number of grid lines between views. A sheet style defines the size of the grid.

The lower left side of this panel sets the beginning point of the array (such as the Lower Left corner); the justification (Centerline) of the sections around the origin point of the sheets; and if the cells are Uniform Per Row, Column (variable width), or All (same width for all).

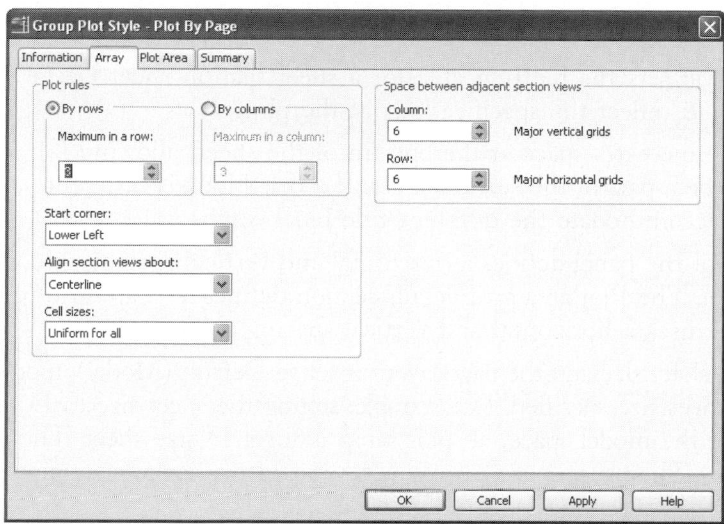

Figure 8.26

Plot Area Panel

The Plot Area panel defines the sheet size for the section views. The sheet style defines the sheet size and it margins. The top left of the dialog box sets the gap between successive sheets (see Figure 8.27).

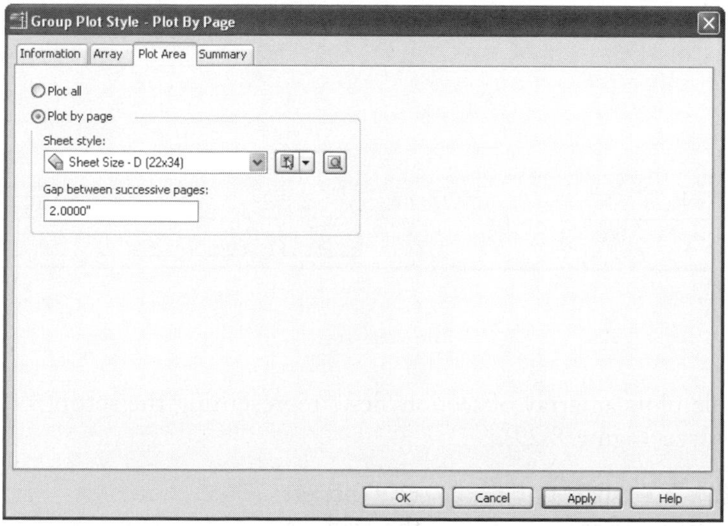

Figure 8.27

SHEET SIZE STYLE

A sheet size style sets the plotting area for a sheet section sheet (see Figure 8.28). The sheet sizes should reflect the specifications of the plotter.

The margins define extra space at the bottom of the sheet, allowing for a border with its information to be a part of the section sheet. If a title block is different, the margins need adjustment to accommodate the different title block shape.

The right side of the panel defines a horizontal and vertical spacing grid for the printable area of the sheet. The Plot area grid details section defines a "snap" grid for the sheet. The group plot style uses a horizontal and vertical spacing value.

When setting a sheet size and the page layout is set to Default (Model), model space must be set up for the sheet size specified. For example, setting the sheet size to D (24x36), the user must make sure the model space tab plot setup is for a D size sheet. The same is true for Default (Layout): The paper space layout must be defined prior to creating the section views.

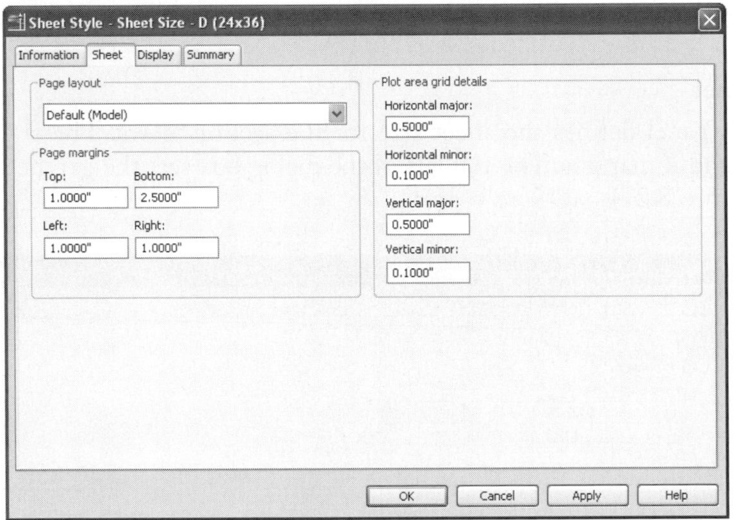

Figure 8.28

PLOT ALL

The plot all style plots an array of section views representing the settings of the Array and Plot Area tabs (see Figure 8.29).

SECTION VIEW BANDS – SET PROPERTIES

When creating section views, the section view and section label styles reference Surface1 and Surface2. Civil 3D assigns the Existing Ground to both variables. Users need to change this association when creating the section views. After setting the values in the Create View or Create Multiple Views commands and selecting an origin point in the drawing, a Section View Bands - Set Properties dialog box appears. Users set the Surface2

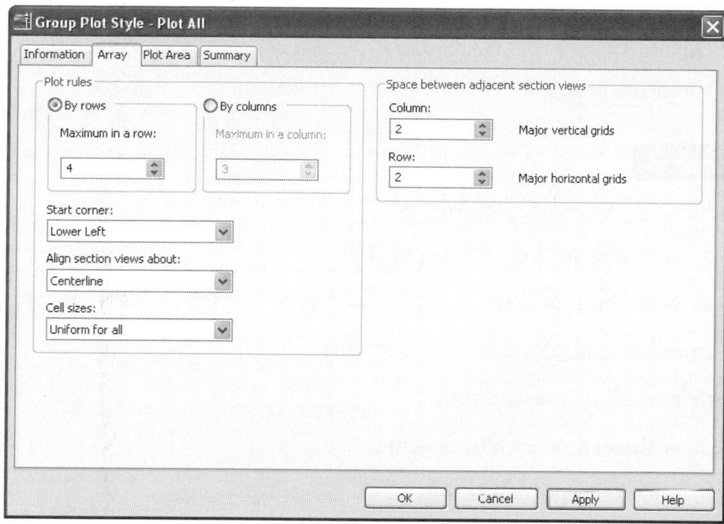

Figure 8.29

variable by clicking in the cell below Surface2 and selecting the appropriate element (see Figure 8.30). The default location for the bands is at the bottom of the section view.

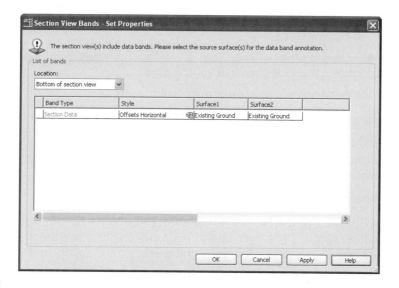

Figure 8.30

ERASING EXISTING SECTIONS

To erase and re-create the existing section views, users will need to resample the data sources. Before re-creating the section views, in Prospector select the Sample Line Group name. Right-click and select Properties..., click the Sections tab, toggle off the checked entries in the Sample column, and click the Resample button at the top right. Click the

Yes button to overwrite sections. After resampling, toggle on the desired Data Sources and click the Resample button. Click the Yes button to overwrite sections. Now the section views can be re-created.

EXERCISE 8-3

When you complete this exercise, you will be:

- Familiar with the settings for paper styles.
- Familiar with the plot page and plot all styles.
- Able to adjust and plot pages of sections.
- Able to plot all of the sections.
- Familiar with section view properties.
- Familiar with section properties.

EXERCISE SETUP

This exercise continues with the drawing from the last exercise. If you did not complete the previous exercise, you can open the drawing *Chapter 8 - Unit 3.dwg*. The file is in the *Chapter 8* folder on the CD that accompanies this textbook.

1. Open the drawing from the previous exercise or the *Chapter 8 - Unit 3* drawing.

REVIEW PAGE SHEET STYLE

When plotting with a page size group style, Civil 3D defines the sheet size, its printable area, and a grid spacing for snapping the cross sections on the page.

1. Click the **Settings** tab to make it the current panel.
2. In Settings, expand the Section View branch until you view the list of styles under the Sheet Styles heading.
3. From the list, click **Sheet Size - D (24x36)**, press the right mouse button, and from the shortcut menu select **Edit...**.
4. Click the **Sheet** tab to view its settings.

   The extra space at the bottom of the sheet allows for a border with its information around the sections. The right side of the panel defines a horizontal and vertical spacing grid for the printable area of the sheet.
5. Click the **Display** tab to view its settings.

   This controls the visibility of the page boundary, its printable area, and the grid.
6. Click the **Cancel** button to exit the dialog box.
7. In Prospector, expand the Corridors branch.

EXERCISES

8. If the Rosewood corridor has the out-of-date icon, select **Rosewood - (1)**, press the right mouse button, and from the shortcut menu select **Rebuild**.
9. Click the **AutoCAD Save** icon to save the drawing.

PLOT ALL

The Plot All method plots an array of sections (see Figure 8.29). The array is four sections wide and has as many rows as necessary to plot all of the sections.

1. Use the AutoCAD Pan and Zoom commands to move the site west.
2. From the Sections menu, select **Create Multiple Views....**

   The Create Multiple Section Views dialog box appears.
3. In the dialog box, near its center, set the Group Plot Style to **Plot All**, and click the **OK** button to continue.
4. Select a point in the lower left of the screen.
5. The Section View Bands - Set Properties dialog box displays.
6. Click the drop-list arrow for **Surface2**, select **Rosewood - (1)**, and click the **OK** button to create the section views.
7. Use the AutoCAD Pan and Zoom commands to view the sections.
8. Click the **AutoCAD Save** icon to save the drawing.

   The only annotation that remains is from the Multipurpose style, All Codes.

This concludes the unit on creating multiple and single section views. Each routine allows you to select which section to draw and what style to use for the section. The next unit covers the Multipurpose styles that annotate the subassembly segments of an assembly.

SUMMARY

- The Create Multiple Section Views command imports section views and sections in an array or page format.
- When creating sections using a sheet size, the page setup for model or paper space must be set to the specified sheet size.
- The Create View command plots one section view at a time.
- The Section View Properties Sections panel assigns, removes, or changes a section's label styles.
- Users need to assign a second surface to the Surface2 variable.

UNIT 4: MULTIPURPOSE CODE STYLES

The section view and section label styles primarily annotate existing ground or, if applied to an earthworks scenario, finished ground. The corridor section annotation of an assembly comes not from section view or section styles, but from code styles from the Multipurpose Style branch of the General branch. The code styles reference a subassembly's points, links, and shapes for their label data. While there may be more than one curb, there is only one code for back-of-curb, flange, and so on. The assignment of label styles to this code creates subassembly labels for all occurrences of this code.

CODE SET STYLES

The Code Set Style - All Codes dialog box lists all the possible points, links, and shapes that can be in an assembly (see Figure 8.31). Each entry has a description, style, label style, and feature line style. The Style column defines how the subassembly's points, links, and shapes display in the drawing. The Label Style column associates each entry with a label style. When assigning a label style, each section will show the label assigned in this dialog box. Even if there are no existing sections, when the user makes a change in the dialog box, the sections update to show the newly assigned style. The last column defines feature line styles for each entry.

CODE LABEL STYLE

A code label style references a point, link, or shape. As a result, when reviewing the contents, the Layout panel focuses on the type of object the style annotates.

MARKER LABEL STYLES

The marker label styles annotate the offset and elevation of a subassembly's point code (see Figure 8.32).

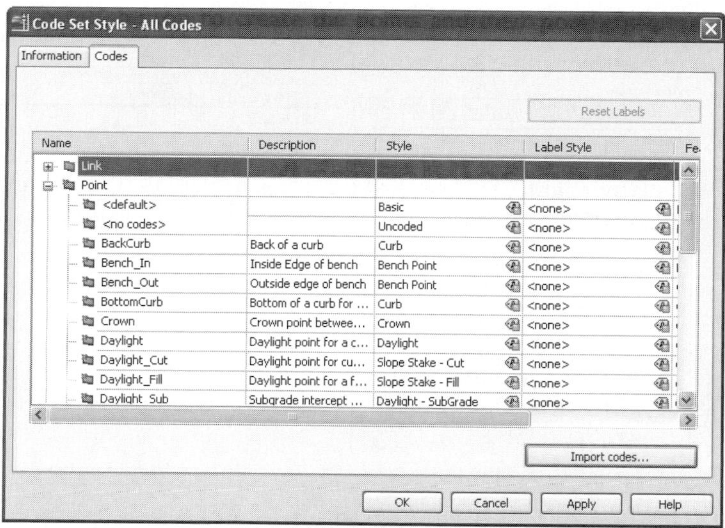

Figure 8.31

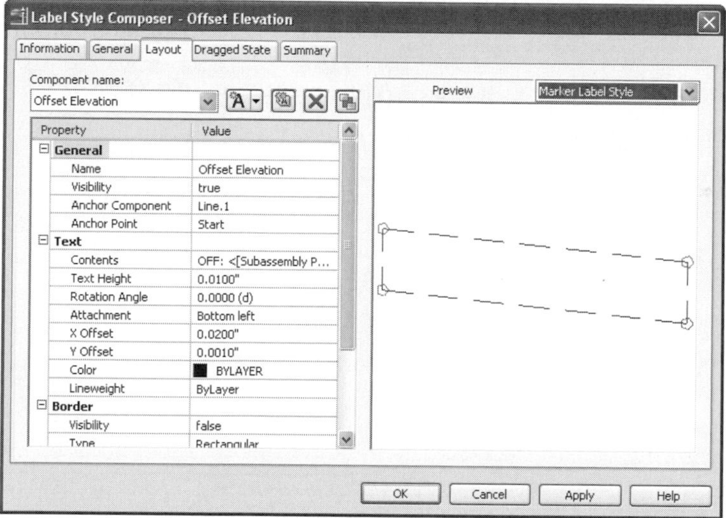

Figure 8.32

LINK LABEL STYLES

A link label style annotates the slope or grade of a subassembly's link (see Figure 8.33).

SHAPE LABEL STYLES

A shape label style annotates a shape in an assembly (see Figure 8.34).

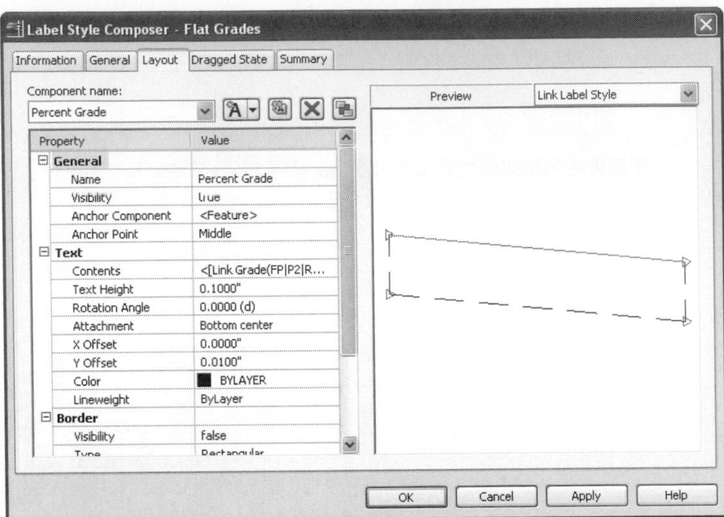

Figure 8.33

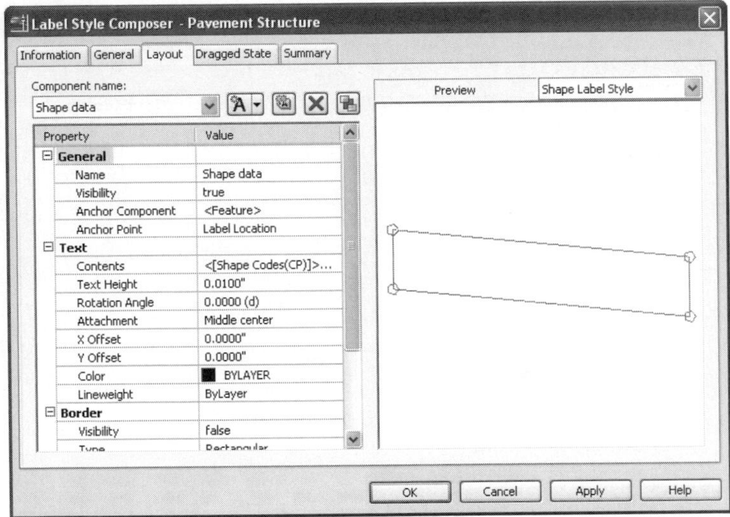

Figure 8.34

EXERCISE 8-4

After completing this exercise, you will be:

- Familiar with the code set label styles.
- Able to create a new code label set.
- Familiar with the code label styles.
- Able to assign a code label style to a code label set.

EXERCISE SETUP

This exercise continues with the drawing from the last exercise. If you did not complete the previous exercise, you can open the drawing *Chapter 8 - Unit 4.dwg*. The file is in the *Chapter 8* folder on the CD that accompanies this textbook.

1. Open the drawing from the previous exercise or the *Chapter 8 - Unit 4* drawing.

REVIEW ALL CODES CODE SET STYLE

The All Codes Set style contains all possible codes, links, and shapes.

1. Click the **Settings** tab to make it current.
2. In Settings, expand the General, then Multipurpose Styles branches until you view the list of code set styles.
3. From the list of styles, select **All Codes**, press the right mouse button, and from the shortcut menu select **Edit…**.
4. Click the **Codes** tab to view its contents.

   The only label is for daylight (Steep Grades).
5. Click the **Cancel** button to exit the dialog box.

MODIFY EXISTING LABELS

The slope label for the daylight slope is too large for the sections and needs adjusting. You also need to resize the text in the Offset and Elevation label style.

1. Click the **Prospector** tab to make it current.
2. In Prospector, expand the Sites branch until you view the list of sections under the SL Collection -1 heading.
3. Scroll through the list of sections until you locate the section for 10+00.
4. Expand the 10+00 branch until you view the list of Section Views under the Section View heading.
5. From the list of section views, select **10+00.00**, press the right mouse button, and from the shortcut menu select **Zoom to**.
6. Use the AutoCAD Zoom and Pan commands to better view the section.

EXERCISES

The text is too large for the section. Adjusting the style definition changes the text in all sections using the style.

7. Click the **Settings** tab to make it current.
8. Expand the General, Labels Styles branch until you view the list of styles for the Link heading.
9. From the list of styles, select **Steep Grades**, press the right mouse button, and from the shortcut menu select **Edit...**.
10. Select the **Layout** tab to make it current.
11. In the Label Style Composer - Steep Grades, Text section, click in the Text Height Value cell, change the size to **0.05**, and click the **OK** button to exit the dialog box.
12. The text changes for the sections.
13. Click the **AutoCAD Save** icon and save the drawing.

EXERCISES

CODE LABEL STYLES

In this exercise you are going to label the station and offset of the edge-of-travelway (ETW). This is done by assigning the the Offset Elevation style for the ETW entry in the Code Set Style - All Codes dialog box (see Figure 8.35).

1. Return to the All Codes style in the General, Multipurpose Styles, Code Set Styles branch.
2. Select **All Codes** from the list, press the right mouse button, and from the shortcut menu select **Edit...**.
3. Scroll through the list and locate the ETW entry.
4. In the All Codes dialog box for the ETW entry, in the Label Style column, click the icon to display the Pick Style dialog box.
5. Click the drop-list arrow, select the **Offset Elevation** style from the list, and click the **OK** button to return to the All Codes dialog box.

   Labels appear on the edge-of-travelway point on the corridor section.
6. Expand the General, Labels Styles branch until you view the list of styles for the Marker heading.
7. From the list of styles, select **Offset Elevation**, press the right mouse button, and from the shortcut menu select **Edit...**.
8. Select the **Layout** tab to make it current.
9. In the Label Style Composer - Offset Elevation, in the Border section Offset Elevation, click in the Background Mask Value cell and change it to **True**.
10. In the Label Style Composer, click the Component Name drop-list arrow, and from the list select **Point Code**.

11. In the Label Style Composer - Offset Elevation, in the Border section for Point Code, click the Background Mask Value cell, and change it to **False**. Click the **OK** button to exit the dialog box.
12. Click the **OK** button to exit.
13. Click the **AutoCAD Save** icon to save the drawing.

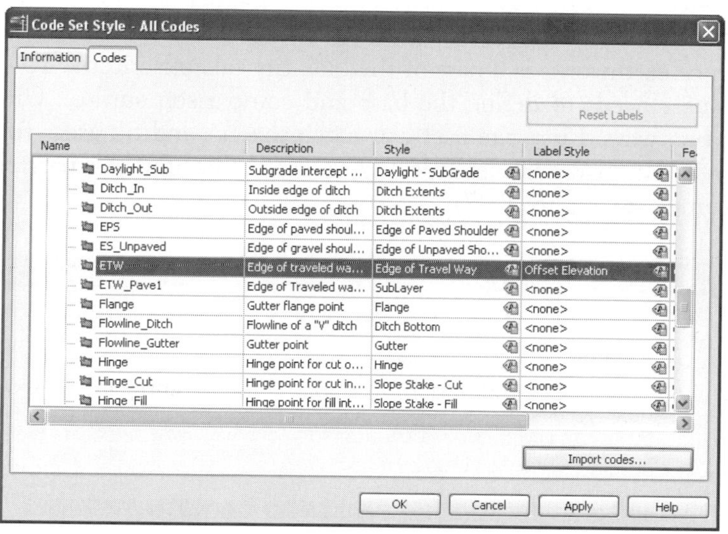

Figure 8.35

This concludes the discussion on labeling corridor components in a section view. The next unit covers the calculation of quantities.

SUMMARY

- The section view and section styles do not label the corridor components.
- The code set style contains the styles for the line work representing the markers, links, and shapes of an assembly.
- A code set style labels the components of a corridor section.

UNIT 5: QUANTITY TAKEOFFS

A final calculation for the corridor and its subassemblies and surfaces are quantity estimates for mass earthworks and material needs. The Volumes routine of the Utilities flyout of the Surfaces menu, as well as a volume surface comparing Existing Ground and Rosewood - Datum, produce an earthworks volume calculation for a roadway design.

However, the quantity takeoff routine creates a formatted report of the earthwork and materials volumes.

Before calculating a volume, the takeoff criteria must be defined. This is done in the Settings, Takeoff Criteria branch.

Users can create multiple material lists for earthwork or material volumes.

EARTHWORKS

The calculation of earthworks is a part of the quantity takeoff criteria. To calculate a mass earthwork, the user needs to define the base and comparison surface. Civil 3D uses this information in two ways. First, the user can hatch the cut and fill areas for sections using the comparison of the two surfaces (see Figure 8.36). Second, the user can calculate an earthworks volume (see Figure 8.37).

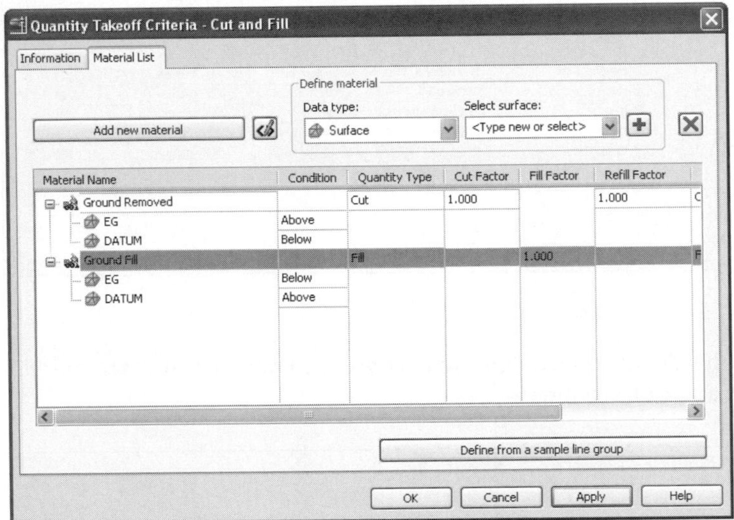

Figure 8.36

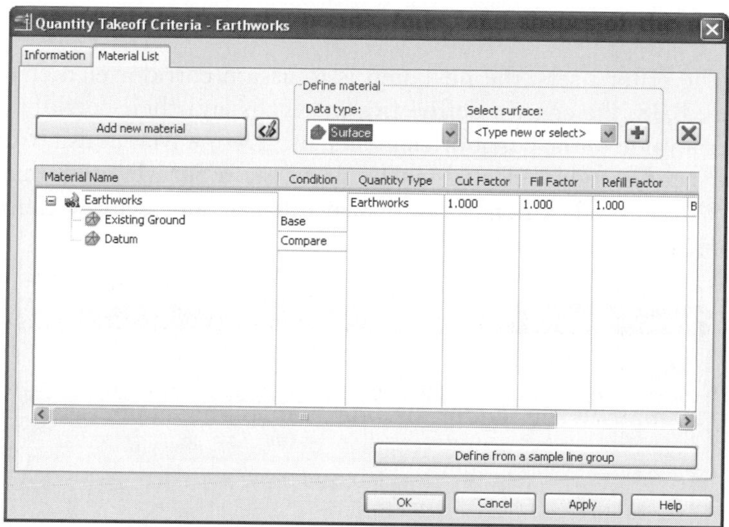

Figure 8.37

MATERIALS

When calculating volumes for materials, the criteria switches to identifying structures (subassemblies) in the corridor sections (i.e., Pave1, Pave2, Subbase, Base, Curb, etc.). See Figure 8.38. When using this criteria set, the volumes are the material needs of the road design.

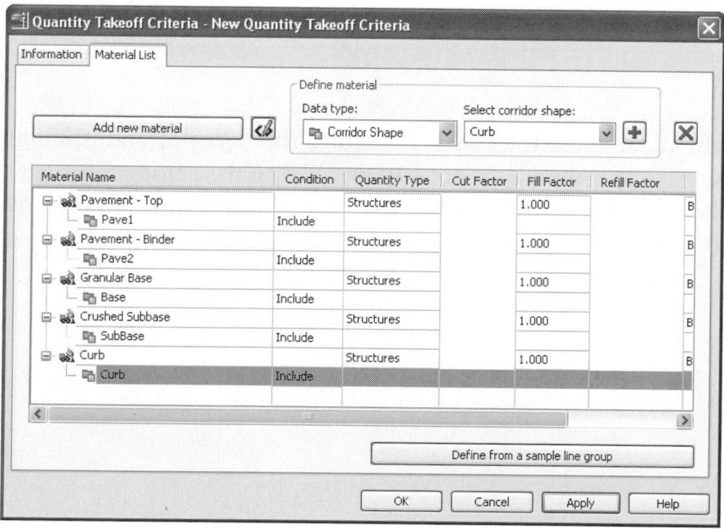

Figure 8.38

QUANTITY TAKEOFF

After defining the criteria sets, the next step is to assign corridor elements to the criteria values. In Figure 8.38, the criteria define five materials and their corridor equivalents. To associate a criteria value to the corridor elements, use Define Materials... from the Sections menu. When this command is selected for the first time, a Set Materials dialog box appears (see Figure 8.39). Click <click here...> to display a list of subassembly elements and select the element from the list.

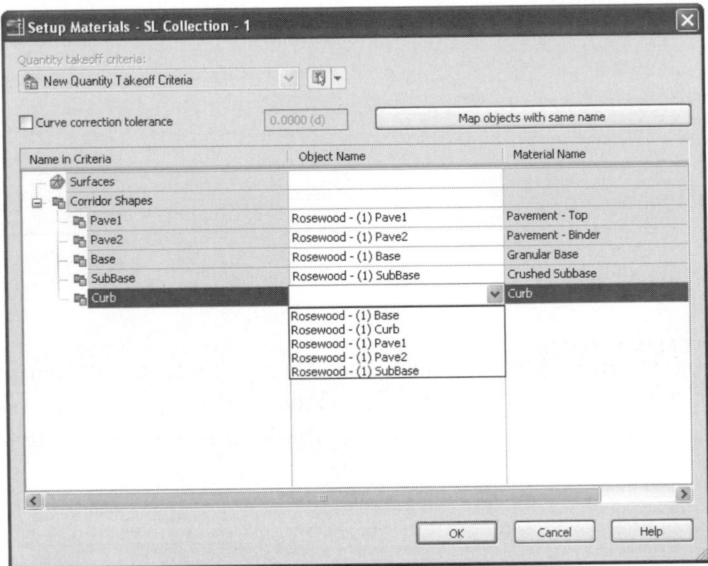

Figure 8.39

When exiting the dialog box, the routine samples the corridor and calculates the volumes. To view the volumes as a report, select Define Materials... from the Sections menu. This command displays the Report Quantities dialog box (see Figure 8.40). In this dialog box, assign the alignment, SLG, material list, and style sheet. Users can also create a LandXML report for the current settings.

Figure 8.40

After setting the values in the Report Quantities dialog box and clicking the OK button, the routine produces a report (see Figure 8.41).

	Crushed Subbase	30.06	0.00	55.66
	Curb	3.23	0.00	5.99
Station: 0+75.000				
	Pavement - Top	4.00	3.70	11.11
	Pavement - Binder	4.00	3.70	11.11
	Granular Base	7.99	7.40	22.20
	Crushed Subbase	30.06	27.83	83.49
	Curb	3.23	2.99	8.98
Station: 1+00.000				
	Pavement - Top	4.00	3.70	14.81
	Pavement - Binder	4.00	3.70	14.81
	Granular Base	7.99	7.40	29.60
	Crushed Subbase	30.06	27.83	111.32
	Curb	3.23	2.99	11.97
Station: 1+25.000				
	Pavement - Top	4.00	3.70	18.51
	Pavement - Binder	4.00	3.70	18.51
	Granular Base	7.99	7.40	37.00
	Crushed Subbase	30.06	27.83	139.15

Figure 8.41

EXERCISE 8-5

After completing this exercise, you will be:

- Familiar with the quantity criteria takeoff.
- Able to create a quantity takeoff earthwork and material report.
- Able to create a quantity takeoff table.

EXERCISE SETUP

This exercise continues with the drawing from the last exercise. If you did not complete the previous exercise, you can open the drawing *Chapter 8 - Unit 5.dwg*. The file is in the *Chapter 8* folder on the CD that accompanies this textbook.

1. Open the drawing from the previous exercise or the *Chapter 8 - Unit 5* drawing.

CUT AND FILL

The first part of the exercise creates cut and fill hatching for each section view.

1. Click the **Settings** tab to make it current.
2. Expand the Quantity Takeoff branch until you view the list of takeoffs under the Quantity Takeoff Criteria heading.
3. From the list, select **Cut and Fill**, press the right mouse button, and from the shortcut menu select **Edit...**.
4. Click the **Material List** tab to view its contents.

   The condition settings are incorrect for Cut and Fill, so they need to change to match Figure 8.36.

5. For EG in the Condition column of Ground Removed, click in the cell to display a drop-list arrow. Click the drop-list arrow, and select **Above**.
6. For Datum in the Condition column of Ground Removed, click in the cell to display a drop-list arrow. Click the drop-list arrow, and select **Below**.
7. For EG in the Condition column of Ground Fill, click in the cell to display a drop-list arrow. Click the drop-list arrow, and select **Below**.
8. For Datum in the Condition column for Ground Fill, click in the cell to display a drop-list arrow. Click the drop-list arrow, and select **Above**.
9. Click the **OK** button to exit the dialog box.

EARTHWORKS

1. From the Quantity Takeoff Criteria list, select **Earthworks**, press the right mouse button, and from the shortcut menu select **Edit...**.
2. Click the **Material List** tab to view its contents.

The criteria compare a base to a comparison surface, just like the Volume routine in Surface/Utilities to define a volume surface.

3. Click the **OK** button to exit the dialog box.

CREATE MATERIAL QUANTITY TAKEOFF CRITERIA

You have five materials in the assembly that have a potential volume: Pave1, Pave2, Base, Subbase, and Curb.

1. In Settings, the Quanitity Takeoff branch, select the **Quantity Takeoff Criteria** heading, press the right mouse button, and from the shortcut menu select **New....**
2. Click the **Information** tab, and for the name of the criteria type **HC3D – Materials**.
3. Click the **Material List** tab to view its contents.
4. In the dialog box, click the **Add New Material** button at the top left to make a new material entry.
5. Click in the Material Name cell and type **Pavement Top** as the name.
6. Click in the Quantity Type cell for Pavement Top; the cell displays a drop-list arrow. Click the drop-list arrow, and select **Structures**.
7. At the top center of the dialog box, to the right of Data Type:, click the drop-list arrow, and select **Corridor Shape** from the list.
8. At the top right of the dialog box, to the right of Select Corridor Shape, click the drop-list, and select **Pave1** from the list.
9. At the top right of the dialog box, click the blue plus sign **+** to add Pave1 to the Pavement Top material name.
10. Using the values in Table 8.1 and repeating Steps 4 through 9, add the remaining materials to the quantity takeoff criteria.

Table 8.1

Material Name	Quantity Type	Data Type	Corridor Shape
Pavement - Binder	Structures	Corridor Shape	Pave2
Granular Base	Structures	Corridor Shape	Base
Crushed Subbase	Structures	Corridor Shape	Subbase
Curb	Structures	Corridor Shape	Curb

When finished, your dialog box should look like Figure 8.38.

11. Click the **OK** button to exit.
12. Click the **AutoCAD Save** icon to save the drawing.

CREATE MATERIAL LISTS

You create material lists that match the quantity takeoff criteria entries to corridor elements.

1. From the Sections menu, select **Define Materials...**.
2. In the Select a Sample Line Group dialog box, set the alignment to **Rosewood - (1)**, the Sample line group to **SL Collection - 1**, and click the **OK** button to continue.

   The Setup Materials dialog box displays. At the top left, the current criteria is Cut and Fill.
3. In the Object Name cell for EG, click **<Click here to set all>**, and from the list select **Existing Ground**.
4. Repeat the previous step for the Datum entry and select **Rosewood - (1) Rosewood - Datum**.
5. Click the **OK** button to compute the materials.

Create a Second Material

1. From the Sections menu, select **Define Materials...**.
2. In the Select a Sample Line Group dialog box, set the alignment to **Rosewood – (1)**, the Sample line group to **SL Collection – 1**, and click the **OK** button to continue.
3. In the Edit Material List dialog box, in the bottom right, click the **Import Another Criteria** button.
4. In the Select a Quantity Takeoff Criteria dialog box, click the drop-list arrow, select **Earthworks**, and click the **OK** button.
5. In the Set Materials dialog box, for Existing Ground, click **<Click here...>** and select **Existing Ground** from the surface list.
6. Repeat the previous step for Datum, but select the **Rosewood - (1) Rosewood - Datum** surface.
7. Click the **OK** buttons to add the materials to the list, and exit the dialog boxes.

   The routine calculates the material values.

Create a Third Material

1. From the Sections menu, select **Define Materials...**.
2. In the Select a Sample Line Group dialog box, set the alignment to **Rosewood – (1)**, the Sample line group to **SL Collection – 1**, and click the **OK** button to continue.
3. In the Edit Material List dialog box, in the bottom right, click the **Import Another Criteria** button.

4. In the Select a Quantity Takeoff Criteria dialog box, click the drop-list arrow, select **HC3D - Materials** from the list, and click the **OK** button to continue.
5. In the Setup Materials dialog box, for Pavement Top, click **<Click here...>**, and select **Rosewood - (1) - Pave1** from the surface list.
6. Repeat the previous step for Pave2, Base, Subbase, and Curb, but select the appropriate subassembly.
7. Click the **OK** button to add the materials to the list.

   The Edit Material List dialog box now contains three material lists, each creating a different volume.
8. Scroll through the materials list in the Edit Material List dialog box.
9. Click the **OK** button to exit the dialog box.

CALCULATE EARTHWORK VOLUMES

You now have the choice of creating a report or table with the current material lists.

1. From the Sections menu, select **Generate Volume Report...**.

   The Report Quantities dialog box displays.
2. In the Report Quantities dialog box, set the Alignment to **Rosewood - (1)**.
3. In the Report Quantities dialog box, set the Material List to **Material List - (2)**.
4. In the Report Quantities dialog box, click the style sheet icon to the right of Select a style sheet, select **Earthwork.xsl**, and click the **Open** button.
5. Click the **OK** button to create an earthwork volume report.
6. If you get a warning about scripts, click the **Yes** button to continue.
7. Click the "**X**" to close the report window.

CALCULATE MATERIAL VOLUMES

1. From the Sections menu, select **Generate Volume Report...**.

   The Report Quantities dialog box displays.
2. In the Report Quantities dialog box, set the Alignment to **Rosewood - (1)**.
3. In the Report Quantities dialog box, set the Material List to **Material List - (3)**.
4. In the Report Quantities dialog box, select the style sheet icon to the right of Select a style sheet, select **Select Material.xsl**, and click the **Open** button.
5. Click the **OK** button to create a materials volume report.
6. If you get a warning about scripts, click the **Yes** button to continue.
7. Click the "**X**" to close the report window.

CREATE A VOLUME TABLE

1. From the Sections menu, the Add Tables flyout, select **Total Volume...**.
2. In the Create Total Volume Table dialog box, set the Alignment to **Rosewood - (1)**, set the Material List to **Material List - (1)**, and click the **OK** button.
3. In the drawing select a point to locate the table in the drawing.
4. Select the **AutoCAD Save** icon to save the drawing.

This ends the unit on calculating volumes from a corridor. The process requires you to define a materials list based on quantity takeoff criteria. After creating the material lists, you create reports or tables displaying the calculated values.

SUMMARY

- Quantity takeoff criteria define the EG and comparison surface for earthworks.
- Quantity takeoff criteria define the number of structures and the subassembly component representing the material.
- A materials list links an entry in the quantity takeoff criteria to a subassembly component.
- After defining a materials list, users can create tables or reports with volume calculations.

This ends the section view and section chapter. The section is the cross-section view of a roadway design. It is an important part of the evaluation and documentation of a design.

CHAPTER 9

Benches, Transitions, and Superelevation

INTRODUCTION

Roadway designs must accommodate a variety of situations. These situations include road widening, cut or fill designs, and sustaining constant speeds. Civil 3D has subassemblies that react to changes in depth above or below the existing ground surface, transitions to widen or narrow lane widths, and transitions to tilt to one side or another to safely maintain speeds through a design.

OBJECTIVES

This chapter focuses on the following topics:

- Creation of Benches When in Deep-Cut Designs
- Creation of a Knuckle Cul-De-Sac Design
- Transition of a Road to Different Lane Widths
- Superelevation of a Road Design Using Alignment Design Parameters

OVERVIEW

Civil 3D uses the parameters of a subassembly, and for transitions it uses additional alignments to create a solution for the situations that will be discussed in this chapter. When designing benches, the subassembly parameters control the height, width, and slope of the bench (see Figure 9.1). If the depth of cut or fill is great, the subassembly will attempt to place a second bench in the search for daylight.

When creating a corridor with transitioning pavement, there needs to be additional alignments (offset alignments) to control the edge-of-travelway. In addition to horizontal alignments, Civil 3D requires a vertical alignment for each horizontal transition alignment. The Create Corridor dialog box is where a user specifies the alignment attachments to the assembly (see Figure 9.2). All of the subassemblies that transition must be attached to the offset alignment.

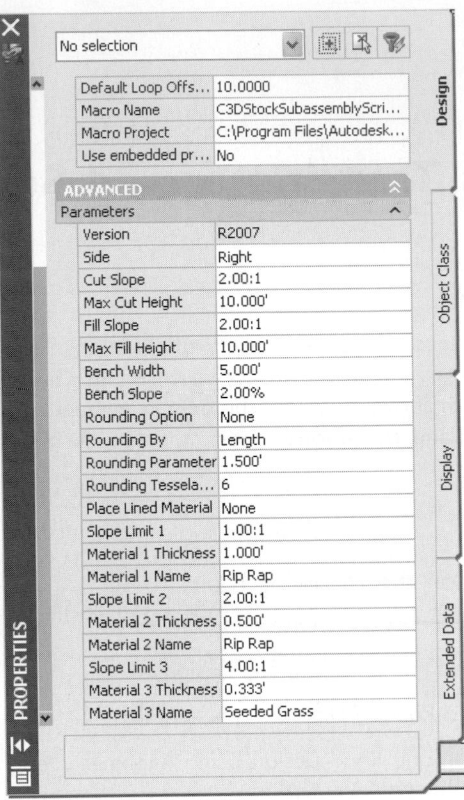

Figure 9.1

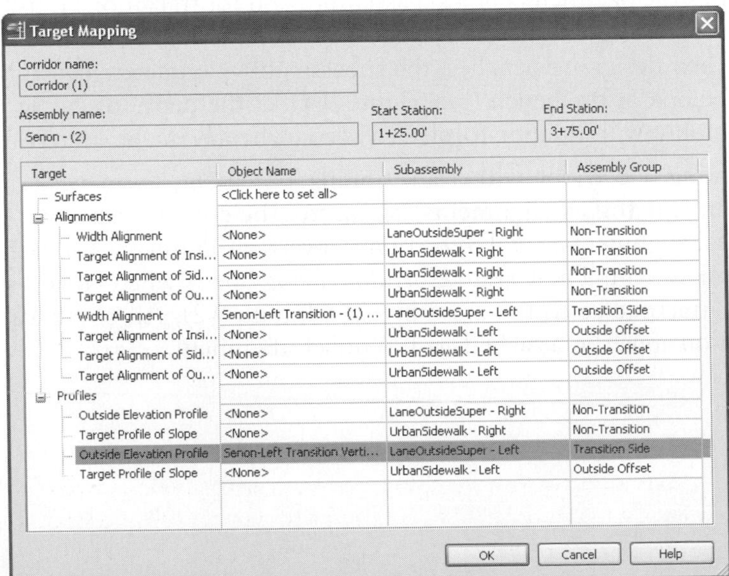

Figure 9.2

When designing superelevations, design speeds and a superelevation table values must be a part of the alignment's properties (see Figure 9.3 and Figure 9.4).

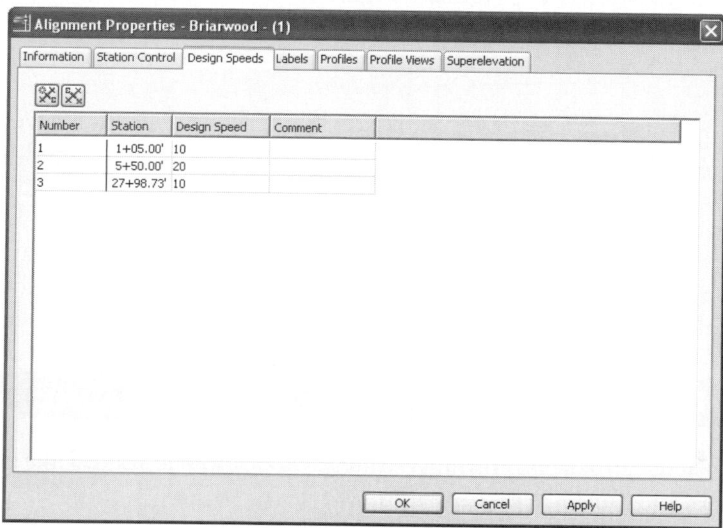

Figure 9.3

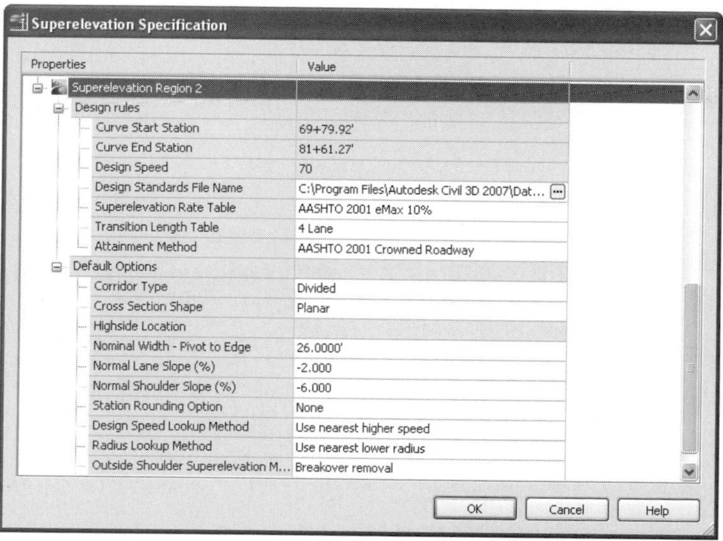

Figure 9.4

UNIT 1

The focus of Unit 1 is the design of benches. Benches are used for roads with deep cuts or large fill amounts. The parameters of the Daylight Bench subassembly define the initial bench. If the design cuts deeper or fills to a higher elevation, multiple benches will appear in the corridor.

UNIT 2

The focus of the second unit is the process of roadway transitioning. Civil 3D requires a second alignment to stretch a pavement segment. The example for this unit is a street knuckle in a subdivision.

UNIT 3

This unit reviews the design of a cul-de-sac. Civil 3D widens the pavement (edge-of-travelway) with additional horizontal alignments (right and left) attached to the assembly.

UNIT 4

Superelevation is a common design method when working with roadways that carry significant traffic, that have higher traveling speeds, and that have safety concerns when encountering turns. The assignment of speeds and the rules for superelevating are the focus of this unit.

UNIT 1: BENCHING

Benching places ledges that pool or move water away from the road. Like ditches, benches generally run parallel to a road design. By adjusting the benching parameters of the DaylightBench subassembly, users can have benches acting as a break in slope that allow water to travel to a ditch cut or away from the road in fill.

The Max cut height parameter of the DaylightBench subassembly (Figure 9.1) determines the occurrence of a bench. If the depth of cut or height of fill exceeds this value (height), the subassembly will create a second bench. For example, if the depth becomes twice the parameter's value, there will be two benches. The subassembly will create a new bench whenever the cut or fill height exceeds the height value.

GENERIC SUBASSEMBLIES

There are many situations where generic links between subassemblies are necessary. For example, there is no daylight subassembly that has a ditch and creates benches. To accomplish this and other design needs, generic links are indispensable.

By adding generic subassemblies to an assembly, users can solve design issues without having to define custom subassemblies. Each type of generic link has a purpose and addresses specific design issues (such as slope, elevation, and offset). An example of using a pair of generic slope and distance links is creating a foreslope to and width of a ditch between a shoulder and a benching daylight subassembly.

EXERCISE 9–1

After completing this exercise, you will:

- Be familiar with design settings for benches.
- Be able to use and set values for generic link subassemblies.
- Create a corridor with benches.

EXERCISE SETUP

You start this exercise by opening the *Chapter 9 - Unit 1.dwg* found in the *Chapter 9* folder of the CD that accompanies this book.

1. Open the drawing *Chapter 9 - Unit 1.*
2. From the File menu, select **Save As...** and save the drawing under a different name.
3. Make sure you can view and select from the Transparent Commands toolbar.

CREATE A MULTIPLE-SURFACE PROFILE

The profile view for the exercise will be located directly under the alignment and contains multiple surfaces.

1. Use the AutoCAD Pan and Zoom commands to move the alignment to the top of the screen.
2. From the Profiles menu, select **Create from Surface....**
3. In the Create Profile from Surface dialog box, hold down the **Ctrl** key, select all of the surfaces, and click the **Add>>** button to place the surfaces on the Profile list.
4. In the dialog box, double-click in the Style cell for Limestone. In Pick Profile Style, click the drop-list arrow, select the **Limestone** style from the list, and click the **OK** button to return to the Create Profile from Surface dialog box (see Figure 9.5).
5. Repeat Step 4 for the Shale profile, assigning it the **Shale** style.

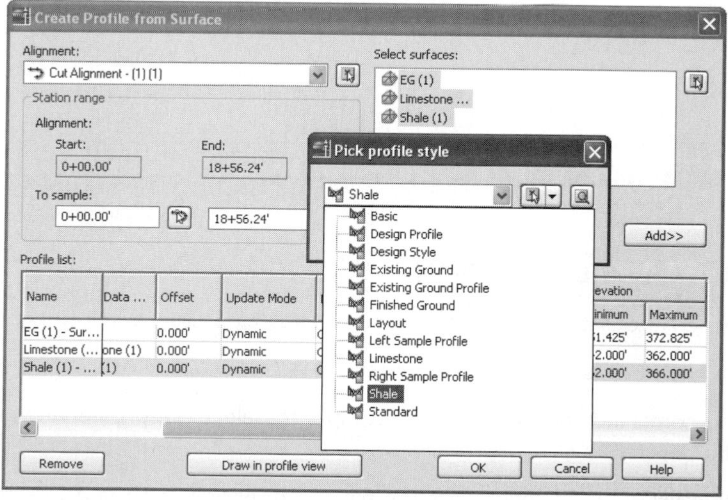

Figure 9.5

6. In the lower middle of the dialog box, click the **Draw in Profile View** button to continue.

7. In the **Create Profile View** dialog box, change the profile type to **Clipped Grid** and click the **OK** button.
8. In the command line, the routine prompts for a location. In the drawing, select a point under the alignment.
9. If necessary, move the profile view to a location similar to Figure 9.6.

   The Event Viewer appears, reminding you to set a Profile2 value.
10. Close the Event Viewer.
11. Click the **AutoCAD Save** icon and save the drawing.

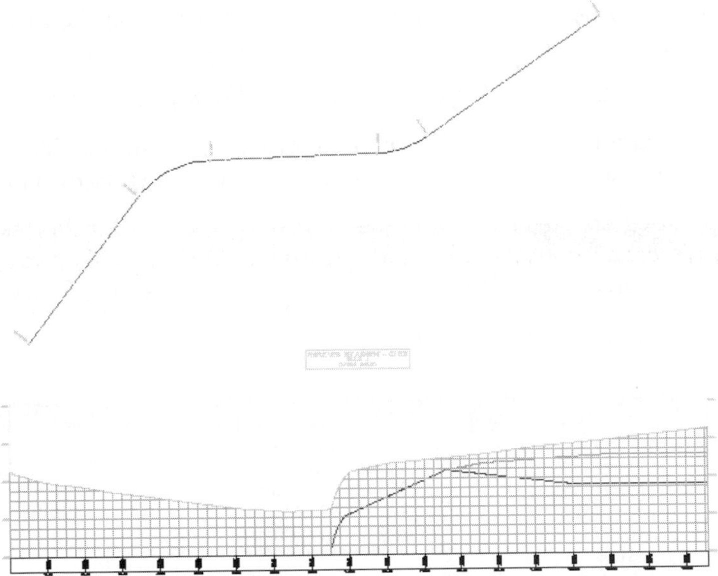

Figure 9.6

CREATE THE VERTICAL ALIGNMENT

1. Use the AutoCAD Pan and Zoom commands to better view the profile.
2. From the Profiles menu, select **Create by Layout...**.
3. In the command line, the routine prompts to select the profile view; select any part of the profile view.
4. After selecting the profile view, the routine displays the Create Profile - Draw New dialog box. For the Name, type **Cut**, leaving the counter, and click the **OK** button to accept the remaining defaults and to continue.
5. In the Profile Layout Tools toolbar, click the drop-list arrow of the leftmost icon and select **Curve Settings...** from the list.

6. In the Vertical Curve Settings dialog box, set the Curve Type to **Parabolic** and the Lengths to **200**.
7. Click the **OK** button to exit the dialog box.
8. Again, click the drop-list arrow of the leftmost icon and select **Draw Tangents With Curves** from the list.

   You may have to zoom in or pan to better see the points.
9. In the drawing, use an intersection object snap, selecting the intersection of the EG surface and the left vertical axes of the profile as the beginning of the vertical design.

   If you are using a running object snap, toggle **OFF** osnap.

   The remaining PVIs are set with the Profile Station Elevation override of Transparent Commands toolbar. Table 9.1 has the stations and their elevation, as does Figure 9.7. The last PVI is the intersection of the EG surface and the right edge of the profile view.
10. From the Transparent Commands toolbar, select **Profile Station Elevation**.

EXERCISES

Table 9.1

PVI Station	Elevation	Curve Length
0+00	Intersection	
6+50	345.00	175.00
11+50	345.00	275.00
18+56.24	Endpoint	

11. In the command line, the override prompts for a profile view; select anywhere on the profile view.
12. In the command line, the prompt asks for a profile station; type **650** as the station and press the **Enter** key.
13. In the command line, the prompt asks for a profile elevation; type **345.00** as the elevation and press the **Enter** key.
14. Repeat Steps 12 and 13, but type the values for station 11+50 and 345.00.
15. Press the **Esc** key to stop the override. Using the endpoint object snap, select the endpoint of the EG surface and the right axes of the profile.
16. Use the AutoCAD Zoom command to view the profile.
17. In the Profile Layout Tools toolbar, click the **Profile Layout Parameters** icon (middle right), then select the **Select PVI** icon (the next button to the left).
18. In the profile view, select a point near the first vertical curve (6+50), change the curve length to **175**, and select a gray color cell to complete the edit.

19. In the profile view, select a point near the second vertical curve (11+50), change the curve length to **275**, and select a gray color cell to complete the edit.
20. Close the Profile Layout Parameters vista.
21. Close the Profile Layout Tools toolbar.

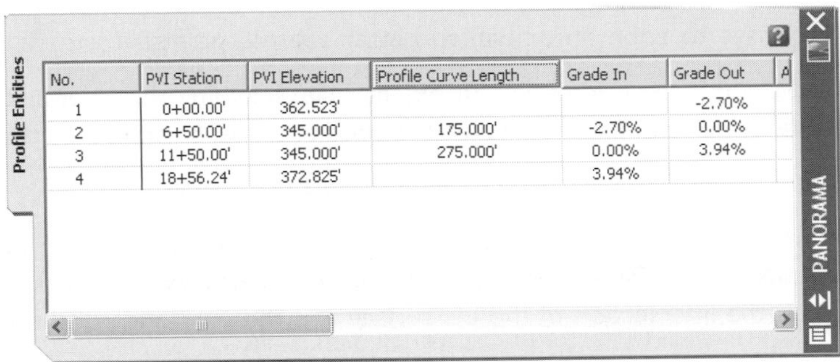

No.	PVI Station	PVI Elevation	Profile Curve Length	Grade In	Grade Out
1	0+00.00'	362.523'			-2.70%
2	6+50.00'	345.000'	175.000'	-2.70%	0.00%
3	11+50.00'	345.000'	275.000'	0.00%	3.94%
4	18+56.24'	372.825'		3.94%	

Figure 9.7

22. Click the **AutoCAD Save** icon and save the drawing.

CREATE AN ASSEMBLY

1. From the Corridors menu, select **Create Assembly....** For the name type **Cut,** leave the counter. Click the **OK** button to accept the defaults, and in the drawing to the right of the profile, select a point to locate the assembly.

ADD SUBASSEMBLIES

The assembly is a combination of the basic lane, a shoulder, two generic links creating the foreslope to a ditch, and the bench daylight subassembly. After placing each subassembly, you should edit the subassembly's properties and assign a more appropriate name. Figure 9.10 is the final configuration for this assembly.

1. From the General menu, select **Tool Palettes Window**.

Basic Lane Subassembly

1. On the palette, click **Imperial - Basic**.
2. From the Imperial - Basic palette, select **BasicLane**. In Properties, the Parameters area, set Side to **Right**, and in the drawing select the assembly to place the lane.
3. In Properties, the Parameters area, change Side to **Left**. In the drawing, select the assembly to place the left basic lane.
4. Press the **Enter** key twice to exit the subassembly routine.
5. If necessary, click the **Prospector** tab to make it current.

6. In Prospector, expand the Subassemblies branch, click the first subassembly entry, press the right mouse button, and from the shortcut menu select **Properties...**.

   It should be the right basic lane.

7. In the dialog box, click the **Information** tab, change the name to **Basic Lane Right**, and click the **OK** button to exit the dialog box.
8. In Prospector, click the second subassembly entry, press the right mouse button, and from the shortcut menu select **Properties...**.
9. In the dialog box, click the **Information** tab and change the name to **Basic Lane - Left**. Click the **OK** button to exit the dialog box.

Shoulder Subassembly

1. From the Imperial – Basic palette, select **BasicShoulder**.

   In Properties, the attachment side should be right.

2. In Properties, the Parameters area, if the attachment Side is **Left**, change it to **Right**. In the drawing, attach the shoulder to the right side of the pavement.
3. Change the attachment Side to **Left** and attach a shoulder to the left side of the roadway.
4. Press the **Enter** key twice to exit the subassembly routine.
5. In Prospector, the Subassemblies branch, click the first shoulder entry, press the right mouse button, and from the shortcut menu select **Properties...**.

   It should be the right basic shoulder.

6. Click the **Information** tab, change the name to **Basic Shoulder - Right**, and click the **OK** button to exit the dialog box.
7. Repeat Steps 5 and 6 and rename the second shoulder.
8. Click the **Information** tab and change the name to **Basic Shoulder - Left**.

Foreslope Generic Link

Adding generic subassemblies allows you to solve design issues without having to define new subassemblies. You will attach two links that act as a foreslope from the shoulder to the beginning of a ditch and the ditch's bottom.

1. In the palette, click the **Imperial - Generic** tab to make current.
2. From the palette, select **LinkWidthAndSlope** from the Generic palette.
3. In Properties, the Parameters area, change the Side to **Right**, the Width to **3**, and the Slope to **–20%** (see Figure 9.8).
4. Attach the link to the right side.
5. In Properties, the Parameters area, change the Side property to **Left** and attach the link to the left side.

EXERCISES

6. Press the **Enter** key twice to exit the subassembly routine.
7. In Prospector, the Subassemblies branch, click the first Link entry, press the right mouse button, and from the shortcut menu select **Properties...**.
8. Click the **Information** tab and change the name to **Ditch Foreslope – Right**. Click the **OK** button to exit.
9. Repeat Steps 7 and 8, renaming the second Link to **Ditch Foreslope - Left**.

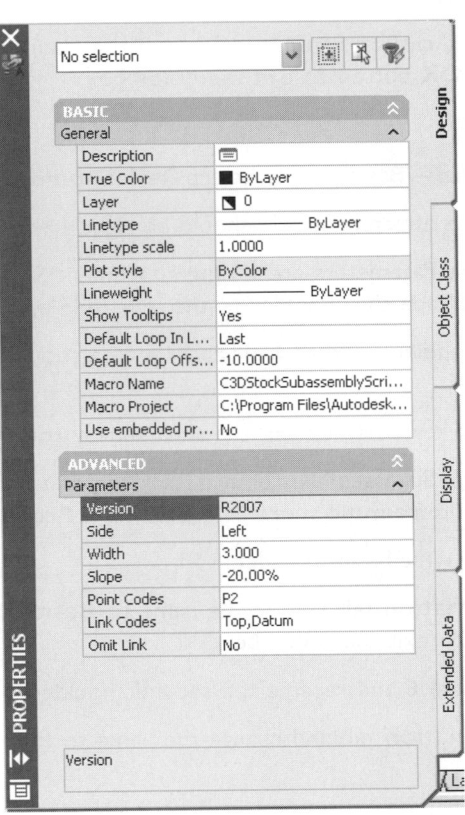

Figure 9.8

Ditch Bottom Generic Link

The Ditch Bottom is a second link that has a slope, distance, and change in elevation.

1. From the Imperial - Generic palette, select **LinkSlopeAndVerticalDeflection**.
2. In Properties, the Parameters area, change the Side to **Right**, the Slope to **1%**, and the Vertical Deflection to **0.025** (see Figure 9.9).
3. In the drawing, attach the link to the right side of the last attached link.
4. In Properties, the Parameters area, change the Side to **Left**, and in the drawing attach the link to the left side of the leftmost link.

5. Press the **Enter** key twice to exit the subassembly routine.
6. In Prospector, the Subassemblies branch, click the first Link entry, press the right mouse button, and from the shortcut menu select **Properties...**.
7. Click the **Information** tab and change the name to **Ditch Bottom - Right** and click the **OK** button to exit.
8. Repeat Steps 6 and 7, renaming the second Link to **Ditch Bottom - Left**.

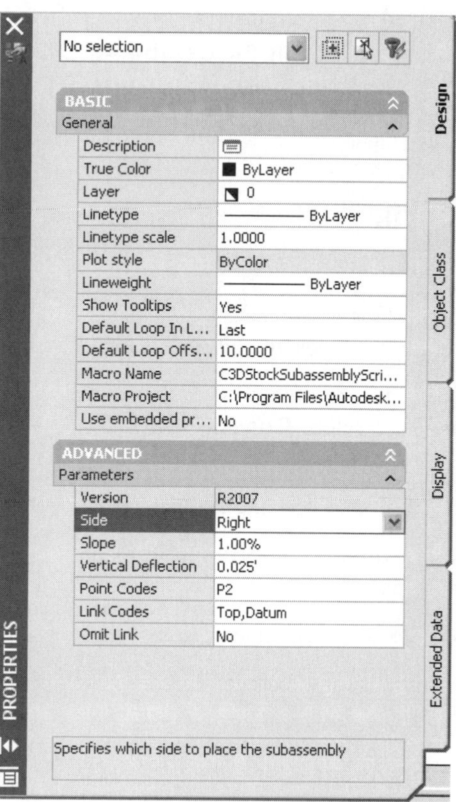

Figure 9.9

Daylight with Bench

1. Change the palette to **Imperial - Daylight**.
2. From the palette, select **DaylightBench**.
3. In Properties, the Parameters area, change the Side to **Right**, the Max Cut and Fill Height to **5**, the Bench Width to **3**, and the Bench Cut and Fill Slopes to **2**.
4. In the drawing, attach the DaylightBench to the right endpoint of the right DitchBottom subassembly (link).

5. In Properties, the Parameters area, change the Side to **Left**, and attach the Daylight-Bench to the left endpoint of the left DitchBottom subassembly (link).
6. Press the **Enter** key twice to exit the subassembly routine (see Figure 9.10).
7. Rename the left and right subassemblies to **Bench - Left** and **Bench - Right**.

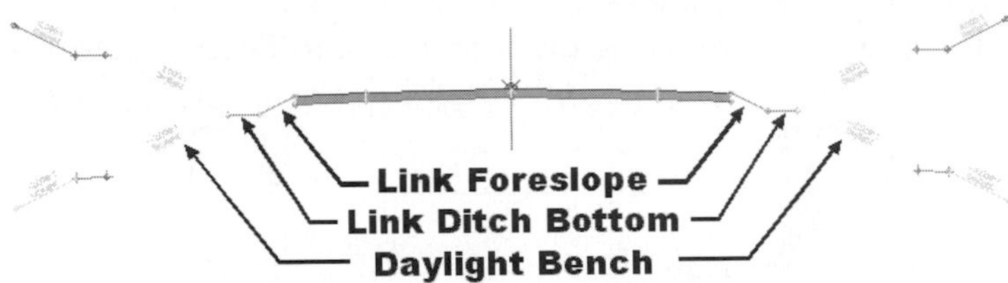

Figure 9.10

CREATE SIMPLE CORRIDOR

1. From the Corridors menu, select **Create Simple Corridor....**
2. In the Create Simple Corridor dialog box, enter **Cut** (leaving in the counter) as the name of the corridor, and click the **OK** button to continue.
3. In the command line, the routine prompts for an alignment; press the **Enter** key. From the alignment list, select **Cut Alignment - (1) (1)**, and click the **OK** button.
4. In the command line, the routine prompts for a profile; press the **Enter** key. Click the drop-list arrow; from the profile list, select **Cut (1)**, and click the **OK** button.
5. In the command line, the routine prompts for an assembly; select the **Cut - (1)** assembly from the screen.

   The Target Mapping dialog box appears.
6. In the Target Mapping dialog box, click in the Surfaces' Object Name cell (**<Click here to set all>**). In the Pick a Surface dialog box, select **EG(1)** (see Figure 9.11).
7. Click the **OK** buttons to exit the dialog boxes.

   Civil 3D calculates the corridor.

VIEW CORRIDOR SECTIONS

1. From the Corridors menu, select **View/Edit Corridor Section**.
2. In the View/Edit Corridor Section routine, press the **Enter** key to select the Cut - (1) corridor, view all of the sections of the corridor, and exit (see Figure 9.12).
3. Click the **AutoCAD Save** icon and save the drawing.
4. From the File menu, select **Close** and exit the current drawing.

EXERCISES

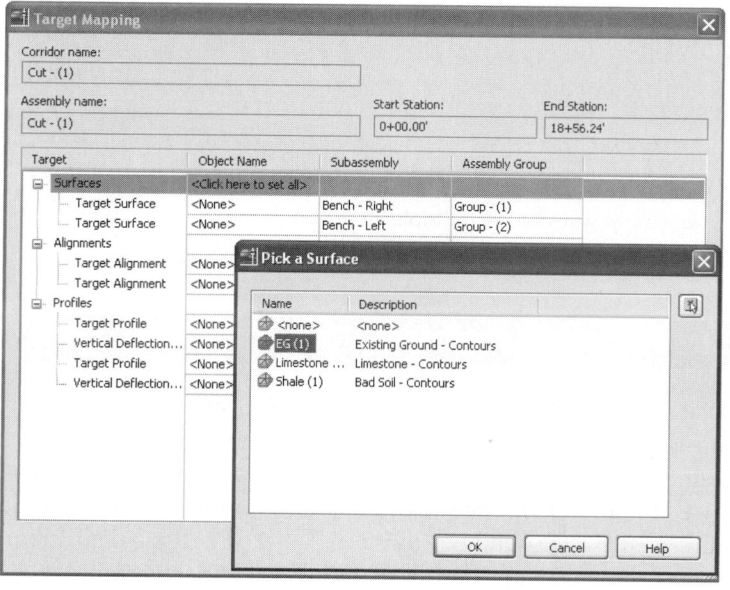

Figure 9.11

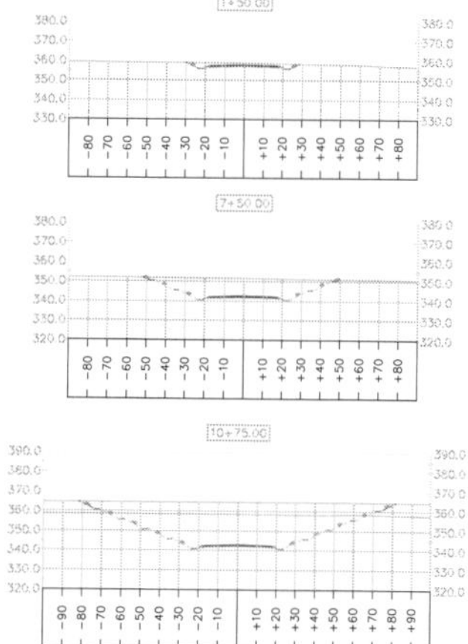

Figure 9.12

This completes the exercise on benching.

SUMMARY

- Generic links create useful subassembly parts without having to define custom subassemblies.
- If the depth of cut or height of fill exceeds the max cut or fill height, the subassembly will create multiple benches.

UNIT 2: SIMPLE TRANSITIONS

Civil 3D has stock subassemblies that stretch according to simple transition design parameters. These parameters address the pavement shape's location and elevation as it transitions to wider or narrower pavement widths. The basic lane transition subassembly uses this simple approach to transitioning (see Figure 9.13). The subassembly's transition value controls its behavior when an alignment affects the subassembly's location. The values for transitioning a basic lane transition are: Hold elevation, change offset; Hold grade, change offset; Change offset and elevation; Hold offset and elevation; and Hold offset, change elevation.

Hold elevation, change offset—The attached offset alignment changes the width of the pavement shape, and the elevation of the edge-of-travelway is held. The cross slope of the pavement changes to maintain the pavement's elevation.

Hold grade, change offset—The pavement grade is held, and as the offset alignment increases the pavement width, the pavement's edge lowers in elevation. If the alignment moves the pavement edge closer to the centerline, the pavement edge will rise in elevation to maintain the grade.

Change offset and elevation—The offset and elevation for the edge-of-travelway is controlled by an offset alignment (horizontal location) and an offset profile (vertical location).

Hold offset and elevation—This parameter makes a subassembly not respond to any attached horizontal or vertical offset alignment.

Hold offset, change elevation—This parameter holds the pavement width as specified, but an offset profile controls the elevation of the edge-of-travelway point.

Sections from an assembly that uses simple transition parameters do have issues. The main issue is the sections on the edge-of-travelway. They are always perpendicular to the centerline, but they are not perpendicular to any arc in the transitional alignment (see Figure 9.14).

What should appear in the sections are curbs and sidewalks perpendicular to the offset alignment's curve (see Figure 9.15).

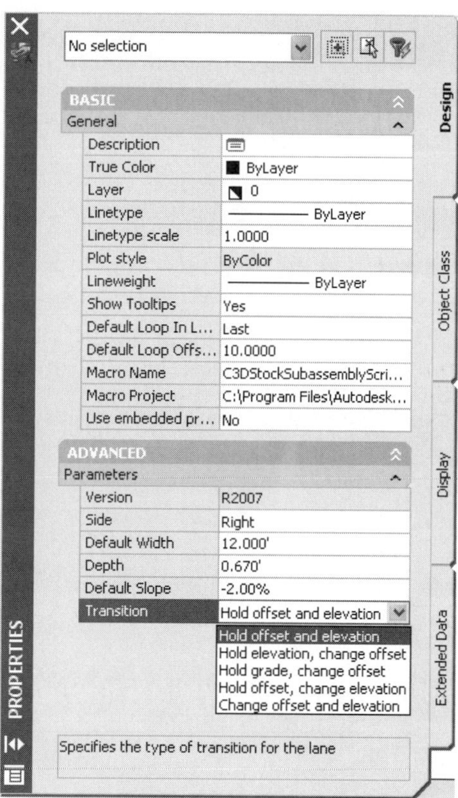

Figure 9.13

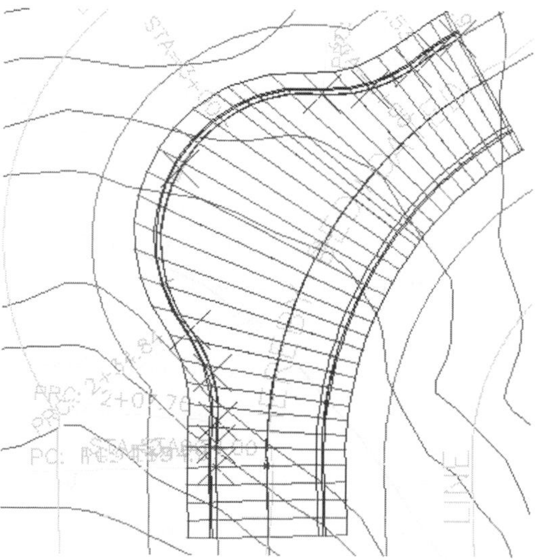

Figure 9.14

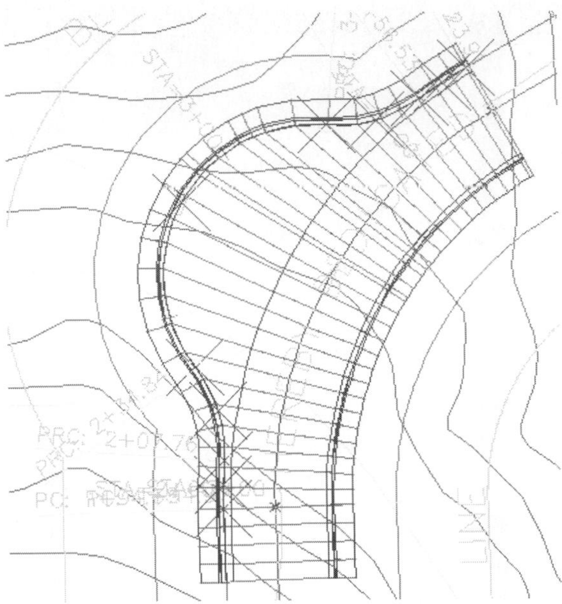

Figure 9.15

For outside subassemblies, the creation of perpendicular sections (curb and sidewalk of Figure 9.15) is from using an assembly offset. The subassemblies attached to an offset are perpendicular to the offset, not the roadway centerline.

The properties of this type of assembly show the subassemblies' offset attachment. The relationship is viewed in the Construction tab of the assembly's properties dialog (see Figure 9.16).

An assembly using an offset alignment must have the outside subassemblies attached to it, not the assembly. When attaching a subassembly, the user must select the offset marker, not the assembly marker. An example of an assembly using an offset is in Figure 9.17.

As mentioned, a transition can also have a vertical alignment that controls the vertical location of a transition point. The basic lane transition does not need a vertical alignment, except when the transition property is set to Change Elevation. When using an offset in an assembly, Civil 3D *requires* an offset vertical alignment to provide elevations for the subassemblies outside of the offset.

Note: All subassemblies, assemblies, and assembly groups should have descriptive names.

When creating an assembly, the user should assign meaningful names to each subassembly. When the subassembly groupings are viewed in the Assembly Properties dialog box, the subassembly names easily identify the important grouping in a potentially complex assembly.

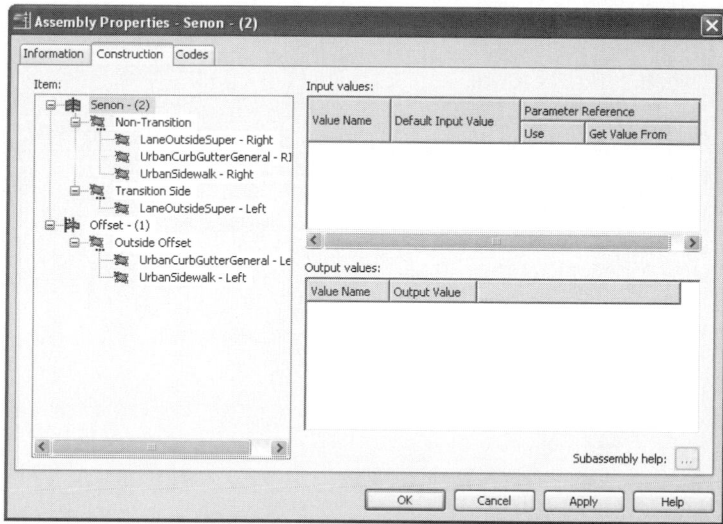

Figure 9.16

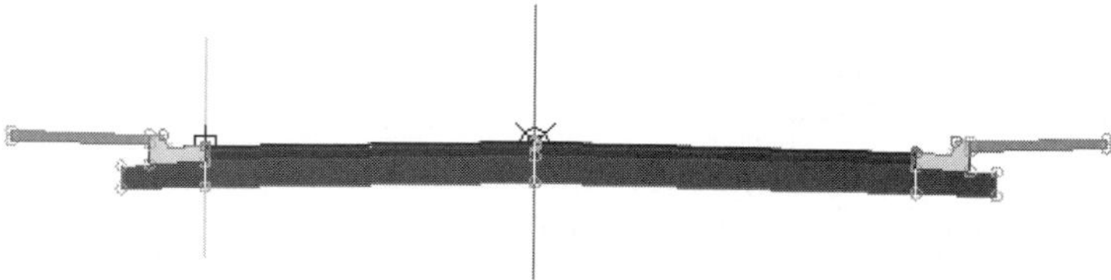

Figure 9.17

This naming rule should apply to any grouping the assembly defines (right, left side, or offset attachments).

IDENTIFYING OFFSET ALIGNMENTS

An assembly offset defines a point of control and, often, an attachment point by a second alignment (transitional alignment). All of the subassemblies that attach to the outside of the offset will be perpendicular to the path of the alignment controlling the offset.

When outside subassemblies are attached, they must be attached to the offset, not the assembly. If the subassemblies are attached to the assembly, the subassemblies remain perpendicular to the centerline alignment. The Assembly Properties dialog box indicates if the subassemblies are attached to the offset or the assembly (see Figure 9.16).

The attachment of offset alignments determines what corridor creation command to use. If creating a simple corridor with the basic lane transition, users can use the Corridor menu pick, Create Simple Corridor. More complex corridors require the use of the Create Corridor menu pick from the Corridor menu. The Create Simple and Create Corridor commands both use a dialog box to set the names of the offset horizontal and vertical alignments (see Figure 9.18).

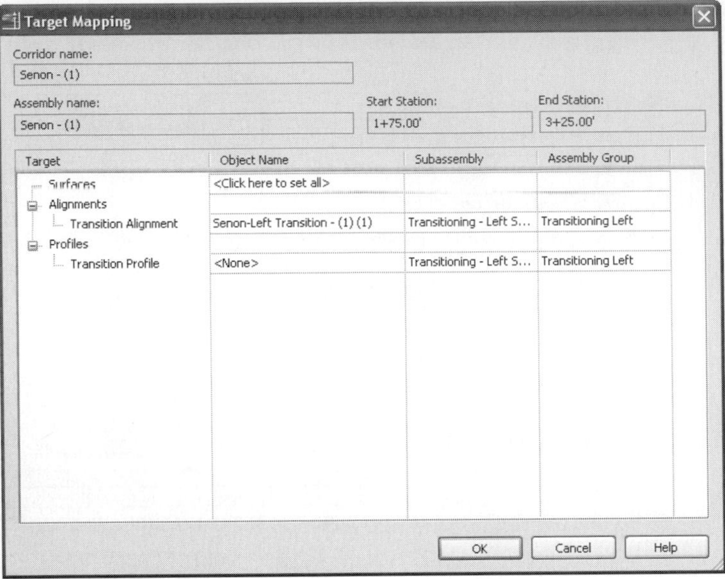

Figure 9.18

CORRIDOR PROPERTIES

After creating the corridor, users can edit the properties of the corridor to add or change attached offset alignments (target name mapping) and/or the frequency to apply assemblies (see Figure 9.19).

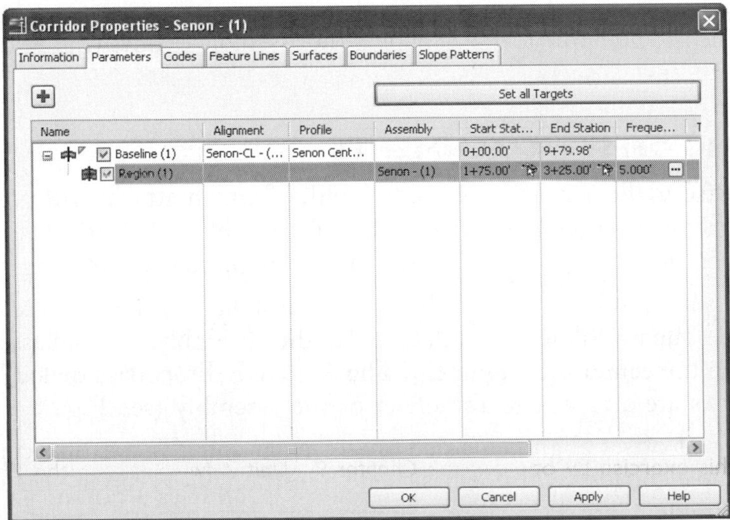

Figure 9.19

CORRIDOR SECTION FREQUENCY

When the corridor is more complex (containing regions, multiple offset alignments, and so on), users should increase the number of sections that create the corridor. This is done in the Frequency to Apply Assemblies dialog box of the Corridor Properties dialog box (see Figure 9.20).

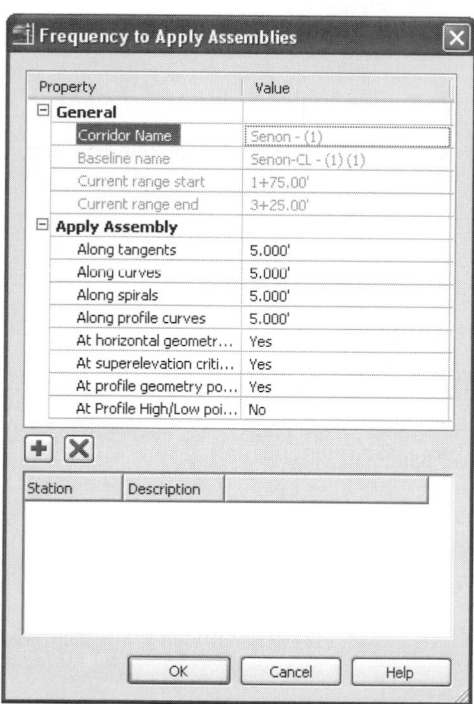

Figure 9.20

EXERCISE 9-2

When you complete this exercise, you will:

- Be familiar with creating a transition assembly.
- Be familiar with the basic lane transition subassembly.
- Be able to transition a lane using an offset horizontal and vertical alignment.

EXERCISE SETUP

You can start this exercise by opening the *Chapter 9 - Unit 2.dwg* found in the *Chapter 9* folder of the CD that accompanies this textbook.

1. Open the drawing *Chapter 9 - Unit 2.*
2. From the File menu, select **Save As...** and save the drawing as **Basic Transition**.

3. If necessary, from the General menu, select **Tool Palettes Window** to display the Civil 3D Imperial tool palette.
4. Click the **Prospector** tab to make it the current panel.

CREATE A BASIC LANE TRANSITION ASSEMBLY

The first assembly is a simple assembly using the basic lane transition, basic curb and gutter, and urban sidewalk subassemblies (see Figure 9.21).

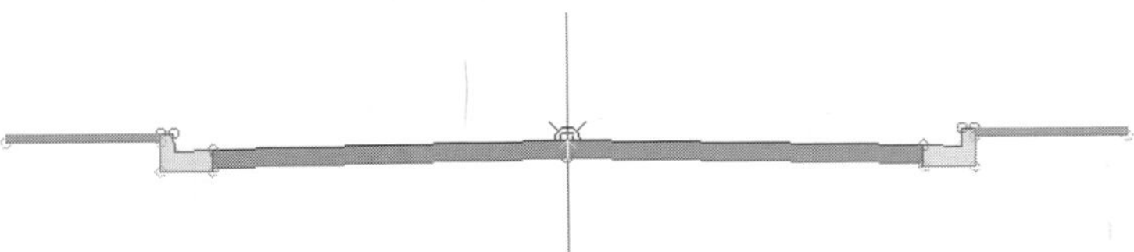

Figure 9.21

Create the Senon Assembly

1. From the Corridors menu, select **Create Assembly....**
2. In the Create Assembly dialog box, type **Senon** as the name of the assembly, leaving the counter. Click the **OK** button, and in the drawing, select a point to the left of the profile.
3. If necessary, zoom and pan to better view the assembly.

Basic Lane - Right Side

1. On the tool palette, select the **Imperial - Basic** tab, and from the palette select **BasicLane**.
2. In Properties, the Parameters area, change the Side to **Right**, set the lane Width to **14**, and in the drawing attach the subassembly to the assembly.
3. Press the **Enter** key twice to exit the routine.
4. In Prospector, expand the Subassemblies branch, click the **BasicLane** subassembly, press the right mouse button, and from the shortcut menu select **Properties....**
5. Click the **Information** tab, change the name of the subassembly to **Non-transitioning - Right Side**, and click the **OK** button to exit the dialog box.

Basic Lane Transition - Left

1. From the Imperial - Basic tool palette, select **BasicLaneTransition**.
2. In Properties, the Parameters area, set the Side to **Left** and the Transition property to **Hold grade, change offset**. In the drawing, attach the subassembly to the assembly and press the **Enter** key twice to exit the routine.

3. Click the **BasicLaneTransition** subassembly in the Subassemblies branch of Prospector, press the right mouse button, and from the shortcut menu select **Properties....**
4. Click the **Information** tab, name the subassembly **Transitioning - Left Side**, and click the **OK** button to exit the dialog box.

Curb, Gutter, and Sidewalk

1. Repeat the process of attaching subassemblies and add a **BasicCurbAndGutter** to the left and right sides of the assembly.
2. Rename the BasicCurbAndGutter to be **Right** or **Left Curb and Gutter** for the appropriate side.
3. Repeat the process of attaching subassemblies and add a **BasicSidewalk** to the left and right sides of the assembly.
4. Rename the Basic Sidewalk to be **Right** or **Left BasicSidewalk**.

ASSEMBLY PROPERTIES

1. In Prospector, expand the Assemblies branch. From the list, select **Senon - (1)**, press the right mouse button, and from the shortcut menu select **Properties....**
2. Click the **Construction** tab to make it current.

   There should be two groups: one for the right side and one for the left side of the assembly.
3. In the dialog box, select the **Group (1)** heading representing the right side of the assembly and press the right mouse button. From the shortcut menu select **Rename**, and rename the group **Non-transitioning**.
4. In the dialog box, select the **Group (2)** heading that represents the left side of the assembly. Press the right mouse button, and from the shortcut menu select **Rename**; rename the group **Transitioning Left**.

   Your Assembly Properties dialog box should now look like Figure 9.22.
5. Click the **OK** button to exit the Assembly Properties dialog box.

CREATE A SIMPLE CORRIDOR

1. From the Corridors menu, select **Create Simple Corridor....**

   The Create Simple Corridor dialog box displays.
2. In the Create Simple Corridor dialog box, change the corridor name to **Senon**, keeping the counter, and click the **OK** button to continue creating the corridor.
3. In the command line, the routine prompts for an alignment. Press the **Enter** key, select from the list **Senon-CL – (1) (1)**, and click the **OK** button to continue.

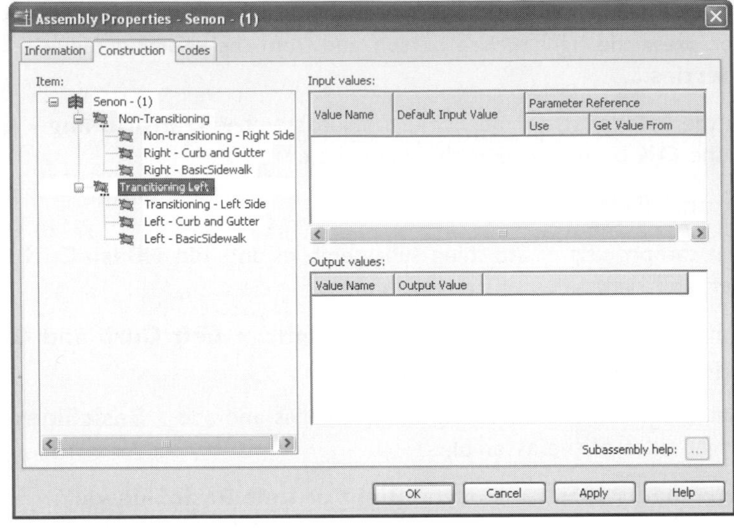

Figure 9.22

4. In the command line, the routine prompts for a profile. Press the right mouse button, click the drop-list arrow, and from the list select **Senon Centerline (1)**. Click the **OK** button to continue.

5. In the command line, the routine prompts for an assembly. Press the **Enter** key, click the drop-list arrow, and from the list select **Senon - (1)**. Click the **OK** button to continue.

   The Target Mapping dialog box displays.

6. In the dialog box, click in the Object Name cell for Transition Alignment; a Pick Horizontal Alignment dialog box appears. Click the drop-list arrow, and select **Senon-Left Transition - (1) (1)** from the list (see Figure 9.23).

   You do not need to set a profile because of the transitioning subassembly parameter (Hold grade, change offset).

7. Click the **OK** button to create the corridor.

8. Click the **AutoCAD Save** icon to save the drawing.

EXAMINE THE CORRIDOR

1. Use the AutoCAD Zoom and Pan commands to view the new corridor.

   The section frequency does not create a good representation of the transition in the knuckle. The section frequency needs to be increased, and you will limit the corridor to a small segment of the alignment.

2. In the drawing, click any part of the corridor, press the right mouse button, and from the shortcut menu select **Corridor Properties...**.

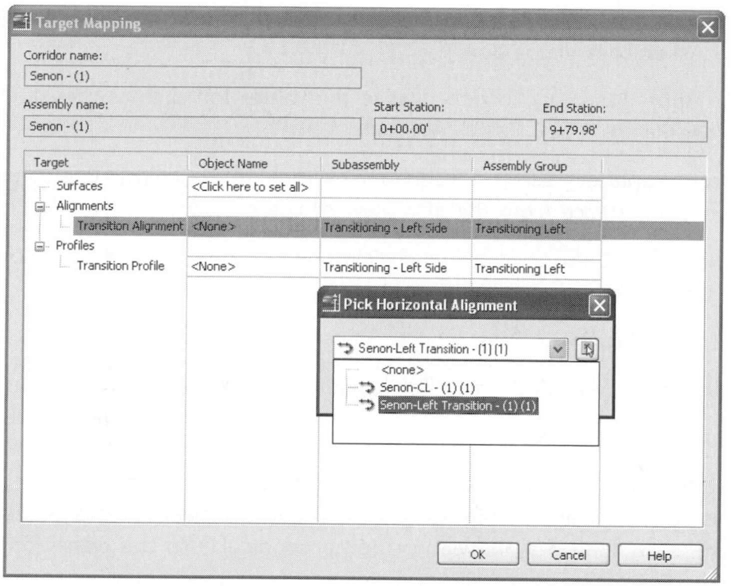

Figure 9.23

3. In the Corridor Properties dialog box, select the **Parameters** tab, and change the Region range from 1+75 (175) to **3+25 (325)**. See Figure 9.24.

   You could graphically select start and end stations by clicking the icon in a station cell and selecting a station value from the plan view of the corridor.

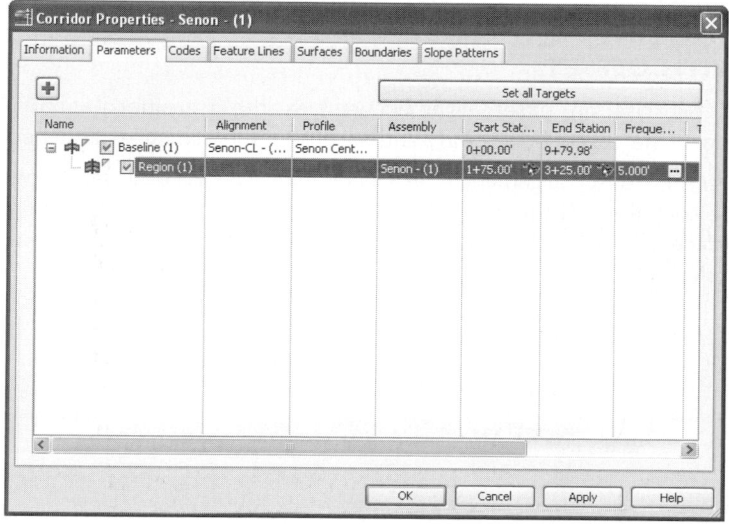

Figure 9.24

4. For Region 1, in the Frequency cell, click the ellipsis to display the Frequency to Apply Assemblies dialog box.
5. In the Apply Assembly section, change the values for each of the Along Entries to **5** feet (see Figure 9.20).

   You can graphically select a frequency by clicking the ellipsis in a value cell and selecting a distance from the plan view of the corridor.
6. Click the **OK** buttons to exit the dialog boxes and to rebuild the corridor with the new section frequency.
7. Click the **AutoCAD Save** icon to save the drawing.
8. Use the AutoCAD Zoom and Pan commands to better view the knuckle.
9. Use the Object Viewer to view the corridor in 3D.

   The transition looks alright. However, the curb and sidewalk are not perpendicular to the roadway edge. When using a basic lane transition in this situation, the sections for the curb and sidewalk will not be perpendicular to the transitioning pavement. Also, notice the constant –2 percent grade from the centerline to the edge-of-travelway for the knuckle. Using these subassemblies is the only control you have over the slope of the pavement.
10. Exit the Object Viewer after viewing the corridor.
11. Open Layer Properties Manager, freeze the layer containing the corridor (**C-ROAD-CORR-Senon – (1)**), and click the **OK** button to exit the dialog box.

    To make the curb and sidewalk perpendicular to the transition curve, the assembly needs two additional alignments: an offset and a vertical.

ASSEMBLY WITH OFFSET

In the previous exercise, an offset alignment defines the transitional control point for the BasicLaneTransition (the edge-of-travelway) and all of the subassemblies attached to that point.

The assembly using an offset alignment is in Figure 9.25. The vertical line between the edge-of-travelway and the curb flange is the offset. You locate the offset by selecting the upper-left endpoint of the PAVE1 surface. The left curb subassembly attaches to the offset. The sidewalk attaches to the top back-of-curb.

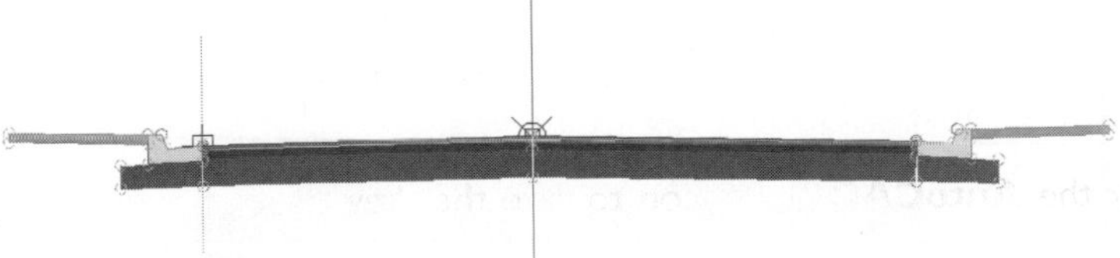

Figure 9.25

CREATE NEW ASSEMBLY

1. Use the AutoCAD Zoom and Pan commands to view the previous assembly.
2. Pan up the assembly so you have a clear working space beneath the previous assembly.
3. From the Corridors menu, select **Create Assembly....**
4. In the Create Assembly dialog box, type **Senon** as the assembly name, leaving the counter. Click the **OK** button, and select a point just below the Senon - (1) assembly.

Create Right Side of Assembly

The right side uses a transitional lane subassembly, but there will be no attached alignment. This causes the Event Viewer to issue warnings about the missing alignment. This is a benign warning and can be ignored.

5. If necessary, from the General menu, select **Tool Palettes Window**.
6. Click the **Imperial - Roadway** tab to make it the current panel.
7. From the tool palette, select **LaneOutsideSuper**; in the Properties palette, the Advanced Parameters section, set the Side to **Right**, change the Width to **14**, and in the drawing, select the assembly.
8. Press the **Enter** key twice to exit the routine.
9. Click the **Imperial - Structures** tab of the tool palette.
10. From the tool palette, select **UrbanCurbGutterGeneral**. In the Properties palette, the Advanced Parameters section, set the Side to **Right**. In the drawing, attach the curb and gutter to the upper-right outside marker (edge-of-travelway) of the right lane.
11. Press the **Enter** key twice to exit the routine.
12. From the same tab of the palette, select **UrbanSidewalk**. In the Properties palette, the Advanced Parameters section, set the Side to **Right**. In the drawing, attach the sidewalk to the upper-right marker (back-of-curb) of the curb and gutter, and press the **Enter** key twice to exit the routine.
13. If necessary, click **Prospector** to make it current.
14. If necessary, expand the Subassemblies branch and change the names of the new subassemblies to indicate they are on the right side of an assembly (e.g., **LaneOutsideSuper - Right**, **UrbanCurbandGutterGeneral - Right**, and **UrbanSidewalk - Right**).
15. Click the **AutoCAD Save** icon to save the drawing.

Assign Left Side Subassembly

1. Click the **Imperial - Roadway** tab of the tool palette to make it current.

2. From the palette, select **LaneOutsideSuper**. In the Properties palette, the Advanced Parameters section, change the Side to **Left**, the Width to **12** feet, and attach it to the assembly. Press the **Enter** key twice to exit the routine.
3. If necessary, expand the Subassemblies branch in Prospector and change the name of the new subassembly to **LaneOutsideSuper - Left**.

Add Offset to the Assembly

1. Use the AutoCAD Zoom command to view the assembly's left half.
2. In the drawing, select the assembly (red vertical line), press the right mouse button, and from the shortcut menu select **Add Offset...**.
3. Use the AutoCAD Zoom command to better view the upper-left end of the PAVE1 shape.
4. Using the **Center** object snap, select the upper, outer PAVE1 marker (red). See Figure 9.25 and Figure 9.26.

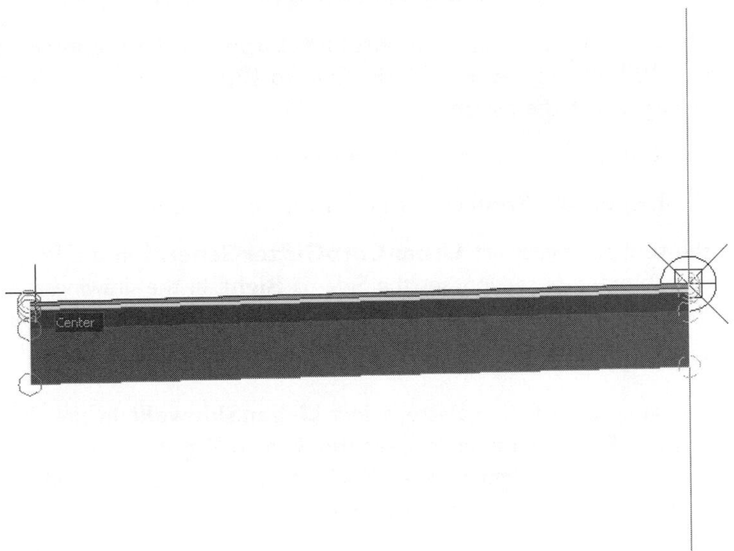

Figure 9.26

An offset marker appears at the location on the assembly as a vertical line (see Figure 9.25).

Attach the Curb and Sidewalk

The left side curb subassembly must be attached to the offset marker, not the assembly. The sidewalk will be perpendicular to whatever alignment the curb is perpendicular to.

1. Click the **Imperial - Structures** tab of the palette to view its contents.
2. From the tool palette, select **UrbanCurbGutterGeneral**. If necessary, in the Properties palette, the Advanced Parameters section, change the Side to **Left**. In the

drawing, select the offset marker (left vertical line) to place the curb and gutter in the assembly, but attached to the offset. Press the **Enter** key twice to exit the routine.

3. From the same palette, select **UrbanSidewalk**. If necessary, in the Properties palette, the Advanced Parameters section, change the Side to **Left**. Select the upper-left marker of the curb (back-of-curb) to place the subassembly, and press the **Enter** key twice to exit the routine.
4. Click the **AutoCAD Save** icon to save the drawing.
5. Click the "**X**" of the tool palette to close the palette.
6. In Prospector, if necessary, expand the Subassemblies branch and change the names of the two new subassemblies to indicate they are on the left side of the assembly (e.g., **UrbanCurbandGutterGeneral - Left**, and **UrbanSidewalk - Left**).
7. Click **AutoCAD Save** icon to save the drawing.

EDIT ASSEMBLY PROPERTIES

1. In Prospector, expand the Assemblies branch, select **Senon - (2)**, press the right mouse button, and from the shortcut menu select **Properties...**.
2. In the dialog box click the **Construction** tab to view its contents.

   The Construction panel indicates the UrbanCurbandGutterGeneral - Left and UrbanSidewalk - Left subassemblies attach to the offset, not the main assembly (see Figure 9.27).

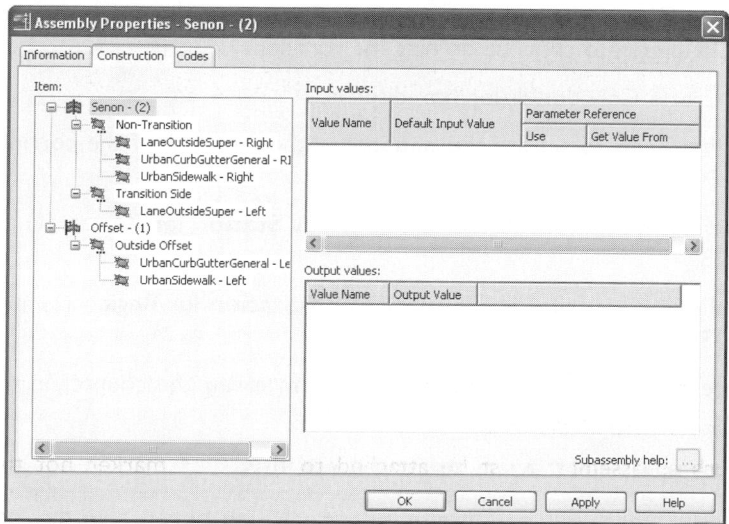

Figure 9.27

Rename the groups in the subassembly using Figure 9.27 as a guide.

3. In the Construction panel, click the group name containing the right side subassemblies, and press the right mouse button. From the shortcut menu, select **Rename** and name the group **Non-Transition Side**.
4. In the Construction panel, click the group name containing the left side LaneOutside-Super subassemblies, and press the right mouse button. From the shortcut menu, select **Rename** and name the group **Transition Side**.
5. In the Construction panel, click the group name (under Offset – (1)) containing the left UrbanCurbGeneral - Left and UrbanSidewalk - Left subassemblies. Press the right mouse button, and from the shortcut menu select **Rename** and name the group **Outside Offset**.
6. Click the **OK** button to exit the Assembly Properties dialog box.
7. Click the **AutoCAD Save** icon to save the drawing.

CREATE A CORRIDOR

1. From the Corridors menu, select **Create Corridor**.
2. In the command line, the routine prompts for an alignment name. Press the right mouse button, and from the alignment list, select **Senon-CL – (1) (1)** and click the **OK** button to continue defining the corridor.
3. In the command line, the routine prompts for a profile name. Press the right mouse button, click the drop-list arrow, and from the profile list, select **Senon Centerline (1)**. Click the **OK** button to continue defining the corridor.
4. In the command line, the routine prompts for an assembly name. Press the right mouse button, click the drop-list arrow, and from the assembly list, select **Senon – (2)**. Click the **OK** button to continue defining the corridor.

   The Create Corridor dialog box displays.

   The first items to set are the stations for the Region 1 of the corridor (see Figure 9.28).
5. In the dialog box, click in the cell for **Start Station** for Region (1) and set its value to **125**.
6. In the dialog box, click in the cell for **End Station** for Region (1) and set its value to **375**.
7. If necessary, replace "Corridor" with **Senon**, leaving the counter, in the Corridor Name area at the top left of the dialog box.

SET OFFSET ALIGNMENT AND PROFILE NAMES

1. In the Alignment cell for Offset - (1), click **<Click here...>**. In the Pick Horizontal Alignment dialog box, select **Senon-Left Transition - (1) (1)** from the list, and click the **OK** button to return to the Create Corridor dialog box (see Figure 9.28).

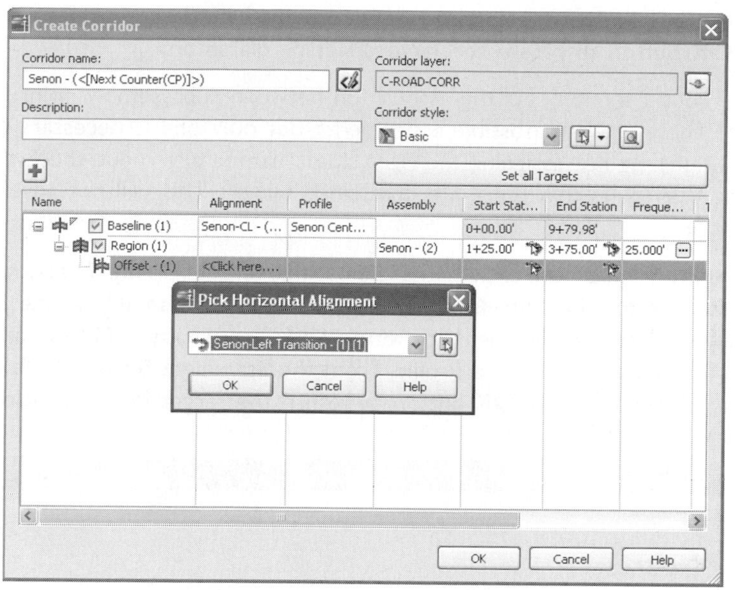

Figure 9.28

The Alignment name appears in the dialog box. The stationing for the offset changes to 1+25 and 3+75, matching the stations of the region, and the Profile column now contains the text "<Click here...>".

2. In the Profile cell for the Offset - (1), click **<Click here...>**. In the Select a Profile dialog box, select **Senon-Left Transition - (1) (1)** as the Alignment. From the Profile list select **Senon-Left Transition Vertical (1)** and click the **OK** button to return to the Create Corridor dialog box.

   This sets the horizontal and vertical alignments for the offset to follow from station 1+25 to 3+75.

SET ALL TARGETS

The Set All Targets dialog box links the subassemblies points to the controlling alignments.

1. In the dialog box, click the **Set All Targets** button at the top right (see Figure 9.29).
2. Widen the Object Name and Subassembly columns so you can see the complete names.

   In the Alignments section, the width for the LaneOutsideSuper - Left is the only item needing an assignment. The width for the right side is set in the subassembly's parameters. By not assigning the right side a named alignment, the Event Viewer will display warnings stating there are unassigned targets. This will not adversely affect the corridor.

3. In the dialog box, the Alignment section, click in the Object Name cell for the **Width Alignment** for the **LaneOutsideSuper - Left** subassembly. In the Pick

Horizontal Alignment dialog box, select **Senon-Left Transition - (1) (1)**, and click the **OK** button to return the Target Mapping dialog box.

The Profiles section sets the association between subassembly points and profiles. Again, there are two possible assignments, but only one is necessary (LaneOutsideSuper - Left). By not assigning the right side a named alignment, the Event Viewer will contain warnings that there are unassigned targets. This will not adversely affect the corridor.

4. In the Profiles section, click in the Object Name cell for the **Outside Elevation-Profile** and for the **LaneOutsideSuper - Left** subassembly. In the Select a Profile dialog box, set the Alignment to **Senon-Left Transition - (1) (1)** and click the drop-list arrow for Select a Profile. From the list, select **Senon-Left Transition Vertical (1)** and click the **OK** button to return the Target Mapping dialog box.

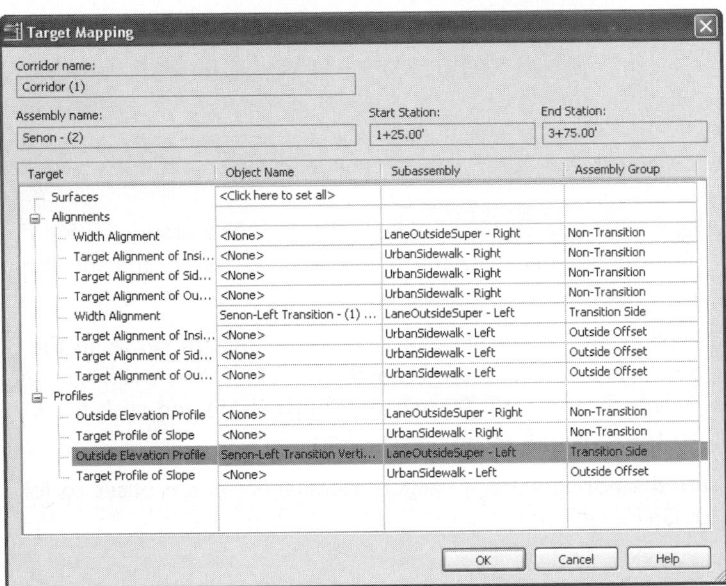

Figure 9.29

5. Click the **OK** button to exit the Target Mapping dialog box and to return to the Create Corridor dialog box.

CHANGE FREQUENCY OF ASSEMBLY IN CORRIDOR

1. In Region (1)'s Frequency cell, click the ellipsis.
2. In the Frequency to Apply Assemblies dialog box, change the Along Tangents and Along Curves frequency to **5** feet, and click the **OK** button to return to the Create Corridor dialog box.
3. Click the **OK** button to create the corridor.

4. Click the **AutoCAD Save** icon to save the drawing.
5. Use the AutoCAD Zoom and Pan commands to view the new Corridor.

   The Curb and Sidewalk sections of the corridor are now perpendicular to the offset alignment's path.
6. Use the Object Viewer to view the corridor in 3D.
7. Exit the Object Viewer after viewing the corridor.
8. Exit the drawing.

This completes the exercise on simple transitions.

SUMMARY

- The basic lane with transition creates simple lane-widening designs.
- The basic lane with transition subassembly has settings that control the transition point's elevation.
- The basic lane with transition needs both an offset horizontal and vertical alignment when the Transition parameter is set to Hold offset, change elevation.
- When using alignment offset, you must have horizontal and vertical alignments controlling the horizontal and vertical location of the offset point.
- The first subassembly to the outside of the alignment's offset point must attach to the offset.

UNIT 3: CREATING A CUL-DE-SAC

Creating a cul-de-sac is similar to transitioning a roadway. In a cul-de-sac, an alignment controls the outside location of the edge-of-travelway and the centerline of the main roadway provides a pavement width (see Figure 9.30). Unlike the previous example, there is no offset alignment.

Having perpendicular sections for the curb and sidewalk around the cul-de-sac means the curb and sidewalk are to the left of the edge-of-travelway and the alignment attaches to the edge-of-travelway (see Figure 9.31).

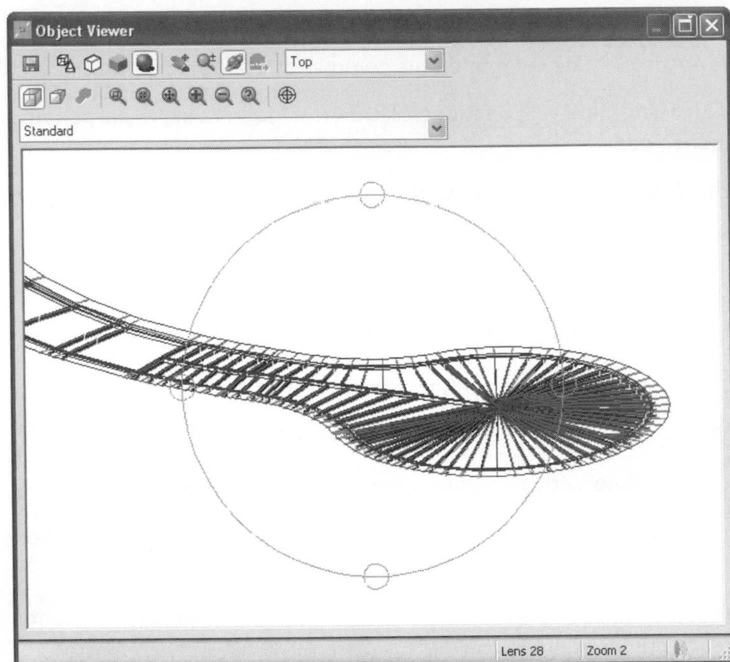

Figure 9.30

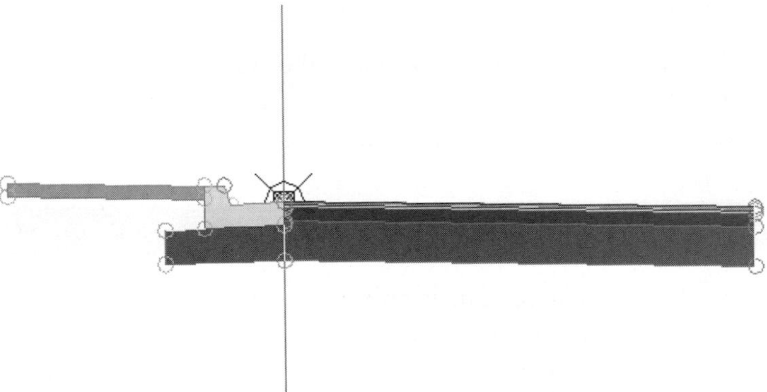

Figure 9.31

CORRIDOR SURFACES

Using more than one region for a corridor introduces issues when creating corridor surface boundaries. When creating a corridor with a single region, the selection is a single feature line from a list (daylight). When using more than one region, the selection of the boundary segments may be an interactive selection process. The location of the selection point of the segment affects the definition of the boundary. When identifying a boundary segment, its definition is sensitive to the nearest endpoint of the selected segment. In the example of a cul-de-sac, the selection of the boundary segments should be in an order that creates a "well-formed" border (i.e., clockwise or counterclockwise around the entire cul-de-sac). See Figure 9.32. Select a feature line that defines the outer limit of the surface and near one of its endpoints, indicative of the direction for the entire boundary.

Not only do users need to be aware of nearest endpoints, but they also need to be aware of the segment's direction. In Figure 9.32, the boundary's direction is along, around, and in the reverse direction of the last segment of the corridor. The reason for the reversed direction is that both sides of the Lorraine – (1) segment go in the same direction (lower to higher stations) and when the boundary comes out of the cul-de-sac and returns to the beginning of Lorraine – (1), it needs to go from higher to lower stations, the reverse direction of its definition.

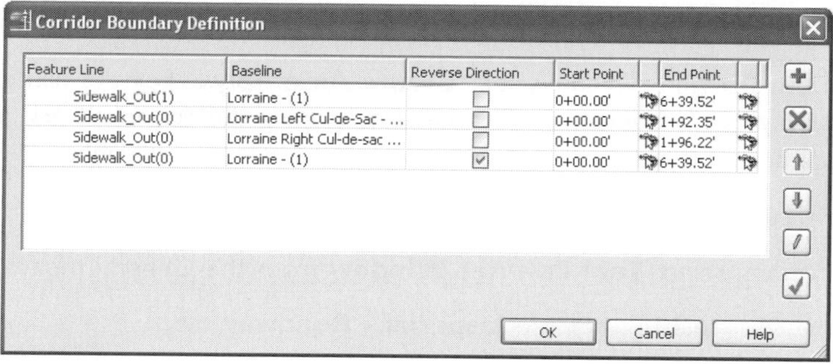

Figure 9.32

EXERCISE 9-3

After completing this exercise, you will:

- Be able to create a cul-de-sac design.
- Be familiar with the Create Corridor command.
- Be able to develop regions within a corridor.
- Create a corridor surface.
- Define a corridor surface boundary interactively.

EXERCISE SETUP

You can start this exercise by opening the *Chapter 9 - Unit 3.dwg* found in the *Chapter 9* folder of the CD that accompanies this textbook.

1. Open drawing *Chapter 9 - Unit 3*.
2. Select **Save As...** from the File menu and save the drawing under a new name.

There are three alignments, one for the baseline of the entire roadway (Lorraine – (1)) and two for the transitions for the cul-de-sac (Lorraine Left and Right Cul-de-sac – (1)). There are two assemblies for the roadway: one that does not transition (stations 0+00 to 5+25) and one that does transition.

The two transitional alignments and their profiles are already in the drawing. You would have had to calculate or determine the profile elevations so they match the main alignment's elevations or reflect the drainage for the area. One way of developing the cul-de-sac design is to use the ETW feature line to determine rough starting elevations. From the Corridors menu, Utilities flyout, you could use the following commands: Polyline from Corridor, Alignment from Corridor, and Profile from Corridor. The Alignment from Corridor routine would prompt you to create a profile.

EXERCISES

CREATE THE TRANSITION ASSEMBLY

When creating the transitional assembly, use Figure 9.31 as a guide.

1. From the View menu, select **Named Views...** and restore **Assembly 2**.
2. From the Corridors menu, select **Create Assembly....** For the name, replace "Assembly" with **Lorraine – Transition**, keeping the counter. Click the **OK** button to exit, and in the drawing select a point to locate the assembly.

Right Side Transition Subassembly

1. If necessary, select **Tool Palettes Window** from the General menu.
2. On the tool palette, click the **Imperial - Roadway** tab.
3. From the Roadway palette, select **LaneOutsideSuper**.
4. In the Properties palette, in the Advanced Parameter section, set the Side to **Right**, and set the Width to **12** feet. In the drawing, select the assembly to place the subassembly, and press the **Enter** key twice to exit the routine.
5. If necessary, click the **Prospector** tab to make it the current panel.
6. Expand the Subassemblies branch and rename the new subassembly **LaneOutsideSuper - Right Transitional**.

Left Side Curb and Sidewalk

1. On the tool palette, click the **Imperial - Structures** tab.
2. From the Structures palette, select **UrbanCurbGutterGeneral**.
3. In the Properties palette, in the Advanced Parameter section, set the Side to **Left**. In the drawing, select the assembly to place the subassembly, and press the **Enter** key twice to exit the routine.
4. From the Structures palette, select **UrbanSidewalk**.
5. In the Properties palette, in the Advanced Parameter section, set the side to **Left**. In the drawing, select the upper back-of-curb marker to place the sidewalk, and press the **Enter** key twice to exit the routine.

6. In Prospector, expand the Subassemblies branch and rename the two new subassemblies **UrbanCurbAndGutterGeneral - Left (Non-Transitional)** and **UrbanSidewalk - Left (Non-Transitional)**.
7. Close the tool palette.
8. Click the **AutoCAD Save** icon to save the drawing.

EDIT ASSEMBLY PROPERTIES

1. Expand the Prospector's Assemblies branch until you view the assembly list.
2. From the list, select **Lorraine – Transition – (1)**, press the right mouse button, and from the shortcut menu select **Properties....**
3. Click the **Construction** tab, and select and rename the group with the right Lane-OutsideSuper to **Transition Subassembly**.
4. Select and rename the group with the Curb and Sidewalk to **Non-Transition Subassemblies**.
5. Click the **OK** button to exit the dialog box.
6. Click the **AutoCAD Save** icon to save the drawing.

CREATE THE NON-TRANSITION CORRIDOR REGION

You will start with a simple corridor, and later you will add complexity to it by editing its properties.

1. From the Corridors menu, select **Create Simple Corridor....** For the name, replace "Corridor" with **Lorraine**, and click the **OK** button to continue creating the corridor.
2. In the command line, the routine prompts for an Alignment. Press the right mouse button, click the drop-list arrow, and from the list, select **Lorraine – (1)**. Click the **OK** button to continue.
3. In the command line, the routine prompts for a profile. Press the right mouse button, click the drop-list arrow, and from the list select **Lorraine CL Vertical (1)**. Click the **OK** button.
4. In the command line, the routine prompts for the assembly. Press the right mouse button, click the drop-list arrow, and from the list select **Lorraine – No Transition – (1)**. Click the **OK** button to continue.

   The routine displays the Target Mapping dialog box. There are no target names to set this time.
5. Click the **OK** button to create the corridor.
6. Use the AutoCAD Zoom and Pan commands to view the corridor.
7. Use the AutoCAD Zoom and Pan commands to view the cul-de-sac from station 5+25 to the end.
8. Click the **AutoCAD Save** icon to save the drawing.

The corridor goes from the beginning to the end of the roadway (center of the cul-de-sac). The non-transitional portion (region) of the corridor should stop at station 5+25.

EDIT THE CORRIDOR PROPERTIES

1. In the drawing, select any line representing the corridor, press the right mouse button, and from the shortcut menu select **Corridor Properties....**
2. Click the **Region (1)** entry to highlight it.
3. Scroll to the right until you view the beginning and ending stations. Click in the cell that represents the ending station of Region (1) and change the ending station to **5+25** (525).
4. Click the **OK** button to create a corridor ending at station 5+25.

   The corridor now ends at station 5+25.

CREATE BASELINE (2)

The control of the corridor around the northern edge of the cul-de-sac passes to a second baseline (centerline). The alignment's vertical controls the edge-of-travelway's vertical location. The Lorraine – (1) alignment stretches the pavement to create the cul-de-sac's paved area.

1. In the drawing, click any line representing the corridor, press the right mouse button, and from the shortcut menu select **Corridor Properties....**
2. If necessary, click the **Parameters** tab to view its contents.
3. In the Parameters panel, click the blue **+** (plus sign) in the upper left of the dialog box to create a new baseline entry.
4. In the Pick Horizontal Alignment dialog box, click the drop-list arrow. From the list, select **Lorraine Left Cul-de-Sac – (1)**, and click the **OK** button to return to the Corridor Properties dialog box.
5. For Baseline (2), in the Profile cell click **<Click here...>**. In the Select a Profile dialog box, click the drop-list arrow. From the list, select **Lorraine Left Transition - Vertical (1)** and click the **OK** button to return to the Corridor Properties dialog box.

CREATE REGION FOR BASELINE (2)

1. In the Corridor Properties dialog box, with the Baseline (2) entry highlighted, press the right mouse button and from the shortcut menu select **Add Region....**
2. In the Select Assembly dialog box, click the drop-list arrow. From the list, select **Lorraine - Transition – (1)**, and click the **OK** button to return to the Corridor Properties dialog box (see Figure 9.33).

   This created Region (1) for Baseline (2) with the selected assembly.
3. Click the **OK** button to create the modified corridor.
4. Use the AutoCAD Zoom and Pan commands to inspect the intersection of the two corridor regions.

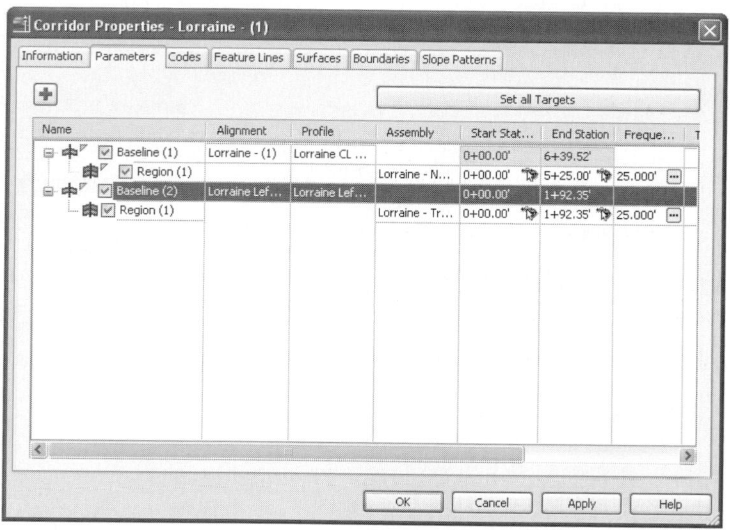

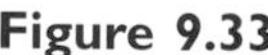
Figure 9.33

The new baseline creates sections following the new alignment and its vertical design. However, the right side of the assembly does not reach the center of the road or cul-de-sac.

To correct the width, you need to assign the centerline of the corridor (Lorraine - (1)) and its vertical to the transition subassembly in the Target Mapping dialog box from Corridor Properties.

TRANSITION LOGICAL NAMES

1. In the drawing, click any line representing the corridor, press the right mouse button, and from the shortcut menu select **Corridor Properties...**.
2. If necessary, click the **Parameters** tab to view its contents.
3. If necessary, click the **Region (1)** entry of Baseline (2) to highlight it, scroll the entry to the right, and click the ellipsis in the Target column.
4. Under Alignments, locate the Width Alignment entry for the Transition Subassembly. Click in the Object Name cell, click the drop-list arrow, and from the list of alignments select **Lorraine – (1)** and click the **OK** button (see Figure 9.34).
5. In the Profiles section, locate the entry Outside Elevation Profile for Transition Subassembly. Click the Object Name cell, make sure the Alignment is Lorraine – (1), and click the drop-list arrow. From the Profile list, select **Lorraine CL Vertical (1)**, and click the **OK** button (see Figure 9.34).
6. Click the **OK** button to exit the Target Mapping dialog box.
7. In the Frequency column for Region (1) of Baseline (2), click the ellipsis, change the Along Tangents and Along Curves sampling rate to **5**, and click the **OK** button to return to the Corridor Properties dialog box.
8. Click the **OK** button to exit the dialog box and to update the corridor.

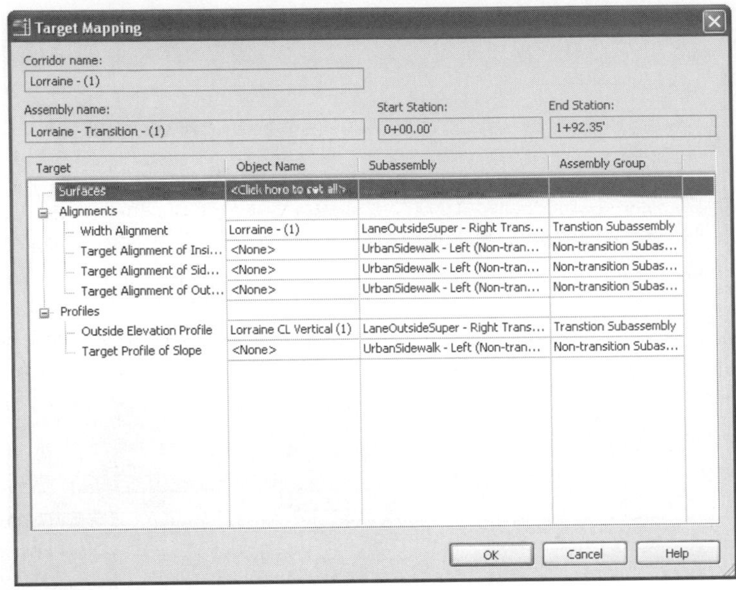

Figure 9.34

9. Use the AutoCAD Zoom and Pan commands to inspect the intersection of the two corridor regions.

   The new baseline and its region widen to the centerline of the roadway and the cul-de-sac.

ADD THE FINAL CUL-DE-SAC REGION

1. In the drawing, click any line representing the corridor, press the right mouse button, and from the shortcut menu select **Corridor Properties...**.
2. If necessary, click the **Parameters** tab to view its contents.
3. Click the blue **+** (plus sign) in the upper left of the dialog box and create a third baseline, **Baseline (3)**.
4. In the Pick Horizontal Alignment dialog box, click the drop-list arrow. From the list, select **Lorraine Right Cul-de-Sac – (1)** and click the **OK** button to return to the Corridor Properties dialog box.
5. For Baseline (3), click in the Profile cell **<Click here...>**. In the Select a Profile dialog box, click the drop-list arrow, and from the list, select **Lorraine Right Transition Vertical (1)**. Click the **OK** button to return to the Parameters panel of Corridor Properties.
6. With Baseline (3) still highlighted, press the right mouse button, and from the short-cut menu select **Add Region...**.

7. In the Select an Assembly dialog box, click the drop-list arrow. From the list, select **Lorraine Transition – (1)**, and click the **OK** button to return to the Corridor Properties dialog box.
8. Scroll the corridor panel to the right until you view the Frequency and Target columns.
9. In the Frequency cell for Region (1) of Baseline (3), click the ellipsis. In the Frequency to Apply Assemblies, change the Along Tangents and Along Curves sampling rate to **5**, and click the **OK** button to return to the Corridor Properties dialog box.
10. In the Target cell for Region (1) of Baseline (3), click the ellipsis.
11. In the Target Mapping dialog box, under Alignments, find the Width Alignment for the Transition Subassembly Assembly Group. Click in the Object Name cell and click the drop-list arrow. From the list of alignments, select **Lorraine – (1)**, and click the **OK** button (see Figure 9.35).
12. In the Profiles section, locate the entry Outside Elevation Profile for the Transition Subassembly. Click in the Object Name cell, make sure the Alignment is Lorraine – (1), and click the drop-list arrow. From the Profile list, select **Lorraine CL Vertical (1)**, and click the **OK** button (see Figure 9.35).

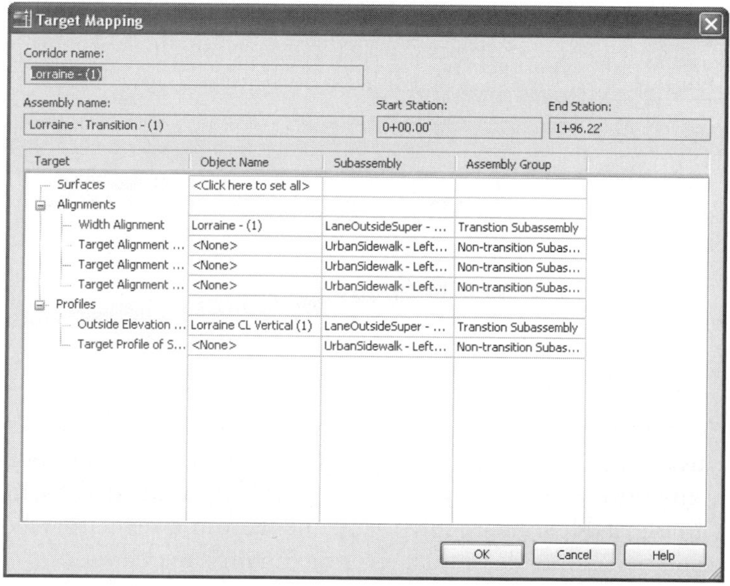

Figure 9.35

13. Click the **OK** buttons to exit the dialog boxes and to update the corridor.
14. Use the AutoCAD Zoom and Pan commands to inspect the intersection of the three corridor regions.

15. Click the **AutoCAD Save** icon to save the drawing.

This ends the exercise on cul-de-sac transitions.

SUMMARY

- One of the keys to cul-de-sacs is defining an assembly and attaching the curb and sidewalk subassemblies to the outside so that they are perpendicular to the alignment around the cul-de-sac arcs.
- The central horizontal and vertical alignments are the width and vertical for the center of the cul-de-sac.

UNIT 4: SUPERELEVATIONS

Superelevations allow a roadway to carry higher speeds around its horizontal curves. To maintain these higher speeds, the roadway design includes subassemblies that rotate (superelevate) the lanes, and possibly the shoulders, toward the center of the horizontal alignment curve. The rotation of the lanes takes advantage of centrifugal force to stabilize a vehicle passing through the curve on the roadway pavement. This design methodology is found in both highway and railway designs (see Figure 9.36).

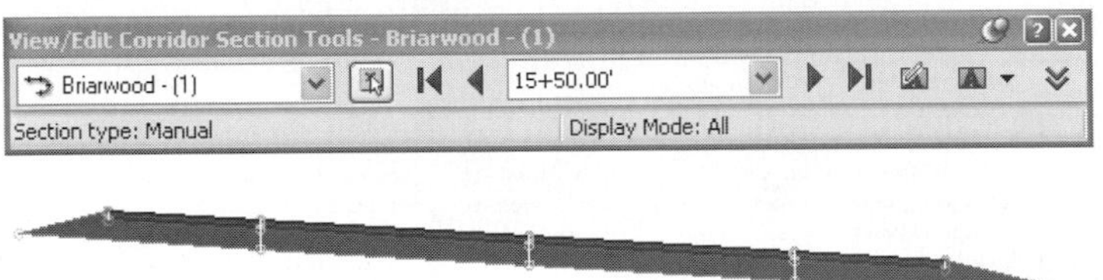

Figure 9.36

The methods of rotating the pavement vary greatly, and there is no uniform standard governing their use or design. The rotation of the pavement can occur about the centerline of the roadway, around its inside or outside lane; or, if it's a divided highway, around the lane edges or centerline. There is usually a document from a regulatory body that defines the necessary distances and angles to achieve maximum superelevation.

The American Association of State Highway & Transportation Officials (AASHTO) publishes documents that define one set of highway design standards. The AASHTO "Green Book" standards are incorporated in Civil 3D as an XML Roadway Design Standards file.

DESIGN STANDARDS FILE

The Design Standards file contains tables with critical superelevation design values. Civil 3D uses the file's values to calculate the roadway cross-section rotation and to check for minimum design values.

The Design Standards file is an XML-based file and can be customized to accommodate differing design standards.

MINIMUM CURVES

A major section of the Design Standards file is the definition of minimum curves for a design based on design speeds and superelevation. The following is an excerpt from the file for a 4 percent superelevation, its design speeds, and the recommended minimum radii. To maintain a higher rate of speed with a 4 percent pavement cross slope, the radius of the road needs to be lengthened to safely handle the greater speeds.

```
<MinimumRadiusTables>
  <!--========================================== -->
  <!-- Defines minimum radii for road type and design speed -->
  <MinimumRadiusTable name="AASHTO 2001 eMax 4%">
        <MinimumRadius speed="15" radius="70"/>
        <MinimumRadius speed="20" radius="125"/>
        <MinimumRadius speed="25" radius="205"/>
        <MinimumRadius speed="30" radius="300"/>
        <MinimumRadius speed="35" radius="420"/>
        <MinimumRadius speed="40" radius="565"/>
        <MinimumRadius speed="45" radius="730"/>
        <MinimumRadius speed="50" radius="930"/>
        <MinimumRadius speed="55" radius="1190"/>
        <MinimumRadius speed="60" radius="1505"/>
```

This minimum radius table defines the smallest radius for a road to maintain the design speed. What if the roadway maintains the same speed, but contains different radii? What amount of superelevation of the total amount allowed does the pavement need to superelevate to safely handle the speed? To answer these questions, Civil 3D references another table that specifies the amount of cross slope based on various curve radii for a fixed speed. The following is an excerpt from the Design Standards file for an urban road with a design speed of 20 and a maximum cross slope of 4 percent (eMax). NC stands for "normal crown" and RC stands for "reverse crown."

```
<DesignSpeed speed="20">
<SuperelevationRate radius="1400" eRate="NC"/>
<SuperelevationRate radius="1200" eRate="RC"/>
<SuperelevationRate radius="1000" eRate="RC"/>
<SuperelevationRate radius="900" eRate="2.1"/>
<SuperelevationRate radius="800" eRate="2.2"/>
<SuperelevationRate radius="700" eRate="2.3"/>
<SuperelevationRate radius="600" eRate="2.5"/>
<SuperelevationRate radius="500" eRate="2.6"/>
<SuperelevationRate radius="450" eRate="2.7"/>
<SuperelevationRate radius="400" eRate="2.9"/>
```

```
<SuperelevationRate radius="350" eRate="3.0"/>
<SuperelevationRate radius="300" eRate="3.2"/>
<SuperelevationRate radius="250" eRate="3.4"/>
<SuperelevationRate radius="200" eRate="3.7"/>
<SuperelevationRate radius="150" eRate="3.9"/>
<SuperelevationRate radius="125" eRate="4.0"/>
```

The maximum rotation of the roadway pavement occurs only at the minimum curve radius for the design speed.

The amount of cross slope changes with the radius of the curve—i.e., the shorter the curve radius, the greater the amount of cross slope needed to maintain the speed. In the previous excerpt, a curve with a radius of 125 has a cross slope of 4 percent and a curve with a radius of 400 will only need a cross slope of only 2.9 percent to maintain the speed.

TRANSITION LENGTHS

A transition length section defines the length of transition for any design speed. The length varies for each speed and radius of curve. For example, a road with a design speed of 20 and a radius of 150 needs 72 feet of transition. If the same road has a radius of 500, it needs only 37 feet of transition.

When setting or selecting standards, Civil 3D will issue a warning about any violations of the standards in the event view.

SUPERELEVATION ATTAINMENT METHODS

Civil 3D supports three basic strategies for superelevating a roadway. The two most common methods are adverse crown removal and planar (uncrowned). The last is a custom formula to calculate critical design values. The following is a brief description of various superelevation attainment methods in Civil 3D.

- *Method 1*—Pivots a crowned roadway about the centerline alignment. Which edge-of-travelway rotates up to remove the crown depends on the direction of the curve. A righthand curve rotates the left side of the road up, and a lefthand curve rotates the right side of the road up.
- *Method 2*—Pivots a crowned roadway around the inside edge of the edge-of-travelway. The direction of the curve defines the inside edge—that is, right turn, right edge; and left turn, left edge. The method forces the outside edge upward to attain the superelevation angle.
- *Method 3*—Pivots a crowned roadway about the outside edge of the edge-of-travelway. The direction of the curve defines the outside edge—that is, right turn, left edge; and left turn, right edge. The method forces the inside edge downward to attain the superelevation angle
- *Method 4*—For divided crown roadways. This method pivots the roadway at the inside of the travelway. The direction of the curve defines the inside edge—that is, right turn, right edge; and left turn, left edge. The method forces the outside edge upward to attain the superelevation angle.

- *Method 5*—For divided uncrowned roadways (planar). This method rotates the entire lane about the inside to create a superelevation. The direction of the curve defines the inside edge—that is, right turn, right edge; and left turn, left edge. The method forces the outside edge upward to attain the superelevation angle.
- *Method 6*—For divided uncrowned roadways (planar). This method rotates the entire roadway about the centerline of the roadway. Which edge-of-travelway rotates up to remove the crown depends on the direction of the curve. A righthand curve rotates the left side of the road up, and a lefthand curve rotates the right side of the road up.
- *Method 7*—For undivided or uncrowned (planar) roadways. This method rotates an uncrowned roadway from its initial slope to an opposing slope. The slopes pivot at the centerline of the roadway or at one of the edges-of-travelway. This is also known as superelevation that opposes a normal slope.
- *Method 8*—For undivided or planar roadways. This method rotates an uncrowned roadway from initial design slope to a greater slope. The slopes pivot at the centerline of the roadway or at one of the edges-of-travelway. This is also known as superelevation that continues a normal slope.

Method 7 is for ramps that start with the roadway sloped in one direction and then has to reverse its slope to make a superelevated curve. This requires a greater length of roadway to accomplish the change of roadway cross slope.

TRANSITION FORMULAS

Civil 3D supports the use of formulas to calculate non-standard superelevation values. The variables available to calculate superelevation parameters are:

- {e} – The superelevation rate (from tables)
- {t} – The superelevation length (from tables)
- {l} – The length of the spiral (from the alignment definition)
- {c} – The normal crown lane slope (from alignment settings)
- {s} – The normal shoulder lane slope (from alignment settings)
- {w} – The nominal width (from alignment settings)

These variables are a part of formulas defining different methods of superelevating a road. Civil 3D supports seven key transition formulas.

LC to FS—Level crown station (LC) to full super (FS) station (runoff)

LC to BC—Level crown station (LC) to beginning of curve (BC)

NC to FS—Normal crown station (NC) to full super (FS) station

NC to BC—Normal crown station (NC) to beginning of curve (BC)

NC to LC—Normal crown station (NC) to level crown station (LC) (Runout)

LC to RC - Level crown station (LC) to reverse crown station (RC)

NS to NC- Normal shoulder station (NS) to normal crown station (NC)

Civil 3D uses the two-thirds rule—i.e., two-thirds of the transition length is along the tangent and the remaining one-third is along the path of the curve.

SHOULDER SUPERELEVATION

Civil 3D supports two methods of superelevating roadway shoulders. The first is breakover removal. This method forces the shoulder slope to match the roadway cross slope before beginning the superelevation of the adjacent lane. This method introduces an additional transitional length because the slope of the shoulder has to rotate up to match the lane slope before beginning the rotation of the transition lane.

The second method is to match lane slopes. In this situation, the shoulder slopes always match the lane slopes.

GENERAL TERMS

When working with superelevations, there are some basic terms to understand:

- *Runout*—The distance over which a lane on one side of the pavement rotates from a normal crown (NC) to no crown (level crown, or LC). Tangent runout distance is another name for runout, because the rotation occurs along the tangent before entering the curve.
- *Runoff*—The distance over which one side of the pavement rotates from a level crown (LC) to full superelevation (FS).
- *Percentage of Runoff*—The runoff percent represents the amount of superelevation that occurs along the tangent before entering the curve. In Civil 3D the default is that two-thirds of the rotation occurs along the tangent and the remainder occurs along the entering curve.
- *E value*—The maximum superelevation rate. The E value is either ft/ft or m/m ratio. A 0.10 E value is a 10 percent grade.

Civil 3D allows users to change the percentage of runoff to a different value in the Design Standards file.

SUPERELEVATION REGIONS

Civil 3D considers each curve a superelevation region. When setting the superelevation parameters, users need to set them for each region (curve) of the alignment.

DESIGN RULES

The first option of the Superelevation Region is the Design rules section (see Figure 9.37). This section identifies the starting and ending station of the curve, the design speed, what Design Standards file to use, the superelevation rate table, the transition length table, and the method of attaining superelevation. The last three settings are tables in the Design Standards file.

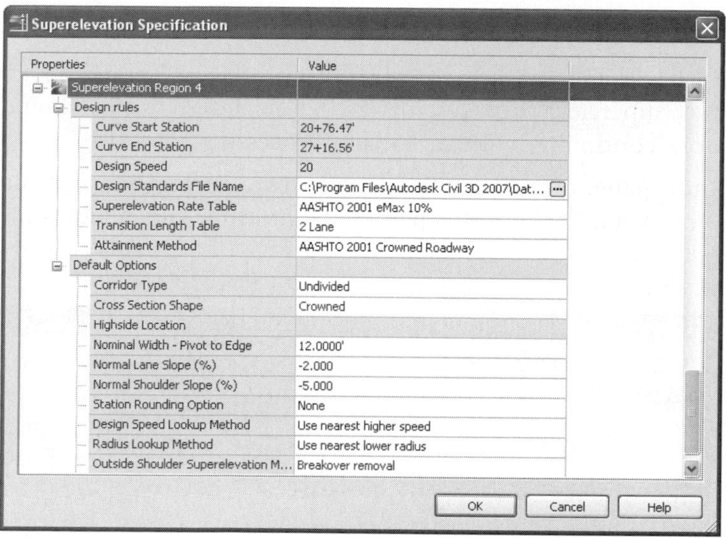

Figure 9.37

DEFAULT OPTIONS

The Default Options section sets corridor, shoulder, and calculation rules. The first part of the section sets the corridor type and cross-section shape (see Figure 9.38). The next three entries set the lane width and grades for the lane and shoulders. The next group of values set which assumptions to use when having to calculate a value when the speed and curve radius is not in the Design Standards file. The last entries specify how to remove the shoulder cross grade.

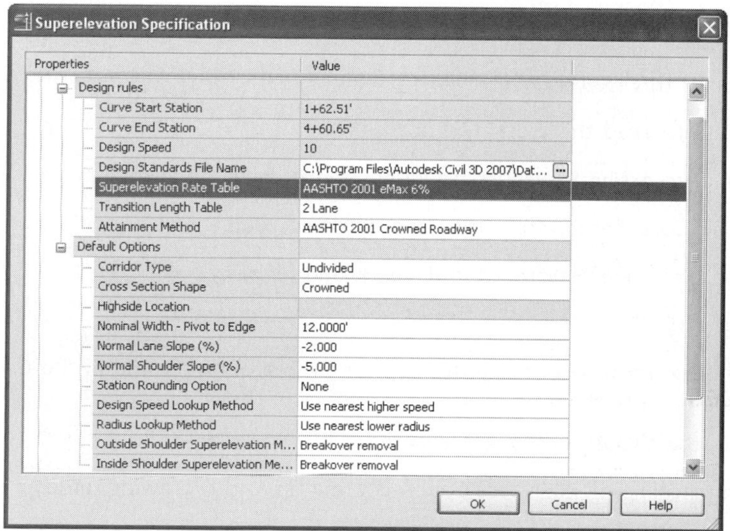

Figure 9.38

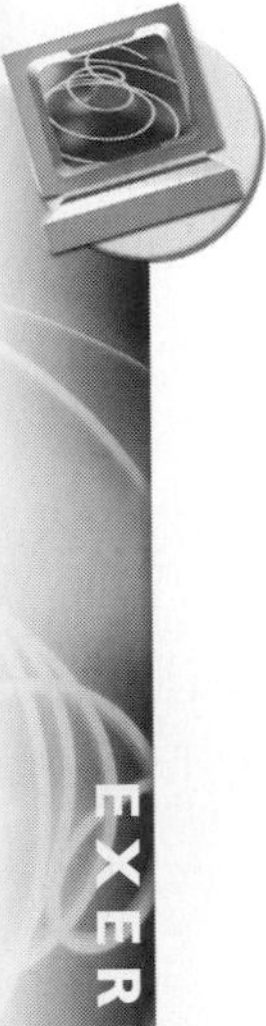

WARNINGS AND ERROR MESSAGES

Civil 3D issues a warning in the Event Viewer if the curve radii and speeds exceed minimums in the Superelevation Design file. These design issues should be dealt with immediately before continuing with the design process.

The Superelevation panel of an alignment's Properties dialog box will indicate problems with a design by using arrows to point to the full superelevation of the region (see Figure 9.39).

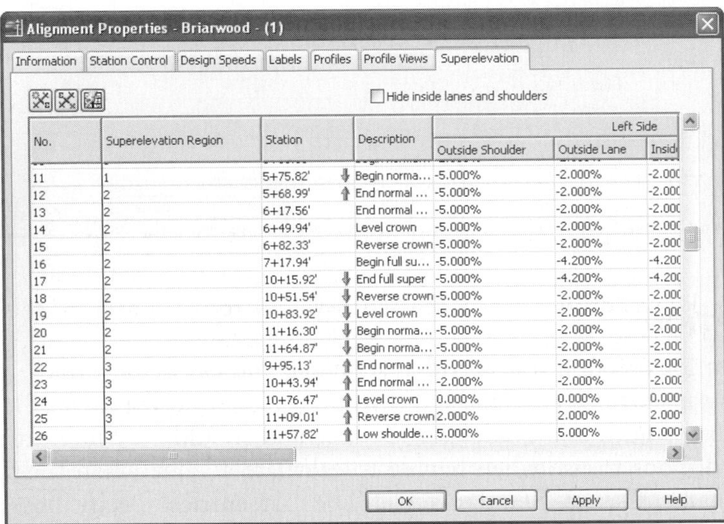

Figure 9.39

EXERCISE 9-4

After you complete this exercise, you will:

- Be able to read the Design Standards file.
- Be able to assign values from Design Standards to an alignment.
- Be able to superelevate a two-lane crowned road.
- Be able to superelevate a four-lane divided highway.

EXERCISE SETUP

You can start this exercise by opening the *Chapter 9 - Unit 4.dwg* found in the *Chapter 9* folder of the CD that accompanies this textbook.

1. Open the *Chapter 9 - Unit 4* drawing.
2. From the File menu, select **Save As...** and save the drawing under a new name.

IMPORT THE EG SURFACE

This exercise uses a single surface, as well as two alignments and their profiles. All of data is in a single XML file. The alignments appear in the drawing after importing them, but the profiles will need to be placed in the drawing with a view containing an existing ground profile.

1. From the File menu, Import flyout, select **Import LandXML....**
2. In the Import LandXML dialog box browse to the *EG.xml* file and click the **Open** button. The file is in the *Chapter 9* folder of the CD that accompanies this textbook.
3. In the Import LandXML dialog box, click the **OK** button to import the file.

   The Event View may display an information warning.
4. Close the Event Viewer.

   The surface should appear on the screen
5. Click the **AutoCAD Save** icon to save the drawing.

IMPORT THE BRIARWOOD ALIGNMENT AND PROFILE

The Briarwood Alignment and Profile are in a second LandXML file. When importing an alignment and profile, Civil 3D will draw the alignment, but will only read the data for the profile and place it in Prospector. You must create a profile view, sampling the EG surface, and when drawing the view you must include the Briarwood vertical alignment.

1. From the File menu, Import flyout, select **Import LandXML....**
2. In the Import LandXML dialog box browse to the *Briarwood.xml* file and click the **Open** button. The file is in the *Chapter 9* folder of the CD that accompanies this textbook.
3. In the Import LandXML dialog box, click the **OK** button to import the file.
4. Use the AutoCAD Zoom and Pan commands to better view the alignment (west central portion of the surface).

   The Briarwood centerline is the small subdivision road at the middle left of the surface.

ASSIGN DESIGN SPEEDS TO BRIARWOOD ALIGNMENT

1. In the drawing, click any segment of the Briarwood Alignment, press the right mouse button, and from the shortcut menu select **Alignment Properties....**
2. Click the **Design Speeds** tab to view the panel.
3. Click the **Add Design Speed** icon to set the first design speed.

   The dialog box disappears and a station jig connects to the alignment. You can select a station with the jig, or you can type in a station on the command line.
4. In the drawing, select a point near station **1+05**.

You return to the Properties dialog box. You can edit the station to an even value and then assign a speed.

5. In the dialog box, edit the station to **1+05** (105) and assign a speed of **10**.

   Instead of selecting a point, you can enter a station value and then, after returning to the Properties dialog box, assign the speed.

6. Repeat the Steps 3 through 5 and assign the speeds at the alignment stations as listed in Table 9-2.

Table 9.2

Station	Speed
1+05	10
5+50	20
27+50	10

7. Click the **Apply** button to accept the design speeds and to remain in the dialog box.

EXERCISES

SET THE SUPERELEVATION PARAMETERS

The alignment's properties maintain the superelevation information and assignments.

1. Click the **Superelevation** tab to view its contents.
2. At the top of the dialog box, toggle **ON** Hide Inside Lanes and Shoulders.

   Briarwood is a two-lane road.

3. In the dialog box, at the top left, click the **Set Superelevation Properties** icon (right of the three icons) to set the superelevation values.
4. Collapse Regions 2, 3, and 4, so only Region 1 is showing.
5. Set the Region 1 and 2 design rules as follows:

   Superelevation Rate Table: **AASHTO 2001 eMAX 6%**

   Transition Length Table: **2 Lane**

   Attainment Method: **AASHTO 2001 Crowned Roadway**

   Use Figure 9.38 as a guide to set the values for the Default Options section of all regions.

6. Set the Region 3 design rules as follows:

   Superelevation Rate Table: **AASHTO 2001 eMAX 8%**

   Transition Length Table: **2 Lane**

   Attainment Method: **AASHTO 2001 Crowned Roadway**

7. Set the Region 4 design rules as follows:

   Superelevation Rate Table: **AASHTO 2001 eMax 12%**

   Transition Length Table: **2 Lane**

   Attainment Method: **AASHTO 2001 Crowned Roadway**

8. Click the **OK** button to set the properties and to return to the Alignment Properties dialog box.

9. Scroll the panel until you view the data for Region 2.

   The blue arrows indicate that some of the elements of the design do not meet minimum AASHTO standards, and the parameters need to be reviewed and possibly changed (see Figure 9.39).

10. Click the **OK** button to exit the dialog box.

11. Click the **AutoCAD Save** icon to save the drawing.

EXERCISES

CREATE THE BRIARWOOD PROFILE

1. From the View menu, use the **Named Views...** to create a view of the screen, and name the view **Briarwood**.

2. Pan the roadway to the right to make open space to the left of the surface.

3. From the Profiles menu, select **Create from Surface....**

   In the Create Profile dialog box, the Briarwood vertical alignment should be already in the Profile list area.

4. In the Create Profile from Surface dialog box, set the alignment to **Briarwood – (1).** At the top right, select the **EG** surface, and click the **Add>>** button at the middle right of the dialog box to add the EG profile to the list (see Figure 9.40).

5. In the dialog box, click the **Draw in Profile View** button at the bottom to call the Create Profile View dialog box.

6. In the middle left of the Create Profile View dialog box, set the Profile Style to **Clipped Grid**, and change the Style for Briarwood Vertical (1) to **Design Style**.

7. Click the **OK** button and, in the drawing, select a point at the left side of the screen.

8. If the Event View displays, close it.

9. If you need to move the profile farther to the left, use the AutoCAD Move command to relocate the profile.

   Your drawing should now look like Figure 9.41.

10. Click the **AutoCAD Save** icon to save the drawing.

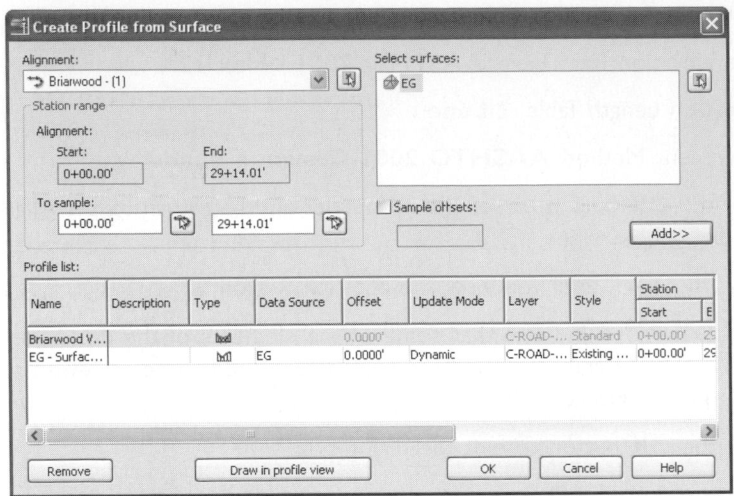

Figure 9.40

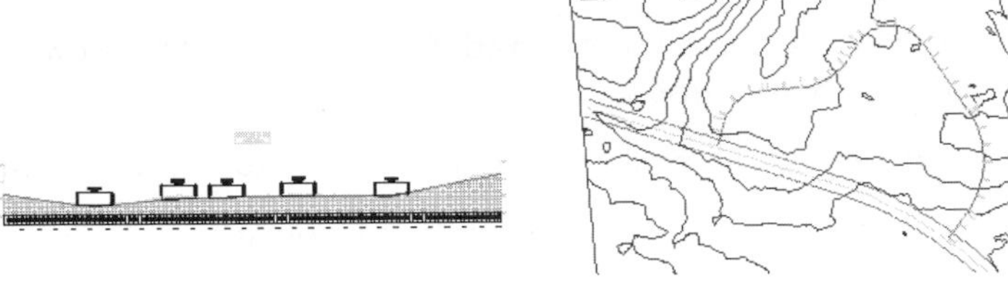

Figure 9.41

CREATE THE ASSEMBLY

See Figure 9.42 for the assembly to create.

1. From the Corridors menu, select **Create Assembly...**.
2. In the Create Assembly dialog box, name the assembly **Briarwood,** leaving the counter. Click the **OK** button and select a point in the drawing near the profile for Briarwood to place the assembly.

ADD SUBASSEMBLIES

1. If necessary, from the General menu, select **Tool Palettes Window**.
2. On the tool palette, click the **Imperial - Roadway** tab.
3. From the palette, select **LaneOutsideSuper**.
4. In the Properties palette, Advanced Parameters, change the Side to **Right**, change the Width to **14** feet, and in the drawing, select the assembly to place the lane.
5. In the Properties, change the Side to **Left**, select the assembly to place the lane, and press the **Enter** key twice to exit the routine.
6. In Prospector, expand the Subassemblies branch to view its list.

7. Select each subassembly and edit its Properties. Rename them **LaneOutsideSuper – Right** and **LaneOutsideSuper – Left**.
8. From the Imperial - Roadway palette, select **ShoulderExtendSubbase**. In Properties, Advanced Parameters, set the Side to **Right**, and in the drawing place the subassembly at the upper-right marker of the pavement's right side.
9. In the Properties, change the Side to **Left**. In the drawing place the subassembly at the pavement's upper-left outer marker and press the **Enter** key twice to exit the routine.
10. In the Prospector's Subassembly branch, select each subassembly and edit its properties. Rename them **ShoulderExtendSubbase – Right** or **ShoulderExtendSubbase – Left**.
11. Click the **AutoCAD Save** icon and save the drawing.

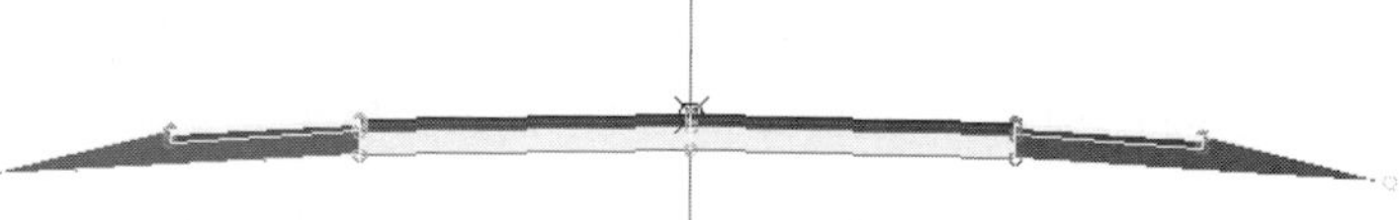

Figure 9.42

EXERCISES

CREATE THE CORRIDOR

1. From the View menu, use the **Named Views...** command to restore the Briarwood view.
2. From the Corridors menu, select **Create Simple Corridor....** Name the corridor **Briarwood,** keeping the counter, and click the **OK** button.
3. In the command line, the routine prompts for an alignment. Press the right mouse button, select **Briarwood – (1)**, and click the **OK** button.
4. In the command line, the routine prompts for a profile. Press the right mouse button, and in the Select a Profile dialog box, click the drop-list arrow. From the list, select **Briarwood Vertical (1)**, and click the **OK** button.
5. In the command line, the routine prompts for an assembly. Press the right mouse button, and in the Select an Assembly dialog box, click the drop-list arrow. From the list, select **Briarwood – (1)** and click the **OK** button.
6. The Target Mapping dialog box displays; there are no mappings to set, so click the **OK** button to create the corridor.

VIEW CORRIDOR SECTIONS

1. From the Corridors menu, select **View/Edit Corridor Section**. In the drawing, select any line representing the Briarwood corridor and press the **Enter** key to view the sections.
2. After viewing the sections, exit the routine and save the drawing.

DIVIDED HIGHWAY CORRIDOR

The divided highway corridor starts at the top left of the site and continues to the middle right of the surface. The LandXML file for this highway is in the *Chapter 9* folder of the CD that accompanies this textbook.

1. Use the AutoCAD Zoom Extents command to view the entire drawing.

IMPORT ROUTE 7380 ALIGNMENT AND PROFILE

1. From the File menu, Import flyout, select **Import LandXML....** Browse to and select the **Route 7380.xml file**, and click the **Open** button to import the file. The file is in the *Chapter 9* folder of the CD that accompanies this textbook.
2. In the Import LandXML File dialog box, click the **OK** button.

   The alignment appears in the northern half of the site.
3. Use the AutoCAD Pan command and pan the site to the left.
4. From the Profiles menu, select **Create From Surface....**
5. In the Create Profile View from Surface dialog box, change the alignment to **Route 7380 – (1)**, select the **EG** surface, and click the **Add>>** button to add the surface profile to the profile list at the bottom of the dialog box. Click the **Draw in Profile View** button at the bottom of the dialog box.
6. In the Select Profiles to Draw area of the Create Profile View dialog box, change the style to **Design Style** for Route 7380 Vertical (1). Click the **OK** button to place the profile to the right of site.
7. If the Event Viewer displays, close it.

EXERCISES

CREATE THE ASSEMBLY

See Figure 9.43 for the assembly.

1. Select **Create Assembly...** from the Corridors menu. Name the assembly **Route 7380,** keeping the counter, and place the assembly near the profile for Route 7380.
2. If necessary, use the AutoCAD Zoom and Pan commands to view the assembly.

ADD SUBASSEMBLIES

1. If necessary, from the General menu, select **Tool Palettes Window**.
2. Click the **Imperial - Roadway** tab.
3. From the palette, select **MedianDepressedShoulderVert**.
4. In Advanced Parameters of the Properties section, change the Hold Ditch Slope to **Hold Ditch at Center, adjust sideslope on high side**. Change the Paved Shoulder Width to **8** feet and change the Unpaved Shoulder Width to **3** feet. In the drawing, select the assembly to attach the subassembly, and press the **Enter** key twice to exit.

   The assembly pivots around the centerline of the median.

5. From the Imperial - Roadway palette, select **LaneOutsideSuper**.
6. In Advance Parameters of the Properties section, change the Side to **Right** and change the Width to **26** feet. In the drawing, select the upper-right marker of the median to place the lane.
7. In Advance Parameters of the Properties section, change the Side to **Left**. In the drawing, select the upper-left marker of the median to place the lane.
8. Press the **Enter** key twice to exit the routine.
9. In Prospector, the Subassemblies branch, edit the properties of both new subassemblies. Name them **LaneOutsideSuper – Right - Divided** and **LaneOutsideSuper – Left - Divided**.
10. From the palette, select **ShoulderVerticalSubbase**.
11. In Advance Parameters of the Properties section, set the Side to **Right**, set the Paved Width to **8**, and set the Unpaved Width to **3**. In the drawing, select the upper-right marker of the pavement to place the subassembly.
12. In Advance Parameters of the Properties section, change the Side to **Left**. In the drawing, place the subassembly at the upper-left outer marker for the pavement, and press the **Enter** key twice to exit the routine.
13. In the Subassemblies branch of Prospector, edit the properties of both new subassemblies, naming them **ShoulderVerticalSubbase – Right – Divided** and **ShoulderVerticalSubbase – Left – Divided**.
14. Click the **Imperial - Daylight** tab.
15. From the palette, select **DaylightStandard**.
16. In Advance Parameters of the Properties section, change the Side to **Right** and attach the subassembly to the right end of the right shoulder.
17. Return to the Properties dialog box. Change the Side to **Left**, and attach Daylight-General to the left side of the assembly.
18. Press the **Enter** key twice to exit the subassembly routine.
19. Edit the Properties for both subassemblies, naming them **DaylightStandard Right – Divided** and **DaylightStandard Left – Divided**.
20. Close the Civil 3D - Imperial tool palette.
21. Click the **AutoCAD Save** icon and save the drawing.

Figure 9.43

EXERCISES

ASSIGN DESIGN SPEEDS

1. Use the AutoCAD Zoom and Pan commands to view the Route 7380 alignment.
2. If necessary, click the **Prospector** tab to make it the current panel.
3. Expand the Sites branch until you view the Route 7380 – (1) alignment.
4. Select the **Route 7380 – (1)** alignment, press the right mouse button, and from the shortcut menu select **Properties...**.
5. Click the **Design Speeds** tab to view its contents.
6. In the Design Speeds panel, click the **Add Design Speed** icon, type the command prompt **10**, and press the **Enter** key.
7. In the Design Speeds panel for station 10, assign **70** as the design speed.
8. Click the **Add Design Speed** icon, type the command prompt **10250,** and press the **Enter** key.
9. If necessary, in the Design Speeds panel, assign **70** as the design speed for station 10250.
10. Click the **Apply** button to set the speeds.

ASSIGN SUPERELEVATION PROPERTIES

1. Click the **Superelevation** tab to view its contents.
2. At the top center of the panel, uncheck the Hide Inside Lanes and Shoulders toggle.
3. Click the **Set Superelevations Properties** icon to set the values for Route 7380 (see Figure 9.44).
4. Set the following values for Superelevations Region 1 and 2:

   Design rules:

   Superelevation Rate Table: **AASHTO 2001 eMax 10%**

   Transition Length Table: **4 Lane**

   Attainment Method: **AASHTO 2001 Crown Roadway**

   Default Options:

   Corridor Type: **Divided**

   Cross Section Type: **Planar**

   Nominal Width - Pivot to Edge: **26**

   Normal Shoulder Slope (%): **-6**

5. Click the **OK** button to set the superelevation parameters for Route 7380.
6. Scroll through the superelevation data and note the stations for maximum superelevation along Route 7380.

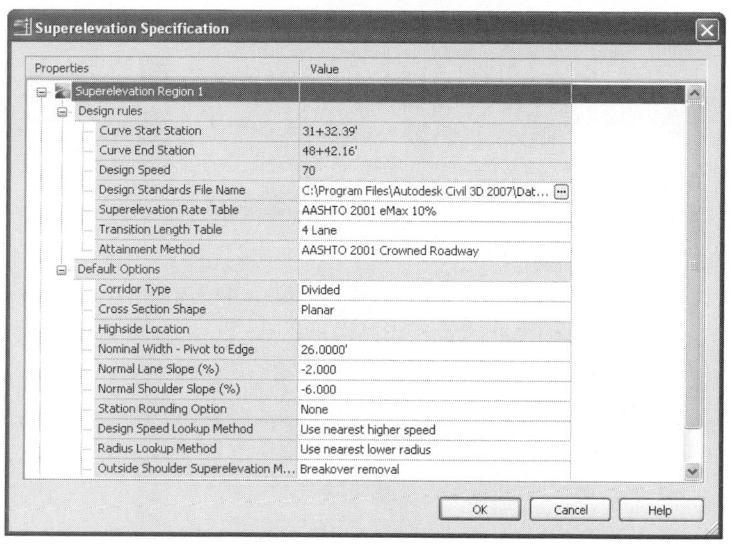

Figure 9.44

7. Click the **OK** button to change the alignment properties.

CREATE THE CORRIDOR

1. From the Corridors menu, select **Create Corridor**.
2. In the command line, the routine prompts for an alignment. Press the right mouse button, select **Route 7380 – (1)**, and click the **OK** button.
3. In the command line, the routine prompts for a profile. Press the right mouse button, and in the Select a Profile dialog box click the drop-list arrow. From the list, select **Route 7380 Vertical (1)** and click the **OK** button.
4. In the command line, the routine prompts for an assembly. Press the right mouse button, and in the Select an Assembly dialog box click the drop-list arrow. From the list, select **Route 7380 – (1)** and click the **OK** button.
5. In the Create Corridor dialog box, name the corridor **Route 7380,** keeping the counter, and click the **OK** button.
6. Click the **Set All Targets** button in the top left of the dialog box.
7. In the Target Mapping dialog box, for Object Name cell of Surfaces, select **<Click here to set all>**. In the Pick a Surface dialog box, select **EG**, and click the **OK** button to return to the Target Mapping dialog box.
8. Click the **OK** buttons to exit and build the divided highway.

   The Event Viewer issues errors for the corridor that do not have names set for transitions.
9. Click the green check mark to hide the Event Viewer.

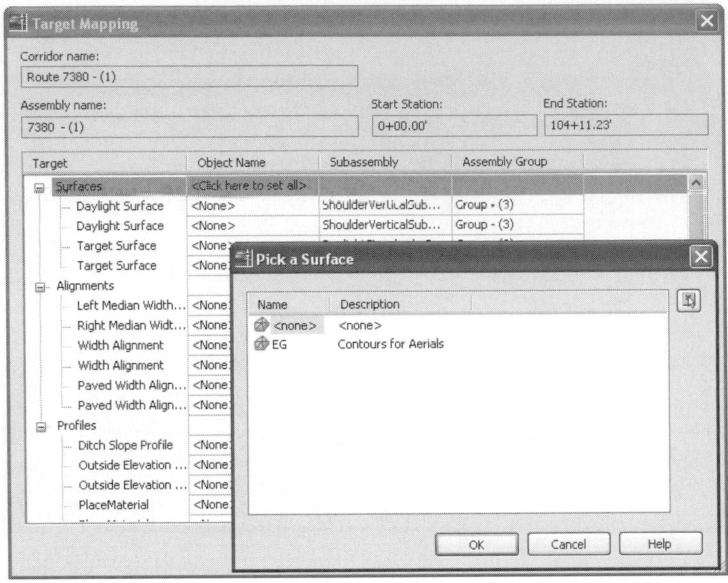

Figure 9.45

10. Click the **AutoCAD Save** icon and save the drawing.
11. Use the AutoCAD Zoom and Pan commands to view the new corridor.

    The maximum superelevation occurs at stations 32+00 to 47+00 and 71+00 to 80+00. Start viewing the first region at station 25+00 and the second region at station 54+00.

12. Select **View/Edit Corridor Section** from the Corridors menu and view the assembly behavior through the curves.

    This completes the exercise on transitions and superelevations. The next chapter focuses on the Civil 3D grading tools.

SUMMARY

- The Design Standards file shipped with Civil 3D reflects the values published in the AASHTO "Green Book."
- The Design Standards file is an XML-based file and can be edited to suit custom needs.
- The Design Standards file contains variables and formulas to accomplish superelevating a roadway.
- Two basic design parameters for superelevation are the design speeds and roadway curve radii.
- Civil 3D evaluates a roadway design based on the alignment and the specified values in the Design Standards file.
- If a curve does not meet the specifications, Civil 3D will issue a warning in the Event Viewer.
- Design speeds and superelevation properties are set in the Alignment Properties dialog box.
- The superelevation properties should be appropriate for the assembly that is creating the roadway corridor.

Grading and Volumes

INTRODUCTION

The Autodesk Civil 3D 2007 grading tools are the least mature of the Civil 3D toolsets. The grading object is a powerful design tool, but its interface is not well developed.

The feature line replaces the 3D polyline of Autodesk Land Desktop and is more powerful because it can contain circular arc segments. Feature lines are the basis for grading solutions. A surface breakline is another use for a feature line. When processing a feature line as surface data, a mid-ordinate variable samples the changing elevations along the curve.

OBJECTIVES

This chapter focuses on the following topics:

- Using the Three Types of Grading Tools in Civil 3D
- Creating Designs Using Point, Feature Lines, and Grading Solutions
- Designing with the Enhanced Grading Object
- Calculating an Earthworks Volume

OVERVIEW

Chapter 4 of this book, Surfaces, introduced some of the basic site design tools (i.e., points, basic feature lines, breaklines, and 3D polylines). The point routines described in this chapter assign elevations to points from surfaces, 3D polylines, or from the interpolation of elevations between controlling points. Figure 10.1 shows the Interpolation routines from the Create Points toolbar.

The Grading menu's grading routines focus on feature lines and the grading object. The grading object sets blending strategies to match an existing surface, projects a surface at a distance using a grade or slope, or projects a surface to a specific depth using a grade or slope. The grading object does not need to have a surface to grade to. It can grade for a distance or to an elevation, relative to or absolute from itself. The grading object can automatically create a surface and calculate a volume.

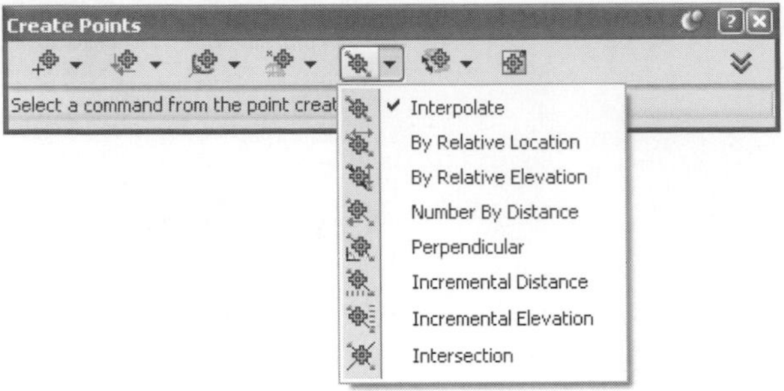

Figure 10.1

The feature line creates the linear shape a grading object uses to create a grading solution. Developing the key feature lines that later create the surface solution is a key step in developing an overall grading solution.

After grading a site or developing a design surface, the next step is to calculate an earthworks volume. The last unit of this chapter describes the process of designing a surface and calculating an earthworks volume.

UNIT 1

The first unit of this section uses the point tools of the Create Points toolbar to create a solution from a series of points. The elevations for some of the points are from surface elevations and point grade/slope interpolations.

UNIT 2

The second unit of the chapter works with the grading object found in the Grading menu. Whether it is a pile, building pad, or grading off the back-of-curb of a parking lot, the enhanced grading object presents the user with several opportunities to quickly evaluate different grading scenarios.

UNIT 3

The third unit focuses on the designing of a second surface with contours, points, and feature line data. After designing the surface, the next step is calculating the earthwork volumes to change the existing surface to the proposed surface.

UNIT 1: GRADING WITH POINTS

The Create Points toolbar contains commands creating new points for a surface representing a grading plan or a proposed site design. Some of the routines assign elevations to points from a surface. Others assign elevations to points by interpolating a grade or slope from a point or between two controlling points. When creating a surface from this point data, the user may need to add breaklines to correctly triangulate the point data. Instead of using point data and breaklines, it maybe simpler to use contours, grading objects, and 3D polylines. When creating a surface from contours, feature lines, and 3D polylines, each object type is a breakline. No matter what object types are surface data, it is necessary to review their effects and resolve any issues.

The Create Points toolbar has three icon stacks with routines for grading: Interpolation, Slope/Grade, and Surface.

INTERPOLATION

The Interpolation icon stack represents several routines that calculate elevations from or between points (see Figure 10.1). Most of the routines require two points (objects or selected). Most routines place points in a direction for a distance at a slope or grade from one point or between two controlling points. To create the new points, the routines may prompt for one or more of the following values: distance, elevation, grade, offset, and slope.

Why would you want to create interpolated points? A wide gap between points along a linear feature, for example a swale, will allow triangles to cross the swale feature. By doing so, the crossing triangle legs indicate the points representing the swale not related and break the assumed constant slope between the swale points. This interpretation of the point data incorrectly represents the related elevations along the swale points. To control this problem, there are two options: to place points between the controlling points of the swale, or to place a breakline between the controlling points of the swale. In either case, the purpose is to make the surface create triangles following the swale data.

INTERPOLATE

The Interpolate routine places points between two existing control points. The elevations of the new points are a straight slope calculation of a change in elevation over a distance. The routine places a specified number of points between two controlling points and each intervening point has a different elevation reflecting a straight slope calculation from the difference in elevation of the controlling points.

RELATIVE LOCATION/RELATIVE ELEVATION

These two routines use a starting and ending point. The new points produced are the interpolated elevations between the starting and ending elevation. The Location routine places points at a distance measured from the first point. The Elevation routine places points along a line from the first to second point whose elevations are calculated at a constant slope or grade. Both routines have an optional offset. If the elevation or distance is greater than the original distance between two points, the routines continue to place points until they reach the distance or elevation.

INTERSECTION

This routine places a point at the intersection of two directions and grades or slopes.

PERPENDICULAR

This routine places a point perpendicular to an object or direction. The elevation of the point is calculated by the distance and grade/slope perpendicular to the intersection with the object.

SLOPE

There are two routines in the Slope icon stack that are similar to and less complicated than the point interpolation routines (see Figure 10.2). The routines are the Slope/Grade - Distance and Slope/Grade - Elevation routines. The Distance and Elevation routines require only one point at which to start, and the second point does not need to be another existing point. When the routine creates the resulting points, it prompts the user to place a point at the end of the distance from the starting point.

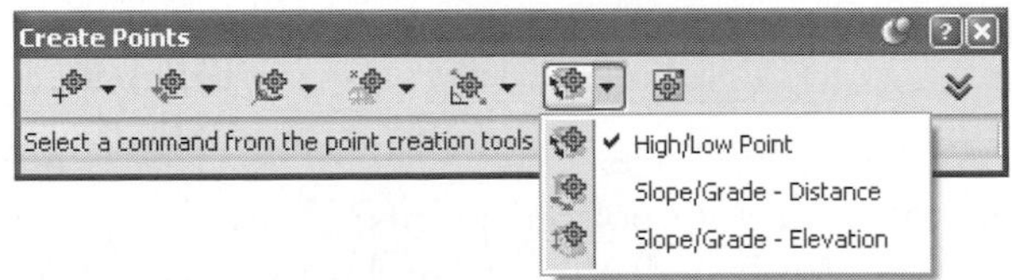

Figure 10.2

SLOPE/GRADE - DISTANCE

The Slope/Grade - Distance routine prompts for a starting point, a direction, a distance, and a slope or grade. The routine prompts for a selected distance and then before proceeding gives the user an opportunity to adjust the distance. The routine then prompts for the number of intermediate points and whether the farthest distance also has a point.

Command:

```
Specify start point: '_PO (Transparent Command: Point Object)
>> Select point object:
Specify start point: (9655.92 8982.44 681.0)
Elevation <681.000'> (Press the ENTER key)
Specify a point to define the direction of the intermediate
    points:
Slope (run:rise) or [Grade] <Horizontal>: 5
Slope: (run:rise): 5.00:1
Grade: (percent): 20.00
Distance: (Select a distance from the select point)
```

The routine responds with the slope and grade, then prompts for the number of intermediate points, and if the endpoint also receives a point.

```
Enter the number of intermediate points <0>: 3
Specify an offset <0.000>: (Press the ENTER key)
Add ending point [Yes/No] <Yes>: (Press the ENTER key)
Command:
```

SLOPE/GRADE - ELEVATION

The Slope/Grade - Elevation routine works much the same as the Distance routine, except that the primary values are elevations. If the routine does not attain the ending elevation over the default distance, it will continue placing points in the direction the user shows the routine. The routine prompts are as follows:

```
Command:
Specify start point: '_PO (Transparent Command: Point Object)
>> Select point object:
Resuming CREATEPOINTS command.
Specify start point: (9655.92 8982.44 681.0)
Elevation <681.000'> (Press the ENTER key)
Specify a point to define the direction of the intermediate
    points:
Slope (run:rise) or [Grade] <Horizontal>: 7
Slope: (run:rise): 7.00:1
Grade: (percent): 14.29
Ending Elevation: 687.5
Enter the number of intermediate points <0>: 4
Specify an offset <0.000>: (Press the ENTER key)
Add ending point [Yes/No] <Yes>: (Press the ENTER key)
Command:
```

Again, if the routine does not attain the user elevation in the distance (the distance between points 1 and 2), the routine continues to place points in the vector direction until reaching the elevation.

SURFACE

These routines require a surface and use the surface's elevations to assign point elevations (see Figure 10.3). If the surface is a terrain surface, the points represent actual surface locations. If the surface is a volume surface, the point elevations represent the difference in elevation between the two surfaces at the location of the points.

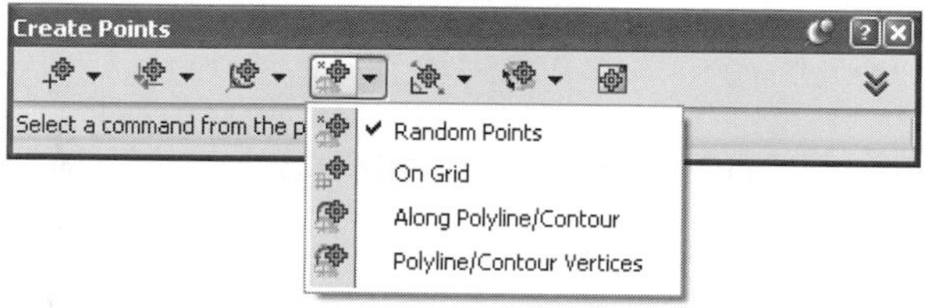

Figure 10.3

RANDOM POINTS

The Random Points routine places a single point into the drawing whose elevation is a surface's elevation at the selected coordinates.

ON GRID

The On Grid routine places points in a user-defined X and Y spacing. The elevations for the points are from the surface at the location grid intersections.

ALONG POLYLINE/CONTOUR

The Along Polyline/Contour routine places a point object at a specified distance along the polyline or contour. This is a measure type of command.

POLYLINE/CONTOUR VERTICES

The Polyline/Contour Vertices routine places a point object at each vertex of the polyline or contour. The elevation of the point is the elevation of the surface at the vertex of the polyline or the elevation of the contour.

EXERCISE 10-1

When you complete this exercise, you will:

- Be able to set points with elevations from a surface.
- Be familiar with placing points between two existing points.
- Be able to place points on a polyline or contour.
- Be able to place points at a distance or elevation.

EXERCISE SETUP

This exercise starts with a new drawing, *Chapter 10 – Unit 1.dwg*. The file is in the *Chapter 10* folder of the CD that accompanies this book.

1. If you are not in Civil 3D, on the Desktop, double-click the **Civil 3D** icon to start the application.
2. Close the initial drawing and Open the *Chapter 10 – Unit 1* drawing file.
3. Use the AutoCAD AutoCAD **Save As...** command to save the file with a different name.

INTERPOLATE

1. From the Points menu, select **Create Points...**.

   The Create Points toolbar displays.
2. On the Create Points toolbar, click the expand toolbar arrow on the right.
3. Expand the Points Creation section, change Prompt For Elevations to **Manual**, change Prompt For Point Names to **None**, change Prompt For Descriptions to **Automatic**, and set the Default Description to **GP** (see Figure 10.4).

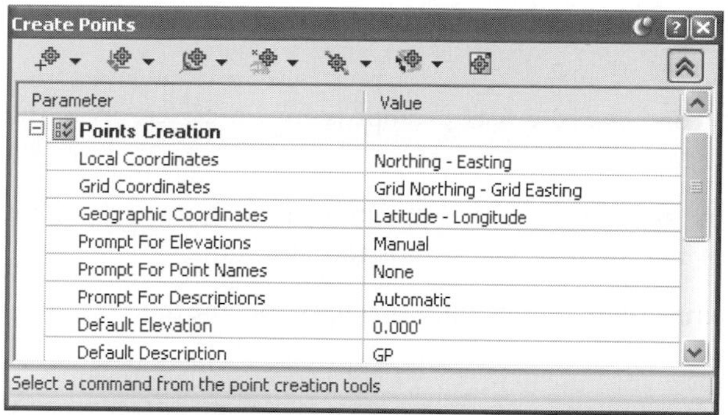

Figure 10.4

4. Click on the up arrow to hide the panel.

   The Interpolate routines use two points between which they calculate elevations for new points. The second point can be an existing or a selected point that you assign an elevation. The new points are between the two selected points and can have an offset. The offset is negative for left and positive for right. You can select two points from the screen, assign them elevations, and continue using the command. You do not need to select a point object for either point.

5. Use the **Manual** routine from the Miscellaneous icon group and select a point to the west of the building points. Assign the point the Elevation **675.00**. The description is automatic. Press the **Enter** key to exit the routine.

   Make sure you can view the Transparent Commands toolbar.

RELATIVE LOCATION

1. From the Create Points toolbar, select the drop-list arrow of the Interpolation icon stack and select **By Relative Location**.

   To select a starting point as a point object, you need to use the Transparent Command of **'PO** or select the **Point Object** icon from the Transparent Commands toolbar.

2. In the command line, the routine prompts for the first point; select the **Point Object** override from the Transparent Commands toolbar. The prompt changes to select point object, and in the drawing select the northernmost point along the building's edge and accept the elevation of 681.00 by pressing the **Enter** key.

3. In the command line, the routine prompts for the second point as a point object; select the new point you just placed in the drawing and press the **Enter** key to accept the elevation of 675.00.

   The routine reports the elevation, distance, and grade between the points and in the drawing shows the selected points, a direction arrow, and a distance jig from the first point.

4. In the command line, the next prompt is for a distance; enter **25** and press the **Enter** key.
5. In the command line, the final prompt is for an offset; press the **Enter** key for no offset distance.

   A new point appears 25 feet away from the first point (building 681.00), and the elevation of the point is an interpolated elevation.
6. In the command line, the routine prompts for another distance; enter **45**, and press the **Enter** key.
7. In the command line, the routine prompts for an offset; press the **Enter** key for no offset.
8. Press the **Esc** key to end the routine.
9. Click the **AutoCAD Save** icon to save the drawing.

INCREMENTAL DISTANCE

This routine works only with the selected points. The focus of this routine is to place new points either at a distance or elevation increment between the first and second points.

1. From the Create Points toolbar, select the drop-list arrow of the Interpolation icon stack and select **Incremental Distance**.

   To select the first and second points as point objects, use the Transparent Command of 'PO or Point Object. Select this from the Transparent Commands toolbar.
2. In the command line, the routine prompts for the first point; select the **Point Object** override from the Transparent Commands toolbar, select the second point down from the north end of the building's edge, and press the **Enter** key to accept the elevation of 681.00.
3. In the command line, the routine prompts for the second point as a point object; select the westerly point you placed in the drawing (elevation of 675.00), and press the **Enter** key to use the elevation of the second as the control.
4. In the command line, the next prompt is for a Distance Between Points; enter **25**, and press the **Enter** key.
5. In the command line, the last prompt is for an Offset; press the **Enter** key for no offset.

   The routine places points between the first and second points and assigns elevations by calculating a constant slope between the two control points.
6. Press the **Esc** key to exit the routine.
7. Erase the new points from the drawing and keep the points along the westerly side of the building.
8. Click the **AutoCAD Save** icon to save the drawing.

SLOPE/GRADE - ELEVATION

This routine creates new points in a direction from an existing or selected point. The elevation is either the point's elevation or one you specify.

1. From the Create Points toolbar, click the drop-list arrow of the Slope icon stack and select **Slope/Grade - Elevation**.
2. From the Transparent Commands toolbar, select the **Point Object** override.
3. In the command line, the routine prompts for the first point as a point object; select the northernmost point object along the building's edge.

   The routine prompts for the next point to set a direction and default distance.
4. In the drawing, select a point to the west.
5. In the command line, the next prompt is for a slope; enter **8** (for 8:1), and press the **Enter** key.

   In the command line, the routine reports the slope and the grade value.
6. In the command line, the next prompt is for the ending elevation; enter **675.00**, and press the **Enter** key.
7. In the command line, the next prompt is for the number of intermediate points; enter **4**, and press the **Enter** key.
8. In the command line, the next prompt is for an offset; press the **Enter** key for no offset.
9. In the command line, the last prompt is about including a point object at the end of the distance; press the **Enter** key to accept.

   The routine creates the new points radiating from the control point and the last point represents the 675 elevation.

   The routine prompts for another direction from the control point.
10. Press the **Esc** key to end the routine.
11. Erase the new points from the drawing and keep the points along the side of the building.
12. Click the **AutoCAD Save** icon to save the drawing.

SURFACE - RANDOM POINTS

The Random Points routine needs a surface to function. There is only one surface in the current drawing, Existing Ground.

1. In Layer Properties Manager, thaw the **C-TOPO-EXISTING GROUND** layer, and click the **OK** button to exit the dialog box.
2. Zoom in the area just west of the points along the building's edge.

3. From the Create Points toolbar, click the drop-list arrow for the Surface icon stack, and select **Random Points** from the list.
4. In the command line, the routine prompts for a surface; in the drawing, select a contour from the surface on the screen to identify the surface.
5. In the drawing, select a few points to set random points, and after placing the points in the drawing, press the **Enter** key twice to exit the routine.

   The routine creates points whose elevations are from the selected surface.
6. Erase the points you just made and leave the points along the building.
7. Click the **AutoCAD Save** icon to save the drawing.

ALONG POLYLINE/CONTOUR

This routine creates points whose elevations are from the selected surface at a measured distance along the path of the polyline.

1. In the status bar, toggle **OFF** Object Snaps.
2. Start the AutoCAD Polyline command and in the drawing, draw two polylines each with multiple vertices.
3. From the Create Points toolbar, click the drop-list arrow to the right of the Surface icon stack and select **Along Polyline/Contour**.
4. In the command line, the routine prompts for a surface; in the drawing, select a contour to identify the surface.
5. In the command line, the routine prompts for a distance; enter **15** as the distance and press the **Enter** key.
6. In the command line, the routine prompts you to select a polyline; in the drawing, select one of the polylines, and press the **Enter** key until you exit the routine.

   The routine creates the points at 15-foot intervals along the polyline, ignoring the polyline vertices.

POLYLINE/CONTOUR VERTICES

This routine creates points whose elevations are from the selected surface at each vertex of the polyline or contour.

1. From the Create Points toolbar, click the drop-list arrow to the right of the Surface icon stack, and select the **Polyline/Contour Vertices** routine from the list.
2. In the command line, the routine prompts for a surface; in the drawing, select a contour to identify the surface.
3. In the command line the routine prompts for you to select a polyline; in the drawing, select the remaining polyline.

   The routine places a point at a vertex. The point's elevation is the surface's elevation at the vertex location.

4. Press the **Enter** key to exit the routine.
5. Erase the points and polylines, keeping the points along the building.
6. Close the Create Points toolbar.
7. From the File menu, select **Close**, exit the drawing, and do not save the changes.

This ends the exercise on points as grading data. The next unit reviews feature lines and grading objects as grading tools.

SUMMARY

- The Surface icon stack of the Create Points toolbar uses a surface to assign elevations to new points.
- The Surface icon stack has routines that create new points on a grid, along a polyline or contour, or randomly.
- The Interpolation icon stack of the Create Points toolbar uses two points to create points.
- The two points used in an Interpolate routine are point objects or points assigned an elevation.
- The new points generally occur between the two control points.
- The Slope icon stack creates new points from a single control point at a slope or grade for a distance or until reaching a specified elevation.
- The Interpolation and Slope routines prompt for an optional offset.

UNIT 2: FEATURE LINES AND GRADING OBJECTS

Civil 3D takes a programmatic approach to solving grading design issues. Traditional grading solves the blending of a design to the existing conditions as offsetting and blending contours or elevations. The Civil 3D grading object can be a linear or closed object and can daylight to a surface or just grade at a distance or a slope. The end result of manipulating the object is a solution using the assigned grades and distances. After creating a solution there is a need to calculate a volume.

The grading object of Land Desktop had its limits. Chief among them was its inability to interact with other grading objects (see Figure 10.5). The most important improvement to the grading object in Civil 3D is its ability to be aware of other grading objects. With this awareness, the grading solution is the intersection of the grades from all objects (see Figure 10.6).

When working through Figure 10.6's grading scenario with just points, one problem is determining the elevations of the intersecting grades. When the user works through the

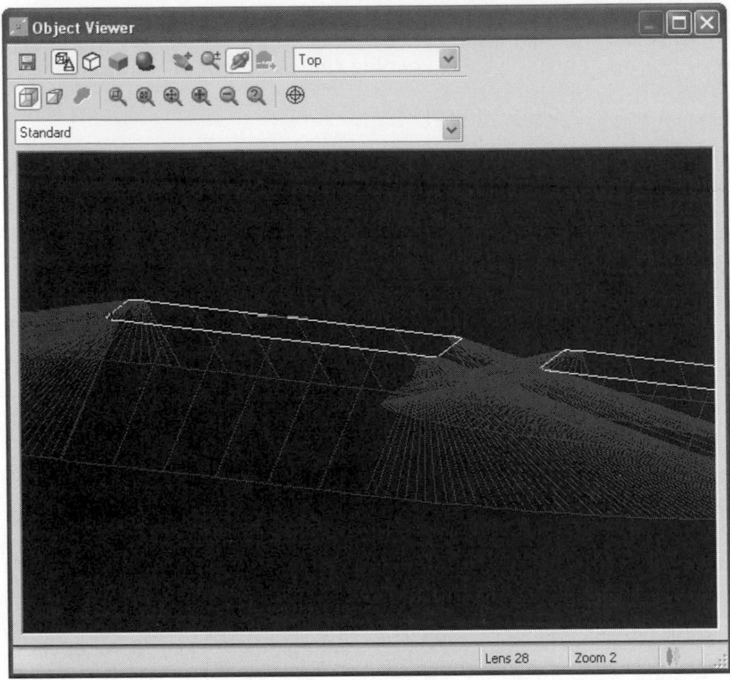

Figure 10.5

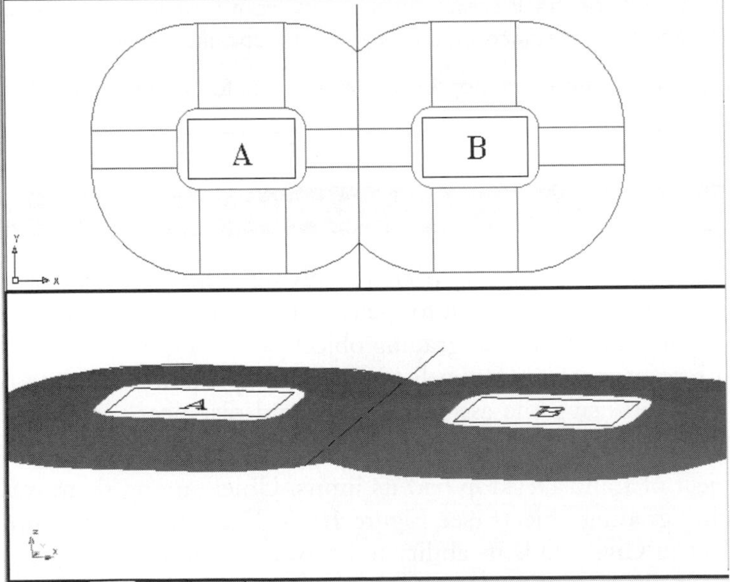

Figure 10.6

same problem, but with multiple pads and varying elevations, points do not present an ease of use solution. The strategy in Figure 10.6 is a simple top-down approach.

Whatever the strategy, control over the grading along the path of the grading object is critical. The grading object allows the user to interactively define segments that have different grades and develop transitions between the various grades (see Figure 10.7).

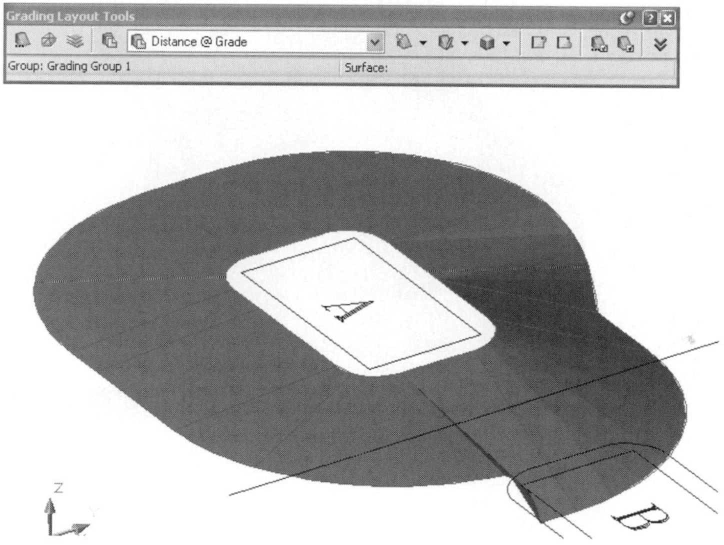

Figure 10.7

Another refinement is the use of a grading solution. A grading solution is automatically a new grading starting point that accepts a new set of grading parameters. In the scenario of building a pond with a bench, the creation of the pond is now a series of grading offsets and slopes (see Figure 10.8).

Another change to grading is the concept of grading groups. A grading group is a collection of grading objects contained with the group's name. This can be a major convenience when working with a large number of grading solutions (like the pond in Figure 10.8 or building pads).

The grading object has a base line (feature line). Select this line to specify an offset side. Once the user establishes the feature and the grading direction, the next set of parameters are the distance and the slope from the grading edge. The distance and slope create a face, grading line, and a target or solution line (see Figure 10.9).

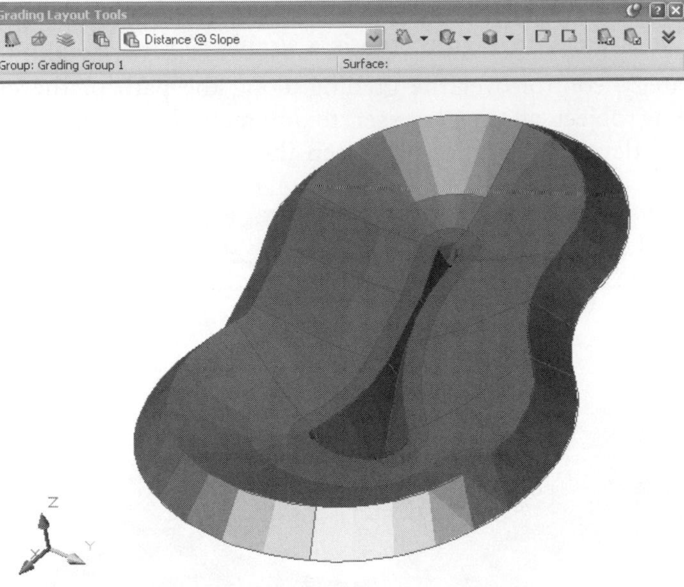

Figure 10.8

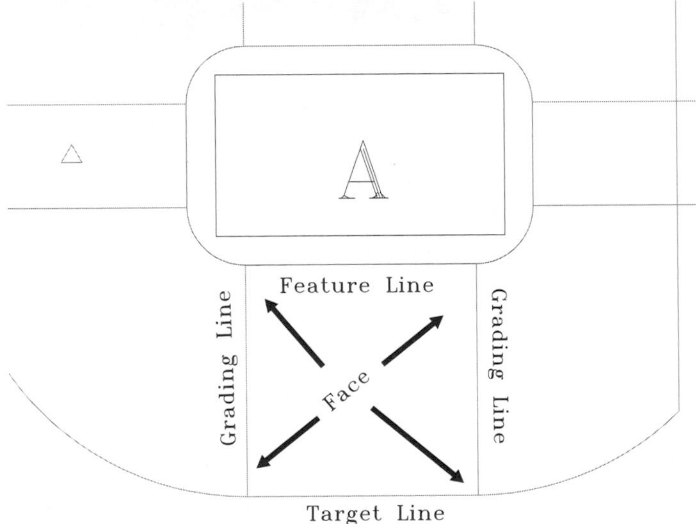

Figure 10.9

EDIT DRAWING SETTINGS

When working with grading objects there will be multiple groups in the drawing. It is good practice to set a modifier for them. This is done in the Drawing Settings dialog box of Settings (see Figure 10.10). It is in the Object Layers panel that the user would set a layer modifier and the value of the modifier.

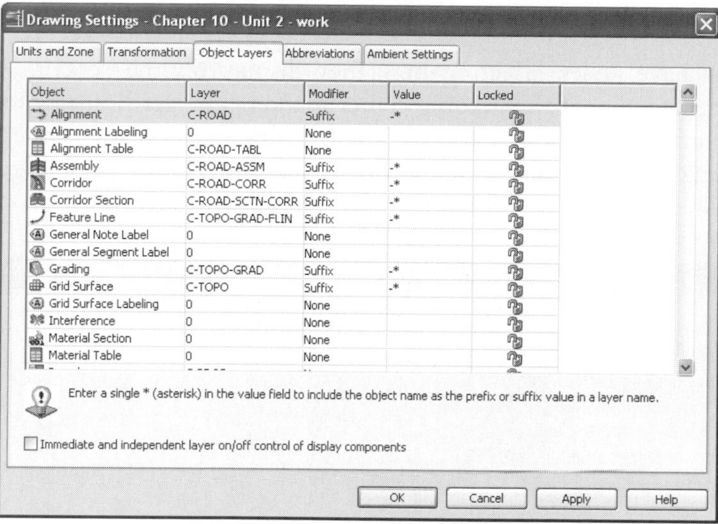

Figure 10.10

EDIT FEATURE SETTINGS

The Edit Feature Settings for grading include the name creation format for the grading groups and grading styles (see Figure 10.11).

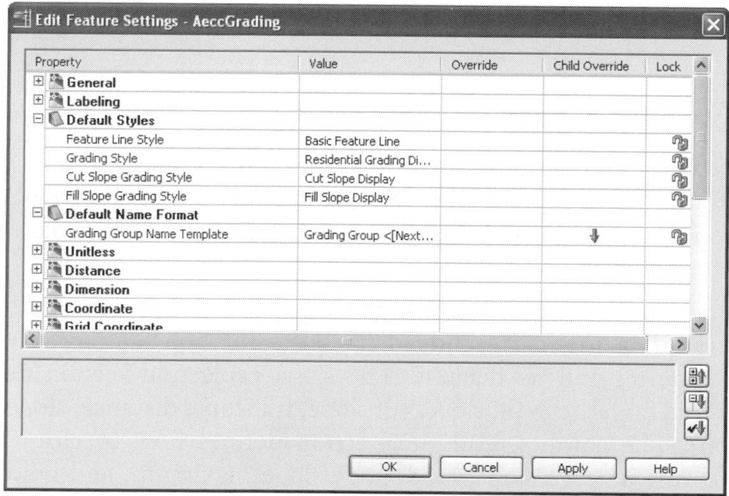

Figure 10.11

GRADING STYLES

Grading objects have styles affecting how they display on the screen. A grading style defines the name and size of the grading center marker. The grading marker is the selector for a grading solution. The Slope Patterns tab identifies an associated slope pattern and the minimum and maximum grades or slopes for the type of grading (see Figure 10.12). The Display tab defines the layers and component visibility when using this style.

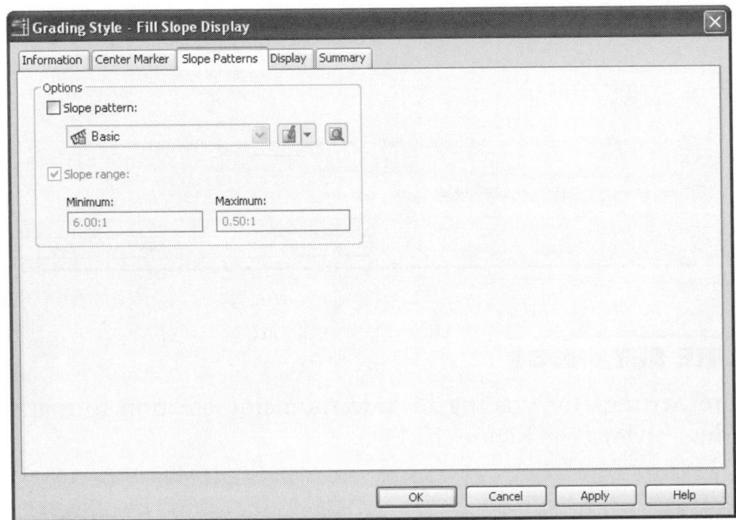

Figure 10.12

GRADING CRITERIA SETS

Grading groups use grading criteria sets. A grading criteria set is an alias for a collection of grading methods (see Figure 10.13). A grading method is a way of determining a solution by a slope or grade by an offset distance, by an elevation (absolute or relative), or by daylighting to a surface.

A grading criteria contains three sections: grading method, slope projection, and conflict resolution. The grading method section defines the method (surface, elevation, relative elevation, and distance) and its defaults. The slope projection section defines how this methods grades to a target (cut/fill slope, cut slope, fill slope, distance, slope, elevation, and relative elevation). The conflict resolution section defines how to clean up overlapping slope projections (use average slope, hold grade/slope maximum, and hold grade/slope as minimum).

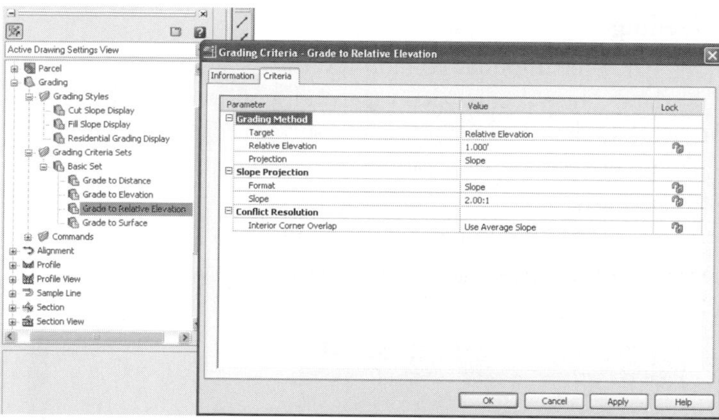

Figure 10.13

FEATURE LINES

Civil 3D draws feature lines or converts them from lines, arcs, and polylines. What is important about a feature line is that it can contain an arc (i.e., a helix or a curb with a radius).

Features lines are also breakline data for a surface. This allows the user to process them with a mid-ordinate value to add data points around any arc segment in the surface data set.

FEATURE LINES TOOLBAR

The Feature Lines toolbar contains all the commands necessary to create, convert, edit, fillet, offset, and join (see Figure 10.14).

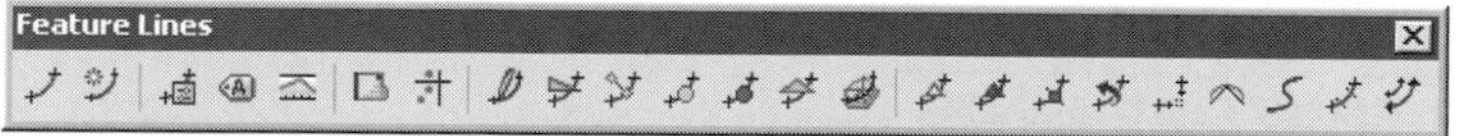

Figure 10.14

The first two icons on the left side of the toolbar draft or convert line work into feature lines. The next three icons display a feature line's properties, label them, or create a quick profile based on a feature line. The next three icons display the Elevation Editor, start a quick elevation edit routine snapping to each editable vertex, or use the command line to edit the feature line vertices.

The middle six icons set the elevation of a feature line by setting a grade/slope between feature line points, referencing another object, inserting or deleting an elevation point (changes the elevation on the feature line but does not break it), inserting a high or low point, or assigning an elevation of a feature line point from a surface.

The remaining icons create new feature line segments or delete existing ones, joins feature lines, reverses their direction, adds fillets to feature lines, fits curves to tessellated sections, smoothes feature lines, weeds out excess vertices, and offsets feature lines.

Feature Line Weeding

There are occasions when feature lines contain too many vertices and need weeding to remove redundant grading data. The Weeding icon (second icon in from the left) allows the user to remove vertices much like the weeding process of surface contour data (see Figure 10.15). The Weed Vertices dialog box contains parameters set to remove vertices. At the bottom of the dialog box, there is a report of how many vertices were removed. The higher the values, the more vertices weeded. As the user adjusts the numbers, if visible, the removed vertices are highlighted in red and the remaining ones in green.

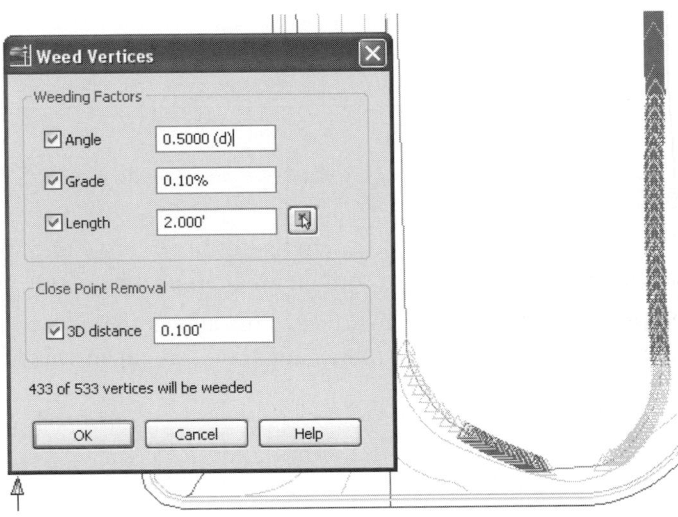

Figure 10.15

CORRIDOR FEATURE LINES

There is another source for feature lines: the corridor. Feature lines are the threads that string together the corridor sections. There is a feature line for each code on a corridor assembly. Export the feature lines in the Corridor menu. An exported feature line can become data for a surface, a polyline, alignment, or profile.

ELEVATION EDITOR

When the user needs to edit the elevations of a feature line, he or she uses the Elevation Editor. The editor is a cell-based panorama that lists all of the elevations on a feature line.

EDIT GRADING

The Grading Editor allows the user to change the parameters of a selected grading object. The Grading Editor lists all of the parameters of the grading solution and allows the user to change their values.

COMMAND SETTINGS

Each grading command displays the current feature settings plus at least one command-specific section. For example, the Create Feature Lines command's Feature Line Creation section determines if the command should delete the original line work and use the converted object layers (see Figure 10.16).

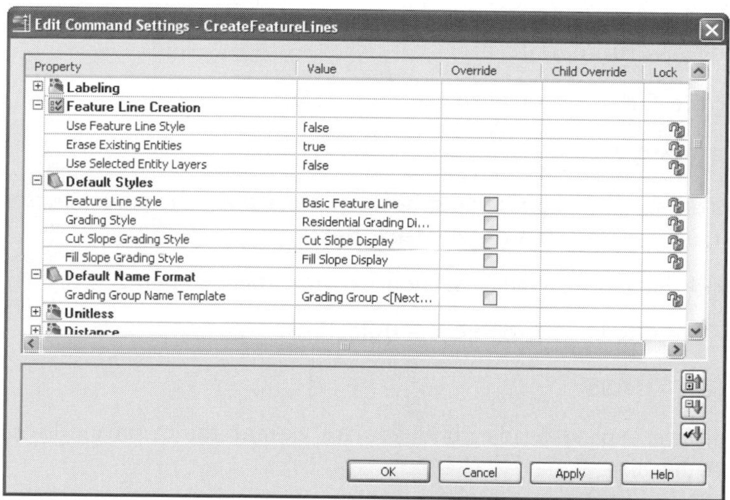

Figure 10.16

EXERCISE 10-2

When you complete this exercise, you will:

- Be familiar with the settings in the Edit Drawing Settings and Edit Feature Settings dialog boxes.
- Adjust a profile PVI and export feature lines from the profile.
- Convert polylines into feature lines and create grading solutions from them.
- Calculate a volume from a grading object.
- Modify feature line elevations.
- Use a feature line as a part of surface data.

EXERCISE SETUP

This exercise starts by opening an existing drawing.

1. Open the *Chapter 10 – Unit 2* drawing from the *Chapter 10* folder of the CD that accompanies this book.

EDIT DRAWING SETTINGS

1. Click the **Settings** tab to make it the current panel.
2. In Settings, select the name of the drawing, press the right mouse button, and from the shortcut menu select **Edit Drawing Settings...**.
3. Click the **Object Layers** tab to view its contents.

   Most objects now have suffix modifiers that append each object with its name (see Figure 10.10).
4. Click the **OK** button to exit the dialog box.

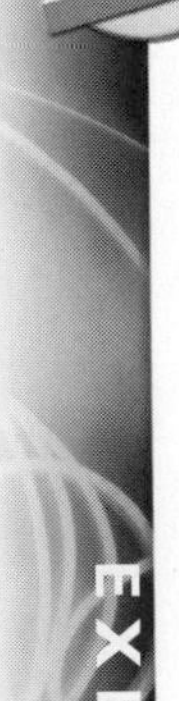

EXERCISES

FEATURE SETTINGS

1. In Settings, select the **Grading** heading, press the right mouse button, and from the shortcut menu select **Edit Feature Settings...**.
2. Expand the various grading sections to view their content.
3. Click the **OK** button to exit the dialog box.

COMMAND SETTINGS

1. In Settings, expand Grading until you are viewing the command list under the Commands heading.
2. From the list, select **CreateFeatureLines**, press the right mouse button, and from the shortcut menu select **Edit Command Settings...**.
3. In the dialog box, expand the Feature Line Creation section to view its settings.

   These settings affect the erasure of converted AutoCAD objects and what layers to use for the new feature lines.
4. Click the **OK** button to exit the Edit Command Settings dialog box.

EXPORTING A FEATURE LINE

There are commands in the Utilities flyout of the Corridors menu to export feature lines from the corridor.

ADJUSTING A PROFILE

1. Click the **Prospector** tab to make it the current panel.
2. In Prospector, expand the Corridors tree, select **93rd - (1)**, press the right mouse button, and make sure Rebuild–Automatic is on.
3. If Rebuild-Automatic is not on, select it to toggle it **ON**.

   The south entrance to the new office complex should have an elevation around 680.35. The road design needs to be close to this value before exporting the feature line.
4. From the View menu, use the Named Views... command to restore the **Entrance** view.

5. Use the AutoCAD ID command and an Endpoint object snap to identify some elevations along the west curb flange line.

   When you place your cursor at the curb flange, a tool tip appears showing the elevation of the curb flange of the corridor and the existing ground surface. The difference between the two surfaces is about 0.35 feet.

6. From the View menu, use the Named Views... command to restore the view **Profile**.
7. In the Profile View, select the **93rd street** profile, press the right mouse button, and from the shortcut menu select **Edit Profile Geometry...**.
8. In the toolbar, click the **Profile Layout Parameters** icon to show the Profile Layout Parameters vista.
9. In the toolbar, click the **Select PVI** icon (to the left of Parameters icon on the right of the toolbar) and select a point near the left PVI (2+10).
10. In Profile Layout Parameters change the PVI elevation to **680.15** and press the **Enter** key to keep the edit (see Figure 10.17).
11. Exit the Profile Layout Parameter dialog box and the Profile Layout Tools toolbar.

    Civil 3D automatically rebuilds the corridor.

12. Restore the **Entrance** view using the Named Views... command of the View menu.
13. Move the cursor to the left entrance curb return. The tool tip should report an elevation around 680.40.

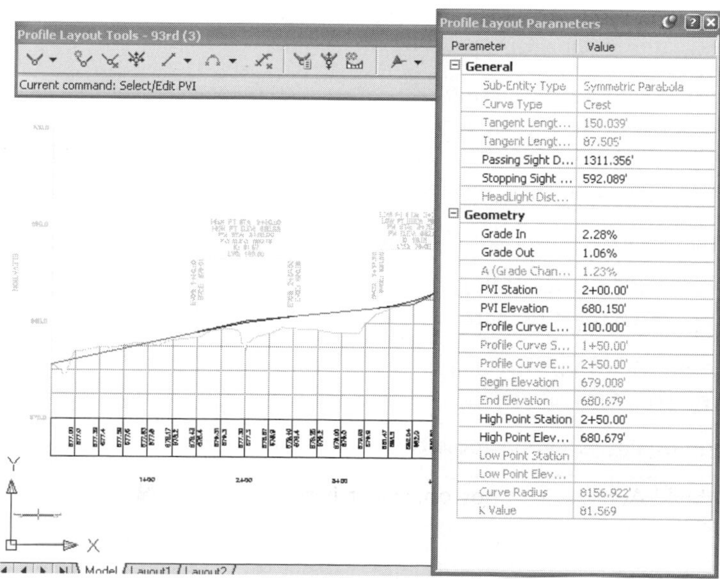

Figure 10.17

EXPORTING A FEATURE LINE

1. In the Layer Properties Manager dialog box, thaw the **C-ROAD-CORR-93rd - (1)** layer, and click the **OK** button to exit.
2. From the Utility flyout in the Corridors menu, select **Grading Feature Line from Corridor**.
3. In the command line, the routine prompts you to select a corridor feature line.
4. In the drawing, select the northernmost blue line of the corridor, and in the Select a Feature Line dialog box, select the **Sidewalk_In** feature line.

   A green line appears in the drawing.
5. In the drawing, export a couple more corridor feature lines. Some feature lines may be on top of one another. If this is the case, just export one of the feature lines.
6. Use the List command, select the object, and review AutoCAD's report about the object.

CREATING DITCHES

The Grading Object is ideal for ditch and pond design.

1. From the View menu, use Named Views... to restore the view **South Ditch**.
2. From the Grading menu, select **Grading Creation Tools....**

   The Grading Creation Tools toolbar displays.
3. On the left side of the toolbar, select the **Set the Grading Group** icon.

   The Create Grading Group dialog box displays.
4. In the Create Grading Group dialog box, do not change the grading group Name, toggle **ON** Automatic Surface Creation, set the Surface style to **_No Display**, toggle **ON** and set the Volume Base Surface to Existing Ground, and click the **OK** button to close the dialog box (see Figure 10.18).

   A Create Surface - Grading Group 1 dialog box appears.
5. In the Create Surface dialog box, enter **Grading Objects** for the surface description, and click the **OK** button to exit the dialog box.
6. In the middle of the Grading Creation Tools toolbar, change the criteria to **Grade to Relative Elevation**.
7. In the Grading Creation Tools toolbar, click the drop-list arrow to the right of the criteria (second icon to the right of the criteria) and select **Create Grading**.
8. In the drawing, select the magenta line representing the top of the southern ditch.
9. In the Create Feature Lines dialog box, toggle **ON** Style, click the drop-list arrow, select the **Grading Ditch** style, and click the **OK** button to continue.

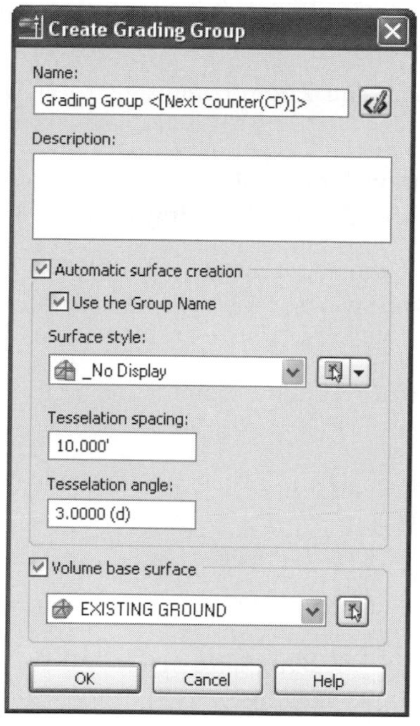

Figure 10.18

10. In the drawing, click in the interior of the ditch and press the **Enter** key to accept applying the criteria to the entire length of the feature line.
11. In the command line, enter **−5** as the relative change in elevation and press the **Enter** key.
12. In the command line, press the **Enter** key to accept a slope parameter, enter **2**, and press the **Enter** key to create the grading.
13. In the drawing, use the AutoCAD transparent pan command and pan to the northern ditch polyline.
14. From the Feature Line toolbar, select the **Weed** icon, select the northern ditch line, review the weeding parameters, and cancel out of the weed command.
15. Repeat Steps 7–12 to create a grading solution for the northern ditch. Use **−6** as a relative distance and **2.5** as the slope to create the ditch.
16. Press the **Enter** key to exit the routine.
17. Click the **AutoCAD Save** icon to save the drawing.

CREATING A POND

Just to the northwest of the building is an outline of a pond. The first series of grading commands will use a grading solution to create a second solution. The interior of the pond will have the floor as an infill to put a floor in the pond.

The event viewer will notify you of several duplicate points throughout this grading exercise. Just close the event viewer panorama.

1. Use the AutoCAD Pan and Zoom commands to view the pond outline to the north of the building.
2. In the Grading Creation Tools toolbar, change the grading criteria to **Grade to Distance** and click the **Create Grading** icon.
3. In the drawing, select the top contour (676) of the pond, in the Create Feature Lines dialog box toggle **ON** Style, set the style to **Basic Feature Line**, and click the **OK** button to continue.
4. In the command line, the routine prompts for a grading side; select a point outside of the polyline.
5. In the command line, apply the grading to the entire length of the polyline and press the **Enter** key to accept.
6. In the command line, the routine prompts for a distance; enter **6** for the distance, and press the **Enter** key to continue.
7. In the command line, the routine prompts for a slope or grade; press the **Enter** key for Slope, for the slope value enter **4**, and press the **Enter** key to solve the grading.
8. Press the **Enter** key to return to the command prompt.
9. In the Grading Creation Tools toolbar, change the grading criteria to **Grade To Surface** and click the **Create Grading** icon.
10. In the drawing, select the outer grading target line and press the **Enter** key to apply the grading to the entire length of the feature line.
11. In the command line, the routine prompts for a cut slope; press the **Enter** key for Slope, enter **4** for the slope value, and press the **Enter** key to continue.
12. In the command line, the routine prompts for a fill slope; press the **Enter** key for Slope, enter **4** for the slope value, and press the **Enter** key to solve the grading.
13. Press the **Enter** key to return to the command prompt.

    The grading solution may intrude on the building.

14. In the Grading Creation Tools toolbar, change the criteria to **Grade to Relative Elevation** and click the **Create Grading** icon.
15. In the drawing, select the pond edge feature line; for the offset side, select a point inside of the pond, and to apply the grading to the entire length of the line press the **Enter** key.
16. In the command line, the routine prompts for a relative elevation; enter **–6** and press the **Enter** key to continue.

17. In the command line, the routine prompts for a slope or grade; press the **Enter** key to enter a slope value, for the slope enter **2.75**, and press the **Enter** key to solve for the interior of the pond.
18. Press the **Enter** key to return to the command prompt.
19. In the Grading Creation Tools toolbar, click the drop-list arrow to the right of Create Grading and select **Create Infill** from the list.
20. In the drawing, click in the center of the pond.
21. Press the **Enter** key to return to the command prompt.

EDIT GRADING

The slope to surface grading solution intrudes on the building.

1. In the drawing, select the grading icon (green triangle), press the right mouse button, and from the shortcut menu select **Grading Editor...**.
2. In the Grading Editor vista, change the fill slope to **2**.

   The grading recomputes.
3. Close the Grading Editor vista.
4. Click the **AutoCAD Save** icon to save the drawing.

VIEWING THE GRADING

1. In the drawing, select the pond grading, press the right mouse button, and from the shortcut menu select **Object Viewer**.
2. After viewing the grading, exit the Object Viewer.

GRADING VOLUME

1. In the Grading Creation Tools toolbar, select the **Grading Volume Tools** icon (fifth icon in from the right).

   This displays a Grading Volume Tools toolbar that reports the volume for the grading group or for selected individuals.
2. If necessary, click the **Entire Group** radio button to get a volume for the all of the grading solutions (both ditches and the pond).
3. Click the red "**X**" to close the toolbars.

   This completes the unit on grading. This portion of the Civil 3D is the least developed. Many of the methods and tools are useful, but it is difficult to complete a grading design with only the tools in Civil 3D.

SUMMARY

- Grading objects are excellent for simple grading scenarios.
- Grading objects understand grading solutions that may intersect and if they do, the resulting grading solution shows the intersection.
- There are some simple grading volume tools.
- A grading group can detach its surface and users can paste the surfaces into an overall surface.

UNIT 3: SITE AND BOUNDARY VOLUMES

The volume calculations are based on the differences in elevations between two surfaces. The surfaces used in the comparison can be any combination of two terrain surfaces types (TIN or Grid). The Grid and TIN volume methods calculate volumes by different algorithms. Each method is directly dependent on the quality of surface data for its results. The better the quality of surface data, the better the resulting volume estimate.

TERRAIN VOLUMES

Civil 3D uses two methods to calculate earthwork volumes. The first method uses the panorama to report the amount of cut and fill material between two specified surfaces (see Figure 10.19). The second method creates a new surface (either a TIN or Grid volume surface) whose properties include the volume between the surfaces (see Figure 10.20).

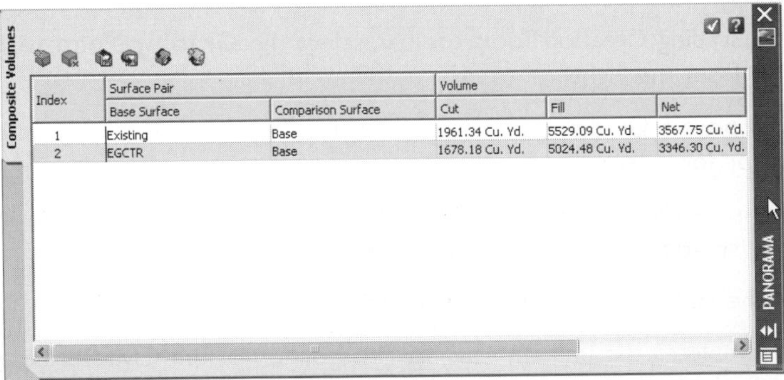

Figure 10.19

Prospector dynamically tracks the status of volume surfaces in the drawing, and uses icons to represent references to the comparison terrain surfaces. If the user attempts to delete a terrain or grid surface that is part of a volume calculation, Prospector will not allow its

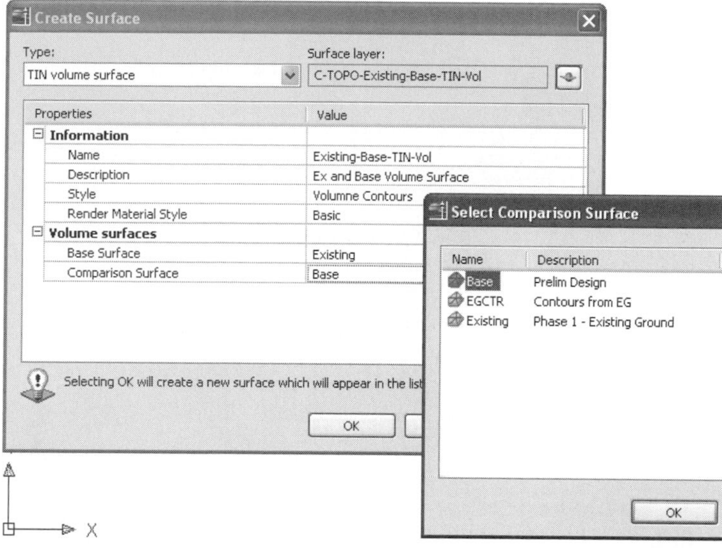

Figure 10.20

deletion. To delete a surface that has a reference (a volume surface), Civil 3D requires the removal of the volume surface before deleting the terrain surface. Prospector's surfaces area indicates if any of the surfaces become out of date because of new surface data or editing.

GRID VOLUME SURFACE

A grid volume surface results from sampling the differences in elevation between two surfaces using a regularly spaced grid in the X and Y directions. There can be a rotation to the sampling grid that reflects a rotation in one or both surfaces. As the grid volume surface samples the two terrain surfaces, it checks for an elevation on both surfaces at each cell corner. If there are elevations from both surfaces at all four corners of a cell, the grid surface assigns the difference in elevation found at each cell corner to the cell corners in the surface (see Figure 10.21).

If the user varies the grid volume surface spacing to smaller or larger cell sizes, the surface reflects the intersection of the two surfaces at that grid spacing. As a result of this, when the user views the properties of the volume surface, it may contain a different volume total from a surface with a different spacing.

The single greatest problem the user faces, outside of bad data, is how often to sample the two terrain surfaces when using the grid volume method. Set the grid spacing to an interval similar to spacing of the terrain surface data. In some cases the terrain data spacing can vary from 50 feet to as small as one-half foot. To sample the site with a 20-foot grid is to sample the interpretations between data with 50-foot spacing and miss the data with one-half foot spacing. If the two surfaces have irregular borders and the sample spacing is too large, the grid volume surface may not include volume data around its border because the spacing doesn't sample the irregular shapes of the terrain surface borders. The grid spacing should sample the greatest amount of intersecting area between the terrain surfaces

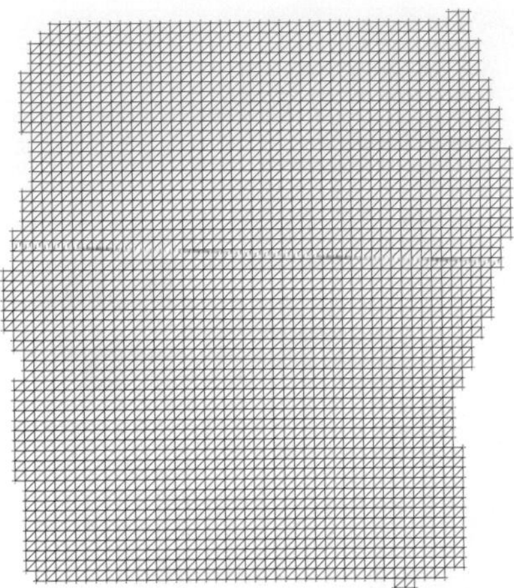

Figure 10.21

with the fewest number of points. This optimum spacing varies based on the size, shape, and relief found in the surfaces. Different sampling intervals in some cases will produce different volumetric results. If the differences in the results are severe, question the spacing of the sampling or the quality of the surfaces.

TIN VOLUMES

Volumes calculated from terrain surfaces containing mostly contour or breakline data limit the effect of the design changes on the volume. The design usually starts by copying the contours of the first surface to another layer. These copied contours become the starting point for the design of the second surface. When the user uses the original contours to design the second surface, the undisturbed contours limit the volume calculations to the first undisturbed contour beyond the design. This is because the volume between all unchanged contours is 0 (zero).

A TIN volume surface is a resulting surface whose triangulation is the combination of the triangulation of the two compared surfaces (see Figure 10.22). At the end of each triangle leg of the volume surface, Civil 3D samples the difference in elevation between the two surfaces. Finally, Civil 3D calculates a earthwork volume from the information within the TIN volume surface.

The elevations on the TIN volume surface are the differences in elevations between the two surfaces. The surfaces' triangles are a composite of the triangles from both terrain surfaces. This type of volume surface is the most comprehensive volume calculation method.

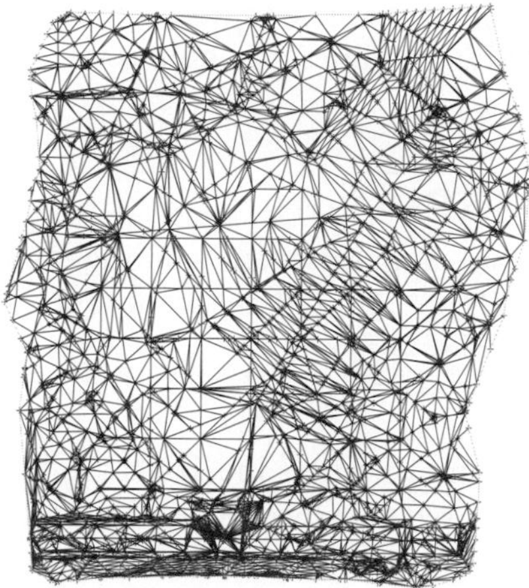

Figure 10.22

BOUNDED VOLUMES

The Bounded Volume utility calculates a volume from a surface for the area defined by a polyline, polygon, or parcel. If the user uses a volume surface, the volume for the boundary is from a comparison between two terrain surfaces. If the user uses this method on a terrain surface, the volume is from an elevation of 0 (zero) to the elevations found on the TIN surface.

DESIGNING SURFACES

The best methods of moderating errors in volume calculations are consistent surface design methodologies, consistent data densities, and an awareness of the strengths and weaknesses of each volume calculation method. What is a good methodology for creating a design and what is the correct density of data are questions open to debate. What the user needs to do is develop a consistent approach to the development of surface data. Evaluate the numbers that a volume routine generates with the "ball park" value manually or mentally figured.

The blending of the second surface design into the first surface is best accomplished by starting with three-dimensional polylines or contours from the first surface. This "seed" surface data is an effective way of creating and blending a design surface into the existing conditions. 3D polylines, feature lines, and contours also aid in the process of designing a new surface. The blending of a design into existing contours is straightforward and areas undisturbed by the design process will produce no volume.

How does one copy a surface? Civil 3D cannot directly copy a surface, so the best method is to paste a surface into a newly defined surface. This action "copies" a surface to the current surface. How do you get the contours out of a Civil 3D surface? To be able to use the contours, explode the surface twice: once to make it a block and a second time to make it polylines with elevation. With polylines the user is then able to trim, draft new, and grade a site.

NEW SURFACE DESIGN TOOLS

The Feature Line Tools toolbar supplies several commands that create data for a design surface. Many of the commands are an implementation of the 3D Polyline commands of Land Desktop.

Stepped Offset

The Stepped Offset routine of the Feature Line Tools toolbar offsets a line, survey figure, 3D polyline, or polyline at a fixed distance. The elevation of the offset object is also adjusted by specifying an amount of change (1, 3, -1, or -3), a grade (25 or -33), or slope (8:1 or -8:1). When entering a slope, it is necessary only to enter the sign (positive or negative) and the run. When entering a grade, enter the value and sign, not the percent sign.

Optionally, the routine can repeat the offset, the Multiple option, if several offsets need to be made with the set grade, slope, or amount of change. The Stepped Offset command in multiple mode assumes the last offset object is next object to be offset.

The snippet below demonstrates how to set the offset distance, the change in elevation, and set the multiple option.

```
Command: Offset layer = Source
Specify offset distance or [Through/Layer]: 8 <Press ENTER>
Select a feature line, survey figure, 3d polyline or polyline to
    offset: <Select the object to offset>
Specify side to offset or [Multiple]: m <Press ENTER>
Specify side to offset: <Select the side to offset>
Specify elevation difference or [Grade/Slope] <0.000>: 1 <Press
    ENTER>
Specify side to offset: <Select the side to offset>
Specify elevation difference or [Grade/Slope] <1.000>: <Press
    ENTER>
```

Draw Feature Line

The Draw Feature Line command drafts a new feature line in the drawing. A feature line can contain a tangent or curve segment. When assigning elevations to a feature line vertex, the user has the choice of a specific elevation, a slope or grade or a difference in elevation between the last and the present vertex, or an elevation from a specified surface.

```
Resuming DRAWFEATURELINE command.
Specify the next point or [Arc]: (5292.54 5351.15 732.4)
Distance 220.546', Grade -0.23, Slope -441.09:1, Elevation 732.
    400'
Specify grade or [SLope/Elevation/Difference/SUrface] <0.00>: e
Specify elevation or [Grade/SLope/Difference/SUrface] <732.400>:
    <Press ENTER>
Specify the next point or [Arc/Length/Undo]:
```

Insert Elevation Point

The Insert Elevation Point routine adds vertices to a feature line whose elevations are a constant grade calculation. The routine physically modifies the feature to include the new vertices. Add the new vertices by a distance (every 10 or 20 units).

Insert High/Low Elevation Point

The Insert High/Low Elevation Point routine creates a new feature line vertex based on intersecting slopes or grades from two adjacent feature line vertices.

```
Command:
Select a feature line, survey figure, parcel line or 3d
    polyline:
Specify the start point: <move cursor to a vertex and select>
Specify the end point: <move cursor to opposite vertex and
    select>
Start Elevation 732.900', End Elevation 732.750', Distance 206.
    176'
Specify slope ahead or [Grade]: 50 <press ENTER>
Specify slope back or [Grade]: 50 <press ENTER>
Select a feature line, survey figure, parcel line or 3d
    polyline: <press ENTER to exit>
Command:
```

VOLUME REPORT TOOLS

There are no Volume Report tools in Civil 3D.

CONTOUR DATA

See the discussion on Contour data in Unit 2 of Chapter 4.

EXERCISE 10-3

This exercise has two parts. The first part is creating a surface whose data is the contours of the existing surface. This is done for two reasons: to compare the calculated volumes between the existing surface (points and breaklines) and a contour data surface and the design surface.

The second part of the exercise creates the design surface. The design of the second surface sets the stage for the volume calculations.

When you complete this exercise, you will:

- Build, evaluate, and calculate volumes.
- Create a surface from contour and 3D polyline data.
- Calculate a volume report (panorama) between two surfaces.
- Create a TIN (Triangular Irregular Network) and grid volume surface.
- Review the statistics of a volume surface.

CREATING A VOLUME SURFACE

This exercise uses the *Chapter 10 – Unit 3* drawing that is in the *Chapter 10* folder of the CD that accompanies this book.

1. If you are not in Civil 3D, double-click the icon to start the application.
2. From the File menu, close the current drawing and do not save it.
3. From the File menu, Open the *Chapter 10 – Unit 3* drawing.

CREATING AN EGCTR SURFACE

The contours for the EGCTR surface are from Existing surface.

1. Click the **Prospector** tab to make it current.
2. In Prospector, select the **Surfaces** heading, press the right mouse button, and from the shortcut menu select **New....**
3. In the Create Surface dialog box, for the name enter **EGCTR**, for the description enter **From EG Contours**, set the Surface Style to **Contours 1' and 5' (Design)**, and click the **OK** buttons.
4. In Prospector, expand the Surfaces branch until you are viewing the Definition branch of the EGCTR surface.
5. In the EGCTR surface's Definition branch, select **Edits**, press the right mouse button, and from the shortcut menu select **Paste Surface....**
6. In the Select Surface to Paste dialog box, select the **Existing** surface.
7. Use the AutoCAD Explode command twice to reduce the EGCTR surface to polylines with elevations.

   The first explode creates a block, and the second creates a set of polylines with a border. Prospector removes the EGCTR surface from the surface list.
8. In the Layer Properties Manager, create a new layer, name it **EGCTR**, and click the **OK** button.
9. In the drawing, select the contours and change them to the **EGCTR** layer.
10. In Prospector, select the Surfaces heading, press the right mouse button, and from the shortcut menu select **New....**
11. In the Create Surface dialog box, for the name enter **EGCTR**, for the description enter **From EG Contours**, set the Surface Style to **Contours 1' and 5' (Design)**, and click the **OK** buttons.
12. In Prospector, expand the Surfaces branch until you are viewing the Definition branch of the EGCTR surface.
13. In the EGCTR surface's Definition branch, select **Contours**, press the right mouse button, and from the shortcut menu select **Add....**

14. In the Add Contour Data dialog box, for the description enter **From EG**, for the Supplementing Distance enter **35**, and click the **OK** button to continue (see Figure 10.23).

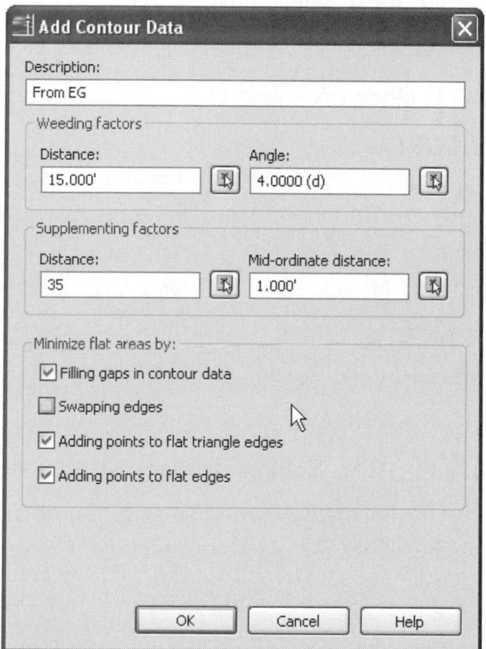

Figure 10.23

15. In the drawing, select all of the contours, but do not select the boundary.
16. In the EGCTR surface's Definition branch, select **Boundaries**, press the right mouse button, and from the shortcut menu select **Add....**
17. In the Add Boundaries dialog box, name the boundary **Outer**, click the **OK** button, select the 3Dpolyline, and press the **Enter** key.
18. In the Layer Properties Manager, freeze the **EGCTR** and **C-TOPO-EGCTR** layers, and click the **OK** button to exit the dialog box.
19. Click the **AutoCAD Save** icon to save the drawing.

CONTOURS FOR THE BASE SURFACE

The Base surface uses the surface paste to create the beginning contours.

1. In Prospector, select the **Surfaces** heading, press the right mouse button, and from the shortcut menu select **New....**
2. In the Create Surface dialog box, for the name enter **Base**, for the description enter **From EG Contours**, set the Surface Style to **Contours 1' and 5' (Design)**, and click the **OK** buttons.

EXERCISES

3. In Prospector, expand the Surfaces branch until you are viewing the Definition branch of the Base surface.
4. In the Base surface's Definition branch, select **Edits**, press the right mouse button, and from the shortcut menu select **Paste Surface....**
5. In the Select Surface to Paste dialog box, select the **Existing** surface.
6. Use the AutoCAD Explode command twice to reduce the Base surface to polylines with elevations.

   The first explode creates a block, and the second creates a set of polylines with a border. Prospector removes the EGCTR surface from the surface list, since the surface is now polyline data.
7. In the Layer Properties Manager, create a New layer, name the layer **BASE**, assign it a color, make Base the current layer, and click the **OK** button to exit layer manager.
8. In the drawing, erase the 3Dpolyline around the contours.
9. In the drawing, select the contours and change them to the Base layer.

MODIFYING THE BASE SURFACE

The site's first editing area is the northeast corner. This area's priority is filling in the swale and moving its contours to the northeasterly edge of the site. If fill is brought into the site and placed in the swale, the elevations within the swale rise moving the lower elevations northeasterly. The effect of filling in the swale is pushing the contours from the site's interior toward its exterior.

To create these changes in the drawing, you need to remove the contours that are in the interior of the swale and redraw new contours at the northeasterly edge reflecting the filling in of the swale (See Figure 10.24).

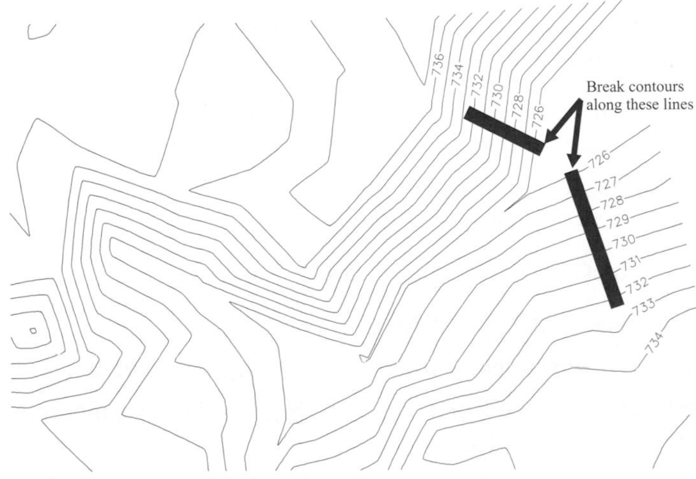

Figure 10.24

Northeast Section

1. Click the **Prospector** tab to make it current.
2. In Prospector, select the **Points** heading to list the points in the preview area.
3. In the preview area, scroll to and select point **111**, press the right mouse button, and from the shortcut menu select **Zoom to**.
4. Use the AutoCAD Zoom command to make your view match what is in Figure 10.24.
5. Using Figure 10.24 as a guide, use the AutoCAD line command to draw the line as indicated in the figure.
6. Using Figure 10.24 as a guide, use the AutoCAD Trim command. In the drawing, select the line just drawn as the cutting edge, press the **Enter** key, and trim the contours on the southerly side of the line to trim away the unwanted portions of the contours.
7. Erase the two lines you used for trimming the contours.

Draw Design Contours

1. In the AutoCAD status bar, toggle **ON** Object Snaps and set it to **Endpoint**.
2. In the drawing and using the AutoCAD Polyline command, select the north end of the 726' contour (the innermost contour), toggle **OFF** the object snap, select a few points to represent the new path of the southerly 726' contour, toggle **ON** object snaps, select the southerly end of the remaining portion of the contour, and press the **Enter** key to exit the polyline command. Use Figure 10.25 as a guide for drawing the new contour.

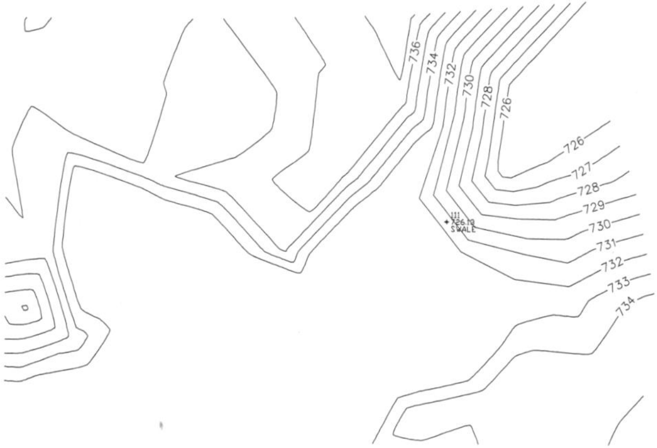

Figure 10.25

3. Check the elevation of the new contour by selecting the polyline, pressing the right mouse button, from the shortcut menu selecting **Properties...**, and reading the elevation value.

   The Elevation value of the polyline should match the elevation of the contour.

4. If the elevation of the new polyline does not match the elevation of the contour, change its elevation to **726** in Properties.

   The engineer wants the new contours to have an 8:1 slope. This means offsetting the contour 8' horizontally and raising its elevation 1'. This is done with the Stepped Offset routine (first icon on the right) of the Feature Line toolbar. The Stepped Offset command in multiple mode assumes the last offset object is the next object to be offset.

5. Use the Stepped Offset routine from the Feature Line toolbar to set the Offset Distance to **8**, select the polyline just drawn, enter **M** to set the Multiple option on, press the **Enter** key, select a point southwest of the polyline, set the Elevation Difference to **1**, continue to pick points to create the remaining five contours with the same Elevation Difference, and press the **Enter** key to exit the command.

   To connect the polylines to the existing contours, activate the grips of the new polylines and contours, and stretch the polylines until they connect to the existing contour segments.

6. In the drawing, activate the grips of the existing contours and offset polylines and stretch the first and last vertices of the new polylines so they connect to the end of the trimmed contours. You may have to zoom in to better see the polylines' and contours' connection points.

7. Click the **AutoCAD Save** icon to save the drawing.

EXERCISES

Northwest Section

The next step is modifying the site's northwest side. You will expand the flat center of the site by erasing the hill contours and by breaking and redrafting other contours in the area (see Figures 10.26 and 10.27).

1. In Prospector, select the **Points** heading to preview the point list.

2. Scroll to and select point **122**, press the right mouse button, and from the shortcut menu select **Zoom to**.

3. Use the AutoCAD Zoom command to make your view match the view in Figure 10.26.

4. In the drawing, use the AutoCAD Erase command to erase the hill contours northwest of the new flat area—the contours around point 122.

5. Using Figure 10.26 as a guide, use the AutoCAD Break command to break the two contours in the drawing at the indicated locations.

6. Using Figure 10.26 as a guide, use the AutoCAD Break command to break off the southern portion of the closed polyline in the drawing.
7. Use the AutoCAD Erase command to erase the depression contour ticks. The easiest method to perform this would be to select a tick, right click and select **Select Similar**, then type in **E** and press the **Enter** key.
8. Press the **F3** key to toggle on Object Snaps.
9. Using Figure 10.27 as a guide, use the AutoCAD Pline command and the Endpoint object snap to redraft the contours connecting the southerly endpoints to the northerly endpoints in the drawing.
10. In the drawing, select the one of the new contours, press the right mouse button, and from the shortcut menu select **Properties....**
11. If necessary, change the contour to the correct elevation.
12. Repeat the previous two steps if necessary to edit the elevation of the remaining contours.
13. Click the **AutoCAD Save** icon to save the drawing.

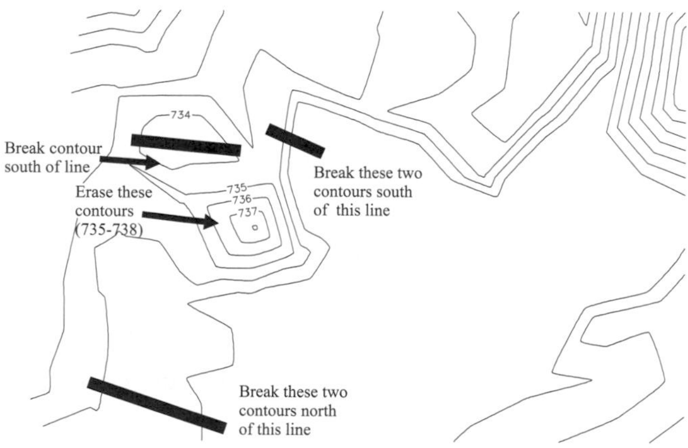

Figure 10.26

Southeast Section

The next area to edit is the southeast side of the site. The flat interior area needs to expand to the southeast (see Figures 10.28 and 10.29). This means lowering the elevations in this area (redrafting the contours toward the southeast to the edge of the surface). You also need to erase the two closed contours.

1. In Prospector, select the **Points** heading to preview the point list.
2. In the preview area, scroll and select point **67**, press the right mouse button, and from the shortcut menu select **Zoom to**.

Figure 10.27

3. Use the AutoCAD Zoom command to make your view of the surface match that of Figure 10.28.
4. Using Figure 10.28 as a guide, use the AutoCAD Erase command to erase the two closed contours and their ticks representing the pond and an adjacent low area, if necessary.
5. Using Figure 10.28 as a guide, use the AutoCAD Break command to break the 733 and 734 contours.
6. Toggle **ON** Object Snaps.
7. Using Figure 10.29 as a guide, use the AutoCAD Pline command to redraft the location of the two contours.
8. In the drawing, select one of the new contours, press the right mouse button, and from the shortcut menu select **Properties...**.
9. If necessary, change the contour to the correct elevation.
10. Repeat the previous two steps if necessary to edit the elevation of the remaining contours.
11. Click the **AutoCAD Save** icon to save the drawing.
12. Use the AutoCAD Zoom Extents command to view the entire surface.

    Your drawing should now like similar to Figure 10.30.

DESIGNING DRAINAGE FOR THE SITE

The problem with the current design is that the site's center is flat and water will not drain. One strategy is draining the site with swales and a small berm. You will create a shallow swale using points from the west side toward the northeast. Figure lines define a second swale and a berm.

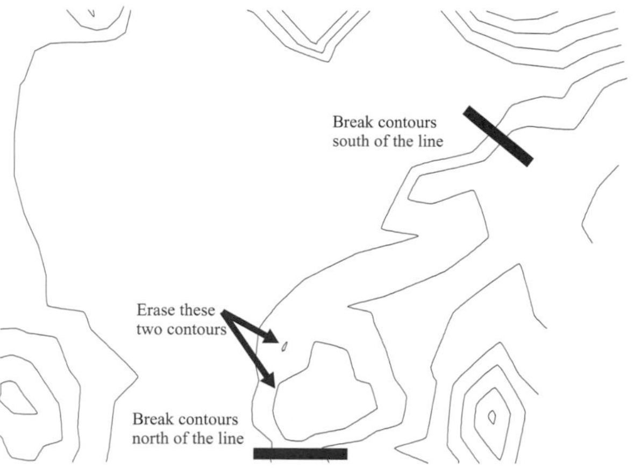
Break contours
south of the line
Erase these
two contours
Break contours
north of the line

Figure 10.28

Figure 10.29

Swale by Points

1. From the View menu, select **Named Views...** and restore the view, Central.
2. Toggle **OFF** Object Snaps.
3. In the Layer Properties Manager, thaw the **V-NODE** layer, and click the **OK** button to exit.
4. Click the **Settings** tab to make it current.
5. In Settings, select the **Points** heading, press the right mouse button, and from the shortcut menu select **Edit Feature Settings....**
6. In the Edit Feature Settings dialog box, expand the Point Identity section, set the Next Point Number to **500**, and click the **OK** button to exit.
7. From the Points menu, select **Create Points....**

Figure 10.30

8. In the Create Points toolbar, select the down arrow at the far right of the toolbar to display the point settings.
9. In the point settings, expand the Points Creation section, and set the following values:

   Elevations: Manual

   Descriptions: Automatic

   Default Description: SWCNTRL
10. From the Create Points toolbar, click the drop-list arrow to the right of the first icon in from the left side, and select **Manual** from the list.
11. Using Figure 10.31 as a guide, place three points, and using Table 10.1 assign the listed elevations.

Table 10.1

Point Number	Elevation
500	732.90
501	732.40
502	732.10

EXERCISES

Figure 10.31

The three points represent the upper,- mid-, and endpoints of a swale. To complete the swale, you need to place points between these control points so the surface correctly triangulates the swale. The Interpolate routine creates these points.

12. In the Create Points toolbar, select the drop-list arrow for the Interpolate icon stack (fifth icon in from the left) and select the **Interpolate** routine.
13. In the command line, the routine prompts for you to select a point object; in the drawing, select point **500**.
14. In the command line, the routine prompts for you to select a second point object; in the drawing, select point **501**.
15. In the command line, the routine prompts for the number of points to place between the two selected points; enter **5** and press the **Enter** key.
16. In the command line, the routine prompts for an offset value; press the **Enter** key to create the interpolated points.
17. Press the **Esc** key to stop the routine and to return to the command line.
18. Repeat Steps 12–17 and create interpolated points between points 501 and 502. Use 3 as the number of interpolated points.
19. Click the **AutoCAD Save** icon to save the drawing.

CREATING A SWALE FROM A FEATURE LINE

You need to make sure you have both the Feature Lines toolbar and the Transparent Commands toolbar on the screen. Use Figure 10.32 as a guide for creating the 3D polyline.

The second swale is a feature line with an elevation of 732.9 at its southern end and the elevation of point 501 (732.4) at its northern end.

1. From the Feature Lines toolbar, select the **Draw Feature Line** icon (first icon on the left).
2. In the Select Style dialog box, toggle **ON** Style, click the drop-list arrow, select from the list of styles **Grading Ditch**, and click the **OK** button to continue.
3. In the command line, the routine prompts for a starting point; in the drawing, select a point near the southerly entrance to the site.
4. In the command line, the routine prompts for an elevation; for the elevation enter **732.90**, and press the **Enter** key to continue.
5. From the Transparent Commands toolbar, select the **Point Object** filter ('PO), and click anywhere on the label of point **501**.
6. In the command line, the routine may prompt for an elevation of <0.00>; if so, enter **E** for elevation, and the routine will echo point 501's elevation (**732.40**). Press **Enter** to accept the elevation (see Figure 10.32).
7. In the command line, the routine prompts to select point object; to end the override press the **Esc** key and then press the **Enter** key.
8. Click the **AutoCAD Save** icon to save the drawing.

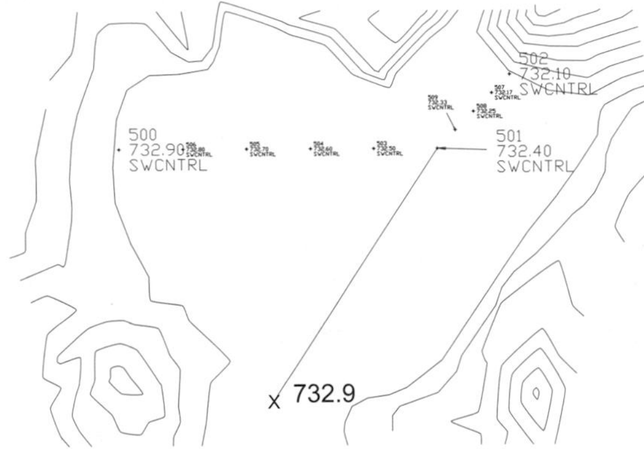

Figure 10.32

The following is the command sequence.

```
Command:
Specify start point: <Select a point in the southern entrance
     area>
Specify elevation or [Surface] <0.000>: 732.9 <Press ENTER>
Specify the next point or [Arc]: '_PO
>>
Select point object: <Select point 501>
Resuming DRAWFEATURELINE command.
Specify the next point or [Arc]: (5292.54 5351.15 732.4)
```

```
Distance 220.546', Grade -0.23, Slope -441.09:1, Elevation 732.
    400'
Specify grade or [SLope/Elevation/Difference/SUrface] <0.00>: e
Specify elevation or [Grade/SLope/Difference/SUrface] <732.400>:
    <Press ENTER>
Specify the next point or [Arc/Length/Undo]:
>>
Select point object: *Cancel* <Press ESC>
>>
Specify the next point or [Arc/Length/Undo]:
Resuming DRAWFEATURELINE command.
Specify the next point or [Arc/Length/Undo]: <Press ENTER>
Command:
```

The feature line has only two vertices: the northern and southern ends. These two points are the only two swale data points for the surface. The Feature Lines toolbar has an Add PI routine for adding vertices to feature lines. The new PI elevations are a straight slope calculation.

Adding Vertices to Swale

1. From the Feature Lines toolbar, select the **Insert Elevation Point** icon (eleventh in from the left) and in the drawing, select the feature line.
2. In the command line, the routine prompts for a distance or increment; enter the letter **I**, press **Enter**, enter **10** feet for the increment, and press the **Enter** key twice to specify the increment and exit the routine.
3. The new elevation points show as blips on the feature line.
4. Click the **AutoCAD Save** icon to save the drawing.

Creating a Berm

The berm sheds water to the point and feature line swale. It has a high point and then is modified with additional elevation points.

1. From the Feature Lines toolbar, select the **Draw Feature Line** icon (first icon on the left).
2. In the Create Feature Lines dialog box, toggle **ON** Style, click the drop-list arrow, select **Grading Ditch** from the list of styles, and click the **OK** button to continue.
3. In the command line, the routine prompts for a starting point; in the drawing, select a point in the southwest between the points and the feature line.
4. In the command line, the routine prompts for an elevation, for the elevation enter **732.90**, and press **Enter** to continue.
5. In the command line, the routine prompts for a second point; in the drawing, select a point between point number 507 and the feature line.
6. In the command line, the routine prompts for an elevation; enter **732.75** as the elevation and press **Enter** twice to set the elevation and exit the routine.

ADDING A HIGH/LOW POINT

The berm needs a high point near its middle. Two positive grades from each end vertex will create a high spot near the middle of the feature line.

1. From the Feature Lines toolbar, select **Insert High/Low Elevation Point** (thirteenth icon in from the left).
2. In the command line, the routine prompts for a feature line, survey figure, parcel line or 3D polyline; in the drawing, select the feature line just drawn.
3. In the command line, the routine prompts to specify the start point; in the drawing, select the southerly end of the feature line just drawn.
4. In the command line, the routine prompts to specify the endpoint; in the drawing, select the northerly end of the feature line just drawn.

   The routine echoes Start Elevation 732.900', End Elevation 732.750', Distance 206.176'
5. In the command line, the routine prompts for specify slope ahead or grade; enter **25** for the slope, and press **Enter**. If the prompt is for grade, enter **S** and press the **Enter** key.
6. In the command line, the routine prompts for specify slope back or grade; for the slope enter **25**, and press the **Enter** key twice to create the high point and to exit the routine.
7. Select the **AutoCAD Save** icon to save the drawing.

Add Vertices to Berm

1. From the Feature Lines toolbar, select the **Insert Elevation Point** icon (eleventh icon from the left) and in the drawing, select the feature line.
2. In the command line, the routine prompts for a distance or increment; enter the letter **I**, press the **Enter** key, for the increment, enter **10** feet, and press the **Enter** key twice to specify the increment and to exit the routine.
3. The new elevation points show as blips on the feature line.
4. Click the **AutoCAD Save** icon to save the drawing.

CREATING THE BASE SURFACE

To calculate a volume you need a second surface. The base surface has points, breaklines, and contour data.

Create Point Group

To assign points to a surface they must belong to a point group.

1. Click the **Prospector** tab to make it current.
2. In Prospector, select the **Point Groups** heading, press the right mouse button, and from the shortcut menu select **New...**.

3. Click the **Information** tab, for the point group Name, enter **Base**.
4. Click the **Include** tab, toggle **ON** With Numbers Matching, and for the range enter **500–520**.
5. Click the **Point List** tab to view the selected points.
6. Click the **OK** button to exit the dialog box.

Creating New Surface

1. In Prospector, select the **Surfaces** heading, press the right mouse button, and from the shortcut menu select **New....**
2. In the Create Surface dialog box, set the surface name to **Base**, for the description enter **Prelim Design**, set the Surface Style to **Border & Triangles & Points**, and click the **OK** buttons to exit.
3. In Prospector, expand the Base branch until you are viewing the data type list under the Definition heading.
4. In the data type list for Base, select **Point Groups**, press the right mouse button, and from the shortcut menu select **Add....**
5. In the Point Groups dialog box, select **Base**, and click the **OK** button.
6. In the data type list for Base, select **Breaklines**, press the right mouse button, and from the shortcut menu select **Add....**
7. In the Add Breaklines dialog box, enter **Breaklines** for the description, set the type to **Standard**, set the Mid-Ordinate distance to **0.01**, click the **OK** button, and in the drawing, select the swale and berm breaklines.
8. In the data type list for Base, select **Contours**, press the right mouse button, and from the shortcut menu select **Add....**
9. In the Add Contour Data dialog box, enter the description of **EG and New**, set the values as shown in Figure 10.33, and when matched, click the **OK** button.
10. In the drawing, select all of the contours, and press the **Enter** key to exit.
11. In Layer Properties Manager, set layer 0 (zero) as the current layer, freeze the layer **Base** and **C-TOPO-FEAT**, and click the **OK** button to exit.

    The new surface is complete.

CALCULATING A VOLUME SUMMARY

1. From the Utilities flyout in the Surfaces menu, select **Volumes....**

   This presents a Composite Volumes vista.

2. In the vista, click the **Create new volume entry** icon (top left of the vista) to create a calculation entry.

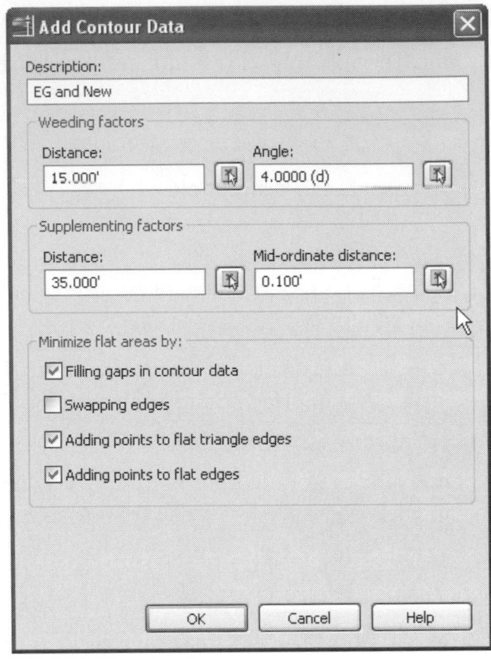

Figure 10.33

3. In the vista, click in the Base Surface cell for index 1 (<select surface>), click the drop-list arrow, and select **Existing** from the list,
4. In the vista, click in the Comparison Surface cell for index 1 (<select surface>), click on the drop-list arrow, and select **Base** from the list, (see Figure 10.34).
5. In the vista, click into the Cut Volume cell for index 1, and the vista calculates a volume.
6. Repeat Steps 2–5 for a second volume calculation using EGCTR and Base.
7. Note down the Cut, Fill, and Net volumes for each comparison.
8. In the Panorama, click the "**X**" at the top of the mast to close the screen.

VOLUME SURFACE STYLE

When creating a volume surface, you assign styles just like any Civil 3D surface. For this exercise, you will define a style showing a 2D view of contour ranges for cut and fill and in 3D as a model.

1. Select the **Settings** tab and expand the Surface collection until you are viewing the list of Surface Styles.
2. In the Surface branch of Settings, select the **Surface Styles** heading, press the right mouse button, and from the shortcut menu select **New...**.
3. In the Surface Style dialog box, select the **Information** tab and enter **Volume Contours** for the name.

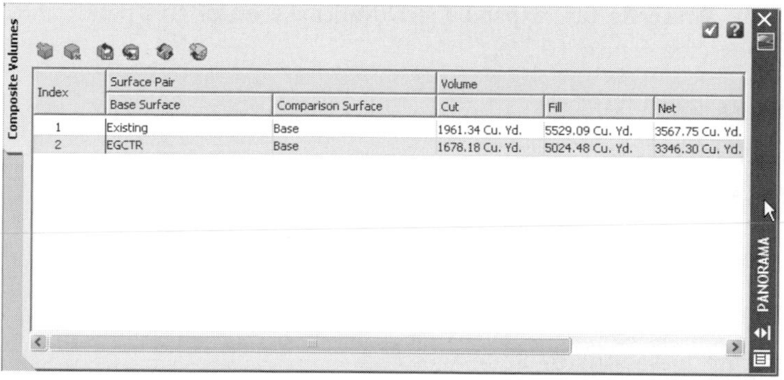

Figure 10.34

4. Click the **Contours** tab, expand the Contour Ranges section and set the following values (see Figure 10.35):

   Group Values by: Equal Interval

   Number of Ranges: 12

   Use Color Scheme: True

   Major Color Scheme: Rainbow

   Minor Color Scheme: Land

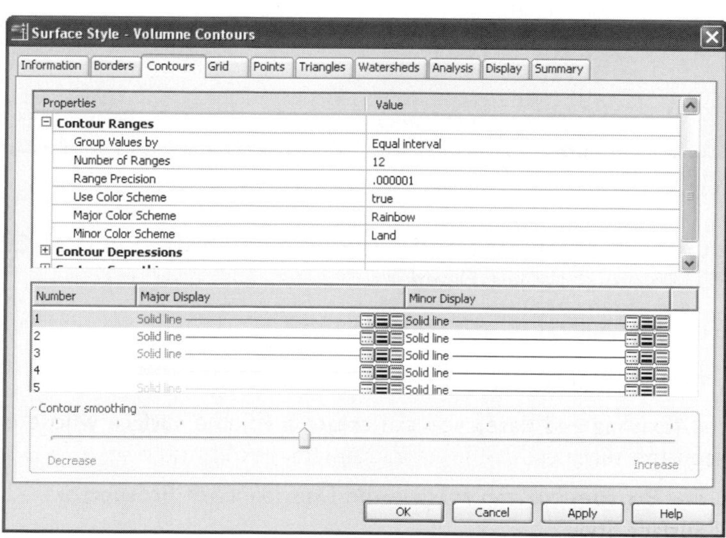

Figure 10.35

5. Click the **Analysis** tab, expand the Elevations tree of this panel, and set the following values (see Figure 10.36):

   Group by: Equal Interval

   Number of Ranges: 12

   Display Type: 3D Faces

   Scheme: Rainbow

   Elevations Display Mode: Exaggerate Elevation

   Exaggerate Elevations by Scale Factor: 4

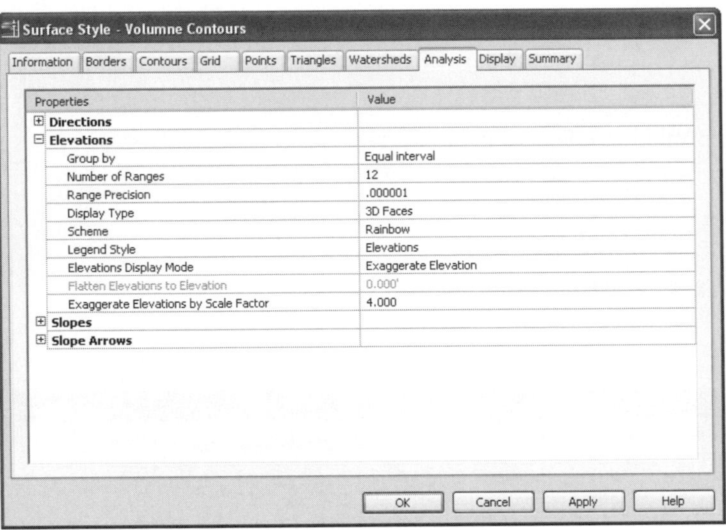

Figure 10.36

6. Click the **Display** tab, set the 2D View to show Border and User Contours, and set the 3D View to show just Elevations.
7. Click the **OK** button to exit the dialog box.

CREATING A TIN VOLUME SURFACE

With two surfaces (Existing and Base), you can create a volume surface whose elevations are the differences between the surfaces.

1. Click the **Prospector** tab to view the Data panel of Prospector.
2. In Prospector, click the **Surfaces** heading, press the right mouse button, and from the shortcut menu select **New...**.
3. In the Create Surface dialog box, set the surface Type to **TIN Volume Surface**, for the name enter **Existing-Base-TIN-Vol**, and enter the remainder of the information found in Figure 10.37.

4. At the bottom of the dialog box, click in the Value cell of Base Surface, click the ellipsis at the right side of the cell, and select the **Existing** surface.
5. Repeat the previous step, but for the Comparison Surface; however, select the **Base** surface.
6. Click the **OK** button to exit the dialog box and to create the Existing-Base-Vol surface.

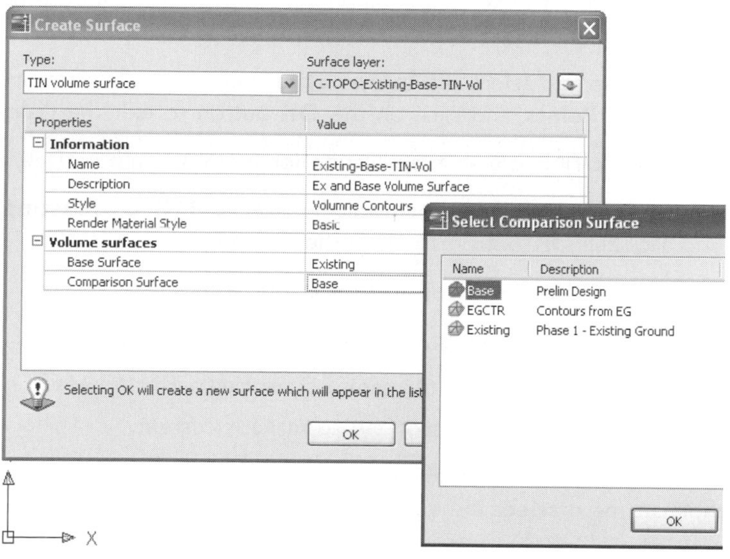

Figure 10.37

SURFACE PROPERTIES

A volume surface's property dialog box reports the amount of cut and fill. The surface contours displays the amounts of surface cut and fill.

1. From the Surfaces list in Prospector, select the **Existing-Base-TIN-Vol**, press the right mouse button, and from the shortcut menu select **Properties....**
2. Select the **Analysis** tab set the Analysis Type to **User-Defined Contours** if necessary, set the Number of Ranges show (**12**), and click the blue **Run Analysis** button.
3. In the Range Details section, change the elevation for contour 6 to **0.00**.
4. Change the Analysis Type to **Elevations** if necessary, set the Number of Ranges show (**12**), and then click the **Run Analysis** button.
5. Select the **Statistics** tab and expand the Volume section to display earthwork calculation.
6. Compare the surface volume to the volume from Surfaces Utilities.
7. Click the **OK** button to exit the Surface Properties dialog box.

The surface contours represent amounts of cut and fill. The style has a 3D view of the amount of cut and fill. The cut areas are the deepest holes and the fill areas are the highest hills.

8. In the drawing, click any surface component, press the right mouse button, from the shortcut menu, select **Object Viewer...**, and view the surface from different angles.
9. Exit the Object Viewer.
10. In the drawing, click any surface component, press the right mouse button, and from the shortcut menu select **Surface Properties...**.
11. In Surface Properties, select the **Information** tab, set the Surface Style to **Border & Triangles & Points**, and click on the **OK** button to exit the dialog box.

    This displays the TIN volume triangulation for a TIN volume surface.
12. Open Layer Properties Manager, freeze the layer **C-TOPO- Existing-Base-TIN-Vol**, and click the **OK** button to close the dialog box.
13. Click the **AutoCAD Save** icon to save the drawing.

CREATING A GRID VOLUME SURFACE

The creation of a Grid Volume surface is the same as creating a TIN Volume surface.

1. If necessary, click the **Prospector** tab to make it current.
2. In Prospector, select the **Surfaces** heading, press the right mouse button, and from the shortcut menu select **New...**.
3. In the Create Surface dialog box, set the surface type to **Grid Volume Surface** and enter the information found in Figure 10.38.
4. In the middle of the dialog box, specify the Grid Spacing to **5** units for X and Y and a rotation of **0** (zero).
5. At the bottom of the dialog box, click in the Value cell of Base Surface, click the ellipsis at the right side of the cell, and select the Existing surface from the list.
6. At the bottom of the dialog box, click in the Value cell of Comparison Surface, click the ellipsis at the right side of the cell, and select the Base surface from the list.
7. Click the **OK** button to exit the dialog box and to create the Existing-Base-Grid-Vol surface.

CREATING USER CONTOURS

To view the cut and fill as contours and elevations, you need to assign the analysis surface style and compute the user contours.

1. From the list of Surfaces in Prospector, select **Existing-Base-Grid-Vol**, press the right mouse button, and select **Properties...** from the shortcut menu.
2. Select the **Analysis** tab, set the Analysis type to **User-Defined Contours** if necessary, set the Number of Ranges to **12** and click the blue **Run Analysis** icon.

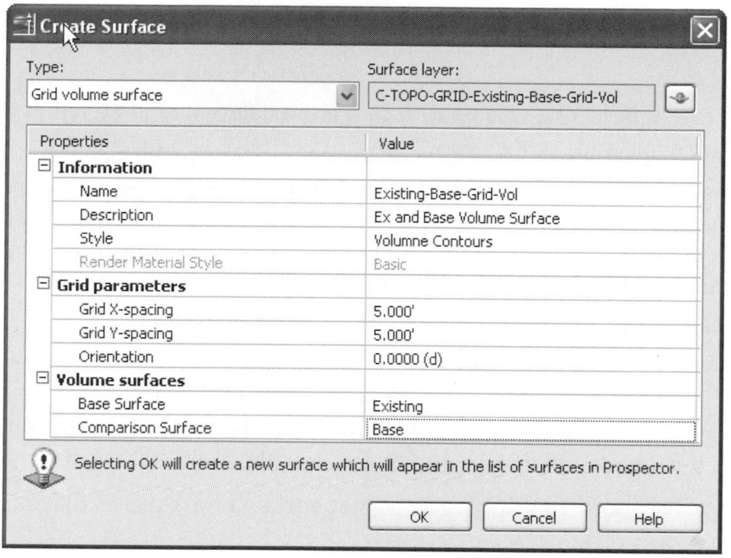

Figure 10.38

3. Set the Analysis to **Elevations** if necessary, set the ranges to **12**, and click the blue **Run Analysis** icon.
4. Select the **Statistics** tab and expand the Volume section to display the cut and fill volumes.
5. Click the **OK** button to exit the Surface Properties dialog box.
6. In the drawing, click any surface component, press the right mouse button, from the shortcut menu select **Object Viewer...**, and view the surface from different angles.
7. Exit the object viewer.
8. Click any surface component, press the right mouse button, and from the shortcut menu select **Surface Properties...**.
9. Select the **Information** tab, change the Surface Style to **Border & Triangles & Points**, and click the **OK** button to view the surface triangulation.

   This shows the grid calculating the surface's volume.
10. Click the **AutoCAD Save** icon to save the drawing.

SUMMARY

- The elevations of a volume surface are the differences in elevation between two terrain surfaces.
- A volume surface can use any combination of terrain and grid surfaces to calculate a volume.
- A property of a volume surface is an overall earthwork volume.
- Analysis surface styles evaluate a volume surface's data.

This ends the chapter on Civil 3D surface design tools and volumes. As mentioned throughout this book, surfaces are fundamental civil data. If a surface is not correct, this error affects many aspects of developing a design solution.

The next chapter is about developing pipe networks. The surface plays a critical role in developing a network design.

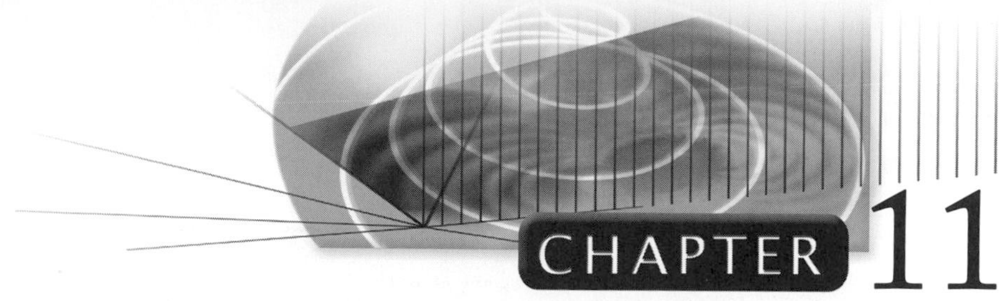

Pipe Networks

INTRODUCTION

Pipe networks are an integral part of a site design solution. The complexity of the piping systems can vary from simple culverts at critical points to several storm and sanitary networks servicing storm water runoff and soiled water for each residential lot. Autodesk Civil 3D 2007 produces a 3D model representing the pipe design solution. The design can be a part of a roadway corridor's sample line group, while a pipe network can be a system of trunk lines with pipe and structures branches that creates a complex web over the extents of the site.

OBJECTIVES

This chapter focuses on the following topics:

- Defining Pipe-Run Specific Settings
- Defining New Pipe Run Structures
- Defining Pipe Runs
- Editing and Analyzing Pipe Run Data
- Annotating Pipe Runs in Plan and Profile View

OVERVIEW

Civil 3D contains tools to create pipe runs or networks of pipes and structures (see Figure 11.1). The pipe design toolset is in the Network Layout Tools toolbar (see Figure 11.2). Most piping designs are a network of interconnected trunk and branch lines. A pipe network can have several trunk lines and each trunk line can have any number of branches. The key is including all of the appropriate trunks and branches with the correct network. For example, the user can define a network with a "Y" shaped trunk line and can later add branches and new trunk lines to the initial "Y" design by editing the network. After completing the edit of the pipe network, Civil 3D lists the network as including the "Y" and all of the newly added pipes and structures.

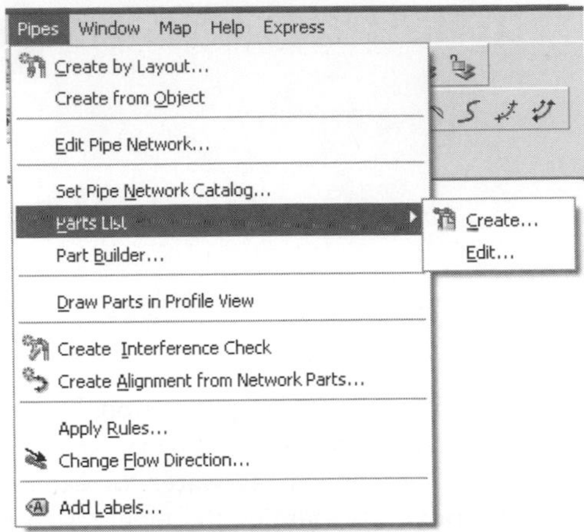

Figure 11.1

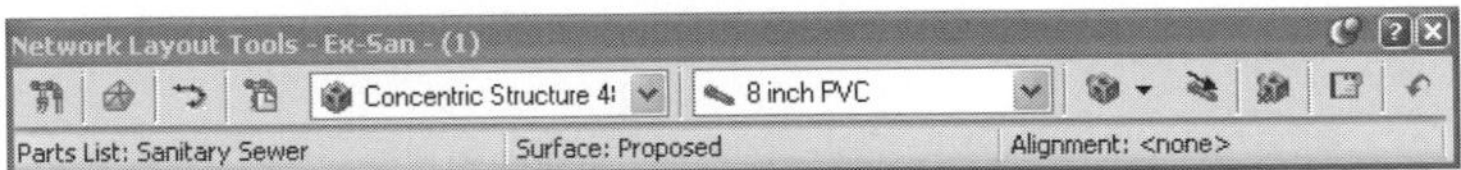

Figure 11.2

The layout routines of the Network Layout Tools toolbar and their resulting pipes and structures depend on a series of settings and styles. These settings and styles govern the type of structure, size of pipe, inverts, and a host of additional values.

Civil 3D does not require an alignment to exist for a pipe network. Each network is a model that stands on its own. If the user wants, he or she can create an alignment and profile from a pipe network. Civil 3D has tools to place all or a selection of pipes and structures in a profile view.

Editing a network is simply selecting a pipe or structure, pressing the right mouse button, and selecting Edit Network... from a shortcut menu. The Edit Network command also occurs in the Pipes menu or in a shortcut menu in Prospector.

When laying out a pipe network, Civil 3D can associate an alignment to the new network (see Figure 11.3). A pipe network can start or end beyond the associated alignment definition. If a pipe run is outside an alignment definition, the user may have to edit the values of the structures and pipes to produce the correct elevations and slopes.

The development, review, and editing of a pipe network design takes place graphically in plan and/or profile view or by modifying values of the pipe network data in a vista editor. If a pipe network does not have an associated alignment, then all of the editing is done in plan view or in the panorama. The user can edit a pipe or structure by three different methods: graphically changing a pipe's or structure's location, editing the properties of a pipe or structure, or by editing the data of pipes and structures in a network vista (the whole network).

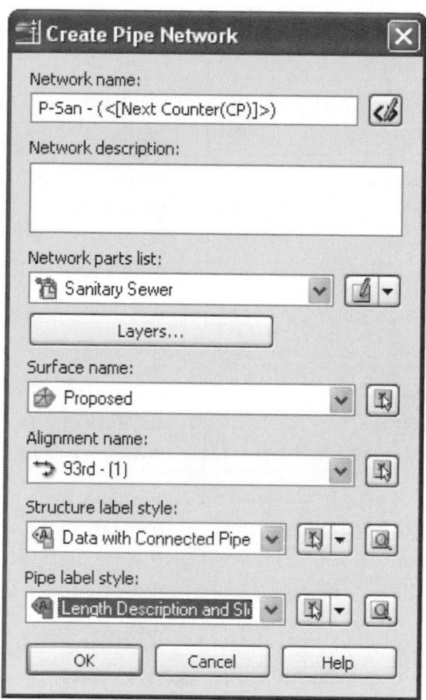

Figure 11.3

One issue to be aware of is the layers for the pipe network. When setting the Network parts list to Sanitary Sewer or Storm Sewer, the layer list does not switch—it stays set to the Storm layers. If the user switches the part list, he or she will need to click the Layers... button in the Create Pipe Network dialog box and set the correct layers for the pipe network to be drafted.

When the user selects a pipe or structure in plan view and presses the right mouse button, a shortcut menu appears. The shortcut menu on the left of Figure 11.4 shows the options that are available to edit the selected pipe, and the shortcut menu on the right shows the options when selecting a structure.

When graphically editing a pipe or structure in plan view, Civil 3D allows moving them, disconnecting them from the network, changing their style, swapping to a new part, or editing the network in a panorama (see Figure 11.4). When the user manipulates a pipe's grip, moving the pipe away from a structure disconnects the pipe from the structure. When the user moves the pipe back to the structure, he or she needs to press the right mouse and select reconnect. If the user is moving both pipes and structures or just the structure, they remain connected after they move. These rules also apply to manipulating structures.

When editing in plan view, place a structure in a pipe run at any point or a pipe between any two structures of the same network. Civil 3D will show a break pipe icon that indicates that a structure will be at that location in the pipe run and a structure connect icon for a pipe connecting to a structure. Pipes and structures of different networks cannot connect to each other.

Figure 11.4

When editing a pipe network in profile view, Civil 3D imports selected pipes and structures or the entire network into the profile view. Within the context of the profile view, the user can graphically change the vertical location of the pipes, change the size of pipes and structures, and swap out structure types.

Editing an entire pipe network takes place in a panorama with vistas of the pipe network. In the vistas, Civil 3D presents the numbers behind the structure and pipe objects. The user can change almost any value for a pipe or structure in these vistas. The biggest issue with the vistas is their size and complexity. The vistas contain numerous cells and they will not display entirely on screen (scroll to see all of the data). Also, an irritant is the placement of data in a vista. Move columns around to view related values that may be separated by several columns. Changes made in the vista do not immediately affect the network; rather, they occur once the user has exited and reentered the vista. Any changes made in the vista are permanent; the user cannot cancel a change or undo it.

Since Civil 3D automates so many of the pipe design tasks, the pipe network and its objects have extensive settings and styles. These settings and styles affect drafting pipes and structures in plan and profile, labeling, structures, and data values. Since there is a difference between the piping and type of structures in sanitary and storm pipe networks, develop two parts lists, each specifying the typical structures and pipes.

Civil 3D includes an extensive catalog of pipe and structure parts. One catalog focuses on storm pipes and another on sanitary pipes and structures. A catalog defines parts that included as standard parts for a typical pipe and structure storm or sanitary network. A parts list is a subset of the catalog listings. A parts list contains descriptions, measurements, rules, and other settings for the plan and profile representations of the structures and pipes in a network.

UNIT 1

Unit 1 focuses on the settings and styles of pipes. As mentioned above, the list of settings and styles is extensive.

UNIT 2

Unit 2 discusses the drafting of pipe networks.

UNIT 3

Unit 3 focuses on reviewing and editing a pipe network's data. Both review and edits can be done in plan or profile view either graphically or from a panorama.

UNIT 4

Unit 4 covers the annotation of the pipe networks when the labeling is not a part of the originally drafted network. It also included pipes in cross sections.

UNIT 1: PIPES SETTINGS AND STYLES

The pipe network portion of Settings uses an extensive list of settings and styles (see Figure 11.5). The Pipe and Structure branches contain settings and values that define shapes, rules, and labeling styles. The Pipe Network section defines lists of parts for typical systems—storm and sanitary—from the definitions in the Pipe and Structure branches.

Since the structures in a storm sewer system are different from those in a sanitary system, it is important to create a separate parts lists of structures and pipes for storm and sanitary systems. The parts list isolates the differences between the storm and sanitary design elements and does not allow them to be mixed in a network design.

The parts list for any one system should include a typical library of structures and pipe sizes. The library should, however, not be limited. A branch of a network may have different structures and pipe sizes than those of a trunk. When working with a storm water system, the branches may be catchbasins rather than simple manhole structures.

When developing parts lists, styles, and settings, it is important to copy them into a template file for use as office standards. This way, every new job starts with the same settings and the user is able to produce drawings with consistency.

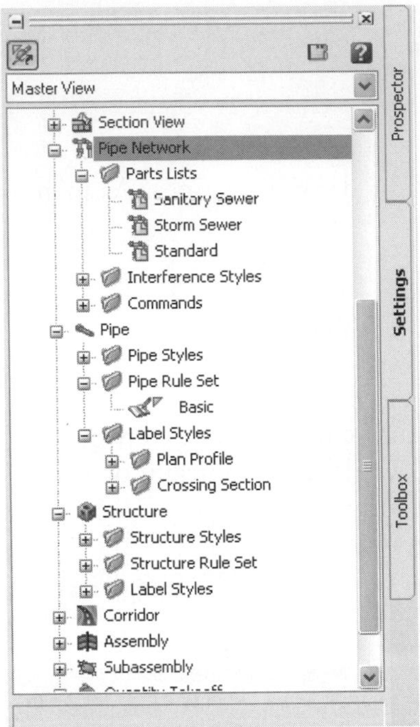

Figure 11.5

EDIT DRAWING SETTINGS

The Edit Drawing Settings dialog box in Settings contains the base layer names for pipe, structure, and network objects (see Figure 11.6). The layer list includes entries for pipe networks in plan, profile, and section views. The problem with the modifier for pipes and structures is that the modifier is each structure and pipe. Civil 3D creates a new layer for each structure and pipe in a pipe network. The pipe network does not have a layer dedicated to just its contents.

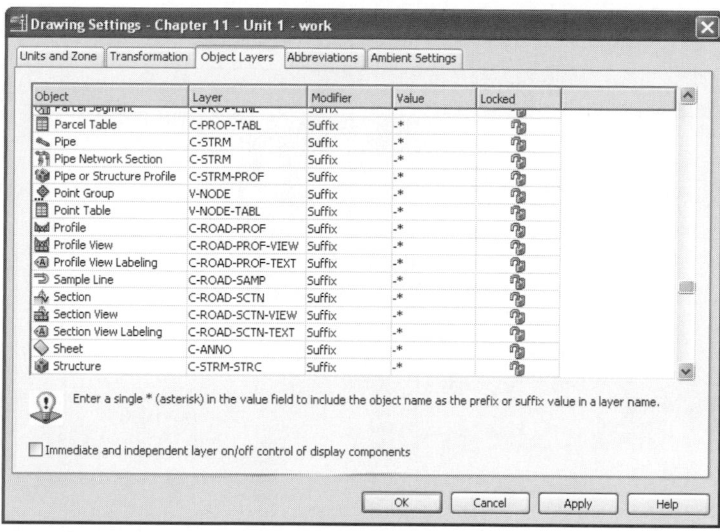

Figure 11.6

PIPE AND STRUCTURE CATALOG

The pipe and structure catalogs supply the basic specs for the parts lists of a pipe network. A catalog populates certain values in the myriad of dialog boxes that appear while developing a parts list for storm and sanitary systems. Unfortunately, there isn't a call in the Civil 3D menu system to view the catalog.

The catalog itself is in the All Users branch of the operating system and is a HTML file that reads XML data files from folders below the catalog (see Figure 11.7). The catalog contains the sizes and other parameters of each type of supported pipe and structure. The only way to add types and parameters is to edit the files in an ASCII editor or the Part Builder.

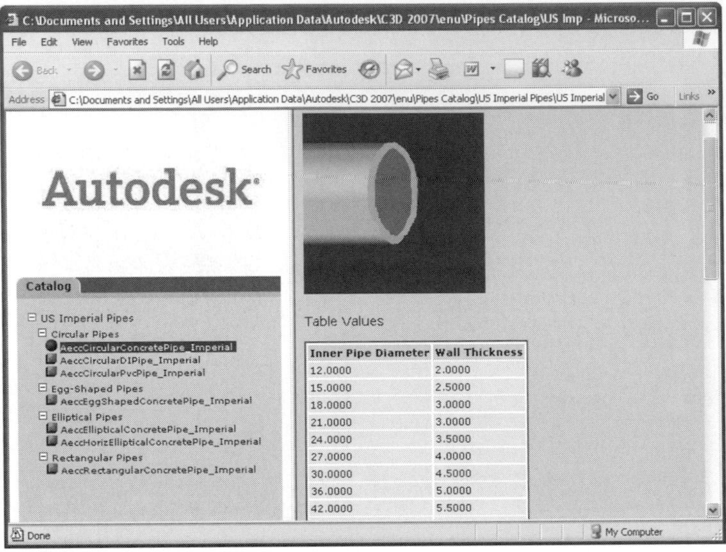

Figure 11.7

PIPE NETWORK PARTS LIST

However, when the user develops or adds to a parts list, the values of the catalog appear as content for the Parts List dialog box (see Figure 11.8). The Add Part Family shortcut menu command calls the Part Catalog dialog box. This dialog box lists the catalog's different categories of parts.

A parts list can include every possible type and size. But like plotting scales, there will be a consistent set of typical pipe sizes and structures for each design project. Civil 3D allows the user to swap a structure part to any size, even those not on the parts list. So, it is relatively easy to handle the exceptions that occur in any project.

As the user develops a parts list, the parts list assigns the basic style for structure and pipe styles, rules, and rendering material. If these need to change, make the change after completing the modification of the parts list. These styles need to be defined prior to assigning them to the members of the parts list. The assignment of styles and rendering material must be made BEFORE using the parts list for a network. Rules affect error reporting, positioning of structures, and other crucial network values. When the user changes the pipes or structures of a parts list, the change affects only those networks created after the change.

After the user toggles on a family in the Part Catalog dialog box and clicks the OK button, the selected family appears as a new group of potential parts for the parts list (see Figure 11.9). When the user selects the new family name, press the right mouse button, and select Add Part Size.... A dialog box appears, showing parameters for a pipe or structure. Some of the parameters for the pipe or structure are parameters read from the catalog and are marked as list in the source column. Other values are calculated or from a table containing a range of values, and identify themselves in the Source column. When

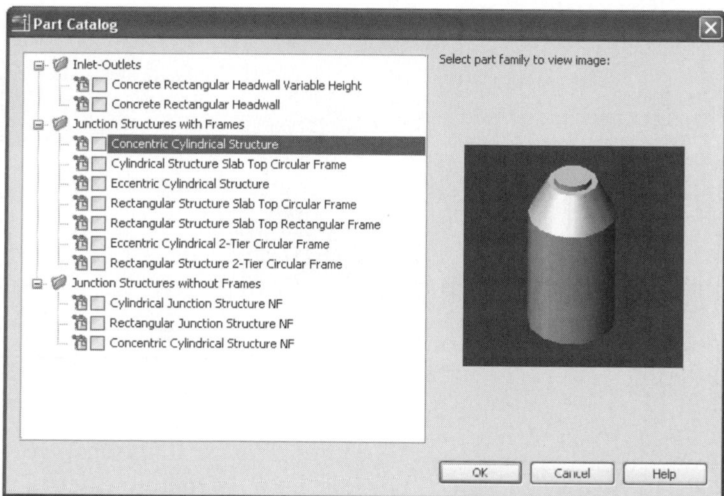

Figure 11.8

clicking into a value cell of a List Source item, clicking the drop-list arrow displays a list of selectable catalog values (see Figure 11.9).

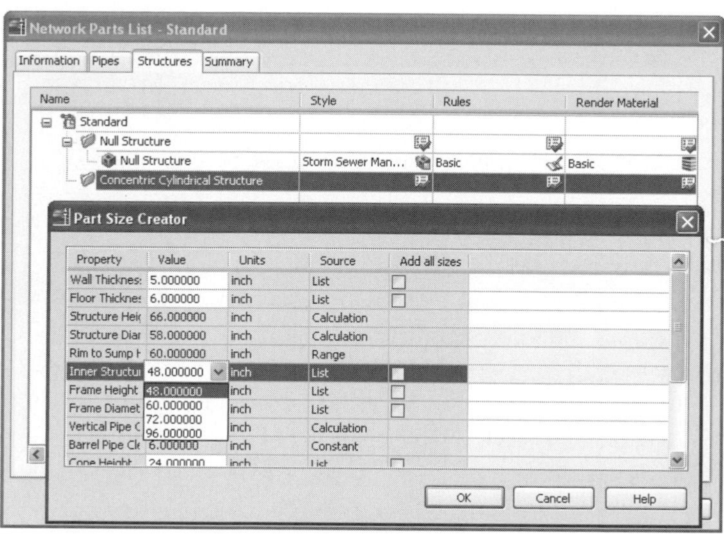

Figure 11.9

The Part Size Creator dialog box has entries for specifying parts: material, frame, grate, and cover part numbers (see Figure 11.10).

As the user adds parts to the parts list, Civil 3D assigns basic styles to structure style, rule, and render material. Figure 11.11 shows a Storm parts list with assigned structure styles. These styles define how the structures appear in plan, profile, and section.

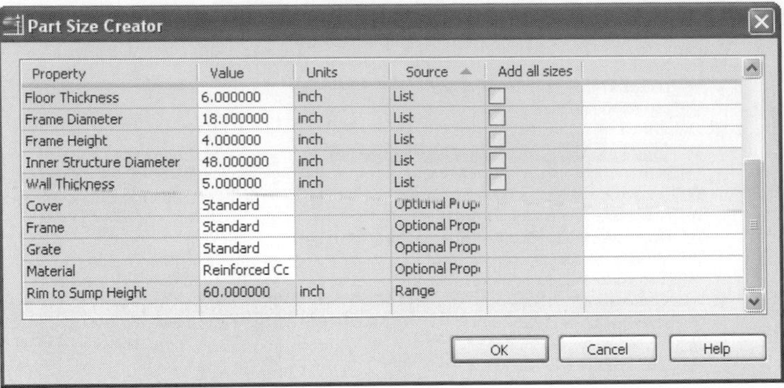

Figure 11.10

There are two ways to assign the styles. The first is to click the icon to the right of the Structure Category in the Style column. This displays a Structure Style dialog box that has a drop list containing all of the structure styles. When the user exits this dialog box, Civil 3D assigns the selected structure type to ALL of the structures in that category. The user can change an individual part in each category by selecting the Style cell for that part. Assigning Rules and Render Materials uses the same method. Selecting a Rule or setting a Render Material at the category level sets the value for all of the instances in the category. Setting a Rule or Render Material for a single member sets the value only for that part.

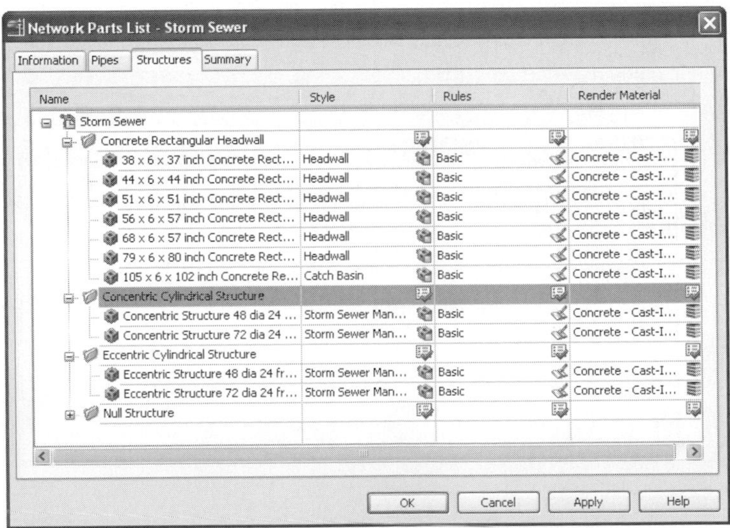

Figure 11.11

PIPE NETWORK EDIT FEATURE SETTINGS

The Edit Feature Settings affect the values found in the styles and commands of the pipe network, pipes, and structures.

The pipe network Edit Feature Settings dialog box sets the default styles for the network parts list and the pipes and structures in the list. Setting the default styles is the function of the Default Styles section of the dialog box (see Figure 11.12). Rather than setting these values for each new part added to a parts list, these settings will populate the entire parts list with these preferred styles. When the user adds a part to the list, these styles are the default styles for the part. The Default Name Format section sets how Civil 3D creates a name for a pipe network, structure, or pipe.

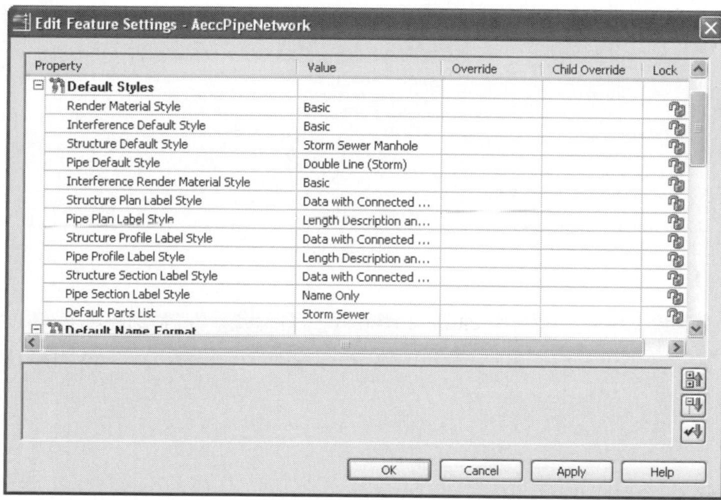

Figure 11.12

The Default Rules section establishes a fundamental rule set for all pipes or structures (see Figure 11.13). However, each pipe and structure can have its own set of rules.

The last section of the Edit Feature Settings for pipe networks affects the location of the labels for pipes and structures in profile and section views. This section refers to an anchor point for the annotation. An anchor point can be above, below, or on the pipe or structure. The location of the anchor point will affect the labeling styles for each object type (see Figure 11.14).

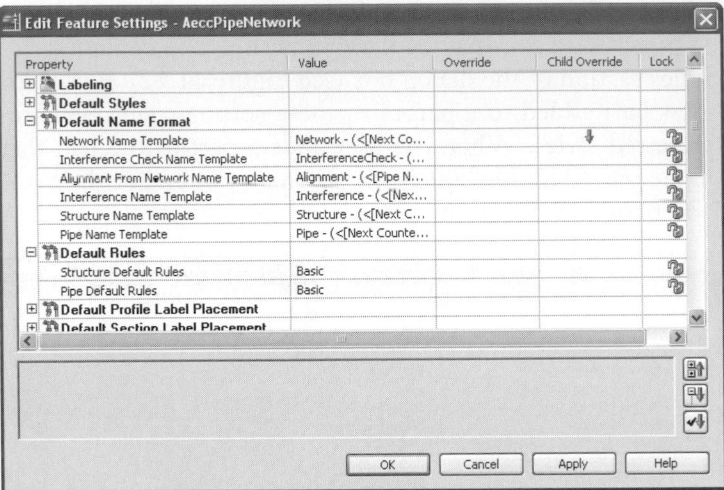

Figure 11.13

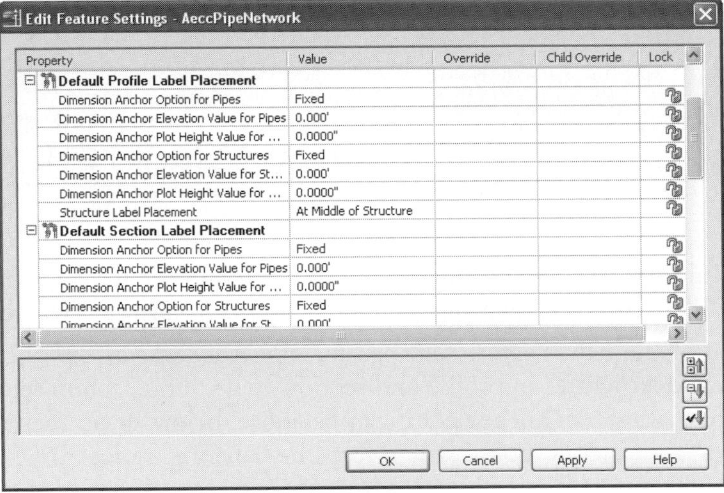

Figure 11.14

PIPE STYLES

Pipe styles define the "look" of the line work for a pipe segment in a pipe network.

PLAN

The settings of the Plan tab define what the pipe shows on the screen when the user looks at a pipe network in plan view (see Figure 11.15). The upper left of the panel defines how the style drafts the walls of the pipe. If the user does not want the specs of the pipe to determine how a pipe appears, he or she can specify an exaggeration factor, a relative factor, or fix the size. The exaggeration factor uses the scale of the drawing to exaggerate the pipe sizes. The relative factor changes the representation of the pipes as the user zooms in or out. The last option is to use the actual size of the pipe.

The upper-right side of the panel defines how Civil 3D drafts the ends of a pipe in plan view. This section contains the same options as the left side, but allows the user to draw the pipe to the interior or exterior wall size.

The bottom portion sets the hatch method for a plan view.

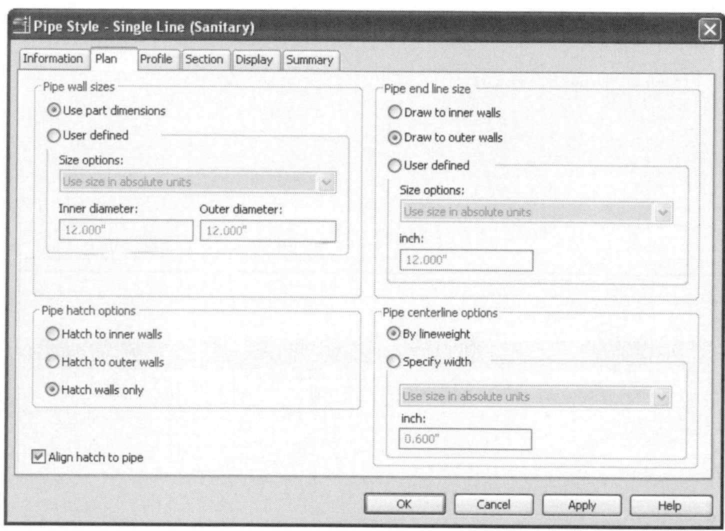

Figure 11.15

PROFILE

The settings for this panel are the same as those in the Plan panel, but apply to a pipe in a profile view (see Figure 11.16). The lower-right section contains hatching options for a crossing pipe.

SECTION

The settings for the Section panel define how to hatch a pipe in a cross section view (see Figure 11.17).

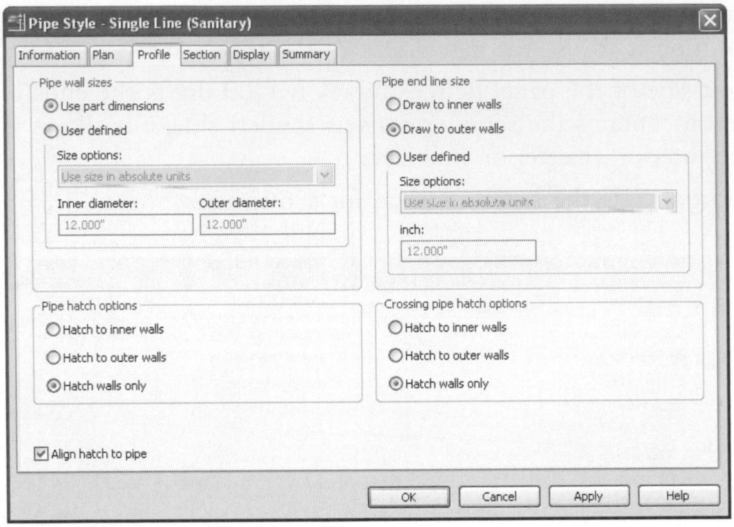

Figure 11.16

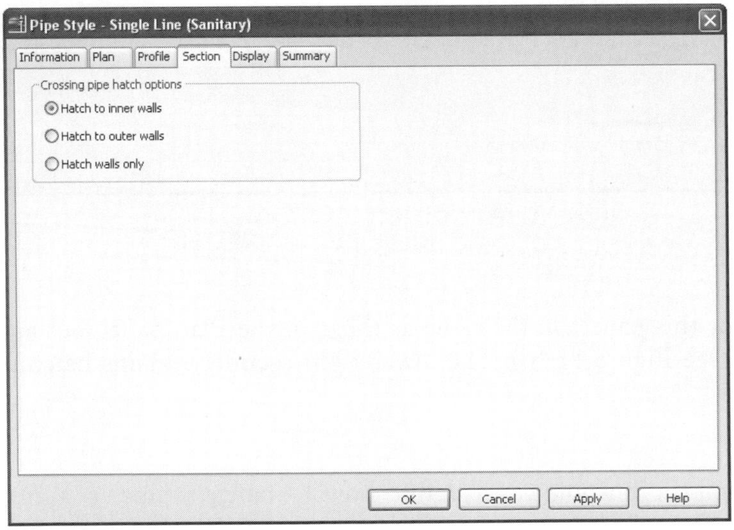

Figure 11.17

DISPLAY

As with any Civil 3D style, the user can set a value for all of the settings, but it is the display panel values that control what is visible in the drawing (see Figure 11.18). When using one of the Civil 3D content templates, there are two pipe styles: single line and double line. It is the styles' visibility settings that differentiate them. The single line style draws the pipe by displaying the centerline of the pipe (left portion of Figure 11.18). The double line style draws two lines representing the interior sides of the pipe (the right portion of Figure 11.18).

Figure 11.18

PIPE RULES

A pipe rules style defines the slope, cover, and segment length rules for pipes (see Figure 11.19). Each size of pipe or family of pipes can have its own set of rules. The Pipe to Pipe Match section allows the user to match pipe inverts by Crown, Centerline, or Invert.

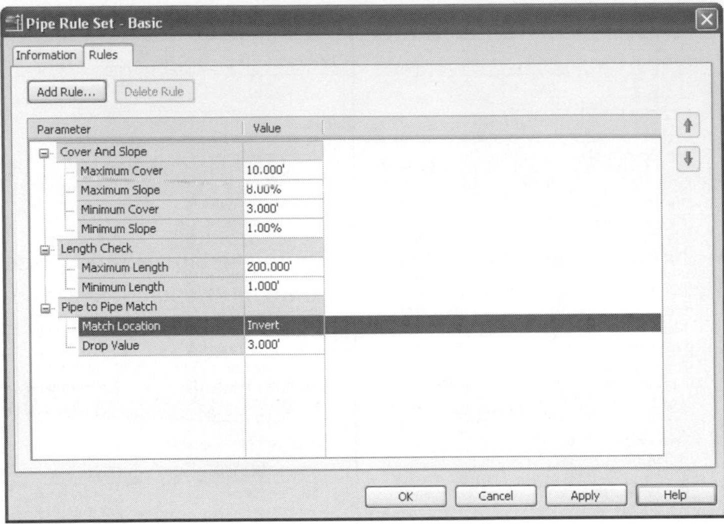

Figure 11.19

PIPE LABELS

When using one of the Civil 3D content drawings, the user has access to several predefined pipe label styles (see Figure 11.20). A pipe label style is the same as any other label style; it is anchored at a point on a pipe, can have more than one content value, and has a dragged state that can be different or the same as the original label. The Text Component Editor defines the label and its formatting.

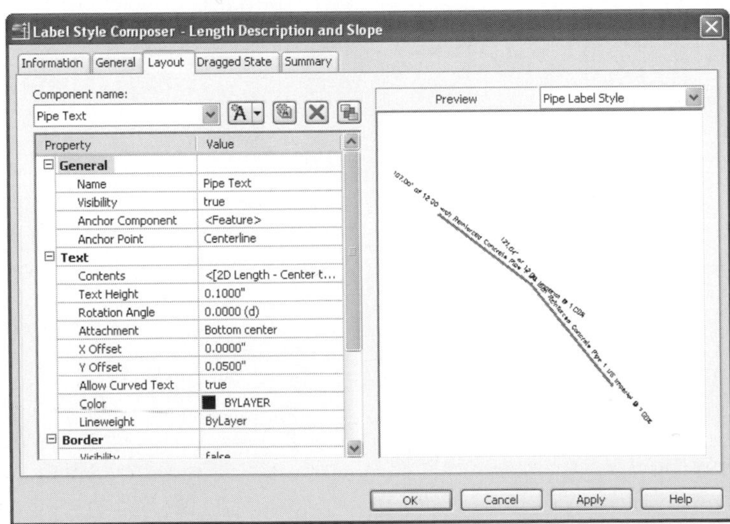

Figure 11.20

STRUCTURE

This family of styles defines the structures that occur between pipe segments.

STRUCTURE STYLE

This branch's styles define the structures' forms for a network.

Model

The Model tab defines structures as shapes from the default catalog (see Figure 11.21). The panel also allows the user to define other simpler shapes to represent the structure: box, cylinder, sphere, and a part-defined shape.

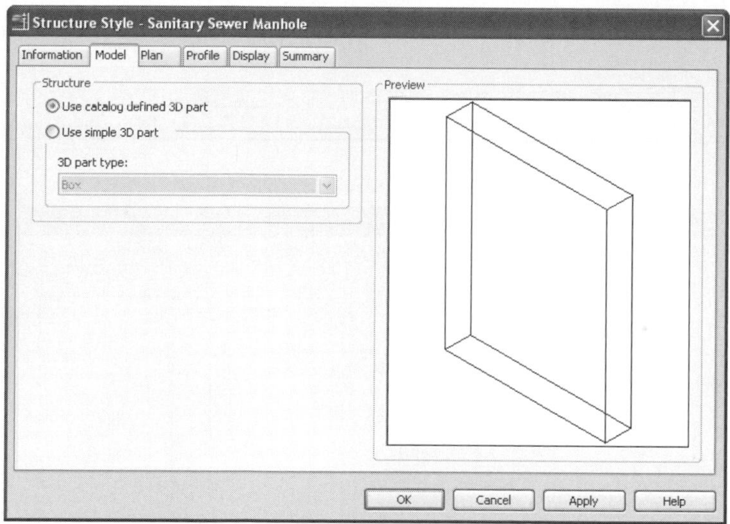

Figure 11.21

Plan

The Plan tab displays a panel defining what symbol to use to identify a structure in plan view (see Figure 11.22). The size options control the size of the symbol in the display. The values for this setting work the same as symbols for point objects. The settings scale a symbol by the drawing scale, a fixed scale, in absolute units, sizes the symbol as a percentage of the screen, or as defined by the catalog parameters.

The Enable Part Masking toggle allows the structure to mask out a connecting pipe.

Profile

Profile tab defines how a structure displays in a profile view (see Figure 11.23). By default, the catalog definition of the structure appears in a profilc view. Other settings in the panel allow the user to define his or her own representation of a structure in a profile view.

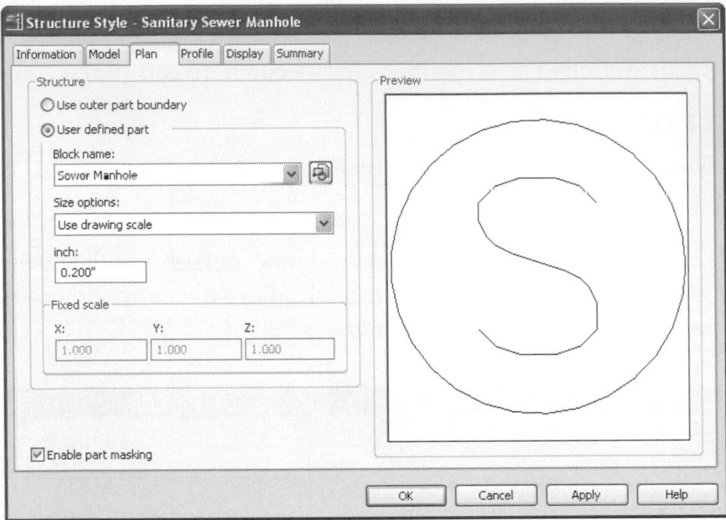

Figure 11.22

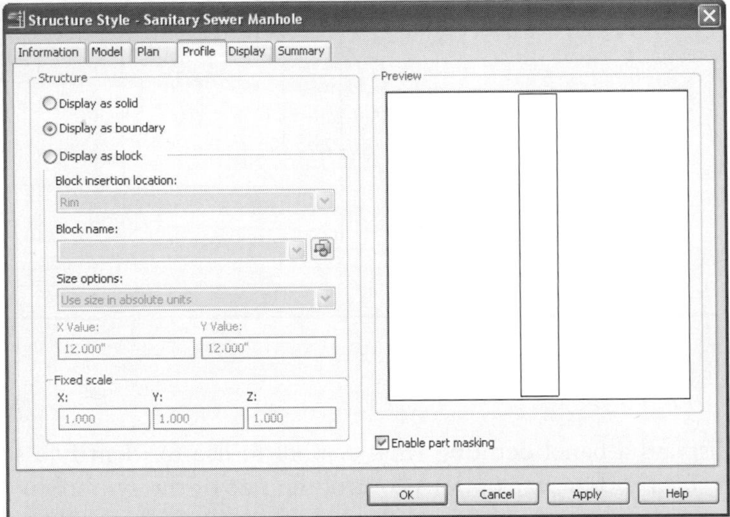

Figure 11.23

Display

The display panel controls what components display and the layers when using a style (see Figure 11.24).

STRUCTURE RULES

A structure rule affects the amount of drop an invert can have across a structure (see Figure 11.25). The user can also set a check for maximum pipe diameter.

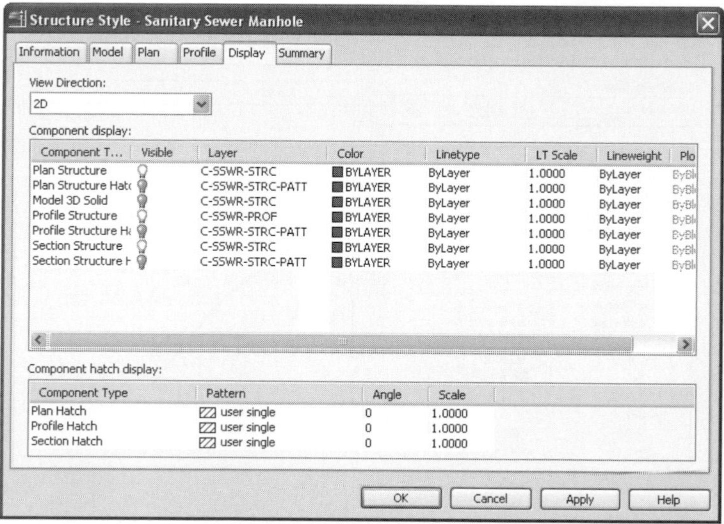

Figure 11.24

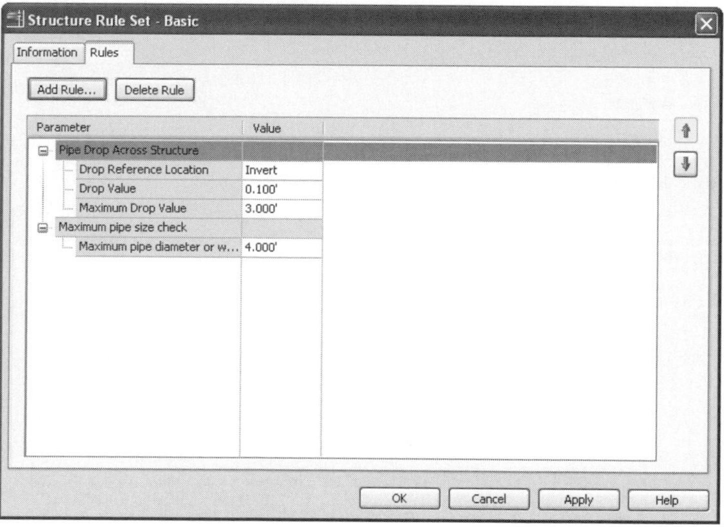

Figure 11.25

STRUCTURE LABELS

These label styles annotate structure information (see Figure 11.26). The Civil 3D content templates provide simple label styles for structures. Develop these label styles in the Label Style Composer interface.

INTERFERENCE STYLES

Pipe networks can check for interferences (colliding pipes). The Interference styles define the markers that locate the interference in the drawing (see Figure 11.27).

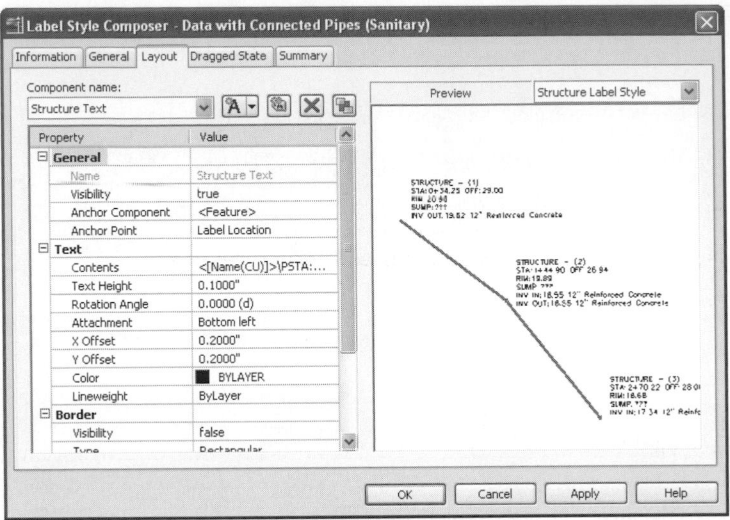

Figure 11.26

Figure 11.27

EXERCISE 11-1

When you complete this exercise, you will:

- Be familiar with the pipe network's Edit Drawing Settings.
- Be familiar with the pipe network's Feature Settings.
- Be able to draft a pipe network.
- Be able to assign materials to pipes and structures.
- Be able to assign pipe rules.
- Be able to add structures to a parts list.

EXERCISE SETUP

This exercise starts with an existing drawing. The drawing *Chapter 11 – Unit 1.dwg* is in the *Chapter 11* folder of the CD that accompanies this book.

1. Double click the Civil 3D icon to start the application.
2. Close the opening drawing.
3. From the File menu use the **Open...** command to open the *Chapter 11 – Unit 1* drawing file.
4. From the File menu use the **AutoCAD Save As...** command to save the drawing under a new name.

EDIT DRAWING SETTINGS

The Edit Drawing Settings dialog box contains several values affecting pipe networks.

1. Click the **Settings** tab to make it the current.
2. At the top of Settings, select the drawing's name, press the right mouse button, and from the shortcut menu select **Edit Drawing Settings....**
3. Click the **Object Layers** tab and view its settings. Your settings should match the values in Figure 11.6.
4. If your settings do not match, make the necessary changes to the pipe values in the dialog box.
5. Click the **OK** button to exit the Edit Drawing Settings dialog box.

EDIT FEATURE SETTINGS

The pipe network Edit Feature Settings values affect both pipes and structures. The function of the values is assigning initial styles and render materials to pipe segments and structures (see Figures 11.12 and 11.13).

1. In the Settings, select the **Pipe Network** heading, press the right mouse button, and from the shortcut menu select **Edit Feature Settings....**
2. Expand the Default Styles and Default Name Format sections.

Civil 3D assigns all the initial styles to pipes or structures from this dialog box. The Default Name Format section sets the naming convention for pipes, structures, and networks. Usually, the name for these objects is the object type followed by a sequential number.

3. In the dialog box, click in the Render Material Style value cell to display an ellipsis and click the ellipsis to display the Render Material Style dialog box.
4. In the Render Material Style dialog box, click the drop-list arrow, select the **Concrete - Cast-In-Place** material, and click the **OK** button to return to the Edit Feature Settings dialog box.
5. Scroll the Pipe Network area of the dialog box and expand the Default Rules and Default Profile Label Placement sections.

   The Default Rules section sets the basic rules for pipes and structures. The Default Profile Label Placement section sets the initial styles for the labels of pipes and structures in a profile view.

6. Click the **OK** button to set the Render Material Style and to exit the Edit Feature Settings dialog box.

PIPE AND STRUCTURE CATALOG

The Pipe and Structure Catalog contains the basic pipe and structure shape and parameters. There is a simple, but poorly documented, interface to add or modify the entries of this file. The command Partbuilder starts an editor to add or modify catalog entries. The catalog is in the All Users path of Documents and Settings (see Figure 11.28). The Catalog is an HTML file located in the US Imperial Pipes and Structures folders: *US Imperial Pipes.htm* and *US Imperial Structures.htm*. You can view the catalog by double clicking on the HTML file (see Figure 11.7).

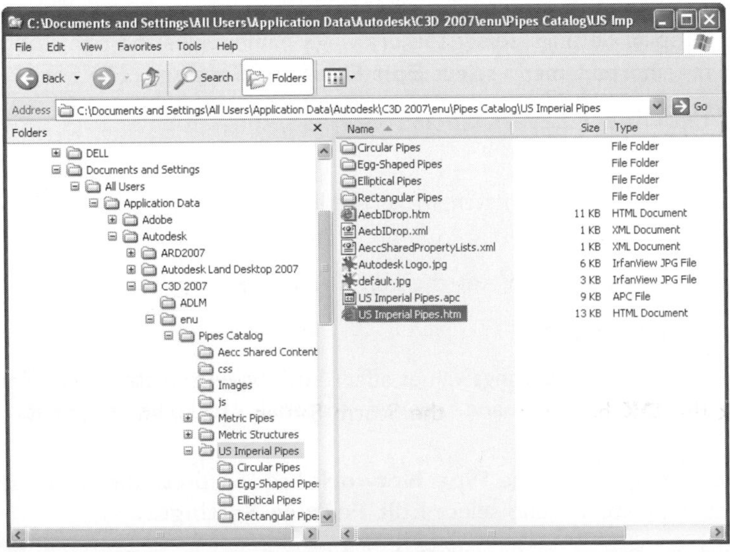

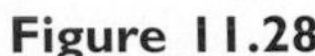
Figure 11.28

1. Start Windows Explorer and make **US Imperial Pipes** the current folder.
2. Double click the *US Imperial Pipes.htm* file.

   Internet Explorer starts and displays the US Imperial Pipes Catalog.
3. Expand the various branches to view the types of pipes and their values.
4. Exit the Catalog.
5. Change to the **US Imperial Structure** folder and double click the *US Imperial Structures.htm* file.

   Internet Explorer starts and displays the US Imperial Structure Catalog.
6. Expand the various branches to view the types of structures and their values.
7. Exit the Catalog.

PARTS LIST

The parts list contains the pipe and structure specs for a typical pipe network. If you are using a content template of Civil 3D, you will have two basic parts lists. You may have to add to the lists or change the part values to those you would use more often in your designs.

Storm Sewer Parts List

1. In Settings, expand Pipe Network until you are viewing a list under the Parts Lists heading.
2. From the list, select the **Storm Sewer**, press the right mouse button, and from the shortcut menu select **Edit...**.
3. Click the **Pipes** tab to view its contents and, if necessary, expand the Concrete Pipe list.

   The panel lists more types of the concrete pipes than you may use in any one project. But over several projects, the list may be complete. You could define a rule for each size and type of pipe. If you did define a rule for each pipe type and size, you will assign the rules here.

   The current style for drafting storm pipes is the double line (storm) style.
4. Click the **Structures** tab to view its contents.
5. If necessary, expand each part family to view its contents.

   Each structure has a descriptive name and has an appropriate object style (see Figure 11.11).
6. Click the **OK** button to exit the Storm Sewer Parts List.
7. Click the **AutoCAD Save** icon to save the drawing file.

EXERCISES

EXERCISES

Sanitary Parts List

1. In the Parts Lists in Pipe Network of Settings, select **Sanitary Sewer** from the list, press the right mouse button, and from the shortcut menu select **Edit....**
2. Click the **Pipes** tab to view its contents and, if necessary, expand the PVC Pipe list to view its contents.

   The panel lists more sizes of PVC pipes than you may use in any one project. But over several projects, the list may be complete. You could define a rule for each size of pipe. If you did define a rule for each pipe size, you will assign the rules here.
3. Click the **Structures** tab to view its contents and, if necessary, expand each cylindrical parts family to view its contents.
4. Click the **OK** button to exit the Sanitary Sewer Parts List.
5. Click the **AutoCAD Save** icon to save the drawing file.

PIPE OBJECT STYLES

1. Expand the Pipe branch until you are viewing a list of styles under the Pipe Styles heading.
2. From the list of styles, select **Single Line (Sanitary)**, press the right mouse button, and from the shortcut menu select **Edit....**
3. Click each tab to view the contents of its panel.
4. Click the **Display** tab to view the Component Visibility settings.

   This style draws the pipe's centerline as the line in the drawing.
5. Click the **OK** button to exit the dialog box.
6. Click **Double Line (Storm)**, press the right mouse button, and from the shortcut menu select **Edit....**
7. Click each tab to view the contents of its panel.
8. Click the **Display** tab to view the Component Visibility settings.

   The pipe's inside diameter draws the two lines.
9. Click the **OK** button to exit the dialog box.

PIPE RULES

1. In Settings, expand the Pipe branch until you are viewing Pipe Rule Set and its list of rules.
2. From the list, select **Basic**, press the right mouse button, and from the shortcut menu select **Edit....**
3. Click the Rules tab to view its content.
4. If necessary, expand each parameter section to view its contents.

In the Cover and Slope section are values for the minimum and maximum slope and cover for pipe. The Pipe Length section defines the shortest and longest pipe section you can have (see Figure 11.19).

5. Click the **OK** button to exit the dialog box.

PIPE LABEL STYLES

Civil 3D labels pipe segments as you draw them or after the design is set. In Unit 4 we will discuss, review, and use these styles to label pipe networks.

STRUCTURE STYLES

Structure Styles define how manholes, flared end sections, catchbasins, etc. of storm and sanitary sewer systems display in plan, profile, and sections. The parameters of the US Imperial Structure catalog define the profile and section shapes.

1. In Settings, expand the Structure branch until you are viewing a list of styles under the Structure Styles heading.
2. From the list of styles, select **Storm Sewer Manhole**, press the right mouse button, and from the shortcut menu select **Edit....**
3. Click the **Model** tab to view its contents.

   The catalog defines the shape, or you can choose a simpler shape.
4. Click the **Plan** tab to view its contents.

   This panel sets the symbol that appears at the structure's location. There are settings affecting the size of the symbol. You can use the actual size of the symbol, exaggerate its size, or make it resize itself each time the display area changes.
5. Click the **Profile** tab to view its contents.

   This panel sets how a structure displays in a profile view.
6. Click the **Display** tab to view the Component Visibility settings.

   This panel controls what components are visible for this structure in plan, profile, or section views.
7. Click the **OK** button to exit the dialog box.

STRUCTURE RULES

1. Expand the Settings tree for Structures until you are viewing the Structure Rule Set branch.
2. Click **Basic,** press the right mouse button, and from the shortcut menu select **Edit....**
3. Click the **Rules** tab to view its contents.
4. If necessary, expand each parameter section to view its contents.
5. Click the **OK** button to exit the dialog box.
6. Click the **AutoCAD Save** icon to save the drawing file.

STRUCTURE LABEL STYLES

Civil 3D can label structures as you draw them or after the design is set. In Unit 4 we will discuss, review, and use these styles on the pipe networks in the current drawing.

This completes the first exercise on Pipe Networks. Civil 3D uses an extensive library of styles, rules, and catalog values to create a pipe network. The next unit creates new pipe networks in the current drawing.

SUMMARY

- Edit Drawing Settings sets the default layers, modifiers, and their values for pipes, structures, and pipe networks.
- The pipe network Edit Feature Settings values define the initial styles, naming format, and labeling values for pipes and structures.
- A catalog provides the basic shapes and specs for pipes and structures.
- A parts list uses the values found in a catalog, but is generally a subset of all possible parts.
- A sanitary pipe has rules affecting slope, cover, and length of pipe segments.
- Civil 3D labels pipes and structures as or after they are created.

UNIT 2: CREATING A PIPE NETWORK

After creating the necessary styles, parts lists, and rules, the next step is creating the pipe networks. The Create Pipe Network command uses a Layout toolbar to create a network and its branches (see Figure 11.2).

The Network Layout Tools toolbar displays the current parts list (storm, sewer, etc.), surface, and alignment (if referenced). The middle portion of the Network Layout Tools toolbar lists the current structure type and pipe size. To the right of the pipe size icons are the drafting mode icons: Structure and Pipes, Pipes only, and Structures only. The Structures Only mode allows the user to locate structures first and then to connect them with pipes later. The user can also draft networks up or down slope. All pipes and structures drafted while in the Create Pipe Network command are members of the current network. If the user exits the command and returns to the command line, her or she must edit the network to add new branches, pipes, or structures.

Civil 3D assigns names for the pipes and structures from the name template in the pipe network's Edit Feature Settings. The user may find it easier to let Civil 3D name the pipes and structure, and then edit their names in the Properities dialog box him- or herself. If the user edits in the Edit Network vista, and if the network is large, the vista will be confusing, showing all of the values in an overly large panorama. Without careful tracking in such a large vista, it is easy to make mistakes.

EXERCISE 11-2

When you complete this exercise, you will:

- Be able to draft a pipe network.
- Be able to set structures and add piping connecting the structures.

EXERCISE SETUP

The exercise continues with the drawing from the first unit of this Chapter. If you did not complete the first exercise, you can open the *Chapter 11 – Unit 2.dwg* file. This file is in the *Chapter 11* folder of the CD that accompanies this book.

1. Open the drawing from the previous exercise or the *Chapter 11 – Unit 2* drawing file.

EXISTING STORM (93RD STREET)

There is an existing storm sewer line near the centerline of 93rd Street (see Figure 11.29). Table 11.1 contains the values for the names and types of the structures for this network. The network starts at the eastern end of 93rd street and should be drawn upslope.

Table 11.1

Manhole #	Type	Material
EX-Stm-01	Concentric 72"	Reinforced Concrete
EX-Stm-02	Concentric 72"	Reinforced Concrete
EX-Stm-03	Concentric 72"	Reinforced Concrete
EX-Stm-04	Concentric 48"	Reinforced Concrete

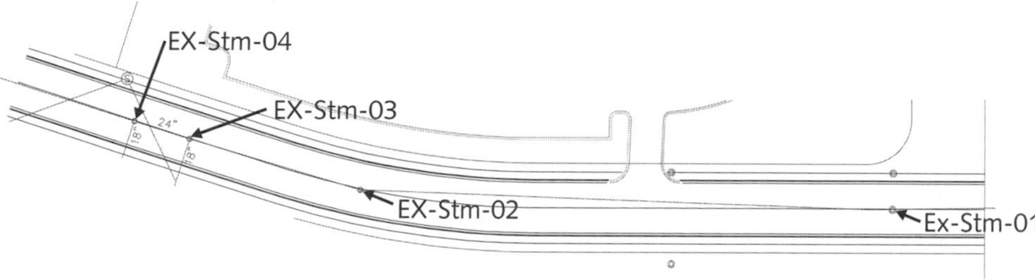

Figure 11.29

This pipe network has several different pipe sizes. The pipe size from EX-Stm-03 to EX-Stm-01 is 36". The pipe between EX-Stm-04 and EX-Stm-03 is 24". Both EX-Stm-04 and EX-Stm-03 have 18" laterals feeding into their barrels. Drawing this pipe network will require changing the drafting mode and pipe sizes.

1. Use the AutoCAD Zoom and Pan commands to view the easterly end and the first manhole of the EX-Stm pipe network.

EXERCISES

The 93rd Street storm sewer uses a special rule, Storm Interceptor. This rule allows the network to be deeper than other storm systems (i.e. parking lot catch basins, etc.). You must set the rule before defining the next network. Otherwise, you have to apply a rule to each existing pipe.

2. If necessary, click the **Settings** tab to make it current.
3. In Settings, select the **Pipe Network** heading, press the right mouse button, and from the shortcut menu, select **Edit Feature Settings...**.
4. If necessary, expand the Default Rules section; for Pipe Default Rules, click in the value cell to display an ellipsis, and click the ellipsis to display the Pipe Default Rules dialog box.
5. In the Pipe Default Rules dialog box, click the drop-list arrow, from the list, select **Storm Interceptor**, and click the **OK** buttons to exit.
6. Click the **Prospector** tab to make it current.
7. From the Pipes menu, select **Create by Layout...**.

   The Create Pipe Network dialog box displays. Use Figure 11.30 as a guide to fill out the dialog box. In the pipe network Edit Feature Settings dialog box, set the default layers to C-STRM. For this pipe network, you do not need to change the layer list.

8. In the Create Pipe Network dialog box, enter **EX-Stm** for the pipe network name, keeping the counter. For the Network parts list select the **Storm Sewer**, set the Surface to **Proposed,** set the Alignment name to **93rd – (1)**, and click the **OK** button to exit the dialog box.

   The Network Layout Tools toolbar appears. Along the toolbar's bottom, Parts List should be listed as Storm Sewer, the Surface should be listed as Proposed, and the Alignment should be listed as 93rd - (1). Use Figure 11.29 as a guide for drafting this network.

9. In the center of the toolbar, click the **Structure** list drop-list arrow, and select from the Concentric Cylindrical Structure parts list the **72" dia Concentric Structure**.
10. In the toolbar, just to the right of center, click the **Pipe Size** drop-list arrow, and select from the list **36 Inch RCP**.
11. In the toolbar, click the **Slope** icon (second to the right of the pipe size) and set the design to **Upslope**.
12. In the toolbar, click the drop-list arrow to the right of the current Drafting Mode (to the right of pipe size), and select from the list **Pipes and Structures**.
13. In the drawing, use the Center object to place the eastern most manhole (EX-Stm-01).
14. In the drawing, also using the Center object snap, locate the EX-Stm-02 and EX-Stm-03 structures.

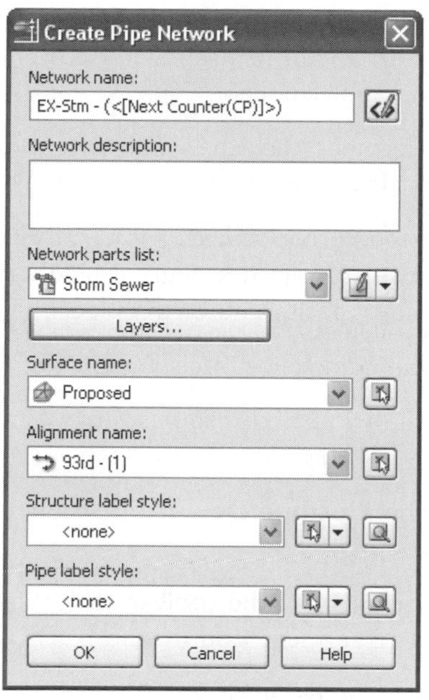

Figure 11.30

15. In the center of the toolbar, click the **Structure** list drop-list arrow, and select from the Concentric Cylindrical Structure parts list **48" dia Concentric Structure**.
16. In the toolbar, click the **Pipe Sizes** drop-list arrow, and from the pipe size list, select **24 Inch RCP**.

    Changing pipe sizes does not interrupt the drafting process.
17. In the drawing, using the Center object snap, select the location of EX-Stm-04.
18. In the toolbar, click the **Pipe Sizes** drop-list arrow and from the pipe size list, select **18 Inch RCP**.
19. In the toolbar, click the **Drafting Mode** drop-list arrow and from the drafting mode list, select **Pipes Only**.

    This breaks the connection to EX-Stm-04. To correctly draft the pipe, you need to reconnect the 18" pipe to EX-Stm-04 before drafting the pipe.
20. In the drawing, place the cursor near EX-Stm-04; a structure connection icon appears. Click near the EX-Stm-04 structure to start drafting the pipe.
21. In the drawing, click a point to the west of the structure as the pipe end.
22. To indicate that you are drafting a new pipe segment connecting to a different structure, click the **Pipe Only** drafting mode icon.

23. In the drawing, place the cursor near EX-Stm-04; the attach structure icon appears. Select a point near the structure and draw an 18" pipe EX-Stm-04 towards the south past the sidewalk polyline line.
24. To start a new pipe segment, click the **Pipe Only** drafting mode icon to place a new 18" pipe segment from EX-Stm-03 towards the south.
25. In the drawing, place the cursor at EX-Stm-03; a structure connection icon appears. Click near the structure to start drawing the pipe and select a second point south of the structure as the opposite end of the pipe.
26. Use the AutoCAD Pan command to view the EX-Stm-01 structure.
27. In the toolbar, click the **Pipe Size** drop-list arrow, and select from the list **36 Inch RCP**.
28. In the toolbar, click the **Upslope/Downslope** icon and set the mode to **Downslope**.
29. To draft a new pipe segment, in the toolbar click the **Pipe Only** drafting mode icon to place a new 36" pipe segment from EX-Stm-01 towards the east.
30. In the drawing, place the cursor at EX-Stm-01; a structure connection icon appears. Click near the structure to start drawing the pipe, and select a second point east of the structure (near the end of the corridor's centerline) as the opposite end of the pipe.
31. Press the **Enter** key to exit the Network Layout Tools toolbar.
32. Click the **AutoCAD Save** icon to save the drawing.

    This simplest way to match names to structures and pipes is by editing the properties of each structure and pipe, rather than trying to use a vista.

33. In the drawing, select each structure and edit its properties, changing each structure's name using the entries in Table 11-1.
34. In the drawing, select each pipe and edit its properties, changing each pipe's name using the entries in Table 11-2.
35. Click the **AutoCAD Save** icon to save the drawing.

Table 11.2

Pipe Name	From/to Manholes	Size
EX-Stm-P01	East end of 93rd to EX-Stm-01	36
EX-Stm-P02	EX-Stm-01 to EX-Stm-02	36
EX-Stm-P03	EX-Stm-02 to EX-Stm-03	36
EX-Stm-P04	EX-Stm-03 to EX-Stm-04	24
EX-Stm-P05	EX-Stm-04 to West end of 93rd	18

Table 11.2 (continued)

Pipe Name	From/to Manholes	Size
EX-Stm-P06	EX-Stm-03 to south of sidewalk	18
EX-Stm-P07	EX-Stm-04 to south of sidewalk	18

CATCHBASIN NETWORK

The last network is a series of catchbasins and pipes with an outfall FES to a pond at the southeast corner of the site. The strategy for drafting this network is first placing an Eccentric 48" structure at the southeast corner of the parking lot next to the detention pond. After placing the manhole, the next step is locating the three catchbasins in the western part of the parking lot. After locating the three western catchbasins, locate a catchbasin in the northeast parking lot. All of these catchbasins connect to the manhole in the southeast. Finally, connect a pipe to the manhole and locate the FES outside of the parking lot in the detention pond (see Figure 11.31 for the placement of the various elements of the catchbasin network).

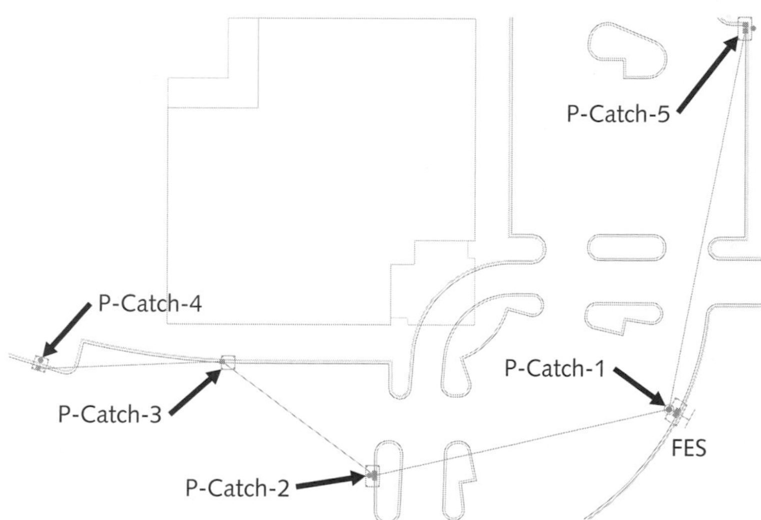

Figure 11.31

Table 11.3 contains the structure names, types, and pipe sizes.

Table 11.3

Structure	Type	Pipe Size	Pipe Name	Material
P-Catch-1	48"x48" Rectangular	12" Concrete	P-Catch-P1	Reinforced Concrete
P-Catch-2	24"x24" Rectangular	12" Concrete	P-Catch-P2	Reinforced Concrete
P-Catch-3	18"x18" Rectangular	12" Concrete	P-Catch-P3	Reinforced Concrete
P-Catch-4	15"x15" Rectangular	12" Concrete	P-Catch-P4	Reinforced Concrete

Table 11.3 (continued)

Structure	Type	Pipe Size	Pipe Name	Material
P-Catch-5	24"x24" Rectangular	12" Concrete	P-Catch-P5	Reinforced Concrete
FES	44"x44" Rectangular Headwall	24" Concrete	P-Catch-P6	Reinforced Concrete

All of the rectangular structures are not in the current parts list for storm sewer. The parts need to be added to the storm sewer parts list.

ADDING A PART FAMILY AND PART TO THE STORM SEWER PARTS LIST

1. Click the **Settings** tab to make it current.
2. In Settings, expand the Pipe Network branch until you are viewing a list under the Parts Lists heading.
3. Click the **Storm Sewer** parts list, press the right mouse button, and select **Edit...** from the shortcut menu.
4. In the Network Parts List - Storm Sewer dialog box, click the **Structures** tab to view its contents.
5. In the Structures tab, select the Storm Sewer heading at the top left, press the right mouse button, and select **Add Part Family...** from the shortcut menu.
6. In the Part Catalog dialog box, toggle on Rectangular Structure Slab Top Rectangular Frame, and click the **OK** button to return to the Network Parts List dialog box.
7. From the list of parts, select **Rectangular Structure Slab Top Rectangular Frame**, press the right mouse button, and select **Add Part Size...** from the shortcut menu.
8. If necessary, change the Inner Structure Width and Length to **15"** and click the **OK** button to create the part. Use Figure 11.32 as a guide.
9. Repeat the previous two steps and add the following sizes: **18x18**, **24x24**, and **48x48**.

   The style for the new structures is storm sewer manhole.

10. In the Part List dialog box, click the **Style** icon to the right of Rectangular Structure Slab Top Rectangular Frame to display the Structure Style dialog box; click the drop-list arrow, from the style list select **Catch Basin**, and click the **OK** button to return to the Parts List dialog box.
11. Click the **OK** button to exit the Storm Sewer Parts List dialog box.

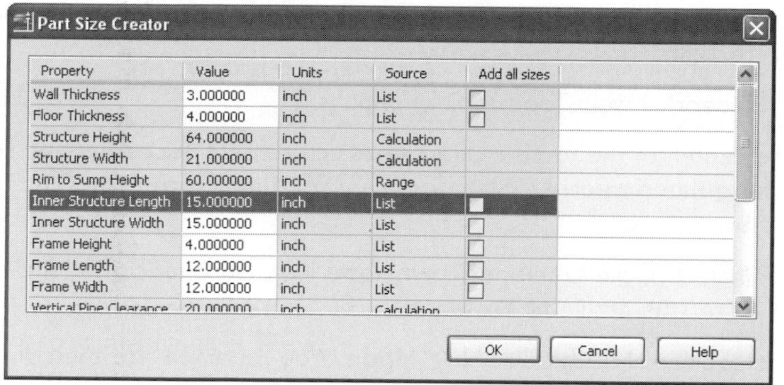

Figure 11.32

DRAFTING THE CATCH BASIN NETWORK

1. Use the AutoCAD Zoom and Pan commands to view the entire parking lot.
2. In the Layer Properties Manager, toggle on the layer PROP-STM-LIN and click the **OK** button to exit.
3. From the Pipes menu, select the **Create by Layout...**.
4. In the Create Pipe Network dialog box, set the name to **P-Catch**, leaving the counter, set the Network parts list to **Storm Sewer**, and set the surface to **Proposed**. Use Figure 11.33 as a guide.

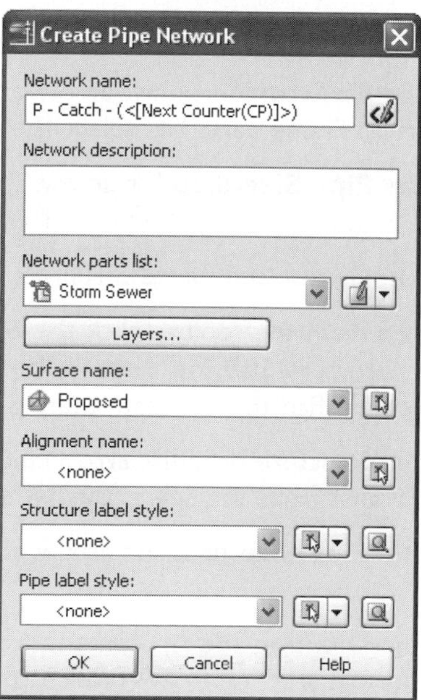

Figure 11.33

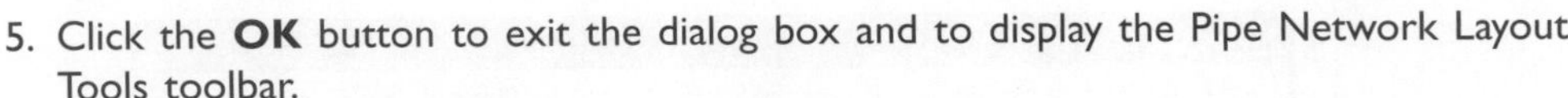

5. Click the **OK** button to exit the dialog box and to display the Pipe Network Layout Tools toolbar.

   The Network Layout Tools toolbar displays.

6. In the center of the toolbar, click the Structure drop list arrow, and select from the **Rectangular Structure Slab Top Rectangular Frame** parts list the **48"x48" Rect Structure**.

7. In toolbar, click the **Upslope/Downslope** icon (second icon to the right of the pipe size) to set the design to **Upslope**.
8. In the toolbar, click the **Drafting Mode** drop-list arrow (to the right of pipe size), and select **Structures Only** from the list.
9. In the drawing, place the Structure at the P-Catch -1 location at the southeast corner of the site. Use Figure 11.31 as a guide for placing the catchbasins.
10. In the toolbar, click the **Structure** drop-list arrow, and from the Rectangular Structure Slab Top Rectangular Frame parts list, select the **24"x24"Rect Structure.**
11. In the toolbar, click the **Pipe Size** drop-list arrow, and from the list of sizes, select **21 inch RCP**.
12. In the toolbar, click the **Drafting Mode** drop-list arrow (to the right of pipe size), and select **Pipes and Structures** from the list.
13. In the drawing, place the cursor near the structure just placed; a connect to structure icon appears. Select the point to attach a pipe.
14. Pan to the west and locate structure P-Catch - 2.
15. In the toolbar, click the **Structure** drop-list arrow, and from the Rectangular Structure Slab Top Rectangular Frame parts list, select the **18"x18" Rect Structure**.
16. In the toolbar, click the **Pipe Size** drop-list arrow, and from the list of sizes, select **15 inch RCP**
17. In the drawing, pan to the west and locate Structure P-Catch - 3.
18. After placing the structure, in the toolbar click the **Structure** drop-list arrow, and from the Rectangular Structure Slab Top Rectangular Frame parts list, select the **15"x15" Rect Structure**. Pan to the west, locating structure P-Catch-4.
19. In the toolbar, click the Structure drop-list arrow, and from the Rectangular Structure Slab Top Rectangular Frame parts list, select the **18"x18" Rect Structure**.
20. In the toolbar, click the **Pipe Size** drop-list arrow, and from the list of sizes, select **21 inch RCP**.
21. In the toolbar, click the **Drafting Mode** icon and from the list, select **Pipes and Structures** to indicate the start of a new branch to the network.
22. If necessary, use the AutoCAD Pan and Zoom to view the P–Catch - 1 structure.

23. In the drawing, place the cursor near P–Catch – 1, and when a connect to structure icon appears, select the point.
24. Use the AutoCAD Pan command to pan north and locate the structure P–Catch - 5.
25. In the toolbar, click the **Structure** drop-list arrow, and from the list in the Concrete Rectangular Headwall part family, select **Headwall 44"x 6"x 44"**.
26. In the toolbar, click the **Pipe Size** drop-list arrow, and from the list of sizes, select **30 inch RCP**.
27. In the toolbar, click the **Downslope/Upslope** toggle and set the design to **Downslope**.
28. In the toolbar, click the **Drafting Mode** icon and from the list select **Pipes and Structures** to start a new branch.
29. In the drawing, place the cursor at P–Catch - 1; when the connect to structure icon appears, select the point and select a second point to the southeast in the detention pond to place the FES.
30. Press the **Enter** key to exit the Create Network routine.
31. Click the **AutoCAD Save** icon to save the drawing.

    This completes the drafting of Pipe Networks in Civil 3D.

EXERCISES

SUMMARY

- A pipe network uses a parts list to determine what structures and pipe sizes are a part of a typical network.
- Adding a branch to an existing pipe network is an edit of an existing network.
- You cannot connect a new pipe network to an existing pipe network.
- The current drafting mode remains active even when changing the structure and/or pipe size.

UNIT 3: PIPE NETWORKS REVIEW AND EDITING

Besides adding branches to an existing pipe network, there are other situations that require edits to a pipe or structure (e.g. changing the rules, resizing a pipe or structure, swapping structures or pipes, changing styles, etc.). Other editing methods are graphically moving, disconnecting and reconnecting elements, and deleting structures and pipes. The results of an interference check may require edits to a pipe network and its components.

When selecting either a pipe segment or a structure and pressing the right mouse button, the user gets a context sensitive shortcut menu (see Figure 11.4). The shortcut menu on the left appears when the user selects a pipe segment. The shortcut menu on the right appears when he or she selects a structure. Both menus can call the Edit Network... command, which is the Pipe Network Layout toolbar.

The Report Manager contains several pipe network reports: Pipes and Structures in html and CSV formats.

NETWORK PROPERTIES

The properties of a pipe network include the original layout settings (see Figure 11.34), what labeling styles annotate the pipe network in a profile (see Figure 11.35), the layers the network uses in section views, and a review of the network statistics.

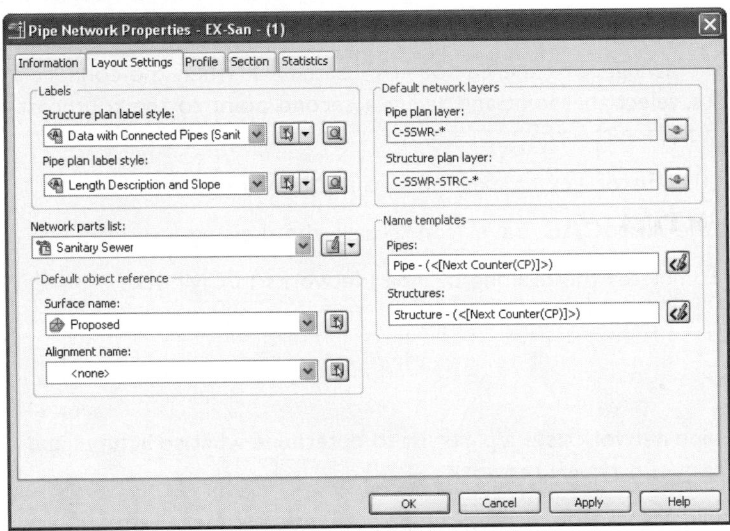

Figure 11.34

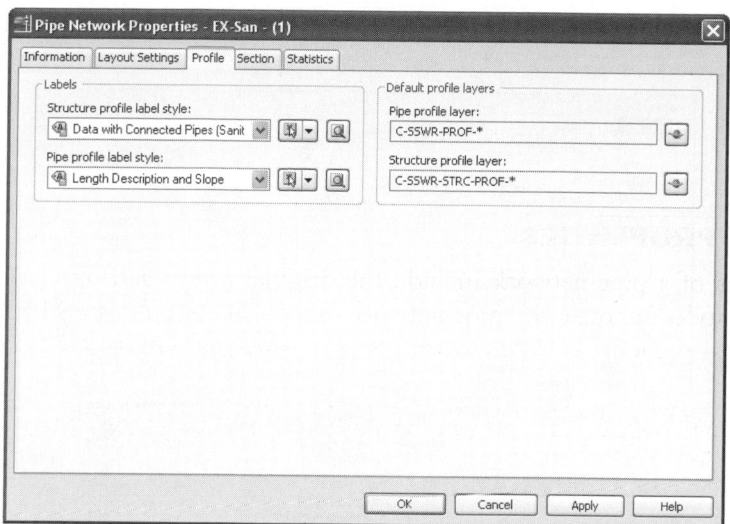

Figure 11.35

PIPE PROPERTIES

When selecting a pipe segment and Pipe Properties... from a shortcut menu, a Pipe Properties dialog box displays with pertinent pipe segment values. The Information tab allows the user to rename and reassign an object style or render material. The Part Properties panel displays all of the information of the part including calculated values (see Figure 11.36). Any value in black print is editable. The Rules panel displays the current rules and how the segment compares to the rule values (see Figure 11.37). If a rule is not met, there is an icon next to the broken rule.

The pipe properties edits are only for the type of pipe selected. The edits do not allow changing pipe types. The best place to change pipe types is in Edit Pipe Networks. In the Network Layout Tools toolbar are commands that erase network elements, provide the ability to switch between parts lists, and provide the ability to draft only pipe segments.

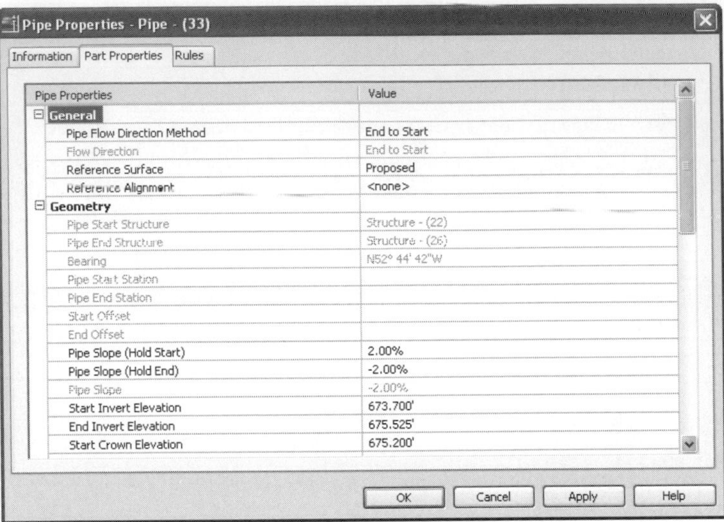

Figure 11.36

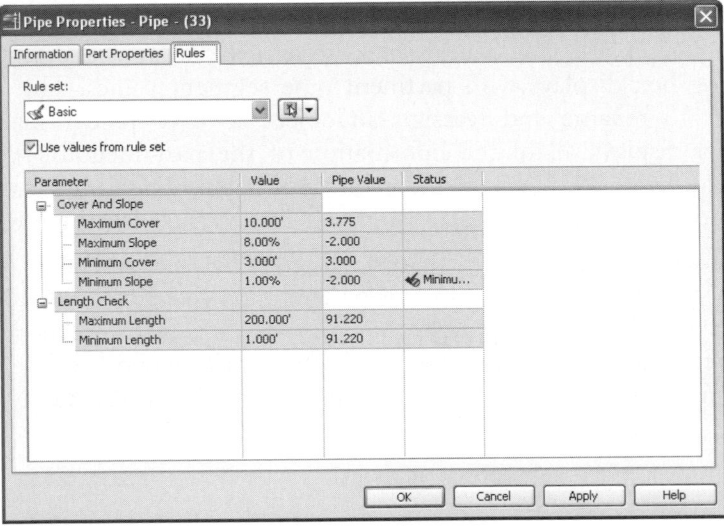

Figure 11.37

STRUCTURE PROPERTIES

The Structure Properties dialog box shows a structure's values. The Information tab allows the user to rename or reassign the object style or render material. The Part Properties panel displays information about the part and any calculated values (see Figure 11.38). Any value in black print is editable.

The Connected Pipes panel displays the pipes connected to the structure, their rules violations, their inverts, and other critical information (see Figure 11.39).

The Rules panel displays the current rules and how the structure compares to their values (see Figure 11.40). If a rule is not met, there is an icon next to the broken rule.

The Structure Properties edits are only for the structure type and they do not allow for structure type swapping. The best place to change the structure types is with the Network Layout Tools toolbar. This toolbar's commands allow the user to erase network elements, provide the ability to switch between parts lists, and provide the ability to draft only structures.

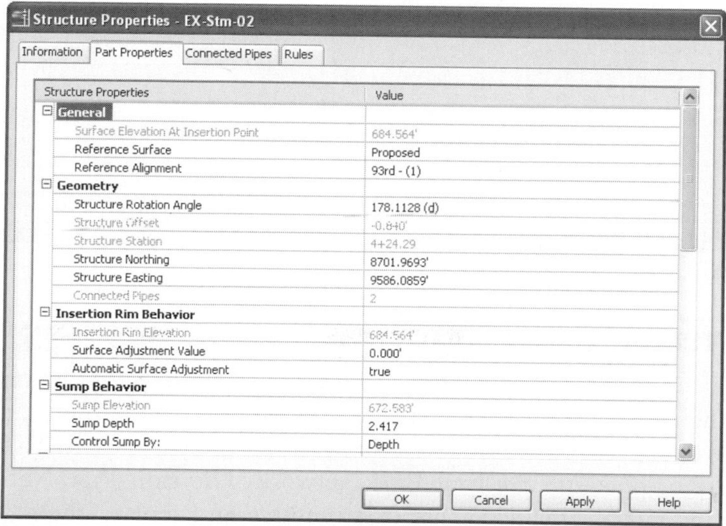

Figure 11.38

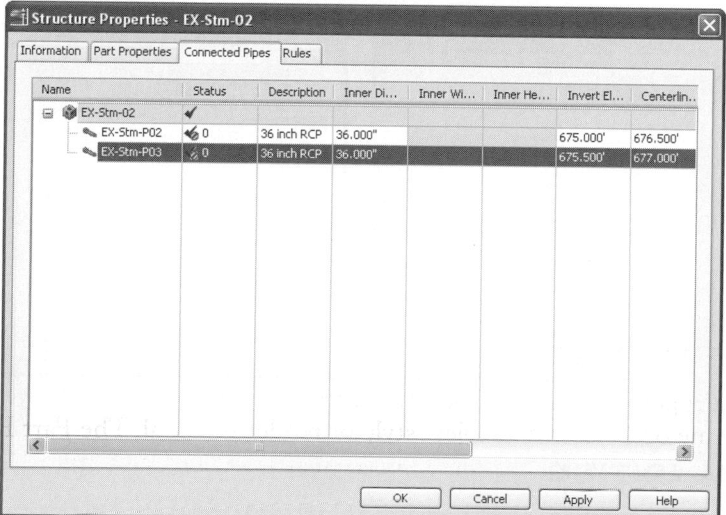

Figure 11.39

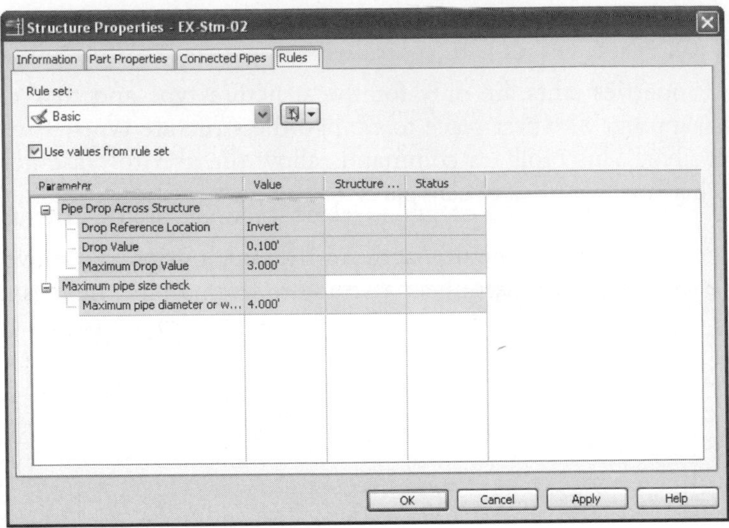

Figure 11.40

SWAP PART: PIPE AND STRUCTURE

The Swap Part command exchanges the currently selected object with another from a list of applicable choices (see Figure 11.41). The user cannot change the type of structure or type of pipe; he or she can only change the sizes within the part family. To change to a new type of structure or pipe, use the Edit Pipe Network. The Edit Pipe Network allows the user to delete individual pipe segments or structures and replace them with parts from different families.

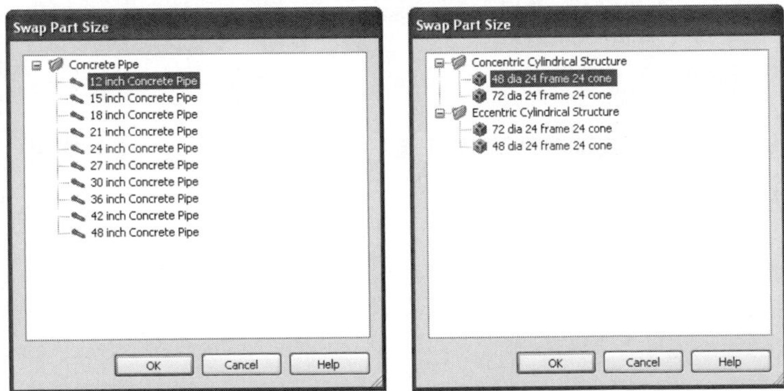

Figure 11.41

GRAPHICAL EDITING

When graphically editing a pipe network, Civil 3D treats the pipes and structures as connected objects. Selecting a structure, activating its grip, and relocating it, the structure drags the attached pipes to its new location.

When the user is grip editing pipes and also wants to move the structure between them (see Figure 11.42), he or she must include the structure in the selection. If the structure is not selected, moving the pipes detaches the pipes from the structure. To attach pipes to a structure, relocate the structure to the vicinity of the pipes and individually reconnect the pipes. Reattach pipes to the structure by clicking on the pipe, pressing the right mouse button, and selecting the connect to part from the shortcut menu. The routine prompts for a connecting structure and in the drawing, select the appropriate structure.

To disconnect a pipe from a structure, activate the pipe's grips and move it away from a structure. If the user wants to disconnect a structure from a pipe, he or she must click the structure, press the right mouse button, select disconnect from part, and select a pipe. If there is more than one pipe, disconnect each pipe and then move the structure, or delete the structure, place a new structure in the desired location, and reconnect the pipes.

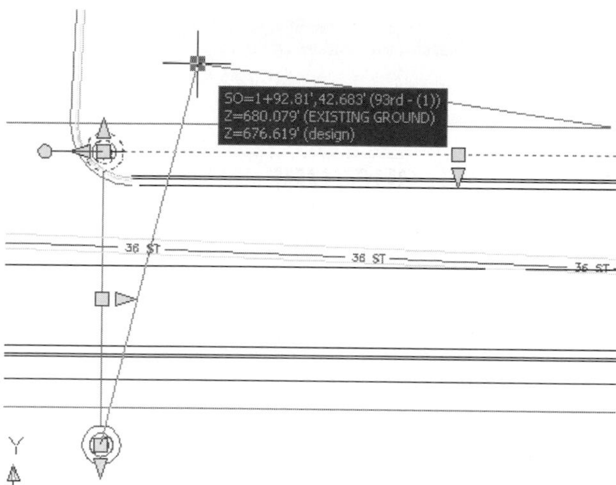

Figure 11.42

CONNECT AND DISCONNECT PART

The Connect and Disconnect Part command allows the user to separate a pipe or structure from an adjacent pipe network object. This is useful when inserting new structures and pipes in existing networks or when graphically editing network components.

PROFILE EDITING

Civil 3D allows the user to edit pipes and structures in a profile view. The pipes and structures that appear in a profile can represent a selection set of or an entire pipe network.

INTERFERENCE CHECKING

When pipes cross there is potential for them to intersect or not have enough vertical separation. The Interference Check routine checks the vertical separation between pipes of different networks or selected pipe segments (see Figure 11.43). The check is a user-specified distance or scale factor and is set by selecting the 3D proximity check criteria button of the Create Interference Check dialog box.

In the Criteria dialog box, set the distance or scale factor for the interference check (see Figure 11.44).

If the user edits any of the pipe networks that participate in an interference check, the interference will show an out-of-date icon. Recalculate the interferences by clicking on the interference and selecting Rerun interference check... from the shortcut menu.

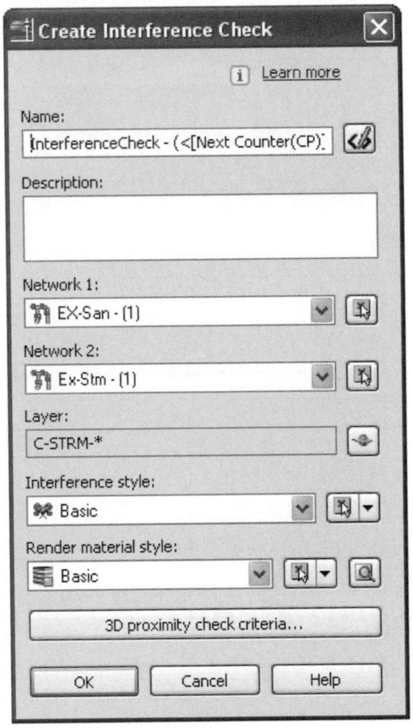

Figure 11.43

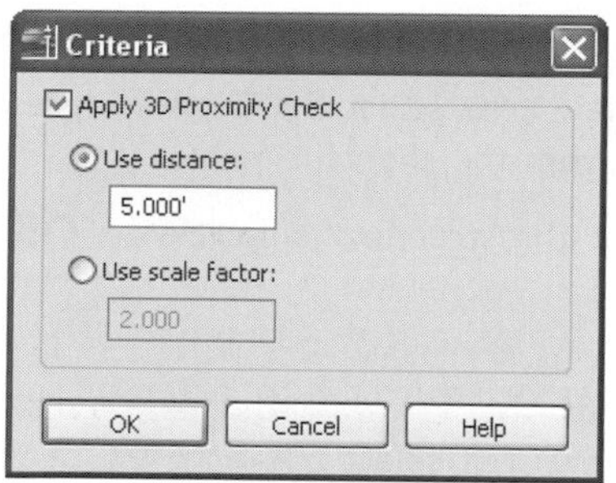

Figure 11.44

EXERCISE 11-3

When you complete this exercise, you will be:

- Able to review and edit the properties of a pipe network.
- Able to graphically edit the location of pipe network segments.
- Familiar with and able to edit the properties of a pipe or structure.
- Familiar with and able to swap out pipes or structures.

EXERCISE SETUP

This exercise continues with the drawing from the previous exercise. If you did not complete the exercise, you can open the *Chapter 11 – Unit 3.dwg* as the file for this exercise. The drawing file is in the *Chapter 11* folder of the CD that accompanies this book.

1. Open the drawing from the previous exercise or the *Chapter 11 – Unit 3* drawing.

PIPE NETWORK PROPERTIES

The Pipe Network Properties dialog box lists the settings active when laying out the network, the profile label styles and layers, section layers, and network statistics.

1. Use the AutoCAD Zoom and Pan commands until you are viewing the EX-Stm - (1) pipe network along 93rd Street.
2. In the drawing, click a pipe or structure, press the right mouse button, and from the shortcut menu select **Network Properties...**.
3. Click the tabs to view the various settings and values in the dialog box.
4. Click the **OK** button to exit the Pipe Network Properties dialog box.

STRUCTURE PROPERTIES

The Structure Properties dialog box displays information about a structure. You can change the type of structure, its name, and many additional values.

1. If necessary, click the **Prospector** tab to make it current.
2. In Prospector, expand the Pipe Networks branch until you are viewing the EX-Stm – (1) pipe network.

3. Click the **Structures** heading to view the list of structures in the preview panel.
4. In Prospector's preview area, select **EX-Stm-01**, press the right mouse button, and select **Properties...**from the shortcut menu,
5. Review the values for the structure and click the **OK** button to exit the dialog box.

SWAP PARTS

Civil 3D allows you to swap pipes or structures to larger or smaller sizes.

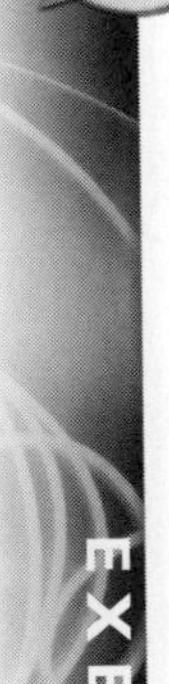

1. In Prospector's preview area, select **EX-Stm-01**, press the right mouse button, and from the shortcut menu, select **Zoom to**.
2. In the drawing at the eastern side of EX-Stm-01, select the pipe (EX-Stm-P1), press the right mouse button, and from the shortcut menu, select **Swap Part...**.

   This displays a pipe list.
3. Select a **48 inch RCP** from the list, and click the **OK** button to exit the dialog box.
4. Click the **AutoCAD Save** icon and save the drawing.

PROFILE REVIEW OF EX-STM - (1)

Civil 3D places part(s) or all of a pipe network into a profile view. The 93rd Street storm system needs to be reviewed in profile.

1. From the Pipes menu, select **Draw Parts in Profile View**.

   The routine prompts for a selection.
2. In the drawing, select a pipe from the EX-Stm-1 network and press the **Enter** key. The prompt changes to Selecting a Profile View.
3. Use the Pan command until you are viewing the profile view and select the **93rd Street Profile View**.

   The EX-Stm-1 network draws in the profile view.

EDIT PIPES IN PROFILE

Using Table 11.4 as a guide, edit each pipe's start and end inverts in the Pipe Properties dialog box.

Table 11.4

Pipe	Start Invert	End Invert
EX-Stm-P01	669.00	668.00
EX-Stm-P02	669.50	675.00
EX-Stm-P03	675.50	679.00
EX-Stm-P04	680.00	681.00
EX-Stm-P05	681.25	682.00

Table 11.4 (continued)

Pipe	Start Invert	End Invert
EX-Stm-P06	681.50	684.00
EX-Stm-P07	681.25	682.00

1. In the 93rd Street Profile view, select the pipe **EX-Stm-P01**, press the right mouse button, and from the shortcut menu select **Pipe Properties....**
2. In the Pipe Properties dialog box, click the **Part Properties** tab, scroll down to the Start and End Invert entries, and using the values in Table 4, change the inverts for pipe EX-Stm-P01.
3. Repeat the previous two steps and edit the inverts for the EX-Stm-1 pipe network (EX-Stm-P02...P5).
4. In the drawing, select any entity representing EX-Stm-1, press the right mouse button, and select from the shortcut menu **Edit Network....**
5. In the Network Layout Tools toolbar, click the **Pipe Network Vistas** icon (right side).
6. Click the **Pipes** tab and edit the start and end inverts for EX-Stm-P06 and EX-Stm-P07.
7. After editing the inverts, close the vista and Network Layout Tools toolbar.
8. After making the edits, click the **AutoCAD Save** icon to save the drawing.

EXERCISES

INTERSECTION BETWEEN EX-STM - (1) AND EX-SAN - (1)

The EX-San – (1) pipe network crosses the EX-Stm – (1) pipe network. The intersection needs to be checked for interferences.

1. Use the AutoCAD Pan and Zoom commands to view the intersection of EX-San – (1) and EX-Stm – (1).
2. If necessary, click the **Prospector** tab to make it current.
3. Expand the Pipe Networks branch until you are viewing the Interference Checks heading.
4. Select **Interference Checks**, press the right mouse button, and from the shortcut menu select **New....**
5. In the command line, the prompt is for an element from the first network; in the drawing, select a pipe from the EX-San – (1) network.
6. In the command line, the prompt is for an element from the second network; in the drawing, select a pipe from the EX-Stm – (1) network.
7. In the Create Interference Check dialog box, click the **3D Proximity Check Criteria** button.

8. In the Criteria dialog box, toggle on Apply 3D Proximity Check, set the distance to 5.0, and click the **OK** button to return to the Create Interference Check dialog box.
9. Click the **OK** button to exit the Create Interference Check.

   The routine finds three interferences.

VIEWING A PIPE NETWORK

Civil 3D creates a model and the 3D Orbit or Object Viewer commands produce a three dimensional view of the selected components.

1. In the drawing, select the interference symbols, pipes, and structures of the two pipe networks, press the right mouse button, and from the shortcut menu select **Object Viewer....**
2. Rotate the model to view the network objects and interferences.
3. Close to exit the Object Viewer.
4. Use the AutoCAD Zoom and Pan commands to view the P-Catch network in the parking lot.
5. In the drawing, select all of the P-Catch elements, press the right mouse button, and from the shortcut menu select **Object Viewer....**
6. Exit the Object Viewer.
7. Click the **AutoCAD Save** icon to save the current drawing.

GRAPHICAL EDITING

Graphically editing a pipe network allows a user to relocate sections or portions of a pipe network. Pipes and structures can be disconnected and reconnected by clicking the network element and selecting Reconnect or Disconnect from the shortcut menu.

1. Use the AutoCAD Zoom and Pan commands to view the western end of the 93rd Street existing Storm and San – (1) pipe networks.
2. In the drawing, select the EX-San – (1) structure and, using its grips, relocate the structure and its pipes.
3. Clear the grips by pressing the **Esc** key.
4. In the drawing, select one of the pipes connected to the EX-San – (1) structure, click the grip nearest the structure, and move the pipe end to a new location.

   The pipe disconnects from the structure.

5. In the drawing, select the pipe just disconnected from the structure, activate the northern grip, move it to the structure, and when the connect to structure icon appears, connect it to the structure.
6. Close the drawing and do not save the changes.
7. Reopen the drawing.

This ends the review of editing of a pipe network. The next unit reviews annotating a network's pipes and structures.

SUMMARY

- To add branches to an existing network, you must edit the network.
- You can not attach a pipe from one network to a structure of a second network.
- The start invert is lowest when drafting upslope.
- The start invert is the highest when drafting downslope.
- When attaching a pipe to a structure, the drafting routine displays a star icon.
- When attaching a pipe to a pipe, the pipe routine displays a break at connection icon.
- If moving structure with pipes, you need only to select the structure. The pipes will relocate to the new structure location.

UNIT 4: PIPE LABELS

The user can label pipes at the same time as he or she drafts a network. The Create Pipe Network dialog box contains settings for the current structure and pipe label styles (see Figures 11.30 and 11.33). The pipe network labeling routine uses the same interface as the surface and parcel labeling routines. However, the label styles are specific to the pipes and structures (see Figure 11.45). The labeling type includes the entire network in plan or profile or single part in plan or profile.

By default pipe labels have plan readability turned off. Structure labels use the plan readability attribute.

All labels appear on the pipe or structure as defined by the style. When labeling single pipes or structures, the routine labels only one object. Reselect the Add button to label a second structure or pipe.

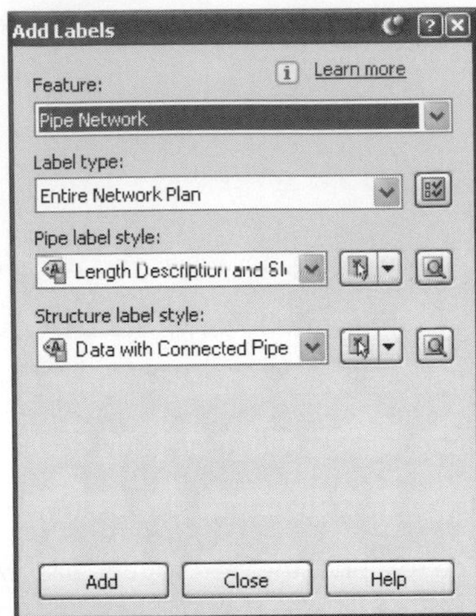

Figure 11.45

STRUCTURE LABEL STYLES

A structure label annotates the specifics of a structure's parameters. A structure label can include the name of the structure, its rim and sump elevation, in and out inverts, clearances, and many more specifics (see Figure 11.46).

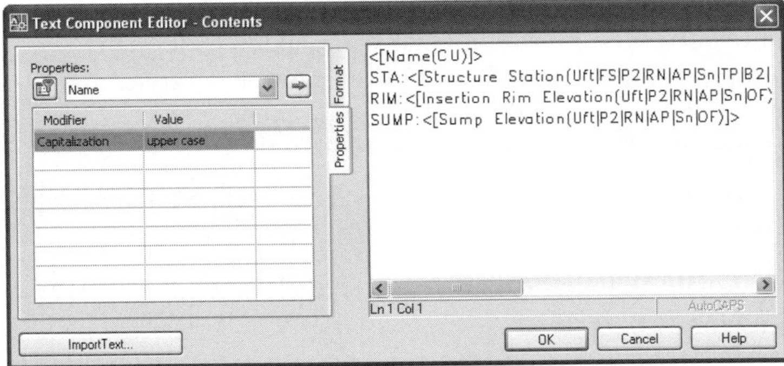

Figure 11.46

PIPE LABEL STYLES

A pipe label annotates specifics of a pipe's parameters. A pipe label can include the name of the pipe, elevation for starting and ending inverts, beginning and ending structure name, cover, and many more specifics (see Figure 11.47).

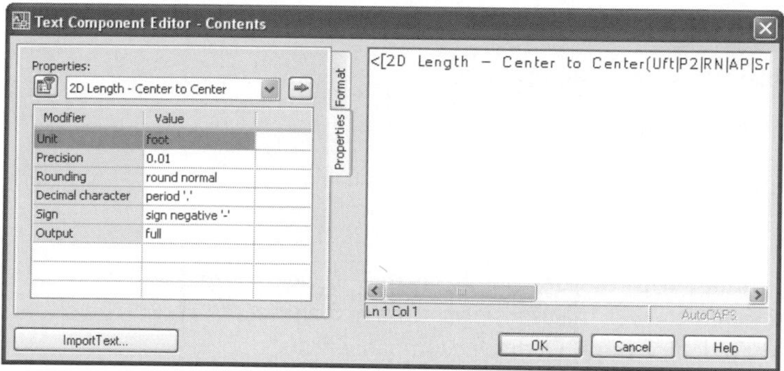

Figure 11.47

PIPES IN SECTIONS

Currently it is difficult to add new data to existing sample line groups. If there is an existing sample line group, it is best just to create a new sample line group that includes the pipes.

The association of pipe and alignment occurs when creating the pipe network (see Figure 11.30 or 11.33). When the user creates the sample line group, all of the pipe networks appear in the Section sampling defaults list at the bottom of the Create Sample Line Group dialog box (see Figure 11.48). Toggle off any pipe network or any other element not wanted in the sections.

After clicking the OK button, it is necessary to add specific stations to the list of sections to add the pipes and structures.

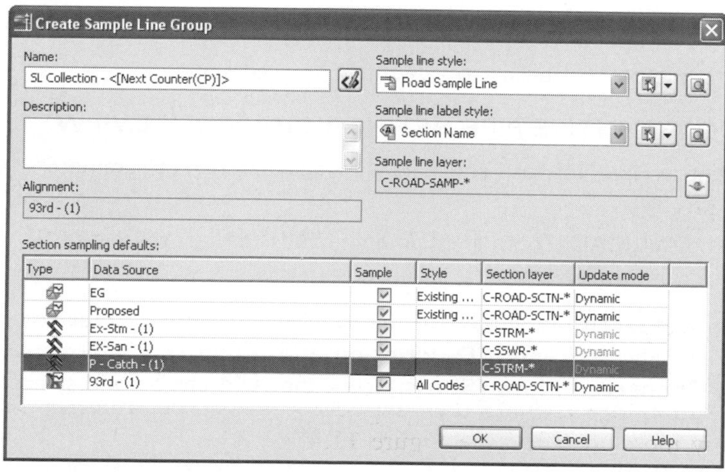

Figure 11.48

EXERCISE 11-4

When you complete this exercise, you will:

- Annotate individual pipes and structures.
- Annotate entire networks in plan and profile.

EXERCISE SETUP

This exercise continues with the drawing from the previous exercise. If you did not complete the exercise, you can open the *Chapter 11 – Unit 4.dwg* as the file for this exercise. The drawing file is in the *Chapter 11* folder of the CD that accompanies this book.

1. Open the drawing from the previous exercise or the *Chapter 11 – Unit 4* drawing.

ANNOTATING AN ENTIRE PIPE NETWORK IN PLAN VIEW

The Add Labels dialog box can annotate an entire pipe network in plan view.

1. Use the AutoCAD Zoom and Pan commands to view the entire 93rd Street storm network in plan view.
2. From the Pipes menu, select **Add Labels....**

   This displays the Add Labels dialog box.
3. In the Add Labels dialog box, set the Label Type to **Entire Network Plan**, set the Pipe Label to **Length Description and Slope**, and set the Structure Label to **Data with Connected Pipes (Storm)**.
4. Click the **Add** button and select a pipe or structure from the 93rd Street existing storm network.
5. Use the AutoCAD Zoom and Pan commands to view the labeling.
6. Drag some of the labeling to view their dragged state.

ANNOTATING AN ENTIRE PIPE NETWORK IN PROFILE VIEW

To annotate an entire Pipe Network in Profile View, set the correct parameters in the Add Labels dialog box.

1. Use the AutoCAD Zoom and Pan commands to view the entire 93rd Street storm network in Profile View.
2. In the Add Labels dialog box, set the Label Type to **Entire Network Profile**, set the Pipe Label to **Length Description and Slope**, set the Structure Label to **Data with Connected Pipes (Storm)**, click the **Add** button, and select a pipe or structure from the 93rd Street existing storm network in the profile view.
3. Use the AutoCAD Zoom and Pan commands to view the labeling.
4. Drag some of the labeling to view their dragged state.
5. Click the **Close** button to exit the Add Labels dialog box.

6. Close the drawing and do not save the labeling.
7. Reopen the drawing file.

ANNOTATING INDIVIDUAL ELEMENTS IN PLAN VIEW

The Add Labels dialog box labels individual pipes and structures in plan view.

1. Use the AutoCAD Zoom and Pan commands to view the EX-San1 structure and pipes at the western end of the 93rd Street storm sewer network.
2. From the Pipes menu, select **Add Labels....**
3. In the Add Labels dialog box, set the Label Type to **Single Part Plan**, set the Pipe Label to **Length Description and Slope**, set the Structure Label to **Data with Connected Pipes (Storm)**, click the **Add** button, and in the drawing, select a structure or a pipe.

   The Add Labels routine labels only one structure or pipe at a time.
4. Label additional pipes and structures by reselecting the Add button and selecting a network pipe or structure.

ANNOTATING SINGLE PARTS IN PROFILE VIEW

The Add Labels dialog box labels individual pipes and structures in Profile View. The Label Type has to be set to Single Part Profile with the appropriate label styles.

1. Use the AutoCAD Zoom and Pan commands to view the profile with the 93rd Street Storm sewer network in the Profile View.
2. In the Add Labels dialog box, set the Label Type to **Single Part Profile**, set the Pipe Label to **Length Description and Slope**, set the Structure Label to **Data with Connected Pipes (Storm)**, click the **Add** button and select a structure or pipe.

   The Add Label routine labels only one structure or pipe at a time.
3. Label additional pipes and structures by reselecting the Add button of the Add Labels dialog box.
4. Click the **Close** button to exit the Add Labels dialog box.

PIPES IN ROAD SECTION VIEWS

1. From the Sections menu, select the **Create Sample Lines....**
2. In the command line, the routine prompts for an Alignment; press the right mouse button, and from the list, select the **93rd – (1)**.
3. In the Create Sample Line Group dialog box, uncheck the P-Catch – (1) pipe network, and click the **OK** button to accept the defaults.
4. In the Sample Line Tools toolbar, select the section creation method of **From Corridor Stations**.

5. In the Create Sample Lines – From Corridor Stations dialog box set the Swath Width to **60** (see Figure 11.49).

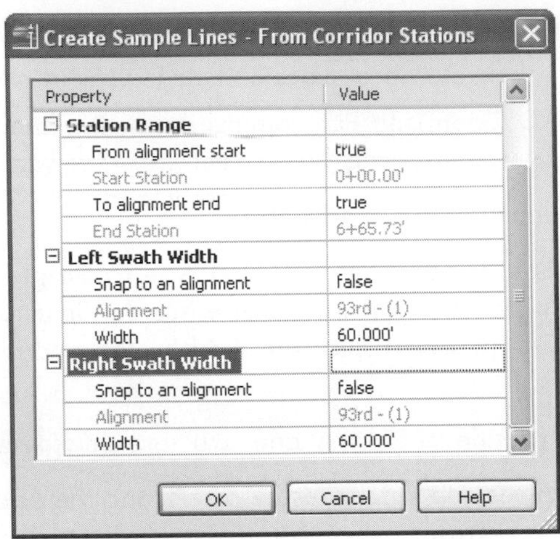

Figure 11.49

6. Click the **OK** button to create section sample lines.

   The routine returns to select a station mode. To create a section centered on a structure, select its station using an AutoCAD Center object snap.

7. Using the station jig, in the drawing, select specific stations representing the pipes and structures, and set the Swath Width to 60' for each new section.

8. Press the **Enter** key after you have completed specifying the stations.

   Note down the sample line stations for the intersections.

9. Use the AutoCAD Pan and Zoom commands to view an empty area in the drawing.

10. If necessary, click the **Prospector** tab to make it current.

11. In Prospector, expand the Corridors branch, select the **93rd – (1)** corridor, press the right mouse button, and from the shortcut menu, select **Rebuild**.

12. From the Sections menu, select **Create View...**.

13. In the Create Section View dialog box, click the drop-list arrow at the middle left for Sample Line, and select one of the sample lines that intersects a structure and/or pipe.

14. Click the **OK** button to create the first section.

15. In the Section View Bands – Set Properties dialog box, click the **OK** button.

16. If necessary, use the AutoCAD Zoom and Pan commands to better view the section.

17. Repeat the Create View command to import additional sections with pipes and/or structures (see Figure 11.50).

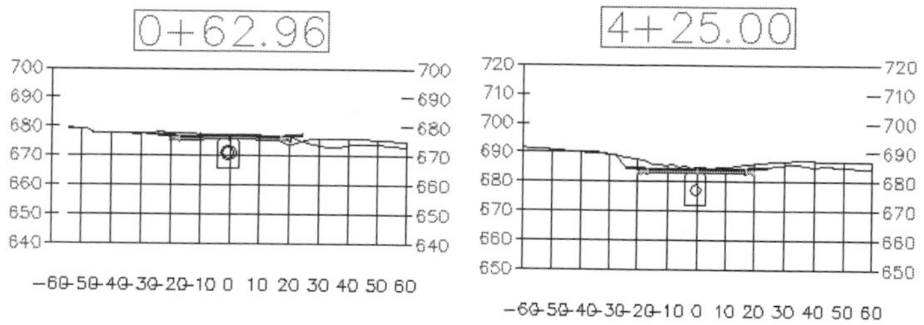

Figure 11.50

Civil 3D provides two basic labels for pipe networks: pipe and structure. You can place the labels as you draft the network or after completing its design.

SUMMARY

- There are two basic label types: pipe and structure.
- The content label styles are simple and do not represent all of the data a label can contain about a pipe or structure.
- When creating a sample line group, the pipe network should exist BEFORE creating the sample lines.
- If you want to add a pipe network to a sample line group, it is best to recreate the sample line group.

This ends the chapter on pipes in Civil 3D. The next chapter goes into the basic methods of sharing data in Civil 3D.

Civil 3D Projects and Vault

INTRODUCTION

The security and timeliness of a project's data is always a concern, as is the sharing of data among design groups or contractors. Initially, Civil 3D provided minimum project support: The drawing was the project. The next step was implementing data shortcuts, or external LandXML files. Other drawings can access these data files and shortcuts and can notify users of an updated version. While effective, they are limited to what can be held in a LandXML file, and there is no control over access. Autodesk Civil 3D 2007 introduces Autodesk Vault technology. This file-based approach to storing data and controlling permissions should make users pause and develop strategies for project data storage (for example, one drawing having all or several smaller drawings, each containing a single piece of data). Other issues with Vault include unknown answers, merging data from local vaults to a corporate vault, and sharing data between company vaults. These issues are not well documented and are not addressed in the implementation documents of Vault.

OBJECTIVES

This chapter focuses on the following topics:

- Data Shortcuts
- Vault

OVERVIEW

Initially, Civil 3D considered a drawing a project, and all of the objects within a drawing represent a project's data. Most of this book views the data in this way. A project considered to be a drawing is a radical departure from the strategy found in the Autodesk Land Desktop product line. In LDT, there is an extensive external folder structure that holds much of the data found in a project; as open and as vulnerable as it is, this is a strategy that has saved many a project's data.

Civil 3D 2006 introduced a new method for sharing data: a data shortcut. A data shortcut is a LandXML file stored in a folder structure, and other drawings reference (use) the data shortcut to create object instances. Civil 3D maintains links between the drawing and the shortcut. If the data shortcut definition changes, Civil 3D notifies the user of the change, and the drawing will need to be updated. A data shortcut expands data availability and ensures some degree of data quality. Data shortcuts were broken in 2007, but they are to be reactivated in a later service pack.

Vault marks the beginning of a more secure method of data storage and sharing. Vault allows administrators to assign basic permission levels to consumers of project data. While much about Vault for the Civil environment is not documented, some basic strategies can make it useful in many offices. The greatest limitations are the number of simultaneous users and the data it stores. The number of simultaneous users is small, about five or six. Most of the time, users will be consuming (referencing) data and not participating as a Vault user. A Civil 3D Vault stores drawings and shares points, surfaces, alignments, profiles, and pipe networks. This list is not the entire list of object types currently implemented in Civil 3D.

UNIT 1

This unit covers creating and using data shortcuts. When referencing shortcut data, Civil 3D uses icons for data states, changes, and references.

UNIT 2

This unit covers use of Vault in a project environment. Topics for this unit include checking out and in, modifying data, and creating references to Vault data.

UNIT 1: DATA SHORTCUTS

A Civil 3D drawing serves as a repository for all project data. Creating shortcuts allows users to share drawing data as project data (see Figure 12.1). Data shortcuts are limited to surfaces, alignments, profiles, and pipe networks. If updating a shortcut (the LandXML file), a user referencing the shortcut's data receives a change notification.

Data shortcuts were broken in Civil 3D 2007, but they will be reactivated in one of the service packs for the product.

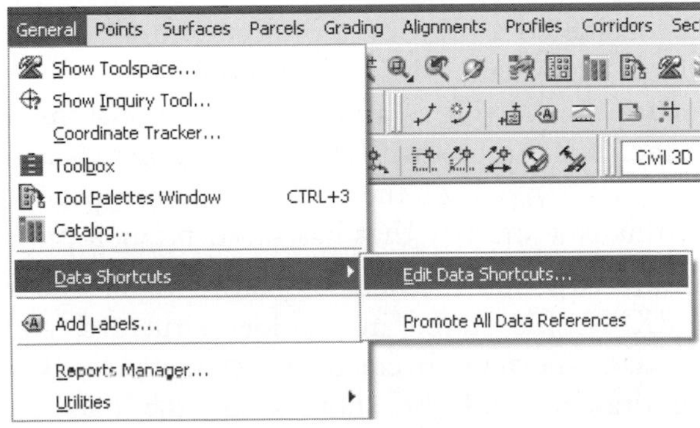

Figure 12.1

CREATING A DATA SHORTCUT

The first step in creating a data shortcut is calling the Edit Data Shortcut vista. The vista contains all of the possible shortcut types and any shortcut defined for the current drawing (see Figure 12.2). If exporting a shortcut, the same vista displays a file name and its location.

If a second drawing creates a reference to a shortcut, the second drawing uses the vista to provide a link to the shortcut files. If a shortcut changes, the user receives a change notification and is prompted to synchronize the data.

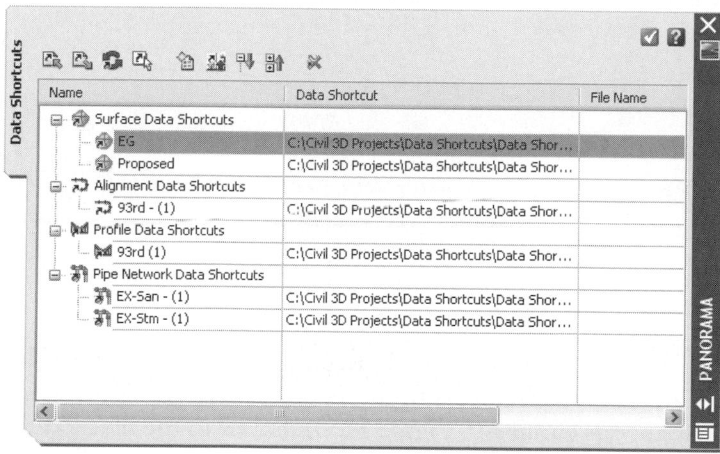

Figure 12.2

Creating a shortcut starts by selecting the Create Data Shortcut icon in the vista, and then in the drawing selecting an object. After creating a shortcut, the vista lists the selected object in the appropriate section with the name of the object instance. The shortcut remains within the drawing until the user selects the Export Data Shortcuts to File Vista icon.

EXPORTING A DATA SHORTCUT

When selecting the Export Shortcut icon, a file dialog box displays, ready to save a LandXML file whose name is the same as the drawing (see Figure 12.3). The best strategy is to store the file at a central location so other users have access to this file with a more descriptive name. Users must export shortcuts for other drawings to reference their data.

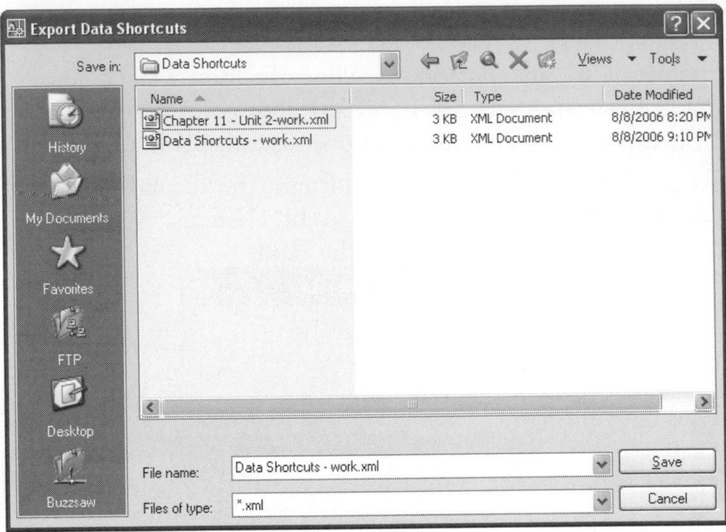

Figure 12.3

IMPORTING A DATA SHORTCUT

When consuming data from an exported shortcut, users use the same Edit Data Shortcuts command, located in the Data Shortcuts flyout menu. After clicking the icon, an Import Data Shortcut dialog box appears (similar to Figure 12.3). Browse to the file's location, select the file, and click Open to read the data to the current drawing. After reading the file, the Data Shortcuts vista lists each shortcut, the drawing and referenced object, the path to the drawing and LandXML files, and the LandXML file. It is critical that these do not move, or the user will have to re-export the LandXML file and re-reference the data file. The Validate Data Shortcuts icon (third icon from the left) validates the paths and files (see Figure 12.4).

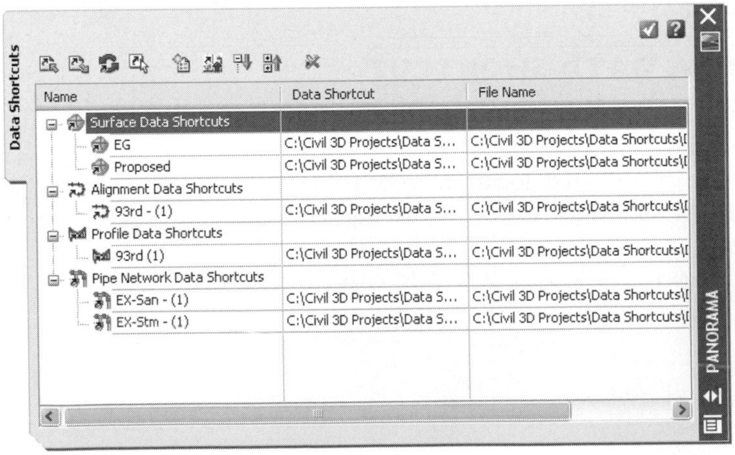

Figure 12.4

REFERENCING AN IMPORTED DATA SHORTCUT

To create a reference to a shortcut, select a shortcut from the list and click the Create Reference icon. The selected shortcut type determines which create object dialog box appears. If selecting an alignment, the Create Alignment Reference dialog box appears (see Figure 12.5). The difference among these create object dialog boxes is that they contain a path to the selected shortcut.

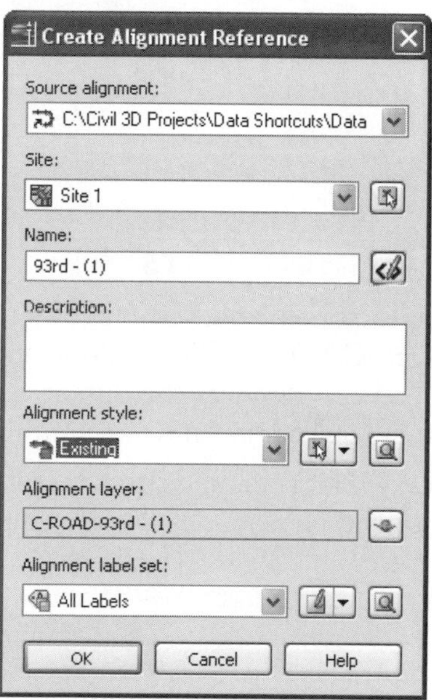

Figure 12.5

After creating a reference to an object, the user can use its data. After creating a reference, Prospector identifies the drawing's object instance as a reference, and the reference object cannot be edited. The object can only be updated by synchronizing it with a changed shortcut file (see Figure 12.6). Figure 12.6 shows the station equation controls as grayed out and not usable.

In addition to the reference icon to the left of the object's name in Prospector, Civil 3D places a reference icon in the AutoCAD tray at the bottom right of the application (see Figure 12.6 and Figure 12.7).

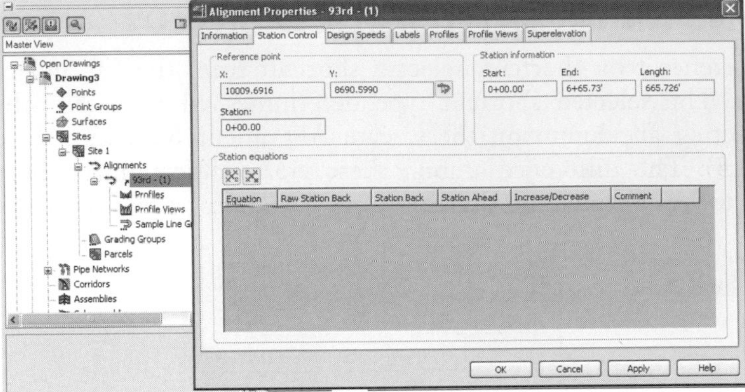

Figure 12.6

SYNCHRONIZING DATA SHORTCUTS

When a data shortcut changes, the out-of-date icon shows to the left of the object's name, and a balloon appears in the AutoCAD tray. The balloon contains a synchronize pick and, when selected, updates the references, removes the out-of-date icons, and leaves a message on the command line as to what objects are now synchronized.

When editing and saving the drawing containing shortcut definitions, Civil 3D knows which drawings reference the data shortcut, and notifies the drawings at the appropriate time.

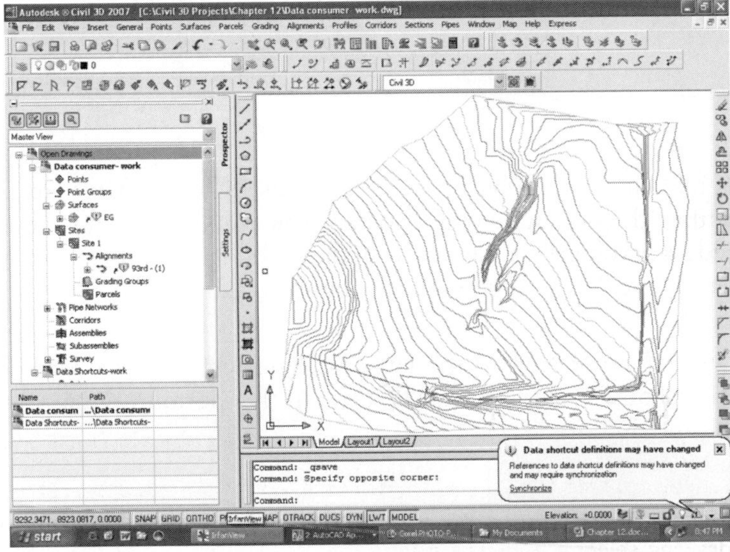

Figure 12.7

PROMOTING A DATA SHORTCUT

When a drawing contains a reference to a shortcut, users cannot edit any of its values. Users can reassign styles and layers, but nothing else. There is the option to promote the reference to a full object; however, this breaks the link to the data shortcut. When promoting a reference to an object, Prospector removes the reference icon from the object listing and no longer notifies the drawing of any changes to the original data.

EXERCISE 12-1

After completing this exercise, you will:

- Be familiar with creating Civil 3D data shortcuts.
- Be familiar with exporting data shortcuts.
- Be familiar with referencing data shortcuts.
- Be familiar with synchronizing drawing and data shortcut values.
- Be familiar with promoting references to drawing objects.

This exercise works with drawings from the CD that accompanies this textbook. The drawings are in the *Chapter 12* folder of the CD.

DATA SOURCE DRAWING

The exercise takes an existing drawing with several Civil 3D objects and creates data shortcuts. After creating the shortcuts, a new drawing references the shortcuts.

1. If not in Civil 3D, start the application by double-clicking the Desktop icon.
2. Close the opening drawing.
3. From the File menu, select **Open...**, browse to the *Chapter 12* folder, and select the *Data Shortcuts* drawing.
4. From the File menu, select **Save As...**, and browse to the Civil 3D *Projects* folder to save the drawing as *Data Shortcuts-work*.

CREATE SURFACE DATA SHORTCUTS

1. Use the AutoCAD Zoom and Pan commands to view the EG contours and the proposed triangulation.
2. From the General menu, Data Shortcuts flyout, select **Edit Data Shortcuts....**
3. In the Data Shortcuts vista, select **Surface Data Shortcuts**. From the vista, click the **Create Data Shortcut by Selection** icon (fourth icon from the left), and in the drawing select an EG contour.
4. Return to the vista and expand the Surface Data Shortcuts section; EG is a listed shortcut.
5. Repeat Steps 3 and 4, but this time select the triangulation of the **Proposed** surface.

EXERCISES

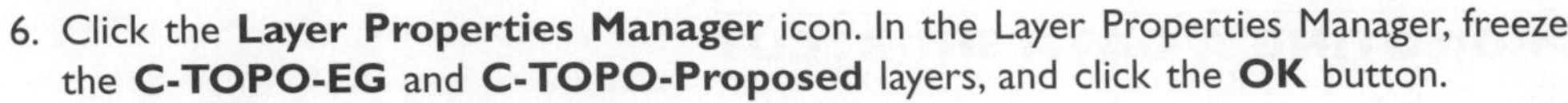

6. Click the **Layer Properties Manager** icon. In the Layer Properties Manager, freeze the **C-TOPO-EG** and **C-TOPO-Proposed** layers, and click the **OK** button.

CREATE PIPE NETWORK DATA SHORTCUTS

1. Use the AutoCAD Zoom and Pan commands to view the southwest corner of the project to view the two pipe networks.
2. In the Data Shortcuts vista, select **Pipe Network Data Shortcuts**. From the vista, select the **Create Data Shortcut by Selection** icon. In the drawing, select a pipe from the **EX-San – (1)** (purple lines).
3. Return to the vista and expand the Pipe Network Data Shortcuts section; Ex-San – (1) is a listed shortcut.
4. Repeat Steps 2 and 3, but this time select the **Ex-Stm – (1)** pipe network (blue lines).

CREATE ALIGNMENT DATA SHORTCUTS

1. Click the Layer Properties Manager icon. In the Layer Properties Manager, thaw the **C-ROAD** layer and click the **OK** button.
2. In the Data Shortcuts vista, select **Alignment Data Shortcuts**. From the vista, click the **Create Data Shortcut by Selection** icon. In the drawing, select a station label for **93rd – (1)**.
3. Return to the vista and expand the Alignment Data Shortcuts section; 93rd – (1) is a listed shortcut.
4. Click the **Layer Properties Manager** icon. In the Layer Properties Manager, freeze the **C-ROAD** layer and click the **OK** button.

CREATE A PROFILE DATA SHORTCUT

1. Use the AutoCAD Zoom and Pan commands to view the profile view of 93rd street at the right of the project.
2. In the Data Shortcuts vista, select **Profile Data Shortcuts**. From the vista, click the **Create Data Shortcut by Selection** icon. In the drawing, in the profile view, select a tangent segment for the **93rd (1)** (red line in the profile view).
3. Return to the vista and expand the Profile Data Shortcuts section; 93rd (1) is a listed shortcut.
4. Click the **AutoCAD Save** icon to save the drawing.

EXPORT DATA SHORTCUTS

1. In the Data Shortcuts vista, click the **Export Data Shortcuts to File** icon (second icon from the left).

2. In the Export Data Shortcuts file dialog box, browse to the Civil 3D *Projects* folder and make a new folder called *Data Shortcuts*. Double-click the *Data Shortcuts* folder, and click the **Save** button to save the LandXML file.
3. Close and save the drawing.

IMPORT DATA SHORTCUTS

1. From the Prospector Master View, expand the Drawing Templates branch. From the list of templates, select **_Autodesk Civil 3D (Imperial) NCS Extended**, press the right mouse button, and from the shortcut menu select **Create New Drawing**.
2. Click the **Settings** tab to make it current.
3. Click the drawing name at the top of Settings, press the right mouse button, and from the shortcut menu select **Edit Drawing Settings....**
4. Click the **Object Layers** tab to make it current.
5. In the Object Layers panel, set the modifier to **Suffix** and the Suffix Value to **-*** for the following objects: Alignment, Alignment Labeling, Corridor, Corridor Section, Profile, Profile View, Profile View Labeling, TIN Surface, and TIN Surface Labeling.
6. Click the **OK** button to exit the dialog box.
7. From the File menu, select **Save As...**, browse to the same Civil 3D *Projects* folder for the *Data Shortcuts - Work* drawing, and save the drawing as *Data Shortcuts - Reference*.
8. From the General menu, the Data Shortcuts flyout, select **Edit Data Shortcuts....**
9. In the Data Shortcuts vista, click the **Import Data Shortcuts From File** icon (the leftmost icon). In the Import Data Shortcuts file dialog box, browse to the Civil 3D *Projects\Data Shortcuts* folder, select the *Data Shortcuts-work.xml* file, and click the **Open** button.
10. The Name, Data Shortcut (object in originating drawing), and File Name (the LandXML file) appear in the Data Shortcuts vista.

EXERCISES

CREATE A DATA SHORTCUT REFERENCE

1. In Surface Data Shortcuts of the Data Shortcuts vista, select **EG**. From the vista, click the **Create Reference** icon (fifth icon from the left).
2. In the Create Surface Reference dialog box, type **EG** as the name, set the style to **Contours 1' and 5' (Background)**, and click the **OK** button to create the surface.
3. The Events vista displays that the surface was created; select the **Data Shortcuts** tab to make it current.
4. Click the **Prospector** tab to make it current.
5. In Prospector, expand the Surfaces branch and note the reference icon to the left of the surface name.

6. Repeat Steps 1 and 2 to create references for the **93rd – (1) alignment**, **EX-Stm – (1)**, and **93rd (1) profile**. The dialog boxes you encounter for repeating Step 2 will be different for each object type.
7. Click the "**X**" to close the close the Data Shortcut panorama.

CREATE A PROFILE VIEW FROM REFERENCE DATA

1. Use the AutoCAD Pan and Zoom commands to move the site to the left side of the screen.
2. From the Profile menu, select **Create from Surface....**
3. In the Create from Surface dialog box, select the **EG** surface and then select the **Add>>** button at the middle right.

   Notice that the 93rd (1) profile is already on the profile list at the bottom.
4. Click the **Draw in Profile View** button, set the Profile View style to **Major Grids**, and click the **OK** button.
5. In the drawing, select a point to the right of the site.
6. In Prospector, expand the Sites branch until you view the Alignments and Profiles branches for 93rd – (1).

   Note that the alignment and profile are references, and the EG profile is from the EG surface reference in the drawing.
7. Click the **AutoCAD Save** icon to save the drawing.

EDIT THE DATA SHORTCUT SOURCE DRAWING

1. Reopen the *Data Shortcuts-work* drawing.
2. In Prospector, under Surfaces, expand the EG branch until you view the data list for the Definition heading.
3. In the data list, select **Edits**; press the right mouse button, and from the shortcut menu select **Raise/Lower Surface**.
4. In the command line, the routine prompts for the amount to adjust the surface. Type **1** and press the **Enter** key to raise the surface 1 foot.
5. Click the **AutoCAD Save** icon to save the drawing.
6. From the Window menu, select the *Data Shortcuts–Reference* drawing.
7. The drawing should show a balloon at the lower right, indicating the drawing has a reference that is out of date. The EG surface should also display an out-of-date icon.
8. Click the **Synchronize** link in the Out-of-Date balloon or in Prospector's Surfaces list select EG, press the right mouse button, and from the shortcut menu select **Synchronize**.
9. Click the **AutoCAD Save** icon to save the drawing.

EXERCISES

PROMOTING A REFERENCE

1. In Prospector, select a reference, press the right mouse button, and in the shortcut menu, select **Promote**.

This creates an object instance in the drawing and breaks the link to the data shortcut. This ends the exercise on data shortcuts.

SUMMARY

- Data shortcuts link an object in a drawing to an external LandXML file.
- When an object in a drawing is a data shortcut and the object changes, all drawings referencing the data shortcut receive an out-of-date balloon notice and an out-of-date icon next to the object.
- You can promote a referenced object to a full object in a drawing.
- When you promote a reference, you break the link to the data shortcut.

UNIT 2: CIVIL 3D VAULT

Civil 3D's approach to data storage, the drawing, is fraught with danger. With a history of frailty, the saving, preserving, and sharing process becomes critical for successfully using Civil 3D. The Vault product is currently the de facto data standard for Autodesk products. Vault is a file-based system and, because of this approach, users need review and a plan for how to store project data safely and efficiently.

The "best" strategy for drawings is still a debatable topic. In a white paper focusing on data storage and Vault, the discussion centers on the object types, data dependencies, and the types of sheets in a submission set. In any discussion, the grouping of dependent object types is the overriding concern, (e.g., a profile needs an alignment and a corridor needs a vertical definition). The question is what mix of objects needs to exist in the drawings that are stored in Vault to gives a team the best access to the data.

The use of Vault data will be a combination of check in, check out, references, and Xrefs. The combination of methods gives data consumers the best data-sharing opportunities in order to execute their design as a part of a team and to keep data secure.

CREATE A PROJECT

Before anything can happen, the user must create a Vault project. Creating a project occurs in the Prospector's Master view. The Master view allows the user to set the working directory as well as create, migrate, and display Vault Explorer.

When selecting New... from the shortcut menu, the Project Properties dialog box appears (see Figure 12.8 and Figure 12.9). The properties of a project are its name, description, the author, and the date.

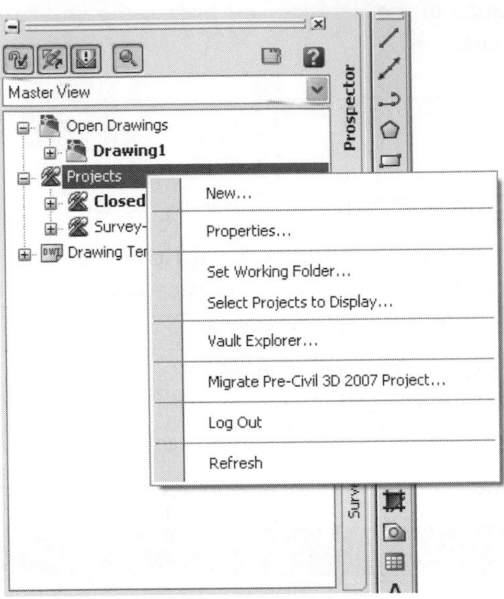

Figure 12.8

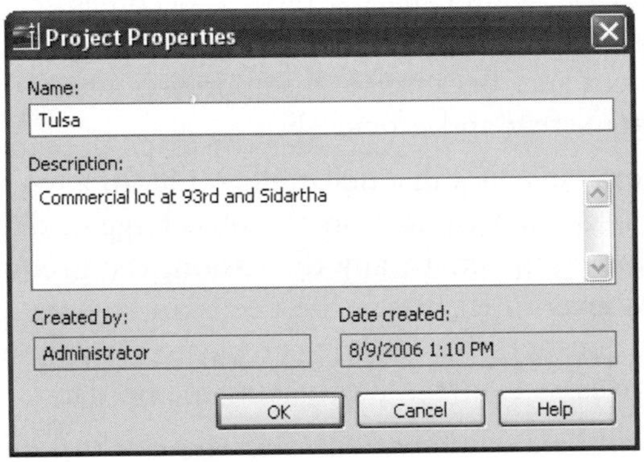

Project Properties
Name:
Tulsa
Description:
Commercial lot at 93rd and Sidartha
Created by:
Administrator
Date created:
8/9/2006 1:10 PM
OK
Cancel
Help

Figure 12.9

When the user clicks the OK button to exit the dialog box, Prospector creates a project folder in the *Working Project* folder (default C:\Civil 3D Projects). This folder contains access databases and other support files (see Figure 12.10).

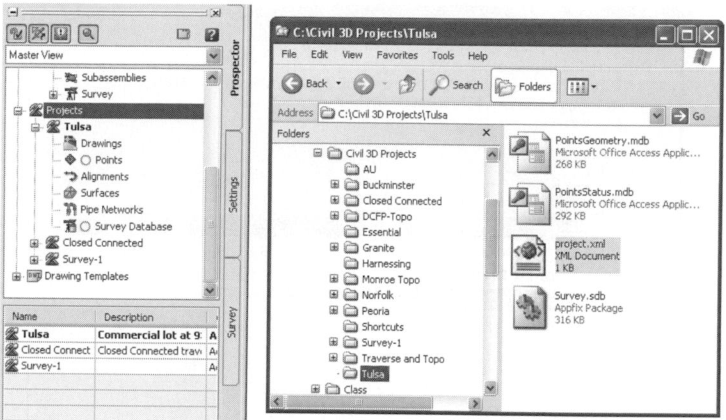

Figure 12.10

ADD DRAWINGS TO A PROJECT

Drawings and data are assigned to a project. When assigning a drawing to a project, the user must decide if he or she wants to retain editing rights or give them up. If the user decides to retain the checked-out status of the drawing or data, the drawing session continues and a copy of the drawing goes to Vault, but the Vault copy is not editable. When you check in and do not retain the checked-out status, anyone with editing status can modify the checked-in file. The decision to check in and to relinquish or keep checked-out status is a toggle in the Add to Project, File Dependencies dialog box (see Figure 12.11). Users who are only data consumers can reference or get read-only copies of the checked-in files. By their user assignment, data consumers cannot edit Vault data.

When checking in a drawing, Civil 3D creates a DWF, automatically saves the drawing, and closes the session. If the user retains the file as checked out, Vault gets a copy of the file and the drawing session continues. If the user "returns unexpectedly to the operating system" or saves and ends the session, when the user returns to the drawing he or she still may be in a checked-out state. If the files are correctly in Vault, the user needs to undo the check-out status. Consult with the Vault administrator before arbitrarily changing a drawing status.

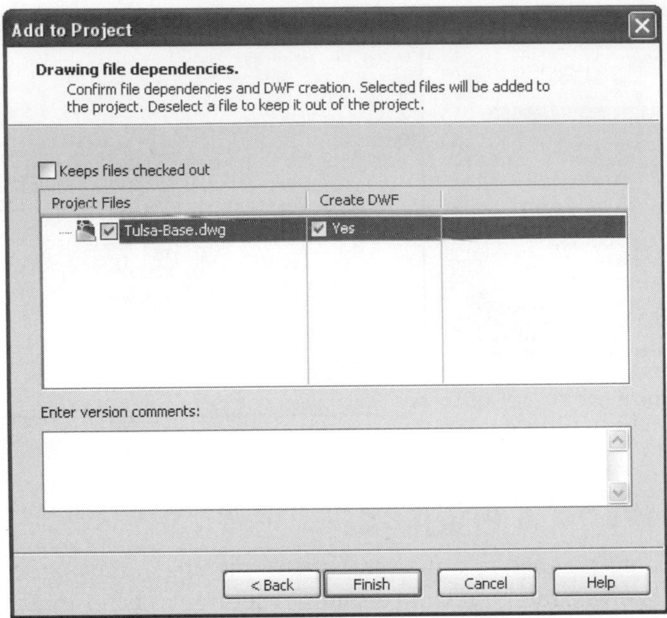

Figure 12.11

ADD TO A PROJECT

To check in a file to Vault, start in Prospector by selecting the drawing's name, click the right mouse button, and from the shortcut menu select Add to Project.... First the user will receive a warning about an automatic drawing save (see Figure 12.12).

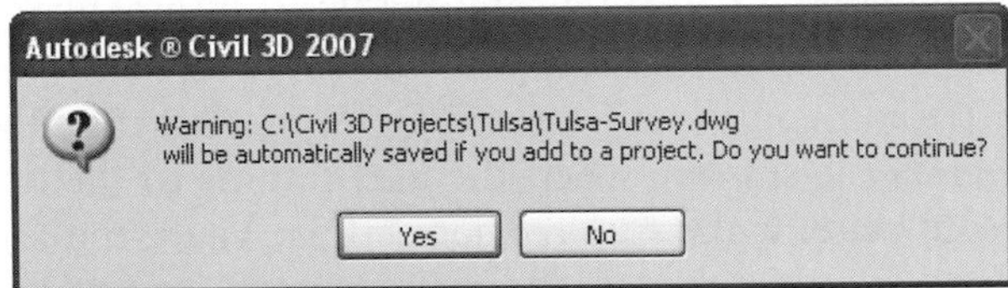

Figure 12.12

Next Civil 3D presents a wizard that goes through the steps necessary to assign a drawing to a project. The first panel asks the user to identify the project (left side of Figure 12.13) and the second panel asks for a drawing file location (right side of Figure 12.13).

Users can define Vault subfolders to store drawings that contain a single data item. For example, a user could define a *Surfaces* folder with and *EG* and *Proposed* subfolders. When selecting a drawing location, the user can identify the appropriate folder for the surface. This also applies to alignments.

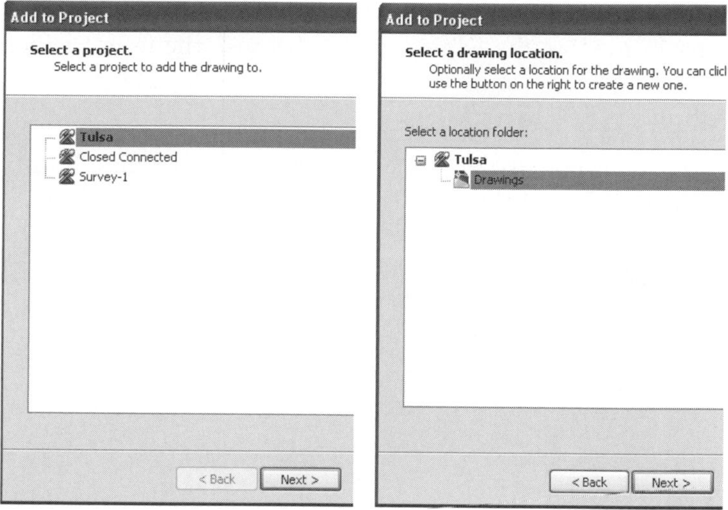

Figure 12.13

The last panel of the wizard is Drawing file dependencies (see Figure 12.14). In this panel, users add a description to the file, create a DWG, and decide whether to check in completely or place a copy in Vault but retain editing privileges. If the drawing uses other files (such as LandXML or LDT surfaces), this panel lists the files and copies them to Vault with the drawing.

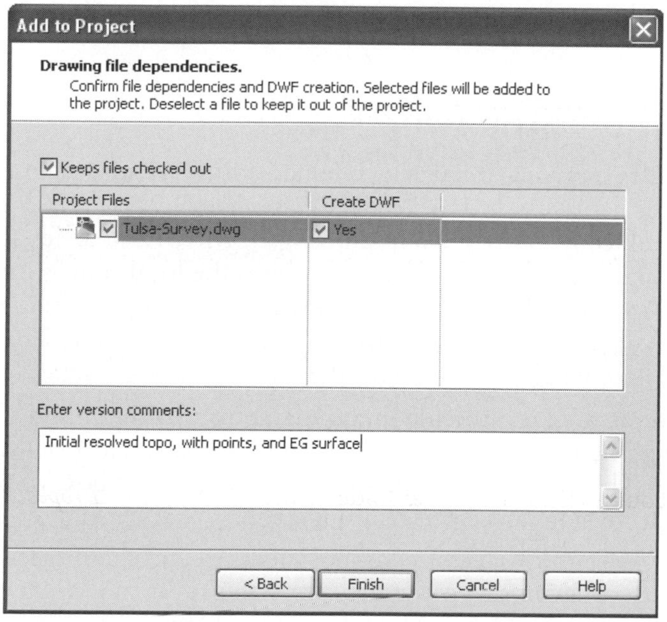

Figure 12.14

If the Keeps files checked out is toggled on, the drawing session remains active. Users must check the drawing in to close the editing session because the user retained editing rights.

DRAWING WITH DEPENDENCIES

When a drawing contains objects that depend on other external files, Vault identifies and by default adds the file as a part of the check-in and check-out process (see Figure 12.15).

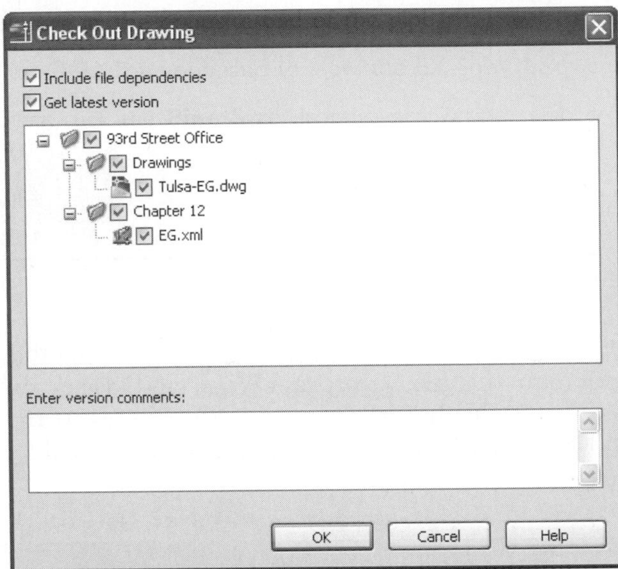

Figure 12.15

CHECK IN AND OUT DRAWINGS

After adding a file to a project and having retained editing rights, the user has two choices. The first option is to save the drawing and exit Civil 3D. In this case, the drawing is retained as a local copy of the Vault drawing. Civil 3D uses icons to indicate this relationship. A white dot with a check mark indicates the local copy is the same as in Vault, and a green circle with a check mark indicates the local copy is newer than the one in Vault. This will happen when the user saves the drawing; save time is newer than the one in Vault. During another drawing session, the user can then check in the file to Vault.

The second option is to check the file in and relinquish editing rights. To check in a file, select the drawing's name in Prospector, right mouse click, and from the shortcut menu select Check In.... This presents the Check In Drawing dialog box, where the user adds a description for the version and creates an updated DWF (see Figure 12.16).

To check out a file, from Prospector's Master View, expand the project until its drawings are in view, select a drawing from the list, press the right mouse button, and from the shortcut menu select Check Out.... The Check Out Drawing dialog box appears. In this dialog box the user identifies what information is desired besides the drawing. If Get

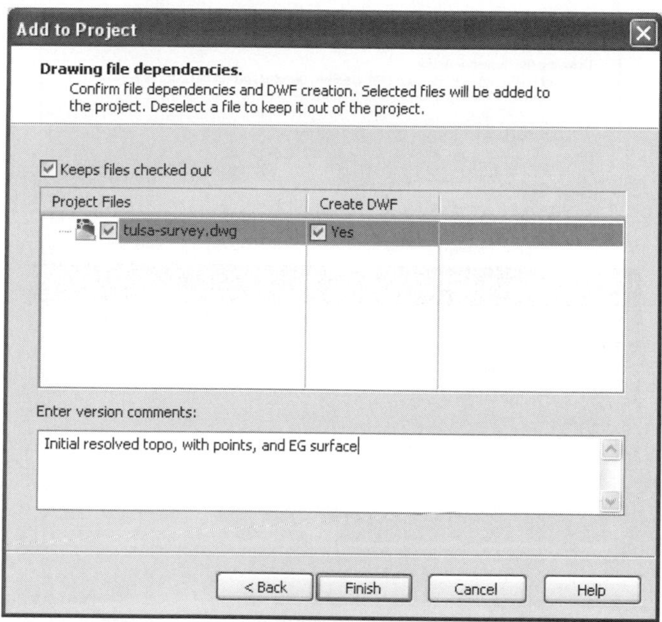

Figure 12.16

Latest version is toggled on, clicking the OK button places a copy of the vault drawing on the user's machine. If Get Latest version is toggled off, it implies the user has a local copy of the drawing to check in to the project. This creates a potential conflict if someone else is using the same process.

Vault icons to the left of the drawing name in both the Projects and Prospector areas indicate the status of the data and drawings. A white circle to the left of the file or data indicates is it available for check out. After checking out the drawing, Prospector shows a white circle with a check mark. If the drawing is saved, the icon changes to a green check mark, indicating the local copy is newer than the Vault copy.

For a Vault project, data is only points, surfaces, alignments, pipe networks, and surveys. The Check In Drawing dialog box changes to show the dependencies within the file (see Figure 12.17). With the data for surfaces from LDT, the LDT DTM data files become dependent files.

In Prospector's Project Drawing area, the check-out drawing is a red DWG icon. When the user places the cursor over the icon, a tooltip notes the file as checked out.

When the user clicks the Next button, the routine displays all of the objects in the drawing that can be Vault data (see Figure 12.18). The Share data panel allows a user to decide if the drawing's data is public to the Vault or if only the drawing can be seen. Users can hide data until it is finalized; then, in another check-in session, the user can mark those items desired to be public in the Vault.

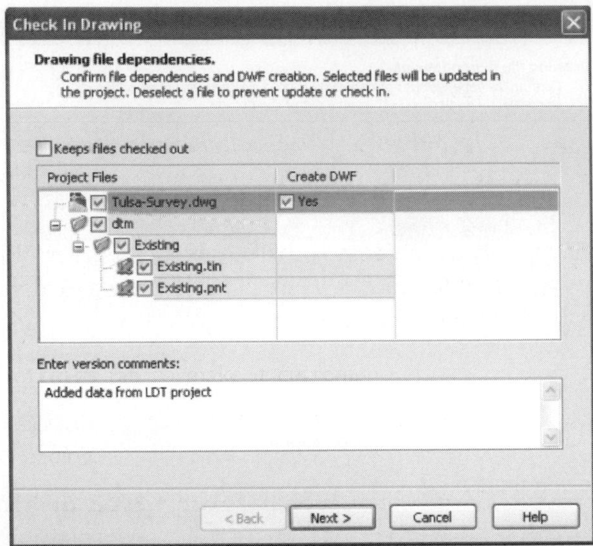

Figure 12.17

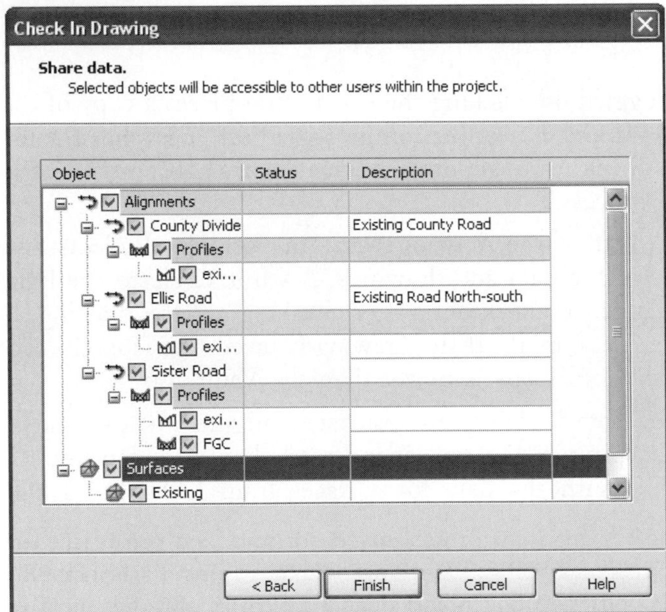

Figure 12.18

REFERENCES

When the user needs a copy of the data for use but does not want to edit the data, the first option is to create a reference to Vault data. Creating a reference to Vault data copies the data locally, marks it as non-editable, and places an icon next the object instance, indicating it as a reference. Vault displays project icons to indicate the status of data: checked out, or available to be checked out.

When a reference is active in a drawing, icons in the Prospector object type list indicate the status of the drawing reference to the Vault data. If the Vault data is more current, the user needs to update the reference to new data.

CREATE A REFERENCE

To create a reference, from Prospector's project data list select the data item, press the right mouse button, and from the shortcut menu select Create reference (see left side of Figure 12.19). When selecting Create Reference from the menu, a Create Surface Reference dialog box displays (see right side of Figure 12.19). Set the style, type a description, and after the OK button is clicked, Civil 3D makes a surface in the drawing and marks it as a reference (see Figure 12.20).

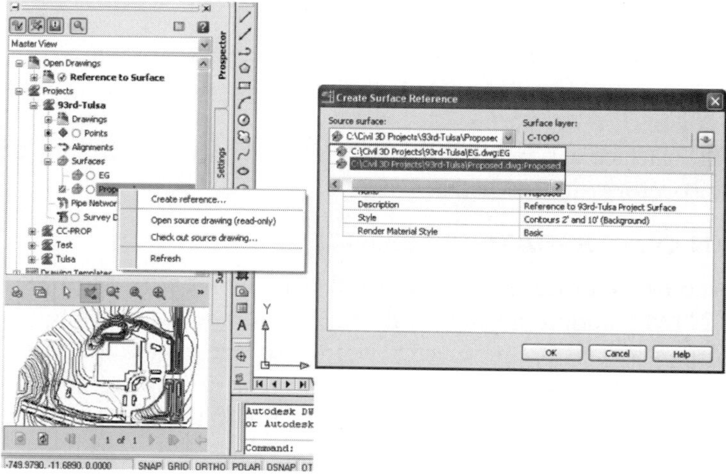

Figure 12.19

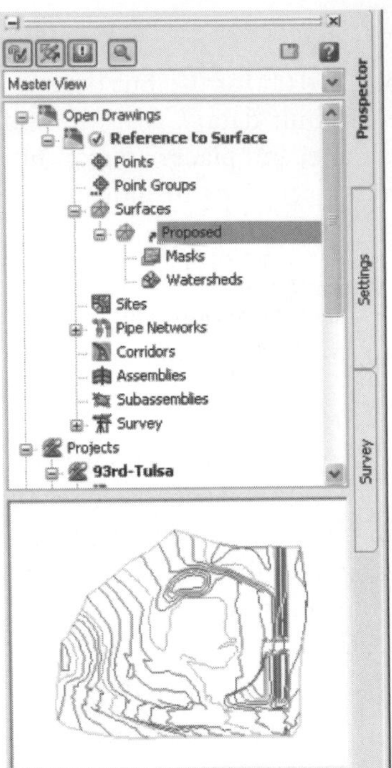

Figure 12.20

After making the surface a Civil 3D Event Viewer displays, notifying the user that the surface is in Prospector's surface list.

UPDATING AN OUT-OF-DATE REFERENCE

If a reference becomes out of date, Prospector displays an icon next to the referenced object (see Figure 12.21). To update an out-of-date reference, select the reference, right mouse click, and from the shortcut menu select Synchronize (see Figure 12.22). This updates the reference and updates the icons. When the user opens a drawing containing a reference that has changed, the data is automatically updated as Civil 3D opens the drawing.

PROMOTING A REFERENCE

Promoting a reference creates an editable object instance in the drawing. Promoting a reference can cause numerous problems. For example, when an alignment is promoted and edited, the values now differ between what is in the drawing and what is in Vault. Vault does not accommodate name conflicts and, if checking the drawing in, Vault will *not* change the alignment to the drawings version.

When the user wants to edit a data item, the best practice is checking out, editing, and checking in.

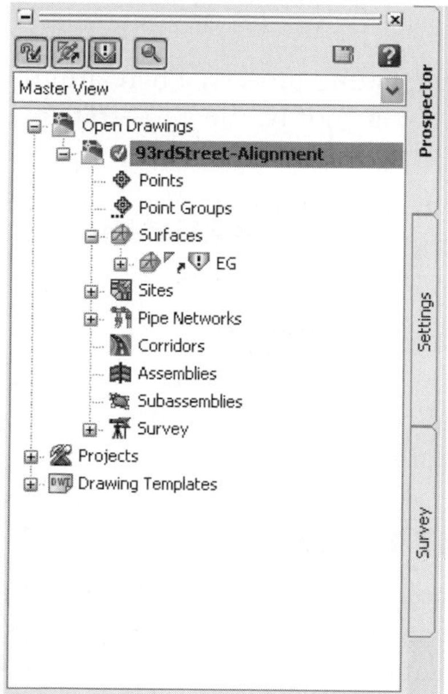

Figure 12.21

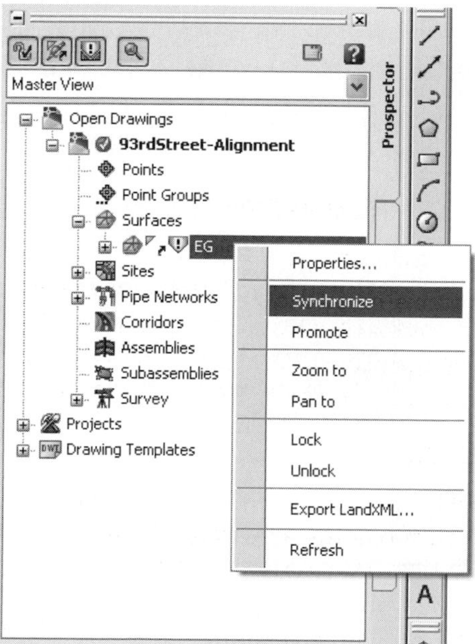

Figure 12.22

XREFS

A reference implies an active updating process. There are times, when the data is finalized or static, that using a reference may not be the best strategy. In Civil 3D there is no direct method of creating an Xref with Vault connections.

In Civil 3D, users must get the latest copy from Vault, save the drawing under a new name, and then attach it as an Xref.

VAULT ICONS

Users need to become familiar with the Vault icon list. Each icon indicates the status of the object as well as the local copy. It would be wise to print the Help page that reviews the Vault icons and their meaning.

VAULT MANAGEMENT TOOLS

The Vault application gives users the ability to manage files and folders of a Vault project. The Vault manager is where an administrator views project file revisions, assigns project files, and copies dependent data (LandXML, DTM, etc.) (see Figure 12.23).

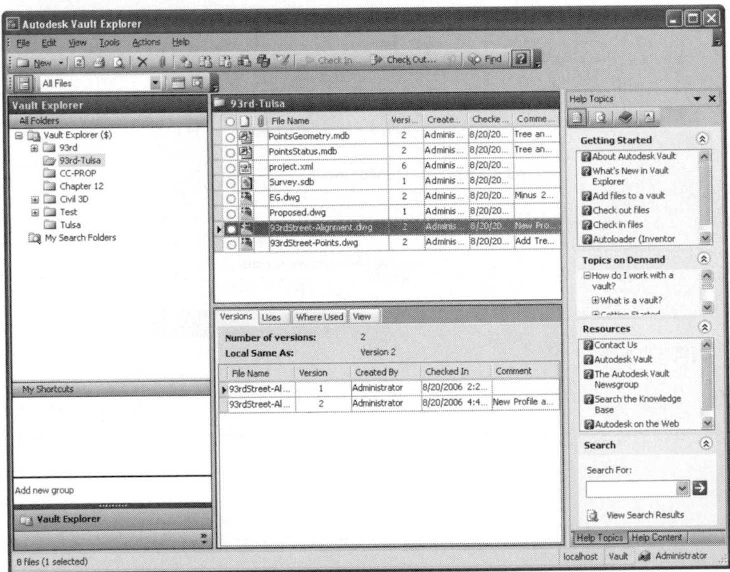

Figure 12.23

If a user wants to add folders to the project structure, that user must be an administrator and must use the administrative tools from the Tools menu.

Other project administration tools are in a shortcut menu. From the Vault Explorer tree at the right, select a project from the list, press the right mouse button, and select the appropriate command from the shortcut menu. These tools include New Folder..., Delete, Check In, Check Out, and so on (see Figure 12.24).

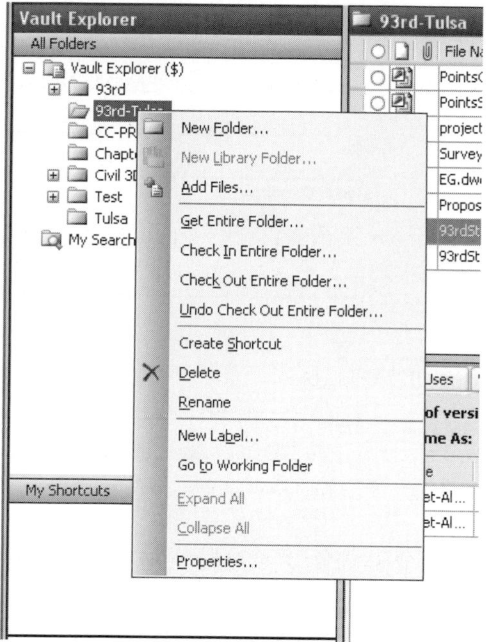

Figure 12.24

VAULT USERS

A Vault administrator creates and manages the user and grouping lists. As explained, the administrative tools are in the Tools menu. Civil 3D has three types of users: Vault administrator, Vault editor, and Vault consumer.

VAULT ADMINISTRATOR

An administrator has full rights to Vault, its files, and its folders. An administrator can create new users, assign them to groups, and assign their roles and permissions.

VAULT EDITOR

A Vault editor has full privileges in Vault, but does not have administrator privileges on the server.

VAULT CONSUMER

A Vault consumer has only read-only access to files and folders.

An administrator defines a user list and, after defining a list of users, assigns them to one or more groups (see Figure 12.25).

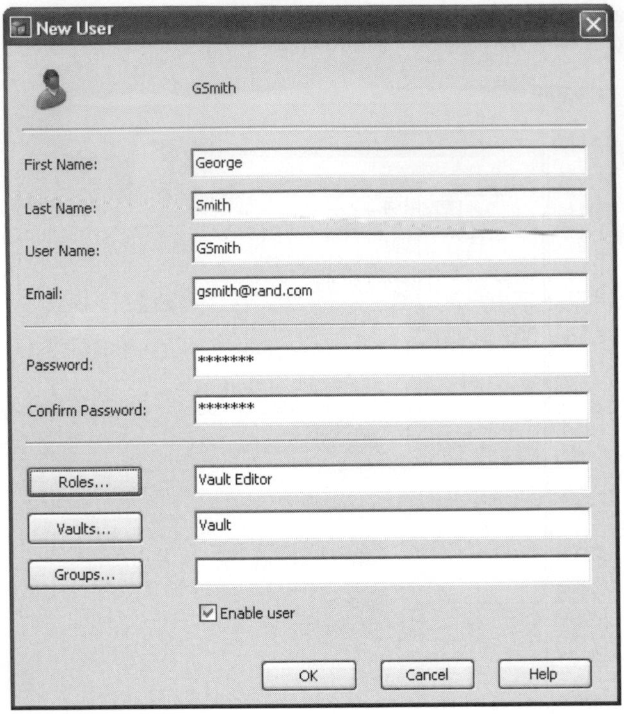

Figure 12.25

GROUPS

After creating a list of users, the next step is to define a group and assign it members (see Figure 12.26). A group contains a list of users who can use the data from the assigned Vault. The administrator should first create the users and then create the group and assign the members (see Figure 12.27).

After creating the group and assigning its members, the Group dialog box shows the member list (see Figure 12.28).

Vault lists all groups and provides administrator access to its values (see Figure 12.29).

Figure 12.26

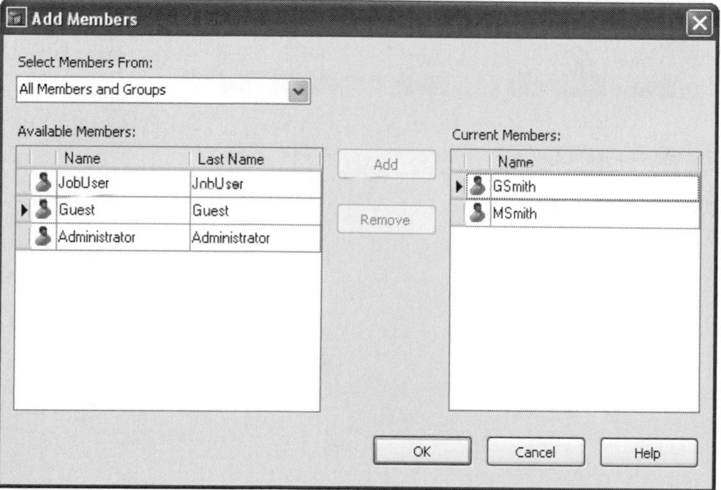

Figure 12.27

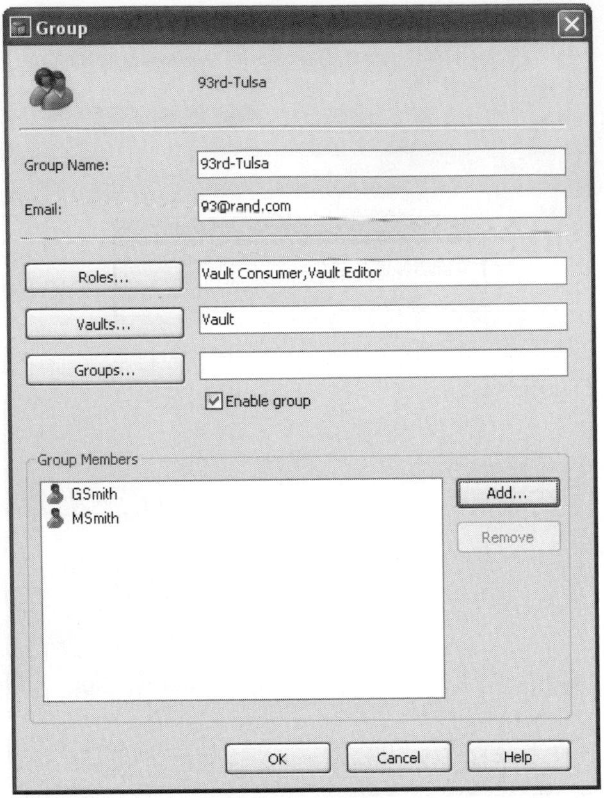

Figure 12.28

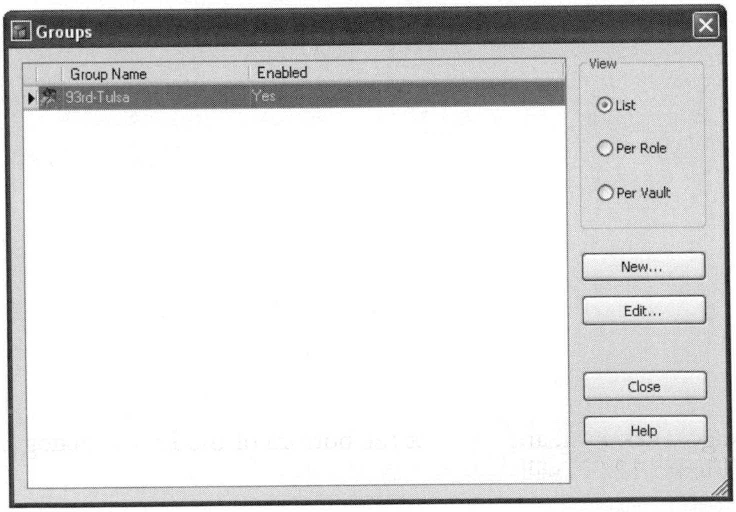

Figure 12.29

VAULT LABELS

A Vault label identifies a point in time of a project (see Figure 12.30). A label can be a rollback target and, when used as such, creates a new file version using the older file. This action restores a project to the revisions that were current when the label was defined. Vault never destroys files; it creates new ones with an incremented revision number.

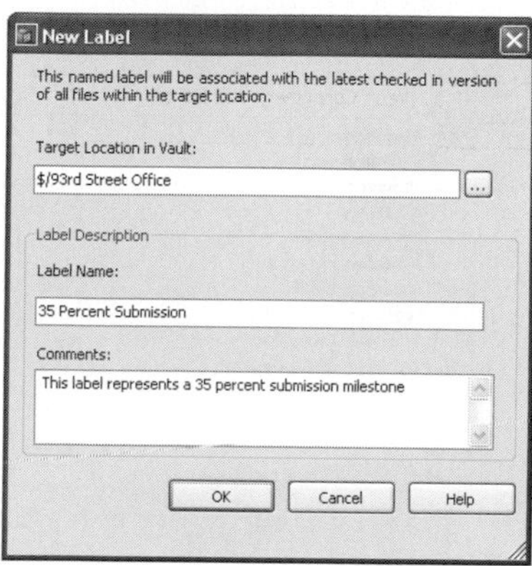

Figure 12.30

An administrator can view all labels active for a project (see Figure 12.31).

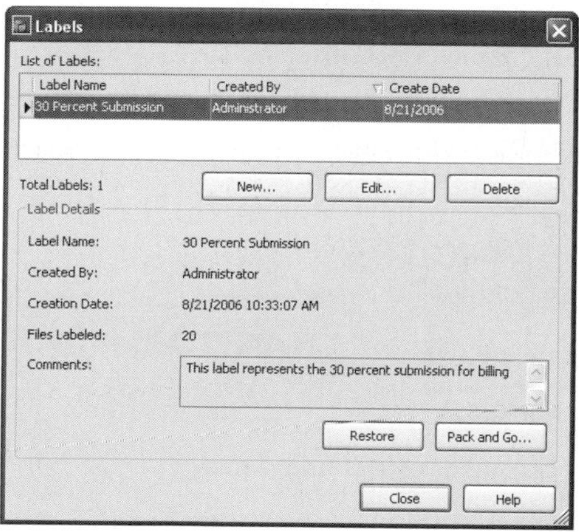

Figure 12.31

After defining a label, a label can generate a pack-and-go assembly of the labeled files. To create the package, click Pack and Go... at the bottom of thc Labels dialog box. The dialog box shown in Figure 12.32 will appear.

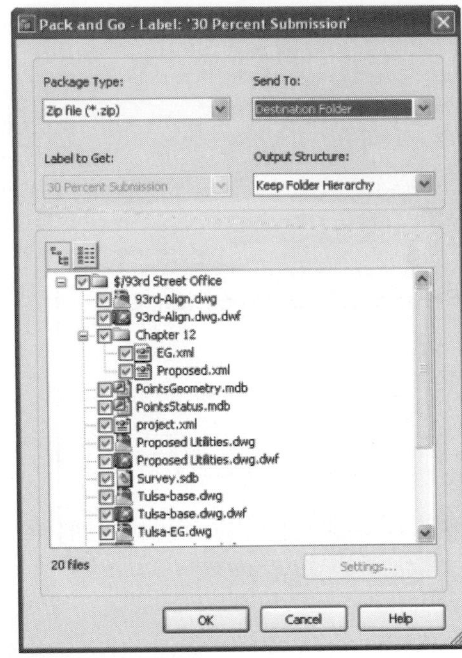

Figure 12.32

LOG IN TO VAULT

To log in to Vault, select the Project heading in Prospector's Master view, press the right mouse button, and select Login. In the Login dialog box, type your User Name and Password and click the OK button.

EXERCISE 12-2

After completing this exercise, you will:

- Create a project.
- Add drawings to a project.
- Check in and check out drawings and data.
- Change data.
- Create and use a reference.

EXERCISE SETUP

You will need to know your Vault login to complete this exercise.

1. If not in Civil 3D, start Civil 3D by double-clicking the Desktop icon.
2. Close and do not save the initial drawing.

3. In Windows Explorer, copy the *93rd* project folder from the *Chapter 12* folder of the CD that accompanies this book and place it in the *Civil 3D* Project folder.

CREATE AN EG DRAWING

1. From the File menu, select **New....** In the Select Template dialog box, browse to the *Chapter 12* folder located on the CD that accompanies this book. From the folder, select the template **HC3D-Extended NCS (Imperial)**, and click the **Open** button.
2. If necessary, click **Prospector** to make it current.
3. From the Surfaces menu, select **Import TIN....**
4. In the Import TIN dialog box, browse to the C:\Civil 3D Projects\93rd\DTM\EG folder, select the **EG.tin** file, and click the **Open** button.
5. This creates a Civil 3D surface from a Land Desktop surface.
6. From the File menu, select **Save As....** In the Save Drawing As dialog box, browse to the C:\Civil 3D Projects\93rd\DWG folder, name the drawing **EG**, and click the **Save** button.

CREATE A PROPOSED SURFACE DRAWING

1. From the File menu, select **New....** In the Select Template dialog box, browse to the *Chapter 12* folder located on the CD that accompanies this book. From the folder, select the template **HC3D-Extended NCS (Imperial)**, and click the **Open** button.
2. If necessary, click the **Prospector** tab to make it current.
3. From the Import flyout of the File menu, select **Import LandXML....**
4. In the Import LandXML dialog box, browse to the C:\Civil 3D Projects\93rd\ folder, select the **Proposed.xml** file, and click the **Open** button.
5. In the Import LandXML dialog box, click the **OK** button to create the surface.
6. From the File menu, select **Save As....** In the Save Drawing As dialog box, browse to the C:\Civil 3D Projects\93rd\DWG folder, name the drawing **Proposed**, and click the **Save** button.

CREATE AN ALIGNMENT DRAWING

1. From the File menu, select **New....** In the Select Template dialog box, browse to the *Chapter 12* folder located on the CD that accompanies this book. From the folder, select the template **HC3D-Extended NCS (Imperial)**, and click the **Open** button.
2. If necessary, click the **Prospector** tab to make it current.
3. From the Import flyout of the File menu, select **Import LandXML....**

4. In the Import LandXML dialog box, browse to the C:\Civil 3D Projects\93rd\ folder, select the **93rd-Align.xml** file, and click **Open**.
5. In the Import LandXML dialog box, click the **OK** button to create the alignment.
6. From the File menu, select **Save As....** In the Save Drawing As dialog box, browse to the C:\Civil 3D Projects\93rd\DWG folder, name the drawing **93rd Street-Alignment**, and click the **Save** button.

CREATE THE 93RD STREET-POINTS DRAWING

1. From the File menu, select **New....** In the Select Template dialog box, browse to the *Chapter 12* folder located on the CD that accompanies this book. From the folder, select the template **HC3D-Extended NCS (Imperial)**, and click the **Open** button.
2. From the File menu, select **Save As....** In the Save Drawing As dialog box, browse to the C:\Civil 3D Projects\93rd\DWG folder, name the drawing **93rd Street-Points**, and click the **Save** button.

CREATE A PROJECT

1. If necessary, in Prospector click the view drop-list arrow and from the list of views, select **Master** view.
2. Prospector should include four open drawings: EG, Proposed, 93rd Street-Alignment, and 93rd Street-Points.

   In the middle of Prospector is the Project Management area. This area lists defined projects and the data.
3. In Prospector's Master view select the **Projects** heading, press the right mouse button, and from the shortcut menu select **Login**.
4. Enter your login values and click the **OK** button to continue.
5. In Prospector's Master view, select the **Projects** heading, press the right mouse button, and from the shortcut menu select **New....**
6. In Project Properties dialog box, type **93rd-Tulsa** as the name, type **Commercial Site with Grading and Pipes** as the description, and click the **OK** button to exit.

   The project is added to the list.
7. In Prospector's Master view, under Projects, expand the 93rd-Tulsa project branch to view the new project.

CHECK IN: EG

1. In Prospector's Master view, from the list of open drawings select **EG**. Press the right mouse button, and from the shortcut menu select **Add to Project....**
2. A Warning dialog box displays; click the **Yes** button to continue.

3. The Add to Project - Select a Project dialog box displays; select **93rd-Tulsa** and click the **Next** button to continue.
4. The Add to Project - Select a Drawing Location dialog box displays. Select the **Drawings** folder of the 93rd-Tulsa project and click the **Next** button to continue.
5. The Add to Project - Drawing File Dependencies dialog box displays. With the EG drawing, EG folders, and surface data files (EG.pnt and EG.tin) toggled off, toggle **OFF** Keep Files Checked Out, and click the **Next** button to continue.

   This copies the dependent data files (EG.pnt and EG.tin) to the Vault project structure.
6. The Add to Project - Share Data dialog box displays; toggle **ON** Surfaces. Type **Final Do NOT Edit** as the surface description, and click the **Finish** button to continue.

   Civil 3D copies the drawing and its dependent file to Vault and closes the EG drawing.

CHECK IN: ADD PROPOSED TO PROJECT

1. In Prospector's Master view, from the list of open drawings select **Proposed**, press the right mouse button, and from the shortcut menu select **Add to Project....**
2. A Warning dialog box displays; click the **Yes** button to continue.
3. The Add to Project - Select a Project dialog box displays; select **93rd-Tulsa** and click the **Next** button to continue.
4. The Add to Project - Select a Drawing Location dialog box displays. Select the **Drawings** folder of the 93rd-Tulsa project and click the **Next** button to continue.
5. The Add to Project - Drawing File Dependencies dialog box displays. With the Proposed drawing and the Proposed.xml file toggled on, toggle **OFF** Keep Files Checked Out, and click the **Next** button to continue.

   This copies the dependent data file (Proposed.xml) to the Vault project structure.
6. The Add to Project - Share Data dialog box displays. Toggle **ON** Surfaces. Type **Proposed - Do NOT EDIT** as the surface description and click the **Finish** button to continue.

   Civil 3D copies the drawing and its dependent file to Vault and closes the proposed drawing.
7. Repeat Steps 1 through 6 to add the 93rd Street-Alignment and Points drawing and its data to the project. Share its data with the project.

CHECK OUT A DRAWING

The current status of the drawings and data is available and ready for check out. The white dots to the left of the items' names in the project listing indicate this status. These icons come from the Vault application (see Figure 12.33).

EXERCISES

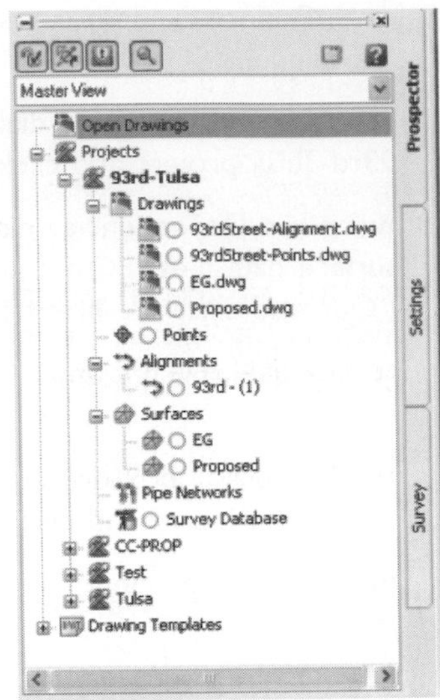

Figure 12.33

1. In Prospector's Master view, expand the 93rd-Tulsa project until you view the list of drawings for the Drawings heading. All of the drawings and their data (surfaces and alignment) are available for check out.

   Check out 93rd Street-Points drawing.

2. For the 93rd-Tulsa project list of drawings, select the **93rd Street-Points** drawing, press the right mouse button, and from the shortcut menu, select **Check Out....**
3. In the Check Out Drawing dialog box, make sure the only item toggled on is Get Latest Version, and click the **OK** button to open the drawing file.
4. In the Project area for 93rd-Tulsa, the icon changes for the drawing as well as the Open Drawing list for the checked-out file (a check with a white background). This icon indicates the drawing is current with the copy in Vault.

ADD POINTS TO THE DRAWING

1. From the Points menu, select **Create Points....**
2. In the Create Points toolbar, click the **Import Points** icon (the last icon on the right) to read an ASCII file that contains point coordinates.
3. If necessary, in the Import Points dialog box set the format by clicking the drop-list arrow and from the list selecting **PNEZD (Comma Delimited)**.
4. In the Import Points dialog box, to the right of Source File, click the yellow folder icon. Browse to the C:\Civil 3D Projects\93rd folder, change the Files of Type to ***.csv**, select the file **MON-TREES-Tulsa.csv**, click the **Open** button, and click the **OK** button to import the points.

5. Click the **AutoCAD Save** icon and save the drawing.
6. The drawing icon in the 93rd-Tulsa project and in the Prospector panel changes to indicate the local copy is newer than the project's copy (green dot with a check mark).
7. In Prospector, Point Groups, select the **_All Points** point group, press the right mouse button, and from the shortcut menu, select **Add Points to Project....**
8. In the Add to Project dialog box, for type **Tree and Boundary Survey Points** as the description, set the Check in Options to **Check In**, and click the **OK** button to add the points to the project (see Figure 12.34).

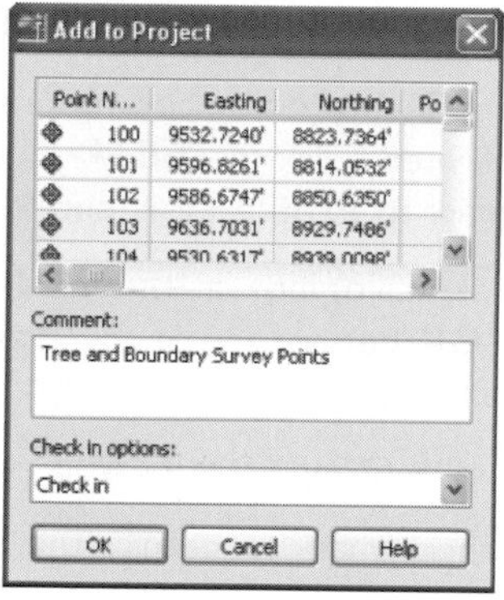

Figure 12.34

9. Click the **AutoCAD Save** icon and save the drawing.
10. In Prospector's Master view, from the Open Drawings list select the 93rd **Street-Points** drawing, press the right mouse button, and from the shortcut menu select **Check In....**
11. In the Check In Drawing dialog box, add **Added Points – Trees and Boundary** to Enter Version Comments, and click the **Finish** button to check in the drawing.

    The points and the drawing are now checked in and a part of the project.

CREATE A DATA REFERENCE

1. In Prospector's Master view, the Project area, select the **93rd Street-Alignment** drawing from the 93rd-Tulsa project list of drawings. Right mouse click and, from the shortcut menu, select **Check Out....**
2. In the Check Out Drawing dialog box, make sure the only item toggled on is Get Latest Version, and click the **OK** button to open the drawing file.
3. In the Project area for 93rd-Tulsa, the icon changes for the drawing as well as the Open Drawing list for the checked-out file (a check with a white background). This icon indicates the drawing is current with the copy in Vault.

4. In Prospector's Master view, from the 93rd-Tulsa project list of Surfaces, select **EG**, right mouse click, and from the shortcut menu, select **Create Reference....**
5. The Create Surface Reference dialog box displays. Set the surface style to **Contours 1' and 5' (Background)** and click the **OK** buttons.
6. In Prospector, expand the Surfaces list to view the EG surface and its reference icon.

CREATE A PROFILE FROM EG FOR THE ALIGNMENT

1. Use the AutoCAD Pan command to move the project to the west so there is clear space to the right of the project for the profile view.
2. From the Profiles menu, select **Create from Surface....**
3. In the Create Profile from Surface dialog box, make sure the **93rd – (1)** alignment is specified and **EG** is highlighted. Click the **Add>>** button at the right middle of the dialog box and click the **Draw in Profile View** button at the bottom of the dialog box.
4. In the Create Profile View dialog box, at the middle left, change the Profile View Style to **Full Grid**, click the **OK** button, and in the drawing select a point to the east of the project for the profile view.
5. An Event Viewer displays, notifying you that you must set a Profile2 after creating it.
6. Click the vista's green check mark to dismiss the panorama.
7. Click the **AutoCAD Save** icon and save the drawing.

   Prospector places a reference icon next to the surface name because the profile view references the surface's data.

MODIFY THE EG SURFACE DRAWING

1. From Prospector's Project area, from the drawing list for the 93rd-Tulsa project, select the **EG** drawing, press the right mouse button, and from the shortcut menu, select **Check Out....**
2. In the Check Out Drawing dialog box, make sure the only item toggled on is **Get Latest Version**, and click the **OK** button to open the drawing file.
3. In Prospector for the EG drawing, expand the Surfaces branch until you view the list of data types under the Definition heading.
4. From the Definition list, select **Edits**, press the right mouse button, and from the list, select **Raise/Lower Surface**.
5. In the command line, the routine prompts for a value. Type **-2** as the value and press the **Enter** key to lower the surface.
6. Click the **AutoCAD Save** icon and save the drawing.
7. In Prospector, select the EG drawing name, press the right mouse button, and from the shortcut menu select **Check In....**

8. In the Check In Drawing dialog box, for the Enter Version Comments add **Lowered EG by 2 feet** and click the **Finish** button to close the drawing.
9. In the 93rd Street-Alignment drawing, in Prospector, expand the Surfaces branch and note the EG surface now has an out-of-date icon.

SYNCHRONIZE A REFERENCE

1. In Prospector's Surfaces list, select **EG**, press the right mouse button, and from the shortcut menu, select **Synchronize**.
2. An Event Viewer appears, notifying you that the surface is up to date.
3. Click the vista's green check mark to dismiss the panorama.
4. Click the **AutoCAD Save** icon and save the drawing.
5. In Prospector, select the 93rd Street-Alignment drawing name, press the right mouse button, and from the shortcut menu select **Check In...**.
6. In the Check In Drawing - Drawing File Dependencies dialog box, for the Enter Version Comments add **New Profile and updated EG** and click the **Next** button.
7. In the Check In Drawing - Share Data dialog box, toggle **ON** Alignments, and click the **Finish** button to check in the drawing and the new data.
8. In Prospector's Project area, expand the 93rd-Tulsa project and view the 93rd Street Alignments branch.

   The branch now includes an EG profile.

EXERCISES

VAULT MANAGER

If you are a Vault administrator, you can edit vault values.

1. In Prospector's Project area, select the **Projects** heading, press the right mouse button, and from the shortcut menu, select **Vault Explorer...**.
2. In Vault Explorer, select the folder 93rd-Tulsa to preview the files.

   Each file has a revision number and a status icon.
3. In the middle panel, select the **93rd Street-Alignment** drawing.
4. The preview at the bottom middle lists the pertinent values for the object.
5. Click the **View** tab to make it current.

   In the preview area is a slider at 2 (the current revision). Slide it to **1** (original version). See Figure 12.35.

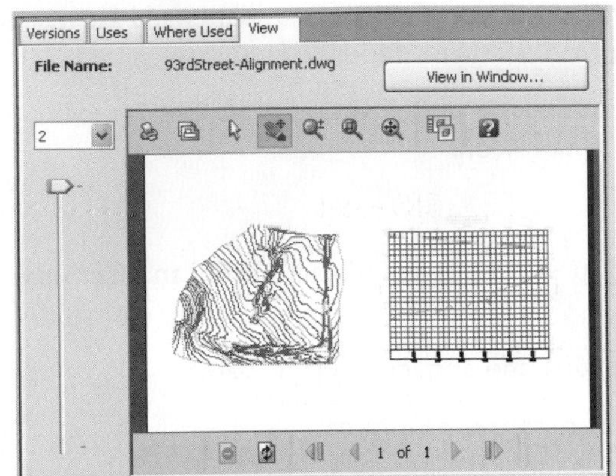

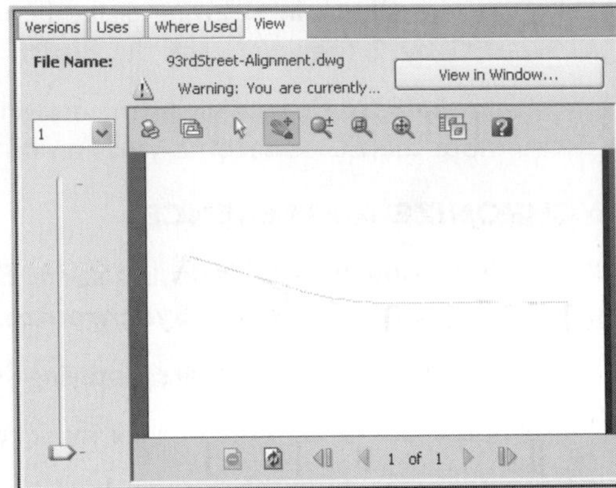

Figure 12.35

6. Close the Vault Explorer application.

   This ends the exercise on Vault.

SUMMARY

- When creating a project, users are defining a Vault project.
- Checking in a drawing can close the current editing session.
- When checking in a drawing that contains data, the user must decide if the data is ready to be shared with other team members.
- If drawing data is marked as not to be shared, a user can check out the drawing and, when checking it back in, can toggle on sharing.
- A Vault administrator creates users and sets their use level.
- Vault labels mark a point in a project's history.
- A Vault label can generate a pack-and-go file and can restore all project files and data to a revision level.

The next two chapters address the survey portion of Civil 3D.

INDEX

A

B

C

D

E

F

K

L

M

Q

R

S

T